PLASTER AND DRYWALL SYSTEMS MANUAL

Third Edition

J. R. GORMAN
SAM JAFFE
WALTER F. PRUTER
JAMES J. ROSE

Published by —
BNI Books, Division of Building News, Inc.
Los Angeles, California 90034

Distributed by —
McGraw-Hill Book Company
New York, St. Louis, San Francisco, Auckland, Bogotá, Hamburg,
London, Madrid, Mexico, Milan, Montreal, New Delhi, Panama, Paris,
São Paulo, Singapore, Sydney, Tokyo, Toronto

INTRODUCTION

Plaster and Drywall Systems Manual is designed to be a complete guide to metal framing, lathing, plastering, drywall and fireproofing installations. It is a modern, authentic and up-to-date reference manual for architects, specifiers, contractors, building officials, inspectors, materials suppliers and other building professionals.

This manual is an outgrowth and expansion of Lathing and Plastering Specifications, authored and promulgated by Western Lath/Plaster/Drywall Industries Association (formerly known as California Lathing and Plastering Contractors Association) in four preceding editions. The expanded manual is intended to provide, under one cover, every bit of information needed by the building professional concerned with lath, plaster, drywall and fireproofing installations.

The WLPDIA specifications have been widely used and referenced as a standard throughout the construction industry for more than two decades. In this expanded manual these specifications, updated to 1988 concepts, appear as Chapter Two and take their place alongside other recognized specifications, standards and codes . . . the combined group representing the most complete and authentic set of standards and specifications on this subject ever published.

Workmanship is of great concern in producing quality lathing, plastering, drywall and fireproofing installations. A tight specification, with adequate inspection, and a qualified contractor insure good workmanship.

ACKNOWLEDGEMENTS

The publishers of this manual gratefully acknowledge the co-operation of a number of the nation's leading specifications, codes and standards organizations in granting reproduction rights on their documents which appear in whole or in part within these pages. Specific credits to the authoring organizations appear at or near where their document or documents have been reproduced.

The editorial group responsible for compilation of this manual is representative of the whole broad spectrum of industry organizations who have an interest in the subject of lath, metal framing systems, plaster, drywall and fireproofing and is listed below:

EDITORIAL ADVISORY COMMITTEE

Charles Beckham, representing Western Lath/Plaster Drywall Industries Association
Ian Hendry, representing Contracting Plasterers' Association of Southern California
Donald MacQueen, representing Gypsum Association
Richard Phillips, representing Structural Engineers Association of Southern California
Robert Welch, representing Stucco Manufacturers Association
Ross Adams, representing Portland Cement Association
Gordon F. Clyde, representing International Conference of Building Officials
John Raeber, AIA, CSI, representing California Council, American Institute of Architects and San Francisco Chapter, Construction Specifications Institute
Richard Blenkinsop, representing Metal Lath/Steel Framing Division of Architectural Metal Manufacturers Association

Library of Congress Cataloging-in-Publication Data

Plaster and drywall systems manual / J. R. Gorman; Jaffe, Sam; W. F. Pruter; J. J. Rose — 3rd ed.
   p.   cm.
  ISBN 0-07-032199-X
   1. Plastering — Handbooks, manuals, etc. 2. Drywall — Handbooks, manuals, etc.
   I. Gorman, J.R.; Jaffe, Sam; Pruter, W.F.; Rose, J.J.
TH8131.P54 1988   88-21532
693.6—dc19   CIP

The editor for this book was Sam Jaffe, the designer was Ramón López and the production supervisor was Richard Berlanga. It was set in Helvetica and Caledonia by BNI Books. Printed by O'Neil Data Systems, Inc. and bound by Stauffer Edition Building Co., Inc.

ABOUT THE AUTHORS

J. R. "DICK" GORMAN

It would be difficult to find a person with a broader acquaintanceship amidst architects, specification writers, lathing and plastering contractors and other professionals involved with lathing and plastering specifications than J. R. "Dick" Gorman.

Dating back to the year 1965 to the present, Gorman has served as director of Technical Services for the Plastering Information Bureau, division of the Southern California Plastering Institute, a non-profit organization dedicated to the upgrading and promulgation of lath and plaster construction in the Southland area.

As such, he has been involved with literally hundreds of lath and plaster jobs in Southern California in terms of specification, inspection, quality control, design and trade promotion.

Prior to that he served as a Captain in the U.S. Corps of Engineers (1956/1961) and architectural representative for Kaiser Gypsum Co. (1962/1965). He received his higher education at Rice University, Houston, Texas, from which he received a BA degree in architecture in 1954 and a BS degree in the same field in 1955.

His numerous industry affiliations include: member, Construction Specifications Institute, Los Angeles and Orange County Chapters; past president, Western Conference of Lathing and Plastering Institutes; past president, International Institute for Lath and Plaster; past vice president, Construction Products Manufacturers Council, Los Angeles Chapter; and affiliate member, International Conference of Building Officials. He also serve as architectural representative and consultant to several manufacturers of construction products on the West Coast.

A resident of the community of Thousand Oaks, California, he enjoys the hobby of gardening with his wife, Alicia, in his spare time and when not involved with the six Gorman offspring.

JAMES J. ROSE

James J. "Jim" Rose is well recognized for his expertise in the field of lathing and plastering, which stems from 13 years of experience as a top flight plastering contractor in the Southern California area and almost three decades of service as executive manager of the Contracting Plasterers Association of Southern California.

He also served as secretary of four other related industry associations during this period.

Today he heads a forensic and technical consulting firm for the construction industry, where his knowledge of lath and plaster is constantly being called upon, particularly as an expert witness in legal actions.

His work as a consultant has taken him as far afield as Germany and Saudi Arabia.

As chief executive officer for the Contracting Plasterers group he was responsible for the full gamut of employer needs, including development of industry standards and specifications, labor management relations including bargaining and arbitration, the operation of various types of insurance group programs, credit information, technical field inspection services, safety programs, seminars, developing and modifying local and model building codes, member communication activities which include training sessions and semi-monthly newsletters. Other services supervised include influencing and interpreting legislation, dealing with public regulatory agencies, and representing the industry on various public service entities as technical consultant.

A native of Berkley, Massachusetts, he received his higher education at two universities, Florida Atlantic University, where he received an MBA in 1983; and University of California at Los Angeles, where he received a BA degree in 1942.

Jim Rose makes his home with his wife, Norma, in the community of Sylmar in the heart of his beloved San Fernando Valley. His favorite hobbies are reading, golf, fishing, checking out installations of his favorite materials — lath, metal framing systems and plaster.

SAM JAFFE

A native of Los Angeles, California, Sam Jaffe grew up in the suburb of Venice, where he attended school and started his career in journalism as editor of the high school weekly newspaper.

During World War II he served in the U.S. Air Force for the three years, with primary duty being the editing of service newspapers. After the War he resumed his career as a journalist, editor and publisher, launching the firm of Building News, Inc. in Los Angeles in 1946, with total orientation to the construction industry.

This firm, which has proliferated over a four decade span and is still very much a leader in the dissemination of industry information and education, started off as publisher of a variety of trade periodicals, but soon branched off into the publication and marketing of technical books.

Today, the total emphasis is upon books. The company publishes around 25 of its own titles, including this manual and markets over 3,000 books on an international basis, with its professional bookstore in West Los Angeles being the "flagship" of the operations.

Among the honors received by Jaffe for his near half-century of literary contribution to the construction industry was his receipt of the 1977 Public Information Award of the California Council, American Institute of Architects.

In 1977, in collaboration with Water Pruter, he conceived the idea of expanding the well-accepted Reference Specifications of the Western Lath/Plaster/Drywall Industries Association into a complete manual/handbook for the related industries. He served as editor of the first and second edition — and now this third edition.

Married long enough to celebrate a golden wedding anniversary next year, he currently makes his home in Beverly Hills, Calif. His offspring includes two daughters and four grand-daughters.

WALTER F. PRUTER

A man whose golfing handicap is 23 has to enjoy the respect of his colleagues. Even more so, in the world of lathing, plastering and drywall the expertise of Walter F. "Walt" Pruter commands the utmost recognition of his peers. For Walt, it's the best of two worlds.

Hard working, hard driving Walt Pruter is one of the busiest men in the industry, trouble-shooting jobs, inspecting work in the field, writing specifications, counseling builders and architects, running trade associations, consulting and researching and writing articles for national trade magazines . . . just to peek at a corner of his daily life. When the pressure lets uₜ he likes to head for the golf course in pursuit of his favorite sport.

Pruter was born in Chicago, Illinois, where he grew up and, in higher education, earned a BS degree in architecture at the Illinois Institute of Technology. One of his first jobs was the U.S. Gypsum Company as an architectural representative in the Midwest, which led to his later career in lath, plaster and drywall. In 1960 he founded the Plastering Information Bureau in Los Angeles, which he operated until 1965, when his colleague, Dick Gorman, took over.

Presently he serves as technical director for the Information Bureau for Lath, Plaster and Drywall in Los Angeles as well as for the Lathing, Metal Furring and Drywall Contractors Association. He is also secretary-treasurer of the International Institute for Lath and Plaster.

He is an affiliate member of the American Institute of Architects, Construction Specifications Institute, International Conference of Building Officials, American Society for Testing and Materials and American Concrete Institute.

He is an accredited author and has written numerous articles for national magazines on the subjects of lathing, steel stud framing, plastering and drywall. With his wife, Carol, he divides his time between two homes in Southern California, one in Glendale and the other in Palm Desert. The Pruter family includes five children.

III

TABLE OF CONTENTS

TABLE OF CONTENTS

CHAPTER ONE

HISTORY
AND
ROMANCE OF
LATH AND
PLASTER

OLD AND THE NEW — Woodcut above depicts plaster work of the Thirteenth Century, highly ornate entrance to the Court of the Lions in The Alhambra, Spain. On opposite page are decorative plaster ceilings in contemporary buildings.

Plaster is a cementitious material or combination of materials and aggregates, which, when mixed with a suitable amount of water (or in some cases, other liquid), forms a plastic mass which, when applied to a surface, adheres to it and subsequently sets or hardens, preserving in a rigid state the form or texture imposed during the period of plasticity. The term "plaster" is used with regard to the specific composition of the material and does not explicitly denote either interior or exterior usage.

In some usages, lath and plaster assemblies contribute sheer values to the building design, and in others, particularly in combination with metal structural studs, lend structural support values.

Traditionally the cementitious material or binder which binds the aggregate particles into a hetero-geneous mass is of three types: gypsum, lime, and portland cement.

While portland cement, strictly speaking, is a relatively modern discovery, forerunners of the same type of material were in use as far back as prehistoric times.

Primitive man, perhaps imitating wasps, birds or beavers, learned early that some natural plastic materials such as mud, lime or adobe, could be manipulated, shaped and applied to rock or brick surfaces. He discovered that some of these hardening, naturally-occuring materials were suitable for weather-proofing, contouring and parging among other usages.

Records do not take us back into that dim primeval past to the dawn of social life when the origin of compounding materials for plastering first occurred. We do not know when the firing of rocks to calcine cementitious ingredients was discovered — whether it was a campfire built over gypsum or lime-

ORNAMENTAL plaster work on the arcades in the Mosque of Ibn-Tulun, Ninth Century, are seen in the woodcut at left. Below, textured acoustical plaster makes beautiful finish for ceiling and curved stairway in contemporary building.

stone rocks which hardened when water fell on the ashes — whether it was shells cast into a fire or ores in a naturally fortuitous combination which set or hydrated.

But we do know that some of the earliest plastering which has remained for us equals or excels in its chemical composition materials in use today. Cliff dwellings, pyramids of Egypt, and structures in Central and South America yield evidences of plastering done 3000-4000 years ago or earlier. While most of the prehistoric plastering was over masonry, there is evidence of wattles or reeds tied with cords providing lath for solid partitions. The Egyptians understood canvas plaster by which thin plaster death masks were made, preserving identical facial features in detail.

Plasters are generally of three types — with gypsum, portland cement (also known as stucco) or lime as the cementitious agent. Gypsum plasters are subject to deterioration in the presence of moisture and samples of primitive plastering using this binder have generally been found in protected areas.

Very early in Greek Architecture we find the use of lime stucco of an exquisite composition, white, fine and thin — often no thicker than eggshell. Some of it has endured weathering better than the stone to which it was applied.

By the time of the Roman Empire the knowledge of blending ingredients and firing cement clinker were well known. Concrete buildings, roads, sewers, viaducts and aqueducts are found in various parts of the world as residues of Roman industry, know-how, and penchance for permanence. Kilns to manufacture cement and other cementitious binders were set up in many areas of the Empire.

Then, as now, the functional utility of plasters, interior and exterior, were well-known. The Biblical Book of Leviticus 14:42 details re-plastering the interior of a house for sanitation and purging a plague. As a sanitary coating it kept out vermin and insects. Plaster is a weatherproofing sealant, to protect against wind, rain and sun.

Plaster is a levelling coat — to make walls and ceilings level or smooth either for their own esthetic appeal or as the base to which coatings, coverings, frescoes or plaster ornamentations may be applied.

In its ornamental usages, plaster is an art form with many styles and types of enrichments. Plaster is a lining for partitions which divide room areas for privacy and diversified functional usages. As a fireproofing coating, plaster has been required by building codes and by law from the time of the Babylonians, following Nero's fire in Rome and in medieval

FLEXIBILITY of plaster as a surfacing material is illustrated by this picture of elevator bay and ceilings with intricate curved planes in Seattle, Washington, department store.

London as notable examples. Fireproofing of metal structural framing is today a large market for the plastering industry. It is an insulative agent for heat transfer control in boats, refineries and many other commercial usages.

As a sound dampener and retarder of intruding noises, both airborne and impact, plaster has historically served well. Now as the world becomes

USE OF PLASTER for ornamentation is limited only to the imagination of the designer. In picture above, graceful Grecian columns and other decorative plaster staff elements embellish this colonial-style home. At left, scratch coat of plaster is applied to metal lath base on prefabricated decorative soffit. Below, artisan puts finishing touches on plaster cast of Seal of the President of the United States.

energy conscious, plaster is found performing well as an insulative material. In its plastic state plaster is shapable and moldable to a great variety of contours, and surface textures and artistic modeling.

Plastered assemblies develop excellent sound attenuation and transfer characteristics not only because of their mass but more effectively when combined with resilient lath mountings or attachments.

In energy conservation, in heating and cooling engineering, plastered assemblies are designed to provide significant savings.

Fireproofing of modern high rise steel framed buildings is achieved most economically and effectively with direct-to-steel sprayed-on fire insulative plasters, cementitious and adhesive.

Plastering of ferro-cement hulls of pleasure and commercial boats and ships is still in its infancy, but a history of successful performance is rapidly being accumulated.

Despite their antiquity, lath and plaster remain modern building materials, giving form and function to the latest design concepts of the most imaginative architectural and engineering creators.

The use of lime as the cementitious binder for plasters, interior and exterior, dates back 25 centuries at least. The burning of lime ore is critical to its eventual hardness and workability characteristics. It should be burned or calcined at as low a controlled temperature as possible. An old Roman law required lime putty to be aged a minimum of three years.

Lime mortars, slow-setting and long-working, were applied in several thin successive coats with time for carbonation (the absorption of carbonic acid gas) before succeeding coats or coatings were applied.

Gypsum, one of the world's most commonly distributed minerals, is frequently found in relatively pure deposits. When ground, burned or calcined and blended with set-controlling additives, it provides a workable cementitious binder.

Gypsum is hydrous calcium sulfate in the dehydrate form. After losing ¾ of its water of combination during calcination, it is termed hemihydrate. Calcined gypsum, when mixed with water, recombines with approximately the same amount of water lost during calcination and reverts to its hard rock-like crystalline form. The rehydration is called setting.

Gypsum plasters are used either neat or aggregated, dependent upon the type of plaster and the designed usage and performance.

The long, needle-like crystals of gypsum, formed in setting, interlock around aggregate particles. The crystals interlace fibers of the paper facing of gypsum lath, thus developing great adhesive attachment. Plaster applied to metal or wood lath forms a mechanical key for attachment. Wood lath is no longer commonly used in the United States.

Several types of aggregates are used. The most common is naturally occuring sand, screened and graded as to particle size distribution. The proper

SAMPLES of well-graded sand used for plaster aggregate after separation into various sizes. (Portland Cement Assn. photo)

SPRAYING lightweight plaster on steel columns for fireproofing.

ONE TYPE of machine used to apply a rock dash or marble-crete finish is seen here. Small pebbles, stones, or marble chips are placed in the hopper and are blown into the surface of the fresh plaster. (Portland Cement Assn. photo)

range of size gradations of sand particles influences plaster strength, workability and tendency to shrink or expand.

Lightweight aggregates may be pumice, exfoliated (popped in a furnace under controlled conditions) perlite or vermiculite or other minerals in their natural or processed state.

In some countries gypsum crushed and graded particles are used as aggregate. Sawdust, wood fibers or crushed coral are also used.

Gypsum plaster possesses the unique characteristic of recalcining on the surface to which relatively low-temperature heat is applied (120-140° F). The greater part of the heat is diverted as steam, resulting from release of the chemically combined water, and little heat is conducted through the gypsum plaster itself.

By varying fineness of grind, types and amounts of admixtures, methods of calcining, etc., several types of gypsum plasters may be produced, some of which may attain compressive strengths well in excess of 10,000 pounds per square inch.

Their use is generally restricted to interior or protected areas because moisture can cause the gypsum plaster to deteriorate.

Lime is made from burning limestone, which is commonly distributed throughout the world's surface. It consists primarily of calcium or magnesium carbonate, or both, and has a wide range of physical and chemical characteristics. Only limestones of extremely high purity (97-99%) are used for producing plasterers' lime. When burned, or calcined, lime gives off carbon dioxide and changes to calcium or magnesium oxide.

Until early in the 20th century lime was the binder in most plaster used on both interior and exterior. Lime plaster returns to its rock-like state over a considerable period of time, after application, by a process of recarbonation in which it absorbs free atmospheric carbon dioxide or carbonic acid gas to replace that lost during calcination. Today lime is not generally used as a basecoat binder but is used as a plasticizer in portland cement masonry and lime-cement mixes. It is also a plasticizing agent in finish plaster mixes.

Portland cement does not occur naturally and is produced primarily from a combination of limestone and an argillaceous or clay-like substance, both of which are abundantly prevalent worldwide. Chemically, it is essentially a combination of various silicates and calcium aluminates. After being properly blended, burned and ground, portland cement, when water is added, starts a complex reaction which ends as a dense hard mass of amorphous gel. The hydration or hardening requires the presence of moisture usually for several days.

Portland cement plaster is used primarily for exterior plastering and for interior usages where wetting, steam or severe dampness are prevalent. It is very durable and relatively unaffected by water and freeze-thaw cycles. Portland cement plaster may be applied to all metallic lath bases as well as to masonry and rough monolithic concrete. It should not be applied to gypsum lath unless a metallic lath has been applied over the gypsum lath.

Portland cement plaster has one characteristic which creates performance problems when not recognized and compensated for: portland cement (and most concretes) shrink slightly during hydration and the subsequent loss of free moisture of mix. Such shrinkage generates stress within the portland cement plaster membrane, particularly when the plaster has

AS ONE workman applies the second exterior coat with the plaster gun, another man screeds or levels off the material to bring it to a true surface. (Portland Cement Assn. photo)

MACHINE APPLICATION of final coat of colored final coat of exterior stucco produces uniform texture and color.
(Portland Cement Assn. photo)

been applied over lath, which often results in cracks and surface distortion.

Other types of stresses which may cause cracking are:

    (a) Stress transfer from the structure
    (b) Thermal shock
    (c) Wind, seismic, vibration or impact stresses.

When the concentration of stresses on the plaster exceeds the maximum strain capacity of the material, ruptures may ensue. The provision of integral stress-relief mechanisms, such as control joints, provides relief from stress build-up and minimizes cracking.

Many new cementitious coatings have been developed in recent years. Epoxies, liquid acrylics, polymers, latexes and elastomeric type binders, among others, are used in special plaster applications.

Pigments are often added to plasters, interior and exterior, for integral color of the finished plaster product, obviating painting or other decorative coatings. Sgraffito is a color-art form in which several layers

A SCARIFIER is used to scratch the base coat of plaster. This forms horizontal ridges that provide maximum mechanical key for the second coat. Expanded metal lath, bottom, with dimples provides proper furring for embedment of the plaster.
(Portland Cement Assn. photo)

FIRST COAT or scratch coat of plaster is troweled over metal reinforcement (stucco netting). This coat must be forced through and behind the reinforcement for complete embedment.
(Portland Cement Assn. photo)

PLASTERER applies the second or brown coat.
(Portland Cement Assn. photo)

FLOATING SKILL is demonstrated here as the plasterer finishes an overhead flat surface. Note the position and pressure of the hand on the float.

of differently colored plaster are laid on and after initial hardening are cut back to desired tints to achieve design effect and color.

The impregnation of color in wet plaster, called fresco, dates back to prehistoric days and reached its flower in medieval Europe. Michelangelo and Da Vinci used this vehicle.

In ancient times vegetable stains, sea shells, sea organisms, natural earth colors and even blood were used to color cementitious mortars. Later on natural mineral ores and chromes were applied. More recently synthetic pigments have come into usage, often with variant results. Mica, colored glitter, and diamond dust may be blown onto the surface of wet plaster to create moving reflective effects.

Other types of specialty plaster applications are:

Skimcoat plaster, which is the application of finish plaster direct to monolithic concrete or masonry bases usually after a bonding agent has been applied.

Veneer or thincoat plaster, which is a high strength gypsum plaster applied over large-size gypsum lath, the paper of which has been treated with a catalyst to induce rapid plaster set.

Ceiling radiant heat plaster, which may be a proprietary formulation or may be sanded gypsum plaster, embeds radiant heat cables, becoming a radiating panel to provide space heating needs.

There are many ·mold stampings, which, when pressed into wet plaster, shape the surface to simulate stone, block or brick. However, any style of masonry or brick may be simulated in wet plaster by carving joints into colored plaster in an artistic pattern.

Marblecrete is an interior exposed aggregate finish in which marble, pebbles or chips of similar material are hand or gun placed into the soft bedding coat or matrix.

There are many other plain and ornamental surface treatments or types of plasters, creating a wide latitude in texture, pattern, color, design and variety.

Fabricating and finishing plaster panels in shops for trucking to jobsites, and fixing into place is a newer technology.

Plaster may be applied to a wide variety of bases, including direct application to concrete, adobe, masonry, block and rock. It is also applied to fabricated structural or non-load bearing bases such as wattles, woven reed, bamboo, burlap, wood lath or foamed plastic sheets. In modern plastering, metallic laths, such as woven wire, expanded metal, or welded wire fabric lath, much of which is paper backed with the lath often crimped for self furring, are commonly used.

Gypsum plaster may be applied to metallic laths or to gypsum sheet lath. Originally holes were punched into gypsum lath to lighten shipping weight. It was commonly thought that the perforations were there to create a mechanical key. However the basic bond of gypsum plaster to gypsum lath is the interlocking of the long needle-like crystals of gypsum into the fiber strands of the lath paper.

In the past three decades the methods of handling the materials of the trade have changed dramatically. Tools and equipment in usage previously had been improved but little in hundreds of years. Mechanical mixers, mortar pumps, power staplers, mechanical trowels, motorized swinging scaffolds, rock embedment guns, metal cutting tools and others have revolutionized handling and application techniques.

Value engineering, in construction, is a study of the long range cost of materials and systems when maintenance, repair, length of service, decorability and other cost elements are factored into the formula. The ultimate utility of plaster is excellent and economical. The potential esthetics and plasticity of form render plaster adaptable for almost any shape, surface texture, or design criteria desired.

Plaster provides acoustical control, fire resistance, thermal insulation, energy conservation, full decorability and is economical to maintain and repair. In form, color, adaptability and permanence it is unexcelled.

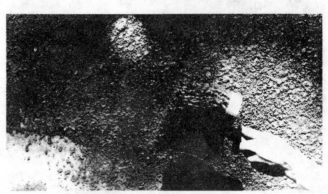

METHOD for applying a dash bond coat with a long bristle brush is illustrated here. (Portland Cement Assn. photo)

CHAPTER TWO

METAL FRAMING
LATHING
FIRE PROTECTION
PLASTERING
EXTERIOR WALL COATINGS
STUCCO
DRYWALL
REFERENCE
SPECIFICATIONS

FOREWORD

To Reference Specifications

These Specifications, constituting the key chapter of this manual, were produced originally in 1958 by the Technical Committee of the California Lathing and Plastering Contractors Association, now known as the Western Lath/Plaster Drywall Industries Association. This organization is the world's largest regional contractors' association in the walls and ceilings industry. Its origin predates World War II and it was re-established in 1946. The Specifications have been updated and reprinted six times prior to this printing.

Intended originally for California architects and specifications writers, the Specifications have been embraced on a national level since their first usage — and are universally recognized as the best technical works ever compiled on the subject of lath, plaster and drywall.

The Specifications are intended to describe **minimum requirements** complying with the appropriate building codes for satisfactory lathing, plastering and drywall construction. Higher standards may be set by the architect and specification writer when he or she believes the project warrants such treatment.

Since it is impossible to cover every design or job condition, these Reference Specifications can only assist in the preparation of specifications for a given project. It is **not intended that an overall reference to these Specifications will relieve the architect from his design responsibilities.**

When referring to the Reference Specifications in any project specification, those paragraph numbers herein which apply to the type of work called for on the project should be included.

Notes are numerous, are important and should be consulted when contemplating use of any paragraph to which they are keyed. Within the notes are to be found the "cautions" and "recommendations," resulting from the combined practical experience of the contractor, journeyman and manufacturer.

Drawings are included in some sections but only where it was felt they would improve the specification writer's knowledge of the system or items described in that section. Where it was felt necessary, the drawings have been referenced by number to the applicable specification paragraphs. Where drawings have been included they will be found within the section, following the indexes specification material.

Drawings do not purport to show any manufacturer's product but are intended to show typical items, accessories and systems available and generally in use in the lathing, plastering and drywall industries.

The update of this 1988 edition represents a consensus of the thinking of the most informed industry professionals.

TABLE OF CONTENTS

To Lathing, Plastering And Drywall Reference Specifications

4. EXTERIOR INSULATION

CLASS PB — TYPE A

5. FIRE PROTECTION (SPRAYED ON)

10. CEMENTITIOUS EXTERIOR WALL COATINGS

11. STUCCO (PLASTER) FINISHES

12. GYPSUM BOARD (DRYWALL) APPLICATIONS

INDEX

To Lathing, Plastering And Drywall Reference Specifications

1. GENERAL REQUIREMENTS

1.1 GENERAL CONDITIONS

1.1.1 General Conditions of the Contract.—As included in the general specifications for this building project, are an integral part of this specification and of the contract. Their contents should be carefully noted. [1]

1.2 GENERAL CONTRACTOR'S RELATED WORK

1.2.1 Temporary Facilities.—The general contractor shall provide, at no cost to subcontractors, temporary stairs and fixed ladders, ramps, runways and hoists; adequate space for storage of materials; temporary closures to control humidity and ventilation; watchman; toilets; adequate light and power; adequate water connections; telephone.

1.2.2 Scaffolding.—The general contractor shall provide all interior and exterior scaffolding and planking over 8' 0" in height. [2]

1.2.3 Floor Protection.—The general contractor shall provide initial protection of all floor areas and maintain for the duration of work. [3]

1.2.4 Heat and Blowers.—The general contractor shall provide necessary temporary heat and blowers to stabilize temperature and control ventilation inside the building. [4]

1.3 SCOPE OF THE WORK

1.3.1 Work Included.—The work includes all labor, materials, and equipment required to complete all lathing and plastering in accordance with the drawings and these specifications and shall include: [5] [6] [7]

1.3.2 Materials Installed.—(Furnished by Others). [8]

1.3.3 Materials Furnished.—(Installed by Others). [9]

1.3.4 Work Not Included. [10]

1.4 SCAFFOLDING & PROTECTION

1.4.1 Scaffolding.—Provide, install and maintain for the duration of the work, scaffolding, staging, trestles and planking necessary for the work of this section.

Scaffolding shall conform with applicable laws and ordinances and shall not unnecessarily interfere with the work of others.

1.4.2 Protection.—Provide, install and maintain all protection required to preserve the work of other trades free from damage due to lathing and plastering operations. When finished materials are installed prior to fireproofing, lathing or plastering, initial protection of such materials shall be provided by the installing subcontractor. [11]

[1] It is desirable to restate here the relative precedence of drawings, specifications and room or finish schedule.

[2] If scaffolding is to be furnished by plastering contractor, see 1.4.1.

[3] Initial protection of floors should be responsibility of general contractor whether or not working surfaces constitute the finish floor or are to receive a covering. This will remove the possibility of duplicate bids for cost-of-protection by general and subcontractor.

[4] Assignment of responsibility for heat and ventilation to general contractor assures control of heat and ventilation on project; prevents duplicate bidding of heating and ventilating costs by general and subcontractor.

[5] If scaffolding is to be furnished by plastering contractor, include paragraph 1.4.1.

[6] Include a concise but comprehensive breakdown of the work involved to assure that bids reflect all work intended to be included in this section. Where work shown on drawings could be interpreted as work of another section, clearly designate responsibility.

[7] If supporting metal framing for other wall and ceiling coverings is to be part of the work of this section, include a statement covering scope.

[8] List items and state under which section or division they are to be furnished.

[9] List items and state under which section or division they are to be installed.

[10] Describe, list and give section in which specified. Use only when necessary.

[11] Provide for initial protection in appropriate section.

1.4.3 Protection By Others. [12] [13]

1.5 PRELIMINARY INSPECTION

1.5.1 Notification of Inspection.—Before proceeding with the work of this section, the contractor shall be notified by the proper authority that required inspections have been made of the work to be covered.

1.5.2 Correction of Unsatisfactory Conditions.—Make a detailed inspection of all areas and surfaces to be enclosed or covered by the work of this section. Report any unsatisfactory condition to the proper authority. Do not proceed with the work of this section until known unsatisfactory conditions have been corrected.

1.5.3 Undetected Faulty Conditions.—Procedure with work in good faith and in accordance with requirements of these specifications will relieve the contractor of responsibility for defects arising in his work due to faults or omissions in work of others, but not detected and reported before work of this section was started and/or completed.

1.5.4 Faulty Materials and Workmanship of Others.—Nothing in this section is to be construed as placing on the lathing and/or plastering contractor final responsibility for faulty materials and workmanship of others.

1.6 PATCHING AND REPAIR OF PLASTER

1.6.1 Surface Defects.—Defects which appear in the work of this section, due to defective workmanship and/or materials furnished and installed hereunder, shall be repaired and refinished with materials and in a manner to meet the requirements of this section.

1.6.2 Damage by Others.—Repair damage done by others to the work of this section. Cost of such repair shall be the responsibility of the damaging party and payment shall be made through the general contractor. [14]

1.7 SPECIFICATION DESIGNATIONS (Lathing & Plastering Materials)

Materials	ASTM	Federal	Other
1.7.1 Accessories, Lathing			
Casing Beads	---	---	SPR[1] R3-60
Control Joints	---	---	2.8.12[2]
Corner Beads	---	---	SPR[1] R3-60
Cornerite	---	---	SPR[1] R3-60
Corner Reinforcement (exterior)	---	---	2.8.5[2]
Drip Screed	---	---	2.8.14[2]
Partition Runners (L or Z shape)	---	---	2.8.8[2]
Partition Terminals (caps)	---	---	SPR[1] R3-60
Screeds (base & parting)	---	---	SPR[1] R3-60
Strip Reinforcement	---	---	SPR[1] R3-60
Window Stools	---	---	2.8.15[2]

[12] Other contractors should provide proper protection for the work of this section. Repair of damage done by other contractors should be made under this section and cost borne by damaging party. See 1.6.2.

[13] Proper precautions must be taken by other contractors to avoid inadvertent covering of their work by lathing and plastering operation (e.g. electrical outlets).

[14] On remodeling of existing buildings, installation of mechanical or electrical work may damage existing plaster. This damage is not usually shown on drawings, or even known before bidding. Repair of such damage should be clearly designated as the responsibility of the trade responsible for damage.

[1]Simplified Practice Recommendations R3-60, U. S. Dept. of Commerce. See pages 84 and 85.
[2]Lathing & Plastering Reference Specifications paragraph number.

Materials	ASTM	Federal*	Other
1.7.2 Aggregate			
Perlite basecoat plaster	C35	—	—
Sand (gypsum plaster)	C35	—	6.6.2[2]
Sand (portland cement plaster) (except gradation)	C897	—	6.6.4[2]
Vermiculite (basecoat plaster)	C35	—	—
Wood Fiber	C28	—	—
1.7.3 Attachments, Lath			
Hanger wire	—	QQ-W-461g & AM-2	—
Nails, Wire and Staples	—	FF-N-105B & AM-2	—
Staples, power driven	—	—	2.2.31[2]
Tie wire	—	QQ-W-461g & AM-2	—
Attachment clips	—	—	2.9.1[2]
1.7.4 Kraft Waterproof Building Paper			
	—	UU-B-790, Feb. 5, 1968	—
1.7.5 Gypsum Materials			
Gypsum Lath, (plain, perforated, insulating, long length)	C37	SS-L-30D	—
Backing board	C442	SS-L-30D (except width)	—
Large size gypsum lath	C37, C442, C588 (except width)	SS-L-30D (except width)	—
Gypsum plasters (neat, ready-mixed, wood-fibered, bond, gauging)	C28	SS-P-00402B	—
Molding plaster	C59	—	—
Keene's cement	C61, C587	SS-C-161A	—
Veneer plaster	—	—	9.5.1[2] —
1.7.6 Lime			
Special finishing hydrate	C206	SS-L-00351	—
Normal finishing hydrate	C6	SS-L-00351	— —
1.7.7 Metal Lath and Wire Fabric Lath			
(For interior gypsum or portland cement plaster)	C847, C1052, C933	QQ-L-101C	SPR[1] R3-60
(For exterior portland cement plaster)	C847, C1032, C933	QQ-L-101C	SPR[1] R3-60
Wire Fabric Lath (woven)	C1032	—	2.6.73[2]
Wire Fabric Lath (Welded)	C 933		
1.7.8 Metal Supports			
Prefabricated studs	—	—	SPR[1] R3-60
Nailable studs	—	—	2.4.2[2]
Screw studs	—	—	2.4.3[2]
Cold rolled channels	—	—	SPR[1] R3-60
Structural studs (50,000 psi)	A245	—	—
(33,000 psi)	—	—	2.5.1[2]

---

[1]Simplified Practice Recommendations R3-60, U.S. Dept. of Commerce
[2]Lathing, Plastering and Drywall Reference Specifications paragraph number.

*Federal Specifications are obtainable in single lots without cost by writing to: Superintendent of Documents, U.S. Government Printing Office, Washington, D.C., 20402.

Materials	ASTM	Federal	Other
1.7.9 Portland Cement			
Regular (Types I, II, III)	C150	SS-C-1960/ GEN and SS-C-1960/3	---
Air-entraining (Types I-A, II-A, III-A)	C175	SS-C-1960/ GEN and SS-C-1960/3	---
Manufactured finish coats	---		6.7.1[2] SMA[3]
Masonry cement	---	SS-C-1960/ GEN and SS-C-1960/1	---
1.7.10 Miscellaneous Materials			
Acoustical plaster or plastic	---	SS-A-118A	---
Bonding agents	---	---	6.10.1[2]
Building paper	---	UU-B-790a	---
Felt, asphalt saturated	---	HH-R-595 & AM-2	---

[2]Lathing, Plastering and Drywall Reference Specifications paragraph number.
[3]Stucco Manufacturers Association, 14006 Ventura Blvd., Sherman Oaks, Calif. 91428.

1.8 REFERENCE DOCUMENTS

1.8.1 Aggregate

ASTM C 35: Inorganic Aggregates For Use in Gypsum Plaster Plaster

1. Scope

These specifications cover those aggregates most commonly used in gypsum plaster, which include perlite, sand (natural and manufactured), and vermiculite. Other aggregates may be employed, provided tests have demonstrated them to yield plaster of satisfactory quality.

2. Definitions

(a) Perlite Aggregate is a siliceous, volcanic glass properly expanded by heat.

(b) Sand Aggregate:

Natural Sand is the fine granular material resulting from the natural disintegration of rock or from the crushing of friable sandstone.

Manufactured Sand is the fine material resulting from the crushing and classification by screening, or otherwise, of rock, gravel or blast furnace slag.

(c) Vermiculite Aggregate is a micaceous mineral properly expanded by heat.

3. Grading

(a) Sieve analysis: See Table to 6.6.2.

(b) For natural or manufactured sand, not more than 50 per cent shall be retained between any two consecutive sieves, nor more than 25 per cent between the No. 50 and No. 100 sieves.

(c) For natural or manufactured sand, the amount of material finer than a No. 200 (74-micron) sieve shall not exceed 5 per cent.

4. Weight of Lightweight Aggregates

(a) The weight of perlite aggregate shall be not less than 7½ nor more than 15 lb. per cu. ft.

(b) The weight of vermiculite aggregate shall be not less than 6 nor more than 10 lb. per cu. ft.

ASTM C 847: Aggregate For Portland Based Plaster

1. Scope

These specifications cover aggregate for use with Portland Cement Plaster.

2. Description

Aggregate for use in portland cement plaster shall consist of natural sand or manufactured sand. Manufactured sand is the product obtained by crushing stone, gravel, or air-cooled iron blast-furnace slag.

3. Grading

(a) See Table to 6.6.4.

(b) The aggregate shall have not more than 50 per cent retained between any two consecutive sieves, nor more than 25 per cent between the No. 50 and No. 100 sieves.

(c) If the fineness modulus varies by more than .20 from the value assumed in selecting proportions for the mortar, the aggregate shall be rejected unless suitable adjustments are made in proportions to compensate for the change in grading.

1.8.2 Gypsum Materials

ASTM C 37: Gypsum Lath

1. Scope

These specifications cover plain, perforated, and insulating types of gypsum lath, designed to be used as a base for the reception of gypsum plaster.

2. Composition

Plain gypsum lath shall consist of an incombustible core, essentially gypsum, with or without fiber but not exceeding 15 per cent of fiber by weight, and surfaced with paper firmly bonded to the core.

3. Weight

The weight of gypsum lath per 1000 sq. ft. shall conform to the following requirements:

Thickness In.	Weight per 1000 sq. ft. (in lbs.)	
	Minimum	Maximum
3/8	1350	1800
1/2	1800	2500

4. Finish

The surfaces of gypsum lath shall be such that they will readily receive and retain gypsum plaster. The corners shall be square with a permissible variation of ¼ in. in the full width of the lath. Gypsum lath shall be free of cracks and imperfections that will render them unfit for use.

NOTE.—Insulating gypsum lath is intended to be installed with the foil surfacing inward, toward the framing members. The foil surface is not adapted to receive or retain plaster.

1. GENERAL REQUIREMENTS

ASTM C 28: Gypsum Plasters

Scope

These specifications cover five gypsum plasters: namely,

Gypsum ready-mixed plaster,
Gypsum neat plaster,
Gypsum wood-fibered plaster,
Gypsum neat plaster, Type R, and
Gypsum gauging plaster for finish coat.

Chemical and Physical Properties:

(a) Gypsum Ready-Mixed Plaster:

1. Description:

Gypsum ready-mixed plaster is calcined gypsum plaster, mixed at the mill with mineral aggregate and other ingredients to control working quality and setting time.

2. Composition:

Gypsum ready-mixed plaster shall contain not more than 3 cu. ft. of mineral aggregate per 100 lbs. of calcined gypsum plaster, to which may be added fiber and material to control working quality and setting time. However, when prepared for application to porous masonry bases, it may contain not more than 4 cu. ft. of mineral aggregate per 100 lbs. of calcined gypsum plaster.

3. Time of Setting:

Gypsum ready-mixed plaster shall have a time of set of not less than 1½ nor more than 8 hours.

4. Compressive Strength:

Gypsum ready-mixed plaster shall have a compressive strength of not less than 400 psi.

(b) Gypsum Neat Plaster:

1. Description:

Gypsum neat plaster is calcined gypsum plaster mixed at the mill with other ingredients to control working quality and setting time. Neat plaster may be fibered or unfibered. The addition of aggregate is required on the job.

2. Time of Setting:

Gypsum neat plaster, when mixed with three parts by weight of standard Ottawa sand, shall set in not less than 2 nor more than 15 hours.

3. Compressive Strength:

Gypsum neat plaster, when mixed with two parts by weight of standard Ottawa sand, shall have a compressive strength of not less than 750 psi.

(c) Gypsum Wood-Fibered Plaster.

1. Description:

Gypsum wood-fibered plaster is a gypsum plaster in which wood fiber is used as an aggregate.

2. Time of Setting:

Gypsum wood-fibered plaster shall set in not less than 1½ nor more than 16 hours.

3. Compressive Strength:

Gypsum wood-fibered plaster shall have a compressive strength of not less than 1200 psi.

(d) Gypsum Neat Plaster, Type R:

1. Description

Gypsum neat plaster is calcined gypsum mixed at the mill with other ingredients to control working quality and setting time. Neat plaster may be fibered or unfibered. The addition of aggregate is required on the job.

2. Time of Setting:

Gypsum bond plaster shall set in not less than 2 nor more than 9 hours.

(e) High Strength Gypsum Base Coat Plaster:

Structo-Base, complyimg with ASTM C 28, Federal specification SS-P-00402B, Type "II" with the added requirement of 2800 psi compressive strength.

(f) High Strength Gypsum Finish Coat Plaster Materials:

C 28, Federal specification SS-P-00402B, Type "V" with the added

(g) Gypsum Gauging Plaster for Finishing Coat:

1. Description:

Gypsum gauging plaster is prepared for mixing with lime putty for the finish coat. It may contain materials to control setting time and working quality.

2. Time of Setting:

Gypsum gauging plaster for the finish coat, when not retarded, shall set in not less than 20 nor more than 40 minutes, and, when retarded, shall set in not less than 40 minutes.

3. Compressive Strength:

Gypsum gauging plaster for finish coat shall have a compressive strength of not less than 1200 psi.

ASTM C 59: Gypsum Molding Plaster

1. Scope:

These specifications cover gypsum molding plaster, a material consisting essentially of calcined gypsum, for use in making interior embellishments and cornices, as gauging plaster, etc.

2. Time of Setting:

Gypsum molding plaster shall set in not less than 20 nor more than 40 minutes.

3. Compressive Strength:

Gypsum molding plaster shall have a compressive strength of not less than 1800 psi.

ASTM C 61: Keene's Cement

1. Scope:

These specifications cover Keene's cement, anhydrous calcined gypsum the set of which is accelerated by the addition of other materials.

2. Time of Setting:

Keene's cement shall set in not less than 20 minutes nor more than 6 hours.

3. Compressive Strength:

Keene's cement shall have a compressive strength of not less than 2500 psi.

1.8.3 Lime

NOTE.—Type S, special finishing hydrate, is differentiated from Type N, normal finishing hydrate, which is covered by the Specifications for Normal Finishing Hydrated Lime (ASTM Designation: C 6), in that a limitation on the amount of unhydrated oxides is specified for Type S hydrate only and the plasticity requirement for Type S hydrate may be determined after soaking for less than 16 hours.

ASTM C 206: Special Finishing Hydrated Lime

Scope:

These specifications cover one type of finishing hydrated lime which is suitable for use in the scratch, brown, and finish coats of plaster, for stucco, for mortar, and as an addition to portland cement concrete. Lime sold under thes specifications shall be designated Type S—special finishing hydrate.

1.8.4 Metal Lath and Wire Fabric Lath

QQ-L-101c—Federal Specifications for Lath, Metal, (and other Metal Plaster Bases).

Scope:

This specification covers metal plaster base for use in interior and exterior construction as plaster bases for interior plaster and exterior stucco.

Types: See paragraph 2.6.2

Requirements:

(a) Material.—All painted metal plaster bases shall be of copper alloy steel. Zinc-coated metal plaster bases shall be of low carbon steel.

(b) Fabrication.—Expanded metal plaster base shall be formed by slitting and then expanding sheet steel. Sheet metal lath shall be formed by perforating sheet steel. Wire base shall be made of wires which cross other wires and are securely attached or woven at all intersections. The ribs of the various rib laths shall be spaced not more than 8 inches on centers.

(c) Paper backing.—

Waterproofed paper. Waterproofed paper for metal plaster base with backings shall be either two sheets cemented together with asphalt, a single sheet impregnated with asphalt, or paper of these types having an asphalt coating on one side. The backing shall completely cover the metal base.

Absorptive paper. Absorptive paper shall not be waterproofed with asphalt, but shall be sized with rosin or starch. For base to receive gypsum, lime and Keene's cement plasters, absorptive paper usually is preferable to waterproofed paper. Metal shall extend at least ½ inch beyond the backing along one side and one end of each sheet.

Selvage. For welded wire fabric, there shall be in the selvage a metal member parallel with the edge of the backing.

(d) Bonding.—Metal plaster base shall provide a mechanical bond and continuous reinforcement, in at least two different directions, for plaster or stucco applied by ordinary plastering operations and it shall permit not less than one-half of the total weight of metal to be embedded fully with a covering of at least 1/16 inch at all points.

(e) Weight of base metal.—Weight of the base metal shall be in accordance with Table V. Weights of expanded-metal and sheet-metal plaster base are for painted or zinc coated material. The weights for base with backing include only the metal on the plaster side.

Weight tolerances. The allowable variations in percentage over and under the nominal weights of metal lath shall be in accordance with mill tolerances in ASTM A 366.

(f) Finish.—Metal plaster base shall be completely covered with a protective coat of paint, unless zinc-coated after fabrication, cut from zinc-coated sheet or fabricated from zinc-coated wire.

NOTE.—For wire gauge, weight and mesh size of woven wire fabric lath see 2.6.73.

1.8.5 Portland Cement

ASTM C 150: Portland Cement

Scope: See paragraphs 6.5.1 and 6.5.3

Basis of Purchase:

The purchaser should specify the type or types desired. When no type is specified, the requirements of Type I-II shall govern.

ASTM C 150: Air-Entraining Portland Cement

Scope: See paragraph 6.5.2

Basis of Purchase:

The purchaser shall specify the type or types. When no type is specified, the requirements of Type I-A shall govern.

ACI Proceedings Title 44-4: Crack Control in Portland Cement Plaster Panels

Method:

Provides unrestrained movement of portland cement plaster panels on wall or ceiling by (1) slip joint at juncture of ceiling and wall, (2) no metal reinforcement carried continuously around internal corners from wall to wall or from ceiling to wall or at other points of potential restraint. This method does away with "the 'restrained' method in which the ceiling and wall slab were fixed together as rigidly as possible with such restraint as was afforded by metal reinforcement and interlocking bond between the different coats of wall and ceiling plaster." With such restraint shrinkage stresses can find relief only in uncontrolled cracks. In the "unrestrained" method "mass movement of the panel is possible without a break or crack in the plaster," except at meeting points at which provision may be made to conceal or fill the controlled separation.

Summary and Conclusions:

The following conclusions have been drawn from the program of study and experimentation conducted by the Bureau of Reclamation of the causes of objectionable cracking in portland cement plaster, and from the results of the method of construction which has eliminated such cracking:

1. The high shrinkage coefficient of cement mortar or plaster is the basic element for consideration in crack eliminations.

Types and Forms	Weights, p.s.y. and openings		
	Small	Medium	Large
Expanded metal lath			
F	2.5, 3.4	1.8, 3.6	1.8, 3.6
SF	2.5, 3.4		
FR	2.75, 3.4		
F3/8R	3.4, 4.0		
F3/4R	0.60, 0.75ᵃ		
FB	2.5, 3.4		1.14, 1.83
SFB			
FRB	2.75, 3.4		
F3/8RB	3.4, 4.0		
Sheet metal lath			
F	4.5		
F3/8R	5.0, 6.3, 7.5 ribs space 8 inches		1.41
Woven or welded wire fabric			
F	2.00, 2.56, 3.26 2 meshes per inch 2.00, 2.56, 3.20, 4.06 2½ meshes per inch		
Welded wire fabric			
SF			
FB			1.14, 1.83
SFB			1.14, 1.83

TABLE V—NOMINAL WEIGHTS OF METAL BACKING

ᵃWeight of Type F3/4R is in p.s.f.

1. GENERAL REQUIREMENTS

2. Most of the objectionable cracking in portland cement plaster is caused by volume change due to shrinkage of the plaster as drying occurs.

3. This shrinkage cannot be avoided when portland cement is used. Since the shrinkage stresses are greater than the tensile strength of the plaster, provision must be made to allow a slab to take substantially its full shrinkage without restraint. A plaster slab of ordinary room size or larger can shrink away from its borders without cracking. If the slab is restrained along its borders or elsewhere, the shrinkage stresses will find relief by cracking. Subsequent shrinkage will tend to open the crack or cracks so formed.

4. The openings or joints formed along the borders of slabs so freed, or at junctures of panels so freed, can be concealed by the use of ornamental or plain beads or strips, by filling the openings with plastics or plaster, or by other means.

5. If drying conditions are properly regulated, the successive coats of portland cement plaster can be placed with complete success at intervals of not over 24 hours. If this is done, no coat need be damp-cured until the last coat is hard enough to receive water without damage. Thereafter, if too-rapid drying is prevented, considerably less water is required than has been customary in the past.

6. Precautions should be taken to provide for slow and uniform drying of the plaster, which will allow the plaster to gain strength and shrink uniformly.

7. The economic and other advantages of completing plaster in 3 to 6 days compared with the 10 to 20 days previously required where portland cement plaster is used, together with the savings effected in the cost of curing and protecting the plaster during the period of drying, are obvious.

Copy of this report—"Crack Control in Portland Cement Plaster Panels" by Bert A. Hall and printed as part of the Proceedings of the American Concrete Institute, in the ACI Journal, Vol. 19, No. 2, October 1947, may be secured by writing: American Concrete Institute, P.O. Box 19150, Detroit, Michigan 48219.

1.8.6 Miscellaneous Materials

UU-B-790 Federal Specification for Building Paper, Vegetable Fibre: (Kraft, Waterproofed, Water Repellent and Fire Resistant)

Scope:

This specification covers building papers composed predominantly of sulphate pulp fibers.

Classification:

The building papers shall be of the following types, grades and styles, as specified:

Type I—Barrier Paper.

Grade A—High water-vapor resistance.

Grade B—Moderate water-vapor resistance.

Grade C—Water resistant.

Grade D—Water vapor permeable.

Style 1—Uncreped, not reinforced.

Style 2—Uncreped, not reinforced, saturated.

Style 3—Creped, one direction, not reinforced.

Style 4—Uncreped, reinforced.

Style 5—Creped one direction, reinforced.

Style 6—Creped two directions, not reinforced.

Style 7—Creped two directions, reinforced.

Type III—Fire-resistant Paper.

Grade F—Water repellent.

GRADE REQUIREMENTS						
Physical Property Requirement	A	B	C	D	E	F
Dry tensile strength; minimum pounds per inch width: both directions	20¹	20¹	20¹	20		
Wet tensile strength; minimum pounds per inch width: machine direction					30	
cross direction					15	
Water resistance hours, minimum	24	16	8	1/6		
Specimen loss, maximum, grams per square inch of surface					0.055	
Weight increase, maximum, grams per square inch						0.025
Water-vapor permeability; grams per square meter per 24 hours: Maximum	4	6				
Minimum			35			
Bursting strength; minimum, points						70
Tearing resistance; minimum, grams						150

¹15-pound only for Style 6; 35-pound may be specified for Styles 1, 2, 3, 4, 5 and 7.

GLOSSARY

NOTE: *Listings ending with (ASTM) were taken from ASTM standards.*

ACCELERATOR — Any material added to gypsum plaster which speeds up the natural set.

ACCESSORIES

Accessories — Linear formed metal, metal and paper, or plastic members fabricated for the purpose of forming corners, edges, control joints, or decorative effects in conjunction with gypsum board and plaster assemblies. (ASTM)

Arch Corner Bead — A corner bead so designed that it can be job-shaped for use on arches.

Arches, Metal — See METAL ARCH.

Base Screed — See SCREEDS.

Casing Bead — Sometimes called a plaster stop, this bead is used where plaster is discontinued; around openings, thus providing a ground; where the plaster adjoins another material; and to form the perimeter of a plaster membrane or panel.

Clips:

Beam Clip — A formed wire section used to attach lath to flanges of steel beams.

Casing Clip — A formed metal section which puts pressure on a casing bead to assure rigid positioning.

Clip for Control of Movement — A flexible, resilient metal section separating the plaster membrane from supports to reduce plaster cracking due to structural movement (and to reduce sound transmission).

Corner Bead Clip — A metal section used, where necessary, to provide an extension for attachment of various types of corner beads.

End Clip — A metal section used to secure ends and edges of gypsum lath.

Furring Clip — A metal section for attaching cross furring to main runners.

Lath Clip (generic) — A metal section to secure lath to supports.

Masonry Wall Clip — See Wall Furring Base Clip.

Metal Base Clip — A formed metal section to which is attached metal base for partitions or walls.

Metal Lath Clip — A formed wire section for fastening metal lath to flanges of steel joists.

Sound Transmission Clip — A flexible, resilient metal clip used to decrease sound transmission through partition and floor assemblies. (Also serves to lessen plaster cracking resulting from structural movement.)

Starter Clip — A metal section used at floor, or initial course of gypsum lath.

Wall Furring Base Clip — A formed metal section used to attach metal base to furred walls.

Concealed Picture Mold — See SCREEDS.

Control Joint (expansion) — A formed metal section limiting the areas of unbroken plaster surfaces to minimize possible cracking due to expansion, contraction, and initial shrinkage in portland cement plaster.

Corner Bead — See METAL CORNER BEAD.

Corner Reinforcement, Exterior — A metal section, usually shaped of wire, for the reinforcement of exterior plaster arrises.

Corner Reinforcement, Interior — Flat or shaped reinforcing units of metal or plastic mesh. See CORNERITE.

Cornerite — Corner reinforcement for interior plastering where the plaster base is not continuous around an internal corner or angle.

Cored Tile or Block — See GYPSUM TILE OR BLOCK. (ASTM)

Fineness Modulus — An empirical factor obtained by adding total percentages of a sample of aggregate retained on each of a specified series of sieves and dividing by 100. The sieve sizes used are: No. 100 (150 μm), No. 50 (300 μm), No. 30 (600 μm), No. 16 (1.18 mm), No. 8 (2.36 mm), No. 4 (4.75 mm), ⅜ in. (9.5 mm), ¾ in. (19.0 mm), 1½ in. (38.1 mm) and larger, increasing in the ratio of 2 to 1. (ASTM)

Inserts:

Furring — A formed metal section which is inserted in concrete or masonry walls for the attachment or support of wall furring channels. See STICKER.

Hanger — A formed metal section which is inserted in concrete members for the attachment of hangers.

Metal Arch — A sheet steel formed arch for use as base (lath) or corner reinforcement at arched openings in partitions.

Metal Corner Bead — Fabricated metal with flanges and nosings at juncture of flanges; used to protect or form arrises.

Metal Partition Base — A fabricated integral metal section which also may serve as a ground for the plaster (attached to framing member or masonry).

Partition Cap — A formed metal section for use at the end of a free-standing solid partition to provide protection of plaster; also used as a stair rail cap, mullion cover, light cove cap, etc.

Picture Mold — See SCREEDS.

Screeds, Metal — See SCREEDS.

Shoe — A formed metal section used in attaching metal studs to floor and ceiling tracks, also the end section of a channel turned to an angle (usually 90°) to permit attachment, generally to other channels.

Track:

Ceiling Track, Ceiling Runner Track or Ceiling Runner — A formed metal section, anchored to the ceiling, into which metal studs for hollow or solid partitions are set; a formed metal section to which lath is attached for studless partitions; a metal channel or angle used for anchoring the partition to the ceiling.

Floor Track, Floor Runner Track or Floor Runner — A formed metal section, anchored to the floor into which metal studs for hollow or solid partitions are set; a formed metal section into which lath is inserted for studless partitions; a wood member into which lath is inserted for studless partitions; a metal channel used for anchoring the partition to the floor.

ACOUSTICAL PLASTER — A finishing plaster designed to correct sound reverberations, or reduce noise intensity.

ADDITIVE — An admixture which is added to a product at the mill during manufacture. See ADMIXTURE.

ADMIXTURE — A material other than water, aggregate or basic cementitious material that is used as an ingredient of plaster and is added to the batch immmediately before or during its mixing for the purpose of improving flow and workability or imparting particular qualities to the mortar.

AGGREGATE — Inert material, used as a filler for mixing with any cementing material. The word used with or in connection with plastering usually means sand, vermiculite or perlite.

AIR ENTRAINMENT — Intentionally introducing into portland cement plaster in its plastic state a controlled number of minute disconnected air bubbles well distributed throughout the mass to improve flow and workability, or to improve other desired characteristics in the mortar.

ALPHA GYPSUM — See GYPSUM.

ALLIGATOR CRACKS — See CHECK CRACKS.

ALL-PURPOSE COMPOUND — A joint treatment compound that can be used as a bedding compound for tape, a finishing compound, and as a laminating adhesive or texturing product.

ANGLE IRON — Metal section sometimes used as main runners in lieu of channels.

ANHYDRITE — The mineral consisting primarily of anhydrous calcium sulfate, $CaSO_4$. (ASTM)

ARCH — Curved top of opening. Structurally designed to carry the overhead load to the side members.

ARCHES, METAL — See Metal Arches under ACCESSORIES.

AREA — The plane surface within certain boundaries.

ARRIS — A sharp edge, forming an external corner at the junction of two surfaces.

ATOMIZER — Device by which air is introduced into material at the nozzle to regulate the texture of machine-applied plaster.

AUTOCLAVED LIME — See LIME.

BAND — A flat moulding.

BANDING — Metal or plastic strapping to secure bundles of gypsum products together in a shipping unit.

BARREL CEILING — A rounded or semi-circular ceiling.

BASE COAT — The plaster coat or combination of coats applied previous to the finish coat.

BASE COAT FLOATING — The finishing act of spreading, compacting, and smoothing of the base coat plaster to a reasonably true plane. (ASTM)

BASE BEAD — See SCREEDS.

BASE SCREED — See SCREEDS.

BEAD — A small round member of a cornice.

BEADED MOULDING — A cast plaster string of beads planted in a moulding or cornice.

BED MOULD OR BED — A flat area in a cornice, designed to have enrichment planted later.

BEDDING COAT — That coat of plaster to receive aggregate or other decorative material of any size, impinged or embedded into its surface, before it sets. (ASTM)

BENCH MARKS — Identification symbols· from which differences in elevations are measured.

BEVEL — A slanted surface.

BINDER — See CEMENTITIOUS MATERIAL.

BLISTERS — Protuberances on the finish coat of plaster caused by application over too damp a base coat, or troweling too soon.

BOND

Chemical Bond — A term, used to describe adherence of one plaster to another or to the base, which implies formation of interlocking crystals or fusion between the coats or to the base.

Mechanical Bond — A term used to describe the physical keying of one plaster coat to another or to the plaster base.

BOND PLASTER — See GYPSUM PLASTER.

BOSS — A Gothic ornament planted at the intersection of mouldings.

BRACKET — A superficial structure usually in angles forming a frame to support lath. Its main purpose is to save material and weight in ornaments or cornices.

BREAK — An interruption in the continuity of a plastered wall or cornice.

BRIDGING — A section sized to fit inside the flanges of studs and channels to stiffen construction.

BROWN COAT — Coat of plaster directly beneath the finish coat. In two-coat work, brown coat refers to the basecoat plaster applied over the lath. In three coat work, the brown coat refers to the second coat applied over a scratch coat. Brown coats are applied with a fairly rough ·surface to receive the finish coat.

BROWN OUT — To complete application of basecoat plastering.

BUCKLES — Raised or ruptured spots which eventually crack, exposing the lath beneath. Most common cause for buckling is application of plaster over dry, broken or incorrectly applied wood lath.

BULL NOSE — This term describes an external angle which is rounded in order to eliminate a sharp corner. Used largely at window returns and door frames. Advantages are ease of cleaning and its durability. Can be made by running with plaster or obtaining a bull nose corner bead with the proper radius.

BUTTERFLIES — Color imperfections on a lime putty finish wall. Large varieties which smear out under pressure of the trowel. Caused by lime lumps not put through a screen; insufficient mixing of the gauging.

BUTTERFLY REINFORCEMENT — Strips of metal reinforcement placed diagonally over the plaster base at the corners of openings before plastering.

BUTTERFLY TIE — See TIES.

CAISSON — A panel sunk below the normal surface in flat or vaulted ceilings.

CALCIUM SULFATE — The chemical compound $CaSO_4$. (ASTM)

CALCINE, CALCINING — To make powdery or to oxidize by removing chemically combined water by action of controlled heat.

CAPITAL OR CAP — The ornamental head of a column or pilaster.

CARRYING CHANNELS — See CHANNELS, CARRYING.

CASE MOULD — Plaster shell used to hold various parts of a plaster mould in correct position. Also used with gelatin and wax moulds to prevent distortions during pouring operation.

CASING BEAD (Sometimes referred to as PLASTER STOP) — See ACCESSORIES.

CASTING PLASTER — See Casting Plaster under GYPSUM.

CASTS — Finished product from a mould. Sometimes referred to as staff. Used generally as enrichments and stuck in place.

CATFACE — Flaw in the finish coat comparable to a pock mark. In some regions basecoat knobs showing through the finish coat are referred to by this term.

CAULK (Sound Caulking) (Finish Caulking) — To seal small openings such as windows and doors in wall or ceiling systems to prevent leakage of sound or to affect a finished appearance and seal between dissimilar materials.

CEILING RUNNER — See Track under ACCESSORIES.

CEILING TRACK — See Track under ACCESSORIES.

CEILINGS — Contact, Furred and Suspended.

(a) **Contact,** as applied to ceiling construction, means that the lath is attached in direct contact with the construction above, without use of runner channels or furring.

(b) **Furred** ceiling construction means that the furring members are attached directly to the structural members of the building.

(c) **Suspended** ceiling means that the furring members are suspended below the structural members of the building.

(d) **Cross Furring** means the furring members are attached at right angles to the underside of main runners or other structural supports.

(e) The term **main runners** denotes the metal channels which are attached to or suspended from the structure above for the support of cross furring.

CEMENT — A material or mixture of materials which, when in a plastic state, possesses adhesive and cohesive properties and which will set in place. Note: The word cement is used without regard to the composition of the materials.

CEMENT PLASTER — (See GYPSUM PLASTER and PORTLAND CEMENT PLASTER).

CEMENTITIOUS MATERIAL — Material binding aggregate particles together into a heterogeneous mass.

CHALK LINE — A straight working line made by snapping a chalked cord between two points.

CHAMFER — A beveled corner or edge.

CHANNELS — Hot or cold-rolled steel used for furring, studs, and in suspended ceilings. Sizes vary according to requirements.

CHANNELS, CARRYING — The heaviest integral supporting member in a suspended ceiling. Carrying channels, or main runners, are supported by hangers attached to the building structure and in turn support various grid systems and/or furring channels or rods to which lath is fastened.

CHANNELS, FURRING — The smaller horizontal member of a suspended ceiling, applied at right angles to the underside of carrying channels and to which lath is attached; the smaller horizontal member in a furred ceiling; in general, the separate members used to space lath from any surface member over which it is applied.

CHASE — A groove in a masonry wall to provide for pipes, ducts or conduits.

CHASE WALL — A partition to enclose mechanical and plumbing systems.

CHECK CRACKS — Cracks in plaster caused by shrinkage, but still bonded to its base.

CHEMICAL BOND — See BOND.

CHIP CRACKS — Similar to check cracks, except the bond has been partially destroyed, causing eggshelling. Sometimes referred to as fire cracks, map cracks, crazing and fire checks, as well as hair cracks.

CLIPS — Special, sometimes patented, devices used to attach various kinds of lath to steel supports.

COAT — A thickness or covering of plaster. See SCRATCH COAT, BROWN COAT, FINISH COAT.

COATING — A layer of finish material which is applied to a surface to decorate, preserve, protect, seal the substrate, or to bridge cracks, which may be sprayed or hand-applied.

COFFERED CEILINGS — Ornamental ceilings made up of sunken or recessed panels.

COMPRESSIVE STRENGTH — The maximum load sustained by a standard specimen of material, when subjected to a crushing force.

COMBINED WATER — The water, chemically held as water of crystallization, by the calcium sulphate dihydrate, or hemihydrate crystal.

CONCAVE — A curved or vaulted surface; the opposite of convex.

CONCEALED PICTURE MOLD — See SCREEDS.

CONSISTENCY (NORMAL) – The number of millilitres of water per 100 g of gypsum plaster or gypsum concrete required to produce a mortar or a slurry of specified fluidity. (ASTM)

CONTACT CEILING — See CEILINGS.

CONTACT FIREPROOFING — See FIREPROOFING.

CONTROL JOINT — See Control Joint under ACCESSORIES.

CONVEX — A protruding rounded surface.

CORE — The gypsum structure between the face and back papers of gypsum board.

CORED TILE or BLOCK – See GYPSUM TILE or BLOCK. (ASTM)

CORNER BEAD — Fabricated metal with flanges and bead at junction of flanges; used to protect arrises.

CORNERITE — Corner reinforcement for interior plastering where metal lath is not continuous.

CORNICE — A moulding with or without reinforcement.

COVE — A curved concave surface.

CRAZE CRACKS — See CHECK CRACKS.

CROSS FURRING — The smaller horizontal members attached at right angles to the underside of main runners or other structural supports. See CHANNELS, FURRING: See FURRING.

CROSS SCRATCHING — See also SCORING. Scratching in two directions of the plaster scratch coat, to provide mechanical bond for the brown coat.

CURE (PORTLAND CEMENT PLASTER OR STUCCO)–(1) To provide conditions conductive to the hydration process of portland cement plaster or stucco.
(2) To maintain proper temperature and a sufficient quantity of water within the plaster to ensure cement hydration. (ASTM)

CURTAIN WALL — A non-load bearing exterior wall supported by the structural elements of a building engineered to withstand designated wind loads.

CUT END — The end of the gypsum lath with the exposed core.

DADO — The lower part of a wall usually separated from the upper by a a moulding or other device.

DARBY — A flat wooden tool, with handles, about 4″ wide and 42″ long; used to smooth or float the brown coat; also used on finish coat to give a preliminary true and even surface.

DASH COAT — See FINISH COAT.

DASH-BOND COAT — A thick slurry of portland cement, sand and water dashed (thrown or blown) on concrete or masonry surfaces by hand or by machine to provide a mechanical bond for the succeeding plaster coat.

DAUB — A glob of adhesive.

DEAD BURNED — Removal of all water content during calcining of gypsum.

DEAD LOAD — The part of the total building load contributed by the structural building elements and materials.

DECIBEL — A unit measure of sound intensity which can be used in expressing sound volume or loudness.

DEFLECTION — The displacement that occurs when a load is applied to a member or assembly.

DEFORMATION — The change in shape of a body brought about by the application of a force. Deformation is proportional to the force within the elastic limits of the material.

DELAMINATION — The separation of the paper plies or surface coverings.

DENSITY – The weight per unit volume of a material. (ASTM)

DENTILS — Architectural terms for small rectangular blocks which are often planted in a series in the bed mould of a cornice.

DEW POINT — The temperature at which air becomes saturated with moisture and below which condensation occurs.

DOPE — Term used by plasterers for additives made to any type of mortar to either accelerate or retard its set.

DOT — A small projection of basecoat plaster placed on a surface and faced out between grounds to assist the plasterer in obtaining the proper plaster thickness and surface plane; occasionally pieces of metal or wood applied to plaster base at intervals as spot grounds to gauge plaster thickness. See SCREEDS, PLASTER.

DOUBLE-UP — When plaster is applied in successive operations without a setting and drying interval between coats.

DRY OUT — Soft, chalky plaster caused by water evaporating before setting.

DRYWALL — See GYPSUM WALL-BOARD.

EFFLORESCENCE — White, fleecy deposit found on the face of plastered walls. Caused by salts in the sand or backing. The process of efflorescing is also referred to as "whiskering" or "salt-petering".

EGG AND DART — Enrichments frequently used in cornices. The design is that of an oval and a dart used alternately. It is said the egg represents life and the dart death.

EGGSHELLING — Refers to the condition of chip-cracked plaster, either base or finish coat. The form taken is concave to the surface and the bond is partially destroyed.

ELASTOMER — Any macromolecular material (such as rubber or a synthetic material having similar properties) that returns rapidly to approximately the initial dimensions and shape after substantial deformation by a weak stress and release of stress.

ELASTOMERIC — Said of any material having the properties of an elastomer, as a roofing material which can expand and contract without rupture.

ELEVATION — Drawing of the interior or exterior vertical sections or sides of a building showing heights, widths, etc.

ENRICHMENTS — Any cast ornament which cannot be executed by a running mould.

EXPANDED METAL — Sheets of metal which are slit and drawn out to form diamond shaped openings. This is used as a metal reinforcing for plaster and termed "metal lath".

EXPANSION JOINT — See CONTROL JOINT.

FACADE — Front elevation of a building.

FACE — The surface designed to be left exposed to view.

FAT — Material accumulated on the trowel during the finishing operation and used to fill in small imperfections. Also a term to describe working characteristics of any type mortar.

FEATHER EDGE — A beveled edge wooden tool, varying in length, used to straighten angles in the finish coat.

FENCE — Term used in cast shops describing a wall of plaster or clay placed around model before pouring material to make the mould.

FIBER

a. Sisal or glass mill-additives to gypsum plaster.

b. Mill-added grained or shredded non-staining wood fiber aggregates in gypsum plaster. Provides fire-rated and dense hard gypsum plaster.

c. Long-length sisal fiber used to affix and reinforce cast ornamental plaster.

FINE AGGREGATE — Sand or other inorganic aggregate for use in plastering. See ASTM C 35 for gypsum plaster and C 144 for portland cement plaster.

FINES — Aggregate particles with a high percentage passing the No. 200 sieve.

FINISH COAT — The final layer of plaster applied over a basecoat or other substrate. (ASTM)

FINISH COAT FLOATING — The finishing act of spreading, compacting and smoothing the finish coat plaster or stucco to a specified surface texture. (ASTM)

FIRE BLOCKING — Intermittent solid cross-framing to retard the spread of flame within the framing cavity.

FIRE-RESISTANCE CLASSIFICATION — A standard rating of fire-resistance and protective characteristics of a building construction or assembly. (ASTM)

FIREPROOFING

Contact — The application of fire-resistive material direct to structural members to protect them from fire damage.

Membrane — A lath and plaster system which is separated from the structural steel members, in most cases by furring or suspension, to provide fire-proofing.

FLASHINGS — Strips of sheet metal or building paper used to waterproof around openings, roof penetrations, etc.

FLAME SPREAD CLASSIFICATION — A standard rating of relative surface burning characteristics of a building material as compared to a standard material. (ASTM)

FLEXURAL STRENGTH — The maximum load sustained by a standard specimen of a sheet material when subjected to a bending force. (ASTM)

FLOAT — A tool shaped like a trowel with handle braced at both ends and wood base or blade; of cork or felt attached to wood, wood or rubber, or formed plastic.

FLOATING

Base Coat Floating — The act of spreading, compacting and smoothing plaster to a reasonably true plane on exterior and interior surfaces.

Finish Coat Floating — The act of bringing the aggregate to the surface to produce a uniform texture.

FLOOR RUNNERS (TRACK) — See Track under ACCESSORIES.

FOAM PLASTER BASE (RIGID TYPE) — A rigid type foamed backing which acts as a plaster base.

FOIL BACK — A gypsum board with a reflective aluminum foil composite laminated to the back surface.

FOIL-BACKED GYPSUM WALL-BOARD — A gypsum wallboard with the back surface covered with a continuous sheet of pure bright finished aluminum foil. (ASTM)

FORCED ENTRY FASTENERS — See POWER DRIVEN FASTENER.

FOUNDATION SCREED — See SCREEDS.

FRAMING — Wood or metal members, such as studs, joists, headers, etc., to which lath is applied.

FREE WATER — All water contained by gypsum board or plaster in excess of that chemically held as water of crystallization. (ASTM)

FRESCO — An art or decorative method consisting of applying a water-soluble paint to freshly spread plaster before it dries.

FURRED CEILING — See CEILINGS.

FURRING — Wall or ceiling construction beyond or below the normal surface plane; designates also the members used in such construction.

FURRING CHANNELS — See CHANNELS, FURRING.

GAUGING — The process of mixing gauging plaster or Keene's cement with lime putty or Type S hydrated lime to control setting time.

GAUGING PLASTER — Specially ground gypsum plaster that mixes easily with lime putty and Type S hydrated lime; available in fast or slow setting formulations.

GELATIN — A product of the packing house, which can be cast into a semi-rigid mould. On account of its flexibility, it is particularly adaptable to moulds containing undercuts, etc.

GESSO — A composition of gypsum plaster whiting, and glue, used as a base for decorative painting.

GLAZING — A condition created by the fines of a machine-dash texture plaster traveling to the surface and producing a flattened texture and shine or discoloration. This may be caused by the basecoat being too wet or the acoustical mortar being too moist. Glazing occurs in hand application when mortar being worked is excessively wet.

GLITTER — A reflective material such as glass, diamond dust or small pieces of variously colored aluminum foil projected into the surface of wet plaster or paint as a decorative treatment.

GRADATION — The particle size distribution of aggregate as determined by separation with standard screen.

GREEN — Wet or damp plaster.

GRILLAGE — A framework composed of main runner channels and furring channels to support ceilings.

GROUNDS — A piece of wood or metal attached to the framing with its exposed surfaces acting as a gauge to determine the thickness of plaster to be applied. Also used by carpenter as a nailing base to support trim.

GROUT — Gypsum or portland cement plaster mortar used to fill crevices or to fill hollow metal frames.

GUSSET — A wood or metal plate affixed over joints (such as truss members) to transfer stresses between members.

GYPSUM

Alpha Gypsum — A term denoting a class of specially processed calcined gypsum having properties of low consistency and high strength.

Calcium Gypsum — A dry powder; primarily calcium sulfate hemihydrate, resulting from calcination of gypsum; cementitious base for production of most gypsum plasters; also called plaster of paris; sometimes called stucco. (ASTM)

Calcium Sulfate – The chemical compound $CaSO_4$. (ASTM)

Casting Plaster — A fast-setting gypsum plaster. See also GYPSUM MOLDING PLASTER.

Edge (of Gypsum Board) – The paper-bound edge as manufactured. (ASTM)

End (of Gypsum Board) – The end perpendicular to the paper-bound edge. The gypsum core is always exposed. (ASTM)

Featured Edge – An edge configuration of the paper bound edge of gypsum board that provides special design or performance. (ASTM)

Gypsum – The mineral consisting primarily of fully hydrated calcium sulfate, $CaSO_4 \cdot 2H_2O$ or calcium sulfate dihydrate (C 22). (ASTM)

Gypsum Backing Board – A ¼ in. to ⅝ in. gypsum board for use as a backing for gypsum wallboard, acoustical tile or other dry cladding. (ASTM)

Water Resistant Gypsum Backing Board – A gypsum board designed for use on walls primarily as a base for the application of ceramic or plastic tile. (ASTM)

Gypsum Board – The generic name for a family of sheet products consisting of a noncombustible core primarily of gypsum with paper surfacing. (ASTM)

Gypsum Concrete – A calcium gypsum mixed with wood chips, or aggregate, or both, used primarily for poured roof decks (C 317). (ASTM)

Gypsum Core Board – A ¾ in. (19.0 mm) to 1 in. (25.4 mm) gypsum board consisting of a single board or factory laminated multiple boards used as a gypsum stud or core in semisolid or solid gypsum board partitions. (ASTM)

Gypsum Formboard – A gypsum board used as the permanent form for poured gypsum roof deck (C 318). (ASTM)

Gypsum Gauging Plaster – A plaster for mixing with lime putty to control the setting time and initial strength of the finish coat. Classified either as quickset or slowset.

Gypsum High Strength Basecoat Plaster – A gypsum cement for use with sand aggregate to achieve high compressive strength plaster.

Gypsum Molding Plaster — A specially formulated plaster used in casting and ornamental plasterwork; may be used neat or with lime.

Gypsum Neat Plaster — A plaster requiring the addition of aggregate on the job. It may be unfibered or fibered (vegetable, or glass fibers).

Gypsum Plaster — Ground calcined gypsum combined with various additives to control the set; used also to denote applied gypsum plaster mixtures.

Gypsum Ready-Mixed Plaster — A term denoting a plaster which is mixed at the mill with a mineral aggregate. It may contain other ingredients to control time of set and working properties. Similar terms are mill-mixed and pre-mixed. Only the addition and mixture of water is required on the job.

Gypsum Sheathing – A gypsum board used a backing for exterior surface materials, manufactured with water-repellent paper and may be manufactured with a water-resistant core. (ASTM)

Gypsum Tile or Block – A cast gypsum building unit. (ASTM)

Gypsum Wallboard – A gypsum board used primarily as an interior surfacing for building structures. (ASTM)

Gypsum Wood-Fibered Plaster — A mill-mixed plaster containing a small percentage of wood fiber as an aggregate, used for fireproofing and high strength.

Low Consistency Plaster — A neat (unfibered) gypsum basecoat plaster especially processed so that less mixing water is required than in standard gypsum basecoat plaster to produce workability. This type plaster is particularly adapted to machine application.

Purity – The percentage of $CaSO_4 \cdot \tfrac{1}{2}H_2O$ in the calcined gypsum portion of a gypsum plaster or gypsum concrete, as defined by Specification C 28, for Gypsum Plasters. The percentage of $CaSO_4 \cdot 2H_2$ in the gypsum or the gypsum portion of fully hydrated, dry, set, gypsum plaster (C 471, C 28). (ASTM)

Ready-mixed Plaster – A calcined gypsum plaster with aggregate added during manufacture. (ASTM)

Setting Time – The elapsed time required for a gypsum plaster to attain a specified hardness and strength after mixing with water. (ASTM)

Synthetic Gypsum – A chemical product, consisting primarily of calcium sulfate dihydrate ($CaSO_4 \cdot 2H_2O$) resulting primarily from an industrial process. (ASTM)

Tapered Edge – An edge formation of gypsum board which provides a shallow depression on the paper-bound edge to receive joint reinforcement. (ASTM)

Veneer Plaster – A specially formulated high-strength plaster for thin-coat application to large-size veneer plaster lath.

Veneer Plaster Base – A gypsum board used as the base for application of a gypsum veneer plaster. (ASTM)

Water-resistant Core – A gypsum board specially formulated to resist water penetration. (ASTM)

GYPSUM LATH

Foil-backed – The same as plain gypsum lath except that in addition, the back surface shall be covered with a continuous sheet of pure bright finished aluminum foil.

Insulating – Same as plain lath except that the back surface is covered with a continuous sheet of pure bright-finished aluminum foil.

Lead-back Lath – Plain gypsum lath to which sheets of lead have been laminated; for shielding from X-rays and other radiation.

Perforated – Same as plain lath except that it has perforations not less than ¾ inch in diameter, with one perforation for not more than each 16 square inches. (No longer manufactured.)

Plain – A sheet or slab having an incombustible core, essentially gypsum, surfaced with paper suitable to receive gypsum plaster. One face may be variously treated; such as mechanical pricking or indenting, or impregation with a catalyst.

Round Edge – A rounded, paper-bound edge formation on gypsum board, commonly used for gypsum lath. (ASTM)

Type "X" – Same as plain lath, except that the core has increased fire-retardant properties to improve its fire-resistive rating.

Veneer Plaster Lath – A large-size base for veneer plasters having an incombustible core, essentially gypsum, surfaced with a special face paper suitable to receive veneer plaster.

GYPSUM PLASTER – See GYPSUM.

GYPSUM OF STRENGTH BASECOAT PLASTER – (See)

HACK – To cut back and roughen a plastered or other surface.

HANGERS – The vertical members which carry the steel framework of a suspended ceiling; or the vertical members which support furring under concrete joist construction; or the wires used in attaching lath directly to concrete joist construction.

HARDENING – The gain of strength of a plastered surface after setting. See SET.

HARDWALL – The term used for base coat plaster. Regionally the term differs; in some cases it refers to sanded plaster, while in others to neat.

HAWK – A flat wood or metal tool 10" to 14" square, with a handle used by the plasterer to carry plaster mortar.

HEADER – A horizontal framing member across the top of a door or window opening in a wall.

HOG RING – A heavy galvanized wire staple applied with a pneumatic gun which clinches it in the form of a closed ring around stud, rod, pencil rod or channel.

HYDRATED LIME – See LIME.

HYDRATION – In portland cement plaster the chemical reaction between water and the cementitious binder, which may be accompanied by a volumetric change or shrinkage.

HYDRAULIC CEMENT – Any cement, such as portland cement, which will set and harden under water. Also refers to a quick-setting expansion-type cement compound which is used to fill cracks and to waterproof.

IMPACT INSULATION CLASSIFICATION (IIC) – A numerical measure of sound transmission through Floor/Ceiling assemblies.

INTERIOR STUCCO – A term (often used interchangeably with the term "interior plaster") designating a finish plaster for walls and ceilings finishing smooth or textured. It is a mechanically blended compound of Keene's cement, lime (Type "S") and inert fine aggregate. Color pigment may be added to produce integrally colored interior stucco. See STUCCO.

JAMB – An upright finished member forming the side of an opening.

JOINING – Point where two mixes on same surface meet.

JOINT PHOTOGRAPHING – The shadowing of the finished joint areas through the surface decoration.

JOINT REINFORCING METAL – Strips of expanded metal, woven or welded wire mesh used to reinforce corners and other areas of plaster and lath. (ASTM)

JOINT REINFORCING TAPE – A type of paper, metal, fabric, glass mesh, or other material, commonly used with a cementitious compound, to reinforce the joints between adjacent gypsum boards. (ASTM)

JOIST – Structural member (wood, steel, other) used as floor, ceiling, or roof framing members.

KEENE'S CEMENT – An anhydrous gypsum plaster characterized by a low mixing water requirement and special setting properties, primarily used with lime to produce hard, dense finish coats. (ASTM)

KEY – The grip or mechanical bond of one coat of plaster to another coat or to a plaster base. It may be accomplished physically by the penetration of wet mortar or crystals into paper fibers, perforations, scoring irregularities, or by the embedment of the lath. See Mechanical Bond under BOND.

LAMINATION – The application of two or more layers of gypsum board.

LASER – A low intensity rotating columnated (relatively non-divergent) light beam construction tool used to establish accurate sightings, plumb and level alignments, vertical and horizontal.

LATH – A base to receive plaster. It is generally secured to framing or furring members. There are five types: gypsum, insulating, metal, wood and wire.

LIGHTWEIGHT AGGREGATE – See VERMICULITE, PERLITE, PUMICE.

LIME – Oxide of calcium produced by burning limestone. Heat drives out the carbon dioxide, leaving calcium oxide, commonly termed "quicklime". Addition of water to quicklime results in chemical changes which are thereafter known as hydrated or slaked lime.

LIME PLASTER – An interior basecoat plaster containing lime, aggregate and sometimes fiber. Lime basecoat plaster is slow-setting. Follow regional

or area practice. It should not be applied to gypsum lath. See PORTLAND CEMENT-LIME PLASTER.

LINE WIRE – See STRING WIRE.

LIVE LOAD – That part of the total load on structural members that is not a permanent part of the structure. It may be variable, as in the case of loads contributed by the occupancy, wind, or snow loads.

LOAD-BEARING PARTITION – A vertical structural interior wall supporting an integral part of the construction above.

LOW CONSISTENCY PLASTER – See GYPSUM.

MACHINE or PUMP – See PLASTERING MACHINE.

MACHINE DIRECTION – The direction parallel to the paper-bound edge of the gypsum board. (ASTM)

MAIN RUNNERS – The heaviest integral supporting member in a suspended ceiling. Main runners (or carrying channels) are supported by hangers attached to the building structure and in turn support furring channels or rods to which lath is fastened.

MARBLE CHIPS – Graded aggregate of maximum hardness made from crushed marble to be thrown or blown onto a soft plaster bedding coat to produce marblecrete.

MARBLECRETE – See FINISH COAT.

MAREZZO – An imitation marble formed with Keene's cement to which colors have been added. Cast on smooth glass or marble beds.

MASKING – Method of affixing paper, plastic or any flexible protective material or coating to protect adjacent work. Particularly used in plastering machine applications.

MASONRY CEMENT – See Standard Specifications for Masonry Cement – ASTM Designation C 91.

MASTIC – A general term usually referring to high viscosity solvent-based adhesives.

MECHANICAL APPLICATION – Application of plaster mortar by mechanical means, generally pumping and spraying, as distinguished from hand placement.

MECHANICAL BOND – See Mechanical Bond under BOND.

MECHANICAL TROWEL – A motor-driven tool with revolving blades used to produce a denser finish coat than by hand troweling.

MEMBRANE FIREPROOFING – See FIREPROOFING.

METAL ARCH — See Metal Arch under ACCESSORIES.

METAL LATH—EXPANDED—Metal lath is of two types, designated as diamond mesh (also called flat or self furring expanded metal lath) or rib. Metal lath is slit and expanded or slit, punched, or otherwise formed, with or without partial expansion, from plain or galvanized steel coils or sheets. Metal lath is coated with rust-inhibitive paint after fabrication, or is made from galvanized sheets.

a. **Diamond Mesh or Flat Expanded Metal Lath** — This term is used to indicate a metal lath slit and expanded from metal sheets or coils into such a form that there will be no rib in the lath; 2.5 lbs., painted; 3.4 lbs., painted or galvanized.

b. **Self-Furring Metal Lath** — A metal so formed that portions of it extend from the face of the lath so that it is separated at least ¼" from the background to which it is attached; painted or galvanized.

c. **Flat Rib Metal Lath** — A combination of expanded metal lath and ribs in which the rib has a total depth of approximately ⅛", measured from top inside of the lath to the top side of the rib, painted.

d. **⅜" Rib Metal Lath** — A combination of expanded metal lath and ribs of a total depth of approximately ⅜", measured from top inside of the lath to the top side of the rib; 3.4 lbs., painted or galvanized; 4.0 lbs., painted.

e. **¾" Rib Metal Lath** — A combination of expanded metal lath and ribs of a total depth of approximately ¾", measured from the top inside of the lath to the top side of the rib, painted.

f. **Paper-Backed Metal Lath** — A factory-assembled combination of any of the preceding defined types of metal lath with paper or other type backing, the assembly being used as a plaster base.

MILDEW — A fungus growth occurring in insufficiently vented and damp surface areas.

MILL-MIXED — See GYPSUM.

MITER — A joint formed by two pieces of material cut to meet at an angle.

MITRE — The joining of two mouldings at their angles.

MODEL — The original from which a mould or copy is made.

MOIST CURE — See CURING.

MORTAR — A material used in a plastic state which becomes hard or set in place. This term is used without regard to the composition of the material

or its specific use or the method of application.

MOULD — Moulds used by plasterers are two in number, namely, running and casting. Running moulds are used for cornice, rails, ribs, moulding, or anything run in place. Casting moulds are used for additional ornamentation that cannot be run in place. Three types are generally used — gelatin, wax and plaster.

MULLION — A vertical division in a frame separating two or more sections.

NAILABLE STUDS — See STUDS.

NAILING CHANNEL — Fabricated from not lighter than 25-gauge steel so as to form slots to permit attachment of lath by means of ratchet-type annular nails—or other satisfactory attachments.

NAIL POPPING — The protrusion of the nail usually attributed to the shrinkage of or use of improperly cured wood framing.

NEAT — Term used to denote plaster material requiring the addition of aggregate. See GYPSUM.

NICHE — Either a curved or square recess in a wall, dependent on the architecture and the use for which it is intended. They are used for housing statutes, vases, telephones, or door chimes.

NOISE REDUCTION COEFFICIENT (NRC) — A relative numerical expression of the ability of a material to absorb sound.

NON-LOAD BEARING PARTITION — A structurally non-essential interior wall assembly for compartmentalizing floor space.

NOZZLE — An attachment at the end of a plastering machine delivery hose, which regulates the fan or spray pattern.

NRC — Noise reduction coefficient is a standard of measurement of sound control in the design of acoustical materials.

OGEE — A curved section of a moulding, partly convex and partly concave.

ON CENTER (o.c.) — The centerline spacing distance between framing members, fasteners or other points of reference.

ORIFICE — Attachment to the nozzle on the hose of plastering machine, of various shapes and sizes, which may be changed to help establish the pattern of the plaster as it is projected onto the surface being plastered.

PARTITION — A wall that subdivides spaces within any story of a building; there are various types of

partitions, commonly known as bearing partition; dwarf partition; fire partition; non-bearing partition; hollow partition; solid partition.

PARTITION CAP — See Partition Cap under ACCESSORIES.

PARTY WALL — A special purpose wall system used to divide compartments for different occupancies. May have requirements for fire and sound.

PENCIL RODS — Mild steel rods of 3/16", ¼" or ⅜" diameter.

PERIMETER RELIEF — Construction detail which allows for building movement. Gasketing materials which relieve stresses at the intersections of wall and ceiling surfaces.

PERLITE — A siliceous volcanic glass properly expanded by heat and weighing not less than 7½ nor more than 15 lbs. per cu. ft., used as a lightweight aggregate in plaster.

PERM — A unit of measurement of Water Vapor Permeance; metric unit, Nanograms per Pascal, second meter squared; U.S. unit, 1 grain per hour/square foot/inch of mercury.

PERMEABILITY — The property of a material to permit a fluid (or gas) to pass through it; in construction, commonly refers to water vapor permeability of a sheet material or assembly and is defined as Water Vapor Permeance per unit thickness. Metric unit of measurement, metric perms per centimeter of thickness.

PERMEANCE (Water Vapor) — The ratio of the rate of water vapor transmission (WVT) through a material or assembly between its two parallel surfaces to the vapor pressure differential between the surfaces. Metric unit of measurement is the Nanograms per Pascal, second meter squared; U.S. unit, 1 grain per hour/square foot/inch of mercury.

PICTURE MOLD — See SCREEDS, GROUNDS.

PILASTER — A projecting square column forming part of a wall.

PINHOLE — A small hole appearing in a cast when the water-stucco ratio has not been accurately measured. Excess water causes pinholes.

PLASTER — A cementitious material or combination of cementitious materials and aggregates that, when mixed with a suitable amount of water, forms a plastic mass which, when applied to a surface, adheres to it and subsequently sets or hardens, preserving in a rigid state the form or texture imposed during the period of plasticity. The term "plaster" is used with regard to the specific composition of the material and does not explicitly denote either interior or exterior use. See GYPSUM, PORTLAND CEMENT PLASTER, LIME PLASTER.

PLASTER BOND – The state of adherence between plaster coats or between plaster and a plaster base, produced by adhesive or mechanical interlock of plaster with base or special supplementary materials. (ASTM)

PLASTER OF PARIS – See CASTING PLASTER. See GYPSUM MOLDING PLASTER.

PLASTERING MACHINE – A mechanical device by which plaster mortar is conveyed through a flexible hose to deposit the plaster in place; also known as a plaster pump or plastering gun. Distinct from "Gunite" machines in which the plaster or concrete is conveyed, dry, through the flexible hose and hydrated at the nozzle.

PLASTIC CEMENT – Portland cement to which small amounts of plastizing agents, not more than 12% by volume, have been added at the mill.

PLASTICITY – Workability and water-retentive characteristics imparted to plaster mortars by such agents as natural cement, lime, asbestos, flour, clays, air-entraining agents or other approved lubricators or fatteners. The duration of mixing time may be a factor in the plasticity of some mortars.

PLASTICIZING AGENT – A product used to increase the flow and/or workability of plaster.

PLATE – The horizontal framing member at the top or base of wall framing.

PLENUM – An enclosed chamber such as the space between a suspended finished ceiling and the floor above.

POLISH – To make plaster finish coat smooth and glossy by troweling.

POPS or PITS – Ruptures in finished plaster or cement surfaces which may be caused by expansion of improperly slaked particles of lime or by foreign substances.

PORTLAND CEMENT – A hydraulic cement produced by pulverizing clinker consisting essentially of hydraulic calcium silicates, and usually containing one or more forms of calcium sulfate as an interground addition. (ASTM)

PORTLAND CEMENT PLASTER – Plaster made with portland cement usually applied in three coats, except when direct to masonry or concrete surfaces.

PORTLAND CEMENT-LIME PLASTER – Portland cement and lime (either Type "S" hydrated lime or properly aged lime putty) combined in proportion as outlined in applicable building code.

POWER DRIVEN FASTENER – A fastener attached to steel, concrete or masonry by a power charge cartridge or by manual impact.

PROCESSED QUICKLIME – See LIME.

PUDDLING – A condition of mechanical dash textures resulting in glazing, texture deviation or discoloration caused by holding the plastering machine nozzle too long in one area.

PUMICE – A lightweight volcanic rock, which, when crushed and graded, may be used as a plaster aggregate.

PUMP – See PLASTERING MACHINE.

PUMPING AGENT – A product used to increase the flow of plaster through hoses during machine applications.

PUTTY – Product resulting from slaking, soaking, or mixing lime and water together.

PUTTY COAT – A troweled finish coat composed of lime putty or Type "S" hydrated lime gauged with gypsum gauging plaster or Keene's Cement. Fine aggregate may be added.

QUICKLIME – See LIME.

RACKING – Lateral stresses exerted on an assembly.

RADIANT HEAT – A heat source which transmits thermal energy by radiation rather than convection. An example is electric resistance wires concealed in a finished ceiling surface.

READY-MIXED PLASTER – See GYPSUM.

REINFORCEMENT – See METAL LATH–EXPANDED, WELDED WIRE FABRIC, WOVEN WIRE FABRIC.

RELIEF – Ornamented prominence of parts of figures above a plane surface.

RELATIVE HUMIDITY – The ratio expressed as a percentage of actual water vapor pressure to the saturation water vapor pressure at a given temperature.

RESILIENT SYSTEMS – See Sound Transmission Clips under ACCESSORIES, CLIPS.

RETARDER – Any material added to gypsum plaster which slows up its natural set.

RETEMPER – Addition of water to portland cement plaster after mixing but before setting process has started. Gypsum plaster must not be retempered.

RETURN – The turn and continuation of a molding, cornice, wall or projection in an opposite or different direction: the

continuation in a different direction of the face of a building, or any member.

REVEAL – The vertical face of a door or window opening between the face of the interior wall and that of the window or door frame, or the like.

RIB LATH – See METAL LATH.

ROCK GUN – A device for throwing aggregate onto a soft bedding coat in applying marblecrete.

RUNNER TRACK – See Track under ACCESSORIES.

RUNNERS – See MAIN RUNNERS.

SADDLE TIE – A specific method of wrapping hanger wire around main runners; also of wrapping tie wire around the juncture of main runner and cross furring.

SAFING – A non-combustible product used at the perimeter of floor and around other penetrations as a fire barrier.

SAND – See AGGREGATE.

SAND FLOAT FINISH – See FINISH COAT.

SCAGLIOLA – An imitation marble. Scagliola is usually precast, using Keene's cement.

SCORING – Grooving, usually horizontal, of portland cement plaster scratch coat to provide mechanical bond for the brown coat. Also a decorative grooving of the finish coat.

SCRATCH COAT – First coat of plaster in three-coat plastering work.

SCREEDS – Long narrow strips of mortar applied horizontally or perpendicularly along a wall or ceiling surface and faced out straight and true to serve as guides for plastering the intervals between them.

SET – The change in mortar from a plastic, workable state to a solid, rigid state.

SETTING TIME – The elapsed time required for a gypsum plaster to attain a specified hardness and strength after mixing with water. (ASTM)

SGRAFFITO – A procedure for decorative purposes generally consisting of two or more layers of differently colored plaster. While still soft, part of the top layer is removed by scratching, exposing part of the base or underlying layer.

SHEAR – (Strengths, Values) – The resistance of an assembly to lateral movement.

SHEET LATH – See METAL LATH.

SHIELDING — Method of protecting adjacent work by positioning temporary protective sheets of rigid material; particularly used for machine applications.

SHIM — To build up low areas; to level or adjust height.

SHOE — A formed metal section used in attaching metal studs to floor and ceiling tracks, also the end section of a channel turned to an angle (usually 90°) to permit attachment, generally to other channels.

SISAL — See FIBER.

SKIM COAT — Last and final coat, referred to as such in some localities.

SLAKING — The act of hydrating quicklime into a putty by the addition of water.

SOFFIT — The underside of an archway, cornice, bead, etc.

SOUND TRANSMISSION CLASS (STC) — A single number rating of the sound insulation value of a partition or wall. It is derived from a curve of its insulation value as a function of frequency; the higher the number, the more effective the sound insulation.

SPAN — Distance between supporting members.

SPLAY ANGLE — Where two surfaces come together forming an angle of more than 90 degrees is referred to as a splay angle.

SPOT GROUND — See DOT.

SPRAY TEXTURE — A surface finish achieved by application of finish coat material with a plastering machine or gun.

STAFF — Plaster casts made in moulds and reinforced with fibre. Usually wired or nailed into place.

STAPLE — A U-shaped metal fastener used to attach building paper, expanded metal, wire or gypsum lath and accessories to framing.

STICKER — A piece of metal channel inserted in concrete or masonry walls for the attachment or support of wall furring channels.

STIFFENER — A horizontal metal shape tied to vertical members (studs or channels) of partitions or walls to brace them.

STRAIGHT EDGE — A flat wooden tool or rod, perfectly true, used to straighten the brown coat or screeds.

STRING WIRE — Wire used on open stud construction, placed horizontally around building to support weatherproofing paper.

STRIP LATH — Metal or wire fabric used over joints of gypsum lath. (Sometimes used to obtain fire rating).

STRIP REINFORCEMENT — See STRIP LATH.

STUCCO — A cementitious mixture used for exterior plaster. Depending upon locality, stucco is the combined base and finish coats or may be the colored finish coat only. The terms is often used interchangeably with "exterior plaster". See INTERIOR STUCCO.

STUCCO NETTING — See WOVEN WIRE FABRIC.

STUDS (METAL, LOAD-BEARING) — Formed from minimum 20-gauge, structural grade strip steel, with punched webs. Widths: 2½", 3¼", 3⅝", 4", 6". Also available in double form in 2½" and 3⅝" widths, to permit attachment of lath by nailing or by other means.

STUDS (METAL, NONLOAD-BEARING).

Prefabricated Types (1⅝", 2½", 3¼", 3⅝", 4" and 6" widths)—

1. 18-gauge channel shapes with perforated webs.

2. Double 7-gauge cold drawn rods welded to a 7-gauge rod bent in diagonal truss design between them. (No longer manufactured).

3. Two 16-gauge angle shapes welded to a 7-gauge rod bent in diagonal truss design between them.

Channel Type — Cold-rolled channels of ¾", 1½" and 2" widths. Hot-rolled channels of ¾", 1", 1½" and 2" (not always available).

Nailing Type — Fabricated from not lighter than 26-gauge steel so as to form slots which permit attachment of the lath by means of specially designed nails, staples, or screws. Widths: 2", 2½", 3¼", 3⅝", 4", and 6". Sometimes called "Nailable." (No longer manufactured).

Screwable Type — Studs fabricated from not lighter than 26 gauge metal with knurled flanges to facilitate easier penetration of self-tapping screws or divergent point staples.

SUCTION — The power of absorption possessed by a plastered surface. Example: The basecoat must have suction in order to absorb the water out of the finish coat.

SUSPENDED CEILINGS — See CEILINGS.

SWEAT OUT — Soft, damp wall area caused by poor drying conditions.

TAPE — A plastic reinforcing mesh or paper used to reinforce angles and to bridge lath joints in Veneer Plastering.

TEMPER — Mixing of plaster to a workable consistency.

TEMPLATE — A gauge, pattern or mould used as a guide to produce arches, curves, and other various work.

TEXTURE — See FINISH COATS.

THERMAL SHOCK — A stress created by an extreme change in temperature that may result in cracking of the plaster which has not yet attained its ultimate strength.

THINCOAT HIGH-STRENGTH PLASTER — See Veneer Plaster under GYPSUM.

THREE-COAT PLASTERING — The application of plaster in three successive coats, leaving time between coats for setting and drying or partial drying.

TIE WIRE — Soft annealed steel wire used to join lath supports, attach lath to supports, attach accessories, etc.

TIES — There are two types used for the attachment of lath: (a) the butterfly tie, which is formed by twisting the wire and cutting so that the two ends extend outward oppositely and (b) the stub tie, which is twisted and cut at the twist. See SADDLE TIES.

TRACK — See Track under ACCESSORIES.

TROWEL FINISH — See FINISH COAT.

TURTLE BACK

1. A term used synonymously with blistering and

2. A term used regionally to denote a small localized area of wind-crazing. See BLISTER.

TWO-COAT PLASTERING — The application of plaster in two successive coats. In two-coat plastering the basecoat is applied in one operation. See DOUBLEUP; See BROWN COAT.

TYPE "S" HYDRATED LIME—See LIME.

TYPE X GYPSUM WALLBOARD – A gypsum wallboard specially manufactured to provide specific fire-resistant characteristics. (ASTM)

TYPE "X" LATH – See GYPSUM LATH.

VAPOR BARRIER – Any material that has a water vapor permeance (perm) rating of one or less. (ASTM)

VENEER PLASTER – See GYPSUM PLASTER.

VERMICULITE – A mineral with the quality of expanding, upon being heated, into a lightweight, porous material, one of whose usages is as aggregate with various types of gypsum plasters.

WADDING – The act of hanging staff by fastening wads made of Plaster of Paris and excelsior or fiber to the casts and winding them around the framing.

WAINSCOT – The lower 3 or 4 feet of an interior wall when it is finished differently from the remainder of the wall.

WASH-OUT—Lack of proper coverage and texture buildup in machine-dash textured plaster caused by the mortar being too soupy.

WASTE MOLD – A precast plaster mold made for the forming of decorative monolithic or cast-in-place concrete. Mold cannot be removed without being destroyed.

WATER ABSORPTION – The amount of water absorbed by a material under specified test conditions commonly expressed as weight percent of the test specimen. (ASTM)

WATERPROOF CEMENT—Portland cement to which waterproofing agents, such as surface repellents, have been added at time of blending materials at the mill.

WATER-REPELLENT PAPER – Gypsum board paper surfacing which has been formulated or treated to resist water penetration. (ASTM)

WATER-RESISTANT CORE – A gypsum board specially formulated to resist water penetration. (ASTM)

WATER VAPOR TRANSMISSION (WVT) – The rate of water vapor flow, under steady specified conditions, through a unit area of a material between its two parallel surfaces and normal to the surfaces. Metric unit of measurement is $1 \text{ g}/24 \text{ h} \cdot \text{m}^2$. See PERMEABILITY, PERMEANCE PERM. (ASTM)

WELDED WIRE FABRIC (INTERIOR or EXTERIOR LATH) – A plaster reinforcement of copper-bearing, soft annealed wire not lighter than 16 gauge, zinc coated, electrically welded at all intersections, forming openings not to exceed 2 x 2 inches; may

have an absorptive paper separator and may have an additional paper backing or foil backing for purposes of waterproofing or insulation; flat or self-furring.

WHITE COAT – See PUTTY COAT. WOVEN WIRE FABRIC; see WELDED WIRE FABRIC.

WIRE CLOTH LATH – A plaster reinforcement of wire not lighter than No. 19 gauge, 2½ meshes per inch and coated with zinc or rust-inhibitive paint. (Not to be used as reinforcement of exterior portland cement plaster.)

WIRE FABRIC LATH – See WOVEN WIRE FABRIC; see WELDED WIRE FABRIC.

WOOD FIBER – See FIBER.

WOOD FIBER PLASTER – See GYPSUM PLASTER.

WORKABILITY – See FAT.

WOVEN WIRE FABRIC – A plaster reinforcement of zinc-coated wire, not lighter than No. 18 gauge when woven into 1-inch openings, or not lighter than No. 17 gauge when woven into 1½-inch openings. Lath may be paper-backed, flat or self-furring.

2. METAL FRAMING

2.1 GENERAL

2.1.1 Wire-Tying Metal Framing.—Use a single strand of No. 16 gauge wire, or a double strand of No. 18 gauge wire for tying metal framing components together. (Detail 29)

For splices use a double wrap tie.

For tying horizontal channels placed at intersecting legs of channel brackets, use a figure-eight tie.

For tying members perpendicular to each other, use a saddle tie.

2.1.2 Screwed Metal Framing.—Use an approved self tapping and/or self drilling screw which provides a minimum $\frac{1}{4}"$ penetration through all metal.

2.1.3 Welded Metal Framing.—Welded connections shall be made by resistance spot or projection welding, fillet welding, or plug welding, and shall be done in accordance with the latest recommended procedures of the American Welding Society. [3]

2.2 NONLOAD-BEARING STUDS

2.2.1 Prefabricated Studs.—Prefabricated studs for nonload-bearing partitions shall be of pressed metal or welded fabrication. They may be cold-formed No. 18 gauge channel shapes with either solid or perforated webs; or double No. 7 gauge minimum, cold-drawn rod chords welded to a minimum No. 7 gauge rod bent in diagonal truss design between chords; or No. 16 gauge steel angles with a single No. 7 gauge rod forming an open-web truss-like pattern between flanges, all points contact-welded. Studs shall be coated with a rust-inhibitive material. (Detail 4)

2.2.2 Nailable Studs.—Not commonly available.

2.2.3 Screw Studs.—For nonload-bearing partitions shall be U- or C-shaped metal channels formed of no lighter than No. 26 gauge galvanized steel and with or without punched stud webs. (Detail 6)

2.2.4 Runner Tracks.—Floor and ceiling runner tracks to which studs are attached shall be the stud manufacturer's regular type for the stud specified. Runner tracks shall be formed from no lighter than No. 26 gauge steel painted with rust inhibitive coating or galvanized. Special runner track as detailed or approved by the architect may be used. [4] (Detail 7)

2.3 LOAD-BEARING STUDS

2.3.1 Prefabricated, Nailable and Screwable.—For load bearing walls or partitions shall be punched or unpunched channel or "C" shapes with minimum 1" flanges. Studs shall be fabricated from not less than the following grades of steel:

    16 ga. and heavier—ASTM A 570—Minimum yield point 50,000 psi
    18 and 20 gauge—ASTM A 570—Minimum yield point 33,000 psi
    All galvanized studs—ASTM A 446 (Detail 8)

2.3.2 Runner Tracks.—Floor and ceiling runner tracks to which load-bearing studs are attached shall have minimum $\frac{7}{8}"$ flanges. Tracks shall be fabricated from minimum 33,000 psi steel. Bridging shall be as shown on the drawings. [5]

2.3.3 Bridging.—For use in load-bearing construction, bridging shall be minimum $\frac{3}{4}"$ cold rolled channel with weld attachment clips at each stud location, or shall be V-bar type weld or screw attached to each stud flange, or shall be as detailed by the manufacturer and approved by the engineer. In all cases bridging shall be adequate to provide lateral support for the stud. Bridging sections located within the stud cavity shall be weld attached.

[4] Special runner track may function both as stud attachment and as ground for plaster.

[5] Architectural drawings must show type of bridging required.

RECOMMENDED / **CONSTRUCTION TECHNIQUE**

Metal Stud Non-Bearing Hollow Partitions

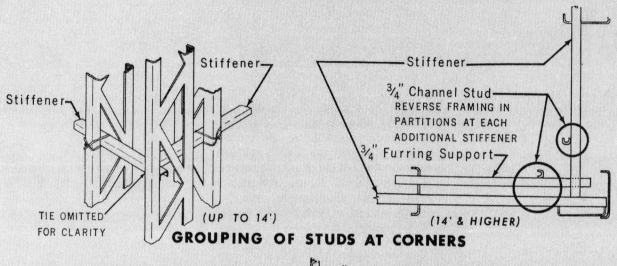

Stiffener

Stiffener

TIE OMITTED FOR CLARITY

(UP TO 14')

Stiffener

¾" Channel Stud
REVERSE FRAMING IN PARTITIONS AT EACH ADDITIONAL STIFFENER

¾" Furring Support

(14' & HIGHER)

GROUPING OF STUDS AT CORNERS

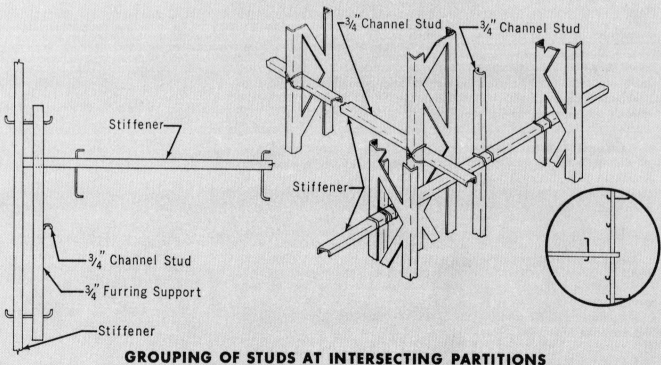

¾" Channel Stud

¾" Channel Stud

Stiffener

Stiffener

¾" Channel Stud

¾" Furring Support

Stiffener

GROUPING OF STUDS AT INTERSECTING PARTITIONS

ERECTION OF STUDS—

Corners and intersections of partitions shall be formed of three prefabricated studs, or a combination of prefabricated studs and ¾-inch channel studs adequately braced. Studs and channels forming internal corners shall be placed 2 inches from the point of partition intersection to allow for tying of lath or cornerite.

2.4 VERTICAL METAL FRAMING
2.5 NONLOAD-BEARING PARTITIONS AND WALLS

2.5.1 Deflection Relief.—When stud partitions abut underside of steel or concrete rigid construction, cut studs short to relieve stress from deflection on framing. [1] [2]

When stud partitions abut steel, concrete or masonry walls or columns, end stud shall not contact abutting surfaces. Secure ends of horizonal stiffeners in such partitions to abutting surface.

2.6 PREFABRICATED STUDS (hollow partitions; walls) (Detail 30)

2.6.1 Erection of Prefabricated Studs.—Align runner tracks accurately to the partition layout at both floor and ceiling and secure to concrete slabs with minimum $\frac{7}{8}''$ powder driven pins, or with $\frac{1}{2}''$ or $\frac{5}{8}''$ concrete stub nails, at not more than 24'' intervals; or wire to the runner or furring channels of ceilings. [4] [5]

Secure each prefabricated stud to runner tracks with two stud shoes at both top and bottom, each wired or crimped to stud; or tack weld stud to the floor runner tracks; or attach to pronged, slotted or snap-in track by inserting end of stud into space provided; or bend stud and attach to floor with concrete stub nail or powder driven pin. [6]

Stiffen all partitions with $\frac{3}{4}''$ channels placed horizontally and not more than 5' apart vertically. Wire stiffeners permanently to the inside of the partition; or secure stiffeners (bridging) in place as recommended by stud manufacturer.

For partitions of unsupported height of 20' or more, horizontal stiffeners shall be $1\frac{1}{2}''$ channels. [7]

Form corners and intersections of partitions of three studs, or a combination of prefabricated studs and $\frac{3}{4}''$ channel studs adequately braced. Place studs and channels forming internal corners 2'' from the point of partition intersection to allow for tying of lath. [8]

2.6.2 Framing Around Door Openings.—Secure two prefabricated studs together and install at each jamb of metal frames, continuous from floor to ceiling. Wire-tie, bolt, weld or otherwise secure stubs to the jamb anchors. Attach jack studs to metal runner track within head of frame, and to runner track or $\frac{3}{4}''$ channel at ceiling. [9] [10]

Wood frames, installed by others, shall have one prefabricated stud at each jamb, continuous from floor to ceiling. Fasten studs to the wood bucks with pairs of 8d nails, four pairs to each jamb.

Place a $1\frac{1}{2}''$ channel horizontal reinforcement in the web openings of the studs 6'' maximum above door openings. It shall extend continuously past the studs alongside the frame and past the next two studs on each side of the opening. Saddle tie or weld the channel to each vertical stud it crosses. [11]

2.7 NAILABLE STUDS (hollow partitions; walls) (Detail 31)

2.7.1 Erection of Nailable Studs.—Align runner tracks accurately to the partition layout at both floor and ceiling and secure to concrete slabs with minimum $\frac{7}{8}''$ powder driven pins, or $\frac{1}{2}''$ or $\frac{5}{8}''$ concrete stub nails, at not more than 24'' intervals; or wire to the runner or furring channels of ceilings. [12]

Secure each nailable stud to runner tracks at both top and bottom by crimping flange or runner track; or by screwing with No. 6 screw; or tack weld bottom of studs to the floor runner tracks; or other methods of fastening studs to floor and ceiling, recommended by stud manufacturer, may be used. [13]

Stiffen all partitions with $\frac{3}{4}''$ channels placed horizontally and not more than 5' apart vertically. Wire stiffeners permanently to the inside of the partition; or secure stiffeners in place as recommended by stud manufacturer.

For partitions of unsupported height of 20'' or more, horizontal stiffeners shall be $1\frac{1}{2}''$ channels. [14]

[1] Slip joint or cushion may also be used to relieve deflection of horizontal construction.

[2] If studs and channel stiffeners in partitions and walls are welded, break up the areas of welded components into panels approximately 24 long to control movement due to temperature change.

[3] Shear panels and other load bearing constructions must have complete details and welding specifications furnished by the structural engineer.

[4] Runner tracks may be lapped at ends and secured with one fastener at lap.

[5] Plenum chamber portions which are to be framed should be clearly designated on drawings.

[6] Where studs extend above suspended ceilings, but not to construction above, each stud should be securely attached to a horizontal $\frac{3}{4}''$ channel placed above the ceiling and along the full length of partition.

[7] Where punched studs do not permit passage of $1\frac{1}{2}''$ channel stiffeners, use two $\frac{3}{4}''$ channels.

[8] Sound insulation of partitions, when a requirement of project, should be specified here. Specify method of reducing sound transference. Specify installation of insulating materials between studs and adjacent construction.

[9] Where heavy or oversize doors are used, attention should be given to developing adequate resistance to impact.

[10] Specify metal door frames to be furnished and installed by others and provide in details that:

[a] Return on flanges of frames shall be of such depth as to permit lath and plaster to enter frames.

[b] Floor anchors in frames shall be of minimum of No. 16 gauge steel, welded to frames and with attachment holes spaced a minimum of 3'' apart. Door frames shall be anchored to floor at floor anchors with bolts, screws or other secure attachment.

[c] Jamb anchors in frames shall be of minimum No. 16 gauge steel, notched or formed to receive studs; or 14-gauge steel 1'' wide strap jamb anchors welded to each flange of frame to which metal studs shall be welded. Jambs shall contain four jamb anchors with one placed near top and one near bottom of each jamb.

[d] Wood buck should be of such size that casing will cover joint between buck and plaster.

[11] If there is not room for $1\frac{1}{2}''$ channel, use two $\frac{3}{4}''$ channels.

[12] Runner track may be lapped at ends and secured with one fastener at lap.

[13] Where studs extend above suspended ceilings, but not to construction above, each stud should be securely attached to a horizontal $\frac{3}{4}''$ channel above the ceiling and along the full length of partition.

[14] On remodeling of existing buildings, installation of mechanical or electrical work may damage existing plaster. This damage is not usually shown on drawings, or even known before bidding. Repair of such damage should be clearly designated as the responsibility of the trade responsible for damage.

Form corners and intersections of partitions of three nailable studs, or a combination of studs and nailable channels, adequately braced. Place studs and channels forming internal corners 2″ from the point of partition intersection to allow for attachment of lath. [15]

2.7.2 Framing Around Door Openings.—Install one nailable stud at each jamb of metal frames, continuous from floor to ceiling. Wire tie, bolt, weld, or otherwise secure stud to the jamb anchors. Attach jack studs to metal runner track within head of frame, and to runner track or ¾″ channel at ceiling. [16] [10]

Wood frames, installed by others, shall have one nailable stud at each jamb, continuous from floor to ceiling. Fasten studs to the wood bucks with pairs of 8d nails, four pairs to each jamb, or other secure attachment.

Place a 1½″ channel horizontal reinforcement in the web openings of the studs 6″ maximum, above door openings. It shall extend continuously past the studs alongside the frame and past the next two studs on each side of the opening. Saddle-tie the channel to each vertical stud it crosses. [17]

2.8 SCREW STUDS (hollow partitions; walls) (Detail 32)

2.8.1 Erection of Screw Studs.—Align runner tracks accurately to the partition layout at floor and ceiling and secure tracks with fasteners spaced at a maximum of 24″ intervals. Plumb and true studs and engage both floor and ceiling tracks. If necessary, studs may be spliced by nesting them with a minimum lap of 8″.

When studs are used in partitions which receive standard lath and plaster, stiffen with ¾″ channels placed horizontally and not more than 5′ apart vertically. Wire stiffeners to inside of studs, or secure as recommended by stud manufacturer. Stiffeners may be omitted when studs receive large size lath and veneer plaster.

2.8.2 Framing Around Door Openings.—Frame openings with single studs at each jamb, or double studs boxed, if heavy doors are used. Studs to run from slab to slab. Bolt or screw studs to the jamb anchors. Attach jack studs to runner track within head of frame and to runner track at ceiling. [18] [10]

Wood frame, installed by others, shall have two screw studs, nested to form a box, continuous from floor to ceiling. Secure studs to wood bucks with pairs of screws or 8d nails, four pairs to each jamb, or other secure attachment.

2.9 DOUBLE CHANNEL STUDS (hollow partitions; walls) (Detail 33)

2.9.1 Erection of Double Channel Studs.—Construct double channel stud hollow partition of a double row of parallel ¾″ channels with spacers or braces.

Align runner tracks or other method of attachment to the partition layout at both floor and ceiling and secure to concrete slabs with minimum ⅞″ powder driven pins, or ½″ or ⅝″ concrete stub nails, at not more than 24″ intervals; or wire the runner track to furring channels or ceilings. [19]

Single piece studs shall not bow when erected because of being forced into place. Where two-piece studs are used, splice by lapping not less than 8″. Interlock flanges and securely wire near each end of splice.

Securely fasten channel studs to runners, to ceilings or to the construction above, or directly to the floor. Install metal base according to the manufacturer's specifications if specified herein.

Stiffen partitions with ¾″ channels placed horizontally and not more than 5′ apart vertically. Wire stiffeners permanently to the inside of both faces of partitions. [20]

Install ¾″ channel spacers, braces or separator clips a maximum of 36″ apart along stiffeners. [21]

[15] Sound insulation of partitions, when a requirement of project, should be specified here. Specify method of reducing sound transference. Specify installation of insulating materials between studs and adjacent construction.

[16] Where heavy or oversize doors are used, attention should be given to developing adequate resistance to impact.

[17] If there is not room for 1½″ channels, use two ¾″ channels.

[18] Where heavy or oversize doors are used, attention should be given to developing adequate resistance to impact.

[19] If ¾″ channel is used as runner track, it should be secured to concrete slab at not more than 30″ intervals.

[20] Prefabricated stud placed horizontally and tied to each row of channel studs may be used as stiffener, omitting cross braces.

[21] Sound insulating double partitions should be specified here and shown on the drawings. Specifications should cover methods of separating materials to avoid sound transference, and installation of insulating materials between studs and/ or frames and adjacent construction.

2.9.2 Framing Around Door Openings.—Install a pair of channel studs at each jamb of metal frames continuous from floor to ceiling. Wire-tie, bolt or weld or otherwise secure studs to the jamb anchors. Attach pairs of channel jack studs to ¾″ channel runner within head of frame, and to runner track or channel runners at ceiling. [22] [10]

Wood frames, installed by others, shall have a pair of channel studs at each side of the opening, continuous from floor to ceiling. Fasten studs to wood bucks with pairs of 8d nails, four pairs to each jamb, or other secure attachment.

Place a ¾″ channel horizontal stiffener 6″ maximum above door openings and tie to inside of one stud in each pair of channel studs. Stiffener shall extend continuousy past the channel alongside the jambs and past the next two pairs of channels on each side of the opening.

2.10 SINGLE CHANNEL STUDS (solid partitions) (Detail 34)

2.10.1 Spacing of Channel Studs (See Table 5-9) [23]

2.10.2 Erection of Single Channel Studs.—Framing for solid partition (with metal lath) shall be a single row of channel studs.

Align runners or other method of attachment to the partition layout at both floor and ceiling and secure to concrete slabs with minimum ⅞″ powder driven pins, or ¾″ concrete stub nails, at not more than 24″ o.c. intervals; or wire the runner tracks or channels to furring channels or ceilings. [24]

Single piece studs shall not bow when erected because of being forced into place. Where two-piece studs are used, splice by lapping not less than 8″. Interlock flanges and securely wire near each end of the splice.

Securely fasten channel studs to runners, to ceiling or to construction above, or directly to floor. Install metal base according to the manufacturer's specifications if specfied herein. [25]

2.10.3 Framing Around Door Openings.—Install one channel stud at each jamb of metal frames, continuous from floor to ceiling. Wire-tie, bolt, weld, or otherwise secure stud to the jamb anchors. Attach channel jack studs to ¾″ channel runner within head of frame, and to runner or channel at ceiling. [26] [10]

Wood frames, installed by others, shall have a channel stud at each side of the opening, continuous from floor to ceiling. Fasten studs to wood bucks with pairs of 8d nails, four pairs to each jamb, or other secure attachment.

Saddle-tie a horizontal ¾″ pencil rod stiffener to each channel stud at 6″ maximum above door opening. It shall be long enough to extend continuously past the studs alongside the jambs and past the next two studs on each side of the opening.

Protect exposed partition ends and tops with a metal partition terminal or corner beads.

2.10.4 Bracing.—Temporarily brace single channel studs on channel side of partition.

2.11 (Reserved)

[22] Where heavy or oversize doors are used, attention should be given to developing adequate resistance to impact.

[23] Spacing of channel studs depends on type and weight of metal lath. Refer to applicable building code.

[24] If ¾″ channel is used as runner track, it should be secured at not more than 30″ intervals.

[25] No horizontal stiffener required in channel stud solid plaster partitions

[26] Do not use heavy or oversize doors in solid plaster partitions.

DETAILS

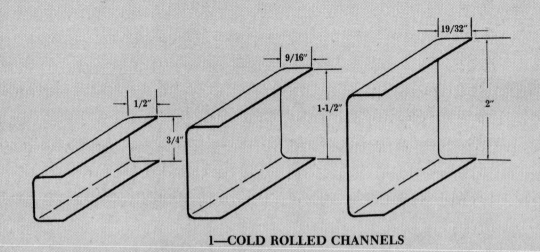

1—COLD ROLLED CHANNELS

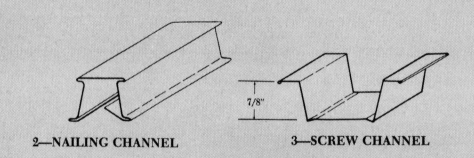

2—NAILING CHANNEL   3—SCREW CHANNEL

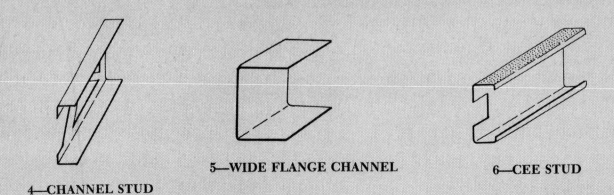

4—CHANNEL STUD

5—WIDE FLANGE CHANNEL

6—CEE STUD

DETAILS

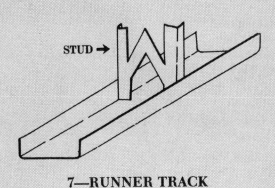

STUD →

7—RUNNER TRACK

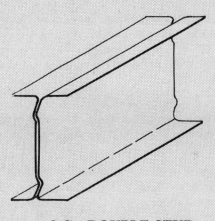

8-A—CHANNEL STUD
(16,18 & 20 ga.)

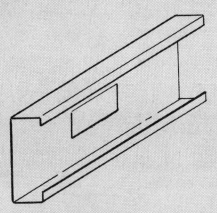

8-B—CEE STUD
(16,18 & 20 ga.)

8-C—DOUBLE STUD

2.12 LOAD-BEARING PARTITIONS AND WALLS   (Detail 36)

2.13 PREFABRICATED AND NAILABLE STRUCTURAL STUDS

2.13.1 Spacing, Size and Height of Load-Bearing Studs.—As indicated on the drawings determined by engineering calculations furnished by the architect.

2.13.2 Erection of Load-Bearing Studs.—Align runner tracks accurately to the partition layout at both floor and ceiling and secure to concrete slabs with minimum 7/8" powder driven pins or 3/4" concrete stub nails at no more than 48" intervals. [34]

Align each stud and make plumb and true and secure to runner tracks at both top and bottom. Methods of fastening studs shall be as recommended by stud manufacturer.

Stiffen or bridge partitions horizontally as required by engineering

[34] Runner track may be lapped at ends and secured with one fastener at lap.

design and secure stiffeners (bridging), as recommended by stud manufacturer.

Form corners and intersections of partitions of three studs, or a combination of studs and channel adequately braced. Place studs forming internal corners not more than 2″ from the point of partition intersection.

2.13.3 Framing Around Door Openings.—Install one load-bearing stud at each jamb of metal frames, continuous from floor to ceiling. Bolt, weld or otherwise secure stud to the jamb anchors. Attach jack studs to metal runner track on interior of head of frame, and to runner track or ¾″ channel at ceiling. [10]

Wood frames, installed by others, shall have one load-bearing stud at each jamb, continuous from floor to ceiling. Fasten studs to the wood bucks with screws or bolts, at four points along each jamb.

2.14 VERTICAL FURRING   (Detail 37-A)

2.14.1 Size, Height and Spacing of Studs.—As indicated on the drawing and as determined by engineering calculations of the architect (See Table 5-7) [34A]

2.14.2 Free-Standing Furring.—Align runners or other method of attachment to the layout at both floor and ceiling and secure to concrete slabs with minimum ⅞″ powder driven pins or ¾″ concrete stub nails at not more than 24″ intervals; or wire runners, or vertical channels to ceiling construction. [37] [38]

Single piece studs shall not bow when erected because of being forced into place. Where two-piece studs are used, splice by lapping not less than 8″. Interlock flanges and securely wire near each end of splice.

Securely fasten studs to runners, to ceiling or to construction above, or directly to floor. If metal base is specified herein, install according to manufacturer's specifications.

Stiffen free-standing furring with ¾″ channels placed horizontally and not more than 4′ 6″ apart, vertically. Wire stiffeners permanently to the unplastered side of each stud.

Erect free standing furring of prefabricated or nailable studs as required for hollow stud partitions. [39]

2.14.3 Braced Furring

2.14.3.1 Bracing for Channel and Prefabricated Studs.—Install minimum ¾″ channel braces between wall and vertical furring at maximum 4′ intervals along horizontal stiffener, or between each stud and the wall, as required at openings. Vertical spacing of braces will depend on height of furring and size of stud. [40] [41]

Secure braces to anchors, or directly to wall, and wire-tie to horizontal stiffeners or to each stud.

[34A] Walls with required minimum heat transmission loss should be specified here and shown on the drawings. Indication should be made of the minimum allowable "R" or "U" value of the wall construction.

[35] Generally may be lathed with either metal or gypsum lath. Walls with required minimum heat transmission loss should be specified here and shown on the drawings. Indication should be made of the minimum allowable "R" or "U" value of the wall construction.

[36] Channel, prefabricated or nailable studs.

[37] Channels or L-shape runners should be used as floor runners. Channel may also serve as plaster ground. Pencil rods or cornerite may be used as ceiling runners.

[38] If ¾″ channel is used as runner track, it should be secured at not more than 30″ intervals.

[39] Use prefabricated or nailable studs for greatest unbraced heights. For erection specification of hollow stud partition, see 3.4 and 3.5.

[40] See Table 5-7 for maximum vertical spacing of braces.

[41] See 3.12.2 and accompanying notes for erection specification for studs.

2.14.3.2 Furring with Vertical Gypsum Lath and Bracing (studless).
—Align metal floor runner, or combination metal clips and base, to wall layout and secure to floor with minimum $7/8''$ powder driven pins or $3/4''$ concrete stub nails or other suitable attachment, spaced at not more than $36''$ intervals. [42] [43] [44]   (Detail 37-B)

Align combination ceiling runner and clip, or a $3/4''$ channel, with the floor runner so that lath will be plumb, and secure with tie wire, or other secure attachment, to the construction above.

Bracing attachments for horizontal stiffeners may be short pieces of $3/4''$ channels driven into masonry joints or securely attached to masonry or concrete surfaces; or special equivalent devices (brackets) securely attached to masonry or concrete surfaces. Space bracing attachments for horizontal stiffeners at not more than $36''$ intervals in both directions, and not more than $4''$ from floor, ceiling, columns, or other abutting construction.

2.15 HORIZONTAL METAL FRAMING

2.15.1 False Beams.—Where false beams are shown on drawings construct with brackets formed of $3/4''$ channel, or screwable furring to size and shape as detailed. Securely attach brackets to cross furring or main runners or to construction above as required.   (Detail 38)

2.16 SUSPENDED CEILINGS

2.16.1 Spacing and Size of Hangers, Main Runners and Cross Furring (See Table 5-8). [47] [48]

2.16.2 Hangers [49]

2.16.2.1 Steel Construction.—Wrap hangers around, insert through, or clip to steel structural supports, to develop the full strength of the hangers. [50]   (Detail 39-A)

2.16.2.2 Concrete Slab Construction.—Secure hangers to steel rebar or mesh reinforcement in concrete slab construction. On existing construction use approved power driven fasteners. [51] [52]   (Detail 39-B)

2.16.2.3 Concrete Joist Construction.—Insert hangers at the bottom of joists and provide with a loop or other deformation to positively enter the concrete, or secure to the reinforcing steel. On existing construction use approved power driven fasteners.   (Detail 39-C)

2.16.2.4 Wood Construction.—Insert wire hangers in holes drilled a minimum of $3''$ above bottom of joists and twist upper end of the hanger three times around itself, or

Loop wire hangers and attach to sides of joists with two $1\frac{1}{2}''$ No. 9 gauge wire staples. Drive staples horizontally or downward, one at the upper end of the loop and another at the point where it crosses itself. [53]   (Detail 39-D)

2.16.3 Main Runners and Cross Furring.—Do not permit main runners or cross furring to come in contact with abutting masonry or concrete walls and partitions; locate a main runner within $6''$ of the paralleling walls to support ends of cross furring; locate hangers to support the ends of main runners.

Where main runners are spliced, the ends shall be overlapped $12''$ (with flanges of channels interlocked) and securely tied near each end of the splice with wire looped twice around the channels.

Saddle-tie wire hangers around the main runners to prevent turning or twisting and to develop the full strength of the hangers. To complete tie, twist hanger at least three times around itself.

[42] Limiting height: 12'0". Minimum plaster thickness: $3/4''$. Not economical for cut-up walls with many openings.

[43] This system consists of two horizontal channel stiffeners placed at third-points between floor and ceiling (or more, dependent on height of furred surface) and attached to the wall with channel or special bracing attachments. Long length lath is attached to the horizontal channels by wire tie or special clips.

[44] Specify which type of metal floor runner.

[47] Where work of mechanical trades is to be framed, it should be clearly designated on drawings.

[48] The term suspended, as applied to ceiling construction, means that the furring members are suspended below the structural members of the building.

[49] Specification paragraphs 3.14.21-24 are for the architect's convenience. Select the one fitting the type of construction involved.

[50] In open web steel joist construction, where light gauge steel decking is used without concrete slab, suitable provision should be made on wide spans for adequate support of intermediate hangers.

[51] Special inserts of at least equivalent strength to the hangers, and to which the hangers can later be attached, may be inserted through or attached to the top of the forms in lieu of anchoring the hangers directly in the concrete.

[52] Where hangers are attached to devices which have been inserted in existing sound concrete construction, such device should develop at least equivalent strength to the hanger. Specify type required by strength and holding power of concrete.

[53] When purlins exceed spacings of 4' o.c., $1\frac{1}{2}''$ screw eyes (size depending on species of wood) should be fastened into heavy wood flooring above, a maximum of 3' o.c. Upper end of hanger should be inserted in screw eye and twisted three times around itself.

Securely saddle-tie or clip the cross furring to the main runners at each crossing, or secure with attachments of equal strength. When furring members are spliced, the ends shall be lapped not less than 8″, with flanges interlocked, and securely wire-tied near each end of the splice.

Form all openings in ceilings and frame all openings for recessed light fixtures or troffers. [53-A]

2.17 FURRED CEILINGS

2.17.1 Spacing and Size of Hangers and Furring. (See Table 5-8). [54]

2.17.2 Steel Joists.—Securely saddle-tie, clip or weld furring members to the underside of steel joists. [55] (Detail 40-A)

2.17.3 Concrete Joists.—In 20″ or 30″ wide form construction, hangers for supporting furring channels or pencil rods against the bottom of concrete joists shall be No. 10 gauge galvanized wire. [55] [56]

(Detail 40-B)

In 20″ wide form construction, place hangers in the center of panels between joists and space not to exceed 36″ on center longitudinally. In 30″ wide form construction place hangers in the center of alternate panels between joists and, in every other panel between joists, place hangers approximately 2″ from each side of joist and space not to exceed 36″ on center longitudinally.

Provide all hangers with a loop or other deformation to positively enter the concrete, or secure to the reinforcing steel. [57]

Erect ¾″ channels parallel to and between the rows of joists. Securely saddle-tie cross furring of ⅜″ round steel pencil rods to the channels at each crossing.

2.17.4 Wood Joists.—Drive 16d common nails in a horizontal position clear through each joist at least 2″ above the bottom edge of joist. [58]

Space nails to conform to the spacing of furring members. Attach channel or pencil rod furring snugly against the bottom edges of joists by securely saddle-tying at nails. (Detail 40-C)

2.18 CONTACT CEILINGS

2.18.1 Steel Joists.—For allowable spans of various types of lath, see Table 47-C of Uniform Building Code in "Codes and Ordinances" chapter of this manual or applicable code. [59 [60] (Detail 41-A)

2.18.2 Concrete Joists.—For allowable spans of various types of lath, see Table 5-9. For attachment of lath see 4.2.11 and 4.2.21 [61] (Detail 41-B)

2.18.3 Wood Joists.—For allowable spans of various types of lath, refer to applicable code. For application and attachment of lath see 4.2. (Detail 41-C)

3.19 WORKMANSHIP

2.19 Metal Framing.—All metal framing under this section shall be true to line, level, plumb, square, curved or as otherwise required.

[53A] Architect shall show detail of framing required for fixtures and troffers and assign responsibility to the mechanical trade involved for locating all openings in the ceiling grillage.

[54] The term furred, as applied to ceiling construction, means that the furring members are attached directly to the structural members of the building.

[55] Hangers are necessary when spacing of joists requires intermediate supports for cross furring.

[56] Specify only the width construction applicable.

[57] Use in specification for either 20″ or 30″ construction.

[58] Two 8d common nails may be used, one on each side of joists, driven diagonally downward to a penetration of at least 1½″ from a point not less than 2″ above the bottom edge of joists.

[59] The term contact, as applied to ceiling construction, means that the lath is attached in direct contact with the construction above, without the use of runner channels or furring.

[60] Select specifictions to fit requirements.

[61] Specify only the width construction applicable.

Plenum chamber partitions which are to be lathed should be clearly designated on drawings and whether on one or both sides.

Where dissimilar materials abut and are to receive plaster as a continuous finish, metal accessory should be placed at the juncture of such materials; or the juncture should be covered with metal lath; or other methods should be specified as a precaution against cracking.

Where lath continues across face of a concrete column or similar structural member, architect should provide relief from transferred stresses.

All wood surfaces wider than 6″ receiving lath other than plain gypsum or paper-backed lath should be covered with water resistant building paper to break bond and avoid water penetration.

Where plaster contacts unplastered construction, break contact, using casing bead or liquid bond breaker. Specify which.

For lathing under portland cement plaster, it is also recommended that architect consult ACI Proceedings, Title 44-4, Crack Control in Portland Cement Plaster Panels. See 1.8.5.

DETAIL 29

TIES AND SPLICES

For Specification See 3.1.1

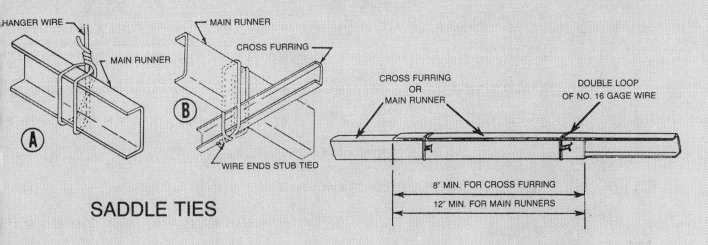

SADDLE TIES

CHANNEL SPLICES

DETAIL 30

PREFABRICATED STUD

Hollow Partition

For Size, Height and Spacing Refer to Manufacturers' Recommendations;
For Erection See 3.4

(1) Ceiling Runner Track

(2) Stud Shoe (wire-tied)

(3) Prefabricated Stud

(4) Door Opening Stiffener

(5) Partition Stiffener

(6) Jack Studs

(7) Metal or Wire Fabric Lath (wire-tied)

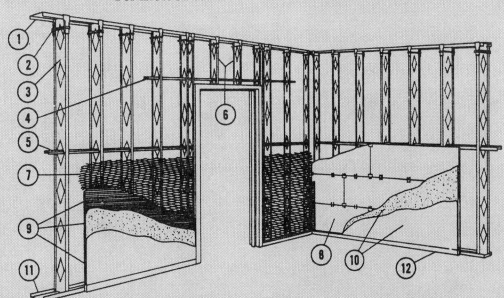

(8) Gypsum Lath (clipped on)

(9) Three Coats of Plaster (Scratch, Brown, Finish)

(10) Two Coats of Plaster (Brown, Finish)

(11) Floor Runner Track

(12) Flush Metal Base Track

DETAIL 31
STEEL STUD
Hollow Partition
For Size, Height and Spacing, Refer to Manufacturers' Recommendations;
For Erection See 3.5

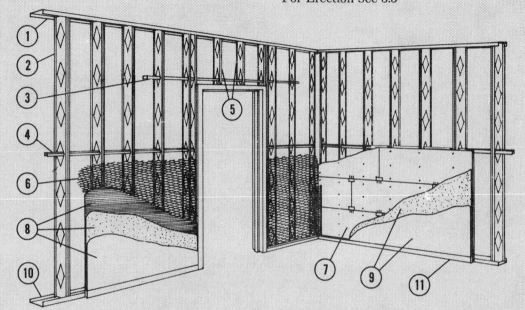

(1) Ceiling Runner Track
(2) Nailable Stud
(3) Door Opening Stiffener
(4) Partition Stiffener
(5) Jack Studs
(6) Metal or Wire Fabric Lath (screwed or wire tied)
(7) Gypsum Lath (nailed, clipped or screwed)
(8) Three Coats of Plaster (Scratch, Brown, Finish)
(9) Two Coats of Plaster (Brown, Finish)
(10) Floor Runner Track
(11) Flush Metal Base

DETAIL 32
SCREW STUD
Hollow Partition
For Size, Height and Spacing, Refer to Manufacturers' Recommendations;
For Erection See 3.6

(1) Ceiling Runner Track
(2) Screw Stud
(3) Door Opening Stiffener[1]
(4) Partition Stiffener[1]
(5) Jack Studs
(6) Metal or Wire Fabric Lath[2] (screwed on)
(7) Gypsum Lath (screwed on)
(8) Three Coats of Plaster (Scratch, Brown, Finish)
(9) Two Coats of Plaster (Brown, Finish)
(10) Floor Runner Track
(11) Flush Metal Base

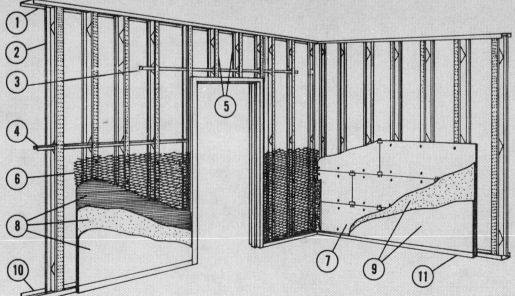

[1] Stiffeners not required in partitions receiving large size lath and veneer plaster.

[2] Check with stud and lath manufacturer for use of metal or wire fabric lath with this type of stud.

DETAIL 33

DOUBLE CHANNEL STUD
Hollow Partition
For Height and Spacing, Refer to Manufacturers' Recommendations;
For Erection See 3.7

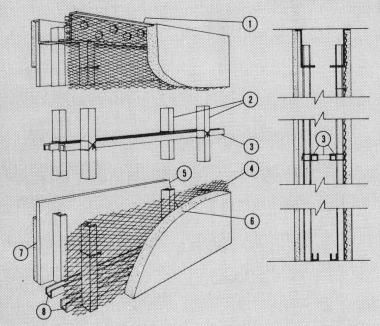

(1) Ceiling Runner
(2) Channel Studs
(3) Partition Stiffener and Channel Spacer
(4) Metal or Wire Fabric Lath (wire-tied)
(5) Gypsum Lath (clipped on)
(6) Three Coats of Plaster (Scratch, Brown, Finish)
(7) Two Coats of Plaster (Brown, Finish)
(8) Floor Runners (channel)

DETAIL 34

SINGLE CHANNEL STUD
Solid Partition
For Stud, Thickness, Height and Length,
Refer to Manufacturers' Recommendations;
For Erection See 3.8; For Plaster Application See 7.6.21 and 7.6.61

(1) Ceiling Runner

(2) Channel Stud

(3) Metal or Wire Fabric Lath (wire-tied)

(4) Plaster

(5) Combination Floor Runner and Screed

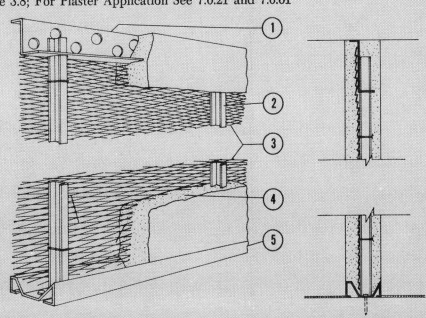

DETAIL 35-A
STUDLESS SOLID PARTITION
For Thickness, Height and Erection See 3.9
For Plaster Application See 7.6.22 and 7.6.62

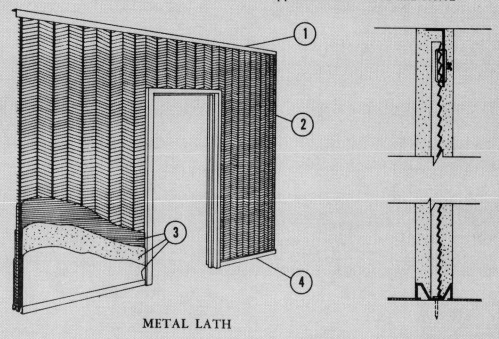

(1) Ceiling Runner
(2) Rib Metal Lath
(3) Plaster
(4) Combination Floor
Runner and Screed

METAL LATH

DETAIL 35-B
STUDLESS SOLID PARTITION
For Thickness, Height and Erection See 3.9
For Plaster Application See 7.6.23

(1) Ceiling Runner
(2) Long Length
Gypsum Lath
(3) Plaster
(4) Combination Floor
Runner and Screed

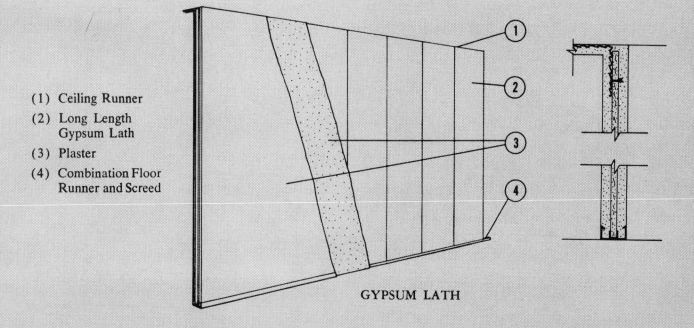

GYPSUM LATH

DETAIL 36

LOAD-BEARING HOLLOW PARTITION
Structural Stud
For Size, Height and Spacing Refer to Manufacturers' Recommendations;

(1) Ceiling Runner Track
(2) Structural Stud (prefabricated)
(3) Structural Stud (nailable)
(4) Jack Studs
(5) Partition Stiffener (bridging)
(6) Metal or Wire Fabric Lath (wire-tied, nailed or stapled)
(7) Gypsum Lath (nailed or stapled)
(8) Three Coats of Plaster (Scratch, Brown, Finish)
(9) Two Coats of Plaster (Brown, Finish)

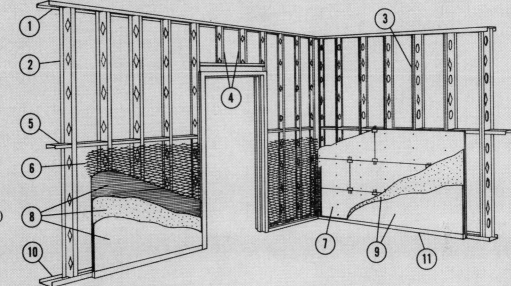

(10) Floor Runner Track   (11) Flush Metal Base

DETAIL 37-A

VERTICAL FURRING
With Studs
For Size, Height and Spacing See Table 5-7; For Erection See 3.12

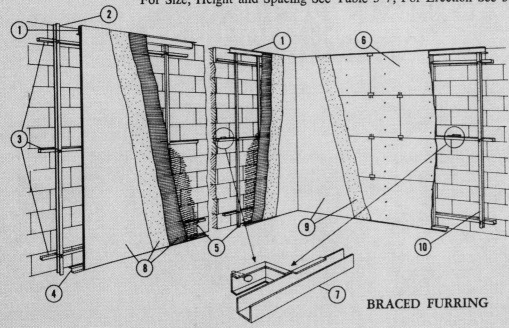

(1) Ceiling Runner[1]
(2) Channel Stud
(3) Horizontal Stiffener[1]
(4) Floor Runner[1]
(5) Metal or Wire Fabric Lath
(6) Gypsum Lath
(7) Bracing
(8) Three Coats of Plaster (Scratch, Brown, Finish)
(9) Two Coats of Plaster (Brown, Finish)
(10) Screw Channel Studs

[1] When ends of studs are secured to runner or directly to construction at floor and ceiling, omit stiffeners at top and bottom.

FREE STANDING FURRING

BRACED FURRING

41

DETAIL 37-B
VERTICAL FURRING
Studless
Limiting Height 12'0"

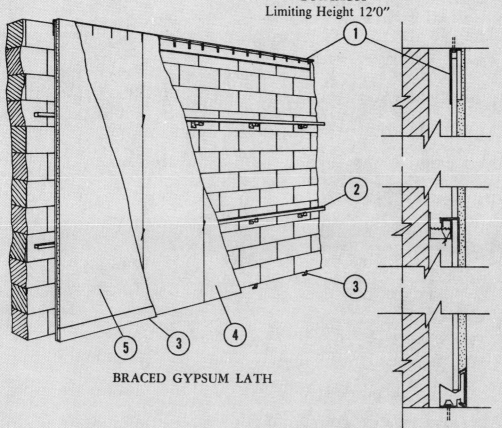

BRACED GYPSUM LATH

(1) Ceiling Runner

(2) Horizontal Stiffener (secured to bracing attachment)

(3) Metal Base and Clips

(4) Gypsum Lath

(5) Three Coats of Plaster (Scratch, Brown, Finish) (Minimum plaster thickness is ¾ inch.)

DETAIL 37-C

COLUMN FURRING
(Alternate Methods)
See 3.12.33 and 3.12.34

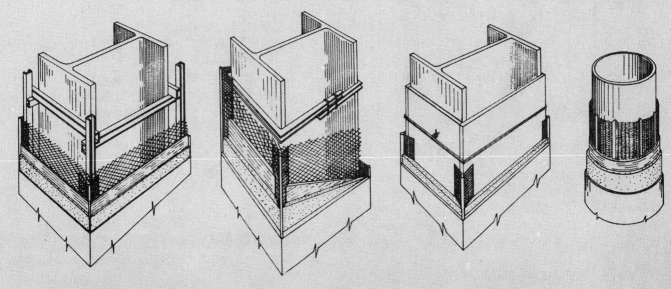

DETAIL 37-D
CONTACT FURRING

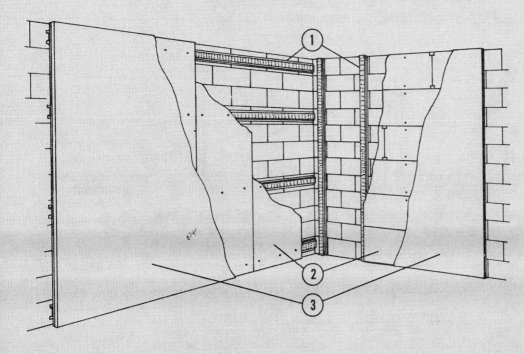

(1) Screw Channel

(2) Gypsum Lath (screwed on)

(3) Two Coats of Plaster (Brown, Finish)

DETAIL 38
FALSE BEAMS
(Alternate Methods)

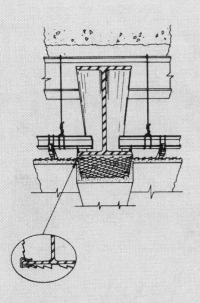

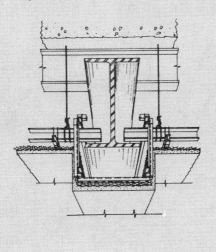

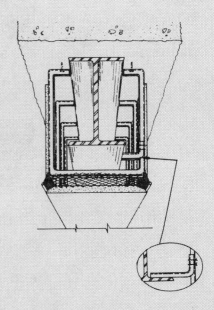

LATH DIRECT TO UNDERSIDE OF BEAM

CHANNEL BRACKETS TO MAIN RUNNERS

CHANNEL BRACKETS TO SLAB

DETAIL 39
SUSPENDED CEILINGS

For Size and Spacing of Components See Table 5-8
For Erection See 3.14

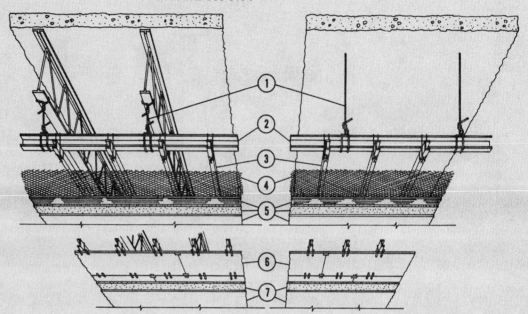

(1) Hanger

(2) Main Runner
Channel

(3) Furring Channel

(4) Metal or Wire
Fabric Lath

(5) Plaster

(6) Gypsum Lath[1]

(7) Plaster

[1] Gypsum lath clip-on systems vary. Lath should be attached as recommended by manufacturer of system.

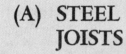

(A) STEEL JOISTS

(B) CONCRETE SLAB

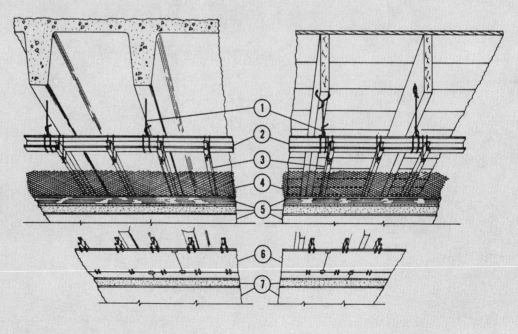

(C) CONCRETE JOISTS

(D) WOOD JOISTS

DETAIL 40
FURRED CEILINGS

For Size and Spacing of Components See Table 5-8

For Erection See 3.15

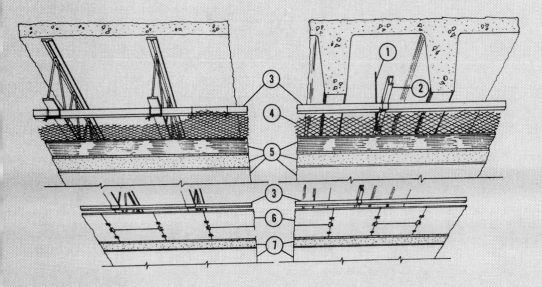

(A) STEEL JOISTS

(B) CONCRETE JOISTS

(1) Hanger
(2) ¾-inch Channel
(3) Cross Furring
(4) Metal or Wire Fabric Lath
(5) Plaster
(6) Gypsum Lath[1]
(7) Plaster

[1] Gypsum lath clip-on systems vary. Lath should be attached as recommended by manufacturer of system.

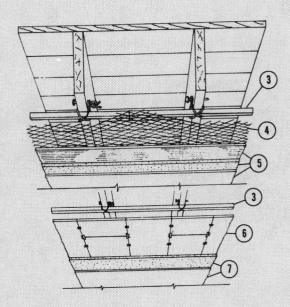

(C) WOOD JOISTS

DETAIL 41
CONTACT CEILINGS

For Size and Spacing of Components See Table 5-8
For Erection See 3.16

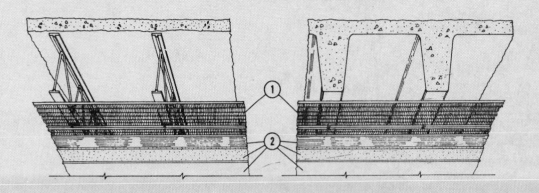

(A) STEEL
JOISTS

(B) CONCRETE
JOISTS

(1) Metal or Wire
Fabric Lath

(2) Plaster

(3) Gypsum Lath

(4) Plaster

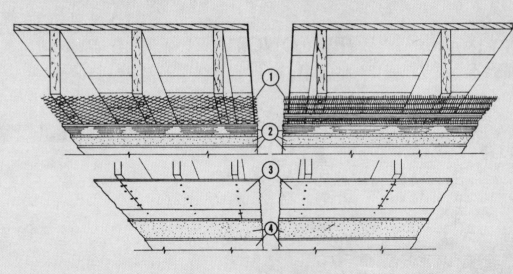

(C) WOOD
JOISTS

3. LATHING
LATHING MATERIALS

3.1 Standard Specifications.—Where published standard specifications are referred to herein, they shall be those of the latest date of adoption. [1]

3.2 WIRE

3.2.1 Hanger and Tie Wire.—Wire shall be galvanized and annealed low carbon steel. The weight of galvanized finish shall be not less than Class I, as set forth in Federal Specifications, QQ-W-461g, Steel Carbon (Round, Bare and Coated).

3.2.2 Gauges of Wire.—All gauges of wire shall be U. S. steel wire gauge (identical to Washburn and Moen).

Wire Gauge	Diameter, In.
No. 20	0.0348
No. 19	0.0410
No. 18	0.0475
No. 16	0.0625
No. 14	0.0800
No. 12	0.1055
No. 10	0.1350
No. 9	0.1438
No. 8	0.1620
No. 7	0.1770

3.2.3 Nails and Staples.—Shall comply with "Federal Specifications for Nails, Wire and Staples," FF-N-105, Amendment 2, except that staple dimensions shall be outside dimensions.

3.2.4 Screws.—Shall be designed specifically for the attachment of metal to metal or gypsum to metal with self tapping point and rust inhibiting coating. Screws shall comply with ASTM C 646.

3.2.4.1 Staples.—Machine or power-driven staples for attaching metal, wire fabric or gypsum lath to wood or metal supports shall be formed from galvanized steel wire having a tensile strength range of 80,000 to 110,000 psi; galvanizing shall be Type I coating [2]

3.3 SHEET METAL MATERIALS

3.3.1 Weight Tolerances.—The allowable variation in percentages over and under specified weights of metal lathing materials and accessories shall be in accordance with customary mill tolerances as established by the American Iron and Steel Institute. Gauges of sheet metal materials for studs and, where hereinafter referred to, of other sheet metal materials and uses shall comply with the following:

Sheet Metal Gauge (U.S. Standard)	Equivalent Thickness (In Approx. Inch Decimals)
No. 12	0.1046
No. 14	0.0747
No. 16	0.0598
No. 18	0.0478
No. 20	0.0359
No. 24	0.0239
No. 25	0.0219
No. 26	

[1] Where any reference is made in these specifications to ASTM, ACI, U.S. Federal Specifications, AIA or CSI, applicable excerpts will be found in Specification Reference 1.8.

[2] Most building code approvals of staples encompass crown widths from 7/16" to 7/8". Dimensions stated in Tables 5-10 and 5-11 are greater than the minimums permitted. When staple lengths are equal, a wider crown will provide improved holding power because of increased support of the material attached. Staple dimensions should always be specified as minimum, to permit the use of increased crown width.

CONSTRUCTION TECHNIQUE

Wire Ties

Several types of ties are used in the installation of metal supports and metal lath:

DOUBLE-WRAP TIE.—Used in tying spliced metal members such as channels or studs, the double-wrap tie is formed by two complete wraps of a single strand of wire around the spliced members at each end of the splice. The tie is completed with a stub tie.

SADDLE TIE (A).—Used in tying hanger wire to main runner channels to support a suspended ceiling, the tie is completed by twisting the hanger wire three times around itself. The saddle tie prevents rotation of the channel, and as the tie tightens it holds the channel firmly.

SADDLE TIE (B).—Generally used in tying cross furring channels to the underside of main runner channels, and horizontal stiffeners in partitions and vertical furring, this type of saddle tie is used to prevent rotation of the furring channel or stiffener. The tie is completed with a stub tie.

FIGURE EIGHT TIE.—Used in attaching a channel runner to channel brackets. Tie is made at intersection of runner and internal angle of each bracket it crosses to secure the runner to both vertical and horizontal legs of bracket. The tie is completed with a stub tie.

STUB TIE.—This is the industry name given to the process of completing any light gage wire tie with (minimum) one and one-half twists of both ends of the tie wire just prior to cutting the twisted ends. The stub tie is used to complete a saddle tie of channel secured to another channel, or to a metal stud, i.e., cross furring (See Detail B); horizontal stiffeners. The stub tie may be used to complete any other simple tie where a standard gage tie wire (No. 16 & No. 18 gage) is used.

BUTTERFLY TIE.—Used almost exclusively in attaching metal lath to metal supports, this is accomplished by a half twist of the two ends of the tie wire forming wings ½ to ¾ in. long, parallel to the direction of the tie, and in opposite directions to each other. Wings must lay up tight against the lath.

NOTE.—Where a single strand of No. 16 gage wire is used to form a tie, double strands of No. 18 gage wire may be substituted, achieving equivalent or greater load capacity.

BUTTERFLY TIE

SAFE LOADS FOR SINGLE STRANDS OF TIE WIRE[1]		
Wire Gage	Safe Load in Lbs.[2]	Estimated Ultimate Strength in Lbs.[3]
18	32	108
16	56	186

(1) A single strand of wire may be a hanger in a suspended ceiling where only one strand supports the load. When a single loop of tie wire is made, two strands of wire support the load. Example: Each 18 gage wire tie attaching metal lath to a channel will support a total load of 2 x 32 or 64 pounds safe load.

(2) Based on a design or working stress of 18,000 psi.

(3) Based on ultimate strength of 60,000 psi.

3.3.2 Accessories.—Metal lathing accessories shall be formed from galvanized sheets of no less than the following gauges: [3]

Corner beads; base screeds;

picture molds; partition runners;

ventilating expansion screeds.................................No. 26 gauge

Casing beads; drip screeds.................................No. 24 gauge

Veneer corner bead, veneer casing and veneer control

joints ....No. 28 gauge

3.4 & 3.5 (Reserved)

3.6 METAL AND WIRE FABRIC LATH

3.6.1 General.—Metal bases for gypsum and portland cement plaster construction shall comply with ASTM C 847—Metal Lath; ASTM C 933—Welded Wire Lath; and ASTM C 1032—Woven Wire Plaster Base.

3.6.2 Types.—Metal plaster base shall be of the following types:

(Detail 9)

Type F—Flat base
Type SF—Self-furring base
Type FR—Flat rib lath
Type F⅜R—Rib lath, ⅜"
Type FR—Flat base with backing
Type SFB—Self-furring base with backing
Type FRB—Flat rib lath with backing
Type F⅜RB—Rib lath, ⅜" with backing
Type F¾R—Rib lath, ¾"

3.6.3 Flat and Self-Furring Diamond Mesh.—Shall weight 2.5, 3.4 lbs. per sq. yd., and shall be expanded from steel sheets, coated with rust-inhibitive paint after fabrication. [7]

3.6.4 Flat Rib Lath.—Shall weigh 2.75, 3.4 lbs. per sq. yd., and shall be expanded from steel sheets, coated with rust-inhibitive paint after fabrication.

3.6.5 Rib Lath.—Shall weigh 3.4 lbs. per sq. yd., and shall be expanded expanded from steel sheets, coated with rust-inhibitive paint after fabrication. [7] [8]

3.6.6 Wire Fabric Lath.—Shall be fabricated of cold drawn steel wire, coated with rust-inhibitive paint after weaving or welding; galvanized after weaving or welding; or shall be fabricated from galvanized wire.

(Detail 10)

3.6.6.1 Welded Wire Fabric Lath.—For interior surfaces and protected horizontal exterior surfaces, with or without absorptive paper separator, shall be fabricated from no lighter than No. 16 gauge wire, with openings not to exceed 2" x 2", welded at all intersections of wire and stiffened continuously, horizontally, at no more than 6" intervals. The lath shall provide metallic reinforcement, in both directions, for the plaster.

[3] In areas where the presence of salt or continued high humidity are likely to adversely affect the exposed portions of metal lathing, accessories should be formed from noncorrosive materials. Specify type.

[6] Paragraphs 2.6.1 through 2.6.74 on metal plaster bases are for the architect's convenience. Specify types required by construction.

[7] Also available with paper backing.

[8] Also available from galvanized sheets.

DETAILS

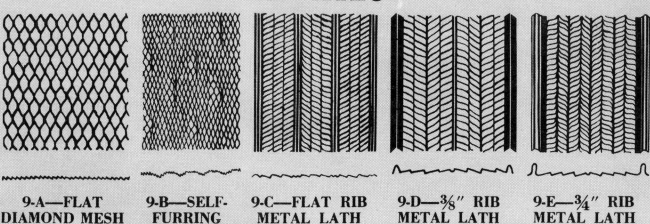

9-A—FLAT DIAMOND MESH METAL LATH **9-B—SELF-FURRING METAL LATH** **9-C—FLAT RIB METAL LATH** **9-D—⅜" RIB METAL LATH** **9-E—¾" RIB METAL LATH**

Note: Not manufactured on West Coast.

3.6.6.2 Welded Wire Fabric Lath With Backing.—For interior and exterior surfaces shall be fabricated from no lighter than No. 16 gauge wire, with openings not to exceed 2″ x 2″, welded at all intersections of wire. Backing shall comply with Federal Specification for Building Paper, UU-B-790, of style and grade described in that specification for the intended use; or backing may be of other waterproofing or insulating material complying with, or exceeding, these requirements. Backing shall be securely attached. The lath shall provide metallic reinforcement, in both directions, for the plaster and shall comply with ASTM C 933.

3.6.6.3 Woven Wire Fabric Lath (Stucco Netting).—For exterior or interior surfaces, with or without backing, shall be of woven steel wire with large openings 1″ minimum, and 1½″ maximum, minimum size of wires No. 20 gauge 1″ openings and No. 17 gauge for 1½″ openings. When manufactured with backing, backing shall comply with Manufacturers Specification for Building Paper of style and grade described in that specification for the intended use; or backing may be other waterproofing or insulating material complying with, or exceeding, these requirements. Backing shall be securely attached. The lath shall provide metallic reinforcement, in both directions, for the plaster and shall comply with ASTM C 933.

3.6.6.4 Paperbacked Woven Wire Fabric Lath (Stucco Netting).— For exterior surfaces, or interior surfaces, with backing, shall be of No. 17 gauge for 1½″ openings. Backing shall comply with Manufacturers Specifications for Building Paper of style and grade described for the intended use, or be of a high absorbent suction paper for gypsum or portland cement plaster for interior or protected locations. Backing shall be securely attached.

3.7 GYPSUM LATH

3.7.1 General—Gypsum lath shall conform to the "Standard Specifications for Gypsum Lath"—ASTM Designation: C 37 (Supplement to ASTM Standards).

3.7.2 Types—Gypsum lath shall be: (Detail 11)

Plain gypsum lath.

Type X lath shall be plain gypsum lath except with a core having increased fire-retardant properties to give higher fire-resistive rating.

Lead-backed lath shall be plain gypsum lath having sheets of lead laminated to one side for control of X-ray transmission.

Veneer plaster lath shall conform to the "Standard Specifications for Gypsum Lath"—ASTM Designation: ASTM C 588 except that dimensions as to thickness, widths and lengths shall not apply.

Insulating gypsum lath shall be plain gypsum lath having aluminum foil laminated to one side to act as a vapor barrier and also as reflective thermal insulation.

3.7.3 Standard Dimensions.—Gypsum lath shall conform in general to the following dimensional requirements:

        Thickness — ⅜, ½, ⅝ or 1 in.
        Width   — 16, 24, or 48 in.
        Length   — 48, or 96 in. (longer lengths available for special uses).

3.8 ACCESSORIES

3.8.1 General.—Metal shapes used as grounds shall be of such size and dimension as to provide for required plaster thickness. [9]

3.8.2 Base or Parting Screeds.—Screeds to divide or separate plaster surfaces from other similar materials shall be fabricated from minimum No. 26 gauge steel sheets.   (Detail 12)

[9] Base screeds, casing beads, ventilation screeds, control joints, etc.

50

3.8.2.1 Terrazzo Screeds.—Screeds to separate plaster from terrazzo shall be non-ferrous metal, or a combination of galvanized steel and non-ferrous metal of design as shown on drawings. [10]

3.8.3 Corner Beads.—Small nose or bull nose corner beads shall be fabricated from minimum No. 26 gauge galvanized steel sheets; veneer plaster corner beads No. 28 gauge; solid flange, expanded flange, or combination of both (or) woven mesh flange. [11]   (Detail 3)

3.8.4 Casing Beads.—Metal casings of various shapes, when used as plaster grounds, shall be formed of minimum No. 24 gauge steel, galvanized; with short or expanded flange. Veneer plaster casing beads shall be formed of No. 28 gauge. [12]   (Detail 4)

3.8.5 Corner Reinforcement (exterior).—External corner (arris) reinforcement for portland cement plaster shall be fabricated from expanded metal with large openings, from welded or woven copper-bearing steel wire of minimum No. 18 gauge, galvanized, or otherwise treated to give superior corrosion resistance. [12]   (Detail 15)

3.8.6 Partition Terminals (Caps).—For solid plaster partitions, shall be made of a minimum No. 16 gauge galvanized curved nose with minimum No. 26 gauge galvanized, expanded wings; (or) minimum No. 24 gauge galvanized square nose, with minimum No. 24 gauge galvanized solid or expanded flange wing.   (Detail 17)

3.8.7 Partition Runners.—For use as tracks in solid plaster partitions, shall be formed or punched from minimum No. 26 gauge galvanized steel. [14]   (Detail 18)

3.8.8 Cornerite.—For reinforcing internal angles over lath, shall be minimum 1.75 metal lath, coated with rust-inhibitive paint, or woven or welded wire fabric of minimum No. 19 gauge wire (or weighing not less than 1.7 lbs. per square yard) galvanized after weaving or welding, unless fabricated from galvanized wire. When shaped for angle reinforcing, it shall have outstanding legs of no less than 2".   (Detail 20)

3.8.9 Strip Reinforcement.—For reinforcing continuous joints of gypsum lath, and joints of dissimilar materials, in the same plane, shall be minimum 1.75 metal lath, coated with rust-inhibitive paint, or woven or welded wire fabric of minimum No. 19 gauge wire (or weighing not less than 1.7 lbs. per square yard) galvanized after weaving or welding, unless fabricated from galvanized wire; minimum width of strip 3".   (Detail 21)

3.8.10 Control Joints.—Shall be fabricated from minimum No. 28 gauge galvanized steel sheets, zinc alloy of PVC vinyl; plain, perforated or expanded flange.

3.8.11 Ventilating Expansion Screed.—Used on horizontal or protected vertical surfaces for ventilation of air spaces under arcades, corridors, canopies, eaves, shall be formed of minimum No. 26 gauge galvanized steel.   (Detail 23)

3.8.12 Drip Screed (Weep Screed).—For use at foundations or horizontal external corner or soffits, shall be formed of minimum No. 26 gauge galvanized steel, zinc alloy, PVC vinyl or welded wire, with or without weep holes. [15]

3.8.13 Window Stools.—When acting as a ground for plaster, shall be formed of minimum No. 16 gauge galvanized steel with square or curved apron, prime coat painted. [13]

[10] Terrazzo screeds with colored plastic grinding strips are available.

[11] Corner beads should be specified on all interior arrises; are not recommended on exterior surfaces under portland cement plaster. See 2.8.5 Corner Reinforcement. Available in zinc alloy.

[12] Available in zinc alloy.

[13] This item no longer manufactured as a stock item and must be custom made.

[14] L or Z shapes. Other types of runner track may function both as stud or lath attachment and as ground for plaster.

[15] Control joints and drip screeds may be of other weight as specified.

3. LATHING MATERIALS

3.9 ATTACHMENT CLIPS (Wire or Sheet Metal)

3.9.1 Clips.—Devices — for attaching framing members to supports, or to each other; for attaching lath to framing members; or for securing lath to lath — shall be formed of galvanized steel wire, or sheet metal, depending on use and manufacturer's requirements. [16]   (Detail 25)

3.10 WEATHER PROTECTION

3.10.1 Building Paper.—Shall comply with Federal Specification UU-B-790 a grade —— [17]

3.10.2 Felt.—Shall comply with Federal Specification for Felt, Asphalt-Saturated, HH-F-191a, and Amendment 2. Felt shall weigh no less than 14 lbs. per 108 sq. ft. [18]

3.10.3 Lath and plaster elements installed in a plane less than 60° from the horizontal require special moisture protection design considerations.

[16] To decrease sound transmission (and dissipate vibration) specify special clips as per manufacturer's design.

[17] Grade B paper has high water resistance and low water-vapor permeability or breathing properties. Grade D paper has low water resistance, and high water-vapor permeability. Specify grade of paper required.

[18] Asphalt saturated felt should be used only where high water-vapor permeability (breathing) is required.

DETAILS

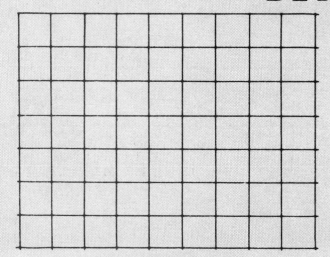

**10-A—PLAIN WIRE
FABRIC LATH**

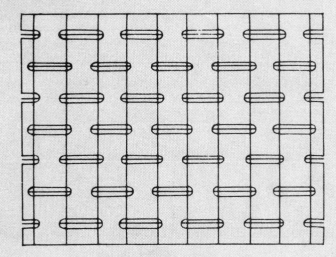

**10-B—SELF-FURRING
WIRE FABRIC LATH**

**10-C—PAPER BACKED
WOVEN WIRE
FABRIC LATH**

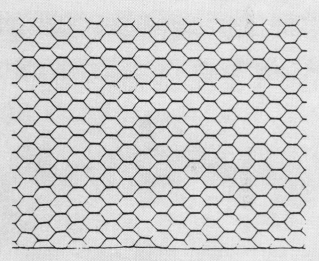

**10-D—WOVEN WIRE
FABRIC LATH
(Also Available Self-Furred)**

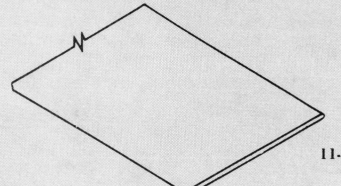

**11-A—PLAIN GYPSUM
LATH**

DETAILS

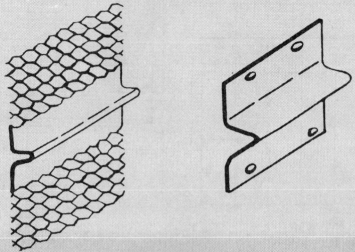

12—BASE OR PARTING SCREEDS

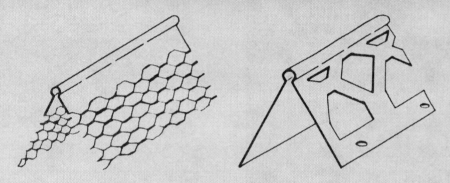

13-A—SMALL NOSE CORNER BEADS

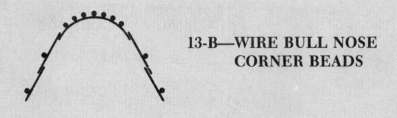

**13-B—WIRE BULL NOSE
CORNER BEADS**

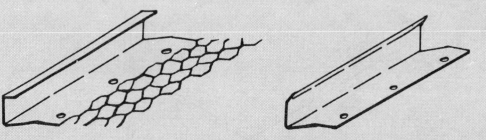

14-A—SQUARE CASING BEADS

DETAILS

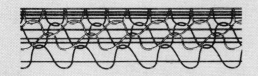

15-A—CORNER REINFORCEMENT (EXTERIOR) WIRE

15-B—CORNER REINFORCEMENT (EXTERIOR) EXPANDED METAL

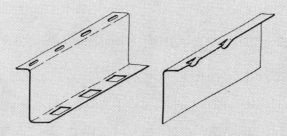

18—PARTITION RUNNERS (Z AND L SHAPE)

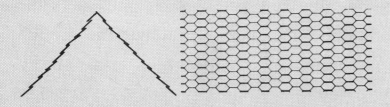

20-A—EXPANDED METAL CORNERITE

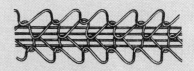

20-B—WIRE CORNERITE

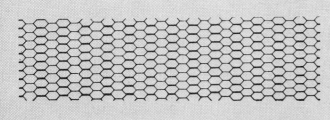

21-A—STRIP REINFORCEMENT (EXPANDED METAL)

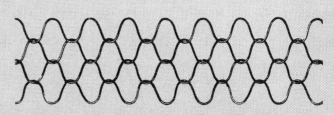

21-B—STRIP REINFORCEMENT (WIRE)

DETAILS

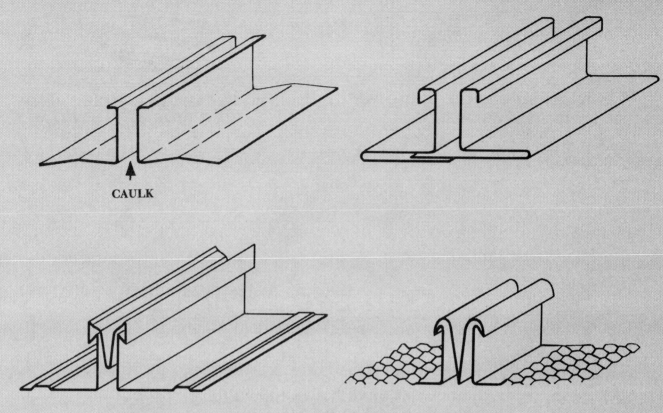

CAULK

22—STRESS RELIEF (CONTROL JOINTS)

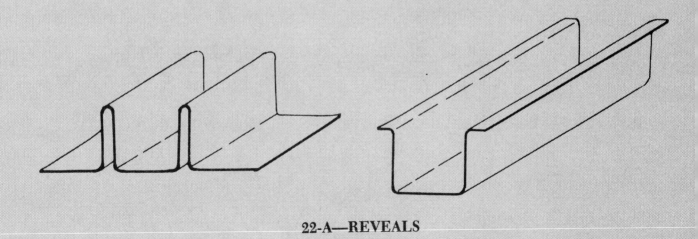

22-A—REVEALS

DETAILS

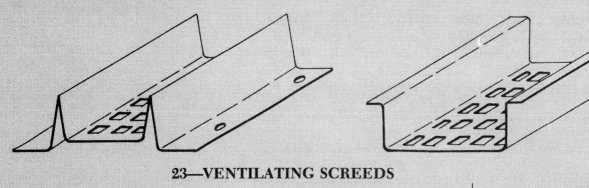

23—VENTILATING SCREEDS

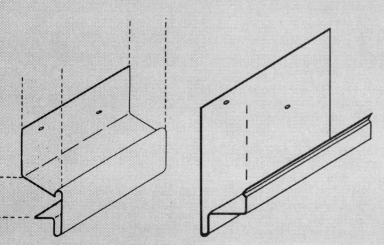

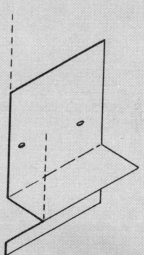

24—DRIP SCREEDS

24-A—WEEP SCREED
(Also available with perforations.
See Recommended Construction Techniques,
pages 92/93).

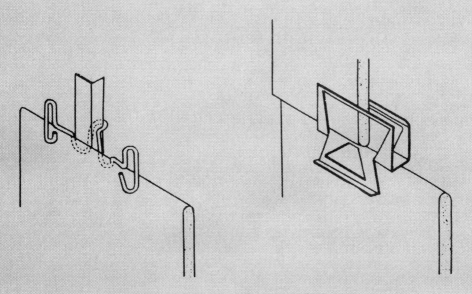

25—GYPSUM LATH ATTACHMENT CLIPS

LATHING APPLICATION

3.11 ATTACHMENT

3.11.1 For Metal Supports.—Wire, nails, screws, and staples shall be of type, minimum size and maximum spacing, along metal supports as required by applicable building code. Special clips shall be as recommended by manufacturer of lath.

3.11.2 For Wood Supports.—Wire, nails, screws and staples shall be of type, minimum size and maximum spacing, along wood supports as required by applicable building code. Special clips shall be as recommended by manufacturer of lath.

3.11.3 For Seismic Stresses (horizontal wood supports).—After attachment of metal or wire fabric lath to horizontal wood supports, drive home a ⅝″ wide, 1½″ long No. 9 gauge, ring shank, hook staple over a 10d nail placed against face of lath not more than 3″ from edge of each sheet. Space staples at not more than 27″ intervals on every joist. Over ⅜″ rib lath, staple may be placed over ribs, omitting 10d nail. [1] [2]

Supplementary Provision For Tying.—Drive 16d common nails in a horizontal position through joists at least 2″ above the bottom edge of joists. Space nails 24″ or 27″ along the sides of alternate joists, spacing depending on the width of lath sheet, and so placed that nails will be not more than 3″ back from edge of each sheet.

In place of 16d common nails driven clear through joists, two 8d common nails may be used; one on each side of joist, driven diagonally downward to a penetration of at least 1½″ from a point not less than 2″ above the bottom edge of joists.

Attach metal lath to each nail (or pair of nails) with a strand of No. 18 gauge wire. Bend the wire into a "U" shape. Double over each leg to form loops and push them through the lath one on each side of the joist, so as to catch the projecting point or head of the nail. Pull the free ends of the wire to tighten the "U" against the lath, and twist the two free ends together beneath the lath. [1] [2]   (Detail 42)

3.12 INTERIOR (application of lath) [3] [4]

3.12.1 Expanded Metal Lath (general application).—Stagger ends of lath to avoid continuous joint on same support. Lap lath at sides not less than ½″ and not less than 1″ at ends. Lap ends of lath at supports. Butt lath into internal angles and reinforce angle with cornerite formed of metal lath. [5] [6] [7] [8]

Attach metal lath to supports at maximum 6″ intervals. [9]

Place an attachment at side laps on supports. On underside of steel joists tie lath with one No. 16 gauge wire or a doubled strand of No. 18 gauge wire.

3.12.1.1 Attachment to Concrete Joists.—On 20″ wide form construction, attach metal lath to the underside of concrete joists by twisting the No. 14 gauge wire hangers as for tie wire, or by clinching other types of hanger wire.

On 30″ wide form construction, attach metal lath to the underside of concrete joists by twisting the No. 14 gauge wire hangers as for tie wire, or by clinching other types of hanger wire. Wire-tie metal lath to the channel runners or power driven fastener.

3.12.1.2 Attachment to Concrete Surfaces. [10]

[1] This construction should be used when the required holding power of nails is not obtainable; and in structures subject to greater-than-ordinary vibration, earthquake stresses or other disturbances.

[2] Specify either hook staple or supplementary tying at option of contractor.

[3] Where work of mechanical trades is framed in (furred) and is to be enclosed with lath and plaster it shall be clearly indicated on the drawings.

[4] Specify metal lath or wire fabric lath with backing behind portland cement thin-set or mortar-set ceramic tile.

[5] For gypsum or portland cement plaster.

[6] Specify type of lath—flat (Type F); flat with backing (Type FB); self-furring (Type SF); self-furring with backing (Type SFB); expanded rib, flat rib or ⅜″ rib. Stud spacing dictates type and weight of lath.

[7] Diamond mesh lath may be bent around internal angles without use of cornerite. Specify when wanted.

[8] When attached to wood supports, nail cornerite only sufficiently to retain position until plastering.

[9] For type, size and spacing of attachments see Table 5-10 (metal supports); Table 5-11 (wood supports).

[10] Where metal lath must be applied directly to large areas of vertical or horizontal concrete surfaces, attachments should be of type sufficient to support the load of the finished material. Metal lath should not be attached to such surfaces with concrete nails. Concrete nails may be used to attach metal lath to concrete columns, beams, pilasters and similar areas. Care must be given to the compressive strength of the concrete in selecting power driven fasteners.

3.12.2 Welded Wire Fabric Lath (general application).—Under portland cement plaster, stagger ends of lath to avoid continuous joints on same supports. Lap all joints at least one mesh. In general, lap ends of lath at supports. Butt lath into internal angles and reinforce angles with conerite reinforcement. [11] [12] [13] [14]

Attach welded wire fabric lath to supports at maximum 6″ intervals. [15]

Place an attachment at side laps on supports. On underside of steel joists, tie lath with one No. 16 gauge wire or a doubled strand of No. 18 gauge wire.

3.12.2.1 Attachment to Concrete Joists.—On 20″ wide form construction, attach welded wire fabric lath to the underside of concrete joists by twisting the No. 14 gauge wire hangers as for tie wire, or by clinching other types of hanger wire.

On 30″ wide form construction, attach welded wire fabric lath to the underside of concrete joists by twisting the No. 14 gauge wire hanger as for tie wire, or by clinching other types of hanger wire. Wire-tie welded wire fabric lath to the $3/4$″ channel runners.

3.12.3 Woven Wire Fabric Lath (stucco netting).—For application under exterior portland cement plaster and interior portland cement or gypsum plaster over wood or metal members. [11] [12] [13] [14] [15] [16]

3.12.4 Gypsum Lath (general application).—Apply gypsum lath, face side out, with the long dimension at right angles to supports. On wood supports, break end joints in each course except that end joints may fall on one support when stripped with 3″ wide metal lath or wire fabric reinforcing. When attached to metal supports with special clips, nails or screws, place ends of lath between supports in alternate courses and clip to adjacent edges. When lath edges are not in moderate contact, cover all joints wider than $3/8$″ with minimum 3″ wide metal lath or wire fabric reinforcing. [17] [18]

Install cornerite at all internal angles fastened sufficiently to retain position during plastering.

Install a minimum 4″x8″ piece of metal reinforcement diagonally at corners of openings larger than 2 sq. ft. [19] [20]

[11] For gypsum or portland cement plaster.

[12] Specify type of lath—flat with backing (or separator paper) (Type FB); self-furring with backing (or separator paper and backing) (Type SFB). Lath should be chosen in conjunction with stud spacing.

[13] Welded wire fabric lath may be bent around internal angles without use of cornerite. Specify when wanted.

[14] When attached to wood supports, nail cornerite only sufficiently to retain position until plastering.

[15] For type, size and spacing of attachments see Table 5-10 (metal supports); Table 5-11 (wood supports).

[16] Where welded wire fabric lath must be applied directly to large areas of vertical or horizontal concrete surfaces, attachments should be of type sufficient to support the load of the finished material. Welded wire fabric lath should not be attached to such surfaces with concrete nails. Concrete nails may be used to attach metal lath to concrete columns, beams, pilasters and similar areas.

[17] For gypsum plaster.

[18] Specify type of lath—plain or perforated, Type X, long length, insulating.

[19] Strip reinforcement may be required on ceilings of wood construction to achieve minimum fire ratings.

[20] All stripping may be omitted when the entire surface is reinforced with not less than 1″ No. 20, U.S. gauge woven wire fabric lath.

DETAIL 42– ATTACHMENT FOR SEISMIC STRESS

Supplementary Tying
(alternate supports)

Flat Expanded Metal Lath
and Earthquake Staple
(each support)

Welded Wire Fabric Lath
and Earthquake Staple
(each support)

Rib Metal Lath and
Earthquake Staple
(each support)

For Specification
See Chapter 4.13

3. LATHING APPLICATION

3.12A STUDLESS PARTITIONS (solid)

3.12A.1 Partition Thickness Height and Length.–Minimum thickness of studless solid partitions shall be 2″. Maximum height shall be 10′.

3.12A.2 Preparation for Erection of Lath.–Secure metal runners or metal base clips or combination runner and screed to the floor to accurately locate the partition; space attachments at maximum 24″ intervals; use fasteners suitable to the floor construction material.

Install L-shape runners, or runners made of cornerite, or runners of equal rigidity so that lath will be plumb and in the center of the partition. Fasten runners to lath ceiling at maximum 12″ intervals with wire ties, staples or suitable clips; to concrete slabs with minimum ⅞″ powder driven pins, or ½″ or ⅝″ concrete stub nails, at no more than 24″ intervals. Ceiling runner shall lap or support lath not less than ½″. [20A]

3.12A.3 Framing Around Door Openings.–For metal lath install one channel stud at each jamb of metal frames, continuous from floor to ceiling. Wire-tie, bolt, weld or otherwise secure stud to the jamb anchors; if frame is of wood, secure channel stud with pairs of 8d nails or other secure attachment. [20B [20C]

For gypsum lath, no additional framing is required around door openings.

3.12A.4 Erection of Lath.

3.12A.5 Metal Lath.–Erect ⅜″ rib metal lath, or lath of equal rigidity, with the long dimension of the sheet vertical. Secure sheets to the ceiling runners with wire ties at 8″ on center and attach to floor runners, metal base clips or anchor in a grouted metal base assembly. Place a wire tie where sheets lap at runners; tie vertical laps at not more than 16″ intervals.

Where studless partitions intersect walls to be plastered, reinforce the angle with cornerite, or bend lath around intersection.

Fasten runners to abutting walls which are not to be plastered and wire the lath to the runner so that the complete assembly is embedded in the plastered partition. [20D]

Protect exposed partition ends and top with a metal partition terminal, or corner beads. [20E]

3.12A.6 Gypsum Lath.–Erect gypsum lath, including the clips and any additional reinforcing, strictly in accordance with the proprietary specifications for the system employed. Wherever possible start a full width at door openings. [20F]

Protect exposed partition ends and top with a metal partition terminal, or corner beads.

3.12A.7 Bracing.–Temporarily brace rib metal lath and gypsum lath studless partitions. [20G]

3.12AA FURRING FOR STUDLESS PARTITIONS

3.12AA.1 Furring for Vertical Gypsum Lath and Bracing (studless).–Align metal floor runner, or combination metal clips and base, to wall layout and secure to floor with minimum ⅞″ powder driven pins or ¾″ concrete stub nails or other suitable attachment, spaced at not more than 36″ intervals. [20H] [20I] [20J]   (Detail 37-B)

[20A] Hollow metal bases may be fully grouted before or after lathing, dependent upon type of base

[20B] Do not use heavy or oversize doors in solid plaster partitions.

[20C] Specify metal door frames shall be of such and installed by others and provide that:

[a] Return on flanges of frames shall be of such depth as to permit lath and plaster to solidly fill frames. Frames shall be carefully plumbed and either anchored to the construction above (struts) or be temporarily braced until the base coat plaster has set.

[b] Floor anchors in frames shall be a minimum of No. 16 gauge steel, welded to frames and with holes in them for attachment, spaced a minimum of 3″ apart. Door frames shall be anchored to floor at floor anchors with bolts, screws or other secure attachment.

[c] Jambs for metal lath solid partitions shall contain four jamb anchors of No. 16 gauge steel formed to permit a channel stud to be tied, bolted or welded to anchors; jambs for gypsum lath solid partitions shall contain inserts for centering lath in frame.

[d] Struts shall not exceed ⅝″ in width.

[20D] Provision should be made for bond breaker and V-joint cut at contact of plaster with unplastered construction. Casing bead may be used

[20E] Electrical outlets should not exceed 1½″ in depth and piping should not exceed 1″ in diameter for 2″ solid partitions. Back of outlets should be coated with liquid bonding agent before plastering

[20F] Specify names of systems and thickness and width of lath.

[20G] Place bracing on rib side of rib lath.

[20H] Limiting height: 12′0″. Minimum plaster thickness: ¾″. Not economical for cut-up walls with many openings.

[20I] This system consists of two horizontal channel stiffeners placed at third-points between floor and ceiling (or more, dependent on height of furred surface) and attached to the wall with channel or special bracing attachments. Long length lath is attached to the horizontal channels by wire tie or special clips.

[20J] Specify which type of metal floor runner

Align combination ceiling runner and clip, or a ¾" channel, with the floor runner so that lath will be plumb, and secure with tie wire, or other secure attachment, to the construction above.

Bracing attachments for horizontal stiffeners may be short pieces of ¾" channels driven into masonry joints or securely attached to masonry or concrete surfaces; or special equivalent devices (brackets) securely attached to masonry or concrete surfaces. Space bracing attachments for horizontal stiffeners at not more than 36" intervals in both directions, and not more than 4" from floor, ceiling, columns or other abutting construction.

Place ¾" channel stiffeners horizontally ¼" minimum and 2¼" maximum from wall and securely wire-tie to attachments or brackets.

Erection of lath: Erect long-length plain gypsum lath vertically. Place bottom of lath in groove of runner or grouted base. Fasten top of lath to ceiling runner with clips, not less than two clips per lath. Use special clip, or wire-tie over a nail, at the edges to each horizontal channel. Apply 3" strip lath full length of joints and below windows.

3.12.3.3A Enclosure of Steel Columns.

—Where columns are shown on drawings to be enclosed. construct enclosure to dimensions as detailed using one of following: [20K]   (Detail 37-C)

(a) **Metal Lath:** Install self-furring metal lath around perimeter of column and wire-tie at laps at not more than 6" intervals; or install ¾" vertical channel furring to channel braces secured to edges of column flanges or otherwise secured to each other, before application of lath. Attach corner beads to provide grounds for required thickness of plaster.

(b) **Gypsum Lath:** Install gypsum lath around perimeter of column and secure with double strands of No. 18 gauge wire around column at not more than 16" intervals. Attach corner beads to provide grounds for required thickness of plaster.

(c) **Gypsum Veneer Lath and Plaster:** Install veneer base around the perimeter of the column and secure to framing with approved screws. Attach veneer corner beads to provide grounds for required thickness of veneer plaster.

3.12.3.4A Enclosure of Round Pipe Columns.

—Where columns are shown on drawings to be enclosed, install self-furring or rib lath around column with long dimension of sheet vertical. Secure with double strands of No. 18 gauge wire around the column at not more than 16" intervals.

3.12.4A Contact Furring   (Detail 37-D)

3.12.4.1A Screw-Channel Furring.

—Align screw-channel with layout at floor and ceiling, and place horizontally or vertically at not more than 24" intervals. Start spacing of screw-channels not more than 2" from floor and ceiling if channels are placed horizontally; or 2" from abutting walls or columns if channels are placed vertically. Cut ends of screw-channels clear of abutting construction. Secure to wall surface with minimum ⅞" powder driven pins or nails placed alternately on opposite flanges spaced at not more than 36" intervals. [46]

3.13 EXTERIOR (application of lath) [21]   (Detail 43)

3.13.1 Control Joints.

—In all portland cement plastered areas, place metal control joints at locations and at spacing shown on drawings. Control joints shall serve as grounds for plaster. [22]

[20K] Where building code requires number of hours of fire resistance, show details of construction on drawings or specify fireproofing requirements, giving authority.

[21] Small amounts of water penetration of exterior lath and plaster, as in other types of wall facing, must be considered in the design stage. While it is possible to design plaster which is highly water resistant, water can find its way through the plastered surface. Where water enters it is usually at (1) windows and doors; (2) plaster separation at control joints; (3) cracks in the plaster at points of structural stress concentration.

Whether or not waterproof paper is used on exterior surfaces, water which penetrates the plaster must be allowed escape down its inner face to ultimately escape at a flashing or weephole. Where exterior plaster on a vertical surface returns onto a soffit, a weep-screed should be placed adjacent to the outer edge of the soffit, or some similar weep-joint should be designed to achieve the same purpose.

On multi-story buildings, if plaster is continuous past floor slabs, metal studs should be offset sufficiently to provide air space between the outer edge of the slab and the inner face of exterior plaster. Floor runners should be open or drilled to allow free flow of water.

On weatherproofed wood frame construction, plaster should not be allowed to bond to concrete slab at grade, or to concrete foundation. Bond prevents the free flow and exit of any water.

On exterior panel and spandrel wall construction, care should be taken to provide waterproof joints at edges.

[22] Control joints to be spaced on centers detailed on drawings as determined by the architect with continuous paper behind.

DETAIL 43
EXTERIOR LATH AND PLASTER

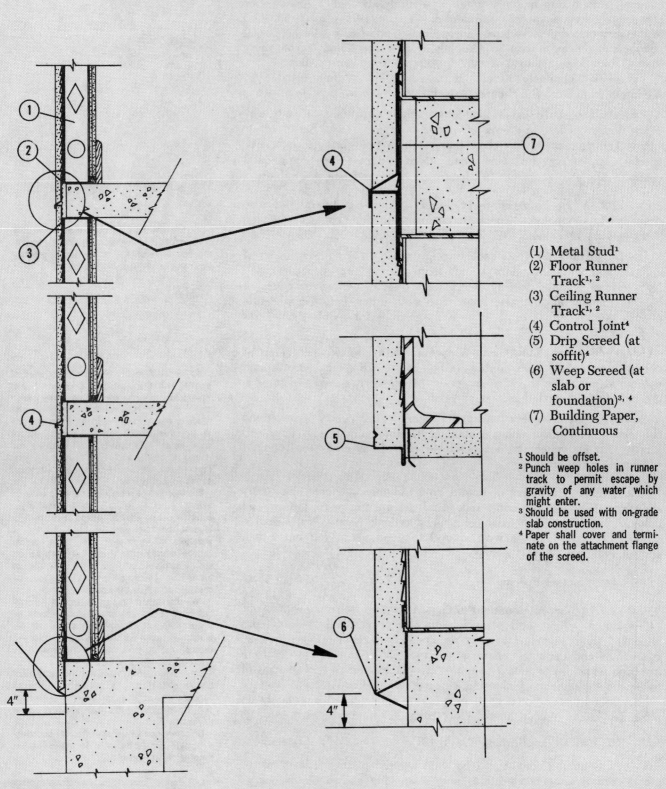

(1) Metal Stud[1]
(2) Floor Runner
    Track[1, 2]
(3) Ceiling Runner
    Track[1, 2]
(4) Control Joint[4]
(5) Drip Screed (at
    soffit)[4]
(6) Weep Screed (at
    slab or
    foundation)[3, 4]
(7) Building Paper,
    Continuous

[1] Should be offset.
[2] Punch weep holes in runner track to permit escape by gravity of any water which might enter.
[3] Should be used with on-grade slab construction.
[4] Paper shall cover and terminate on the attachment flange of the screed.

(A) OPEN METAL FRAME CONSTRUCTION

DETAIL 43
EXTERIOR LATH AND PLASTER

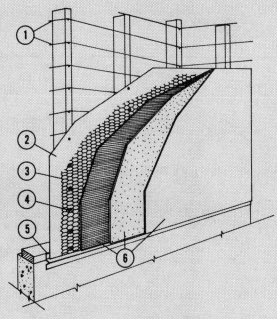

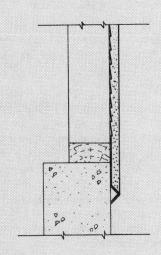

(B) OPEN WOOD FRAME CONSTRUCTION

(1) Wire Backing[1]
(2) Building Paper[1, 6]
(3) Wire Fabric Lath[1]
(4) Approved Fasteners[2]
(5) Weep Screed[3, 6]
(6) Three Coats of
Plaster (Scratch,
Brown, Finish)

[1] Paper-backed wire fabric lath may be used omitting separate wire and paper.
[2] Self-furring lath may be used.
[3] Should be used with on-grade slab construction.
[4] Use Control joints wherever possible.
[5] Section 4708 of the Uniform Building Code requires two layers of approved paper over solid wood sheathing.
[6] Paper shall cover and terminate on the attachment flange of the weep screed.

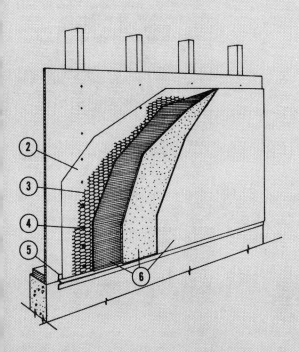

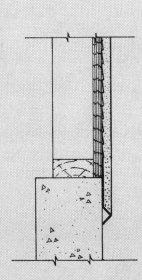

(C) SHEATHED WOOD FRAME CONSTRUCTION

3.13.2   Expanded Metal Lath (general application).—Apply metal lath with the long dimension at right angles to supports. Stagger ends of lath to avoid continuous joint on same support. Lap lath at sides not less than ½" and not less than 1" at ends. Lap all ends at supports. At internal angles, butt lath and reinforce angle with cornerite. Wire-tie cornerite along edges or secure along edge with other attachments. At external angles (arrises) cut ends of lath and reinforce arris with corner reinforcement designed to provide solid plaster corner. [23] [24] [25]

Attach metal lath to supports at maximum 6" intervals. [26]

Place an attachment at side laps on supports. Place a wire tie at side laps between supports.

On vertical surfaces, furr lath over solid wood backing at least ¼". Furring attachments or self-furring lath may be used.

3.13.3   Weather Protection.—Over open or solid wood backing frame construction to which metal lath is to be applied, apply water resistant paper to exterior vertical surfaces. Apply paper to supports and lap upper courses over lower courses not less than 2"; lap foundation at least 2"; lap vertical joints at least 6" [27] [29] [31] [32]

Double paper should be installed over solid wood backing.

3.13.4   Welded Wire Fabric Lath (general application).—Stagger ends of lath to avoid continuous joint on same support. Lap all joints at least one mesh. Lap all ends at supports. At external angles (arrises) cut ends of lath and reinforce arris with corner reinforcement of a design which provides a solid plaster corner. [28]

Attach welded wire fabric lath to supports at maximum intervals. [30]

Place an attachment at side laps on supports. Place a wire tie at side laps between supports.

On vertical surfaces, furr lath over open or solid wood backing frame construction at least ¼". Furring attachments or self-furring lath may be used.

3.13.4.1   Weather Protection.—Over wood frame construction to which welded wire fabric lath is to be applied, apply water resistant paper to vertical surfaces with the long dimension at right angles to supports and lap upper courses over lower courses not less than 2"; lap foundation at least 2"; lap vertical joints at least 6"; on open frame construction lap vertical joints to next support. [29] [31] [32]

Over solid backing there should be a 3" minimum vertical lap.

3.14.4   Woven Wire Fabric Lath (stucco netting) (general application).

3.14.4.1   Wire Backing (string wire).—Over open wood frame construction, attach wire of not less than No. 18 gauge to vertical surfaces, stretched taut horizontally and spaced at not more than 6" intervals. [33] [34] [35]

3.14.4.2   Weather Protection.—Over all wood frame construction to which woven wire fabric lath is to be applied, apply water resistant paper to supports and lap upper courses of paper over lower courses not less than 2"; lap paper over foundation at least 2"; lap vertical joints of paper at least 6"; on open frame construction lap vertical joints to next support. [27] [36]

[23] For portland cement and portland cement-lime plaster.

[24] Specify type of lath—flat (Type F); flat with backing (Type FB); self-furring (Type SF); self-furring with backing (Type SFB); expanded rib, flat rib or ⅜" rib. Stud spacing dictates type and weight of lath.

[25] Diamond mesh lath may be bent around internal and external angles (arrises) without use of corner reinforcement. Specify when wanted.

[26] On vertical surfaces, over solid backing using furring attachments or self furring lath.

[27] Where floor runner track or mud sill is placed on finish floor level of ongrade concrete slab, provide means for escape of water which may penetrate a surface such as a weep screed.

[28] Lath may be bent around both internal and external angles (arrises) without use of corner reinforcement. Specify when wanted.

[29] Where weep screeds occur, paper shall cover and terminate on the attached flange of the weep screed. (See next page).

[30] For type, size, and spacing of attachments see Table 5-10 (metal supports); Table 5-11 (wood supports).

[31] Where floor runner track or mud sill is placed on finish floor level of on-grade concrete slab, provide means for escape of water which may penetrate surface.

[32] Welded wire fabric, woven wire fabric and expanded metal lath with water resistant paper backing may be applied over metal framing in lieu of water resistant paper.

[33] String wire backing should be nailed only sufficiently to achieve taut condition until application of first coat of plaster. It should not be overly taut, or nailed too often.

[34] Where lath on vertical surfaces is to extend between rafters or other similar projecting members, specify solid wood backing, installed by others, to provide support for lath and attachments.

[35] Woven wire fabric lath which contains horizontal wire stiffening and water resistant paper backing may be used over open frame construction, omitting application of separate string wire backing and water resistant paper.

[36] Where floor runner track or mud sill is placed on finish floor level of on-grade concrete slab, provide means for escape of water which may penetrate surface.

RECOMMENDED | CONSTRUCTION TECHNIQUE

PENETRATION FLASHING RECOMMENDATIONS

FOR PORTLAND CEMENT PLASTER WALL AND PENETRATING FIXTURES

Section 1707(b) of the Uniform Building Code states *"Exterior openings exposed to the weather shall be flashed in such a manner as to make them waterproof."* The following procedure is recommended to achieve this intent in the flashing of penetrations to include, but not limited to, windows, doors, vents, etc.

Penetration Flashing Material

Material for flashing shall be barrier coated reinforced flashing material and shall provide for 4-hour minimum protection from water penetration when tested in accordance with ASTM D-779. Flashing material shall carry continuous identification. Sealant shall be Butyl to comply with Fed. Spec. TT-S-1657.

Application

To flash penetrations, a strip of approved flashing material at least nine inches wide must be applied in weatherboard fashion around all openings. Apply the first strip horizontally immediately underneath the sill, cut it sufficiently long to extend past each side of the window, door, or vent, so that it projects beyond the vertical flashing to be applied. (Detail

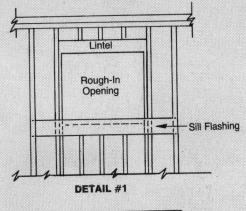

DETAIL #1

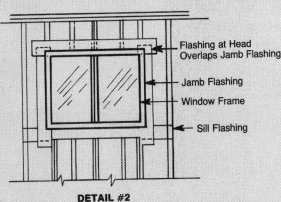

DETAIL #2

#1) Fasten the top edge of the first segment to the wall, but do not secure the body and lower edge of the first horizontal strip, so the weather resistant building paper applied later may be slipped up and underneath the bottom flashing in weatherboard fashion. In the case of low-set windows, apply approved paper the full height from the bottom of the plate line to the bottom of the window sill when the window is flashed.

Next, apply the two vertical side sections of flashing. Cut the side sections sufficiently long to extend the width of the flashing above the top of the window and the same distance below the window. Apply the side sections over the bottom strip of flashing. (Detail #2)

The penetrating fixture then is installed by pressing the nailing flange positively into a continuous bead of sealant which extends around the bottom and vertical perimeter of the inserted fixture.

NOTE: The continuous bead of sealant that is applied to the underneath side of the nailing flange of windows, doors, and vents is not to be construed as a substitute for flashing.

Apply the top horizontal section of flashing last, overlapping and sealed against the full height of the outer face of the top nailing flange with a continuous bead of sealant. Cut the top piece of flashing sufficiently long so that it will extend to the outer edge of both vertical strips of side flashing. (Detail #3)

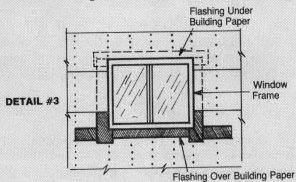

DETAIL #3

Installation of Exterior Plaster weather resistant paper underlayment to complete acceptable penetration flashing.

Commence at the bottom of the wall and overlapping the weep screed flange lay the approved weather resistant paper up the wall, overlapping 2″ min. in weatherboard fashion. Be sure that A is placed under the sill strip flashing.

NOTE: These recommendations are based on extensive experience and are not in any way intended to imply guarantees.

3. LATHING APPLICATION

3.14.4.3 Application of Lath.—Lap all joints at least one mesh but not less than 1", and lap all upper courses of lath over lower courses; lap all ends at supports. At external angles (arrises) cut ends of lath and reinforce arris with corner reinforcement of a design which provides a solid plaster corner. [37] [38]

Attach woven wire fabric lath to supports at maximum 6" intervals. [39]

On vertical surfaces, furr lath over wood frame construction at least ¼".

3.15 CORNER BEADS AND OTHER ACCESSORIES

3.15.1 Attachment.—Attach metal plastering accessories to supports or to plaster base, using shims if necessary, so as to provide true grounds for plaster. Wire-tie, nail or staple accessories to supporting surface sufficiently to hold accessory in place during plastering. [40] [41]

3.16 ACCESS PANELS

3.16.1 Location and Sizes.—Provide and set metal access panels, complete with frames and hardware, in the following quantities, sizes and type, to be installed in plaster walls and ceilings; the exact location is determined by the drawings. [42]

Provide........panels, type........, size........x........in plastered ceilings
Provide........panels, type........, size........x........in plastered ceilings
Provide........panels, type........, size........x........in plastered ceilings
Provide........panels, type........, size........x........in plastered walls
Provide........panels, type........, size........x........in plastered walls
Provide........panels, type........, size........x........in plastered walls

3.16.2 Construction.—Construct frames of steel not lighter than No. 16 gauge and with minimum No. 24 gauge expanded or perforated metal wings, designed to finish flush with plaster. Panels shall be flush panel type, constructed of steel not lighter than No. 18 gauge, and have concealed spring hinges with screw-driver operated latch. Panels shall be removable by taking pin out of hinge leaf. Finish shall be factory-applied prime coat. [43]

3.16.3 Installation.—Set access panel frames in place at locations shown, straight and true. Securely anchor frames to adjacent furring or structural framing in accordance with manufacturer's recommendations.

3.17 WORKMANSHIP

3.17.1 Application of Lath.—Securely attach all lath and related materials so as to provide a proper base and/or reinforcement for plaster. Securely attach accessories so as to keep them plumb, straight or level as required, and to provide required thickness of plaster.

[37] Lath may be bent around both internal and external angles (arrises) without use of corner reinforcement. Specify when wanted.

[38] Lath may be interrupted at each control joint. Where control joints are spaced closer than 12', lath may be broken at alternate joints.

[39] For type, size, and spacing of attachments see Table 5-11 (wood supports).

[40] Architect should make clear who is to furnish and install metal shapes of special design and gauge of metal, particularly those which act both as plaster grounds and as attachments for work of others.

[41] When metal window stools serve as grounds for plaster, place installation in lathing section.

[42] Architect should assign responsibility for location of access panels in acoustical tile ceilings over plaster base to acoustical tile contractor.

[43] Chrome finish panels should be furnished and installed by others.

TABLE 3-1 — VERTICAL FURRING

VERTICAL FURRING MEMBER	UNBRACED[1]				BRACED[1]			
	STUD SPACING[2]				STUD SPACING[2]			
	24″	19″	16″	12″	24″	19″	16″	12″
	Maximum Furring Heights				Maximum Vertical Distance Between Braces			
¾″ Channel	6′	7′	8′	9′	5′	5′	6′	7′
1½″ Channel	8′	9′	10′	12′	6′	7′	8′	9′
2″ Channel	9′	10′	11′	13′	7′	8′	9′	10′
2″ Prefab. Stud	8′	9′	10′	11′	6′	7′	8′	9′
2½″ Prefab. Stud	10′	11′	12′	14′	8′	9′	10′	11′
3¼″ Prefab. Stud	14′	16′	17′	20′	11′	13′	14′	16′

[1]When vertical furring is braced, height is unlimited.   [2]For types of lath and allowable spacing of supports, see Table 3-2.

TABLE 3-2 — TYPES OF LATH—MAXIMUM SPACING OF SUPPORTS

TYPE OF LATH[3]	Minimum Weight (psy), Gauge & Mesh Size	VERTICAL			HORIZONTAL	
		WOOD	METAL		Wood or Concrete	Metal
			Solid Plaster Partitions	Other		
Expanded Metal Lath (Diamond Mesh)	2.5	16″	16″	12″	12″	12″
	3.4	16″	16″	16″	16″	16″
Flat rib Expanded Metal Lath	2.75	16″	16″	16″	16″	16″
	3.4	19″	24″	19″	19″	19″
Stucco Mesh Expanded Metal Lath	1.8 and 3.6	16″[3]	—	—	—	—
⅜″ Rib Expanded Metal Lath	3.4	24″	—	24″	24″	24″
	4.0	24″	—	24″	24″	24″
Sheet Lath	4.5	24″	—[4]	24″	24″	24″
¾″ Rib Expanded Metal Lath (Not manufactured in West)	5.4	—	—[4]	—	36″[5]	36″[5]
Wire Fabric Lath — Welded	1.95 lbs., 11ga., 2″ x 2″	24″	24″	24″	24″	24″
	1.4 lbs., 16 ga., 2″ x 2″	16″	16″	16″	16″	16″
	1.4 lbs., 18 ga., 1″ x 1″[6]	16″[3]	—	—	—	—
Wire Fabric Lath — Woven[3]	1.4 lbs., 17 ga., 1½″ Hex.[3]	24″	16″	16″	24″	16″
	1.4 lbs., 18 ga., 1″ Hex.[3,6]	24″	16″	16″	24″	16″
⅜″ Gypsum Lath (plain)	—	16″	—	16″[7]	16″	16″
(Large Size)	—	16″	—	16″[7,8]	16″	16″
½″ Gypsum Lath (plain)	—	24″	—	24″	24″	24″
(Large Size)	—	24″	No supports; Erected vertically	24″	24″	16″
⅝″ Gypsum Lath (Large Size)	—	24″	No supports; Erected vertically	24″	24″	16″

[1]For Fire-resistive Construction, conform to local building code. For Shear-resisting Elements, conform to local building code.

[2]Metal lath and wire fabric lath used as reinforcement for portland cement plaster shall be furred out away from vertical supports at least ¼ inch. Self-furring lath meets furring requirement. Exception: Furring is not required on steel supports having a flange width of 1 inch or less.

[3]Wire backing required on open vertical frame construction except under expanded metal lath and paperbacked wire fabric lath.

[4]May be used for studless solid partitions.

[5]Contact or furred ceilings only. May not be used in suspended ceilings.

[6]Woven wire or welded wire fabric lath, not to be used as base for gypsum plaster, without absorbent paperbacking or slot-perforated separator.

[7]Span may be increased to 24 inches on vertical screw or approved nailable assemblies.

[8]When attached to screw or nailable stud or channel in vertical construction, may be increased to 24″; for veneer plaster.

67

TABLE 3-3

TYPES OF LATH-ATTACHMENT TO METAL SUPPORTS

MINIMUM SIZE & MAXIMUM SPACING OF ATTACHMENTS[1, 2]

TYPE OF LATH	TIE WIRE: Galvanized; Soft Annealed — Gauge	TIE WIRE: Max. Spacing — VER	TIE WIRE: Max. Spacing — HOR	NAILS: Round Steel Wire; Ring Shank — Gauge	NAILS: Round Steel Wire; Ring Shank — Head	NAILS: Round Steel Wire; Ring Shank — Penetration	NAILS: Max. Spacing — VER	NAILS: Max. Spacing — HOR	SCREWS (Power Driven): Flat Head Self-drilling Self-tapping Galvanized, or blued, or painted	SCREWS: Max. Spacing — VER	SCREWS: Max. Spacing — HOR	STAPLES (Power Driven): Galvanized Steel Wire; Round Wire; Flat Wire Notched;* Chisel-point; Divergent-point; Divergent-chisel-point — Type of Support	Leg	Wire Gauge	Crown	STAPLES: Max. Spacing — VER	STAPLES: Max. Spacing — HOR
Expanded Metal Lath[3]	18 ga.	6″	6″	12½	Head[3]	⅝″	6″	6″	¾ in. No. 6 Head[3]	6″	6″	nlb[4] lb[5]	⅝″ 1″	16 FN 14	9/16″ ⅞″	6″ 6″	6″ 6″
⅜ inch Rib Metal Lath[3]	18 ga.	6″	6″	12½	Head[3]	⅝″	6″	6″	¾ in. No. 6 Head[3]	6″	at rib	nlb[4] lb[5]	15/16″ 1⅜″	16 FN 14	9/16″ ⅞″	at rib	
Wire Fabric Lath[3]	18 ga.	6″	6″	12½	Cap Nail[3]	⅝″	6″	6″	¾ in. No. 6 with washer	6″	6″	nlb[4] lb[5]	⅝″ 1⅛″	16 FN 14	9/16″ ⅞″	6″ 6″	6″ 6″
⅜ inch Gypsum Lath[6]	—	—	—	12½	¼″	⅝″	6″	6″	¾ in. No. 6	12″	6″	nlb[4] lb[5]	15/16″ 1⅛″	16 FN 14	9/16″ ⅞″	6″ 6″	6″ 6″
(Large size)	—	—	—				8″[8]	8″[8]		12″[7]	8″	nlb[4] lb[5]	15/16″ 1⅛″	16 FN 14	9/16″ ⅞″	7″ 7″	7″ 7″
½ inch Gypsum Lath[6]	—	—	—	12½	¼″	⅝″	5″	5″	1 inch No. 6	12″	6″	nlb[4] lb[5]	1⅛″ 1¼″	16 FN 14	9/16″ ⅞″	5″ 5″	5″ 5″
(Large size)	—	—	—				8″[8]	8″[8]		12″[7]	8″	nlb[4] lb[5]	1⅛″ 1¼″	16 FN 14	9/16″ ⅞″	7″ 7″	7″ 7″
⅝ inch Gypsum Large Size Lath	—	—	—	12½	¼″	⅝″	8″[8]	8″[8]	1 inch No. 6	12″[7]	8″	nlb[4] lb[5]	1⅛″ 1¼″	16 FN 14	9/16″ ⅞″	6″ 7″	6″ 7″

[1]Except where indicated all attachments, other than tie wire, are for nonload bearing supports. All attachments are minimum outside dimensions; all spacings are approximate; attachments shall be placed at metal lath laps crossing supports, and approximately two inches (2″) from gypsum lath edges which cross supports.

[2]For wire and sheet metal clips and attachments see 2.9; install as directed by lath manufacturer.

[3]Wire and nail or screw head shall engage minimum two strands or the rib of metal lath; or minimum one strand of wire lath; washers may be used if necessary.

[4]nlb = Non load bearing.

[5]lb = Load bearing.

*Approved—Round wire staples without notches may be used with nailable studs designed to receive a conventional nail.

[6]Receives one-half inch (½″) gypsum plaster. When attached to prefabricated open web, and nailable studs, end joints shall generally fall between supports; end joints may occur on supports if attachment is provided with washer, diamond clip, or similar device engaging minimum three-eighth inch (⅜″) on each sheet of lath. When attached to screw studs, end joints may occur on, or between, supports.

[7]Nine inch (9″) spacing of screws on edges which fall on supports.

[8]For other than veneer plaster, space attachments six inches (6″) maximum.

FN = Flat wire notched.

TABLE 3-4

TYPES OF LATH-ATTACHMENT TO WOOD AND METAL[1] SUPPORTS

TYPE OF LATH	NAILS[2,3] Type and Size	MAXIMUM SPACING[5] Vertical (In Inches)	MAXIMUM SPACING[5] Horizontal (In Inches)	SCREWS[3,6] MAX. SPACING[5] Vertical (In Inches)	SCREWS[3,6] MAX. SPACING[5] Horizontal (In Inches)	STAPLES[3,4] Round or Flattened Wire: Wire Gauge No.	STAPLES Crown (In Inches)	STAPLES Leg[7] (In Inches)	STAPLES MAX. SPACING[5,6] Vertical (In Inches)	STAPLES MAX. SPACING[5,6] Horizontal (In Inches)
1. Diamond Mesh Expanded Metal Lath and Flat Rib Metal Lath	4d blued smooth box 1½ No. 14 gauge $\frac{7}{32}$" head (clinched)[7] 1" No. 11 gauge $\frac{7}{16}$" head, barbed 1½" No. 11 gauge $\frac{7}{16}$" head, barbed	6 6 6	– – 6	6	6	16	¾	⅞	6	6
2. ⅜" Rib Metal Lath and Sheet Lath	1½" No. 11 ga. $\frac{7}{16}$" head, barbed	6	6	6	6	16	¾	1¼	At Ribs	At Ribs
3. ¾" Rib Metal Lath	4d common 1½" No. 12½ gauge ¼" head 2" No. 11 gauge $\frac{7}{16}$" head, barbed	At Ribs	– At Ribs	At Ribs	At Ribs	16	¾	1⅝	At Ribs	At Ribs
4. Wire Fabric Lath[9]	4d blued smooth box (clinched)[8] 1" No. 11 gauge $\frac{7}{16}$" head, barbed	6 6	– –	6	6	16	¾	⅞	6	6
	1½" No. 11 gauge $\frac{7}{16}$" head, barbed 1¼" No. 12 ga. ⅜" head, furring 1" No. 12 gauge ⅜" head	6 6 6	6 6			16	$\frac{7}{16}$[9]	⅞	6	6
5. ⅜"Gypsum Lath	1⅛" No. 13 gauge $\frac{19}{64}$" head, blued	8[10]	8[10]	8[10]	8[10]	16	¾	⅞	8[10]	8[10]
6. ½"Gypsum Lath	1¼" No. 13 gauge $\frac{19}{64}$" head, blued	8	8[10] 6[11]	8[10]	8[10] 6[11]	16	¾	1⅛	8[10]	8[10] 6[11]

[1]Metal lath, wire lath, wire fabric lath and metal accessories shall conform with the provisions of U.B.C. Standard No. 47-4.

[2]For nailable nonload-bearing metal supports, use annular threaded nails or approved staples.

[3]For fire-resistive construction, see Tables No. 43-B and No. 43-C. For shear-resisting elements, see Table No. 47-I. Approved wire and sheet metal attachment clips may be used. (Tables referred to are from Uniform Building Code, 1979 Edition)

[4]With chisel or divergent points.

[5]Maximum spacing of attachments from longitudinal edges shall not exceed 2 inches.

[6]Screws shall be an approved type long enough to penetrate into wood framing not less than ⅝ inch and through metal supports adaptable for screw attachment not less than ¼ inch.

[7]When lath and stripping are stapled simultaneously, increase leg length of staple ⅛ inch.

[8]For interiors only.

[9]Attach self-furring wire fabric lath to supports at furring device.

[10]Three attachments per 16-inch-wide lath per bearing. Four attachments per 24-inch-wide lath per bearing.

[11]Supports spaced 24 inches o.c. Four attachments per bearing per 16-inch-wide lath. Five attachments per 24-inch-wide lath per bearing.

4. EXTERIOR INSULATION

Note: Following is an outline specification for the two primary varieties of exterior insulation and finish systems, developed and adopted by the Exterior Insulation Manufacturers Association (EIMA), the national industry promotional and educational organization for this industry, in June, 1984, and revised in April 1987.

It is reproduced in this manual with permission of EIMA.

The Exterior Insulation Manufacturers Association, Inc. has prepared these generic specifications for use as guidelines when specifying Exterior Insulation and Finish Systems. (Hereinafter referred to as EIFS or System). Specific installation instructions and procedures for each particular system must be obtained from the manufacturer supplying the materials. EIMA does not endorse any particular system or make any recommendations in regard to a specific manufacturer's products or application procedure. It is the sole responsibility of the manufacturer to provide all parties with specific recommendations regarding their product application and limitations.

Note: This specification is intended as a guide for experienced specification writers only, and is not intended to be used verbatim as an actual specification. It must be integrated into and coordinated with the procedures of each architectural firm and the requirements of the specific project. It is intended solely to cover the exterior insulation portion of the building and assumed that the exterior insulation finish system is being applied to a sound substrate.

EXTERIOR INSULATION AND FINISH SYSTEM
CLASS PB (POLYMER BASE) — TYPE A (EXTERNALLY REINFORCED)

4.1 GENERAL

4.1.1 Description.

A. Provide all labor, materials and equipment necessary to install the EIFS. The system consists of insulation board, adhesive/base coat, reinforcing fabric, finish coat and accessories as recommended by the wall system manufacturer.

B. Related work specified elsewhere:

1. Light Gauge Metals Section 05400.
2. Sheathing Section 06100.
3. Unit Masonry Section 04200.
4. Concrete Section 03300.
5. Sealants Section 07900.

C. Terms/Definitions.

1. EIFS or System. Comprised of insulation board, adhesive/base coat, reinforcing fabric, finish coat and applicable accessories that interact together to form an energy efficient exterior wall.

2. Insulation Board. A component of the system of a specific type and density that functions to reduce heat flow through the wall and serves as the surface to receive the adhesive/base coat.

3. Adhesive/Base Coat. Adheres the insulation board to the substrate and also performs as the base to embed the reinforcing fabric and receive the finish coat.

4. Reinforcing Fabric. Balanced, open mesh glass fabric properly treated for compatibility with other materials of the system, which functions to strengthen the system.

5. Finish Coat. Functions as the weathering surface. It shall be a factory mixed finish supplied by the system manufacturer in various colors, finishes, and can be textured.

6. Accessories. Products, such as corner beads and stops that may be utilized in conjunction with the different systems.

7. Primers. Products utilized as adhesion intermediaries between the system and the substrate, as recommended by some system manufacturers.

8. Mechanical Fasteners. Utilized to fasten the insulation board to the substrate as recommended by some system manufacturers.

9. Wet Material. The adhesive/base coat and finish coat components of the system in its initial states.

4.1.2 Quality Assurance.

A. Applicator Requirements.

1. Application of the system must be by an applicator approved by the manufacturer.

B. Code Approval.

1. The system shall be recognized for the intended use by the applicable building codes.

C. Details.

1. Conform with system manufacturer's current published typical details and specific recommendations for the project.

4.1.3 Submittals.

A. The applicator shall submit evidence with the bid that he is an approved applicator of the system manufacturer with whom the project has been bid.

B. Samples.

1. The applicator shall, before the project commences, provide the owner/architect with a sample of suitable size of the system as required for each color and/or texture to be utilized on the project for approval.

2. Each sample shall be prepared using the same tools and techniques as required for the actual application.

3. An approved sample shall be available and maintained at the application site.

4.1.4 Product, Delivery, Storage and Handling.

A. Deliver all materials supplied by the system manufacturer in original, unopened packages with legible manufacturer's identification and labels intact.

B. Store all products supplied by the wall system manufacturer in a cool, dry place out of direct sunlight, protected from weather and other damage. In addition, wet materials shall be stored at a temperature not less than 40°F at all times.

4.1.5 Job Conditions.

A. Weather/Environmental Conditions.

1. Application of the system shall not take place during inclement weather unless appropriate protection is employed.

2. Installation of wet materials in temperatures less than 40°F shall take place only if supplementary heat is provided and is maintained for a minimum of 24 hours after application of the wet materials.

B. Protection.

1. Protect surrounding areas and surfaces during the application of the wall system.

2. The system shall be protected when work ceases for the day or when an area is completed so that water will not infiltrate behind the system.

4.1.6 Coordination/Scheduling.

A. The work in this section requires close coordination with related sections and trades.

B. The tops of all walls must immediately be covered with either the final trim or temporarily protected to prevent water infiltration behind the system. The cap flashing shall be installed as soon as possible after the finish coat has been installed.

C. All sealants shall be installed in a timely manner.

4.2 PRODUCTS

4.2.1 Manufacturers.

A. The following manufacturer(s) are approved for the project:

_____

B. All products shall be obtained from the selected system manufacturer or its approved supplier.

4.2.2 Materials.

A. Adhesive — shall be compatible with the insulation board, substrate and reinforcing fabric.

B. Portland Cement — ASTM C 150, as approved by the wall system manufacturer.

C. Insulation Board — shall comply with the following EIMA guidelines:

1. Expanded Polystyrene (EPS):

(a) Type ASTM C 578; Federal Specifications HH-I-524C Type (I or II); and tested in accordance with ASTM E 84 or UL 723, with a flame spread of less than 25 as required by the model building codes. If required by Model Building Code or other Agency Code, the insulation board shall be inspected by an independent third party inspection agency which shall certify that the insulation board meets the foregoing specifications. Third party inspecting agencies should be listed by a model code or other applicable code as approved to perform such inspection and certification.

(b) Insulation board shall be aged; if air dried, insulation board shall be aged for not less than six weeks in block form prior to cutting and shipping. Other methods of aging shall be equivalent to six-week aged (air dried) insulation board and certified as equivalent by insulation board manufacturer.

(c) Variations in tolerances in dimensions of insulation board shall be minimized.

(d) All insulation board shall be labeled on each package to provide information required by model code or other applicable codes and the appropriate model code recognition.

(e) Thickness shall be _____ or as noted on drawing.

D. Reinforcing Fabric — shall be balanced, open mesh, glass fiber fabric properly treated or compatibility

with other materials of the system.

E. Base Coat — shall be compatible with the insulation board and reinforcing fabric.

F. Primers — may be required by the system manufacturer as adhesion intermediaries.

G. Finish Coat — shall be factory mixed. Type, color and texture to be approved by the architect/owner.

H. Mechanical Fasteners — may be required by the system manufacturer.

I. Accessories — shall be as per system manufacturer's recommendations.

J. Water — clean and potable.

K. Sealants — shall be listed by the system manufacturer and approved by the sealant manufacturer for the selected system.

4.2.3 Performance Characteristics.

A. General Physical Properties—The system shall meet or exceed the following performance standards when tested in accordance with the following methods:

1. Accelerated Weathering Test: ASTM G 23-69 test, or ASTM G 26-77. Operating light and water exposure apparatus for exposure of nonmetallic materials. Testing period: 2,000 hours.

Test specimen: The sample shall be a 1″ exterior insulation system mounted on gypsum board of ½″ thickness and cured for a minimum of 28 days. The sample size shall be suitable for test equipment. Results: No cracking, flaking or deleterious effects.

2. Wind Driven Rain. Test: Federal Standard TTC 555B.

Test specimen: The sample shall be a 1″ exterior insulation system mounted on gypsum board of ½″ thickness and cured for a minimum of 28 days.

Results: no visible leaks for dampness throughout to the rear face and less than 90 gram weight increase.

3. Moisture Resistance. Test: Federal Standard 141 A, Method 6201. Test period: 14 days exposure.

Test specimen: The sample shall be a 1″ exterior insulation system mounted on gypsum board of ½″ thickness and cured for a minimum of 28 days. Results: No deleterious effects.

4. Salt Spray Resistance. Test: ASTM B 117 Salt Spray (Fog) Testing. Testing period: 300 hours.

Test specimen: The sample shall be a 1″ exterior insulation system mounted on gypsum board of ½″ thickness and cured for a minimum of 28 days. Results: No deleterious effects.

5. Absorption—Freeze. Test: The specimen subjected to four days underwater soak followed by 60 cycles as follows: −10°C for two hours and +20°C for two hours.

Test specimen: Three samples 4″ x 8″ of 1″ exterior insulation system, coated on all sides with the manufacturer's base coat, reinforcing fabric and finish coat. The sample shall cure for a minimum of 28 days.

Test period: 60 freeze/thaw cycles. Results: No visible change.

6. Mildew Resistance Test: MIL Standard 810 B, Method 508.

Test specimen: The sample shall be sample of

finish coat only on clean 2″ x 2″ glass substrate. The finish coat shall be applied according to the manufacturer's printed instructions. Specimens shall be cured for a minimum of 28 days. Results: No mildew growth supported.

7. Abrasion Resistance. Test: Federal Standard 141 A, Method 6191; ASTM D 968.

Test specimen: The sample shall be a 1″ exterior insulation system mounted on gypsum board of ½″ thickness and cured for a minimum of 28 days. Results: No cracking, checking or loss of film integrity after 500 liters of sand.

B. Fire Test Performance.

1. Surface Burning. Test: ASTM E 84 or UL 723.

Test specimen: Three samples mounted on ¼″ cement asbestos board. The board shall consist of base coat, reinforcing fabric, and finish coat applied in coverage rates consistent with manufacturer's current application instructions. Results: The flame spread and the smoke development shall be less than 25.

2. Full Scale Diversified Fire Testing. Test: ASTM E 108 modified for vertical walls.

Test specimen: Two sample panels the size to meet test requirements. The sample shall consist of a minimum of 4″ of insulation mounted on gypsum board of ½″ thickness and cured for a minimum of 28 days. Results: The system shall not contribute to significant vertical or horizontal flamespread.

C. Structural Testing.

1. Impact Resistance. Test: EIMA Test Methods and Standard 101.86. Refer to Appendix A.

2. Negative Wind Load/Full Scale Testing. Test: ASTM E 330 positive and negative air pressure.

Test specimen: The sample shall be 4′ x 4′ and consist of studs, sheathing and 1″ exterior insulation system. Results: Withstand wind loads required by local building codes, as follows ——————————

4.3 EXECUTION

4.3.1 Installation.

A. Installation shall be performed strictly in accordance with system manufacturer's current published instructions.

4.3.2 Substrate.

A. The substrate shall be a type approved by the system manufacturer.

B. The substrate shall be free of planar irregularities greater than ¼″ and shall be sound and free of foreign substances.

C. Unsatisfactory conditions shall be corrected before the application of the system.

4.3.3 Application.

A. Mixing—All materials requiring preparation in the field shall be labeled with complete mixing instructions. All such instructions shall be followed by the applicator.

B. Method of Attachment—Shall be approved by the system manufacturer.

1. Adhesive Method—The adhesive shall be applied to the insulation board as per the system manufac-turer's recommendations or, if none, by one of the following methods:

(a) The adhesive shall be applied to the entire surface of the insulation board using a notched trowel. Notches shall be ½″ x 12″ and spaced no more than 1″ apart.

(b) The adhesive shall be applied to the entire perimeter of the insulation board in a ribbon fashion that is two inches wide by ⅜″ thick. Dabs of approximately 4″ in diameter in the same thickness are then applied 8″ on center over the remainder of the board.

2. Adhesive and Mechanical Fastener Method.

(a) Adhesive shall be applied using the ribbon and dab method, notched trowel methods or as per system manufacturer's current published instructions.

(b) Mechanical fasteners shall be of a type and spacing as per system manufacturer's current published instructions.

C. Installation of insulation shall be as per system manufacturer's current published instructions:

1. The application of the insulation board shall commence at the base of the wall from a temporary or permanent line or support.

2. When the adhesive has been applied to the back of the insulation board, it shall be installed by sliding it into place until it abuts adjoining insulation board.

3. Pressure shall then be applied over the entire surface of the insulation board so as to achieve uniform contact and high initial grab. The insulation board should be checked occasionally for proper contact to the substrate. Proper contact has been achieved when a piece of insulation board is removed and the adhesive is adhered approximately equally to both the substrate and the insulation board.

4. The insulation boards are to be applied in a running bond pattern, with staggered vertical joints and interlocking EPS boards at the inside and outside corners.

5. All insulation boards shall be butted tightly. Any gaps greater than ⅛ of an inch that may occur between the insulation boards shall be filled with insulation. Gaps shall not be filled with adhesive or any other non-insulating material.

6. The adhered insulation board shall be allowed to remain undisturbed for 24 hours prior to proceeding with the installation of the base coat/reinforcing fabric. Where mechanical fasteners are used, follow wall system manufacturer's current published instructions.

D. Applying Standard Reinforcing Fabric and Base Coat.

1. Prior to base coat/reinforcing fabric application, all insulation board irregularities greater than 1/16″ must be rasped flush.

2. Apply the base coat to the entire surface of the insulation board to the thickness specified by the system manufacturer.

3. Fully embed the reinforcing fabric in the wet base coat troweling from the center to the edge of the reinforcing fabric so as to avoid wrinkles. The reinforcing fabric shall be continuous at all corners and lapped a minimum of 2½ inches at the mesh joints.

E. Impact Layers.

1. Higher impact performance may be accomplished by:

(a) Multiple layers of reinforcing fabric.

(b) Incorporating higher impact resistant fabrics.

2. All areas requiring higher impact performance shall be detailed on the plans. The application shall be installed in accordance with manufacturer's recommendations.

F. Applying Finish.

1. The base coat/reinforcing fabric application must be allowed to set prior to the application of the finish coat.

2. Apply the finish coat to the dry base coat maintaining a wet edge at all times so as to obtain a uniform appearance. The thickness of the finish coat shall be as per wall system manufacturer's current published instructions.

3. The texture and color of the finish shall be as specified and as per the approved sample. All mechanics applying and texturing the finish shall utilize the same equipment and techniques to achieve uniformity.

4.3.4 Job Site Clean-Up.

A. All excess wall system materials shall be removed from the job site by the wall system applicator.

B. All surrounding areas where the wall system has been applied shall be left free of debris and foreign substances.

EXTERIOR WALL INSULATION AND FINISH SYSTEM SPECIFICATIONS
CLASS PM (Polymer Modified Mineral Base) — TYPE A (Externally Reinforced)

4A.1 GENERAL

4A.1.1 Description.

A. Provide all labor, materials and equipment necessary to install the EIFS. The system shall be a field applied or prefabricated wall covering consisting of insulation board, glass fabric reinforcing mesh or corrosion resistant stucco netting, mechanical anchors, fiber reinforcement, polymer modified cementitious base coat and choice of color and textures in finish coats: Cementitious finish with colored acrylic sealer, synthetic finish with integral color, or exposed aggregate finish.

B. Related Work Specified Elsewhere:

1. Light Gauge Metals Section 05400.

2. Sheathing Section 06100.

3. Unit Masonry Section 04200.

4. Concrete Section 03300.

5. Sealants Section 07900.

6. Carpentry . . . wood studs and sheathing.

C. Terms and Definitions.

1. EIFS or System. Comprised of insulation board, mechanical anchors, reinforcing mesh and base and finish coats that are tested and proven compatible to form an exterior insulated wall covering and finish system.

2. Insulation Board. A component of specific type and density that functions as a thermal envelope to reduce heat flow through the wall and serves as a substrate for uniform application of base and finish coats.

3. Mechanical Anchors. Pre-drilled, power driven or air driven fasteners, clips, screws and washers used to affix EIFS system to masonry, concrete, or metal or wood stud framed substrates.

4. Reinforcing Fabric.

(a) Balanced, open mesh glass fabric properly treated for compatibility with other materials of the system, which functions to strengthen the system.

(b) Stucco netting (see manufacturer) woven wire lath, one inch, 20 gauge galvanized.

5. Base Coat. Polymer modified cementitious coating, troweled or sprayed on insulation board over reinforcing mesh. Base coat is $1/8$ to $1/4$ inch in thickness and contains fiber reinforcement.

6. Finish Coat.

(a) Polymer modified cementitious coating, which is spray or trowel applied to the base coat.

(b) Synthetic finish, pre-mixed, integrally colored, rolled, sprayed or trowel applied.

(c) Exposed aggregate embedded into cementitious base coat matrix.

7. Sealer. Factory colored, acrylic based film-forming coating, spray or roller applied to cementitious finish coats.

8. Accessories. Products such as expansion joints, control joints, corner beads, stops, etc., utilized as recommended by system manufacturer.

9. Adhesives. Adhesives used to affix insulation board to substrate as per EIFS manufacturer's recommendations.

4A.1.2 Quality Assurance.

A. Applicator Requirements.

1. Must be trained and approved in the installation of the system.

2. Must be approved in writing by the manufacturer prior to the bid process.

B. Insulation Board Manufacturer.

1. Board manufacturer must be approved by system manufacturer and must be under a written agreement to produce insulation board in accordance with system manufacturer's specification. Board shall be Code approved and listed by Underwriters Laboratories labeled with manufacturer's pertinent information.

C. Code Approvals. The system shall be recognized for the intended use by applicable Building Codes.

4A.1.3 Submittals.

A. Applicator must submit evidence with the bid

4. EXTERIOR INSULATION

that he is currently an approved applicator of the system.

B. Samples.

1. The applicator shall, before the project commences, provide the architect/owner with samples for approval. Samples shall be of the system and of suitable size as required to accurately represent each color and texture to be utilized on the project.

2. Each sample shall be prepared using the same tools and techniques as required for the actual application.

3. An approved sample shall be available and maintained at the job site.

4A.1.4 Product Delivery, Storage and Handling.

A. Deliver to job site all materials in unopened, undamaged containers, clearly marked and identified with manufacturer's name and description of contents.

B. Store all products supplied by the system manufacturer in a cool, dry place out of direct sunlight, protected from weather and other damage. In addition, store all wet materials at a temperature not less than 40°F at all times. Store insulation board away from open flame.

4A.1.5 Conditions of Installation.

A. Weather/Environment Conditions.

1. Application of finish coats shall not take place during inclement weather unless appropriate protection is employed.

2. Installation of wet materials shall be at an ambient temperature of 40°F and rising and maintained above 40°F for 24 hours after application of coating.

3. Supplemental heat must be furnished for applications in less than 40°F.

B. Protection of Work.

1. Protect surrounding areas and surfaces during the application of the system.

2. The system shall be protected when work ceases for the day or when an area is completed so that water will not infiltrate behind the system.

4A.1.6 Coordination and Scheduling.

A. Installation of system shall be coordinated with other construction trades.

B. Tops of walls must be immediately covered to avoid water infiltration. Copings or flashing shall be installed as soon as possible after finish coat to protect system.

C. All sealants to be installed in a timely manner.

D. Sufficient manpower and equipment must be employed to ensure a continuous operation, free of cold joints, scaffolding lines, etc.

4A.2 PRODUCTS

4A.1.1 Manufacturers.

A. The following manufacturers are approved for this project.

1. _____

2. _____

B. All system components shall be obtained from the system manufacturer or its approved supplier.

4A.2.2 Materials.

A. Thermal Insulation Board, ASTM C 578 Type IV.

Rigid extruded foam plastic insulation board shall have high density, smooth extruded, natural skinned surface and shall be free of facing films, papers or foils.

Typical properties:

1. Minimum compressive strength 25 psi, tested in vertical direction in accordance with ASTM D 1621-73.

2. Maximum water absorption 0.3% by volume when testing in accordance with ASTM C 272-76. Insulation board shall also meet physical properties required by ASTM C 578 for Type IV thermal insulating board.

For manufacturers allowing other types of rigid cellular polystyrene, see manufacturer's specification.

B. Reinforcing Fabric—shall be balanced, open mesh, glass fiber fabric properly treated for compatibility with other materials of the system, constructed of continuous filament yarns produced in accordance with ASTM D 578.

C. Base Coat—shall be a fiber reinforced cementitious coating consisting of polymer modifiers, Portland Cement (Type I), fine aggregates and fiber reinforcement. It shall be compatible with insulation board and reinforcing fabric.

D. Fiber Reinforcement—loose fill chopped fiber strands compatible with cementitious base coat as per manufacturer's requirements.

E. Finish Coat—To be specified from among:

1. Cementitious coatings with polymer modifiers, Portland Cement (Type I), and fine aggregates in accordance with ASTM C 897 and manufacturer's requirements. Color and texture to be approved by architect/owner.

Color Sealer—for cementitious finish shall be acrylic based, factory mixed sealer, compatible with finish coat. Color to be approved by architect/owner.

2. Synthetic finish (factory blended and integrally colored). Color and texture to be approved by architect/owner.

3. Exposed aggregate finish. Aggregate type, color and base coat tint to be approved by the architect/owner — size of aggregate not to exceed twice base coat thickness.

F. Fasteners—shall have the necessary pull out, tensile and shear strength as installed to resist design loads imposed on the system.

1. Screws shall be corrosion resistant steel drill screws meeting the requirements of ASTM C 1002. Screw threads shall be adequate to pull screw head below the surface of the foam, when used with the manufacturer's washer attachment.

2. Nails, Staples or Pins—hand or power driven, corrosion resistant, used with manufacturer's washer attachment.

3. Expansion Anchors—expandable sheath with corrosion resistant hammer drive pin, pre-drilled and used with manufacturer's washer attachment.

G. Accessories—conventional plaster trim accessories, J-mold, drip channel, expansion joints, control joints and corner beads, of exterior grade vinyl, zinc or galvanized metal. Ground size to be compatible with coating thickness.

H. Sealant—as approved for compatibility by the system manufacturer.

4A.2.3   Performance Characteristics.

A. General Physical Properties—the system shall meet or exceed the following performance standards when tested in accordance with the following methods:

1. Accelerated Weathering. Test: ASTM G 23-81 — Method 1 or ASTM G 53-81.

Standard practice for operating light and water exposure apparatus for exposure of nonmetallic materials. Test period: 2,000 hours.

Test specimen: The sample shall be manufacturer's standard finish, one inch thick system, and cured for 28 days. Result: No cracking, flaking, peeling, blistering or other deleterious effects.

2. Wind Driven Rain. Test: Federal Specification—TT C 555B.

Specimen: Sample shall be one inch exterior insulation system with manufacturer's standard finish 8″ x 16″ cured for 28 days. Result: No visible leaks, no blistering, cracking or wear of the finish coat and less than 75 grams weight gain.

3. Salt Spray Resistance. Test: ASTM B 117 Salt Spray (Fog) Testing. Test period: 300 hours.

Test specimen: Sample shall be one inch exterior insulation system with manufacturer's standard finish, 8″ x 16″ cured for 28 days. Result: No deleterious effects.

4. Absorption Freeze–Thaw. Test: ASTM C 67-81.

Specimen: Sample shall be one inch exterior insulation system with manufacturer's standard coating on one face cured for 28 days, 3″ x 5″.

The specimen is subject to 50 cycles consisting of 20 hours of freezing at 16°F and four hours of thawing in water of 75°F ± 10 degrees. Result: No visible damage—negligible weight gain.

5. Mildew Resistance. Test: mil Standard 810B, Method 508.

Specimen: Sample shall be coating only on 2″ x 2″ glass slides. The finish coat shall be applied according to manufacturer's printed instructions. Specimen shall be incubated for a minimum of 28 days. Result: No growth supported.

6. Abrasion Resistance. Test: ASTM D 968-81, Method A (Falling Sand Abrasion Test).

Specimen: Sample shall be a one inch exterior insulation system with manufacturer's standard finish and cured for 28 days. Result: No cracking, checking or loss of film integrity after 500 liters of sand.

7. Water Vapor Transmission (coating only). Test: ASTM E 96 Water Method, Procedure B.

Specimen: Facing only, consisting of mesh, base and finish coats, standard thickness. Result: Minimum—greater than insulation permeability. Maximum—7.5 perms.

8. Water Vapor Transmission (extruded polystyrene). Test: ASTM E 96 Water Method, Procedure B.

Specimen: Insulation only, one inch thickness. Result: 1.0 perm.

B. Fire Test Performance.

1. Surface Burning Characteristics. Test: ASTM E 84.

Specimen: Finish mounted on three ¼″ cement asbestos boards 20¼″ x 96″, applied at coverage rates consistent with manufacturer's current application instructions. Result: Flame spread index—less than 25. Smoke developed value—less than 25.

2. Full Scale Diversified Fire Test. Test: ASTM E 108 Modified for Vertical Walls.

Specimen: Two panels sized to meet the test requirements (6′ x 10′). One panel shall have the facing removed from an area 4″ high x 24″ wide, centered 2′ above the bottom of the panel, so that the foam is exposed. The sample shall consist of a minimum of 4″ of insulation mounted on gypsum board of ½″ thickness and cured for a minimum of 28 days.

Result: The system shall not contribute to significant vertical or horizontal flame spread. Finish shall prevent fire involvement of the insulation core. No fall-off of coating.

C. Structural Testing.

1. Impact Loading Test. Test: ASTM E 695, 30 lb. Impact Mass.

Method: Twelve impacts by 30 lb. lead shot mass standard test bag, swung as a pendulum, from 6″ to 6′ drop heights at 6″ increments.

Specimen: Sample shall be 2″ exterior insulation system mounted on ½″ gypsum sheathing, mounted to 24″ c. to c. 18 gauge steel studs. Result: No cracking, no denting.

2. Negative Wind Load Full Scale Testing.

Test: ASTM E 330 Positive and Negative Air Pressure. Test load shall be maintained for five minutes as required by ASTM C 72-80.

Specimen: The sample shall be a minimum of 4′ x 4′ and consist of studs, sheathing and one inch exterior insulation system in accordance with manufacturer's latest application instructions. Result: Withstand wind load requirements of local building codes.

4A.3   EXECUTION

4A.3.1   Installation.

A. Installation shall be in accordance with system manufacturer's current written instructions.

4A.3.2   Substrate.

A. The substrate shall be a type approved by the system manufacturer.

B. The substrate shall be free of planar irregularities greater than ¼″ and shall be sound.

C. Unsatisfactory conditions shall be corrected before the application of the system.

4A.3.3   Application.

A. Installation shall conform to the manufacturer's written instructions.

1. Insulation board, reinforcing mesh and mechanical anchors.

(a) Boards shall be placed from a level base line with vertical joints to be staggered and butted tightly.

Board joints shall be interlocked at corners. Surfaces of adjacent boards shall be flush. Boards shall be affixed to substrate or structural members with mechanical fasteners. (Fastening pattern shall be according to manufacturer's written instructions with spacing not less than one fastener per two square feet.) For adhesive attachment, see manufacturer's printed instructions.

(b) Reinforcing glass fabric shall be installed according to manufacturer's written instructions.

(c) Installation and placement of control/expansion joints shall be in accordance with manufacturer's written instructions, in areas not to exceed 150 square feet and 20 feet lineally.

(d) Sealant requiring application prior to base and finish coat application shall be placed in accordance with the system manufacturer's written instructions to clean, dry surfaces.

2. Base and Finish Coat.

(a) Base coat shall be mixed in accordance with manufacturer's written instructions.

(b) Apply coating tightly to insulation surface, by hand trowel or spray, over reinforcing mesh, to achieve bond and contact. Apply to a uniform thickness as specified by the system manufacturer. Insulation surface shall be entirely free of deleterious material. Base coat shall be applied to level out surface areas and to fill joints smooth with adjacent areas. Base coat surface shall be completed and textured in a manner that will be acceptable for finish coat.

(c) Finish coat shall be mixed and applied to a thickness in accordance with manufacturer's written instructions. A wet edge should be maintained to obtain a uniform appearance. Texture and color shall be specified and matched to the approved sample. All mechanics applying finish and texture shall utilize the same equipment and techniques to achieve uniformity.

4A.3.4 Job Site Clean-Up

A. All excess wall system materials shall be removed from the job site by the wall system applicator.

JOINT SEALANT SPECIFICATION FOR EXTERIOR INSULATION AND FINISH SYSTEMS (EIFS) CLASS PB AND PM

1. PRODUCTS

A. Sealant shall meet Federal Specification TT-S-00227E, Type II, Class A and ASTM C 920, Type M, Grade NS, Class 25.

1. Sealant recommended and listed by the EIFS system manufacturer and approved by the sealant manufacturer for the selected EIFS system.

B. Primer.

1. As per sealant manufacturer's recommendation to be compatible with recommended sealant and EIFS system.

C. Back-Up Material.

1. Closed-cell polyethylene rod approved by sealant manufacturer.

2. Bond-breaker tape approved by sealant manufacturer.

2. PREPARATION

Must be done in a good and workman-like manner which meets the following minimum requirements or standards. The specific manufacturer's latest instructions must be followed.

A. Inspection.

1. Examine surfaces to be caulked for:

(a) Defects or coatings on the substrate that will adversely affect the execution and quality of work.

(b) Deviations beyond allowable tolerances for the installation of sealants.

2. Work shall not be started until usatisfactory conditions are corrected.

B. Joint Design.

1. Minimum size of joint shall be ½" or four times the anticipated movement, whichever is greater.

2. Minimum caulking joint, ½" x ¼"; depth of joint shall not exceed width of joint. For joints larger than ½" x ½", depth of the sealant shall be not more than ½".

3. Maximum joint size — approximately 2" width by ½" depth in a single application.

C. Surface Preparation.

1. All areas where the system meets dissimilar materials or in areas where the system has a joint through itself shall have a sealant joint. The EPS board shall be cut back from the dissimilar adjoining material a minimum of ½" to form a sealant joint. Prior to sealing, all EPS board edges shall be covered as per system manufacturer's specification. Allow this application to dry before applying sealant materials.

2. (a) Joints to be caulked shall be clean, dry and free of dust. Loose mortar or other foreign materials shall be removed.

(b) All EPS edges shall be coated with base coat, reinforcing fabric and finish or as per manufacturer's recommendation. Caulk shall not come in direct contact with EPS board or reinforcing fabric.

3. All joints which have EIFS panels on both sides or are abutting similar porous construction materials must have sealant Primer applied as per listed sealant manufacturer's recommendation. Primer shall be completely dry before applying sealant.

4. All anodized aluminum surfaces shall be wiped clean to remove temporary organic protective coatings or surface contamination. After drying, surfaces are to be primed per sealant manufacturer's specifications. Primer shall be allowed to dry prior to applying sealant.

5. Clean ferrous metals of all rust, mill scale and coatings by wire brush, or grinding. Remove oil and grease and temporary protective coatings with a high performance cleaner approved by the sealant manufacturer.

6. Precast or masonry joint surfaces shall be wire brushed, then air blown clean. The joint surfaces must be free of form release agents or retarders, or must be tested

on project site for compatibility.

7. Sealants shall not be applied to masonry joints where a water repellent or masonry preservative has been applied prior to caulking. Waterproofing treatments shall be applied after caulking when specified.

8. Clean plastic by chemical cleaners or other means which are not harmful to substrates or leave residues capable of interfering with adhesion of joint sealant. The indiscriminate use of solvent for cleaning joints cannot only be ineffective but, in the case of plastics, could damage the surface. Recommendations should be obtained from the sealant manufacturer.

9. Do not caulk joints until they are in compliance with requirements of the sealant and EIFS system manufacturers, the detail as shown on the drawings, and the specific requirements of other sections of the specifications.

10. All joint surfaces composed of EIFS panels shall have no voids, cracks, and be of the proper depth requirements.

D. Joint Backing:

1. Joint backing should be used to control the depth of joint to recommended thickness. Where depth of joint will not permit use of joint backing, a bond-breaker tape must be installed to prevent three-sided adhesion.

2. Joint backing shall be a closed-cell polyethylene rod or other backing material to prevent adhesion to sealant material. Select a size to allow for 25% compression of the backing when inserted into the joint.

3. APPLICATION

Must be performed in a good and workman-like manner which meets the following minimum requirements or standards. The specific manufacturer's latest instructions must be followed.

A. Apply sealant with a gun with proper size nozzles. Use sufficient pressure to fill all voids and joints solid to the back-up material.

B. Surface of sealant shall be a full smooth bead, free of ridges, wrinkles, sags, air pockets and embedded impurities.

C. After all joints have been completely filled, they shall be neatly tooled to eliminate air pockets or voids, and to provide a smooth, neat appearing finish.

D. Protect the adjacent system from excess sealant material. Use masking tape where required to prevent contact of sealant with adjoining surfaces which otherwise would be permanently stained or damaged by such contact or by cleaning methods required to remove sealant smears. Remove tape immediately after tooling without disturbing joint seal.

E. Job Site Clean-Up.

1. All excess materials shall be removed from the job site by the sealant applicator.

2. All surrounding areas where the joint sealant has been applied shall be left free of debris and foreign substances.

5. FIRE PROTECTION (SPRAYED ON)

Because of the rapid rate at which the lathing, plastering and drywall industries' fire research activity provides new test data, it is almost impossible to publish a complete listing of fire resistance ratings that is up to date.

Also, interpretation and recognition of fire test data varies among building departments and other governmental agencies which enforce fire protective standards of construction.

Unforunate as this may be, it is a recurrent fact of life to the architect, engineer and contractor who must keep abreast of the most recent data and its degree of acceptance by enforcement agencies.

In addition to the publication of nationally recognized authorities which list fire resistive ratings, most manufacturers' associations publish complete lists of fire resistive assemblies tested with one or more of their members' products.

It is recommended that during preparation of preliminary drawings and specifications, the architect consult the building department having jurisdiction, national fire resistive listing authorities, or manufacturers of lath and plaster fireproofing products, for specific information on "approved" tested assemblies.

Standard lathing and plastering construction, with little exception, provides the architect with an almost limitless choice of fire protection methods.

Standard lathing and plastering construction will usually meet fire protection standards. These have been adquately covered in other sections of these reference specifications.

In this section are recommended specifications for the application of sprayed-on fire protection.

FIRE RESISTANCE RATINGS

The fire resistance rating of a wall or partition or floor and ceiling construction is essentially the time in hours the construction will remain in place and prevent temperatures on the unexposed side from exceeding a certain amount (250°F. higher than the starting temperature—max. 325°F.—when the construction is exposed to the standard test fire, simulating what might be anticipated in a fire in an actual structure).

With masonry walls and partitions the critical feature is usually the temperature rise on the unexposed surface.

With columns, beams and girders the critical feature is usually the ability to carry the load . . . usually dependent on the heat insulating value of the protective covering and its ability to stay in place during the fire exposure.

With incombustible floor construction the critical feature has usually been the temperature rise on the top unexposed surface in view of adequate protection provided on the underside.

The use of perlite or vermiculite aggregate in place of sand in plaster increases its resistance to fire.

. . . ratings of plaster facing on walls and partitions are not applicable to similar finishes on ceilings . . . plaster may fall from a ceiling considerably before it will fall from a vertical surface.

In fires in buildings the greatest number of deaths are caused by exposure to flame and hot or toxic gases, suffocation by smoke, deprivation of oxygen, and panic. Restrictions and limitations applicable to fire safety found in local building codes are designed primarily to prevent the spread of fire from one area to another, and to limit the use of potentially dangerous flammable surface materials and so safeguard public health and welfare.

Flame spread requirements are set up by building codes with the intention of regulating the use of finishing materials with respect to the rapidity with which flames spread over their surfaces. Tests made in accordance with ASTM Specification E 84, commonly known as the Tunnel Test, are nationally accepted. From them classifications are reported by Underwriters Laboratories Inc.

To assist the designer most building codes include a division or chapter which states . . . "for the purpose of determining time periods of fire resistance, as required by the provisions of this Code, materials of construction shall be assumed to have the fire-resistive ratings given herein . . ." and these are detailed in tables showing the fire-resistive periods for "Protection of Structural Elements," for "Walls and Partitions" and for "Various Floors, Ceilings and Roofs."

5.1 SPRAYED-ON FIRE PROTECTION

5.1.1 Scope.—Furnish and apply sprayed-on fire protection as shown on the drawings and specified in this Section. [1]

5.1.2 Fire-resistive Ratings.—Apply sprayed-on fire protection material so as to provide the fire-resistive ratings indicated on the drawings or set forth herein.

5.1.3 Test Criteria.—Fire ratings and thickness of fire protection material proposed for use on the structural members and decking as indicated shall be established only by certified tests made by a testing laboratory recognized by the building department or state fire inspection agency having jurisdiction over the project. [2]

5.1.4 Workmanship.—Apply material in strict accordance with manufacturer's printed instructions and any additional requirements specified herein. Use experienced applicators and equipment of the type recommended by the material manufacturer.

5.1.5 Coordination.—Advise general contractor prior to the application of fire protection materials. All other trades must have completed the installation of all items such as clips, sleeves, hangers, clamps, and supports of all kind for work suspended from, attached to, or passing through construction required to receive sprayed-on fire protection. No mechanical services such as ducts, pipes, conduit, etc., shall be installed prior to application of sprayed-on fire protection materials.

5.1.6 Cleanup.—After completion, clean up all rubbish resulting from this work and remove it from the premises. Leave clean all finish surfaces adjacent to sprayed areas.

5.2 MATERIALS

5.2.1 Fire Protection Material.—Shall be a factory-mixed blend of inorganic insulators and binders prepared for fire-protection purposes. Material may be cementitious type or fiber type.

Tested in accordance with ASTM E 84, material shall have a flame spread rating of 15 or less.

Material shall have been tested for fire resistance in accordance with the applicable requirements of ASTM E 119.

Deliver fire-protection material to the job in unopened factory-sealed bags, with each bag properly identified and bearing the Underwriters Laboratories' re-examination label. Keep dry until ready for use.

5.2.2 Accessory Materials.—Provide bonding adhesives, sealers and other materials for use with the approved fire-protection material as required by manufacturer.

5.3 APPLICATION

5.3.1 Temperature and Ventilation.—Apply sprayed-on fire protection material under temperature and ventilation conditions recommended by manufacturer. Temporary ventilating devices, or closures for openings, if needed, shall be provided before application of fireproofing. [3]

5.3.2 Surface Preparation.—Surfaces to receive fire protection shall have been cleaned by others of all dirt, grease, rust, scale, loose paint, and any other extraneous substances which might prevent permanent adhesion of fire protection. [4] [5]

[1] These specifications are recommended as a part of the plastering section, covering the machine application of fire protection materials directly to structural steel, roof and floor decks, beams and girders. When supplemented by the general conditions, plans, details and schedules, the specification should provide the contractor with sufficient information to establish costs and to construct the work properly.

[2] Indicate whether building department or state agency has jurisdiction.

[3] Responsibility for temperature and ventilation control should be assigned to prime contractor to avoid duplication of charges in bids of general and subcontractors. See 1.2.4.

[4] Architect may wish to require that manufacturer certify that surfaces have been inspected and are suitable to receive sprayed-on fire protection.

[5] Since control of surfaces prior to application of fire protection material is not possible, architect should provide that cleaning of surfaces be done by general contractor.

5.3.3 Bonding Adhesive.—Where required by manufacturer, apply bonding adhesive on all surfaces to receive sprayed-on fire protection material in accordance with manufacturer's printed instructions.

5.3.4 Fire Protection Material.—Application of fire protection material shall be in accordance with manufacturer's specifications.

5.3.5 Surface Finish.—Where structural members are indicated to receive a surface finish, wrap or cage with metal or wire fabric lath, and plaster with approved fire resistive plaster. [6]

[6] Architect should indicate on drawings or in specifications.

5.3.6 Repair damage to sprayed fire-protection materials from failure of other trades to install hangers, inserts, anchors, built-ins, and similar items, or from negligence; trades responsible for damages, as determined by the general contractor, shall pay costs for repairing the damage.

5.3.7 When testing of fireproofing materials is required by the governing code, architect, or owner, the cost for such testing shall be paid for by the owner.

CALIFORNIA STATE FIRE MARSHAL'S LISTING SERVICE

An important reference guide to fire-resistive materials and systems is the approval service operated by the Office of the State Fire Marshal, State of California.

The State Fire Marshal tests materials and systems for specific uses and issues approvals for one year.

Approved products and systems are each described in printed statements which contain general descriptions of the subject products and systems and defines their fire-resistive qualities.

These approval sheets are compiled into a loose-leaf binder titled "California State Fire Marshal Listing Service."

Copies of this book may be obtained from the Office of The State Fire Marshal, 7171 Bowling Drive, Suite 600, Sacramento California, 95823, telephone (916) 427-4178. Because the price varies from time to time it is necessary to telephone and inquire on the price before ordering.

Any product or system is eligible for testing by the State Fire Marshal by submission of an application form and payment of the required fee.

Approval by the State Fire Marshal is not a necessary prerequisite for the use of a particular material or system in every jurisdiction in California — but in many cases it is mandatory and in every case the approval is a definite benefit.

6. PLASTERING MATERIALS

6.1 GENERAL

6.1.1. Standard Specifications.—Where published standard specifications are referred to herein, they shall be those of the latest date of adoption. [1]

6.2 QUALITY AND USE OF MATERIALS

6.2.1 Quality.—The materials provided and installed shall be new and of the quality required by this Section.

6.2.2 Use of Gypsum Plaster.—Gypsum plaster may be used on any type of base for fire resistive or structural purposes, but shall not be used where it will be subjected to alternate wetting and drying or continuous moisture exposure, or for exterior plaster, except in protected areas defined by building codes or building departments.

6.2.3 Use of Portland Cement Plaster.—Portland cement plaster or Portland cement-lime plaster may be used for any purpose in any location but shall not be applied over gypsum lath, without metal lath and paper being installled first.

6.2.4 Delivery of Materials.—Deliver materials so as to insure uninterrupted progress of the work.

6.2.5 Containers.—Deliver all manufactured materials in the original packages or containers bearing the name of the manufacturer and brand.

6.2.6 Protection of Materials.—Keep all cementitious materials dry until used. Keep materials off the ground, under cover, and clear of damp walls or other damp surfaces.

6.3 GYPSUM PLASTERS

6.3.1 Gypsum Plasters.—Shall be neat (or) mill-mixed (or) wood fibered (or) bond (or) gauging, gypsum plaster (as specified for their particular use) complying with the "Standard Specifications for Gypsum Plasters," ASTM Designation: C 28. [3]

6.4 LIME

6.4.1 Hydrated Lime.—Both dolomite and high calcium, used in the preparation of lime putty, shall be a standard brand conforming to the "Standard Specifications for Special Hydrated Lime," ASTM Designation: C 206.

6.4.2 Lime Putty.—Shall weigh no less than 83 lbs. per cubic foot and shall be made of hydrated lime or quicklime.

[1] Where any reference is made in these specifications to ASTM, ACI, U.S. Federal Specifications, AIA or CSI, applicable excerpts will be found in Specification Reference 1.8.

[2] On exterior eave overhangs or soffits, or interior locations, portland cement plaster may be applied over gypsum lath backing when plaster is reinforced as for other solid backing.

[3] For definitions and uses of gypsum plasters see Glossary, Specification Reference 1.9.

6. PLASTERING MATERIALS

6.5 PORTLAND CEMENT

6.5.1 Portland Cement.—For plaster shall conform to "Standard Specifications for Portland Cement," ASTM Designation: C 150, Type I, Type II or Type III. [4]

6.5.2 Air-entraining Portland Cement.—Shall conform to "Standard Specifications ASTM Designation: C 175, Type I-A, Type II-A or Type III-A.

6.5.3 Plastic Cement.—Shall meet the requirements of "Standard Specifications for Portland Cement," ASTM Designation: C 150, Type I or Type II except in respect to the limitations on insoluble residue, air-entrainment, and additions subsequent to calcination. Plasticizing agents may be added to portland cement Types I and II in the manufacturing process, but not in excess of 12 percent of the total volume.

When plastic cement is used, no lime or other plasticizer may be added to the cement plaster at the time of mixing.

6.6 AGGREGATES

6.6.1 Gypsum Basecoat Plaster Aggregates.—Shall comply with "Tentative Specifications for Inorganic Aggregates for Use in Interior Plaster," ASTM Designation: C 35. [6]

6.6.2 Grading.—Aggregate for gypsum basecoat plaster, except as provided elsewhere in these specifications, shall be graded within the following limits:

Sieve Size	Percentage Retained on Each Sieve					
	Perlite by Volume		Vermiculite by Volume		Sand by Weight	
	Max.	Min.	Max.	Min.	Max.	Min.
No. 4	0	—	0	—	0	—
No. 8	5	0	10	0	5	0
No. 16	60	10	75	40	30	5
No. 30	95	45	95	65	65	30
No. 50	98	75	98	75	95	65
No. 100	100	88	100	90	100	90

6.6.3 Gypsum-Lime Smooth Finish Aggregate.—Aggregates for use in gypsum-lime smooth finish, such as silica sand, perlite, etc., shall be graded within the following limits:

Sieve Size	Percentage Retained on Each Sieve			
	Perlite by Volume		Sand by Weight	
	Max.	Min.	Max.	Min.
No. 20	0	—	0	—
No. 30	10	0	10	0
No. 100	100	40	100	40
No. 200	100	70	100	70

[4] Type I is for use in general plastering. Should always be specified with addition of plasticizers for workability.

Type II should be used when the plaster will be exposed to moderate sulphate action (alkali).

Type III should be specified or approved for use when high early strength is desireable.

[6] Sand, wherever it is available, should be washed natural sand.

6.6.4 Portland Cement Plaster Sand.—Except as provided elsewhere in these specifications, shall be clean and well graded from coarse to fine, meeting the requirements of ASTM C 897, except gradation of sand shall conform to the following requirements:

Sieve Size	Percent Retained (by Weight) On Each Sieve	
	Max.	Min.
No. 4	0	0
No. 8	10	0
No. 16	40	10
No. 30	65	30
No. 50	90	70
No. 100	100	95

6.7 PROPRIETARY AND SPECIAL PURPOSE PLASTERS

6.7.1 Manufactured Finishes (exterior and interior).—When factory mixed shall comply with the general requirements of the "Specifications and Standards for Manufactured Stucco Finishes," prepared by the Stucco Manufacturers Association, 14006 Ventura Blvd., Suite 204, Sherman Oaks, CA 91403, or with requirements of applicable building codes. [7] [8]

6.7.2 Acoustical Plaster.—Shall develop a noise reduction co-efficient of .... Texture and color shall closely approximate sample in architect's office or in job offfice. Upon completion of application, the lathing and plastering contractor may be required to submit to the architect a dated and signed Certificate of Compliance, showing name of contractor, name and address of project, and certifying that the factory-manufactured acoustical plaster has been applied in accordance with installation specifications set forth by the architect and the manufacturer.

6.7.3 Veneer Plaster. [10]

6.7.4 Marblecrete.—Shall be an exposed aggregate finish consisting of natural or integrally colored aggregate, partially embedded in a white or colored bedding coat of portland cement or portland cement-lime plaster. Marblecrete shall be applied over portland cement plaster base coats, (or) concrete (or) masonry surfaces. [11]

6.8. ADDITIVES AND ADMIXTURES (for portland cement plaster).

6.8.1 Lime. [12]

6.8.2 Other Additives and Admixtures.—Plasticizing or air-entraining agents, when used, shall not reduce the compressive strength of the mortar more than 15% below the strength of the mortar without the plasticizing or air-entraining. Other additives and admixtures, when used, shall not reduce the compressive strength, tensile strength, flexural strength, impact strength, or resistance to abrasion of the mortar below the strength of the mortar without the additive or admixture. Use additives or admixtures only upon the approval of the architect or as specified herein. [13]

[7] For material specification of various manufactured finishes see Specification Reference No. 10, Stucco Finishes.

[8] Interior and exterior finishes may be colored in the field, although it is difficult to accomplish uniformity of color. When desired, architect should specify the addition of pure mineral oxides guaranteed by manufacturer.

[9] Deleted.

[10] See Specification Reference No. 9.

[11] Specify which type of base. See Specification Reference No. 10, Stucco Finishes, for Marblecrete specification.

[12] For material specification see 6.4.

[13] Specify type of admixture required. When plasticizing or air-entraining agents are used, the amounts will vary with the agent used and the degree of plasticity or air-entraining desired. The kind and amount of plasticizing or air-entraining agent necessary for proper workability should be determined in advance of starting the job. The smallest amount needed should be used to secure the desired plasticity or air-entraining.

6.9 WATER

6.9.1 Mixing Water.—Shall be clean, fresh, suitable for domestic consumption, and free from such amounts of mineral and organic substances as would affect the set of gypsum plaster. [14]

[14] Water containing salt or alum, or water in which tools have been washed, accelerate the set. Water containing organic or vegetable matter may retard the set.

6.10 BONDING AGENTS

6.10.1 Liquid Bonding Agents—Surface Applied.—Bonding agents shall be a resinous emulsion which will provide bond for gypsum, lime, portland cement, or acoustical plaster finishes to gypsum or portland cement, plaster, concrete, masonry, wood, steel, painted or unpainted, old or new surfaces. When properly cured and dried bond tensile strength and bond compressive shear strength shall be as outlined in Military Specification MIL-B-19235C(YD), Table I, when tested in accordance with 4.3.4.1 and 4.3.4.2 of that same specification. Minimum tensile strength shall be 60 p.s.i. and minimum compressive shear strength shall be 300 p.s.i. [15]

[15] Bonding agents should not be used as waterproofing agents.

6.10.2 Liquid Bonding Agents—Integral.—Bonding agents shall be a non-reemulsifiable resinous emulsion which will provide bond for gypsum, lime, portland cement, or acoustical plaster finishes to gypsum or portland cement, plaster, concrete, masonry, wood, steel, painted or unpainted, old or new surfaces. When mixed according to manufacturers' recommendations, properly cured, dried, and tested samples shall have the following minimum strengths: [16]

[16] Manufacturer of bonding agent should be consulted for proper water/ bonding agent ratio.

Tensile Strength	400	ASTM C 190
Compressive Strength	2500	ASTM C 109
Compressive Shear Strength	175	Fed Spec. MMM-B-350a
Flexural Strength	900	ASTM C 348
Brick Bond Strength	300	ASTM C 321

6.11 COATINGS

6.11.1 Waterproof Coatings.—Waterproof coatings shall meet or exceed the requirements as set forth in Federal Specification TT-P-0035 when tested in accordance with that specification with regard to Impact Resistance 4.4.5, Resistance to Wind-Driven Rain 4.4.7, and Accelerated Weathering 4.4..8. Coating shall also pass 3000 litre sand abrasion (Fed. Spec. TT-P-14lb, Method 619.1). Percent of absorption shall be less than 2.75 percent (ASTM C 67). Waterproof coatings must show no adverse effects after 5% salt solution exposure, 300 hours (ASTM B 117). Waterproof coatings must exhibit resistance to fungus growth, 21 days (Fed. Spec. TT-P-29b). Waterproof coatings shall show no evidence of chalking, checking, scaling, peeling, or blistering after 4000 hours exposure Atlas Twin Arc (ASTM D 822 and ASTM C 23).

7. PLASTERING TECHNIQUES

7.1 PREPARATION FOR PLASTERING

7.1.1 Temperature and Ventilation.—When the prevailing outdoor temperature at the building site is below 40°F. do not apply interior plaster unless a uniform temperature of not under 40°F. has been, and continues to be, maintained in the building prior to the application of plaster, while plastering is being done, and until plaster is dry. Prevent concentration of heat on plaster areas near the heat source.

For exterior plaster, do not apply when prevailing outdoor temperature is below 40°F. If freezing is expected, do not apply plaster beyond period of day necessary to allow hydration. [1] [2]

7.1.2 Masonry Surfaces.—On which suction must be reduced shall be wet down sufficiently before plastering operations start. [3] [4] [5]

7.1.3 Poured Concrete Surfaces (preparing bond).—Give concrete surfaces a dash bond coat of portland cement plaster or treat with a liquid bonding agent. [6] [7] [8]

7.1.3.1 Dash Bond.—Shall be a portland cement plaster composed of 1 part cement to 1½ parts graded sand by volume, mixed to a mushy consistency. Forcibly dash plaster on the surface, leave undisturbed and keep damp at least 24 hours following its application. Allow to cure (harden) before application of plaster. [9]

7.1.3.2 Bonding Agent.—Surface-applied or integral bonding agent should be used in strict accordance with manufacturer's direction.

7.1.3.3 Roughening of Concrete (by others).—If, in the opinion of the plastering contractor, concrete surfaces are not sufficiently rough and clean to provide proper bond for plaster, they shall be cleaned and adequately roughened by other than the plastering contractor.

7.1.4 Examination of Surfaces to be Plastered.—Examine all construction, grounds and other accessories, to insure that finished plaster surfaces will be true to line, level, plumb, square, curved, or as otherwise required without requiring excessive thickness of plaster.

7.2 SEQUENCE OF OPERATIONS

7.2.1 Interior.—Complete all plastering in rooms and spaces where prefabricated acoustical units are to be installed. In rooms having wainscots of materials other than plaster, apply finishing coat of plaster on walls above wainscot before wainscot work is completed.

Installation of hangers and sleeving for the work of all trades shall be completed before application of plaster fireproofing. Complete all plaster fireproofing to structural steel members or to steel decking before installation of plumbing, mechanical and electrical work, and before installation of suspended ceiling systems.

[1] Responsibility for temperature and ventilation control should be assigned to prime contractor to avoid duplication of charges in bids of general and subcontractors. See 1.2.4.

[2] To minimize the possibility of plaster cracking because of structural movements caused by thermal changes, the building should be permitted to become adjusted to continuous, uniform heating conditions for at least 48 hours prior to the start of interior plastering.

[3] Specify in Masonry Section that surfaces to be plastered shall be turned over to the plastering contractor cleaned of all dust, loose particles and free of other foreign matter. Since masonry units do not usually provide sufficient mechanical bond, specify dash bond method (see 7.1.31), or liquid bonding agent (see 7.1.32) to be applied at option of contractor. Over glazed units specify attachment of metal or wire fabric lath.

[4] Plaster will not satisfactorily bond to masonry surfaces to which bituminous compounds have been applied. Do not specify.

[5] In areas of heavy rainfall, gypsum plaster should not be applied directly to interior surfaces of exterior masonry construction. Such surfaces may be treated with a dash coat and a scratch coat of portland cement plaster to which an approved waterproofing admixture has been added; and left with a good mechanical key. After curing, apply gypsum plaster brown coat. Under extreme conditions furring may be necessary.

[6] Specify either 7.1.31 or 7.1.32 at option of contractor.

[7] Over rough formed surfaces dash bond coat may be used. Over plywood or metal formed surfaces, liquid bonding agent is advisable. When form release compound is used, specify one that is compatible with the bonding agent.

[8] Specify in Concrete Section that concrete surfaces to be plastered shall be turned over to the plastering contractor cleaned of all dust, loose particles, parting and similar compounds, and other foreign matter.

[9] The use of dressed (smooth) lumber, metal and plywood forms, their oiling or greasing, and the vibrating of concrete produces concrete surfaces so smooth that roughness necessary for mechanical bond of plaster is absent.

7. PLASTERING TECHNIQUES

7.3 THICKNESS

7.3.1 Gypsum Plaster Thickness (See Table 8-1) [10]

7.3.2 Portland Cement Plaster Thickness (See Table 8-1)

7.3.3 Solid Plaster Partition Thickness (See Table 5-5 for Channel Solid [11]

7.4 PROPORTIONS

7.4.1 Gypsum Plaster Basecoats.—Proportion neat gypsum plaster to mineral aggregates as shown in Table 8-2.

7.4.1.1 Wood Fiber Plaster.—Use without the addition of aggregate except as otherwise specified. On unit masonry surfaces, add sand to wood fiber plaster in proportions of not more than 100 lbs. of plaster sand to 100 lbs. of plaster. [12]

7.4.1.2 Bond Plaster.—Use without aggregate over concrete surfaces. May not be readily available. [13]

7.4.2 Gypsum Finish Coats. [14] [15] [16]

7.4.2.1 Keene's Cement-Lime Smooth Finish (Keene's Cement Coat).—Proportion by weight: 1 part of Keene's cement, not more than 1½ parts of dry hydrated lime, and ½ part fine white sand. [17] [18]

7.4.2.2 Gauging-Lime Smooth Finish (putty coat).—Proportion by volume, 1 part of gauging plaster, and not more than 3 parts of lime putty. [19] [20]

Where the basecoat plaster aggregate is perlite or vermiculite, include not more than 1 cubic foot of fine aggregate such as fine silica sand.

7.4.2.3 Keene's Cement-Lime Sand Float Finish.—Proportion by weight; 1 part of Keene's cement to not more than 3 parts of dry hydrated lime (or an equivalent amount of lime putty) and maximum 4 parts of graded sand. [21] [22]

7.4.2.4 Machine Dash Finish.—Proportion by weight; 1 part Keene's cement, 1 part dry hydrated lime (or an equivalent amount of lime putty) and not more than 3 cubic feet of lightweight aggregate. [23] [24]

7.4.2.5 Manufactured Gypsum Finishes.—Proportioned smooth, sand float and machine dash finishes either natural or colored, in accordance with the manufacturer's directions. [25] [26] [27]

7.4.3 Portland Cement and Portland Cement Lime Plaster Basecoats. [28] [29]

---

[10] Where fire resistive rating is required, thickness of plaster should be determined from local building code.

[11] For studless solid plaster partition thickness, see 3.9.1.

[12] Wood fiber plaster containing a maximum 100 lbs. of plaster sand is recommended over all plaster bases in areas not exposed to frequent wetting, but where completed plaster may be subjected to severe usage, making a high-strength basecoat desirable.

[13] Concrete surface shall be rough enough to assure initial, mechanical bond. See 7.1.3 for preparing bond.

[14] Humidity and temperature changes require adjustment in amount of gauging plaster or Keene's cement in finish coat. Consultation on proportions between the plastering contractor and the architect will assure best results.

[15] All the proportioned finishes in this specification are suitable for texturing, except that additional sand may be added to smooth finishes when a texture is specified. Specify texture by name (see Specification Reference No. 10) or by manufacturer's number where finish is proprietary.

[16] Finish plaster proportions contained herein are of standard hardness. When surfaces will be subjected to extreme abuse, special finishes should be specified. Consult CLPCA on proportioning.

---

[17] Standard Hardness.

[18] This mix is equivalent to 100 lbs. of Keene's cement to not more than:
  3 (50 lb.) bags of hydrated lime or
  1½ barrels (270-300 lbs.) of lime putty or 3⅜ cu. ft. of lime putty or
  26 gallons of lime putty.
Keene's cement and lime are usually mechanically mixed.

[19] Standard Hardness.

[20] This mix is equivalent to 100 lbs. of gypsum gauging plaster to not more than:
  4 (50 lb.) bags of hydrated lime or
  4½ cu. ft. of lime putty or
  35 gals. of lime putty.

[21] Standard Hardness.

[22] Specify maximum sand sieve size.

[23] Soft.

[24] Sometimes called "simulated acoustic," this finish should be limited to ceilings and wall areas over 7'6" from floor.

[25] Standard Hardness.

[26] Specify maximum sand sieve size for float finishes.

[27] Integrally colored finishes should be specified by manufacturer's color number and keyed to finish schedule. If color selection has not been made when specifications are written, some indication should be given of the depth of color (light, medium, dark) to permit accurate bidding.

[28] Do not add plasticizing agents to portland cement lime plaster.

[29] More aggregate is permitted in portland cement-lime plaster since the lime is considered a cementitious material, not a plasticizer. Proportions are considered optimum. For more information see ASTM C 926-81.

7.4.3.1 Portland Cement Basecoat Plaster.—Proportion portland cement to mineral aggregates as shown in Table 8-3. Cement may be either in standard, plastic or gun plastic portland cement or types specified in this Section. When plastic portland cement is used, no plasticizing agents shall be added at job site. [30] [31] [33]

7.4.3.2 Portland Cement-Lime Basecoat Plaster.—Proportion portland cement and lime to mineral aggregate as shown in Table 8-3. Cement must conform to ASTM C 150/C 175.

7.4.3.3 Plasticizers.—From 10 to 20 lbs. of dry hydrated lime (or an equivalent amount of lime putty) may be added as a plasticizing agent to each sack of portland cement in standard portland cement basecoat plaster. When added to mortar on the job site, limit the amount of plasticizing agents (other than lime) in standard portland cement basecoat plaster to the smallest amount necessary for proper workability, or as recommended by manufacturer. [32] [33] [34]

Do not add plasticizing agents to portland cement-lime plaster or to plastic cement mixes.

7.4.4. Portland Cement Finish Coats. [37]

7.4.4.1 Portland Cement-Lime Smooth Finish.—Proportion by weight; 1 part standard portland cement, not more than 1 part of dry hydrated lime (or an equivalent amount of lime putty) and not more than 2½ parts of graded aggregates. [38] [39]

7.4.4.2 Portland Cement-Lime Sand Float Finish.—Proportion by weight; 1 part standard portland cement, not more than ½ part dry hydrated lime (or an equivalent amount of lime putty) and no more than 3 parts of graded aggregate. [40] [41]

7.4.4.3 Portland Cement-Lime Dash Finish.—Proportion by weight; 1 part standard portland cement, not more than ½ part dry hydrated lime (or an equivalent amount of lime putty) and not more than 1 part #20 mesh, and 1 part #16 mesh graded aggregate. [42] [43]

7.4.4.4 Portland Cement Tunnel Dash.—Proportion by weight; one part standard portland cement, not more than ½ part dry hydrated lime (or equivalent amount of lime putty) and not more than one part #16 mesh and one part #12 mesh graded aggregate.

7.4.4.5 Manufactured Portland Cement Finishes.—Proportion smooth, sand float and machine dash finishes, either natural or colored, in accordance with the manufacturer's directions. [44]

7.5 MIXING

7.5.1 Gypsum Plaster Basecoats.—Thoroughly mix plaster and aggregate in the proportions specified herein, with only sufficient water to attain proper consistency for application. Proper consistency for machine applied gypsum plaster may be determined by slump test. Material for slump test shall be taken from nozzle of plastering machine hose. The maximum allowable slump shall be 3″ using a 2″ x 4″ x 6″ slump cone.

7.5.2 Gypsum Finish Coats.—Thoroughly mix both job-mixed and proprietary finishes with water to the proper consistency for application.

[30] Specify types of cement under materials.

[31] In areas of known alkali or sulfate conditions, exterior plaster should stop above grade.

[32] Recommended proportions to one sack cement:

Sand (vol.)	Lime (lbs.)
3 to 3½ parts	15
4 to 4½ parts	20

[33] Equal parts of standard portland and plastic cements may be used with good results.

[34] Obtain list of approved air-entraining and/or plasticizing agents from local building departments. Specify brands.

[37] See Specification Reference No. 10 for textures.

[38] **Very Hard Finish.**

[39] Use only on interior surfaces. Smooth portland cement finishes over portland cement basecoat plaster tend to crack because of extremely high, surface tensile stress. The use of high lime content in portland cement plaster will produce a very hard finish, but with reduced surface tensile stress. Wherever possible, a fine sand float finish should be specified.

[40] **Very Hard Finish.**

[41] Specify maximum sand sieve size. Plastic cement may be used instead of standard if lime content is reduced proportionately.

[42] **Very Hard Finish.**

[43] When heavy dash textures are desired, so specify and request samples.

[44] Integrally colored finishes should be specified by manufacturer's color number and keyed to finish schedule. If color selection has not been made when specifications are written, some indication should be given of the depth of color (light, medium, dark) to permit accurate bidding.

7. PLASTERING TECHNIQUES

7.5.3 Portland Cement Plaster Basecoats.—Thoroughly mix plaster and aggregate in the proportions specified herein, with only sufficient water to attain proper consistency for application. Proper consistency for machine applied portland cement plaster may be determined by slump test. Material for slump test shall be taken from nozzle of plastering machine hose. The maximum allowable slump shall be 2½″ using a 2″ x 4″ x 6″ slump cone. [45]

7.5.4 Portland Cement Finish Coats.—Thoroughly mix both job-mixed and proprietary finishes with water to the proper consistency for application.

7.5.5 Lime Putty.—Prepare lime putty in accordance with the printed directions of the manufacturer of the lime. [46]

7.6 APPLICATION

7.6.1 Gypsum Plaster Basecoats (general)

7.6.1.1 Two-Coat Work.—Apply basecoat with sufficient material and pressure to bond to gypsum lath and masonry. Bring the plaster out to grounds and straighten to true surface. [47]

7.6.1.2 Three-Coat Work.—Apply scratch coat with sufficient material to form good keys or bond on lath. Cover the lath well and cross scratch. [48]

Apply brown coat to set scratch coat, bring out to grounds, and straighten to a true surface. [49]

7.6.1.3 Over Concrete.—After surface has been prepared to assure bond, apply gypsum basecoat plaster in one or two coats, as required to bring out to grounds, and straighten to a true surface. [50] [51]

7.6.1.4 Plaster Screeds.—On metal lath, or wire fabric lath, place plaster screeds wherever permanent grounds are too far apart to serve as guides for rodding. Establish true surface plaster screeds with a rod before plaster has set (hardened).

7.6.2 Gypsum Solid Plaster Partitions (basecoat application).

7.6.2.1 Single Channel Stud Solid Partitions (metal lath).—Apply scratch coat to the lathed side of channel with sufficient material to form good keys. Cover the lath well and cross scratch. After scratch coat has set (hardened), apply backup coat to the channel side in (one or) two coats, bring plaster out to grounds, and straighten to a true surface. Maintain temporary bracing until backup coat has set (hardened). After the backup coat has set, apply brown coat over the face of scratch coat on lath side, bring out to grounds and straighten to a true surface. Extend plaster to floor or fill metal bases, and fill door frames solid. [52]

When plastered from one side only, apply scratch coat to brick layer of lath and allow to set (harden). Apply second coat and bring out to face of channel and allow to set. After attachment of metal lath over second coat, apply successive coats of plaster to required thickness. [53]

7.6.2.2 Studless Metal Lath Solid Partitions (basecoat application).—Apply scratch coat on flat side of lath opposite bracing with sufficient material to form good keys. Cover the lath well and cross scratch. Apply brown and backup coats with one of the following alternate methods:

(a) Apply backup coat to braced side of metal lath. Apply brown coat over the face of scratch coat.

(b) Apply brown coat over the face of scratch coat. Apply backup coat on the braced side of the partition.

Bring backup and brown coats out to grounds and straighten to a true surface. Extend plaster to floor or fill metal bases, and fill door frames solid. [54] [55]

[45] Also applies to Portland Cement-Lime plaster.

[46] Putty may be used immediately after combining water and lime, or following a soaking period, as recommended by the manufacturer.

[47] Two-coat work is standard application procedure over gypsum lath and masonry. It may not be used over metal or wire fabric lath.

[48] Three-coat work is standard application procedure over metal lath and wire fabric lath.

[49] Scratch coat or brown coat need not be completely dry before applying succeeding coat.

[50] For preparation of concrete surfaces see 7.1.3.

[51] Gypsum bond plaster, if specified, should be applied in accordance with manufacturer's directions.

[52] Sequence: 1. Scratch. 2. Back-up coats. 3. Brown. 4. Finish.

[53] See Specification Reference No. 12 — Construction Techniques No. 2.

[54] Before plastering, backs of electrical outlets and similar surfaces should be coated with liquid bonding agent or fiberglass reinforcement may be used.

[55] Sequence: 1. Scratch. 2. Back-up (or Brown). 3. Brown (or Back-up). 4. Finish.

RECOMMENDED / CONSTRUCTION TECHNIQUE

Solid Partitions Plastered from One Side

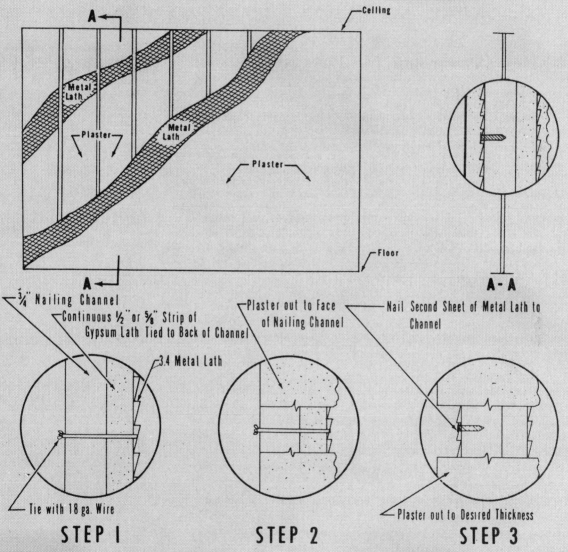

STEP 1   STEP 2   STEP 3

This assembly is recommended for use in shafts and in areas requiring fire protection where it is possible to plaster only from one side of the partition. The fire resistance of selected solid plaster assemblies using one plane of lath is shown in table as listed in the Uniform Building Code, 1976 Edition. The assembly recommended here uses two planes of lath and may be considered to possess equivalent fire resistivity.

ERECTION: Securely place nailing channels to provide adequate support for lath. Place continuous strips of gypsum lath spacer on the back side of the stud. Attach metal lath over the gypsum strips and wire-tie to the studs. Apply scratch and brown coats of plaster out to face of studs and allow to set. Nail second plane of lath to nailing channel studs with annular nails and plaster to required thickness.

CONSTRUCTION	TOTAL PARTITION THICKNESS	
	2 hr.	1 hr.
Incombustible studding with metal or wire lath, approved vermiculite-gypsum or approved perlite-gypsum plaster	2½ in.	2 in.
Incombustible studding with metal or wire lath, neat wood fiber gypsum plaster	2 in.	
Incombustible studding with metal or wire lath		2 in.

7.6.2.3 Studless Gypsum Lath Solid Partitions (basecoat application).—Apply approximately ⅜″ thick scratch coat to both sides of gypsum lath with sufficient pressure to form a good bond. Cross scratch and allow to set (harden) and partially dry. Apply brown coat to side opposite bracing and allow to set and partially dry. Remove temporary bracing and apply brown coat to braced side. Bring brown coats out to grounds and straighten to a true surface. Extend plaster to floor or fill metal bases and fill hollow metal door frames solid. [56]

7.6.3 Gypsum Finish Coats. [57]

7.6.3.1 Smooth Finishes.—Apply finish plaster to an approximate thickness of ⅛″ and fill out to a true even plane. Trowel well to a smooth finish, free from blemishes.

7.6.3.2 Float Finishes.—Apply finish plaster to an approximate thickness of ⅛″ and uniformly float to a true plane. [58]

7.6.3.3 Textured Finishes.—Apply finish plaster to an approximate thickness of ⅛″. Bring to a true plane and surface and then texture to meet specifications. [59]

7.6.3.4 Skim Coat Plaster Finish.—Apply finish plaster over concrete base. (Note to specifier: This application will improve finish quality of concrete, but cannot be used to straighten concrete surfaces.)

7.6.4 Acoustical Plaster.—Apply acoustical plaster over a gypsum plaster brown coat, or other base approved by manufacturer of acoustical plaster. Apply in strict accordance with manufacturer's directions. [60]

7.6.5 Portland Cement Plaster Basecoats (application).—Apply scratch coat with sufficient material to form good keys on metal lath or wire fabric lath. Embed and fill all spaces of lath and score horizontally. Do not apply brown coat sooner than 24 hours on interior surfaces and 48 hours on exterior surfaces. Apply brown coat to scratch coat, bring out to grounds, straighten to a true surface, float and compact and leave sufficiently rough to assure adequate bond for finish. On exterior vertical surfaces, cold joints in brown coat shall not occur over cold joints in scratch coat. Surface shall be free from imperfections which may reflect in the finish coat. Portland cement plaster shall be moist cured per Table E-3. [61]

7.6.5.1 Over Masonry and Concrete.—After surface has been prepared to assure bond, apply portland cement plaster scratch and brown coats as specified herein. A liquid hardener agent shall be applied to properly prepared surface. Thickness up to and including ⅝″ can be applied. Thickness grades require self furred metal lath. [62] [64]

7.6.6 Portland Cement Solid Plaster Partitions.

7.6.6.1 Single Channel Stud Solid Partitions.—Apply portland cement plaster in the same sequence as for gypsum plaster application on this partition. There shall be a minimum curing period of 24 hours between coats on interior plaster, and 48 hours on exterior plaster. [63]

7.6.6.2 Studless Metal Lath Solid Partitions.—Apply basecoats in the same sequence as for gypsum plaster application on this partition assembly. There shall be a minimum curing period of 24 hours between coats. [64]

7.6.7 Portland Cement Finish Coats (application).

7.6.7.1 Portland Cement-Lime Float Finishes.—Apply finish plaster to an approximate thickness of ⅛″ minimum and uniformly float to a true even plane.

[56] Sequence: 1. Scratch both sides. 2. Brown one side. 3. Brown other side. 4. Finish.

[57] The finish coat may be applied to a partially dry basecoat or to a dry basecoat which has been dampened to control suction.

[58] **Float-texture is governed by maximum sieve sizes of sand. A fine float finish may be obtained with 20 to 30 sieves and a coarse finish with 16 to 20 sieves.**

[59] For special textures see Specification Reference No. 10.

[60] Specify thickness, texture, color and NRC and names of approved acoustical plaster manufacturers.

[61] Specify both scratch and brown coats under thin-set ceramic tile; scratch coat only under mortar-set tile.

[62] For preparation of masonry and concrete surfaces, see 7.1.2 and 7.1.3. If dash-bond coat is applied, scratch coat may be omitted and lesser thickness of basecoat plaster applied in one operation.

[63] See 7.6.21 for sequence of gypsum plaster application.

[64] See 7.6.22 for sequence of gypsum plaster application.

7.6.7.2 Portland Cement-Lime Textured Finishes.—Apply finish plaster to an approximate thickness of ⅛″ minimum. Texture to achieve finish as specified herein. [66]

7.6.7.3 Portland Cement-Lime Dash Finishes.—Machine apply finish plaster in two coats evenly and uniformly. Apply the first coat to provide the texture pattern. Apply the second coat to obtain uniformity in color and texture. When practical, complete both coats on the same day, the second following the first when the latter is sufficiently firm to receive the second coat. [67] [68]

7.6.8 Curing Portland Cement Plaster.—After application, each basecoat of portland cement shall contain sufficient moisture for at least 24 hours on interior and 48 hours on exterior to assure hydration. If necessary to moisten plaster, avoid soaking and apply water in fine fog spray. Moist curing for interior portland cement plaster not required except where job conditions indicate it necessary.

Do not apply plaster to surfaces which contain surface water. [69]

7.7 CONTROL JOINTS

7.7.1 Control joints will be placed as shown on the drawings.
See Sec. 4.3.1.

7.8 SAMPLES

7.8.1 Finish Samples.—Prepare samples on the job showing finish texture and/or color for architect's approval before start of application of finish coats on project. Finish work shall approximate in texture and color the sample approved. [71]

7.9 SPECIAL WORK

7.9.1 Ornamental Work.—Run or cast all work in accordance with the details and models and leave level, true and even. Miter all corners and leave sharp, true and clean.

7.9.2 Marblecrete.—See Stucco Finishes, Specification Reference No. 11.

7.9.3 Veneer Plaster.—See Specification Reference No. 9.

7.9.4 Back-Plastered Exterior Walls.—Apply scratch coat to exterior face of metal lath with sufficient material to form good keys and cover well, and then scratch horizontally.

After scratch coat has set firm and hard and is partially dry, apply back-plaster coat between vertical supports to interior face of lath to an approximate thickness of ½″. Surface need not be straightened and shall remain moist for 24 hours before application of lath to interior face of support.

7.10 WORKMANSHIP

7.10.1 Interior and Exterior.—Finish all plaster, interior and exterior, true and even, within ¼″ tolerance in 5′, without imperfections which can be attributed to the lathing and plastering contractor's work and materials. Plumb, curve, or level and square plaster with adjoining work (which itself is plumb, level and square) and form a proper foundation for wood moldings, trim, paint and other finishing materials.

7.11 CLEANUP

7.11.1 Final Cleanup.—Upon final completion of work in an area, remove rubbish, debris, scaffold and tools and leave area clean of all surface plaster. Where the type of floor covering requires further cleaning, such cleaning shall be the work of others. [72]

[65] Smooth finishes may be applied directly to concrete surfaces to which a liquid bonding agent has been applied. This application will improve finish quality of concrete, but cannot be used to straighten concrete surfaces.

[66] For special textures see Specification Reference No. 10.

[67] Dash finishes may be applied directly to concrete surfaces if the surface is sufficiently free of imperfections. Dash finishes will not conceal form joints, rock pockets, etc. If such a condition is anticipated, concrete should be either sacked or a leveling coat of plaster specified.

[68] On areas where it is impractical to machine apply dash finishes, finish may be broom dashed by hand.

[69] Application of water for curing of finish coats is not usually required.

[70] An approved curing compound which permits painting may be applied to smooth finishes to prevent excessive loss of moisture.

[71] Specify when samples are desired. Also specify size of sample.

[72] Initial protection of floors and their maintenance should be specified under general contractor's work to avoid duplication of cost of protection in bidding by general and subcontractors.

RECOMMENDED / CONSTRUCTION TECHNIQUE

Water Penetration of Exterior Plaster Walls

(Wood Frame Construction)

Water penetration of all types of wall facings must be considered in the design stage. While it is possible to design plaster of varying density, with attendant degrees of water resistance, portland cement plaster should not be assumed to be waterproof. Water may find its way through the plastered surface, particularly when wind-driven, heavy rainfall occurs within a concentrated period. Constant use of sprinklers or prolonged flooding of the plaster surface may cause the same condition. Water may enter through plaster; through plaster separation alongside windows, doors and control joints; cracks in plaster at points of structural stress concentration; shrinkage cracks.

PREVENTION

It is essential when designing exterior plaster walls, to provide the wall with a self-weeping characteristic.

A water-resistant building paper is usually required by building codes under exterior plaster on residential construction. When properly applied, the paper provides adequate additional protection, and water which penetrates the plaster will run down the exterior face of the paper and seek exit at the plate or mudsill. If the construction at the mudsill is designed to permit the water to escape, no water entry to the inside will occur from this source.

Where the building is constructed with studs placed directly on an on-grade concrete floor slab system, and the plaster membrane is continuous over the face of the slab, no effective method of draining moisture to the outside exists.

One popular method of accomplishing this is through use of a drip screed installed at the juncture of the mudsill and the concrete slab. This system and an alternate system are shown below (check your building department for the approved method):

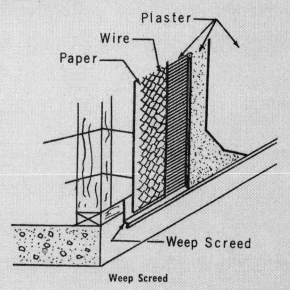

Weep Screed

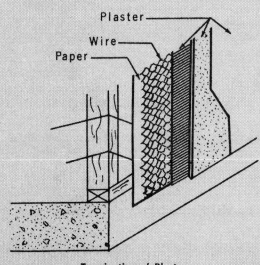

Termination of Plaster

RECOMMENDED/CONSTRUCTION TECHNIQUE *(Continued)*

FINDING SOURCE

Where foundation drip screeds have not been installed, and water penetration has occurred, it must not be presumed that the source of entry is the portland cement plaster. It is necessary to eliminate other potential sources before deciding on the method of correction. The only sure way is to test the suspected wall is with a water hose. The hose test should always be planned to eliminate or to define a given area of a wall as the source:

    1. Test area with no openings (below windows).
    2. Test window openings individually.
    3. Test door opening individually.

Possible sources of water entry:

a. No flashing.
b. Improperly installed flashing.
c. Inadequately sealed window and door frames.

d. Gas meter pipes.
e. Meter boxes.
f. Vents.
g. Leaking roof.

Where the portland cement plaster has been determined as the source of water entry, it is usually because of the condition shown below:

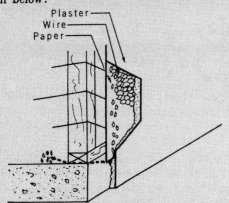

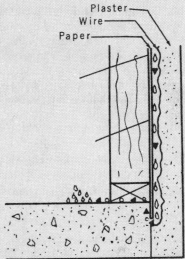

Water penetrating wall is entrapped by bond of plaster to slab. Water finds escape beneath mudsill and backs into inside of house.

CORRECTION

If the source of water is through the plaster as described above, the correction is relatively simple—provide some means of escape to the outside. The most effective methods are the two shown below:

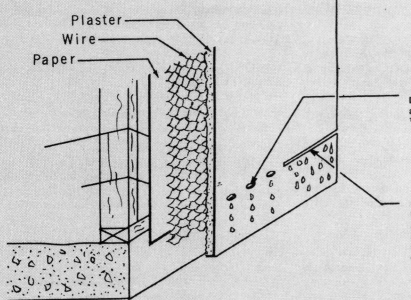

Drill weep hole 12″ o.c. at a point just above the building paper, and at least ½″ to ¾″ below top of concrete floor slab.

Using a power saw with a carborundum blade, cut plaster back to building paper at same point as for drilling.
It is recommended that a trench or some other drainage system be provided at base of plaster when this procedure is used.

RECOMMENDED | # CONSTRUCTION TECHNIQUES

APPLICATION OF DIFFERENT TYPES OF PLASTER

PLASTIC CEMENT PLASTER

Scratch coat: 1 part plastic cement, 4 parts sand.

Brown coat: 1 part plastic cement, 5 parts sand.

Application:

Scratch coat: Apply to a minimum ⅜″ thickness to embed and cover lath. Score lightly in a horizontal direction only. Keep moist for at least two days. Apply brown coat after scratch coat has been in place two days.

Brown coat: Dampen, but do not saturate, the scratch coat. Apply portland cement plaster to bring thickness of scratch and brown coats to a minimum ¾″. Float and rod to a level plane with a maximum variation of ¼″ under a 10-ft. straightedge placed at any point on plaster surfaces. Keep brown coat moist for two days, and allow to hydrate an additional five days before application of the finish coat.

Alternate method option permissible. *(See Notes)*

Caveats: Do not add other ingredients to straight plastic cement mixes, such as admixtures, water repellents, waterproofers, lime and the like.

Plastic cement is manufactured to entrain air (nominally 19% to 23%). Plastic cement mixes must be moist-cured the same as for any portland cement mix.

Plastic cement is manufactured expressly for stucco application and is available only in the far West.

MASONRY CEMENT AND REGULAR CEMENT PLASTER

Scratch coat: ½ part masonry cement, ½ part regular cement Type I or II, 4 parts sand.

Brown coat: ½ part masonry cement, ½ part regular cement Type I or II, 5 parts sand.

Application:

Scratch coat: Apply to a minimum ⅜″ thickness to embed and cover lath. Score lightly in a horizontal direction only. Keep moist for at least two days. Apply brown coat after scratch coat has been in place two days.

Brown coat: Dampen, but do not saturate, the scratch coat. Apply portland cement plaster to bring thickness of scratch and brown coats to a minimum ¾″. Float and rod to a level plane with a maximum variation of ¼″ under a 10-ft. straightedge placed at any point on plaster surfaces. Keep brown coat moist for two days and allow to hydrate an additional five days before application of the finish coat.

Alternate method option permissible. *(See Notes)*

Caveats: Masonry cement is a specialty cement manufactured for masonry mortar (ASTM C 91). Strength development with masonry cement is lower than with regu-

lar portland cements. The proportion of actual portland cement in masonry cement is low compared to other specialty cements.

It does not follow that masonry cement mixes provide better performance than regular stucco mixes over concrete and masonry block.

PLASTIC CEMENT AND REGULAR CEMENT PLASTER (Half-and-half)

Scratch coat: ½ part plastic cement, ½ part regular cement Type I or II, 4 parts sand.

Brown coat: ½ part plastic cement, ½ part regular cement Type I or II, 5 parts sand.

Application:

Scratch coat: Apply to a minimum ⅜″ thickness to embed and cover lath. Score lightly in a horizontal direction only. Keep moist for at least two days. Apply brown coat after scratch coat has been in place for two days.

Brown coat: Dampen, but do not saturate, the scratch coat. Apply portland cement plaster to bring thickness of scratch and brown coats to a minimum of ¾″. Float and rod to a level plane with a maximum variation of ¼″ under a 10-foot straightedge placed at any point on plaster surfaces. Keep brown coat moist for two days, and allow to hydrate an additional five days before application of the finish coat.

Alternate method option permissible. *(See Note)*

Caveats: If necessary, admixtures can be used with half-and-half mixes, but are not recommended.

Half-and-half mixes are more appropriate for machine-application than regular cement mixes, and this mix formulation has provided satisfactory performance on most types of buildings.

MASONRY CEMENT AND REGULAR CEMENT PLASTER

Scratch coat: 1 part masonry cement, 1 part regular cement Type I or II, 4 parts sand.

Brown coat: 1 part masonry cement, 1 part regular cement Type I or II, 6 parts sand.

Application:

Scratch coat: Apply to a minimum ⅜″ thickness to embed and cover lath. Score lightly in a horizontal direction only. Keep moist for at least two days. Apply brown coat after scratch coat has been in place two days.

Brown coat: Dampen, but do not saturate, the scratch coat. Apply portland cement plaster to bring thickness of scratch and brown coats to a minimum ¾″. Float and rod to a level plane with a maximum variation of ¼″ under a

tion only. Keep moist for at least two days. Apply brown coat after scratch coat has been in place two days.

Brown coat: Dampen, but do not saturate, the scratch coat. Apply portland cement plaster to bring thickness of scratch and brown coats to a minimum ¾″. Float and rod to a level plane with a maximum ¼″ variation under a 10-foot straightedge placed at any point on plaster surfaces. Keep brown coat moist for two days, and allow to hydrate an additional five days before application of the finish coat.

Alternate method option permissible. *(See Notes)*

Caveats: PRF is a proprietary admixture which has proved its value in contributing to superior strength development in portland cement plaster, minimizing cracking, and eliminating efflorescence. It has a long and honorable record of use and has increased compressive strengths of plaster used for prison facilities to 7 ksi in seven days.

PORTLAND CEMENT - GLASS FIBER ENHANCED ACRYLIC-MODIFIED BROWN COAT PLASTER

Scratch coat: 1 part portland cement Type I or II, 3½ parts sand, not more than 10 lbs. lime for workability, 2 lbs. glass fiber (½″).

Brown coat: 1 part portland cement Type I or II, 4 parts sand. Brown coat is mixed with a solution comprised of one part acrylic compound such as Acryl 60 and three parts of water.

Application:

Scratch coat: Apply to a minimum ⅜″ thickness to embed and cover lath. Score lightly in a horizontal direction only. Keep moist for at least two days. Apply brown coat after scratch coat has been in place two days.

Brown coat: Apply acrylic-modified portland cement plaster to bring thickness of scratch and brown coats to a minimum ¾″. Float and rod to a level plane with a maximum variation of ¼″ under a 10-foot straightedge placed at any point on plaster surfaces. Allow brown coat to air-cure.

Caveats: Glass fibers provide an optimum scratch coat. Acrylic-modified brown coat improves tensile and flexural properties to resist cracking.

Because acrylic-modified portland cement applications harden by film formation from the exterior surface to the back plane of plaster, the brown coat should not be rodded as this can damage necessary film formation.

A finish coat over the acrylic-modified brown coat should also be acrylic-modified or an acrylic polymer finish.

FINISH COATS OF PLASTER

★

Integrally-colored manufactured stucco finish applied to ⅛′ thickness. Apply with sufficient material and pressure to bond to and conceal the brown coat.

★

Job-mixed stucco finish: 1 part portland cement, 1 part hydrated lime, 3 parts sand. White or gray cement may be used.

★ ★

Manufactured stucco applied in a two-step operation with a tight trowel coat followed by application to required thickness.

★ ★ ★

Manufactured stucco applied with a tight trowel coat followed by application to required thickness. When stucco is thoroughly dry, a fog spray of stucco wash is applied with an air sprayer.

★ ★ ★ ★

Manufactured stucco finish mixed with a solution comprised of one part acrylic compound such as Acryl 60 or equal, and three parts water.

★ ★ ★ ★

Acrylic polymer finish with the following minimum properties: Tensile strength, 150 psi; % elongation under maximum stress, 160%; % elongation at break, 360%; low temperature flex at −35°F. 18° bend over ⅛″ mandrel; weatherometer test with no change at 5,000 hours.

NOTES

Alternate method option: Where job conditions are such that rapid application of base coats is feasible, the brown coat may be applied as soon as the scratch coat has attained sufficient rigidity to accept it without damage to the scratch coat. In such instances, the brown coat is moist-cured as outlined in building codes and ASTM C 926 (job mixed portland cement-based plaster).

Masonry cement (ASTM C 91): Masonry cement is defined as: "3.1 cement, masonry - a hydraulic cement for use in mortars for masonry construction, containing one or more of the following materials: portland cement, portland blast-furnace slag-cement, portland-pozzolan cement, natural cement, slag cement, or hydraulic lime; and in addition usually containing one or more materials such as hydrated lime, limestone, chalk, calcareous shell, talc, slag, or clay, as prepared for this purpose."

Textures and finishes: There are some heavy textures which can best be executed in the brown coat such as simulated adobe, heavy Spanish, and others. The brown coat should be applied to accommodate execution of this kind of texture. Some building code jurisdictions permit paint as an alternate finish over base coats of plaster. This is a "two-coat" system and effectively "short-changes" the user. Three-coat plaster applications offer the best performance.

Base coats and finish coats: Do not choose a finish with a lesser rating than a base coat.

Fly ash: Fly ash is an excellent addition to portland cement mixes. The fly ash must be low carbon.

10-ft. straightedge placed at any point on plaster surfaces. Keep brown coat moist for two days, and allow to hydrate an additional five days before application of the finish coat.

Alternate method option permissible. *(See Notes)*

Caveats: In some areas of the country, this mix has provided satisfactory performance in all kinds of climates. The common denominator appears to be the applicator whose application techniques result in satisfactory performance.

Masonry cement is not manufactured for stucco applications, however, and strength developments are not as high as with regular cements.

This is a rich mix in that there is a high proportion of cement to aggregates.

PORTLAND CEMENT/LIME PLASTER

Scratch coat: 1 part regular cement Type I or II, 1 part hydrated lime Type S, 8 parts sand.

Brown coat: 1 part regular cement Type I or II, 1 part hydrated lime Type S, 9 parts sand.

Application:

Scratch coat: Apply to a minimum ⅜″ thickness to embed and cover lath. Score lightly in a horizontal direction only. Keep moist for at least two days. Apply brown coat after scratch coat has been in place two days.

Brown coat: Dampen, but do not saturate, the brown coat. Apply portland cement plaster to bring thickness of scratch and brown coats to a minimum ¾″. Float and rod to a level plane with a maximum variation of ¼″ under a 10-foot straightedge placed at any point on plaster surfaces. Keep brown coat moist for two days, and allow to hydrate an additional five days before application of the finish coat.

Alternate method option permissible. *(See Notes)*

Caveats: Portland cement/lime plaster requires a longer mixing time than portland cement and sand mixes. They provide good workability and easier machine application. They are not popular with plasterers generally. Portland cement/lime formulations offer good suction for uniform color of finish coats. Ultimate strengths cannot equal those of regular portland cement plaster.

PORTLAND CEMENT — REGULAR PLASTER PLASTER

Scratch coat: 1 part portland cement Type I or II; 3½ parts sand, not more than 10 lbs. lime, Type S, for workability.

Brown coat: 1 part portland cement Type I or II; 4 parts sand, not more than 10 lbs. lime, Type S, for workability.

Application:

Scratch coat: Apply to a minimum ⅜″ thickness to embed and cover lath. Score lightly in a horizontal direction only. Keep moist for at least two days. Apply brown coat after scratch coat has been in place two days.

Brown coat: Dampen, but do not saturate, the scratch coat. Apply portland cement plaster to bring thickness of scratch and brown coats to a minimum ¾″. Float and rod to a level plane with a maximum variation of ¼″ under a

10-foot straightedge placed at any point on plaster surfaces. Keep brown coat moist for two days, and allow to hydrate an additional five days before application of the finish coat.

Alternate method option permissible. *(See Notes)*

Caveats: Maintain a low cement/water ratio to minimize volume change (shrinkage). Cement content should not be increased on the premises that more cement leads to better performance. For machine-application, a low carbon fly ash is recommended. *(See Notes)*

PORTLAND CEMENT - GLASS FIBER ENHANCED

Scratch coat: 1 part portland cement Type I or II, 3½ parts sand, not more than 10 lbs. lime Type S for workability, 2 lbs. glass fibers (½″).

Brown coat: 1 part portland cement Type I or II, 3½ parts sand, not more than 10 lbs. lime Type S for workability, 2 lbs. glass fibers (½″).

Application:

Scratch coat: Apply to a minimum ⅜″ thickness to embed and cover lath. Score lightly in a horizontal direction only. Keep moist for at least two days. Apply brown coat after scratch coat has been in place two days.

Brown coat: Dampen, but do not saturate, the scratch coat. Apply portland cement plaster to bring thickness of scratch and brown coats to a minimum ¾″. Float and rod to a level plane with a maximum variation of ¼″ under a 10-foot straightedge placed at any point on plaster surfaces. Keep brown coat moist for two days, and allow to hydrate an additional five days before application of the finish coat.

Alternate method option permissible. *(See Notes)*

Caveats: Glass fibers have unquestionably proved their value in enhancing performance of portland cement plaster. Greater tensile and compressive strengths and significantly reduced cracking are major features of glass fiber enhancement.

Glass fibers are added to the mixer after it has been operating for at least two minutes to minimize glass fiber breakage. AR or E glass may be used.

PORTLAND CEMENT—ADMIXTURE ENHANCED PLASTER

Scratch coat: 1 part portland cement, Type I or II; 3½ parts sand, 3 oz. PRF as manufactured by Gibco, Inc., Tulsa, OK.

Brown coat: 1 part portland cement, Type I or II, 4 parts sand, 3 oz. PRF as manufactured by Gibco, Inc., Tulsa, OK.

Application:

Scratch coat: Apply to a minimum ⅜″ thickness to embed and cover lath. Score lightly in a horizontal direc-

> **Remember that plaster membranes are thin and as such they tend to lose water more rapidly than thicker installations of portland cement materials. It is essential that evaporation of water from the mix applied to walls and ceilings be compensated for by prompt moist-curing.**

8. PLASTERING TABLES

TABLE 8-1

THICKNESS OF PLASTER

PLASTER BASE	FINISHED THICKNESS OF PLASTER FROM FACE OF LATH, MASONRY, CONCRETE	
	Gypsum Plaster	Portland Cement Plaster
Expanded Metal Lath	5⁄8″ minimum[2]	5⁄8″ minimum[2]
Wire Fabric Lath	5⁄8″ minimum[2]	3⁄4″ minimum (interior)[3] 7⁄8″ minimum (exterior)[3]
Gypsum Lath	1⁄2″ minimum	
Gypsum Veneer Base	1⁄16″ minimum	
Masonry Walls[4]	1⁄2″ minimum	1⁄2″ minimum
Monolithic Concrete Walls[4, 5]	5⁄8″ maximum	7⁄8″ maximum
Monolithic Concrete Ceilings[4, 5]	3⁄8″ maximum[6, 7, 8]	1⁄2″ maximum[7, 8]

[1]For Fire-resistive Construction, conform to local Building Code.
[2]When measured from back plane of expanded metal lath, exclusive of ribs or self-furring lath, plaster thickness shall be 3⁄4-inch minimum.
[3]When measured from face of support or backing.
[4]Because masonry and concrete surfaces may vary in plane, thickness of plaster need not be uniform.
[5]When applied over a liquid bonding agent, finish coat may be applied directly to concrete surface.
[6]Approved acoustical plaster may be applied directly to concrete, or over base coat plaster, beyond the maximum plaster thickness shown.
[7]On concrete ceilings, where the base coat plaster thickness exceeds the maximum thickness shown, metal lath or wire fabric lath shall be attached to the concrete.
[8]An approved skim coat plaster 1/16 inch thick may be applied directly to concrete.

TABLE 8-2

GYPSUM PLASTER PROPORTIONS

NUMBER OF COATS	COAT	PLASTER BASE OR LATH	MAXIMUM VOLUME AGGREGATE PER 100# NEAT PLASTER [1, 2] (CUBIC FEET)	
			Damp Loose Sand[3]	Perlite or Vermiculite[3]
Two-Coat Work	Basecoat	Gypsum Lath	2½[4]	2½[4]
	Basecoat	Masonry	3[4]	3[4]
Three-Coat Work	First Coat	Lath	2[5]	2[5]
	Second Coat	Lath	3[5]	3[5]
	First & Second Coat	Masonry	3[5]	3[5]

[1]Wood fibered gypsum plaster may be mixed in the proportions of 100 pounds of gypsum to not more than one cubic foot of sand where applied on masonry or concrete.
[2]For Fire-resistive Construction, conform to local Building Code.
[3]When determining the amount of aggregate in set plaster, a tolerance of 10 per cent shall be allowed.
[4]Combinations of sand and lightweight aggregate may be used, provided the volume and weight relationship of the combined aggregate to gypsum plaster is maintained.
[5]If used for both first and second coats, the volume of aggregate may be two and one-half cubic feet.

TABLE 8-3

PORTLAND CEMENT PLASTER						
COAT	VOLUME CEMENT	MAXIMUM WEIGHT (OR VOLUME) LIME PER VOLUME CEMENT[2]	MAXIMUM VOLUME SAND PER VOLUME CEMENT[3]	APPROXIMATE MINIMUM THICKNESS[4]	MINIMUM PERIOD MOIST CURING	MINIMUM INTERVAL BETWEEN COATS
First	1	20 lbs.	4	⅜″[5]	48[6] Hours	48[7] Hours
Second	1	20 lbs.	5	1st and 2nd Coats total ¾″	48 Hours	7 Days[8]
Finish	1	1[9]	3	1st, 2nd and Finish Coats ⅞″	—	8
PORTLAND CEMENT-LIME PLASTER[10]						
COAT	VOLUME[11] CEMENT	MAXIMUM VOLUME LIME PER VOLUME CEMENT	MAXIMUM VOLUME SAND PER COMBINED VOLUMES CEMENT AND LIME	APPROXIMATE MINIMUM THICKNESS[4]	MINIMUM PERIOD MOIST CURING	MINIMUM INTERVAL BETWEEN COATS
First	1	1	4	⅜″[5]	48[6] Hours	48[7] Hours
Second	1	1	4½	1st and 2nd Coats total ¾″	48 Hours	7 Days[8]
Finish	1	1[9]	3	1st, 2nd and Finish Coats ⅞″	—	8

[1]Exposed aggregate plaster shall be applied in accordance with Section 10.10. Minimum overall thickness shall be ¾ inch.

[2]Up to 20 pounds of dry hydrated lime (or an equivalent amount of lime putty) may be used as a plasticizing agent in proportion to each sack (cubic foot) of Type I and Type II Standard portland cement in first and second coats of plaster.

[3]When determining the amount of sand in set plaster, a tolerance of 10 percent may be allowed.

[4]See Table No. 8-1.

[5]Measured from face of support or backing to crest of scored plaster.

[6]The first two coats shall be as required for the first coats of exterior plaster, except that the moist curing time period between the first and second coats shall be not less than 24 hours and the thickness shall be as set forth in Table 8-1. Moist curing shall not be required where job and weather conditions are favorable to the retention of moisture in the portland cement plaster for the required time period.

[7]Twenty-four hours minimum interval between coats of interior portland cement plaster. As an alternate method of application, the second coat may be applied as soon as the first coat has attained sufficient rigidity to receive the second coat. When using this method of application calcium aluminate cement up to 15 per cent of the weight of the portland cement may be added to the mix. Curing of the first coat may be omitted and the second coat shall be cured as set forth in this Table.

[8]Finish coat plaster may be applied to interior portland cement base coats after a 48-hour period.

[9]For finish coat plaster, up to an equal part of dry hydrated lime by weight (or an equivalent volume of lime putty) may be added to Types I, II and III Standard portland cement.

[10]No additions of plasticizing agents shall be made.

[11]Type I, II or III Standard portland cement.

9. VENEER PLASTERING

9.1 DESCRIPTION

9.1.1 Veneer Plaster Construction.—Shall consist of large size gypsum lath attached to wood or metal supports as specified, and which shall receive a thin overall monolithic plaster coating (finish) applied in one or more coats to a thickness of 1/16" to 1/8". Minimum overall thickness of lath and veneer plaster shall be 1/2".

9.2 LATHING MATERIALS [1]

9.2.1 Gauges of Wire (See 2.2.2)

9.2.2 Nails & Staples (See 2.2.3)

9.2.3 Power Driven Staples (See 2.2.31)

9.2.4 Nailing Channels (See 2.3.4)

9.2.5 Screw Channels (See 2.3.5)

9.2.6 Nailable Studs nlb (See 2.4.2)

9.2.7 Screw Studs nlb (See 2.4.3)

9.2.8 Structural Nailable Studs lb (See 2.5.1)

9.2.8.1 Runner Track (See 2.5.2)

9.2.8.2 Bridging (See 2.5.3)

9.3 GYPSUM LATH

9.3.1 General.—Gypsum lath shall conform to the "Standard Specifications for Gypsum Lath—ASTM Designation C 588."

9.3.1.1 Types of Lath.—Gypsum lath for veneer plaster shall be plain, (or) Type "X", (or) insulating, gypsum lath as specified herein, or as dictated by fire resistance requirements. Face side of lath shall have a special paper designed for application of veneer plaster. [2] [3] [4]

9.3.1.2 Dimension of Lath.—Gypsum lath for veneer plaster systems shall meet the following nominal dimensional requirements:

Thickness: ⅜", ½", or ⅝" (greater thickness may be used)
Width: 48" (width may vary slightly to meet job requirements)
Length: 96" or longer to meet job requirements

9.4 ACCESSORIES

9.4.1 General.—Metal shapes used as grounds for veneer plaster shall be of such a size as to provide for required thickness.

9.4.2 Corner Beads.—For use at all external corners shall be formed of minimum .015 inch thick zinc coated steel, or other approved material, having minimum ⅞" wings, with minimum ⅜" diameter holes.

9.4.3 Casings.—Used to provide a finished edge at window and door jambs, at openings, at partition terminals, and at intersections with other other materials, shall be formed of minimum .015 inch thick zinc coated steel, or other approved material.

[1] Paragraphs 9.2.1 through 9.2.82 cover metal framing materials which can be used to construct metal partitions and walls, or ceiling grillages to which veneer lath and plaster are applied.

[2] See 2.7.2.

[3] Type "X" (special fire retardant) designates gypsum lath for veneer plaster complying with these specifications, that provides fire retardant rating at least equal to those obtained with the same thickness of Type "X" gypsum wallboard, when tested in accordance with requirements of the Method of Fire Tests of Building Construction Materials (ASTM Designation: E-119).

[4] Consult manufacturers for independent test data on assembly particulars, materials, and ratings for specific type of construction.

DETAIL 50
VENEER PLASTER

For Veneer Plaster Construction See Specification Reference 9

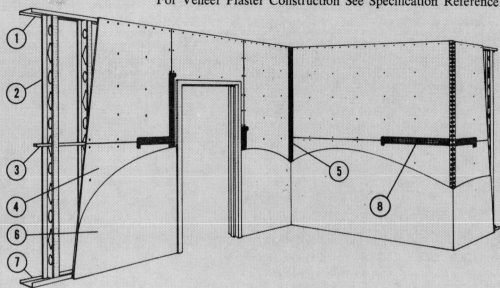

(1) Ceiling Runner Track
(2) Metal Stud (nailable or screw)
(3) Horizontal Stiffener[1]
(4) Large Size Lath.
(5) Angle Reinforcement
(6) Veneer Plaster 1/16 to 1/8 inch thick)
(7) Floor Runner Track
(8) Joint Reinforcement

[1] Stiffener is omitted with screw studs.

(A) METAL STUD CONSTRUCTION

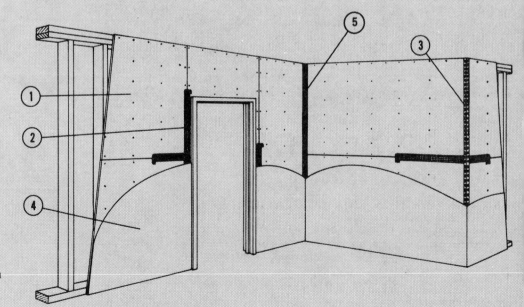

(1) Large Size Lath
(2) Joint Reinforcement
(3) Corner Bead
(4) Veneer Plaster (1/16 to 1/8 inch thick)

(B) WOOD STUD CONSTRUCTION

9.4.4 Partition Bases.—Recessed, flush type, or reveal type base for veneer plaster partitions or other walls, shall be formed of minimum No. 26 gauge steel, galvanized or coated with a rust inhibiting paint. Clips and splice plates shall be the manufacturer's regular type for the base used.

9.4.5 Joint Reinforcement.—Strip reinforcement used to reinforce joints of lath base for veneer plaster shall be glass fiber mesh a minimum width of 2"; or perforated cross fibered paper tape, or other approved material of equal strength. [5]

9.4.6 Metal Trim Staples.—Staples to secure corner beads or casings shall be flattened (galvanized) steel with wire minimum 9/16" legs; for securing joint reinforcement mesh, leg length shall be minimum ¼".

9.5 PLASTERING MATERIAL

9.5.1 Veneer Plasters.—Shall be a proprietary specially formulated high-strength gypsum base plaster for hand or machine application to large size lath or other properly prepared surfaces. Veneer plaster shall be formulated for application as a thin monolithic basecoat plaster over which a finish may be applied, or may be formulated for application as a finish plaster and shall conform to ASTM C 587. It shall have a setting time of from twenty to ninety minutes and a minimum compressive strength of 1500 psi. Setting time of plaster shall be controlled at the time of manufacture, or by introduction of, or contact with a setting agent, as recommended by the manufacturer.

9.5.2 Standard Plaster Finishes.—For application over veneer basecoat plaster shall be gypsum-lime smooth finish, (or) Keene's Cement-Lime float finish (or) machine dash finish; or approved manufactured regular gypsum finishes. [6]

9.5.3 Water.—For mixing with veneer plaster shall be clean, fresh, suitable for domestic consumption, and free from such amounts of mineral or organic substances as would affect the set of the plaster.

9.5.4 Bonding Agent.—(See 6.10). [7]

9.6 METAL FRAMING [8]

9.6.1 Metal Framing Members.—Shall be straight, true, and properly aligned so that plane of lath at edges and ends is not offset.

Where metal studs extend above suspended ceilings each stud shall be securely attached to a horizontal ¾" channel placed above the ceiling and along the full length of the partition.

9.7 WOOD FRAMING (by others) [9] [10]

9.8 LATH ERECTION AND APPLICATION

9.8.1 Application of Large Size Lath.—Apply large size lath for veneer plaster in either a vertical or horizontal direction. All ends and edges of the lath shall fall on supports except when edge joints are at right angles to supports.

On wood frame construction apply lath first to the ceiling and then to walls. On metal framing, lath may be applied in any sequence. Fit ends and edges of lath as close together as permitted by framing, but do not force into place. At external corners butt and fit lath so as to provide solid corner. Stagger end joints when lath is applied across supports. Place joints on opposite sides of partitions on different supports. Wherever possible, do not place joints at corners of door and window frames.

[5] Other types of material may be used to reinforce joints in lath and to strengthen plaster over joints, provided they are approved by the architect and the building department which has jurisdiction.

[6] See 7.4.22; 7.4.23; 7.4.24; 7.4.25.

[7] Bonding agents are sometimes used to bond veneer plaster to concrete or masonry interior surfaces. Use only on recommendation of manufacturer.

[8] For erection of metal studs and ceiling grillages see Specification Reference No. 3; or consult manufacturer.

[9] Wood framing by others should meet the minimum requirements of HUD and local building codes. Framing member should be straight, true, of uniform dimension, and properly aligned, and should have a moisture content not in excess of 15% at the time of the gypsum large size lath application. Bowed or twisted studs or joists should be straightened by others.

[10] For spacing of supports for large size lath see Table 5-9.

Attach lath from center to edges and ends, pressing the lath firmly against the supports. Place attachments approximately ⅜″ from edges of lath. Set attachment flush with the surface of the lath but do not break paper.

Internal vertical and horizontal angles on wood frame construction may be floated by not attaching lath to the supports in the angles; where supports are at right angles to the internal angle, attach lath approximately 8″ away from angle.

Cut lath to fit electrical outlets, pipes or other required openings.

9.8.2 Application of Joint Reinforcement.—Apply joint reinforcement to all joints either by stapling or by embedding in veneer plaster as recommended by manufacturer. Do not overlap reinforcement at intersections. Secure joint reinforcement by one of the following methods:

Stapling: Staple reinforcement at ends on each side of joint and at maximum 24″ intervals along reinforcement on alternate sides of joint. Staple reinforcement on one side only of vertical angles, and on ceiling side only of horizontal angles. [11]

Embedding: Embed reinforcement mesh or tape in veneer plaster at all joints before application of base or finish coat. If finish plaster is to be machine applied, plaster at joints shall be free of trowel marks or ridges. [11]

9.8.3 Application of Accessories.—Install corner beads at all external angles, set tight against lath and attach by nail, staple or by crimping. Attach accessories at not more than 12″ intervals. [12] [13]

9.9 VENEER PLASTERING

9.9.1 Mixing.—Mix veneer plaster in strict conformity with recommendations of manufacturer.

9.9.2 Application.—Apply veneer plaster by hand or machine to a minimum thickness of 1/16″ as directed by manufacturer, and as required to achieve the specified finish. Finish shall be (a) smooth, (b) trowel texture, (c) spray texture, as indicated on drawings or on room finish schedule. [14] [15] [16]

[11] Where a fire rated assembly is specified the joint treatment should be as specified by the manufacturer of the veneer plaster system.

[12] Approved adhesives may also be used.

[13] Where other metal trim is required for protection of edges at windows and openings and at intersections with other materials, etc., so indicate on drawings.

[14] Specify finish treatment required. Machine applied textures are light stipple (sand finishes); depth of trowel texture is limited by thickness of veneer plaster.

[15] When acoustic-type texture (non-rated) finish is applied over veneer plaster on ceilings prior to plastering lath on walls, the wall lath should be protected from overspray so as not to affect bond and setting time of plaster.

[16] For ASTM Standard C 843, Standard Specification for Application of Gypsum Veneer Plaster, see chapter on Industry-Wide Reference Standards and Specifications.

10. CEMENTITIOUS EXTERIOR WALL COATINGS

10.1 DESCRIPTION

10.1.1 Cementitious Exterior Wall Coatings.—One coat proprietary exterior plaster systems a minimum of ⅜″ thick consisting of products containing various combinations of sand, cement, lime, fibers, admixtures, and water that are applied over woven wire fabric or metal lath to insulation boards, fiberboards, wood sheathing or other similar substrates. Color may be added to the plaster mix at the factory by the manufacturers or on the job by the plastering contractor in accordance with the manufacturer's instructions. An ⅛″ conventional stucco finish with color may be used as a finish coat. [1]

10.2 MATERIALS

10.2.1 Cement.—Type I or II portland cement complying with UBC Standard No. 26.1. [2]

10.2.2 Sand.—Must be clean and free from deleterious amounts of loam, clay, silt, soluble salts and organic matter [3]. Sampling and testing must comply with ASTM C 987.

10.2.3 Fibers.—Chopped fibers are proprietary items included by each manufacturer.

Description of type of materials (A/R & E type polypropelene, acrylic), fiber length, purpose, etc. should be specified (see specific manufacturer used).

10.2.4 Admixtures.—Proprietary ingredients which are added by each manufacturer to improve the quality of his product.

10.2.5 Lime.—Per ASTM specification or UBC Standard No. 47-16 (if used). [4]

10.2.6 Insulation Board.—Expanded polystyrene (EPS) insulation board with a normal density of 1.5 pounds per cubic food, a Class I flame spread rating and a smoke-developed rating not to exceed 450. Boards are 1″-1½″ thick and have ⅜″ high tongue with compatible groves for horizontal joints. All boards must be manufactured under a recognized quality control program. (UBC 1712 and ICBO ES acceptance criteria for foam plastics). Each polystyrene foam plastic insulation board is to be identified in accordance with its respective ICBO ES or NES evaluation report. Additionally the board density must be noted.

10.2.7 Attachments.—All fasteners must be specifically described and be corrosion resistant.

10.2.7.1 Gauges of Wire.—(See 2.2.2)

10.2.7.2 Nails and Staples.—(See 2.2.3)

10.2.7.3 Power Driven Staples.—(See 2.2.3.1)

10.2.8 Wire Fabric Lath.—Minimum No. 2Ø gauge, 1 inch galvanized steel woven wire fabric or 1.75 expanded metal lath (must be galvanized). Lath must be self-furred or furred. Unfurred lath is permitted over polystyrene boards with no sheathing behind them.

10.2.9 Gypsum Sheathing Board.—Water resistant gypsum sheathing complying with UBC Standard No. 47-10.

10.2.10 Fiberboard.—Minimum ½″ asphalt impregnated fiberboard complying with UBC Standard No. 25.24 as a regular density sheathing.

[1] These systems are proprietary systems and reference should be made to manufacturers and their appropriate code approvals as required.

[2] Type I is for use in general plastering. It is usually specified with additions of plasticizers for workability.

Type II is for use where the plaster will be exposed to moderate sulfate action (alkali).

[3] Sand wherever it is practicable should be washed natural sand.

[4] For material specifications, see 6.4.

10. CEMENTITIOUS EXTERIOR WALL COATINGS

10.2.11 Plywood.—Minimum ⅜″ exterior grade plywood for studs 16″ on center and minimum ½″ exterior grade plywood for studs spaced 24″ on center. Plywood will comply with UBC Standard No. 25-9.

10.2.12 Caulking.—Acrylic latex caulking material complying with ASTM C 834.

10.2.13 Weather Resistive Barrier.—Minimum grade "D" Kraft building paper complying with UBC Standard No. 17-1 or asphalt saturated rag felt complying with UBC Standard No. 32-1. Must comply with 1707 of the Uniform Building Code.

10.2.14 Accessories.—All trim screeds and corner reinforcement must be either zinc coated or galvanized metal or plastic.

10.2.15 Water.—Shall be clean, fresh, suitable for domestic consumption and free from such amounts of mineral and organic substances as would effect of the plaster.

10.3 APPLICATION [5]

10.3.1 General.—The exterior cementitious coating is applied by hand troweling or machine spraying in one coat to a minimum ⅜ inch thickness. The wire fabric lath must be embedded in the minimum coating thickness and therefore cannot be exposed. Finish coat, if required, must be applied within 48 hours after the base coat unless the latter is sprayed/brushed with a bonding treatment complying with UBC Standard No. 47-1 or an acrylic bonding adhesive is added to the finish coat stucco mix prior to the finish coat application. Fasteners for lath must penetrate 1 inch minimum into wood studs. Flashing, corner reinforcement, metal trim and weep screeds must be installed as shown in details. The coating is applied at ambient air temperatures ranging from 40 degrees F. to 110 degrees F. by contractors approved by the manufacturer.

10.3.2 Application Over Open Framing.—*Insulation Board:* The weather resistive barrier is placed over open wood studs spaced 24 inches on center, maximum. The insulation board is then placed horizontally with tongue faced upward and is temporarily held in place with galvanized staples or roofing nails. Vertical but joints must be staggered a minimum of one stud space from adjacent courses and occur directly over studs. Wire fabric lath is then applied tightly over the polystyrene board and fastened through the board to studs with No. 11 gage galvanized roofing nails or No. 16 gage galvanized staples spaced six inches on center with a minimum one inch penetration into the studs. Staples must have a minimum crown width of ½ inch. Stapling is permitted only in southern pine, coast Douglas fir, or larch wood species. Care must be taken to avoid over-driving fasteners. The lath is applied with 1½ inch laps at all joints. Application to steel studs is similar except that Type S screws are installed at six inches on center. Screws must penetrate studs at least ¼ inch. Wall bracing in accordance with Section 2517 (g) 3 of the Code or acceptable alternate is required. Outside wall corners and parapet corners are covered with metal corner reinforcement, and weep screeds are installed at the bottom in accordance with Section 4706 (e) of the Code.

Galvanized metal 1⅜ inch J-shaped trim pieces are installed at other areas where foam is exposed. At windows and doors, butting J trim metal edges must be caulked. Holes for hose bibs, electrical panels, and other penetrations (except those caused by fasteners) of substrate surfaces must also be caulked.

10.3.3 Application Over Solid Backing.

(a) *Fiberboard.* Minimum ½ inch-thick fiberboard sheathing is installed directly over wood studs spaced 24 inches on center, maximum. The fiberboard is temporarily held in place with corrosion-resistant staples or roofing nails. A weather-resistive barrier of two layers of building paper is applied

[5] These systems are normally proprietary in general. Following application instructions should be superseded by published manufacturers instruction and code approvals.

over the fiberboard prior to lath or optional insulation board. The furred lath is then attached to studs through the sheathing with fasteners and spacing as described for insulation board in Table 25-Q of the UBC. All walls must be braced in accordance with the Code. Exposed sheathing edges are protected with screeds.

(b) *Gypsum Sheathing.* Minimum ½ inch-thick, water-resistant gypsum sheathing may be installed directly on wood studs at a maximum of 24 inches on center in a manner similar to fiberboard. Gypsum sheathing is fastened in accordance with Table No. 47-G of the UBC. A weather-resistive barrier is required over the gypsum sheathing prior to installation of the metal lath and coating. Minimum ½ inch EPS may be installed over the sheathing prior to furred or self-furred lath and coating.

(c) *Plywood.* Minimum ⁵⁄₁₆″ inch-thick plywood is installed directly to wood studs spaced 16 inches on center, maximum. Where studs are spaced 24 inches on center, minimum ⅜″ inch-thick plywood must be used. The weather-resistive barrier, consisting of two layers of Type D building paper or approved equal, wire fabric lath and coating are applied as described for fiberboard.

10.4 FIRE RESISTIVE CONSTRUCTION.—All assemblies shall be tested in accordance with UBC Standard 43-1 (ASTM E 119-83). [6]

10.4.1 One Hour Fire Assembly.—1. (a) *Interior Face.* One layer of ⅝ inch-thick Type X gypsum wallboard, water-resistant backer board of veneer base is applied parallel at right angles to the interior face of 2 by 4 wood studs spaced 24 inches on center maximum. The wallboard is attached with 6d coated nails, 1⅞ inch long with ¼ inch-diameter head, at 7 inches on center to studs, plates, and blocking. All wallboard joints must be backed with wood framing.

(b) *Exterior Face.* One layer of minimum ⅝-inch-thick Type X water-resistant gypsum sheathing, 48 inches wide is applied parallel to studs with No. 11 gage galvanized roofing nails, 1¾ inches long with ⁷⁄₁₆ inch or ½-inch-diameter heads at 4 inches on center at board edges and 7 inches on center at intermediate studs. Or staples may be used per Section 43-B of the UBC. The sheathing is fastened to top and bottom plates at 7 inches on center. A weather-resistive barrier is required over the sheathing. The furred or self-furred lath and wall coating are then applied as described in this specification.

10.4.2 Other One-Hour Configurations.—If properly tested and approved by the ICBO Evaluation Service, other assemblies will be acceptable.

10.5 INSPECTION.—Building department inspection will be required on wire lath prior to application of plaster. [7]

10.6 CONTROL JOINTS.—Control joints must be installed as specified by the architect, designer, builder or exterior coating manufacturer. [8]

10.7 CURING.—Plaster shall be moist cured as specified by the material manufacturer. [9]

10.8 IDENTIFICATION.—The factory prepared mix is delivered to the job site in *water-resistant* bags with labels bearing the following information:

1. Name and address of manufacturer and appropriate evaluation report number.
2. Identification of product.
3. Weight of packaged mix.
4. Storage instructions.
5. Maximum amount of water and sand that may be added.

[6] Appropriate code jurisdiction must be consulted to determine acceptability of fire assemblies.

[7] Local code jurisdiction must be consulted to determine amount of inspection required. Inspection criteria can vary widely from city to city.

[8] Control joints must be drawn on the elevations. General statements such as 10' on center, 100 sq.ft., etc., should not be allowed.

[9] Moist curing requirement will vary greatly by local conditions. Contact manufacturer, code authorities, etc.

11. STUCCO (PLASTER) FINISHES

11.1 GENERAL REQUIREMENTS

11.1.1 Packaging.—Each manufacturer shall package manufactured stucco in sealed, multi-wall bags bearing his name, brand, weight and color identification.

11.1.2 Additives.—Only clean water shall be added to manufactured stucco.

11.2 EXTERIOR STUCCO [1]

11.2.1 Uses.—Over any properly prepared portland cement base (plaster, concrete or masonry).

11.2.2 Materials.—A packaged blend of portland cement (ASTM C 150), hydrated lime (C 206 or C 207) and properly graded quality aggregate, with or without color.

11.3 MARBLECRETE BEDDING COAT [2]

11.3.1 Uses.—As a bedding coat to receive exposed aggregate.

11.3.2 Materials.—A packaged blend of portland cement (ASTM C 150), hydrated lime (C 206 or C 207) and properly graded quality aggregate, with or without color.

11.4 PORTLAND CEMENT STUCCO PAINT [3]

11.4.1 Uses.—Compensates for inclement weather, variation in base coat thickness, and other conditions leading to mottled colors. Color fog coating (spray application) is recommended as standard procedure to achieve color uniformity.

11.4.2 Materials.—A packaged blend of portland cement (ASTM C 150), hydrated lime (C 206 or C 207) and color.

11.5 ACOUSTIC-TYPE FINISH (Exterior) [4]

11.5.1 Uses.—Over any properly prepared portland cement base, such as plaster, concrete or masonry where texture of interior acoustic ceilings must be matched, or where such a texture is desired on exterior horizontal surfaces. Specify only thickness required to achieve desired texture.

11.5.2 Materials.—A packaged blend of portland cement (ASTM C 150), hydrated lime (C 206) and vermiculite or perlite aggregate (C 35).

11.6 INTERIOR COLORED STUCCO [5]

11.6.1 Uses.—A gypsum-lime base finish for use over gypsum plaster base coats where a colored finish is desired.

11.6.2 Materials.—A packaged blend of Keene's cement (ASTM C 61), hydrated lime (C 206) and properly graded quality aggregate, with color.

This specification reference presupposes that basic lathing and plastering specifications have already been prepared for the project and relates only to materials and application methods.

[1] Special Recommendations:
[a] For color control add measured amounts of water required to maintain uniform consistency for type of texture specified.
[b] Light pastels are recommended with float textures. Darker colors may be specified with dash or troweled textures.
[c] Clean mixer thoroughly between color changes.
[d] Curing is necessary under hot, dry or windy conditions. When required, fog lightly with water the day following application.

[2] Special Recommendations:
[a] For color control add measured amounts of water required to maintain uniform consistency.
[b] Clean mixer thoroughly between color changes.
[c] Curing is necessary under hot, dry or windy conditions. When required, fog lightly with water the day following application.
[d] Specify penetrating sealer or non-penetrating glaze to enhance colors and protect surface.

[3] Special Recommendations:
[a] Mix material by adding water slowly and stirring until a thick paste is formed. Allow to stand for 10 minutes. Add more water to obtain smooth flowing mixture slightly thicker than milk for hand application, water consistency for spraying. Water proportion must remain constant to produce uniform color.
[b] Hand application requires a large brush. Machine application is by pressure spray equipment; one coat for a color similar to existing surface, two when specifying different color.
[c] Except during damp weather, surface shall be dampened slightly 12 hours after completion and re-dampened at intervals until it hardens.

[4] Special Recommendations:
[a] Mix in a mechanical mixer until fluffy (at least 10 minutes).
[b] Specify two coats applied by plaster machine. Spray second coat to uniform texture when visible moisture has left surface.

[5] Special Recommendations:
[a] For color control add measured amounts of water required to maintain uniform consistency for type of texture specified.
[b] Light pastels are recommended with a smooth finish. Darker colors may be specified with float, dash or troweled textures.
[c] Clean mixer thoroughly between color changes.

11.6.3 Properties.—When tested in accordance with ASTM C 472, interior colored stucco shall have a minimum compressive strength of 500 psi.

11.7 INTERIOR WHITE COAT

11.7.1 Uses.—Over gypsum plaster base coats where a smooth white finish is desired. Specify standard interior white coat for residential use, commercial type for commercial buildings and hospitals.

11.7.2 Materials.—A packaged blend of Keene's cement (ASTM C 61) or gypsum gauging plaster, hydrated lime (C 206) and properly graded quality aggregate, without color.

11.7.3 Properties.—When tested per ASTM C 472, standard interior white coat shall have a minimum compressive strength of 500 psi and commercial type shall have a minimum compressive strength of 600 psi.

11.8 ACOUSTIC-TYPE FINISH (Interior) [6]

11.8.1 Uses.—Over gypsum plaster basecoats and prepared concrete surfaces. Material is designed to simulate texture of acoustic plaster. Specify only the thickness required to achieve desired texture. Use only on horizontal surfaces, or on portions of vertical surfaces, which are at least 7′ 6″ from floor.

11.8.2 Materials.—A packaged blend of Keene's cement (ASTM C 61), hydrated lime (C 206) and properly graded quality aggregate.

11.9 MARBLECRETE

11.9.1 Description.—Marblecrete shall consist of exposed natural or integrally colored aggregate, partially embedded in a natural or colored bedding coat of portland cement plaster.

11.9.2 Location.—Apply marblecrete in areas where shown on the drawings, or called for in the finish schedule or in these specifications. [7]

11.9.3 Materials.—Lathing and plastering materials for marblecrete shall be standard lathing and plastering materials. [8]

11.9.3.1 Aggregate.—For marblecrete finish shall consist of marble chips or pebbles, and shall be clean and free from harmful amounts of dust and other foreign matter. [9]

Size of marblecrete aggregate shall conform to the following grading standard:

Chip Size (Number)	Passing Screen (Inches)	Retained on Screen (Inches)
0	⅛	1/16
1	¼	⅛
2	⅜	¼
3	½	⅜
4	⅝	½

NOTE.—Chips larger than No. 4 must be applied manually. Use only for random accent.

Aggregate shall be blended by sizes in the percentages of graded chips called for in the specified sample.

11.9.3.2 Liquid Bonding Agent.—See 6.10.1. [10]

[6] Special Recommendations:
[a] Mix in a mechanical mixer until fluffy (at least 10 minutes).
[b] Specify two coats applied by plaster machine. Spray second coat to a uniform texture when visible moisture has left surface.

[7] Specify in lathing section: control joints, casing beads, parting screeds, or other metal sections to define panels as shown on drawings, and to serve as grounds for marblecrete. Marblecrete panels should not exceed 120 square feet, or 11 feet in any direction. Specify total ground thickness and show on drawings.

[8] e.g. portland cement; lime; sand; gypsum, etc.

[9] Other natural aggregate such as quartz, cinders, sea shells, or integrally colored manufactured aggregate, such as crushed glass, china, ceramics, may be specified, provided they are weather resistant, permanent in color, moderately hard (3 or more on the MOH scale) and are compatible with the bedding coat.

[10] Liquid bonding agent may be applied to plaster basecoat before application of bedding coat. If bonding agent is desired, so specify; or permit use at option of contractor.

11.9.3.3 Sealer.—Waterproofing shall be a clear penetrating liquid; (or) glaze shall be a clear non-penetrating liquid. Sealer shall be non-staining and shall resist deterioration from weather exposure.

Apply as directed by the manufacturer. [11]

11.9.4 Bases.—Apply marblecrete over (a) concrete, (b) masonry, (c) portland cement plaster basecoat, (d) gypsum plaster basecoat. [12]

11.9.4.1 Concrete and Masonry.—Give masonry and poured concrete surfaces which are to receive marblecrete a dash bond coat of portland cement plaster; or treat with a liquid bonding agent. [13]

Apply plaster using one of the following optional methods: [14]

(1) Apply bedding coat over dash bond coat or liquid bonding agent and double back to required thickness.

(2) Apply brown (leveling) coat over dash bond coat or liquid bonding agent and straighten with rod and darby before applying bedding coat.

11.9.4.2 Basecoat Plaster.—Over metal or wire fabric lath apply basecoat plaster in one of the following optional methods:

(1) Apply scratch coat and allow to set. Apply brown (leveling) coat minimum ⅜" thick and straighten with rod and darby. Leave rough and allow to set before applying bedding coat.

(2) Apply same as above except that bedding coat may be applied as soon as brown coat is firm enough to support bedding coat without sagging, sliding, or otherwise affecting bond.

(3) Apply scratch coat minimum ½" thick using double-back method, cover lath completely and straighten with rod and darby. Scratch horizontally and allow to set before applying bedding coat. Minimum overall thickness of scratch and bedding coat shall be 1".

Over gypsum lath apply bonding agent and brown coat to a minimum ⅜" thickness, straighten with rod and darby, leave rough and allow to set before applying bedding coat.

11.9.5 Bedding Coat.—Factory prepared bedding coat shall be a portland cement and lime plaster with properly graded quality aggregate, with or without color.

Job proportioned bedding coat shall be composed of one part portland cement, one part Type S lime, and maximum three parts of graded white or natural sand by volume. [15]

11.9.5.1 Mixing.—Mix manufactured and job-proportioned bedding coat with only sufficient water to attain proper consistency for application and embedment of aggregate.

11.9.5.2 Thickness.—The thickness of the bedding coat shall be determined by the maximum size of the aggregate specified and shall conform to the following: [16]

Bedding Coat Thickness	Aggregate Size (Max.)	
(Minimum—Inches)	Number	Inches
⅜	#0	⅛
⅜	#1	¼
⅜	#2	⅜
⅜	#3	½
½	#4	⅝
80% of aggregate dimension	Larger than #4	

[11] Sealer is recommended to improve and retain color and cleanliness of marblecrete.

[12] Specify those bases applicable to project.

[13] See 7.1.2 Masonry Surfaces; 7.1.3 Poured Concrete Surfaces.

[14] Check with local building officials for method of application approved. All building codes require minimum overall thickness of exterior portland cement plaster to be ⅞ inch, gypsum plaster minimum ½ inch.

[15] If bedding coat is gypsum plaster, proportion 100 pounds neat gypsum plaster and maximum 200 pounds of graded white sand.

[16] Overall thickness will depend on thickness of basecoat plaster plus thickness of bedding coat.

11.9.6 Application of Marblecrete.—Apply bedding coat to proper thickness and straighten to a true, reasonably smooth surface with rod and darby. Allow bedding coat to take up until it attains the proper consistency to permit application of aggregate.

Apply the aggregate to the bedding coat, starting at the perimeter of a panel area and working towards the center. Tamp lightly and evenly to assure embedment of the aggregate and to bring surface to an even plane. Overspray aggregate must be thoroughly rewashed and completely dry before reusing.

11.9.7 Curing.—Portland cement marblecrete shall retain sufficient moisture for hydration (hardening) for 24 hours minimum. Where weather conditions require, keep marblecrete damp by spraying lightly.

11.9.8 Murals.—The architect or his representative shall transfer design pattern shown on large scale drawings to base which receives marblecrete mural. The transferred design shall serve as guide for the application of flexible or rigid metal separating screeds. Artist's rendering, and color and aggregate key, shall be provided contractor at time of bidding.

For pictures of stucco (plaster) colors and textures, ee 16-page color supplement following this chapter.]

12. GYPSUM BOARD (Drywall) APPLICATIONS

12.1 GYPSUM BOARD MATERIALS

12.1.1 General.

12.1.1.1 Standard Specifications.—Where published standard applications are referred to herein, they shall be those of the latest date of adoption. [1]

12.2 MATERIALS

12.2.1 Gypsum Board Panels.—In lengths as long as practical to minimize the number of adjoins, gypsum panels shall be: [2]

12.2.1.1 Regular Gypsum Board.—Mill fabricated gypsum panel composed of a fire-resistant gypsum core encased in a heavy natural finish paper on the face side and a strong liner paper on the back side. [ASTM C 36]

12.2.1.2 Type X (Special Fire-Resistant) Gypsum Board.—Gypsum board manufactured to provide a greater degree of fire-resistance than regular gypsum board. [ASTM C 36]

12.2.1.3 Foil-Back Gypsum Board.—Gypsum panels made by laminating aluminum foil to the back surface. Vapor permeability shall not exceed .30 perms.

12.2.1.4 Water Resistant Gypsum Board.—Gypsum board with a gypsum core that has been treated with special asphalt composition. The green multi-layered face and back paper are chemically treated to combat penetration of moisture. [ASTM C 630]

12.2.1.5 Gypsum Coreboard.— A one inch thick, 24″ wide gypsum board encased in strong gray paper on both sides. To be used in laminated gypsum board construction. [ASTM C 442]

12.2.1.6 Exterior Gypsum Ceiling Board (Soffit Board).—A weather and sag resistant board designed for the weather protected side of eaves, canopies and carports. Covered with water repellent face paper suitable for painting. [ASTM C 931] [3]

12.2.2 Fasteners and Accessories.

12.2.2.1 Screws.—For attaching gypsum board to steel or wood framing. Standard Specification for Steel Drill Screws for the Application of Gypsum Board. [ASTM C 1002]. Type of screw as recommended by the manufacturer of the gypsum wallboard.

12.2.2.2 Nails.—Standard Specification for Nails for Application of Gypsum Wallboard. [ASTM C 514]. Special nails for pre-decorated gypsum board should be as recommended by the pre-decorated gypsum board manufacturer.

12.2.2.3 Cornerbead, Control Joints and Edge Trim.—Shall be made from corrosion-resistant steel or plastic designed for its intended use. Flange shall be free of grease contaminants or materials which could adversely affect bond of joint treatment or decoration.

12.2.2.4 Joint Reinforcing Tape and Joint Compound.—Standard Specification for Joint Treatment Materials for Gypsum Wallboard Construction. [ASTM C 475]. Same brand as the manufacturer of the gypsum board.

[1] Where any reference is made in these specifications to ASTM, U.S. Federal Specifications, AIA or CSI, applicable excerpts will be formed in Specification Reference 1.8.

[2] Provide suitable fascia and molding around the perimeter to protect the gypsum board from direct exposure to water. Unless protected by metal or other water stops, place the edges of the board not less than 1/64 inch away from abutting vertical surfaces.

[3] Specify minimum acceptable thickness, ⅜, ½ or ⅝ inch. Thickness determined by framing spacing and fire or sound resistance.

12.3 DELIVERY, IDENTIFICATION, HANDLING AND STORAGE

12.3.1 All materials should be delivered in original packages, containers or bundles bearing brand name, applicable standard designation, and name of manufacturer or supplier for whom product is manufactured.

12.3.2 All materials should be kept dry, preferably by being stored inside building — under roof. Where necessary to store gypsum board products and accessories outside, they should be stacked above ground, properly supported on a level platform and fully protected from weather and direct sunlight exposure.

12.3.3 Gypsum board should be neatly stacked flat with care taken to prevent sagging or damage to edges, ends and surfaces.

12.4 APPLICATION OF SINGLE LAYER GYPSUM BOARD

12.4.1 In general apply gypsum board to ceilings first, then to walls.

Note: For specific information in laminated gypsum wallboard construction and shaftwall systems refer to manufacturer's printed instructions.

12.4.2 To minimize end joints, use panels of maximum practical lengths. Gypsum board at openings shall be located so that no joint will align with edges of opening unless control joints will be installed at these locations. Stagger end joints in successive courses with joints on opposite sides of a partition placed on different studs.

12.4.3 All cut edges and ends of the gypsum board shall be smoothed to make a neat joining.

12.4.4 Space the fasteners when used at edges of boards not more than one inch from the edges and not less than ⅜ inch from the edges and ends of gypsum board (except where floating angles are used). Perimeter fastening into partition plate or track at the top and bottom is not required except where the fire rating, structural performance and other special conditions require such fastening.

12.4.5 Fasteners should be driven so that the heads are slightly below the plane of the face paper. Avoid fracturing the face paper or damaging the core. Hold the panel in firm contact with the framing while driving the fasteners. Install fasteners in the field of the board, first working towards ends and edges.

12.4.6 Install trim at all external and internal angles formed by the intersecting of gypsum board surfaces or gypsum board with other surfaces, Apply corner bead to all vertical and horizontal external corners in accordance with the manufacturer's printed instructions. Fasteners to be installed six inches o.c. nominally or attach trim with a crimping tool. Attach gypsum board control joint with 9/16" type "G" staples spaced not over 6" apart in each flange. Cut end joints square and align for a neat fit.

12.5 APPLICATION OF EXTERIOR GYPSUM CEILING BOARD (SOFFIT BOARD)

12.5.1 Framing should be no more than 16 inches for ½-inch thick gypsum ceiling board or 24 inches for ⅝-inch trick gypsum ceiling board. Long dimension of the board should be perpendicular to the framing.

12.5.2 End joints must fall on supports with 1/16 to ⅛ inch space between butted ends of boards. Fasten boards to supports with screws spaced 12" o.c. or nails spaced 8" o.c.

12.6 FINISHING OF GYPSUM BOARD

12.6.1 Finish all face panel joints and internal angles with materials which meet ASTM C 475 or those materials recommended by the gypsum board manufacturer.

12.6.2 No finishing operation shall be done until interior temperature has been maintained at a minimum of 50 degrees F for a period of at least 48 hours and thereafter until the compounds have completely dried.

12.6.3 Keep gypsum panels free of dirt or other foreign matter which could cause lack of bond. Fill all dents and gouges. Set the fasteners below the plane of the board. All joints shall be even and true. Keep the board tight against the framing.

12.6.4 Tape shall be properly applied either by applying compound to joint (buttering), pressing in tape, and wiping off excess compound or by mechanical tool designed for the process.

12.6.5 Apply the second coat with sufficiently wide tools to extend the compound at least three inches beyond the joint center. Draw down to a smooth even plane. After drying or setting, sand as needed to eliminate any high spots or excessive compound.

12.6.6 Third coat when required should be applied with tools which will permit feathering of joint treatment edges approximately six inches from center of joint.

12.6.7 After drying final coat shall be lightly sanded to leave a smooth, even surface covering the joint. Caution should be taken to not raise the nap of the paper when sanding.

12.6.8 Cover fastener heads with three successive coats each applied in a different direction except if special finish tool is used.

12.6.9 Finish corner beads, control joints and other trim with successive coats of joint treatment compound as recommended by the compound manufacturer, so as to make the trim piece flush with the board surface.

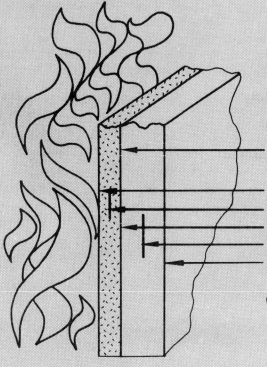

HOW GYPSUM RETARDS HEAT TRANSMISSION

AFTER TWO HOUR EXPOSURE TO HEAT FOLLOWING ASTM E 119 TIME-TEMPERATURE CURVE

Vertical line represents plane of calcination at depth of about 2″. Temperature never greatly exceeds 212 F. behind plane of calcination.

Temperature of exposed surface = 1900 F.

Temperature 1″ from exposed face = 950 F.

Temperature 2″ from exposed face = 220 F.

Temperature 4″ from exposed face = 180 F.

Temperature at back surface = 130 F.

(Data from Underwriters Laboratories, Inc.)

PLASTER TEXTURES

DEPICTED in this book are specimens of plaster textures which are in most common use today. The term "plaster" is used generally to describe material prepared with either portland cement or gypsum. The term "stucco" used herein describes a factory-prepared, integrally colored finish which over the years, has come to be used to describe all colored portland cement finishes.

Each of the pictures shown herein was photographed from the same distance of approximately 10 feet and depicts a one square foot panel. Plaster, applied in a plastic state, may create a great variety of design configurations or texture patterns.

Colored plaster is used from the Gulf of Mexico to the northern border of the United States, and beyond. It has been a popular finish in Canada for many years. All types of buildings from single family homes to high rise structures utilize colored plaster.

There are many reasons for this wide acceptance. Since it lasts the life of the building, redecoration is necessary only when a change in color is desired. Nevertheless, stucco is one of the most economical materials a builder can specify. Marblecrete, for example, produces a highly attractive surface at a fraction of exposed aggregate concrete's cost. The architect is limited only by his or her imagination in choice of texture treatments.

Designers, builders and contractors can pick from a wide variety of textures on the following pages. It is recommended that the contractor prepare samples before construction to insure that all parties agree on the finish to be produced.

Color selection should be made at the time specifications are written. Special colors not included in the manufacturer's current color chart usually require an additional charge. Deeper colors call for additional pigment and are therefore higher priced.

Textures illustrated aren't intended to show the whole range of possibilities. They are the most commonly used. Identifying names are those by which the texture is commonly referred to in the trade. It should be noted that a designer, contractor or journeyman may have differing concepts of such terms as Spanish, Monterey or English. Suggested application procedures are included as general, abbreviated guidelines to the production of each texture.

The textures shown can be used on either exteriors or interiors. Interior plaster is formulated with gypsum rather than cement. Generally speaking, heavier textures with deeper relief are used on the outside.

Mill-mixed colored stucco is produced under controlled conditions in modern plants to insure uniformity from batch to batch.

A House In Palos Verdes, California

Architect: B.O.A. Architects
               San Pedro, California

Photograph: Wayne Thom Photography

12″

Light Lace

Suggested Application Procedures

1. Trowel, float or dash on a first coat to completely cover base.
2. When surface moisture leaves, trowel apply light second coat in random directions.
3. Knock down surface lightly with trowel.

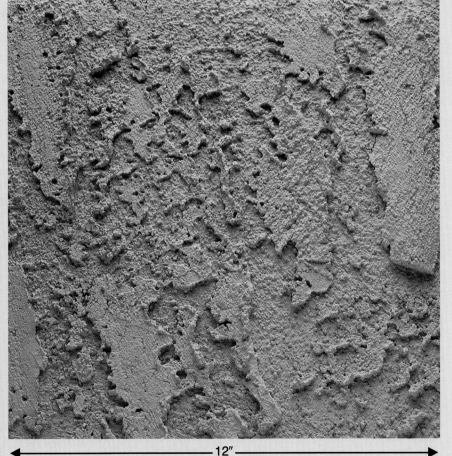

Heavy Lace

Suggested Application Procedures

1. Trowel, float or dash on a first coat to completely cover base.
2. When surface moisture leaves, trowel apply second coat in random directions.

12″

Light Dash

Suggested Application Procedures

1. Apply a first dash coat to produce complete color coverage.
2. Apply a second dash coat for texture depth and uniformity when first coat is dry, using a plaster mix of thinner consistency.
3. Use proportionately more atomizing air at the gun nozzle.

12″

Medium Dash

Suggested Application Procedures

1. Apply a first dash coat to produce complete color coverage.
2. Apply a second dash coat for texture depth and uniformity when first coat is dry.
3. Use a medium amount of atomizing air at the gun nozzle.

12″

← 12″ →

Heavy Dash

Suggested Application Procedures
1. Apply a first dash coat to produce complete color coverage.
2. Apply a second dash coat for texture depth and uniformity when first coat is dry, mortar to be of a relatively stiff consistency.
3. Use relatively less atomizing air at the gun nozzle and lower water ratio of the plaster.

← 12″ →

Tunnel Dash

Suggested Application Procedures
1. Apply first dash coat to produce complete color coverage.
2. When surface moisture leaves — or on second day — apply a second heavy texture coat.
3. Use low atomizing air and reduce water ratio of the plaster.

Knockdown Dash

Suggested Application Procedures

1. Apply first dash coat in thin consistency to produce complete color coverage.
2. Apply a coarse second dash coat for texture depth and uniformity, allowing some of first coat to show through.
3. Trowel lightly after moisture leaves surface.

12″

Monterey

Suggested Application Procedures

1. Trowel on a first coat, leaving relatively smooth.
2. Apply second coat in a random texture, using overlapping strokes of the trowel.

12″

← 12″ →

Combed

Suggested Application Procedures

1. Apply finish coat in sufficient thickness to accommodate depth of grooves without exposing base (brown) coat.
2. Rod and darby, leaving surface reasonably straight and true.
3. Using a strip as a guide, comb surface vertically (or horizontally) with a template, formed to achieve pattern detailed on drawings.

Note: Special mix required.

Deep Relief

Suggested Application Procedures

1. Trowel on a heavy texture coat.
2. After surface moisture is absorbed, apply heavy second coat, leaving it rough under the trowel.

← 12″ →

Fine Sand Float

Suggested Application Procedures

1. Trowel on a finish coat and double back with a second application. Plaster mix is to be formulated with a 30-mesh aggregate or a blend of 20-30.

2. Using circular motion, rub surface with float to achieve uniform pattern, bringing sand particles to surface. An absolute minimum of water should be used in floating.

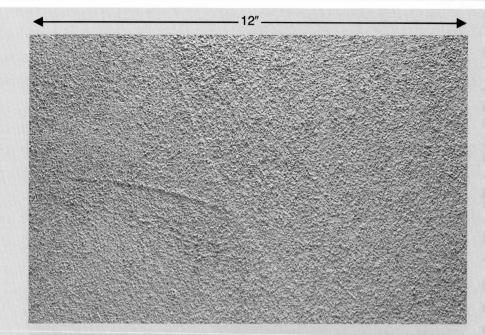

Medium Sand Float

Suggested Application Procedures

1. Trowel on a finish coat and double back with a second application. Plaster mix is to be formulated with 20-mesh aggregate.

2. Using circular motion, rub surface with float to achieve uniform pattern, bringing sand particles to surface. An absolute minimum of water should be used in floating.

Heavy Sand Float

Suggested Application Procedures

1. Trowel on a finish coat and double back with a second application. Plaster mix is to be made with coarse aggregate or relatively coarse blend.

2. Using circular motion, rub surface with float to achieve uniform pattern, bringing sand particles to surface. An absolute minimum of water should be used in floating.

← 12″ →

Scraped

Suggested Application Procedures

1. Apply finish coat approximately ¼ in. thick and allow to take up until surface moisture leaves.
2. Scrape vertically with a steel joint rod or trowel held at right angles to the plane of the wall. Remove sufficient material to leave a torn surface, free from smooth spots and joinings.

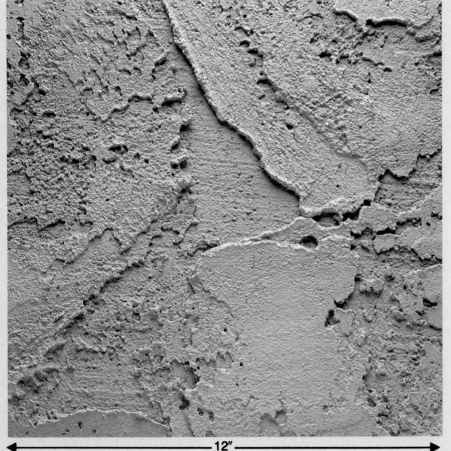

Spanish

Suggested Application Procedures

1. Trowel on a first coat, leaving relatively smooth.
2. Apply second coat in random texture, using overlapping strokes of the trowel.

← 12″ →

California

Suggested Application Procedures

1. Trowel on a first coat to completely cover base.
2. Apply a thin texture coat with trowel in a random pattern, overlapping strokes.
3. Flatten higher areas with a trowel.

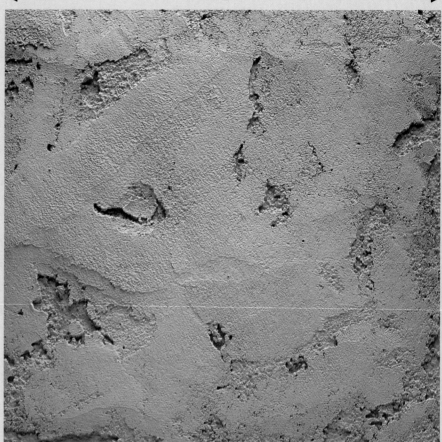

Arizona

Suggested Application Procedures

1. Trowel on a heavy texture coat.
2. After surface moisture is absorbed, apply heavy second coat, leaving it rough under the trowel with small area texture pats.

Frieze

Suggested Application Procedures

1. Trowel on a first coat using doubleback method, and rake with a coarse brush or broom.
2. Splatter dash sparingly with dash broom, using mortar of fairly stiff consistency to partially cover the surface.
3. After moisture leaves surface, trowel down high spots, retaining general pattern of dash texture.

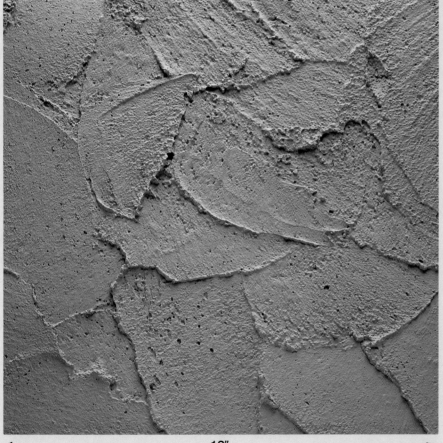

English

Suggested Application Procedures

1. Trowel on a first coat to completely cover base.
2. Using a rounded trowel, apply a thick texture coat with short strokes in varying directions, leaving a rough, irregular pattern.

Trowel Sweep

Suggested Application Procedures

1. Trowel on a first coat to completely cover base.
2. Apply a second coat with fan-shaped strokes, lapping each other so as to form narrow, high ridges where mortar flows over the toe of the trowel.

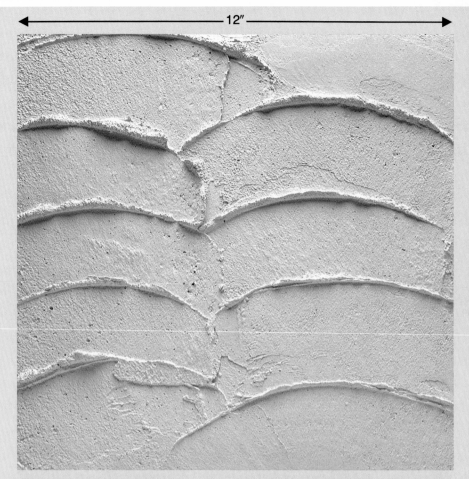

12″

Web

Suggested Application Procedures

1. Trowel on a first coat and broom lightly in varying directions, using a sweeping motion.
2. Using a trowel, apply a texture coat in strips approximately 2″ x 6″, forming a more or less rectangular pattern.
3. Trowel surface lightly.

12″

← 12″ →

Briar

Suggested Application Procedures

1. Trowel on first coat to completely cover base.
2. Apply a texture coat, holding trowel at angle to surface. With heel of trowel serving as a pivot point, produce fanlike ridges in radiating pattern.
3. Flatten higher areas with trowel.

Brick

Suggested Application Procedures

1. Trowel on a first coat of mortar-colored joint material.
2. Trowel on a second coat of brick-colored stucco material.
3. Comb or broom the surface with a coarse fiber brush or broom to achieve desired grain. A light troweled texture may also be applied.
4. Using brick template or straightedge, rake joints to depth required to expose mortar joint material.

← 12″ →

Marblecrete

Suggested Application Procedures

1. Apply bedding coat to proper thickness.
2. Straighten with rod and darby, leaving surface reasonably smooth.
3. Apply aggregate to bedding coat.

← 12″ →

Simulated Timber

Suggested Application Procedures

1. Spread finish coat plaster to desired panel texture.
2. Lay on a narrow band (i.e. 6″ to 8″ wide) of same material in pattern of half-timber.
3. Cut shallow groove on each side of simulated timber.
4. Lightly trowel face of simulated timber to relatively smooth surface.
5. If peg marks are desired, lightly press large screw head near ends of simulated timber.

← 12″ →

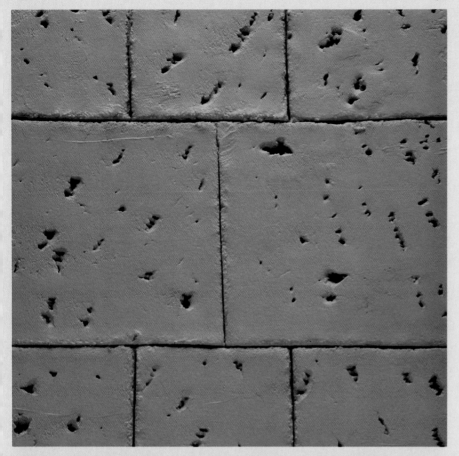

12"

Travertine

Suggested Application Procedures

1. Apply travertine finish coat ¼ in. to ⅜ in. thick over damp base.
2. Rod and darby.
3. Lay out travertine design as shown on drawings. To simulate joints, stamp in lines with a joint rod or rake out with raking tool.
4. With spring steel or a wire brush pick out small portions of surface to imitate indentations of travertine stone.
5. Water trowel smooth, retaining indentations.

12"

Imagination

Plaster textures are only limited by the imagination of the designer and the talents of the craftsman. This is an example of a texture achieved by the square end of a small tool.

Rock 'n Roll

Suggested Application Procedures

1. Trowel on a first coat to completely color base coat.
2. Apply a second coat of a specially prepared finish plaster containing selected size pebbles to achieve desired texture.
3. The action of the trowel or float rolls the pebbles to create miniature troughs in a linear or circular pattern.

Glacier

Suggested Application Procedures

1. Trowel on a first coat to completely cover base.
2. Float the surface to raise the aggregate.
3. Trowel on a light texture coat over the floated surface.

CHAPTER THREE

INDUSTRY-WIDE
REFERENCE
STANDARDS
AND
SPECIFICATIONS

BY WAY OF EXPLANATION . . .

Included in this chapter are the latest versions of other major reference standards and specifications that anyone involved with drawing plans or writing specifications for lath, plaster, fireproofing and drywall installations would be concerned with.

METAL LATH/STEEL FRAMING DIVISION OF NATIONAL ASSOCIATION OF ARCHITECTURAL METAL MANUFACTURERS

This organization is the national trade association of the major manufacturers of metal lath and steel framing. Its *"Specifications For Metal Lathing and Furring"* are the reconized standard for the industry and are reproduced verbatim in this chapter with the permission of the association, located at 600 So. Federal St., Chicago, Illinois.

AMERICAN SOCIETY FOR TESTING AND MATERIALS

The American Society for Testing and Materials, founded in 1898, is a scientific and technical organization formed for "the development of standards on characteristics and performance of materials, products, systems and services; and the promotion of related knowledge." It is the world's largest source of voluntary consensus standards. The Society operates through more than 110 main technical committees. These committees function in prescribed fields under regulations that ensure balanced representation among producers, consumers and general interest publications.

ASTM consensus standards are universally recognized and those affecting lath, metal framing systems plaster and drywall installations are reproduced in this chapter. All ASTM standards which follow are reprinted by permission of the American Society for Testing and Materials, copyrighted.

Important updates on a number of these standards have been made in recent months prior to the printing of this manual, which reflects the latest updated versions of the standards. These updates have been promulgated and voted upon by ASTM's Committee C 11.03 on Gypsum and Related Building Materials and Systems.

In addition to construction specifications ASTM also publishes testing specifications for various building materials. The major ASTM specifications for testing lath, plaster and drywall materials appear in Chapter Four, *Requirements, Tests and Standards.*

American Society for Testing and Materials is located at 1916 Race Street, Philadelphia, Pennsylvania 19103, telephone (215) 299-5400.

specification writing for the construction industry, known both in electronic and printed copy forms. The Spectext Housing and Urban Development (HUD). These standards, (214) 363-6747. Copies priced at $14.00 each, may be obtained Conference of Building Officials; the Los Angeles City Building Code, distributed by Building News, Inc. and the

CONSTRUCTION SPECIFICATIONS INSTITUTE

The Construction Specifications Institute (CSI), headquartered in Alexandria, Virginia, is the national society of professional specifiers for the construction industry. Through committees of specialists in all areas of construction CSI in years past authored and published a series of specification documents for all types of construction installations which were universally recognized as the most authorative documents of their type in existence.

In more recent years CSI has joined forces with three other professional groups in a new coalition established for the purpose of developing uniformity and upgrading specification writing for the construction industry known as the Engineers Joint Contract Document Committee (EJCDC).

The old CSI specifications have been merged into the new uniform specifications marketed by this group under the namestyle of Spectext, which is available to users both in elections and printed copy from. The Spectext specifications applicable to the plaster, lath and drywall industries are reproduced in this manual with the permission of EJCDC.

Specifications writers may utilize these specifications by photocopying directly from this manual; or by purchase of copies from The Construction Specifications Institute.

U.S. DEPARTMENT OF HOUSING AND URBAN DEVELOPMENT (HUD)

All housing units designed to be eligible for FHA mutual mortgage insurance must conform to standards written and promulgated by the U.S. Department of Housing and Urban Development. These standards, known as HUD Minimum Property Standards (MPS), are published in three separate volumes: Volume 1, *MPS for One and Two Family Dwellings*, 4900.1; Volume 2, *MPS for Multifamily Housing*, 4910.1; and Volume 3, *MPS for Care-Type Housing*, 4920.1.

Each of these volumes contains a brief section on lath, plaster and drywall installations which is reproduced in this chapter.

Whereas in former years the MPS were equivalent to detailed building codes, they are today very brief and sketchy standards. HUD, for the most part, follows local building codes in establishing criteria for FHA mutual mortgage insurance.

However, HUD also publishes a more detailed guidance document, known as the *Manual of Acceptable*

BY WAY OF EXPLANATION — *(Cont.)*

Practices, 4930.1. This manual goes into more depth and provides guidelines for all types of construction which will qualify for FHA insurance. Those sections from this manual pertaining to lath and plaster installations are also reproduced in this chapter.

The MPS are sold by the Superintendent of Documents, U.S. Government Printing Office, Washington, D.C. 20402. The purchase price includes the MPS, its quarterly revisions for three years and mailing charges. Prices are: 4900.1, $10.25; 4010.1, $8.95; 4920.1, $7.45; and 4930.1, $22.15.

OTHER INDUSTRY-WIDE REFERENCE STANDARDS AND SPECIFICATIONS

In addition to the industry-wide standards and specifications reproduced in this chapter, attention is directed to the following documents, which are widely referenced and highly regarded in the industry:

LATH AND PLASTER SELECTION DATA AND ASSEMBLIES

The newest edition of this book, published in May, 1977, contains excellent tables, design data and details on lath and plaster partitions, furring, ceilings, beams, columns, grounds and components.

The tables present the essential details of dimension, weight, mix and laboratory tests of various lath and plaster assemblies. They are intended to facilitate the selection of the system best suited for the individual job by detailing the qualities of each assembly with appropriate illustrations.

This book was edited and published by the Texas Lathing and Plastering Contractors Association and the Texas Bureau For Lathing and Plastering, 6500 Green-field Avenue, Suite 460, Dallas, Texas 75206, telephone (214) 363-6747. Copies priced at $400 each, may be obtained by writing the publishers.

FIRE RESISTANCE DESIGN MANUAL

This manual, published by the Gypsum Association, allows the user to quickly and easily determine the essential performance characteristics of a wide range of fire resistant construction assemblies. Comparison of these characteristics enables him to be more accurate in meeting particular design requirements. It also provides authoritative data for building officials, fire fighting and insurance personnel.

Assemblies in this book use gypsum products. Thus, they provide fire resistant walls and partitions, floor-ceilings, columns, beams and roof decks. They are classified according to use and fire resistance ratings, with walls, partitions and floor-ceilings further classified with sound transmission class (STC) ratings and impact isolation classification (IIC) ratings. Structural height limitations of nonload-bearing partitions are also included.

The assemblies are divided into the above mentioned categories: walls and partitions, floor-ceilings, columns, beams and roof decks. They are then listed according to their fire resistance rating starting at one hour and in descending order of their sound classifications.

The *Fire Resistance Design Manual* is referenced in: the BOCA Basic Building Code, published by the Building Officials and Code Administrators International; the Uniform Building Code, published by the International Conference of Buuilding Officials; the Los Angeles City Building Code, distributed by Building News, and the Standard Building Code, published by the Southern Building Code Congress.

Single copies may be obtained free by sending written requests to Gypsum Association, 1603 Orrington Avenue, Evanston, Illinois 60201, telephone (213) 491-1744.

SPECIFICATIONS FOR METAL LATHING AND FURRING

NOTE: These specifications are reproduced with permission of their publisher, the Metal Lath/Steel Framing Division of the National Association of Architectural Metal Manufacturers (NAAMM). A new, updated version of these specifications is now being prepared by NAAMM and may be obtained in the Spring of 1989 by writing this organization at 600 So. Federal Street, Chicago, Illinois 60605.

These specifications are a guide for the specification writer, inspector and contractor. They provide detailed information about many metal lath and plaster assemblies, not all of which may be incorporated into a single job. In such instances, certain sections may be omitted, but Section One, General Specifications, is necessary to all other sections and must be retained.

Advisory notes are indicated by letters beside the text where applicable. Other notes, under Design Assumptions and General Notes, are intended to provide background information only.

It should be observed that application techniques do not offer, nor does job performance require, precision in measurement of the spacing of attachments, furring, studs, etc. Reasonable tolerances should be permitted.

DESIGN NOTES

A Wall-ceiling junctures or intersections of partitions may be treated either as restrained construction or as unrestrained construction. The discontinuity of unrestrained construction is particularly advantageous where differential movement may be expected between intersecting plaster membranes.

B Control joints should be designed and installed in a manner to permit the plaster membrane to expand and contract freely without inducing excessive stress in the plaster. (See Section Ten for specific information on material selection and installation.)

C Where non-load bearing partitions or furring abutt the underside of a concrete slab or run parallel to and abutt concrete beams or steel members, a slip or cushioning joint should be provided so that loads will not be transferred to the partition or furring. For specifications and details of specific types of slip joints consult appropriate manufacturer's literature. Common joint construction generally includes track, runners, clips and casing bead used in combination with one or more sealing materials such as elastic caulking, compressible fiberboards and vinyl gaskets.

D Concrete stub nails (for attaching track, runners, studs, etc.) should be a minimum of 1/2" long and should penetrate concrete at least 3/8". If other fasteners (powder drives, expansion drives, etc.) are used, they should provide attachment as strong as the concrete stub nails.

E Plaster should be applied in five coats. To prevent bowing during erection, the following sequence should be followed:
   1. apply scratch coat on side opposite temporary braces, allow it to set and partially dry;
   2. next apply brown coat to the side scratched. When it has set, remove temporary brace and apply back-up coat and finish coats.

F Metal door frames should be specified that have jamb anchors which are notched or formed to receive a stud and are sized to fit snugly within the interior of the jamb. Jamb anchors should be made of a minimum of 16 gage steel and extend no more than 3/4" out of the back of the jamb. One anchor should be located near the top of the jamb (within 2"), another spaced approximately 12" from the top, another located at the floor, and the rest spaced no more than 24" apart. (4 jamb anchors and a floor anchor at each side).

To preclude spalling and cracking of the plaster, the frame should be grouted solidly.

G Wood casings should be specified that lap over plaster at least 1".

H Electrical boxes greater than 1½" in depth should not be specified for 2" solid partitions.

J Corners and intersections of partitions should be formed of three studs, or a combination of prefabricated studs and 3/4" cold-rolled channels, adequately braced. Studs and channels forming internal corners should be placed 2" from the point of partition intersection to allow for tying of cornerite or lath.

If channels or prefabricated steel studs are attached to each other by welding, the areas of welded components should be broken up into panels to control movement during temperature changes.

K Metal base should be filled with grout after studs have been installed and either before or after metal lath has been wire tied to the studs.

L Plaster should be applied in five coats. To prevent bowing during erection, the following sequence should be followed:
   1. apply scratch coat on lath side (opposite side of studs), allow it to set and partially dry;
   2. next apply back-up coat on channel side and allow it to set;
   3. then apply brown coat on lath side, then finish coats on each side.

M To obtain maximum results from the sound insulating double partition, the following should be observed.
   1. In braced construction, do not cross brace except at mid-height since every connection between partition faces reduces the sound isolating quality. (Mid-height cross braces reduce the sound transmission loss potential by about 2 db.).
   2. Be sure that stiffeners do not connect both faces of the partition because they will provide direct sound paths.
   3. Be sure that service lines (ducts, piping, etc.) are not tied to the studs (or furring, etc.). If they are, they can cause a partition face to act as a "loud speaker" for mechanical noises.
   4. Install lath and plaster behind all openings in the plaster face (i.e. behind electrical boxes, cabinets, etc.).
   5. Do not install telephones, door bells and other noise sources on partition.
   6. Be careful that windows and doors do not provide an air path for sound to by-pass the entire partition.
   7. A split, or double door frame should be used. Door frames should be installed according to the manufacturer's directions and cork or other insulation should be used to provide a sound deadening casing.

Design Notes continued on page 125

SECTION ONE

GENERAL SPECIFICATIONS

1.1 MATERIALS

1.11 Metal Lath—expanded, rib or sheet, with or without paper backing —shall be fabricated from steel and given a protective coating of paint after fabrication, or shall be fabricated from galvanized steel.

1.12 Channels shall be cold roll-formed from 16 gage steel and given a protective coating of paint.

1.13 Prefabricated steel studs shall be fabricated as channels from 18 gage steel with punched openings along entire length of web (between flanges). They shall be given a protective coating of paint, or fabricated from galvanized steel.

1.14 Wire shall be galvanized, annealed steel wire. The weight of galvanized finish shall be not less than for Class 1, as set forth in Federal Specification QQ-W-461, "Wire, Steel, Carbon (Round, Bare and Coated)". Wire used for ties shall not be smaller than 18 gage.

1.15 Ceiling runner shall be fabricated from not lighter than 28 gage steel; painted or galvanized; formed into an "L", "Z", or channel; slotted, perforated or pronged; and shall be capable of holding studs or lath securely in place. Also, 2" x 2" cornerite, or track may be used as a ceiling runner.

1.16 Floor runner shall be fabricated from not lighter than 28 gage steel; painted or galvanized; formed into a "Z" or channel; slotted, perforated, or pronged; and shall be capable of holding studs or lath securely in place. Also, track may be used as a floor runner.

1.17 Track shall be fabricated from not lighter than 24 gage steel; painted or galvanized; formed to hold studs securely in place with slots, prongs, or with wire ties.

1.18 Staples shall be manufactured from round, semiround or flat 16 gage wire.

1.2 INSTALLATION

1.21 Install metal lath with its long dimension across the supports; rib metal lath with ribs against the supports. Attach metal lath to supports at 6" o.c. Wire tie metal lath at side laps not to exceed 9" o.c.

1.22 Lap diamond mesh and flat rib metal laths at sides at least ½". Lap ⅜" rib lath by either nesting edge ribs or lapping sides ½". Lap all lath at least 1" and nest major ribs at ends of sheets. Ends should lap over supports.

A 1.23 Where diamond mesh metal lath is used on ceiling and walls or partitions, bend lath into corner and lap at least 6" onto the adjoining lath; or butt lath into corner and apply cornerite over the abutting laths.

Where rib and sheet metal laths are used on ceiling or walls and partitions, butt lath into corner and apply cornerite over the abutting laths.

A 1.24 Where diamond mesh metal lath is used on abutting walls or partitions (i.e. internal corners), start lath one stud away from corner and bend it into corner and carry it onto the adjoining surface; or butt lath into corner and apply cornerite over the abutting laths.

Where rib and sheet metal laths are used on abutting walls or partitions (i.e. internal corners), butt lath into corner and apply cornerite over the abutting laths.

A 1.25 Where abutting surface is not to be plastered, attach metal lath to the abutting surface so that lath edge and attachments will be embedded in the plaster.

1.26 Where unrestrained construction is desired, install casing bead or control joint along plaster edge.

1.27 Install metal plaster accessories to plaster line (use shims if necessary) and attach accessory (to metal lath, gypsum lath, wood, concrete, masonry, etc.) by wire-tying, nailing or stapling through expanded wings or through holes provided in accessory. Attachments should be strong enough to hold accessory in place during plastering.

INDEX

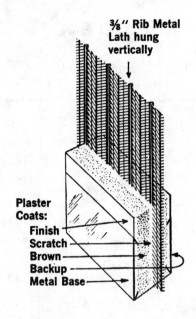

⅜" Rib Metal Lath hung vertically

Plaster Coats:
Finish
Scratch
Brown
Backup
Metal Base

Type of Lath	Weight of Lath Lb. Per Sq. Yd.	WALLS AND PARTITIONS			CEILINGS	
		Wood Studs	Solid Partitions	Steel Studs, Wall Furring, etc.	Wood or Concrete	Metal
Diamond Mesh (flat expanded)	2.5	16	16	13½	12	12
	3.4	16	16	16	16	16
Flat Rib	2.75	16	16	16	16	16
	3.4	19	24(3)	19	19	19
⅜" Rib (1) (2)	3.4	24	(4)	24	24	24
	4.0	24	(4)	24	24	24
¾" Rib	5.4	—.	(4)	24(5)	36(6)	36(6)
Sheet Lath (2)	4.5	24	(4)	24	24	24

Table title: **MAXIMUM SPACING OF SUPPORTS FOR METAL LATH (Inches)**

NOTE:

Weights are exclusive of paper, fiber, or other backing.

(1) 3.4 lb., ⅜" Rib Lath is permissible under Concrete Joists at 27" c.c.

(2) These spacings are based on a narrow bearing surface for the lath. When supports with a relatively wide bearing surface are used, these spacings may be increased accordingly, and still assure satisfactory work.

(3) This spacing permissible for Solid Par-

titions not exceeding 16' in height. For greater heights, permanent horizontal stiffener channels or rods must be provided on channel side of partitions, every 6' vertically, or else spacing shall be reduced 25%.

(4) For studless solid partitions, lath erected vertically.

(5) For interior wall furring or for application over solid surfaces for stucco.

(6) For contact or furred ceilings only.

SECTION TWO

SOLID PARTITIONS — STUDLESS

2.1 MATERIAL SELECTION

B 2.11 Metal lath shall be of the type and weight for the partition height as set forth in the following table:

LATH TYPE	LATH WEIGHT (lbs./sq. yd.)	MAXIMUM PARTITION HEIGHT
⅜" Rib	3.4	8'4"
⅜" Rib	4.0	10'4"
¾" Rib	.60 lb./sq. ft.	12'4"
Sheet	4.5	10'4"

2.12 Partition thickness shall be a minimum of 2".

2.2 INSTALLATION

C 2.21 Attach runner:

D 1. to concrete, with concrete stub nails, powder driven fasteners or expansion drives, 12" o.c.

2. to steel, with wire ties, bolts or welds, 12" o.c.

D 2.22 Attach metal base clip 16"

o.c. with concrete stub nails, powder driven fasteners or expansion drives.

2.23 Erect metal lath so that long dimension of sheet is vertical. Erect lath so that a minimum of 5/8" of plaster may be applied to the flat side of the lath.

2.24 Attach metal lath to ceiling runner with wire ties 8" o.c. Attach metal lath:

1. to floor runner with wire ties 8" o.c., or

2. to metal base clip with wire ties, or

3. by setting metal lath into a grouted metal base assembly.

2.25 Wire tie side lap of sheets at each end and at 12" o.c. along the lap.

E 2.26 Brace partition on rib side at mid-height for partition height 8'4" or under, or at third-points for partition heights over 8'4", until the

brown coat on the unbraced side has been applied and has set.

F, G 2.27 Lath at door frames by installing 3/4" channel stud at each jamb. Install stud:

1. at metal door frame, by wire-tying, bolting or welding to all jamb anchors:

2. at wood door buck, by either wire-tying to pairs of 8d nails driven into buck at 2' o.c., or by using 1" wood screws 2' o.c.

Insert metal lath as far as possible into the re-entrant space of a metal frame. Notch lath to pass around jamb anchors.

2.28 Place two 3/4" channel studs back-to-back at cased opening and partition end. Install partition cap or corner beads to protect exposed end of partition.

H 2.29 Cover back of outlet or switch box with metal lath.

SECTION THREE

SOLID PARTITIONS — CHANNEL STUDS

3.1 MATERIAL SELECTION

B   3.11 Metal lath shall be of the type and weight for the spacing of supports as set forth in the following table:

LATH TYPE	LATH WEIGHT (lbs./sq. yd.)	MAXIMUM SUPPORT SPACING
Diamond Mesh	2.5	16"
Diamond Mesh	3.4	16"
Flat Rib	2.75	16"
Flat Rib	3.4	24"

3.12 Channel studs shall be of the size for the height of the partition as set forth in the following table:

CHANNEL STUD	MAXIMUM PARTITION HEIGHT
¾" C.R.C.	16'
1½" C.R.C.	24'

3.13 Partition thickness shall be appropriate for the height, and length (distance between columns, walls or other vertical structural members) of the partition as set forth in the following table:

THICKNESS	MAXIMUM PARTITION HEIGHT	MAXIMUM[1] PARTITION LENGTH
1½"[2]	8'6"	No restriction on partitions under 10' high
2"	12'	24'
2¼"	14'	28'
2½"	16'	24'
2¾"	18'	27'
3"	20'	20'
3¼"	24'	24'

[1]for greater length, increase thickness by 20%.
[2]channel web shall be turned parallel with and adjacent to the lath.

3.14 Horizontal stiffeners shall be ¾" cold-rolled channels.

3.2 INSTALLATION

C   3.21 Attach runner:

D   1. to concrete, with concrete stub nails, powder driven fasteners or expansion drives, 12" o.c.;

2. to steel, with wire ties, bolts or welds, 12" o.c.

D   Attach clip to floor with concrete stub nail.

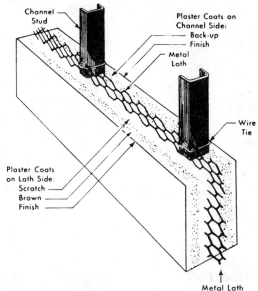

Metal Lath and Plaster Solid Partition with Channel Studs

J   3.22 Extend channel stud to a metal lath ceiling or to the construction above the ceiling. Where channel stud extends to ceiling:

1. use runner to hold stud in place, or

2. punch stud through ceiling lath and wire tie to a horizontal ¾" channel extending the full length of the partition.

Where channel stud extends to the construction above the ceiling use a runner to hold stud in place.

K   Attach channel stud to floor with runner, clip or by direct attachment.

Attach channel stud:

1. to runner, by setting stud into hole or prongs in runner or by wire tying stud to runner;

2. to clip, by setting stud into hole or prongs in clip or by wire-tying stud to clip.

Attach channel stud directly to the floor:

D   1. by bending lower end of stud and nailing with concrete stub nail, or

2. by setting stud into a hole cut in concrete floor.

Cut studs to proper length (so that they will not bow).

3.23 Splice channels by lapping, with flanges interlocked, at least 12". Wire tie each end of splice with at least two loops of No. 16 gage wire.

L   3.24 Install horizontal stiffeners 6' apart where required. For partition having unsupported height: from 0' to 20', no stiffeners are required; over 20', place stiffener so that back of channel (web) is against studs. Saddle tie stiffeners to each stud.

F, G   3.25 Lath at door frames by installing a channel stud at each jamb:

1. at metal door frame, by wire-tying, bolting or welding to all jamb anchors;

2. at wood door buck, by either wire-tying to pairs of 8d nails driven into wood buck at 2' o.c., or by using 1" wood screws 2' o.c.

Install jack studs 16" o.c. by attaching ¾" channel studs to top of door frame and bracing the channel side with a ⅛" x 1¼" flat bar not more than 6" above the frame and extending the brace at least one full stud space beyond the opening.

Insert metal lath as far as possible into the re-entrant space of a metal frame. Notch lath to pass around jamb anchors.

3.26 Place two channel studs back-to-back at cased opening and partition end. Install partition end cap or corner beads to protect exposed end of partition.

H   3.27 Cover back of outlet or switch box with metal lath.

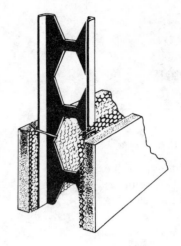

SECTION FOUR

HOLLOW PARTITIONS — PREFABRICATED STEEL STUDS

4.1 MATERIAL SELECTION

B 4.11 Metal lath shall be of the type and weight for the spacing of supports as set forth in the following table:

LATH TYPE	LATH WEIGHT (lbs./sq. yd.)	MAXIMUM SUPPORT SPACING
Diamond Mesh	2.5	13½"
Diamond Mesh	3.4	16"
Flat Rib	2.75	16"
Flat Rib	3.4	19"
⅜" Rib	3.4	24"
⅜" Rib	4.0	24"
Sheet	4.5	24"

4.12 Prefabricated steel studs shall be of the size and spacing for the height of the partition as set forth in the following table:

MAXIMUM PARTITION HEIGHTS

OVERALL[1] PARTITION THICKNESS	STUD SIZE	STUD SPACING[2] 24"	19"	16"	12"
3⅛" or 3⅜"	1⅝" or 1⅞"	7'	8'	9'	10'
3½"	2"	8'	9'	10'	14'
4"	2½"	9'	14'	15'	18'
4¾"	3¼"	13'	18'	21'	22'
5½"	4"	16'	20'	22'	25'
7½"	6"	20'	24'	26'	29'

[1]plaster thickness is ⅝" from finished face to face of lath (¾" from finished face to face of stud). For ⅜" rib lath, partition thickness is ½" greater than shown.

[2]for partition length exceeding 1½ times partition height, reduce allowable height by 20%.

4.13 Horizontal stiffeners shall be 3/4" cold-rolled channels.

4.2 INSTALLATION

C 4.21 Attach track:

D 1. to concrete, with concrete stub nails, powder driven fasteners, or expansion drives, 24" o.c.

2. to steel, with wire ties, bolts or welds, 24" o.c.

J 4.22 Extend prefabricated steel stud to a metal lath ceiling or to the construction above the ceiling. Where prefabricated steel stud extends to ceiling:

1. use track to hold stud in place, or

2. punch stud through ceiling lath and wire tie stud to a horizontal 3/4" channel extending the full length of the partition.

Where prefabricated steel stud extends to the construction above the ceiling:

1. use track to hold stud in place, or

2. use piece of stud bent at a right angle to hold stud in place.

Attach stud:

1. to track, by welding or by wire-tying a pair of stud shoes to the stud;

2. to pieces of bent stud, by interlocking the flanges and wire-tying each side.

Attach prefabricated steel stud to the floor with track, or by direct attachment. Attach stud:

1. to track, by welding, peening, or by wire-tying a pair of stud shoes to the stud;

2. to pronged, slotted, or snap-in track, by inserting stud into space provided.

D Attach prefabricated steel stud directly to the floor by bending end and nailing to floor with a concrete stub nail.

Cut studs to proper length (so that they will not bow).

4.23 Install horizontal stiffeners 4'6" apart. For partition having unsupported height: from 8' to 18', place stiffeners on alternate interior sides; over 18', place on each interior side. Saddle tie stiffeners to each stud.

F, G 4.24 Lath at door frames by installing prefabricated steel studs at each jamb. Install stud:

1. at metal door frame, by wire-tying, bolting or welding stud to all jamb anchors;

2. at wood door buck, by either wire-tying to pairs of 8d nails driven into buck at 2' o.c. or by using 1" wood screws 2' o.c.

For metal door frames install an additional steel stud. Install second stud:

1. by wire-tying or welding back-to-back to the first stud, or

2. by locating stud 2" away from first stud and providing suitable attachments between the two studs.

Install prefabricated steel studs as jack studs 16" o.c. over the head of door frame by:

1. attaching prefabricated steel stud to floor track located within the interior of frame. Reinforce jack studs with a 3/4" cold-rolled channel stiffener located not more than 6" above the frame and extending two full stud spaces beyond the opening, or

2. attach prefabricated steel stud to top of door frame and brace each interior face of the prefabricated steel stud with a 3/4" cold-rolled channel located not more than 6" above the frame and extending at least two full stud spaces beyond the opening.

Insert metal lath as far as possible into the re-entrant space of a metal frame. Notch lath to pass around jamb anchors.

SECTION FIVE

HOLLOW PARTITIONS — CHANNEL STUDS

5.1 MATERIAL SELECTION

5.11 Metal lath shall be of the type and weight for the spacing of supports as set forth in the following table:

LATH TYPE	WEIGHT (lbs./sq. yd.)	MAXIMUM SUPPORT SPACING
Diamond Mesh	2.5	13½"
Diamond Mesh	3.4	16"
Flat Rib	2.75	16"
Flat Rib	3.4	19"
⅜" Rib	3.4	24"
⅜" Rib	4.0	24"
Sheet	4.5	24"

5.12 Studs shall be a double row of 3/4" cold-rolled channels with separation and spacing for the height of the partition as set forth in the following table:

MAXIMUM PARTITION HEIGHTS

OVERALL¹ PARTITION THICKNESS	STUD² SEPARATION	STUD SPACING³ 24"	19"	16"	12"
3½"	½"	9'	12'	13'	14'
3¾"	¾"	12'	14'	15'	17'
4½"	1½"	14'	16'	17'	20'
5"	2"	16'	18'	20'	23'
5½"	2½"	18'	21'	23'	26'

¹Plaster thickness is ⅝" from finished face to face of lath (¾" from finished face to face of stud). For ⅜" rib lath, partition thickness is ½" greater than shown.

²Separation is the clear distance between rows of studs.

³For partition length exceeding 1½ times partition height, reduce allowable height by 20%.

5.13 Horizontal stiffeners shall be not less than 3/4" cold-rolled channels.

5.14 Cross braces shall be not less than 3/4" cold-rolled channels.

5.2 INSTALLATION

5.21 Attach runner:

D
1. to concrete, with concrete stub nails, powder driven fasteners, or expansion drives, 12" o.c.
2. to steel, with wire ties, bolts or welds, 12" o.c.

D Attach clip to floor with concrete stub nail.

J 5.22 Extend channel stud to a metal lath ceiling or to the construction above the ceiling. Where channel stud extends to ceiling:

1. use runner to hold stud in place, or
2. punch stud through ceiling lath and wire tie stud to a horizontal 3/4" channel extending the full length of the partition.

Where channel stud extends to the construction above the ceiling use a runner to hold stud in place.

Attach channel stud to floor with runner, clip or by direct attachment. Attach channel stud:

1. to runner, by setting stud into hole or prongs in runner or by wire-tying stud to runner;
2. to clip, by setting stud into hole or prongs in clip or by wire-tying stud to clip.

Attach channel stud directly to the floor:

D
1. by bending lower end of stud and nailing to floor with concrete stub nail, or
2. by setting stud into a hole cut in concrete floor.

Cut studs to proper length (so that they will not bow).

5.23 Splice channels by lapping, with flanges interlocked, at least 12". Wire-tie each end of splice with at least two loops of No. 16 gage wire.

5.24 Install horizontal stiffeners 4'6" apart. For partition having unsupported height: from 8' to 18', place stiffeners on alternate interior sides; over 18', place on each interior side (in pairs). Saddle tie stiffeners to each stud.

5.25 Install cross braces 4'6" apart along each pair of studs. If stiffeners are installed in pairs, cross braces may be placed 4' apart between stiffeners instead of between each pair of studs. Saddle tie cross braces to each stud, or when cross bracing between stiffeners, saddle tie each end of cross brace to a stiffener. If a prefabricated metal stud is used as a horizontal stiffener and it is tied to both sides of the partition, no additional cross braces are necessary.

F, G 5.26 Lath at door frames by installing two channel studs (four per opening) at each side of frame. Install stud:

1. at metal door frame, by wire-tying, bolting or welding stud to all jamb anchors;
2. at wood door buck, by either wire-tying to pairs of 8d nails driven into buck at 2' o.c., or by using 1" wood screws 2' o.c.

Install jack studs 16" o.c. by attaching channel studs to top of door frame and bracing interior side with 3/4" cold-rolled channel not more than 6" above the frame and extending the brace at least two full stud spaces beyond the opening.

Insert metal lath as far as possible into the re-entrant space of a metal door frame. Notch lath to pass around jamb anchors.

SECTION SIX

SOUND INSULATING DOUBLE PARTITIONS

6.1 MATERIAL SELECTION

B 6.11 Metal lath shall be of the type and weight for the spacing of supports as set forth in the following table:

LATH TYPE	LATH WEIGHT (lbs./sq. yd.)	MAXIMUM SUPPORT SPACING
Diamond Mesh	2.5	13½"
Diamond Mesh	3.4	16"
Flat Rib	2.75	16"
Flat Rib	3.4	19"
⅜" Rib	3.4	24"
⅜" Rib	4.0	24"
Sheet	4.5	24"

6.12 Studs for **unbraced** hollow partitions shall be of the type, size and spacing for the height of the partition as set forth in the following table:

MAXIMUM PARTITION HEIGHTS

STUD TYPE	STUD SIZE	STUD SPACING 24"	19"	16"	12"
Cold-Rolled Channel	¾"	6'	7'	8'	9'
Cold-Rolled Channel	1½"	8'	9'	10'	12'
Cold-Rolled Channel	2"	9'	10'	11'	13'
Prefab. Steel Stud	2"	8'	9'	10'	11'
Prefab. Steel Stud	2½"	10'	11'	12'	14'
Prefab. Steel Stud	3¼"	14'	16'	17'	20'

6.13 Studs for **braced** hollow partitions shall be of the type, size and spacing for the height of the partition as set forth in the following table:

MAXIMUM PARTITION HEIGHTS

STUD TYPE	STUD SIZE	STUD SPACING 24"	19"	16"	12"
Cold-Rolled Channel	¾"	10'	11'	12'	14'
Cold-Rolled Channel	1½"	13'	15'	16'	18'
Cold-Rolled Channel	2"	15'	17'	18'	21'
Prefab. Steel Stud	2"	13'	15'	16'	18'
Prefab. Steel Stud	2½"	16'	18'	20'	23'
Prefab. Steel Stud	3¼"	23'	26'	28'	32'

6.14 Horizontal stiffeners shall be ¾" cold-rolled channels.

M 6.2 INSTALLATION

C 6.21 Attach track:

D 1. to concrete, with concrete stub nails, powder driven fasteners, or expansion drives, 24" o.c.

2. to steel, with wire ties, bolts or welds, 24" o.c.

6.22 Erect studs in two rows so that each row is in direct contact with only one partition face.

J 6.23 Extend prefabricated steel stud to a metal lath ceiling or to the construction above the ceiling. Where prefabricated steel stud extends to ceiling:

1. use track to hold stud in place, or

2. punch stud through ceiling lath and wire tie stud in place.

Where prefabricated steel stud extends to the construction above the ceiling:

1. use track to hold stud in place, or

2. use piece of stud bent at a right angle to hold stud in place.

Attach stud:

1. to track, by welding or by wire-tying a pair of stud shoes to the stud;

2. to pieces of bent stud, by interlocking the flanges and wire-tying each side.

K Attach prefabricated steel stud to the floor with track, or by direct attachment. Attach stud:

1. to track, by welding, peening, or by wire-tying a pair of stud shoes to the stud;

2. to pronged, slotted, or snap-in track, by inserting stud into space provided.

D Attach prefabricated steel stud directly to the floor by bending end and nailing to floor with a concrete stub nail.

Cut studs to proper length (so that they will not bow.

J 6.24 Extend channel stud to a metal lath ceiling or to the construction above the ceiling. Where channel stud extends to ceiling:

1. use runner to hold stud in place, or

2. punch stud through ceiling lath and wire tie stud to a horizontal ¾" channel extending the full length of the partition.

Where channel stud extends to the construction above the ceiling use a runner to hold stud in place.

K Attach channel stud to floor with runner, clip or by direct attachment.

Attach channel stud:

1. to runner, by setting stud into hole or prongs in runner or by wire-tying stud to runner;

2. to clip, by setting stud into hole or prongs in clips or by wire-tying stud to clip.

Attach channel stud directly to the floor:

D 1. by bending lower end of stud and nailing to floor with concrete stub nail, or

2. by setting stud into a hole cut in concrete floor.

Cut studs to proper length (so that they will not bow).

6.25 Splice channels or prefabricated steel studs by lapping, with flanges interlocked, at least 12". Wire tie each end of splice with at least two loops of No. 16 gage wire.

6.26 Install horizontal stiffeners 4' apart on each inside face of the partition. Saddle tie horizontal stiffeners to each stud.

6.27 Saddle tie cross braces between mid-height stiffeners every 3'. Do not cross brace except at mid-height.

F, G 6.28 Lath at door frames by installing two channel studs at each side of the frame. Install stud:

1. at metal door frame, by wire-tying, bolting or welding stud to all jamb anchors;

2. at wood door buck, by either wire-tying to pairs of 8d nails driven into buck at 2' o.c., or by using 1" wood screws 2' o.c.

Install jack studs 16" o.c. by attaching channel stud to top of door frame and bracing interior side with ¾" cold-rolled channel not more than 6" above the frame and extending the brace at least two full stud spaces beyond the opening.

Insert metal lath as far as possible into the re-entrant space of a metal door frame. Notch lath to pass around jamb anchors.

SECTION SEVEN

VERTICAL FURRING

7.1 MATERIAL SELECTION

7.11 Metal lath shall be of the type and weight for the spacing of supports as set forth in the following table:

LATH TYPE	LATH WEIGHT (lbs./sq. yd.)	MAXIMUM SUPPORT SPACING
Diamond Mesh	2.5	13½"
Diamond Mesh	3.4	16"
Flat Rib	2.75	16"
Flat Rib	3.4	19"
⅜" Rib	3.4	24"
⅜" Rib	4.0	24"
Sheet	4.5	24"

7.12 Studs for **unbraced furring** shall be of the type, size and spacing for the height of the assembly as set forth in the following table:

MAXIMUM FURRING HEIGHTS

FURRING TYPE	FURRING SIZE	24"	19"	16"	12"
Cold-Rolled Channel	¾"	6'	7'	8'	9'
Cold-Rolled Channel	1½"	8'	9'	10'	12'
Cold-Rolled Channel	2"	9'	10'	11'	13'
Prefab. Steel Stud	2"	8'	9'	10'	11'
Prefab. Steel Stud	2½"	10'	11'	12'	14'
Prefab. Steel Stud	3¼"	14'	16'	17'	20'

7.13 Studs for **braced furring** shall be of the type, size and spacing for the **distance between braces** (furring spans) as set forth in the following table:

MAXIMUM FURRING SPANS

FURRING TYPE	FURRING SIZE	24"	19"	16"	12"
Cold-Rolled Channel	¾"	5'	5'	6'	7'
Cold-Rolled Channel	1½"	6'	7'	8'	9'
Cold-Rolled Channel	2"	7'	8'	9'	10'
Prefab. Steel Stud	2"	6'	7'	8'	9'
Prefab. Steel Stud	2½"	8'	9'	10'	11'
Prefab. Steel Stud	3¼"	11'	13'	14'	16'

7.14 Horizontal stiffeners shall be not less than 3/4" cold-rolled channels.

7.15 Braces shall be not less than 3/4" cold-rolled channels.

7.2 INSTALLATION

C 7.21 Attach track:

D 1. to concrete, with concrete stub nails, powder driven fasteners, or expansion drives, 24" o.c.

2. to steel, with wire ties, bolts or welds, 24" o.c.

J 7.22 Extend prefabricated steel stud to a metal lath ceiling or to the construction above the ceiling. Where prefabricated steel stud extends to ceiling:

1. use track to hold stud in place, or

2. punch stud through ceiling lath and wire tie stud in place.

Where prefabricated steel stud extends to the construction above the ceiling:

1. use track to hold stud in place, or

2. use piece of stud bent at a right angle to hold stud in place.

Attach stud:

1. to track, by welding or by wire-tying a pair of stud shoes to the stud;

2. to pieces of bent stud, by interlocking the flanges and wire-tying each side.

Attach prefabricated steel stud to the floor with track, or by direct attachment. Attach stud:

1. to track, by welding, peening, or by wire-tying a pair of stud shoes to the stud;

2. to pronged, slotted, or snap-in track, by inserting stud into space provided.

D Attach prefabricated steel stud directly to the floor by bending end and nailing to floor with a concrete stub nail.

Cut studs to proper length (so that they will not bow).

J 7.23 Extend channel stud to a metal lath ceiling or to the construction above the ceiling. Where channel stud extends to ceiling:

1. use runner to hold stud in place, or

2. punch stud through ceiling lath and wire tie stud to a horizontal 3/4" channel extending the full length of the partition.

Where channel stud extends to the construction above the ceiling use a runner to hold stud in place. Attach channel stud to floor with runner, clip or by direct attachment.

Attach channel stud:

1. to runner, by setting stud into hole or prongs in runner or by wire-tying stud to runner;

2. to clip, by setting stud into hole or prongs in clip or by wire-tying stud to clip.

Attach channel stud directly to the floor:

D 1. by bending lower end of stud and nailing to floor with concrete stub nail, or

2. by setting stud into a hole cut in concrete floor.

Cut studs to proper length (so that they will not bow).

7.24 Install horizontal stiffeners no more than 4'6" apart and at least 1/4" away from the wall. Saddle tie stiffeners to each stud and to braces or to anchors with three loops of No. 18 gage wire.

7.25 Install braces 2' apart (horizontal spacing) between horizontal stiffeners and the wall or between each stud and the wall (vertical spacing established in 7.13). Wire tie braces to anchors, inserted or driven into wall, with three loops of No. 18 gage wire, or insert or drive braces into wall.

G, 7.26 Lath at door frames by in-
N stalling casing bead at door frame to separate the plaster surface from frame.

SECTION EIGHT

METAL LATH ON WOOD STUDS

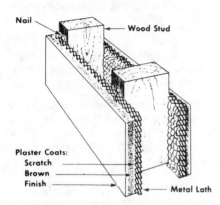

Nail

Wood Stud

Plaster Coats:
Scratch
Brown
Finish

Metal Lath

8.1 MATERIAL SELECTION

B   8.11 Metal lath shall be of the type and weight for the spacing of supports as set forth in the following table:

LATH TYPE	LATH WEIGHT (lbs./sq. yd.)	MAXIMUM SUPPORT SPACING
Diamond Mesh	2.5	16"
Diamond Mesh	3.4	16"
Flat Rib	2.75	16"
Flat Rib	3.4	19"
3/8" Rib	3.4	24"
3/8" Rib	4.0	24"
Sheet	4.5	24"

8.12 Attachments for metal lath shall be 4d common nails, 1" roofing nails with 7/16" heads, or 1" No. 14 gage wire staples, or power-driven 7/8" 16 gage wire staples with 3/4" crown.

8.2 INSTALLATION

8.21 Attach metal lath to each wood stud with nails or staples 6" o.c., driven at least 3/4" into the stud.

G   8.22 Install attachments as follows:

1. common nails, drive in half way and bend over to engage at least three strands of diamond mesh lath or bend over the ribs of flat and 3/8" rib laths;

2. roofing nails, drive in so as to engage at least two strands of diamond mesh lath or drive thru the ribs of flat and 3/8" rib laths;

3. staples, drive in so as to engage at least two strands of diamond mesh lath or drive over the ribs of flat and 3/8" rib laths.

SECTION NINE

METAL LATH ON EXTERIORS (STUCCO)

9.1 MATERIAL SELECTION

B   9.11 Metal lath shall be of the appropriate type and weight for the assembly type (vertical or horizontal) and spacing of supports as set forth in the following table:

VERTICAL ASSEMBLIES (WALLS)

LATH TYPE	LATH WGT. (lbs./sq. yd.)	MAXIMUM SUPPORT SPACING Wood Studs	Solid[1] Surfaces	Metal Studs
Diamond Mesh	3.4	16"	16"[2]	16"
Flat Rib	2.75	16"	16"[2]	16"
Flat Rib	3.4	19"	19"[2]	19"
3/8" Rib	3.4	24"	24"	24"
3/8" Rib	4.0	24"	24"	24"
Sheet	4.5	24"	24"	24"
3/4" Rib	.60 lbs./sq. ft.	......	24"	......

HORIZONTAL ASSEMBLIES (SOFFITS)

LATH TYPE	LATH WEIGHT (lbs./sq. yd.)	MAXIMUM SUPPORT SPACING Wood/ Concrete Supports	Metal Supports
Diamond Mesh	3.4	16"[3]	16"[3]
Flat Rib	2.75	16"[3]	16"[3]
Flat Rib	3.4	19"[3]	19"[3]
3/8" Rib	3.4	24"	24"
3/8" Rib	4.0	24"	24"
Sheet	4.5	24"	24"

[1]Attach metal lath to solid surfaces with nails or staples spaced 6" apart in a line running across the short dimension of the sheet. The maximum support spacing as set forth in this table is the distance between lines of supporting nails.

[2]Special furring devices are required or a self-furring lath shall be used.

[3]Furring required only if support surface is more than 1⅝" wide.

9.12 Attachments for metal lath shall be appropriate for the type assembly and construction as follows:

Vertical Assemblies (Walls)

1. Steel supports . . . attachments for securing metal lath shall be at least 18 gage wire.

2. Wood supports . . . attachments for securing metal lath or expanded stucco mesh shall be galvanized nails which provide for a penetration into the supports of at least 1" or power-driven galvanized staples which provide at least 3/4" penetration.

Horizontal Assemblies (Soffits)

1. Concrete . . . attachments for securing metal lath shall be at least 14 gage wire hairpins, hooks, or loops. They shall be provided with a loop or other deformation to positively enter the concrete or shall be secured to the reinforcing steel.

2. Steel supports . . . attachment for securing metal lath shall be at least one loop of 16 gage wire or two loops of 18 gage wire.

3. Wood supports . . . attachments for securing metal lath shall be not less than No. 11 gage barbed roofing nails which provide for a penetration into the supports of at least 1⅜" or power-driven galvanized staples which provide at least 3/4" penetration into the supports.

9.13 Building paper, where applied separately from lath, shall be waterproofed paper, meeting Federal Specifications UUB-790 and shall weigh not less than 14 lbs. per 108 square feet.

9.14 Paper backing, supplied as an integral part of the lath, shall be waterproofed paper meeting Federal Specifications UUB-790.

9.2 INSTALLATION On Wood Frame Construction

9.21 Install furring devices (except for self-furring lath) to provide clearance for "plaster-keys" to form.

9.22 Stagger end laps. Start metal lath one stud away from corner and bend around corner, or use cornerite. Carry lath over foundation at least 2".

9.3 INSTALLATION On Sheathed Construction or Overcoating

9.31 Fasten building paper, in horizontal layers, to the sheathing or surface being overcoated, with each strip lapping the strip below at least 3". (This is not necessary where lath with integral waterproof paper backing is used.)

9.32 Install furring devices (except for self-furring lath) to provide clearance for "plaster-keys" to form.

9.33 Stagger end laps. Return lath around corners or use cornerite. Carry lath over foundation at least 2".

9.4 INSTALLATION On Steel Frame Construction

9.41 Install furring devices (except for self-furring lath) to provide clearance for "plaster-keys" to form. (Not necessary where bearing area of support is less than 1⅝" wide.)

9.42 Stagger end laps. Return lath around corners or use cornerite. Carry lath over foundation at least 2".

SECTION TEN

CONTROL JOINTS

10.1 MATERIAL SELECTION

10.11 Control joints shall be pairs of casing beads mounted back-to-back or specially designed metal plastering accessories (expansion joints).

10.12 Material shall be galvanized steel or zinc alloy.

10.2 INSTALLATION

10.21 Install control joints with attachment only to the edges of abutting sheets of lath, so that lath is not continuous or tied across the joint.

10.22 Where control joints are placed parallel to framing members, install joints so that none is more than 4" away from a framing member.

10.23 In gypsum plaster ceilings, install control joints to create panels no larger than 2,400 sq. ft., with no dimension exceeding 60 ft.

10.24 For exterior portland cement plaster, install control joints to create panels no larger than 144 sq. ft. with no dimension exceeding 18 ft. or a length to width ratio of 2½ to 1.

10.25 Install control joints at all locations where panel sizes or dimensions change. Joints shall extend the full width or height of the plaster membrane.

DESIGN NOTES

(Continued from page 116)

Door frames should not be supported by the vertical furring.

Flashing should be of the non-corrosive type and should be located as follows:

1. at the tops of all openings
2. along the sides of all openings
3. where trim projects under copings and brick sills
4. at intersections of walls and roofs
5. under built-in gutters
6. around all roof openings
7. and at all other places where flashing can be used to prevent water from getting behind the stucco.

Flashing should be installed prior to lathing.

To minimize stucco cracks caused by structural deflections, the following provisions should be required in the appropriate specifications (i.e. carpentry, etc.);

1. In wood frame construction, require corners to be diagonally braced, studs to be doubled at corners and around openings, and trusses to be provided over openings more than 4' wide.
2. In steel frame construction, require studs and furring to be securely fastened to the structural frame.

Metal lath should be used as reinforcement over all surfaces (old brick, block, sheathing, etc.) that do not provide a satisfactory bond for stucco. Special consideration must be given to attachment of the lath to these surfaces. Each type of construction must be designed according to its own characteristics.

Where intermediate supports are desired to provide for appropriate furring or lath spans, the following should be added to the specifications:

"___.1___ Intermediate supports shall be of the appropriate size and spacing (center to center distance between intermediate support and adjacent supports) for the distance between hangers (spans) as set forth in the following table:

MAXIMUM ALLOWABLE SPANS

INTERMEDIATE SUPPORT	SPACINGS				
	48"	36"	24"	16"	12"
⅜" Pencil Rod	.....	.....	.....	2'	2'6"
¾" C.R.C.	.....	.....	2'	3'	4'
1½" C.R.C.	.....	2'	3'	4'	4'
2" C.R.C.	3'	4'	5'	.....	.....
	4'6"	5'	7'		

"___.1___ Hangers for intermediate supports shall be of the type and size for the area of the ceiling supported as set forth in the following table: (insert table 12.14)

"___.2___ Wire Hanger Installation for Intermediate Supports (insert hanger installation specification as set forth in 12.21)"

R Where greater-than-ordinary vibrations are anticipated the following additional supports for standard nailing should be specified:

"10.22 Supplementary Tying for Contact Ceiling on Wood Joists

"Drive 16d common nails in a horizontal position through joists at least 2" above the bottom edge of joists. Space nails 24" or 27" o.c. along the sides of alternate joists, (the spacing depending on the width of metal lath sheet used), and so placed that when the sheets of metal lath are in position the nails will be not more than 3" back from the edge of each sheet. In place of 16d common nails driven through joists, two 8d common nails may be used, one on each side of joist, driven diagonally downward to a penetration of at least 1½", from a point not less than 2" above the bottom edge of joists.

"Attach metal lath to each nail (or pair of nails) with a loop of No. 18 gage wire. Bend the wire into a "U"-shape. Double-over each leg to form loops and push them through the lath one on each side of the joist, so as to catch the projecting point or head of the nail. Pull the free ends of the wire to tighten the "U" against the lath, and twist the two free ends together beneath the lath."

SECTION ELEVEN

CONTACT CEILINGS

11.1 MATERIAL SELECTION

B 11.11 Metal lath shall be of the
Q type and weight for the spacing of supports as set forth in the following table:

LATH TYPE	LATH WEIGHT (lbs./sq. yd.)	MAXIMUM SUPPORT SPACING
Diamond Mesh	2.5	12″
Diamond Mesh	3.4	16″
Flat Rib	2.75	16″
Flat Rib	3.4	19″
⅜″ Rib	3.4	24″
⅜″ Rib	4.0	24″
Sheet	4.5	24″

11.12 Attachments for metal lath shall be appropriate for the type of construction as follows:

Concrete . . . attachments for securing metal lath to the underside of concrete joists shall be wire hairpins, hooks or loops. They shall be provided with a loop or other deformation to positively enter the concrete, or shall be secured to the reinforcing steel.

Attachments shall be not less than No. 14 gage wire when twisted as a tie, and not less than No. 10 gage wire when struck over. They shall be spaced at not more than 5″ o.c. along the joists.

Steel . . . attachments for securing metal lath to the underside of steel joists shall be not less than one loop of No. 16 gage wire or two loops of No. 18 gage wire.

Wood . . . attachments of diamond mesh and flat rib metal laths to the underside of wood joists shall be not less than No. 11 gage barbed roofing nails with 7/16″ heads, 1½″ long, or power-driven staples 7/8″ long with 3/4″ crown. For 3/8″ rib and sheet laths, nails shall be at least 2″ long and power-driven staples at least 1¼″ long.

11.2 INSTALLATION

11.21 Attach metal lath:

1. to concrete joist by twisting wire hairpins or loops under the lath, or by striking wire over (10 gage minimum);

2. to steel joist, by twisting loops under lath;

R 3. to wood joist, by driving roofing nail or power-driven staple home.

SECTION TWELVE

FURRED CEILINGS

12.1 MATERIAL SELECTION

B, 12.11 Metal lath shall be of the
Q type and weight for the spacing of supports (furring) as set forth in the following table:

LATH TYPE	LATH WEIGHT (lbs./sq. yd.)	MAXIMUM SUPPORT SPACING
Diamond Mesh	2.5	12″
Diamond Mesh	3.4	16″
Flat Rib	2.75	16″
Flat Rib	3.4	19″
⅜″ Rib	3.4	24″
⅜″ Rib	4.0	24″
Sheet	4.5	24″

12.12 Furring shall be of the type and spacing for the span between supports (joists) as set forth in the following table:

	MAXIMUM ALLOWABLE SPANS SPACINGS			
FURRING	24″	19″	16″	12″
⅜″ Pencil Rod		2′	2′	2′6″
¾″ C.R.C.	3′0″	3′6″	4′	4′

12.13 Attachments for furring shall be appropriate for the type of construction as follows:

Concrete . . . attachments for securing furring to the underside of concrete joists shall consist of wire hairpins, hooks or loops. They shall be provided with a loop or other deformation to positively enter the concrete or shall be secured to the reinforcing steel. Attachments shall be not less than No. 14 gage wire.

Steel . . . attachments for securing furring to the underside of steel joists shall be not less than one loop of No. 16 gage wire or two loops of No. 18 gage wire.

Wood . . . attachments for securing furring to the underside of wood joists shall be not less than No. 18 gage wire.

12.2 INSTALLATION

12.21 Attach furring:

1. to concrete joists, by twisting wire hairpins or loops under the furring;

2. to steel joists, by saddle-tying;

3. to wood joists, by saddle-tying to:

   a. a 16d common nail driven horizontally through the joist at least 2″ above the bottom, or

   b. a pair of 8d common nails (one on each side) driven diagonally downward to a penetration of at least 1½″ from a point at least 2″ above the bottom.

SECTION THIRTEEN

SUSPENDED CEILINGS

13.1 MATERIAL SELECTION

13.11 Metal lath shall be of the type and weight for the spacing of supports (cross-furring) as set forth in the following table:

LATH TYPE	LATH WEIGHT (lbs./sq. yd.)	MAXIMUM SUPPORT SPACING
Diamond Mesh	2.5	12"
Diamond Mesh	3.4	16"
Flat Rib	2.75	16"
Flat Rib	3.4	19"
3/8" Rib	3.4	24"
3/8" Rib	4.0	24"
Sheet	4.5	24"

13.12 Main runner shall be of the size and spacing for the distance between hangers (runner spans) as set forth in the following table:

MAXIMUM ALLOWABLE RUNNER SPANS

MAIN RUNNERS	48"	42"	36"	30"	24"
3/4" C.R.C.			2'	2'6"	2'6"
1½" C.R.C.		3'	3'6"	4'	4'6"	5'
2" C.R.C.	4'6"	5'	5'	6'	7'

SPACINGS

13.13 Cross-furring shall be of the type and spacing for the distance between runners (furring spans) as set forth in the following table:

MAXIMUM ALLOWABLE FURRING SPANS

CROSS-FURRING	24"	19"	16"	12"
3/8" Pencil Rod	....	2'	2'	2'6"
3/4" C.R.C.	3'0"	3'0"	3'6"	4'

SPACINGS

13.14 Cross-furring shall be saddle-tied to main runners with not less than No. 16 gage wire.

13.15 Hangers shall be of the type and size for the area of ceiling supported as set forth in the following table.

TYPE HANGER	MAXIMUM AREA SUPPORTED (sq. ft. per hanger)
3/16" Pencil Rod	25
1" x 3/16" Flat	25
1¼" x 1/8" Flat	25
12 Gage Wire	8
10 Gage Wire	12
9 Gage Wire	12.5
8 Gage Wire	16
7 Gage Wire	17.5
6 Gage Wire	25

13.2 INSTALLATION

13.21 Wire Hanger Installation

Attach wire hanger to the construction above the desired ceiling. For each of the installations described below, execute the attachment in a manner which will insure the development of the full hanger strength.

Attach wire hanger to concrete by:

1. securing the wire hanger to the steel reinforcement, or

2. by securing the wire hanger to special inserts, or

3. by looping end and embedding it in the concrete. (Not recommended for use with lightweight aggregate concrete.)

Attach wire hanger to steel by:

1. wrapping the wire hanger around or through the steel member, or

2. by bolting the wire hanger to the steel member, or

3. by clipping the wire hanger to the steel member, or

4. by tying the wire hanger to a drop clip.

Attach wire hanger to wood by:

1. inserting it through hole drilled 3" or more above bottom of wood joists, and twisting end three times around itself, or

2. by tying top of hanger to a 30d nail, driven to a penetration of 3" or more from a point 5" or more from the bottom of the joist, or

3. by stapling top of hanger to joist with four 1½" No. 9 gage wire staples (three near upper end and fourth to hold loose end.)

Attach wire hanger to the suspended member by saddle-tying the lower end of the wire hanger to the member in a manner which will restrain twisting and turning of the member and which will insure development of the full hanger strength.

13.22 Strap Hanger Installation

Attach strap hanger to the construction above the desired ceiling. For each of the installations described below, execute the attachment in a manner which will insure the development of the full hanger strength.

Attach strap hanger to concrete by:

1. bending the strap around the steel reinforcement, or

2. by bending 3½" of the end and embedding the bend 2¼" or more within the concrete, or

3. by bolting to special inserts.

Attach strap hanger to steel by:

1. bolting it to the steel, or

2. by wrapping the hanger around or through the steel member and bolting or tying the end back to the strap, or

3. by bolting to special clips.

Attach strap hanger to wood by spiking through strap into joist or by bolting through holes in joist and strap.

Attach strap hanger to suspended member by bolting through holes punched in the strap 3/8" or more from the end. Bolts for attaching strap hangers shall have at least a 3/8" diameter.

13.23 Locate a hanger within 6" of ends of main runners.

13.24 Locate a main runner within 6" of walls to support ends of cross-furring.

13.25 Permit no part of the suspension grillage (main runners or cross furring) to come into contact with abutting walls or load bearing partitions.

13.26 Where main runners are suspended from joists, and where spacings will permit, run main runners tranverse to joists.

SECTION FOURTEEN

BEAM BOXES

14.1 MATERIAL SELECTION

14.11 Metal lath shall be of the type and weight for the spacing of supports as set forth in the following table:

LATH TYPE	LATH WEIGHT (lbs./sq. yd.)	MAXIMUM SUPPORT SPACING
Diamond Mesh	3.4	13½"
Flat Rib	2.75	12"
Flat Rib	3.4	19"
⅜" Rib	3.4	24"
⅜" Rib	4.0	24"
Sheet	4.5	24"

14.12 Beam box brackets shall be formed from 3/4" cold-rolled channels. If the beam box brackets support the lath directly (i.e. without furring), they shall be spaced in accordance with the type and weight of lath as set forth in the table above. If the beam box brackets support furring, they shall be spaced in accordance with the furring type as set forth in the following table:

FURRING TYPE	BRACKET SPACING
¼" Pencil Rod	19"
¾" C.R.C.	36"

14.13 Stiffeners for the corners of beam box shall be 3/4" cold-rolled channels.

14.14 Attachments shall be not less than the following:

1. for tying brackets to structural members or ceiling grillage, 14 gage wire saddle tie or six loops of 18 gage wire;
2. for tying stiffeners to brackets, one loop of 18 gage wire;
3. for tying furring to brackets, one loop of 16 gage wire or two loops of 18 gage wire;
4. for tying lath to supports, one loop of 18 gage wire.

14.15 Supplementary hangers for supporting ends of ceiling grillage which support beam box brackets shall be of the appropriate size for the plaster area supported by the beam box as set forth in the following table:

HANGER SIZE (WIRE)	AREA TO BE SUPPORTED (i.e., one-half the total area supported by one beam box bracket.) — (in sq. ft.)
12 gage	8
10 gage	12
9 gage	12.5
8 gage	16
7 gage	17.5
6 gage	25

14.2 INSTALLATION

14.21 Install supplementary hangers within 6" of where the beam box bracket is attached to the ceiling grillage. Attach hanger to the member supporting the beam box bracket in the same manner as the hangers on the rest of the ceiling grillage are attached.

SECTION FIFTEEN

COLUMN FURRING

15.1 MATERIAL SELECTION

B   15.11 Metal lath shall be 3.4 lb. per square yard diamond mesh for columns made of rolled structural steel shapes (ie. "WF" or "I").

15.12 Furring shall be:

1. integral furring (i.e. self-furring or rib lath), or
2. 3/4" cold-rolled channels.

15.2 INSTALLATION

15.21 For columns made from rolled structural steel shapes:

1. erect self-furring metal lath so that it neatly fits the column and wire tie at laps 6" o.c.; **or**
2. install cold-rolled channels as furring brackets around column, and attach metal lath to furring brackets and wire tie laps at 6" o.c.; or
3. install vertical furring (in accordance with Section Seven) and attach metal lath to furring.

15.22 For columns made of steel pipes, erect rib metal lath with long dimension of sheet vertical. Bend metal lath around column, with ribs against the column. Attach metal lath to column with loops (around the column) of 18 gage wire at 16" o.c.

SECTION SIXTEEN

METAL LATH REINFORCEMENT FOR PLASTER BASES OTHER THAN METAL LATH

16.1 MATERIAL SELECTION

16.11 Corners . . . the plaster at all internal vertical or horizontal corners shall be reinforced with cornerite not less than 4" wide (2" on each surface). Plaster at all external corners shall be reinforced with galvanized, expanded wing or perforated corner bead.

16.12 Lath Joints . . . the plaster over all lath joints that extend parallel with the supports for more than 16" shall be reinforced with stripite not less than 4" wide.

16.13 Openings . . . the plaster around all openings of more than two square feet shall be reinforced with strips of metal lath not less than 6" by 12". Metal lath shall be located at the corners of the openings.

16.14 Chases . . . the plaster over all chases or similar openings shall be reinforced by covering the chase, etc., with metal lath.

16.15 Solid Surfaces . . . the plaster over all solid surfaces which do not provide adequate mechanical or chemical bond (i.e. portland cement over gypsum or any plaster over painted surfaces) shall be reinforced by covering the surface with self-furring metal lath.

16.16 Door Frames . . . the plaster at door frames that are continuous from floor to ceiling, in masonry partitions, shall be reinforced by covering the frame from the top of the opening to the ceiling with a strip of metal lath at least 12" wide.

16.17 Electrical Outlet . . . the plaster over the back of an electrical outlet box shall be reinforced by covering the back of the box with metal lath.

16.2 INSTALLATION

16.21 Install metal lath as follows:

1. where metal lath is applied over another plaster base (reinforcement of opening corners, over door frames, etc.), attach metal lath lightly to material being reinforced.
2. where metal lath is applied over open spaces (chases, etc.), firmly attach metal lath on 6" centers along each edge support with an attachment suitable to the support.

DESIGN ASSUMPTIONS AND GENERAL NOTES

001 Selection of metal lath is usually determined by the: assembly orientation (vertical or horizontal), support spacings, and bearing surface of the supports. A guide to selection is shown in the table* in each assembly section. However, when the overall assembly size reaches a point where the rigidity is significantly reduced, the allowable support spacings should be reduced accordingly.

*If lath is used, other than that produced by members of the MLA, caiculations should be made independently of tables, graphs and charts given in this publication.

002 The physical properties of studs, as given by the manufacturer, are used for design purposes. Several types of studs are not symmetrical in the plane of loading and require consideration of shear center phenomena. Normal structural design procedures are usually adequate. Design analysis used in conjunction with this specification has been made on minimum properties and is therefore acceptable for use with any Metal Lath Association member's products.

003 The physical properties of channels produced by members of the Metal Lath Association are given in the following table*. Channels are unsymmetrical about their minor axis and require consideration of shear center phenomena. When channels are spliced, the entire length is considered to act as a continuous member. Fiber stress is usually limited to 18,000 psi. When consideration of warping stress is included, local overstress of 20% is allowed.

ELEMENTS OF COLD-ROLLED CHANNELS(a)

Depth	Flange Width	U.S. Standard Gage	Weight per 1000 ft.	S (Inch³)	I (Inch⁴)
¾"	7/16"	16	300 lbs.	.0200	.0075
1½"	7/16"	16	475 lbs.	.0538	.0404
2"	19/32"	16	590 lbs.	.1005	.1005

NOTE: (a) Properties S and I are computed about the strongest axis of the channel.

*If channels are used, other than those produced by members of the MLA, calculations should be made independently of tables, graphs and charts given in this publication.

004 Recommended wire hanger gages are shown in tables in the appropriate assembly sections. These recommendations are based on a ceiling load of 10 lbs. per square foot, and the ultimate strength of the wire hanger and the grillage members. For loadings exceeding these recommendations, attention should be given to local failure of the grillage member. Vibra-

tional loads have not been considered, and if they are anticipated, appropriate design modifications are required.

Gage	Cross-Sectional Area in Sq. Inch	Max. Area Supported Per Hanger (a)
12 U.S.S.	.0087	8 sq. ft.
10 U.S.S.	.0143	12 sq. ft.
9 U.S.S.	.0173	12.5 sq. ft.
8 U.S.S.	.0206	16 sq. ft.
7 U.S.S.	.0246	17.5 sq. ft.
6 U.S.S.	.0290	25 sq. ft.
5 U.S.S.	.0337	25 sq. ft.

NOTE: (a) Multiply individual figures by 10 lbs. per sq. ft. (average weight of ceiling) to determine the approximate load to be supported by each hanger. Allowable areas are based on spacings suitable to ultimate strength of wire hangers and grillage members.

005 Fire resistance ratings are given to assemblies by building codes and are based on fire endurance tests. A listing of test results for metal lath and plaster assemblies are set forth in ML/SFA technical literature, (available free of charge: write to ML/SFA, 221 N. La Salle Street, Chicago, Ill. 60601.)

006 Variations in weights and thicknesses of products fabricated from cold-rolled carbon steel sheets shall not exceed permissible variations given in the Standard Specification for General Requirements for Carbon and High-strength Low-alloy steel, Hot-rolled Strip, Hot-rolled sheets, and Cold-rolled Sheets, ASTM A568.

007 Plaster load is assumed to be ten pounds per square foot; however, three coat portland cement plaster is heavier and its use requires appropriate design modification.

008 Deflection of a plastered assembly is usually limited to 1/360 of the smallest span, assuming that the plaster load produces the deflection. Spans are also limited to prevent a panel from losing its plaster when the plasterer trowels the adjacent panel.

009 Normally, plaster systems are designed to be structurally independent of the main supporting structure. (i.e. whenever possible, prevent structural loads from being transferred to the plaster assembly, especially reversing or vibrational loads.)

010 Corrosion is not a normal design criterion. However, when certain conditions exist, an undesirable galvanic corrosion may take place. These special corrosive conditions require special design.

Designation: C 841 – 87

Standard Specification for

INSTALLATION OF INTERIOR LATHING AND FURRING[1]

This standard is issued under the fixed designation C 841; the number immediately following the designation indicates the year of original adoption or, in the case of revision, the year of last revision. A number in parentheses indicates the year of last reapproval. A superscript epsilon (ε) indicates an editorial change since the last revision or reapproval.

This specification has been approved for use by agencies of the Department of Defense and for listing in the DoD Index of Specifications and Standards.

1. Scope*

1.1 This specification covers the minimum requirements for interior lathing and furring for full thick gypsum plastering. Other materials may be used provided that their physical characteristics and durability under conditions of usage are at least equal in performance to those described.

NOTE 1—To secure desirable results, this specification should be coordinated with the provisions of Specification C 842.

1.2 Where a specific degree of fire resistance is required for plastered assemblies and constructions, details of construction shall be in strict compliance with official reports of fire tests conducted in recognized testing laboratories in accordance with Methods E 119 (see Annex A1.2).

1.3 Where a specific degree of sound control is required for plastered assemblies and constructions, details of construction shall be in compliance with official reports of tests conducted in recognized testing laboratories in accordance with the applicable sound tests of Method E 90, Test Method C 423, or Method E 492 (see A1.3).

1.4 General information regarding matters of a contractual nature concerning lathing and furring is contained in Annex A1. Technical information relating to lathing and furring materials is provided in Annex A2. Erection data for door frames installed in lath and plaster hollow partitions using prefabricated steel studs or channel studs is provided in Annex A3.

2. Referenced Documents

2.1 *ASTM Standards:*
A 641 Specification for Zinc-Coated (Galvanized) Carbon Steel Wire[2]

C 11 Definitions of Terms Relating to Gypsum and Related Building Materials and Systems[3]
C 37 Specification for Gypsum Lath[3]
C 423 Test Method for Sound Absorption and Sound Absorption Coefficients by the Reverberation Room Method[4]
C 473 Methods for Physical Testing of Gypsum Board Products and Gypsum Lath[3]
C 842 Specification for Application of Interior Gypsum Plaster[3]
D 3678 Specification for Rigid Poly (Vinyl Chloride) (PVC) Interior-Profile Extrusions[5]
E 84 Test Method for Surface Burning Characteristics of Building Materials[6]
E 90 Method for Laboratory Measurement of Airborne-Sound Transmission Loss of Building Partitions[4]
E 119 Methods for Fire Tests of Building Construction and Materials[6]
E 492 Method of Laboratory Measurement of Impact Sound Transmission Through Floor-Ceiling Assemblies Using the Tapping Machine[4]

* A Summary of Changes section appears at the end of this specification.

3. Definitions

3.1 *contact ceiling*—a ceiling in which the one half of the total weight of the metal.

[1] This specification is under the jurisdiction of ASTM Committee C-11 on Gypsum and Related Building Materials and Systems, and is the direct responsibility of Subcommittee C11.03 on Specifications for Application of Gypsum and Other Products in Assemblies.
Current edition approved May 15, 1987. Published June 1987. Originally published as C 841 – 76. Last previous edition C 841 – 81.
[2] *Annual Book of ASTM Standards,* Vol 01.06.
[3] *Annual Book of ASTM Standards,* Vol 04.01.
[4] *Annual Book of ASTM Standards,* Vol 04.06.
[5] *Annual Book of ASTM Standards,* Vol 08.04.
[6] *Annual Book of ASTM Standards,* Vol 04.07.

lath is attached in direct contact with the construction above, without the use of main runners or cross furring.

3.2 *furred ceiling*—a ceiling in which the furring used for the support of the lath is attached directly to the structural members of the building.

3.3 *suspended ceiling*—a ceiling in which the main runners and cross furring are suspended below the structural members of the building.

3.4 *main runners*—the runners that are attached to or suspended from the construction above for the support of cross furring.

3.5 *cross furring*—a ceiling assembly in which the furring members are attached at right angles to the underside of the main runners or structural supports for support of the lath.

3.6 *furring (used in vertical construction)*—the spacer elements used to maintain a space between the structural elements and the lath.

3.7 *primary members*—the members (main runners or structural supports) to which the cross furring is attached.

3.8 *metal plaster bases*—expanded metal lath, sheet metal lath, welded and woven wire lath.

3.9 *face side (gypsum lath)*—the side opposite the paper cover seam laps.

3.10 *surface transition*—the change in width of a surface, usually as a result of penetrations (doors, windows) or architectural demands.

4. Delivery of Materials

4.1 All materials shall be delivered in the original packages, containers, or bundles bearing the brand name and manufacturer's (or supplier's) identification.

5. Storage of Materials

5.1 All materials shall be kept dry, preferably by being stored inside. Where necessary to be stored outside, materials shall be stacked off the ground, supported on a level platform, and protected from the weather and surface contamination.

5.2 Materials shall be neatly stacked flat with care taken to avoid damage to edges, ends, or surfaces.

6. Materials

6.1 *Wire*—Gages of wire hereinafter referred to shall be in accordance with the following:

Wire Gage (U.S. Steel Wire Gage)[A]	Diameter, in.
No. 20	0.0348
No. 19	0.0410
No. 18	0.0475
No. 17	0.0540
No. 16	0.0625
No. 14	0.0800
No. 13	0.0915
No. 12	0.1055
No. 11	0.1205
No. 10	0.1350
No. 9	0.1483
No. 8	0.1620

[A] Identical to Washburn and Moen gage.

NOTE 2—Allowable variations in percentages for specified weights or for thicknesses attributed to gage numbers shall be in accordance with current mill tolerances as established by the American Iron and Steel Institute for all steel products included in this specification.

6.2 *Gypsum Lath*—See Specification C 37.

6.3 *Expanded Metal Lath*—Diamond mesh, rib or sheet lath, with or without paper backing, shall be fabricated from steel sheet of the weights and configuration indicated in Table 1. Lath shall be given a protective coating of paint after fabrication, or shall be fabricated from zinc-coated (galvanized) steel.

6.4 *Wire Lath:*

6.4.1 *Welded Wire Lath,* with or without absorptive paper separator and with or without paper backing, shall be fabricated from copper-bearing, cold-drawn, galvanized steel wire not less than No. 16 gage, with openings not to exceed 2 by 2 in., welded at all intersections of wire and stiffened continuously, parallel to long dimension of the lath at not more than 6-in. (152-mm) intervals.

6.4.2 *Woven Wire Lath,* fabricated from copper-bearing, cold-drawn, galvanized steel wire, with minimum openings of 1 in., maximum openings of 2¼ in. and a maximum area of 4 in.² (26 cm²) for a single opening. Minimum size of wires shall be No. 18 gage for 1-in. (25.4-mm) openings, No. 17 gage for 1½-in. (38.1-mm) openings, and No. 16 gage for 2-in. (50.8-mm) openings.

6.5 *Paper or Other Backing*—When used on metal plaster bases, backing shall be securely held in place by or attached to the metal plaster base. The backing shall permit full embedment, in at least ⅛ in. (3.2 mm) of plaster, of at least one half of the total length of the strands and one half of the total weight of the metal.

6.6 *Accessories*—Corner beads, base screeds, control joints, concealed picture molds, and similar items shall be steel with a protective coating of paint after fabrication or fabricated from zinc-coated (galvanized) steel, or high impact poly (vinyl chloride) (PVC) plastic in accordance with Specification D 3678. Steel shall be not less than 0.015 in. (0.38 mm) thick uncoated. Plastic shall be not less than 0.030 in. (0.76 mm) thick. All accessories shall have perforated or expanded flanges or clips shaped so as to permit complete embedment in the plaster, to provide means for accurate alignment, and to secure attachment of the accessory to the underlying surface. Accessories shall be designed to receive or to permit application of the specified plaster thickness.

6.6.1 Casing beads shall be steel, plastic, or aluminum. Steel shall be not less than 0.0172 in. (0.437 mm) thick, uncoated. PVC plastic shall not be less than 0.030 in. (0.88 mm) thick, high impact in accordance with Specification D 3678. Aluminum shall be not less than 0.50 in. (13 mm) clear plastic coated. Accessories shall be designed for the intended use.

6.6.2 Control joints may be formed by using a single prefabricated member or may be fabricated by installing casing beads back to back with a flexible barrier membrane behind casing beads or inserted into the casing beads. The separation spacing shall be not less than ⅛ in. (3.2 mm) or as required by the anticipated thermal exposure range.

6.7 *Cornerite*—minimum 1.75 lb/yd² (0.059 kg/m²) expanded metal lath, galvanized, or given a protective coating of paint after fabrication, or woven or welded fabric of minimum No. 19 gage wire (or weighing not less than 1.7 lb/yd²) fabricated from galvanized wire. When shaped for angle reinforcing, it shall have outstanding legs of not less than 2 in. (50.8 mm).

6.8 *Channels*—Hot-rolled or cold-rolled steel, free of rust with a coating of rust inhibitive paint, of the following minimum weights per thousand linear feet:

Sizes, in.	Hot-Rolled, lb	Cold-Rolled, lb
¾	300	300
1	410	…
1½	1120	475
2	1260	590

NOTE 3—Channels used in areas subject to corrosive action of salt air shall be galvanized.

6.9 *Wire*, in accordance with Specification A 641, Class 1 coating (galvanized), soft temper.

6.10 *Steel*, for rod and flat (strap) hangers shall be mild steel, zinc or cadmium plated or protected with a rust-inhibiting paint.

6.11 *Clips*—Devices for attaching framing members to supports or to each other; for attaching lath to framing members; or for securing lath to lath, shall be formed of zinc-coated (galvanized) steel wire or sheet, depending on use and manufacturer's requirements.

7. Installation

7.1 *General Requirements for Application of Lath and Accessories*:

7.1.1 Apply all lath with the long dimension at right angles to supports, unless otherwise specified.

7.1.2 Use corner beads to protect all external corners and to establish grounds. Attach them to substrate in such a manner as to ensure proper alignment during application of plaster. (See Annex A2 for additional information.)

7.1.3 Install all metal accessories in such a manner that flanges and clips provided for their attachment are completely embedded in the plaster.

7.1.4 *Control Joints*—Install expansion and contraction (control) joints in ceilings exceeding 2500 ft² (232 m²) in area and in partition, wall, and wall furring runs exceeding 30 ft (9 m). The distance between ceiling control joints shall not exceed 50 ft (15 m) in either direction. Install a control joint where the ceiling framing or furring changes direction. The distance between control joints in walls or wall furring shall not exceed 30 ft. Install a control joint where an expansion joint occurs in the base exterior wall. Wall or partition height door frames may be considered as control joints.

7.1.5 Isolate, with casing bead or other suitable means, non-load bearing partitions from load bearing members, to avoid transfer of structural loads to partitions.

7.2 *Application of Gypsum Lath to Supports*:

7.2.1 *General*—Apply gypsum lath so that vertical joints do not occur nearer than one full stud space from edges of openings in walls or partitions.

7.2.2 Apply gypsum lath with the face side out. End joints shall fall on different supports in alternate courses, or apply the lath so that end joints are continuous on one support. In the latter case, cover the continuous end joints with 3-in. (76-mm) wide strips of metal lath or welded or woven wire fabric, and offset or stagger the long edge joints of lath in alternate courses. In all cases, butt the lath together.

7.2.3 When gypsum lath is attached on ceilings by clips providing edge support only, use only plain gypsum lath and three-coat plaster. Where clips are used, install them in accordance with the manufacturer's directions. Where clips are used to attach gypsum lath to horizontal wood supports, attach the clips to the sides of the supports.

7.2.4 Provide cornerite on gypsum lath at all internal angles, and lightly nail or staple it to the lath and not to the framing members, except do not use cornerite in unrestrained construction, or where the drawings or specifications make other provisions for the treatment of internal angles.

7.2.5 Where gypsum lath surfaces intersect or where joint surfaces are to be plastered without lathing (such as masonry), install a casing bead at the intersection or joining instead of using cornerite or metal lath stripping attached to the lathing structure.

7.2.6 Diagonally strip gypsum lath at corners of doors, windows, or other openings with self-furring diamond mesh metal lath or wire lath, not less than 6 in. (152 mm) wide by 12 in. (305 mm) long.

7.2.7 Use metal lath stripping to cover chases and similar breaks in continuity or horizontal or vertical surfaces that are to receive plaster. Extend stripping a minimum of 3 in. (76 mm) on all sides of the openings.

7.3 *Application of Gypsum Lath to Wood Supports*:

7.3.1 Except where required otherwise, for certain fire-resistant construction, securely attach gypsum lath to wood supports with nails or staples conforming to Table 2.

7.3.2 Drive the nails so that the face of the head is flush with the face of the gypsum lath, and not closer than ⅜ in. (9.5 mm) from the

edges of lath. Machine-drive staples with the crown parallel to the nailing members. The crown shall bear lightly against, but not cut into, the face of the lath.

7.4 *Application of Gypsum Lath to Metal Supports*:

7.4.1 Attach gypsum lath to horizontal or vertical metal supports with clips, staples, screws, or nails, or a combination thereof.

7.4.2 Drive screws so that the face of the head is flush with the face of the gypsum lath and not less than ⅜ in. (9.5 mm) from the edges of the lath. Machine-drive staples with the crown parallel to the framing members. The crown shall bear tightly against, but not cut into, the face of the lath.

7.5 *Studless Solid Partitions (Gypsum Lath and Plaster)*:

7.5.1 Gypsum lath shall be plain, ½ in. (12.7 mm) thick, 24 in. (610 mm) wide, and in floor-to-ceiling lengths not to exceed 12 ft (3.7 m).

7.5.2 Space the anchors for securing wood floor runners to the floor at not more than 24 in. (610 mm) on center. Anchors shall penetrate floor surface to a depth of not less than ⅝ in. (15.9 mm). The width of such runners shall correspond to the overall partition thickness and their upper surface shall be grooved parallel to the length of the runner in the center to a depth of at least ½ in. (12.7 mm) and to a width to accommodate snugly the thickness of the lath.

7.5.3 Securely anchor metal floor runners, bases, and clips at not more than 24 in. (610 mm) on center. Runners shall be designed to hold the lower edge of the lath securely in position and to assure specified minimum thickness of plaster on both faces.

7.5.4 Align metal ceiling runners, to which the lath may be clipped or wire tied with the floor runners to ensure plumb installation of the lath, and firmly secure them to the ceiling construction.

7.5.5 Gypsum lath shall be of such length as to allow not less than ¼ in. (6.4 mm) nor more than 1-in. (25.4-mm) top clearance in the ceiling runner. Erect gypsum lath vertically so as to engage the ends in or to the floor and ceiling runners.

7.5.6 In erecting lath, align the vertical edges in accordance with the printed directions of the manufacturer of the partition system.

7.5.7 Use temporary bracing. Erect the bracing, and attach the lath to the braces in accordance with the printed directions of the manufacturer of the partition system.

7.6 *Application of Metal Plaster Bases to Supports*:

7.6.1 *General*—The spacing of supports for the type and weight of metal lath used shall conform to the requirements of Table 1.

7.6.2 *Attachments for Metal Plaster Bases to Wood Supports*:

7.6.2.1 Attach diamond mesh expanded metal lath, flat rib expanded metal lath, and welded wire lath to horizontal wood supports with 1½ in., No. 11 gage, ⁷⁄₁₆ in. head, barbed, galvanized, or blued roofing nails driven flush; and attach to vertical wood supports with 6d common nails, or 1-in. roofing nails driven to a penetration of at least ¾ in. (19 mm) or 1-in. No. 14 gage wire staples driven flush.

7.6.2.2 Attach ⅜-in. (9.5-mm) rib expanded metal lath and sheet lath to horizontal and vertical wood supports with nails to staples ⅜ in. longer than required (7.6.2.1) to provide at least 1⅜-in. (34.9-mm) penetration into horizontal wood supports, and ¾-in. (19-mm) penetration into vertical wood supports.

7.6.2.3 When used on vertical wood supports, bend common nails over to engage at least three strands of lath, or over a rib-for-rib lath.

7.6.2.4 Other methods of attachment which afford carrying strength equal to or greater than those described in 7.6.2.1 through 7.6.2.3 may be used.

7.6.3 *Attachments for Metal Plaster Bases to Metal Supports*:

7.6.3.1 Except as provided in 7.6.3.2, securely attach all metal plaster bases to metal supports with No. 18 gage wire ties or by other means of attachment which afford carrying strength and resistance to corrosion equal to or superior than that of the wire.

7.6.3.2 Attach rib metal lath to open web steel joists by single ties of galvanized, annealed steel wire, not lighter than No. 18 gage, with the ends of each tie twisted together 1½ times.

7.6.4 *Attachments for Metal Plaster Bases to Concrete Supports*—Attach rib metal lath to concrete joists by loops of galvanized, annealed steel wire, not lighter than No. 14 gage, with the ends of each loop twisted together.

7.6.5 *Spacing of Attachments for Metal Plaster Bases*—Space attachments for securing metal plaster bases to supports not more than 7 in. (178 mm) apart for diamond mesh and flat rib laths and at each rib for ⅜-in. (9.5-mm) rib lath.

7.6.6 *Lapping of Metal Plaster Bases*:

7.6.6.1 Secure side laps of metal plaster bases to supports, and tie between supports at intervals not to exceed 9 in. (229 mm).

7.6.6.2 Lap expanded and sheet metal lath ½ in. (12.7 mm) at sides, or edge ribs may be nested. Lap welded wire lath one mesh at sides and ends. Lap expanded and sheet lath 1 in. at ends or ends may be nested. When end laps occur between supports, lace the ends of sheets of all metal plaster bases or adequately tie with No. 18 gage, galvanized, annealed steel wire.

7.6.7 *Procedure for Application of Metal Plaster Bases to Supports*:

7.6.7.1 Generally, apply metal plaster bases to ceilings first.

7.6.7.2 Stagger the ends of sheets wherever possible.

7.6.7.3 Where furred or suspended ceilings greater than 50 ft (15 m) in either direction or 2500 ft² (232 m²) in area butt into or are penetrated by columns, walls, beams, or other elements, the following procedure shall be used: Abutt the sides and ends of the ceiling lath at the horizontal internal angles and terminate the lath at a casing bead, control joint, relief casing, or similar device designed to keep the sides and ends of the ceiling lath and plaster free from the adjoining vertically oriented elements. Do not use cornerite instead of casing beads, etc., at the internal angles between such ceilings, walls, or partitions. Lath shall not be continuous through control joints but shall be stopped at each side.

7.6.8 Ends of sheets of diamond mesh metal lath and welded wire lath on partitions not abutting structural walls, columns, or floor-ceiling slabs may be either bent into or around vertical corners and carried on to at least one support away from the corner, or the ends of the sheets of such lath may be butted into corners when cornerite is applied over the abutting laths. Butt rib metal lath or sheet lath into corners and apply cornerite over the abutting laths.

7.6.9 Begin and terminate the ends of metal plaster bases of load-bearing walls and partitions abutting structural walls, columns, or floor-ceiling slabs for each surface at the internal angles; do not use cornerite at such angles, but provide either casing beads or control joints at the extreme edges of the abutting surfaces to keep the sides or ends of the lath and plaster free and clear of the abutting surfaces.

7.7 *Studless Solid Partitions (Metal Lath and Plaster)*:

7.7.1 Metal plaster bases used for studless solid partitions shall have the sides (long dimension) of the sheet vertical. Wire-tie lath to ceiling runners and ceiling lath by ties spaced not more than 8 in. (203 mm) apart and suitably anchor to the floor runners or base. At vertical internal and external corners, metal plaster shall be bent and returned 6 in. (152 mm) on abutting surfaces. Lap rib metal lath ½ in. (12.7 mm) at sides, or nest outside ribs. Wire-tie side laps at intervals not exceeding 12 in. (305 mm).

7.7.2 Use temporary bracing as specified in 7.5.7.

7.7.3 Where partitions abutt load-bearing walls, columns, beams, floor-ceiling slabs, or other structural elements, the applicable provisions of 7.6.9 shall apply.

7.8 *Suspended and Furred Ceilings (Metal Lath and Gypsum Lath Construction)*:

7.8.1 *Hangers and Inserts*—Hangers shall be of ample length.

7.8.1.1 Secure hangers (without inserts) in concrete to steel reinforcement of slabs or otherwise embedded in concrete so as to develop full strength.

7.8.1.2 Inserts shall develop the full strength of the hangers which are attached to them.

7.8.2 *Hangers for Suspended Ceilings Under Wood Constructions*—Hangers shall conform with the requirements of Table 3 both as to size and maximum area to be supported, except as modified in this section. Hangers shall be attached to supports by one of the following methods:

7.8.2.1 Insert into holes drilled a minimum of 3 in. (76 mm) above bottom of joists and with the upper end of the hanger twisted three times around itself.

7.8.2.2 Three 12d nails driven on a downward slant into sides of joists with not less than 1¼-in. (31.8-mm) penetration and at least 5 in. (127 mm) from bottom edges, not over 36 in. (914 mm) on center, and with the upper end of the hanger in each case twisted three times around itself.

7.8.2.3 Four 1½-in. (38.1-mm), No. 9 gage wire staples driven horizontally or on a downward slant into sides of joists, three near the

upper end of the loop and the fourth to fasten the loose end.

7.8.2.4 Where spacing of supports is more than 4 ft (1.2 m) on center, 1½-in. (38.1-mm) No. 0 screw eyes, or equivalent, spaced no more than 3 ft (0.9 m) on centers may be screwed into heavy wood flooring so that the supported area will be not more than 9 ft² (0.[ ] m²). Insert the upper end of the wire hanger in the screw eye and in each case twisted three times around itself.

7.8.2.5 Spike flat hangers to sides of joists with two 12d nails driven through holes drilled in the hanger and clinch at least 3 in. (76 mm) above the bottom of the joists.

7.8.3 *Minimum Size for Hangers*—Gage of wire hangers, diameter of rod hangers, and sizes of flat hangers shall be as specified in Table 3.

7.8.4 Where 1 by ³⁄₁₆-in. (25.4 by 4.8-mm) flat inserts and hangers are used, ⁷⁄₁₆-in. (11 mm) diameter holes shall be punched on the center line at the lower end of inserts and upper end of hanger to permit the attachment of the hangers to the insert.

7.8.5 Holes in both inserts and hangers shall not be closer than ⅜ in. (9.5 mm) from the ends.

7.8.6 Flat steel hangers shall be bolted to by ³⁄₁₆-in. (25.4 by 4.8-mm) inserts with ⅜-in. (9.5-mm) diameter round-head stove bolts.

7.8.7 The nuts of bolts shall be drawn up tight.

7.8.8 Saddle-tie wire or rod hangers or wrap them around main runners so as to prevent turning or twisting of the runners and to develop the full strength of the hangers and the runners. Attach smooth or threaded rod hangers to inserts and runners with special attachments.

7.8.9 Bolt the lower ends of flat hangers to the main runners, or bend them tightly around runners and carry up and above the runner and bolt to the main part of the hanger. Bolt shall be ⅜-in. (9.5-mm) diameter, round-head stove bolts.

7.9 *Main Runners*:

7.9.1 Minimum sizes and maximum spans and spacings of main runners for the various spans between hangers or other supports shall be as specified in Table 3.

7.9.2 Provide a clearance of not less than [ ] in. (25.4 mm) between the ends of main runners and abutting masonry or reinforced concrete walls, partitions and columns, except where special conditions require that main runners shall be let into abutting masonry or concrete construction, provide a clearance within such constructions of not less than 1 in. from the ends and not less than ⅛ in. (3.2 mm) from the tops and sides of the runners.

7.9.3 Locate a main runner within 6 in. (152 mm) of the paralleling walls to support the ends of the cross furring. The ends of main runners shall be supported by hangers located not more than 6 in. from the ends.

7.9.4 When main runners are spliced, overlap the ends not less than 12 in. (305 mm) (with flanges of channels interlocked) and securely tie near each end of the splice with double loops of No. 16 (or double loops of twin strands of No. 18) gage, galvanized wire.

7.10 *Cross Furring*:

7.10.1 Minimum size and maximum spans and spacings of various types of cross furring for various spans between main runners and supports shall be as specified in Table 3.

7.10.2 Securely saddle-tie cross furring to main runners with No. 16 gage galvanized wire, or a double strand of No. 18 gage galvanized wire or with special galvanized clips, or equivalent attachments.

7.10.3 When cross furring members are spliced, overlap the ends not less than 8 in. (203 mm) (with flanges of channels interlocked) and securely tie near each end of the splice with double loops of No. 16 gage galvanized wire or twin strands of No. 18-gage galvanized wire.

7.10.4 Do not let cross furring into nor come in contact with abutting masonry or reinforced concrete walls or partitions except that, where special conditions require that cross furring shall be let into abutting masonry or concrete construction, the applicable provisions of 7.9.2 shall apply.

7.10.5 Grillage (main runners and cross furring) shall be interrupted at control and expansion joints.

7.11 *Metal Furring for Walls*:

7.11.1 Attachments for furring shall consist of nails driven securely into concrete or into masonry joints, short pieces of ¾ channels used as anchors driven into masonry joints, or other devices specifically designed as spacer elements. Space them horizontally not to exceed 2 ft (0.6 m) on centers. Space them vertically in accordance with horizontal stiffener spacing so that they project from the face of the wall to permit ties to be made.

7.11.2 Where dampproofing has been damaged in installation of attachments, repair the dampproofing with the same material before proceeding with the installation of the furring.

7.11.3 Horizontal stiffeners shall be not less than ¾-in. (19-mm) hot-rolled or cold-rolled channels, spaced not to exceed 54 in. (1372 mm) on centers vertically, with the lower and upper channels not more than 6 in. (152 mm) from the floor and ceiling, respectively, and not less than ¼ in. (6.4 mm) clear from wall face. Securely tie them to attachments with three loops of No. 18-gage, galvanized, soft-annealed wire, or equivalent devices.

7.11.3.1 Use special devices, securely attached to concrete or masonry, instead of horizontal members, for support of vertical members.

7.11.4 Vertical members shall be not less than ¾-in. (19-mm) hot-rolled or cold-rolled channels, in accordance with requirements in Table 1. Saddle-tie members to horizontal members with three loops of No. 18 gage galvanized soft-annealed wire, or equivalent devices, at each crossing, and securely anchor to the floor and ceiling constructions. If furring is not in contact with the wall, install channel braces between horizontal stiffeners and the wall, spaced 2 ft (0.6 mm) apart (horizontal spacing).

7.12 *Workmanship*—Erect metal construction furring and lathing so that finished plaster

surfaces will be true to line, level, plumb, square, curved, or as required to receive specified plaster thickness.

TABLE 1 Types and Weights of Expanded Metal Lath, Sheet Lath, or Welded Wire Lath and Corresponding Maximum Permissible Spacing of Supports[A]

Type of Metal Plastering Base	Minimum Weight of Metal Base, lb/yd² (kg/m²)	Maximum Permissible Spacing of Supports Center for Center, in. (mm)				
		Walls (Partitions)			Ceilings	
		Wood Studs	Solid Partitions	Steel Studs Wall Furring, etc.	Wood or Concrete	Metal
Expanded metal lath:						
Diamond mesh	2.5 (0.08)	16 (406)	16 (406)	13.5 (343)	12 (305)	12 (305)
	3.0 (0.1)	16 (406)	12 (305)	12 (305)	12 (305)	12 (305)
	3.4 (0.12)	16 (406)	16 (406)	16 (406)	10 (254)	16 (406)
Flat rib	2.5 (0.08)	16 (406)	12 (305)	12 (305)	12 (305)	12 (305)
	2.75 (0.09)	16 (406)	16 (406)	16 (406)	16 (406)	16 (406)
	3.0 (0.1)	16 (406)	16 (406)	16 (406)	16 (406)	13 (330)
	3.4 (0.12)	19 (482)	24 (610)	19 (482)	19 (482)	19 (482)
⅜-in. (9.5-mm) rib	3.0 (0.1)	19 (482)	N/A[D]	16 (406)	16 (406)	16 (406)
	3.4 (0.12)	24 (610)	N/A	19 (482)	19 (482)	19 (482)
	3.5 (0.12)	24 (610)	N/A	19 (482)	19 (482)	19 (482)
	4.0 (0.14)	24 (610)	N/A	24 (610)	24 (610)	24 (610)
¾-in. (19-mm) rib	5.4 (0.18)	24 (610)	N/A	24 (610)	36 (914)	36 (914)
Sheet lath	4.5 (0.15)	24 (610)	N/A	24 (610)	24 (610)	24 (610)
Wire lath welded	1.4 (0.05)[B]	16 (406)	16 (406)	16 (406)	16 (406)	16 (406)
	1.95 (0.07)[C]	24 (610)	24 (610)	24 (610)	24 (610)	24 (610)

[A] If paper-backed lath is used, limit to lath having an absorbent and a perforated slotted paper separator only.
[B] Welded wire paper-backed lath, 16-gage.
[C] Welded wire, paper-backed lath, 16-gage face wire.
[D] Not applicable.

TABLE 2 Nails and Staples[A] for Attaching Gypsum Lath to Horizontal and Vertical Wood Supports

		Maximum Spacing, in. (mm)			Minimum Gage Requirements, in. (mm)			
Width of Lath	Thickness of Lath	Distance Between Supports	Number of Attachments per Bearing	Approximate Spacing c to c of Attachments	Length of Leg	Depth of Support Penetration	Diameter of Flat Head or Blued Nails or Crown Width of Staples[A]	Gage of Shank of Nails or Staples[A]
16 (406)	⅜ (9.5)	16 (406)	4 (102)	5 (127)	1⅛ (28.6) 1 (25.4)[A]	1¾ (44) ⅝ (15.9)[A]	¹⁹⁄₆₄ (7.5) ⁷⁄₁₆ (11)[A]	13 (330) 16 (406)[A]
24 (610)	⅜ (9.5)	16 (406)	6 (152)	4½ (114)	1⅛ (28.6) 1 (25.4)[A]	¾ (19) ⅝ (15.9)[A]	¹⁹⁄₆₄ (7.5) ⁷⁄₁₆ (11)[A]	13 (330) 16 (406)[A]
16 (406)	½ (12.7)	24 (610)	4 (102)	5 (127)	1¼ (32) 1⅛ (28.6)[A]	¾ (10) ⅝ (15.9)[A]	¹⁹⁄₆₄ (7.5) ⁷⁄₁₆ (11)[A]	13 (330) 16 (406)[A]
24 (610)	½ (12.7)	24 (610)	6 (152)	4½ (114)	1¼ (32) 1⅛ (28.6)[A]	¾ (19) ⅝ (15.9)[A]	¹⁹⁄₆₄ (7.5) ⁷⁄₁₆ (11)[A]	13 (330) 16 (406)[A]

[A] Galvanized staples.

TABLE 3 Suspended and Furred Ceilings, Minimum Sizes for Wire, Rod, and Rigid Hangers; Minimum Sizes and Maximum Spans and Spacings for Main Runners; and Minimum Sizes and Maximum Spans and Spacings for Cross Furring

NOTE—1 in. = 25.4 mm; 1 ft² = 0.093 m²

HANGERS		
	Maximum Ceiling Area Supported, ft²	Minimum Size of Hangers
Hangers for Suspended Ceilings	12.5	9 gage wire
	16	8 gage wire
	18	³⁄₁₆-in. diameter, mild steel rod[A]
	20	⁷⁄₃₂-in. diameter, mild steel rod[A]
	22.5	¼-in. diameter, mild steel rod[A]
	25.0	1 by ³⁄₁₆-in. mild steel flat[B]
Attachments for Tying Runners and Furring Directly to Beams and Joists		
For Supporting Runners		
Single Hangers Between Beams[C]	8	12 gage wire
	12	10 gage wire
	16	8 gage wire
Double Wire Loops at Beams or Joists[C]	8	14 gage wire
	12	12 gage wire
	16	11 gage wire
For Supporting Furring Without Runners[C] (Wire Loops at Supports)		
Types of Support:		
Concrete	8	14 gage wire
Steel	8	16 gage (2 loops)[D]
Wood	8	16 gage (2 loops)[D]

SPANS AND SPACINGS OF MAIN RUNNERS[E,F]

Minimum Size and Type	Maximum Span Between Hangers or Supports, in.	Maximum Center to Center Spacing of Runners, in.
¾ in., 0.3 lb/ft, cold- or hot-rolled channel	24	36
1½ in. 0.475 lb/ft, cold-rolled channel	36	48
1½ in., 0.475 lb/ft, cold-rolled channel	42	42
1½ in., 0.475 lb/ft, cold-rolled channel	48	36
1½ in., 1.12 lb/ft, hot-rolled channel	48	54
2 in. 0.59 lb/ft, cold-rolled channel	60	48
2 in., 1.26 lb/ft, hot-rolled channel	60	60
1½ by 1½ by ³⁄₁₆-in. angle	60	42

SPANS AND SPACINGS OF CROSS FURRING[E,F]

Minimum Size and Type	Maximum Span Between Runners or Supports, in.	Maximum Center to Center Spacing of Cross Furring Members, in.
¼-in. diameter pencil rods	24	12
³⁄₈-in. diameter pencil rods	24	19
³⁄₈-in. diameter pencil rods	30	12
¾ in., 0.3 lb/ft, cold- or hot-rolled channel	36	24
	42	19
	48	16
1 in., 0.410 lb/ft, hot-rolled channel	48	24
	54	19
	60	12

[A] It is highly recommended that all rod hangers be protected with a zinc cadmium coating.
[B] It is highly recommended that all flat hangers be protected with a zinc or cadmium coating or with a rust-protective paint.
[C] Inserts, special clips, or other devices of equal strength may be substituted for those specified.
[D] Two loops on No. 18 gage wire may be substituted for each loop of No. 16 gage wire for attaching steel furring to steel or wood joists.
[E] These spans are based on webs of channels being erected and maintained in a vertical position.
[F] Other sections of hot- or cold-rolled members of equivalent beam strength may be substituted for those specified.

ANNEXES

A1. GENERAL INFORMATION

A1.1 The work includes all labor, materials, services, equipment, and scaffolding required to complete the lathing and furring of the project in accordance with the drawings and specifications.

A1.2 Details of construction to achieve required fire resistance may be obtained from official reports of fire tests conducted at recognized fire testing laboratories in accordance with Methods E 119.

A1.3 Details of construction to achieve required sound control may be obtained from official reports of tests conducted at recognized sound testing laboratories in accordance with the applicable sound tests of Method E 90, Method C 423, or Method E 492.

A1.4 Scaffolding shall be constructed and maintained in strict conformity with applicable laws and ordinances.

A1.5 The work shall be coordinated with the work of other trades.

A1.6 Surfaces and openings shall be examined before furring or lathing are applied thereto, the proper authorities shall be notified, and unsatisfactory conditions shall be corrected prior to applicating of furring or lathing.

A2. TECHNICAL INFORMATION

A2.1 Ventilation above lath and plaster ceilings located under attics, roofs, or similar unheated spaces shall be adequately ventilated by providing effective cross ventilation for all spaces between roof and top floor ceilings by screened louvers or other approved or acceptable means.

A2.1.1 The ratio of total net free ventilating area to ceiling area shall be not less than $L/150$, except that the ratio may be $L/300$ provided:

A2.1.1.1 A vapor barrier having a transmission rate not exceeding 1 perm is installed on the warm side of the ceiling, or

A2.1.1.2 At least 50 % of the required ventilation area is provided by ventilators located in the upper portion of the space to be ventilated (at least 3 ft (0.9 m) above eave or cornice vents) with the balance of the required ventilation area provided by eave or cornice vents.

A2.1.2 Attic space that is accessible and suitable for future habitable rooms or walled-off storage space shall have at least 50 % of the required ventilating area located in the upper part of the ventilated space as near the high point of the roof as practical and above the probable level of any future ceilings.

A2.2 Attachments for support of wood subpurlins or wood furring strips by purlins, joists, or trusses, where ceiling lath is attached to subpurlins. Wood subpurlins or furring strips to be furnished and installed by others and to which lath is to be attached for ceiling construction shall be secured to supporting purlins, joists, or trusses by adequate nailing, lag screws, or wire loops. In schools, theaters, hospitals, nursing homes, shopping centers, and generally in all buildings of public assembly, where wood subpurlins or wood furring strips are used for the attachment of plastering bases on ceilings, attachment of the wood subpurlins, or wood furring to supporting purlins, joists, or trusses should be by lag screws, galvanized wire loops, or similar positive means (see 7.3).

A2.3 *Floating Angle Construction*—Except where required otherwise for certain fire-resistive construction, nails or staples occurring less than 6 in. (152 mm) from internal angles may be omitted in the application of gypsum lath, provided each piece of lath is nailed to at least two supports. Where lath abuts masonry, the angle should not be "floated" and a plaster stop or casing bead shall be used.

A2.3.1 Gypsum lath, for floating angle construction, shall be applied to ceiling areas first, so that the ceiling lath will be supported in the angles by the upper edge of the top course of lath applied to walls or partitions.

A2.4 *Wood Furring*—For the attachment of wall finishes, wood furring shall be not less than ¾ in. (19 mm) by 1½ in. (38.1 mm) (actual size) where applied to solid backing such as masonry. Wood furring attached to supports, spaced not more than 24 in. (610 mm) on centers, shall be not less than 1½ by 1½ in.

A3. DESIGN AND ERECTION OF HOLLOW DOOR METAL FRAMES FOR PREFABRICATED STEEL STUD AND CHANNEL STUD, LATH AND PLASTER HOLLOW PARTITIONS

A3.1 Hollow metal frames shall be fabricated of not less than 16-gage shop-primed steel, and shall be designed to provide space for studs, lath, and plaster. Returns on flanges of door frames shall not prevent free intrusion of lath and plaster into the frames.

A3.2 Insert-type floor anchor clips shall be at least 16-gage steel, and welded to the flanges of the frames at the floor and provided for two floor anchors spaced a minimum of 3 in. (76 mm) apart. If the frame is of such size that suitable anchorage cannot be provided, the clip shall extend beyond the frame to provide the required anchorage. At least one of the holes in the insert shall be located within the frame. Where the frame is supplied with inserts that do not permit the use of two anchors, supplementary clips may be used provided they fit snugly in the frame and are fastened to the floor through the welded clip.

A3.3 Jamb anchor insert clips shall be of 16-gage steel minimum, notched or formed to receive the stud, of a size to fit snugly within the interior of the jamb, with no part extending more than ¾ in. (19 mm) beyond the back of the jamb, located near the top of the jamb, approximately 12 in. (305 mm) from the top, and a maximum of 24 in. (610 mm) on centers and securely attached to the frame.

A3.4 Jam anchors shall be so formed and installed as to prevent independent movement of any component of the frame cross section in the clip itself, and shall be so formed that when a stud is fastened to the anchor clip there can be no movement of the stud and frame independent of each other. The clips shall be of such design as to permit the first stud to be placed within or immediately adjacent to the door frame. If jamb anchors that are notched or formed to receive metal studs are not available, jamb anchors that adhere to all other requirements may be used, provided they are welded or bolted to the stud.

A3.5 Door frames shall be erected with spreaders in place, plumbed, aligned, and temporarily braced at specific locations.

A3.6 Frames shall be anchored at the floor with two anchors spaced at least 3 in. (76 mm) apart. Where attachment is to concrete floors, powder-driven or expansion-type anchors shall be used.

A3.7 *Solid Plaster Partitions with Steel Studs*:

A3.7.1 Steel studs for solid plaster partitions shall be of such size and number and so located as to provide backing at all corners. Steel studs shall be set to the required dimensions, properly aligned, made plumb and true, securely anchored to the floor and ceiling construction, and temporarily braced, if necessary.

A3.7.2 Where studs do not fail at edges of openings, extra studs shall be added. Such studs shall be securely anchored to the door frames.

A3.7.3 Space above headings of openings extending to the ceiling shall be reinforced by installing jack studs 16 in. (406 mm) on centers over the heads of the openings.

A3.8 *Hollow Metal Frames* (for Prefabricated Steel Stud, Channel Stud, Lath and Plaster Hollow Partitions):

A3.8.1 Metal studs shall be installed at each jamb of the steel frame by wiretying, bolting, or welding them to jamb anchors in the frame. The edges of the studs shall be inserted into the notches or other devices provided in the jamb anchor insert clips.

A3.8.2 The first stud on either side of the frame shall be fastened to suitable anchorage at the top and bottom of the partition in accordance with the instructions of the stud manufacturer. A second metal stud shall be placed not more than 2 in. (50 mm) away from the first stud, using adequate means of attachments between the two; or, two metal studs shall be attached to jamb anchors and securely fastened back to back.

A3.8.3 Where metal door frames have an adjustable strut equal in stiffness to a stud, running from each jamb to the ceiling, one metal stud shall be installed at each jamb in the manner previously noted.

A3.8.4 At the head of the metal frame, a section of metal floor track shall be placed horizontally within the head section, and fastened to the jamb studs with wire ties, screws, or welds. Frames over 3. in. (813 mm) wide shall be provided with anchor insert clips spaced not over 24 in. (610 mm) on centers.

A3.8.5 Metal jack studs shall be installed over the head of the frame, spaced 16 in. (406 mm) on centers maximum, and shall be attached to the metal floor track. They shall be reinforced with a ¾-in. (19-mm) cold rolled channel stiffener placed inside the partition and not more than 6 in. (152 mm) above the frame. Floor track may be omitted if one ¾-in. cold-rolled channel stiffener is placed horizontally on each interior face of the studs and not more than 6 in. above the frame. Channel stiffeners shall extend at least two full stud spaces beyond the opening and shall be saddle-tied to each stud they cross.

A3.8.6 Lath shall be inserted into jambs as far as possible and notched to pass jamb anchors, except where frames are required to be plaster grouted full.

A3.8.7 Where partition assemblies require the installation of control joints over the head of door frames to relieve stresses within the assemblies, such control joints may be installed in the following manner:

A3.8.7.1 Install two metal jack studs over the head of the frame, each stud approximately 3 in. (76 mm) from the center of the span with a vertical lath joint occurring between the two studs. Tie either a pair of casing beads edge to edge, or prefabricated control joint, onto the lath, centered over the parallel with the lath joint.

A3.8.7.2 Where it is aesthetically desirable to locate control joints at the outer edge of the jamb, the procedure outlined in A3.8.7.1 may be followed except that supporting members shall consist of a metal jack stud and a partition stud.

Summary of Changes

This section identifies the location of changes to this specification that have been incorporated since the last issue. Committee C-11 has highlighted those changes that affect the technical interpretation or use of this specification.

(1) Section 7.2.3 was deleted.

LATHING ACCESSORIES

The following accessories are the most common used in the function of lathing. These two pages are reproduced from the Eighth Edition of *Ramsey/Sleeper Architectural Graphic Standards* with permission of the publisher, John Wiley & Sons. This 854-page book is the "bible" of design professionals, containing design details for every conceivable usage in construction. Copyright 1988 by John Wiley & Son, Inc.

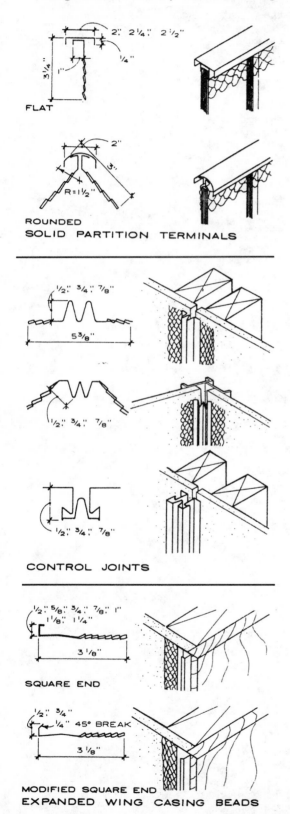

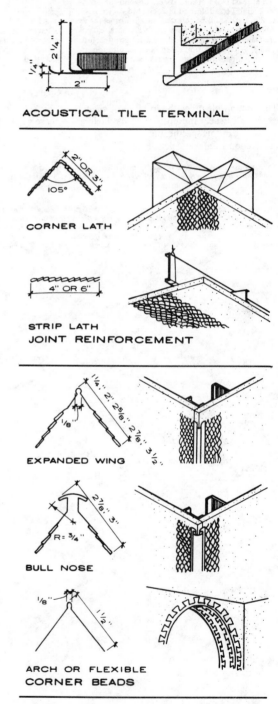

GENERAL NOTES

1. Certain accessory items are available in high impact PVC plastic and can be utilized with stucco, interior veneer, and conventional plaster items. Stock color is white. Special colors available on request from manufacturer.
2. Extruded aluminum shapes used mostly for stucco are available in a variety of anodized finishes.

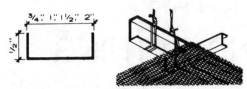

COLD ROLLED CHANNEL

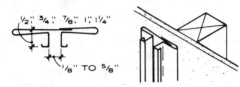

EXPANSION JOINT

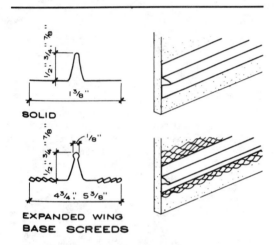

SOLID

EXPANDED WING BASE SCREEDS

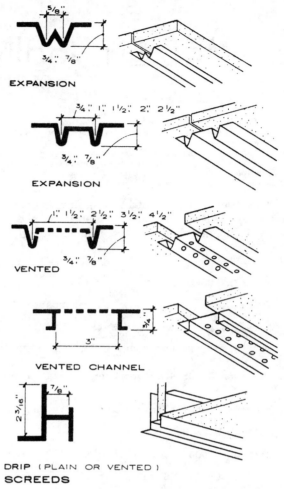

EXPANSION

EXPANSION

VENTED

VENTED CHANNEL

DRIP (PLAIN OR VENTED)
SCREEDS

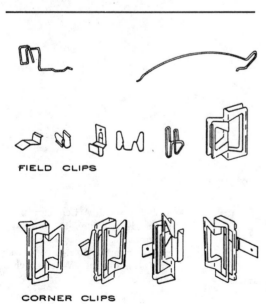

FIELD CLIPS

CORNER CLIPS

NOTE: OTHER CLIP TYPES ARE AVAILABLE

MISCELLANEOUS

CLIPS FOR GYPSUM LATH SYSTEM

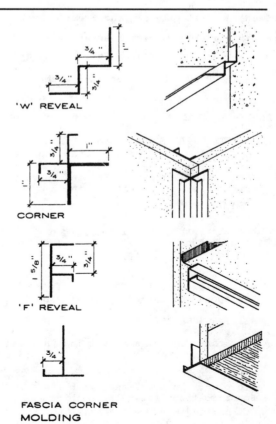

'W' REVEAL

CORNER

'F' REVEAL

FASCIA CORNER
MOLDING

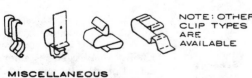

SPECTEXT SPECIFICATION FOR

METAL FURRING AND LATHING
SECTION 09206

This section includes metal furring, framing and lathing usually associated with wet applied gypsum or cement plaster finish. This section includes proprietary and descriptive type specifications. Use only one specifying type to avoid any conflicting requirements. Edit accordingly.

PART 1   GENERAL

1.01   WORK INCLUDED

A.   [Wall] [bulkhead] [and] [furred space] framing.

B.   Metal lathing for wet plaster finish.

1.02   WORK INSTALLED BUT FURNISHED UNDER OTHER SECTIONS

A.   [Section 08305 - Access Doors: Metal access panels.]

B.   Section [_____ - _____]: [Mechanical] [Electrical] components built into work of this Section.

> 1.02 List sections which furnish products for installation under this section. When a product will be furnished "by Others," outside the contract, it is considered to be furnished "by Owner."

1.03   RELATED WORK

A.   Section 05400 - Cold Formed Metal Framing.

B.   Section 08111 - Standard Steel Doors and Frames: Installation of door frames.

C.   Section 09111 - Metal Stud Framing System.

D.   Section 09210 - Gypsum Plaster.

E.   Section 09220 - Portland Cement Plaster.

F.   Section 09260 - Gypsum Board Systems.

1.04   REFERENCES
A.   ANSI/ASTM C841 - Installation of Interior Lathing and Furring
B.   ANSI/ASTM C847 - Metal Lath.
C.   FS QQ-L-101 - Lath, Metal, (and other Metal Plaster Bases).
D.   ML/SFA (Metal Lath/Steel Framing Association) - Specifications for Metal Lathing and Furring.

1.05   SYSTEM DESCRIPTION

A.   Fabricate vertical wall and furred space framing to limit finish surface to [1/180] [_____] deflection under lateral point load of [100] [_____] lbs ([445] [_____] N).

B.   Fabricate horizontal ceiling and soffit framing to limit finish surface to [1/360] [_____] deflection under superimposed dead loads [and wind uplift].

> 1.05 Use this Article carefully; restrict descriptions to total system parameters, performance or design requirements. Do not repeat statements made in "Work Included" Article.

1.06   QUALITY ASSURANCE

A.   Applicator: Company specializing in metal furring and lathing work with [_____] years [documented] experience.

B.   Perform work in accordance with [ANSI/ASTM C841.] [ML/SFA - Specifications for Metal Lathing and Furring.] [_____.]

> 1.06B Many agencies offer manuals of practice for installing furring and lathing materirals. Select an appropriate document or agency and edit this section accordingly.

1.07   REGULATORY REQUIREMENTS

   A.   Conform to [applicable] [_____] code for fire rated
         assemblies in conjunction with Section [09210] [_____] as
         follows:
         1.   Fire Rated Partitions: Listed assembly by [UL] [FM]
              No. [_____.]
         2.   Fire Rated [Ceiling] [and] [Soffits]: Listed assembly
              by [UL] [FM] No. [_____.]
         3.   Fire Rated Structural Column Framing: Listed assembly
              by [UL] [FM] No. [_____.]
         4.   Fire Rated Structural Beam Framing: Listed assembly by
              [UL] [FM] No. [_____.]

1.08   SUBMITTALS

   A.   Submit product data under provisions of Section [01300.]
         [01340.]

   B.   Provide product data on furring and lathing components,
         structural characteristcs, material limitations and
         finish.

   C.   Submit samples under provisions of Section [01300.]
         [01340.]

   D.   Submit [two] [_____] samples [_____x_____] inch
         ([_____x_____] mm) in size illustrating component design,
         material, and finish.

   E.   Submit manufacturer's installation instructions under
         provisions of Section [01300.] [01340.]

1.08B   Use the following two paragraphs for submission of physical samples.

1.08E   When manufacturers' instructions for specific installation requirements are utilized, carefully edit Part 3, Execution, requirements to avoid conflict with those instructions.

1.09   COORDINATION

   A.   Coordination work of this Section with installation of
         hollow metal frames.

   B.   [Coordinate work with installation of] [Install] metal
         access panels [and rigidly secure in place.] Refer to
         Section 08305.

   C.   Coordinate the installation of bucks, anchors, blocking,
         and electrical and mechanical work which is to be placed
         in or behind framing, furring, and lathing. [Allow such
         items to be installed [during framing.] [after framing is
         complete.]]

   D.   Coordinate furring and framing with installation of metal
         studding of Section [05400.] [09260.]

PART 2   PRODUCTS

2.01   ACCEPTABLE MANUFACTURERS

   A.   [_____.]

   B.   [_____.]

   C.   [_____.]

   D.   Substitutions: Under provisions of Section [01600.]
         [01630.]

2.01   If only one manufacturer is acceptable, list in Article 2.02 and delete this Article; if more than one, list in this Article. If product substitution procedure is used, include Paragraph D. Refer to TAS Sections 09205, 09207 and 09209.

2.02   FRAMING MATERIALS

   A.   Furring Channels: Formed steel; minimum [25] [_____] gage
         ([0.5] [_____] mm) thick, [3/8] [_____] inch ([10] [_____]
         mm) deep x [3/4] [_____] inch ([19] [_____] mm) high;
         length as required.

2.02   Edit the following descriptive specifications for any conflicts with manufacturers' products specified above.

B. Main Ceiling Channels: Formed steel; minimum [18] [_____]
    gage ([1.2] [_____] mm) thick, [3/4] [_____] inch ([19]
    [_____] mm) deep x [1-1/2] [_____] inch ([38] [_____] mm)
    high; length as required.

C. Resilient Channels: Formed steel; minimum [25] [_____]
    gage ([0.5] [_____] mm) thick; size and length as
    required, [serrated] [_____] face, [flatened 'Z']
    [hat shaped] [_____] profile.

2.02C Resilient channels can be used when an acoustic attenuation feature is required.

D. Hangers: Galvanized steel, of size and type to suit
    application, to rigidly support ceiling components in
    place, with maximum deflection as indicated.

2.02D At exterior soffit locations, hanger wires will not resist wind uplift loads. Consider using channels described in the "lateral bracing" paragraph below by spanning between rigid supports. In humid locations, consider specifying a monel or aluminum wire.

E. Lateral Bracing: Formed steel; minimum [16] [_____] gage
    ([1.5] [_____] mm) thick; size and length as required.

F. Casing Bead: Formed sheet steel; minimum [25] [_____] gage
    ([0.5] [_____] mm) thick; thickness governed by plaster
    thickness; maximum possible lengths; [expanded metal]
    [solid] flanges, with [square] [bullnosed] [bevelled]
    [_____] edges.

2.02F The following accessory listings describe components manufactured in steel. These items are also available in extruded plastic and zinc for special applications.

G. Control [and Expansion] Joint Accessories: Formed sheet
    steel; minimum [25] [_____] gage ([0.5] [_____] mm) thick;
    [accordian] [_____] profile, 2 inch (50 mm) [expanded
    metal] [solid] flanges each side.

2.02G Control and expansion joint accessories are available in a variety of designs, sizes and shapes. Beware of joint components that are relatively inflexible; they can cause adjacent plaster work to crack just beyond the joint.

H. Anchorage and Fastening Device: Approved devices of type
    and size to suit application; to rigidly secure ceiling
    furring members in place.

2.03 LATHING MATERIALS AND ACCESSORIES

A. Metal Lath: [ANSI/ASTM C847;] [FS QQ-L-101;] [flat diamond
    [self-furring] mesh] [flat rib, 3/8 inch (10 mm) high]
    [rod-ribbed, 3/4 inch (19 mm) high] [stamped sheet] [paper
    interwoven lath]; of weight to suit application.

B. Corner Mesh: Formed sheet steel; minimum [26] [_____] gage
    ([0.5] mm) thick; [perforated] [expanded] flanges shaped
    to permit complete embedding in plaster; minimum [2]
    [_____] inch ([50] [_____] mm) size.

C. Strip Mesh: Expanded metal lath, minimum [26] [_____] gage
    ([0.5] [_____] mm) thick; [2] [_____] inch ([50] mm) wide
    x [24] [_____] inch ([600] [_____] mm) long.

D. Anchorages: Tie wire, nails, screws and other metal
    supports, of type and size to suit application; to rigidly
    secure lathing materials in place.

E. Polyethylene Sheet: Clear, 6 mil (0.15 mm) thick.

2.04 FINISHES

A. Framing Materials: [Galvanized.] [Rust inhibitive primer.]

B. Hangers, Anchors, and Fastening Devices: [Galvanized.]
    [_____.]

C. Lath Materials: [Galvanized.] [Rust inhibitive primer.]

PART 3   EXECUTION

3.01   INSPECTION

A. Verify that surfaces conditions are ready to receive work.

B. Beginning of installation means acceptance of [existing surfaces.] [substrate.]

3.02   WALL AND FURRED SPACE FRAMING

A. [Erect wall furring by directly attaching to [concrete masonry] [concrete] [_____] walls.]

B. Erect furring channels [horizontally.] [vertically.]. Secure in place on alternate channel flanges at maximum [24] [_____] inches ([600] [_____] mm) o.c.

C. Spacing furring channels maximum [16] [24] inches ([400] [600] mm) on center, not more than [4] [_____] inches ([100] [_____] mm) from [floor and ceiling lines.] [abutting walls.] [_____.]

D. Erect resilient channels at maximum [24] [_____] inches ([600] [_____] mm) on center. [Place joints over framing members.]

E. Establish control [and expansion] joints with specified joint device.

****** [OR] ******

F. Establish control [and expansion] joints with [back to back] casing beads [set [1/4] [_____] inch ([6] [_____] mm) apart]. Set both beads over [6] [_____] inch ([150] [_____] mm) wide strip of polyethylene sheet for air seal continuity.

3.03   CEILING AND SOFFIT FRAMING

A. Install furring to height indicated. Erect after above ceiling work is complete. Coordinate the location of hangers with other work.

B. Install ceiling furring independent of walls, columns, and above ceiling work. Securely anchor hangers to structural members or embed in structural slab. Space hangers to achieve deflection limits indicated.

C. Space main carrying channels at maximum [72] [_____] inch ([1 800] [_____] mm) centers; not more than [6] [_____] inches ([150] [_____] mm) from wall surfaces. Lap splice securely.

D. Securely fix carrying channels to hangers to prevent turning or twisting and to transmit full load to hangers.

E. Place furring channels perpendicular to carrying channels, not more than [2] [_____] inches ([50] [_____] mm) from perimeter walls, and rigidly secure. Lap splice securely.

F. Reinforce openings in suspension system which interrupt main carrying channels or furring channels with lateral channel bracing. Extend bracing minimum [24] [_____] inches ([600] [_____] mm) past each opening.

G. Laterally brace suspension system.

H. Erect resilient channels at maximum [_____] inches ([_____] mm) on center. Rigidly secure in place.

I. Establish contraction, control [, and expansion] joints with specified joint device.

3.02 Verify that by specifying the following component spacings, the specified performance criteria described in Article 1.05 is not compromised.

J. Establish contraction, control [, and expansion] joints with [back to back] casing beads [set [1/4] [_____] inch ([6] [_____] mm) apart.] Set both beads over [6] [_____] inch ([150] [_____] mm) wide strip of polyethylene sheet for air seal continuity.

3.04 LATHING MATERIALS

A. Apply metal lath taut, with long dimension perpendicular to supports.

B. Lap ends minimum [one] [_____] inch ([25] [_____] mm). Secure end laps with tie wire where they occur between supports.

C. [Lap sides of diamond mesh lath minimum [1-1/2] [_____] inches ([38] [_____] mm). Nest outside ribs of rib lath together.]

D. Attach metal lath to wood supports using nails at maximum [_____] inches ([_____] mm) on center.

E. Attach metal lath to metal supports using [tie wire] [_____] at maximum [6] [_____] inches ([150] [_____] mm) on center.

F. Attach metal lath to [concrete] [concrete masonry] using wire [hair pins,] [hooks,] [or] [loops]. Ensure that anchors are securely attached to backup surface and spaced at maximum [24] [_____] inches ([600] [_____] mm) on center.

G. Continuously reinforce internal angles with corner mesh, except where the metal lath returns 3 inches (75 mm) from corner to form the angle reinforcement; fasten at perimeter edges only.

H. Place [beaded] [bullnosed] external angle with mesh at corners; fasten at outer edges only.

I. Place strip mesh diagonally at corners of lathed openings. Secure rigidly in place.

J. Place [4] [_____] inch ([100] [_____] mm) wide strips of metal lath centered over junctions of dissimilar backing materials. Secure rigidly in place.

K. Place casing beads at terminations of plaster finish. Butt and align ends. Secure rigidly in place.

L. Establish contraction, control [,and expansion] joints with specified joint device.

M. Establish contraction, control [,and expansion] joints with [back to back] casing beads [set [1/4] [_____] inch ([6] [_____] mm) apart.] Set both beads over [6] [_____] inch ([150] [_____] mm) wide strip of polyethylene sheet for air seal continuity.

N. Place control joints vertically above each top corner of door [and glazed] frames to [6] [_____] inches ([150] [_____] mm) above ceiling line.]

3.05 TOLERANCES

A. Maximum Variation from True Lines and Levels: [1/8 inch in 10 feet] [_____] ([3 mm in 3 m] [_____]).

B. Maximum Variation from True Position: [1/8] [_____] inch ([3] [_____] mm).

3.06 SCHEDULE

3.05J Exterior soffits require careful attention to thermal and structural movement and shrinkage caused by curing of plaster. Jointing of cement plaster work should not exceed 3 to 4 feet (1 to 1.2 m) without scribed contraction joints specified in the relevant cement plaster section. Control joints in this work should not exceed 12 feet (3.5 m). Joints constructed of back to back casing beads can be filled with a sealant capable of flexible joint movement. Expansion joints are usually associated with building expansion joint lines and require special consideration. Select one of the following paragraphs and coordinate with the drawings.

3.04L Coordinate the following paragraph with the statements under Paragraphs 3.02 E and F and 3.03 I and J. Do not repeat requirements.

3.06 Provide a schedule to identify fire rating assembly, classifications as applicable to this section, varying finish requirements, or any other special conditions.

DRAWING COORDINATION CONSIDERATIONS

These Drawing Coordination Considerations describe Drawing related items that should be considered with **SPECTEXT Section 09111.**

Drawing Terms: The generic terms in the adjoining column (A thru H) should be used in drawing notes to ensure definitions parallel the specification section. If more than one type of an item indicated is required, indicate on drawings and in the specification section as "Type A," "Type B," etc.

Coordination: The adjoining items (A thru J) require special attention when coordinating drawings and specifications, to ensure parallel information and references.

Caution: The adjoining item A indicates that caution should be observed prior to incorporating into drawings or specifications.

A. ANCHORAGE DEVICES: Fasteners used to attach runners or studs to building structural system or supporting components.
B. FASTENERS: Spiral threaded screws for attaching studs to runners, furring to studs, and splicing studs.
C. FURRING: Channel shaped steel components used for interruptions to stud framing. Furring members are fastened or wire tied into position.
D. METAL BACKING: Sheet metal strips of various thicknesses appropriate to attachment function requirements.
E. METAL STUDS: Vertical 'C' or channel formed steel components, cut to a length slightly less than nominal stud height to accommodate structural building movement and deflection, and fitted into horizontal runners. Studs must be sized to suit height of partition and to accommodate any live and dead loads anticipated (such as wall hung cabinets or fixtures).
F. RUNNERS: Horizontal channel shaped steel components, fitted to the floor and overhead structure or ceiling, configured to receive the vertical studs.
G. SEALANT: Non-hardening compound used to acoustically seal stud framing system between occupied spaces. Sealant is placed between runners and substrate surfaces.
H. STRUCTURAL STUDS: Vertical 'C' or channel shaped steel components, load bearing, sized to accommodate live and dead loads anticipated.

A. Detail size and location of sheet metal backing or wood blocking within stud framing system for anchoring or supporting fixtures or equipment.
B. For detailing assistance in the correct placement of acoustic sealant, refer to literature provided by the stud and partition manufacturers and the Gypsum Association.
C. Coordinate the physical configuration of insulation and sealant locations for acoustic separations. If two or more partition acoustical ratings are required, indicate acoustic values on Drawings. If the specification section indicates an acoustic rating for all stud framed partition assemblies, do not indicate insulation thickness on Drawings.
D. Detail the stud positions at exterior wall locations and openings to permit attachment of the air and vapor barrier materials. Coordinate these details with the specification requirements of SPECTEXT Section 07190.
E. Coordinate and schedule various partition types and heights and acoustic and fire rated wall assemblies, on the Drawings.
F. Ensure stud thickness is adequate to accommodate required insulation thickness.
G. At light switch locations, ensure double studding at door jambs can be achieved.
H. Coordinate anticipated deflection of structural members when detailing stud top runner attachment. Detail a vertical slip connection with the studding.
I. Indicate horizontal bracing, if applicable.
J. Detail special framing conditions where fire rated partitions enclose or pass-by structural framing or joists.

A. Do not use studs for framing horizontal soffits, horizontal furring, or for irregular framing where studs cannot be securely and rigidly fixed into runners at each end of the stud. Use furring components for this purpose.

143

SPECTEXT SPECIFICATION FOR

METAL STUD FRAMING SYSTEM
SECTION 09111

This section includes light gage non-axial load bearing metal stud framing, usually 20 gage or lighter, including framed integral openings and bracing. This section is usually used for exterior wall non-load bearing framing infill, designed to resist wind/suction loads. This section can be edited to provide stud materials for interior partition systems.

Load bearing stud framing for structural conditions is included in SPECTEXT Section 05400. This section can be supplemented with performance criteria statements, provided conflicting prescriptive statements are edited or deleted accordingly.

PART 1   GENERAL

1.01   SECTION INCLUDES

A.   Formed metal stud framing [at [_____] locations.]

B.   Framing accessories.

1.02   PRODUCTS FURNISHED BUT NOT INSTALLED UNDER THIS SECTION

A.   Section [_____-_____]: [Anchors] [_____] for attaching work of this Section.

1.02 List sections which specify installation of products furnished in this section and indicate specific items.

1.03   PRODUCTS INSTALLED BUT NOT FURNISHED UNDER THIS SECTION

A.   Section [_____-_____]: Anchors for support of [_____.]

B.   Section [_____-_____]: [Window] [_____] anchors.

1.03 List sections which furnish products for installation in this section and indicate specific items.

1.04   RELATED SECTIONS

A.   Section 05400 - Cold-Formed Metal Framing.

B.   Section [_____-_____]: Metal fabrications attached to stud framing.

C.   Section 06114 - Wood Blocking and Curbing: Rough wood blocking within stud framing.

D.   Section 07190 - Vapor and Air Retarders.

E.   Section 07213 - Batt and Blanket Insulation: Insulation within stud framing.

F.   Section 07900 - Joint Sealers.

G.   Section 09206 - Metal Furring and Lathing.

H.   Section 09210 - Gypsum Plaster.

I.   Section 09220 - Portland Cement Plaster.

J.   Section 09260 - Gypsum Board Systems: Metal studs for partitioning.

K.   Section 10616 - Demountable Gypsum Board Partitions: Metal studs for partitioning.

1.04 Include only reference standards that are to be indicated within the text of this section. Refer to TAS Sections 09205, 09207 and 09209. Edit the following, adding and deleting as required for project and product selection.

1.05   REFERENCES

A.   ASTM A525 - General Requirements for Steel Sheet, Zinc-Coated (Galvanized) by the Hot-Dip Process.

B.   ANSI/ASTM A591 - Steel Sheet, Cold-Rolled, Electrolytic Zinc-Coated.

C.   ASTM C645 - Non-Load (Axial) Bearing Steel Studs, Runners (Track) and Rigid Furring Channels for Screw Application of Gypsum Board.

1.05 List reference standards that are included within the text of this section. Refer to Section 01090 for abbreviations and addresses of standards agencies and associations. Edit the following as required for project conditions. Refer to TAS Sections 05400, 09205 and 09260.

D. ASTM C 754 - Installation of Steel Framing Members to Receive Screw-Attached Gypsum Wallboard, Backing Board, or Water-Resistant Backing Board.

E. FS TT-P-645 - Primer, Paint, Zinc-Chromate, Alkyd Type.

F. GA 203 - Installation of Screw-Type Steel Framing Members to Receive Gypsum Board.

1.06   SYSTEM DESCRIPTION

A. Metal stud framing system for exterior wall infill, with exterior sheathing specified in Section [_____], [_____] type insulation specified in Section [_____], interior gypsum board specified in Section [_____], [and] [_____].

B. Metal stud framing system for interior walls, with [[_____] type acoustic insulation specified in Section [_____],] gypsum board specified in Section [_____], [and] [_____.]

C. Maximum Allowable Deflection: [1/180] [1/270] [1/360] [_____] span.

D. Design system to accommodate construction tolerances, deflection of building structural members, and clearances of intended openings.

1.07   SUBMITTALS

A. Submit shop drawings under provisions of Section [01300.] [01340.]

B. Submit shop drawings [of prefabricated work] indicating component details, [stud layout,] [framed openings,] [anchorage to structure,] [type and location of fasteners,] and accessories or items required of other related work.

C. Describe method for securing [studs] [_____] to tracks, [splicing,] and for blocking and reinforcement to framing connections.

D. Submit product data under provisions of Section [01300.] [01340.]

E. Submit product data describing standard framing member materials and finish, product criteria, load charts, limitations, [and] [_____.]

F. Submit manufacturer's installation instructions under provisions of Section [01300.] [01340.]

1.08   QUALITY ASSURANCE

A. Perform work in accordance with [GA 203] [and] [ASTM C754.]

B. Maintain one copy of [each] document on site.

1.09   MOCK-UP

A. Provide mock-up of [exterior wall infill] [_____] under provisions of Section [01400.] [01405.]

B. Coordinate construction of mock-up with related Sections.

C. Provide analysis of mock-up under provisions of Section [01400.] [01410.]

D. When accepted, mock-up will demonstrate minimum standard for the Work. Mock-up may [not] remain as part of the Work.

1.10   PRE-INSTALLATION CONFERENCE

A. Convene [one] [___] week prior to commencing work of this Section, under provisions of Section [01200.] [01210.]

B. Discuss construction of mock-up with related Sections.

1.06   Use this Article carefully; restrict paragraph statements to describe components used to assemble this system. Do not repeat statements made in Article 1.01, Section Includes.

1.06C   The deflection limits in Paragraph C relate directly to exterior wall systems, and the nature, type and exposure of the finish to be installed over the metal stud framing work. Deflection limits will vary with type of surface finish, lateral wind/suction loads, span of studs, and weather exposure. Smooth finish interior plasters may require a deflection limit of 1/270 or less.

1.07   Do not request submittals if Drawings sufficiently describe the products of this section or if proprietary specifying techniques are used. The review of submittals increases the possibility of unintended variations to Drawings, thereby increasing the Specifier's liability. Shop drawings may be required for shop fabricated framing or special details. Metal framing component manufacturer's product data is usually acceptable to illustrate standard framing details.

1.07F   When manufacturer's instructions for specific installation requirements are referenced in Part 3 Execution, include the request 1.07F for submittal of those instructions. Edit the Part 3 statements to avoid conflict with those instructions.

1.09   Use this Article for full sized erected assemblies required for review of construction, coordination of work of several sections, testing or observation of operation.

1.10   Include this Article when specifying a mock-up.

1.11   SEQUENCING AND SCHEDULING

    A. Sequence work under the provisions of Section [01005.] [_____.]

    B. Sequence work with other work directly affected by this Section.

    C. Coordinate work under provisions of Section [01005.] [01040.] [01041.]

    D. Coordinate the work of [related Sections.] [_____.]

PART 2   PRODUCTS

2.01   MANUFACTURERS

    A. [_____] Product [_____.]

    B. [_____] Product [_____.]

    C. [_____] Product [_____.]

    D. Substitutions: Under provisions of Section [01600.] [01630.]

2.02   STUD FRAMING MATERIALS

    A. Framing System Components: ASTM C645.

****** [OR] ******

    A. Studs: [ASTM A525, galvanized to [G90] [___] coating class,] [ANSI/ASTM A591, electrogalvanized,] non-load bearing rolled steel, channel shaped, punched for utility access, [as scheduled.] [as follows:]
        1. Width: [1-5/8] [2-1/2] [3-1/2] [4] [6] [___] inches ([40] [60] [90] [100] [150] [___] mm).
        2. Thickness: [25] [27] [___] gage ([0.56] [0.45] [___] mm).

    B. Runners: Of same material and finish as studs, bent leg retainer notched to receive studs [with provision for crimp locking to stud.] [Ceiling runners with extended legs.]

    C. Furring and Bracing Members: Of same material and finish as studs, thickness to suit purpose.

    D. Fasteners: [GA 203.] [Self-drilling, self-tapping screws.] [_____.]

    E. Metal Backing: [20] [___] gage ([0.9] [___] mm thick) [galvanized] steel for reinforcement of [_____.]

    F. Anchorage Devices: [Power driven.] [Powder actuated.] [Drilled expansion bolts.] [Screws with sleeves.] [_____.]

    G. [Acoustic Sealant: As specified in Section [09260.] [_____.]

    H. Primer: FSTT-P-645, for touch-up of galvanized surfaces.

2.03   FABRICATION

    A. Fabricate assemblies of [framed sections] [_____] to sizes and profiles required; with framing members fitted, reinforced, and braced to suit design requirements.

    B. Fit and assemble in largest practical sections for delivery to site, ready for installation.

1.11 Use this Article to identify special coordination with other specialty sections such as mechanical or electrical, or special components, integral with this section, such as grab bars, or accommodating wall switches in double studding.

2.01 In this Article, list the manufacturers acceptable for this project. If product substitution is allowed, include Paragraph D. Refer to SPEC-DATA II Building Products Directory.

2.01D Edit the following descriptive specifications to identify project requirements and to eliminate any conflict with manufacturers' products specified above. Refer to TAS Sections 05400, 09205 and 09260.

2.02 Edit the following text recognizing wind and suction loads applied to exterior wall assemblies and its effect on member size and spacing. Studs used in the exterior and in potentially corrosive environments should be galvanized.

2.02A If multiple material gages or stud widths are required, schedule this criteria at the end of this section.

2.03 Delete the following paragraphs when shop fabricated assemblies are not required.

PART 3   EXECUTION

3.01   EXAMINATION

    A.   Verify that conditions are ready to receive work.

    B.   Verify field measurements are as [shown on Drawings.] [instructed by the manufacturer.]

    C.   Verify that rough-in utilities are in proper location.

    D.   Beginning of installation means installer accepts existing conditions.

3.02   ERECTION

    A.   Align and secure top and bottom runners at [24] [___] inches ([600] [___] mm) oc. [Place [one] [two] [___] beads of [acoustic] sealant between runners and substrate.] [Achieve air seal between runners and substrate in conjunction with Section [07190.] [_____.]]

3.02A Edit Paragraph A for sealant or air seal between runner and substrate as appropriate.

    B.   Fit runners under and above openings; secure intermediate studs at spacing of wall studs.

    C.   Install studs vertically at [12] [16] [24] [___] inches ([300] [400] [600] [___] mm) oc. [Place [one] [two] [___] beads of [acoustic] sealant between studs and adjacent vertical surfaces.] [Achieve air seal between studs and adjacent vertical surfaces in conjunction with Section [07190.] [_____.]]

3.02C Check span capabilities of stud with intended sheathing material to determine spacing. Edit Paragraph C for sealant or air seal between studs and adjacent vertical surfaces as appropriate.

    D.   Connect studs to tracks using [crimping] [clip and tie] [fastener] [_____] method.

    E.   Stud splicing [not] permissible. [Splice studs with 8 inch (200 mm) nested lap, secure each stud flange with [flush head screw.] [crimped indentation.] [_____.]]

    F.   Construct corners using minimum three studs.

    G.   Double studs at wall openings, door and window jambs, and not more than 2 inches (50 mm) each side of openings.

    H.   Brace stud framing system and make rigid.

    I.   Coordinate erection of studs with requirements of [door and window frame] [_____] supports and attachments.

    J.   Align stud web openings.

    K.   Coordinate installation of bucks, anchors, and blocking with electrical and mechanical work to be placed in or behind stud framing.

    L.   Blocking: [Secure wood blocking to studs.] [Secure steel channels to studs.] [Install blocking for support of [plumbing fixtures,] [toilet partitions,] [wall cabinets,] [toilet accessories,] [hardware,] [and] [_____].]

3.02L Wood blocking can be placed in vertical studs at door jambs to assist in securing the door frame.

    M.   Extend stud framing to ceiling only. Attach ceiling runner securely to [acoustical ceiling track] [ceiling framing] in accordance with [manufacturer's instructions.] [details indicated.]

    M.   Refer to Drawings for indication of partitions extending to ceiling only and for partitions extending through ceiling to structure above. Maintain clearance under structural building members to avoid deflection transfer to studs. [Provide extended leg ceiling runners.]

    N.   Coordinate placement of insulation in multiple stud spaces made inaccessible after stud framing erection.

3.02N The following paragraph identifies the need to make a similar statement in the appropriate insulation section for coordination purposes.

3.03   TOLERANCES

    A.   Maximum Variation From True Position: [___] inch ([___] mm).

    B.   Maximum Variation of any Member from Plane: [___] inch ([___] mm).

3.04   SCHEDULE

    A.   Exterior Wall Infill (Main Floor): 6 inch nominal stud width, wind load of 18 lb/sq ft, sheathed for aggregate finish specified in Section 09825.

    B.   Exterior Wall Infill (Typical Floors): 4 inch nominal stud width, sheathed for metal cladding construction specified in Section 07461.

    C.   Interior Partitions (Typical Floors): 4 inch studs at demising walls and 3-1/2 inch studs elsewhere, gypsum board finish both faces, typical.

    D.   Interior Acoustic Partitions: 3-1/2 inch studs, batt insulation, 45 STC required.

3.03 Include this Article when tolerances are important to the nature of the project. Tolerances required will vary depending on the nature of subsequent construction of finish. Flat, smooth surface finishes will require rigid erection tolerances.

3.04 Provide a schedule when framing member nominal dimensions vary, locations require identification, design loads vary with location, or special conditions require identification. The following examples may assist in developing such a schedule.

Designation: C 842 – 85

Standard Specification for
APPLICATION OF INTERIOR GYPSUM PLASTER[1]

This standard is issued under the fixed designation C 842; the number immediately following the designation indicates the year of original adoption or, in the case of revision, the year of last revision. A number in parentheses indicates the year of last reapproval. A superscript epsilon (ε) indicates an editorial change since the last revision or reapproval.

1. Scope

1.1 This specification covers the minimum requirements for full-thickness (as specified in Table 6) gypsum plastering on gypsum, metal, masonry, or monolithic concrete bases designed or prepared to receive gypsum plaster.

NOTE 1—To secure desirable results, this specification should be coordinated with the provisions of Specification C 841.

1.2 Where a specific degree of fire resistance is required for plastered assemblies and construction, details of construction shall be in strict accordance with official reports of fire tests conducted in recognized testing laboratories in accordance with Methods E 119 (see Annex A1).

1.3 Where a specific degree of sound control is required for plastered assemblies and construction, details of construction shall be in compliance with official reports of tests conducted in recognized sound testing laboratories in accordance with the applicable sound tests of Method E 90, Test Method C 423, or Method E 492 (see Annex A1.2).

2. Applicable Documents

2.1 *ASTM Standards:*

C 5 Specification for Quicklime for Structural Purposes[2]

C 11 Definitions of Terms Relating to Gypsum and Related Building Materials and Systems[2]

C 28 Specification for Gypsum Plasters[2]

C 35 Specification for Inorganic Aggregates for Use in Gypsum Plaster[2]

C 59 Specification for Gypsum Casting and Molding Plaster[2]

C 61 Specification for Gypsum Keene's Cement[2]

C 206 Specification for Finishing Hydrated Lime[2]

C 423 Test Method for Sound Absorption and Sound Absorption Coefficients by the Reverberation Room Method[3]

C 631 Specification for Bonding Compounds for Interior Plastering[2]

C 841 Specification for Installation of Interior Lathing and Furring[2]

E 90 Method for Laboratory Measurement of Airborne-Sound Transmission Loss of Building Partitions[3]

E 119 Methods of Fire Tests of Building Construction and Materials[4]

E 492 Method of Laboratory Measurement of Impact Sound Transmission Through Floor-Ceiling Assemblies Using the Tapping Machine[3]

3. Definitions

3.1 *catfaces*—blemishes or rough depressions in the finish coat caused by variation in base coat thickness.

3.2 *coat*—a thickness or layer of plaster applied over a surface in a single application.

3.2.1 *scratch coat*—the first coat of plaster applied over a lath or other substrate.

[1] This specification is under the jurisdiction of ASTM Committee C-11 on Gypsum and Related Building Materials and Systems and is the direct responsibility of Subcommittee C11.03 on Specifications for Application of Gypsum and Other Products in Assemblies.

Current edition approved April 26, 1985. Published June 1985. Originally published as C 842 – 76. Last previous edition C 842 – 79.

[2] *Annual Book of ASTM Standards.* Vol 04.01.
[3] *Annual Book of ASTM Standards.* Vol 04.06.
[4] *Annual Book of ASTM Standards.* Vol 04.07.

3.2.2 *base coat*—the sum of the scratch and brown coats or the total coats in place prior to application of finish coats.

3.2.3 *brown coat*—the second coat of plaster applied in either two-coat or three-coat work.

3.2.4 *finish coat*—the last coat of plaster applied in either two-coat or three-coat work.

3.2.5 *two-coat work*—the application of the scratch and brown coats from the same mix with no time allowed for setting of the scratch coat before the brown coat is applied (also called double-back).

3.2.6 *three-coat work*—the application of plaster in three successive coats, leaving time between coats for setting or drying, or both, of the plaster.

3.3 *smooth-trowel finish*—a finish resulting from steel troweling with a minimum of water after the plaster has become firm. A smooth finish free of trowel marks, blemishes, or other imperfections.

3.4 *texture finish*—a finish resulting from (*1*) trowel application followed by floating or texturing of the surface with any of a variety of tools using a minimum of water or (*2*) machine application which may or may not be hand textured. The surface should be textured uniformly, free of slick spots or other blemishes.

3.5 *metal bases*—expanded metal, welded or woven wire, or punched sheet metal plaster bases.

3.6 Other terms relating to gypsum shall be as defined in Definitions C 11.

4. Delivery of Materials

4.1 All manufactured materials shall be delivered in the original packages, containers, or bundles bearing the brand name and manufacturer (or supplier) identification.

5. Protection of Materials

5.1 Plasters and other cementitious materials shall be kept dry until used; they shall be stored off the ground, under cover, and away from walls with condensation and other damp surfaces. Metal products shall be protected, while stored, against rusting.

6. Environmental Conditions

6.1 *Temperatures*—When the ambient outdoor temperature at the building site is less than 40°F (4.4°C), maintain a continuous uniform temperature of not less than 40°F and not more than 80°F (26°C) in the building for

a period of not less than 1 week prior to the application of plaster (Note 2), while the plastering is being done, and for a period of at least 1 week after the plaster is set, or until the plaster is dry. Distribute the heat evenly in all areas, and use deflective or protective screens to prevent concentrated or uneven heat on the plaster near the heat source.

NOTE 2—The building should be permitted to become adjusted to the uniform temporary heating conditions for at least 1 week prior to the start of plastering to minimize the possibility of plaster cracking due to structural movements caused by thermal changes. Blasts of cold air through openings should be avoided.

6.2 *Ventilation*—Provide ventilation to remove the water in excess of that required for hydration of the plaster, immediately subsequent to its application and set (Note 3). See Annex A1.3 for provisions for ventilating portions of underside of roofs enclosed with lath and plaster ceilings.

NOTE 3—In glazed buildings ventilation shall be accomplished by keeping windows open approximately 2 in. (50 mm) top and bottom or side-pivoted windows approximately 4 in. (100 mm), to provide air circulation. For enclosed areas of buildings that lack natural ventilation, provisions shall be made to exhaust the moist air to the outside by mechanical means such as temporary circulators, exhaust fans, or the air conditioning system. A portion of the water used in mixing gypsum plaster is necessary for the chemical reaction that rehydrates and hardens the plaster. The excess amount required for mixing and application must be exhausted from the plaster after set has taken place. Free circulation of air is necessary to avoid "sweat-outs." Vent enclosed areas lacking natural ventilation by use of temporary mechanical means. If glazed sashes are not in place and the building is subjected to hot dry winds, openings shall be screened with cheesecloth or similar materials. Free circulation of hot dry winds over the face of freshly applied plaster causes "dry-outs" by removing water from the plaster before the setting action is completed.

7. Materials

7.1 *Gypsum Plasters* (Note 4)—The following plasters shall conform to Specification C 28:

7.1.1 Ready-mixed.

7.1.2 Neat.

7.1.3 Wood-fibered.

7.1.4 Gauging for finish coat.

7.2 *Gypsum Casting and Molding Plasters*—Specification C 59.

7.3 *Gypsum Keene's Cement*—Specification C 61. (See Note 4.)

NOTE 4—Gypsum plasters are not recommended for use in exterior locations or interior "wet" areas.

7.4 *Lime:*

7.4.1 *Special Finishing Hydrated Lime*—Specification C 206, Type S.

7.4.2 *Quicklime for Structural Purposes*—Specification C 5.

7.5 *Lime Putty*—Lime putty exceeding .8 weight % of unhydrated magnesium oxide shall not be used for finish coat plaster.

7.6 *Aggregates:*

7.6.1 *Aggregates for Base Coat Plaster*—Specification C 35. Fine sand having rounded particle shape, uniform in size (frequently called "quicksand") shall not be used for base coat plastering.

NOTE 5—Such material (quicksand) produces a weak plaster, causes workability problems, and may result in separation of gypsum plaster and aggregate when pumping through a plaster machine.

7.6.2 *Aggregates for Finish Coat Plasters*—Specification C 35, except that gradation shall be within the limits specified in Table 1.

7.6.3 Sand for job-mixed lime putty-gypsum gauged, sand float finish (see 9.4.5) shall be graded within the limits specified in Table 2.

7.6.4 Vermiculite for job-mixed finish coat plaster shall weigh not less than 7 nor more than 12 lb/ft³ (34 to 59 kg/m³) and shall be graded within the limits specified in Table 3. In all other respects vermiculite shall meet the requirements of Specification C 35.

7.7 *Water*—Water used in mixing and finishing plaster shall be clean, fresh, suitable for domestic water consumption, and free of such amounts of mineral or organic substances as would affect the set, the plaster, or any metal in the system.

NOTE 6—Water containing salt or alum, or water in which tools have been washed, accelerates the "set" and may cause efflorescence. Water from stagnant pools and wells frequently contain organic or vegetable matter which may retard the "set," cause staining, or interfere with the bond.

7.8 *Metal Products*—Metal products shall have a corrosion protective coating.

7.9 *Bonding Compounds*—Specification C 631.

8. Surface Preparation

8.1 *Examination*—All surfaces, such as unit masonry, monolithic concrete, lathing of all types, and accessories that are to receive plaster shall be examined by the plastering contractor or his representative before the plaster is applied. The proper authorities shall be promptly notified of all unsatisfactory conditions. Do not apply plaster until unsatisfactory conditions are corrected.

8.1.1 Do not apply plaster over any surface of unit masonry or monolithic concrete that has been coated with materials such as bituminous compounds or form release compounds that are detrimental to the plaster or act as a deterrent to bond or adhesion. Only a furred or self-furring metal plaster base shall be attached to such surfaces to receive the plaster.

8.1.2 Carefully examine metal or plastic accessories, such as beads, screeds, molds, and wood grounds to see that they are straight, plumb, level, square, and true to the required angles and curves as the case may require. Securely attach such accessories to the substrate so as to permit embedment of flanges, before plaster application is started.

8.2 *Conditioning of Surfaces:*

8.2.1 *Masonry Surface*—Immediately prior to plaster application, the suction of which must be reduced, wet down the masonry surface, until satisfactory suction is obtained (Note 7). There shall be no free water visible on the surface at the time of plaster application.

NOTE 7—This is particularly important if wood-fiber plaster or other high plaster-low aggregate ratio material is used.

8.2.2 *Gypsum Lath*—Do not wet down gypsum lath and metal plaster bases.

8.2.3 *Monolithic Surfaces*—On solid bases such as cast-in-place or precast concrete, where bond to plaster depends on its ability to absorb water (suction) and surface roughness (mechanical key), obtain the bond by one of the following methods:

8.2.3.1 The surface shall be sufficiently rough or shall be roughened to ensure mechanical key by sand-blasting, wire brushing, acid wash, or chipping. The surface shall be completely free of grease, oil, form release compounds, dirt, dust, paint, laitance, efflorescence (Note 8), or other foreign matter that would inhibit suction, mechanical keying, or cause staining.

NOTE 8—Laitance and efflorescence can be removed by washing first with a 10 % solution of commercial hydrochloric acid (muriatic acid) and water and then with clean water to remove all traces of acid. Grease, oil, and form release compounds shall be removed with cleaning agents compatible with surface and subsequent plastering.

NOTE 9—The use of dressed (smooth) lumber, metal, plaster and plywood forms, their oiling, greasing, or both, use of form release compounds on such forms, and the vibration of concrete produce concrete surfaces so dense or smooth, that the mechanical bond necessary for plaster is absent. Lumber, oiled forms, and exposed reinforcement metals may cause staining. Check concrete surface: nonstaining, disappearing-type release agents should be used that have proven to be compatible with plaster, on surfaces to be plastered without use of metal reinforcement. Depressions in concrete surface, deeper than ¹/₈ in. (3.2 mm) shall be patched flush to the surface with compatible materials prior to plaster application. Fins or protrusions extending more than ¹/₁₆ in. (1.6 mm) from the concrete surface shall be removed. Protrusions of ¹/₁₆ in. or less shall be feathered out with compatible materials prior to plaster application.

8.2.3.2 Use bonding compounds to bond plaster to solid bases having low or questionable suction. Apply the bonding compounds in accordance with the manufacturer's printed directions. Do not use bonding compounds in assemblies given a fire resistance rating based on a full-scale Methods E 119 fire test, unless the specific bonding agent was used in the tested assembly. Do not use bonding compounds that are water re-emulsifiable in areas of a building where cyclic or continuous exposure to very humid or wet conditions, or in which a dew point condition may occur in the plaster (Note 10). Do not use bonding compounds instead of metal plaster base over monolithic surfaces that require more than ³/₈ in. (9.5 mm) for horizontal and ⁵/₈ in. (15.9 mm) for vertical surfaces to achieve required lines or planes. Protect applied bonding compound from adherence of airborne dirt and dust where immediate covering of the compound with plaster will not occur.

NOTE 10—Dew point conditions occur frequently in such buildings as commercial laundries, natatoriums, and other wet areas such as showers, hydro-therapy, and washing rooms.

8.2.3.3 Where bond cannot be obtained over the entire surface by any of the methods specified in 8.2.3.1 or 8.2.3.2, then use metal lath. The metal lath used shall be self-furring or be furred out, not less than ¹/₄ in. (6.4 mm)

from the surface of the concrete.

8.3 *Grout:*

8.3.1 *Metal Bases*—Grout metal bases for solid partitions with plaster. Screed the plaster grout approximately ¹/₄ in. (6.4 mm) below the top edge of the base, and in such fashion as to form a center groove that will permit the lath to extend ³/₄ in. (19 mm) below the top edge of the base, or plaster grout may be placed after installation of lath.

8.3.2 *Metal Frames*—Fill the jambs of hollow metal frames solid with plaster grout.

9. Mix Design

9.1 *Mixing*—Unless specifically approved, do not use hand mixing. Where either mechanical or hand mixing is used, make provision for disposal of waste cleaning water.

9.1.1 Do not use frozen, caked, or lumpy material. Do not retemper or use material that has partially set. Clean mixing boxes, tools, and mechanical mixers and free of all set and hardened materials after mixing each batch. Mix each batch separately.

9.1.1.1 Accelerate the plaster, if necessary, to provide a setting time of not more than 4 h beginning with the addition of the mixing water to the batch of dry plaster.

9.1.2 *Hand Mixing:*

9.1.2.1 Mixing boxes shall be watertight. Where mixing is done in a building, provide water-resistive protection under and around the mixing boxes and water containers.

9.1.2.2 *Gypsum Neat Plaster*—Add aggregate at the job. Mix plaster and aggregate dry to a uniform color, pull to one end of the box, hoe and chop into water at the other end, and thoroughly mix to the desired consistency.

9.1.2.3 *Gypsum Ready-Mixed Plaster*—Do not add aggregate. Place the plaster at one end of the box, hoe into water at the other end, and thoroughly mix to the desired consistency.

9.1.2.4 *Gypsum Wood-Fibered Plaster*—Where used without aggregate, place the plaster at one end of the box, hoe into water at the other end, and thoroughly mix to the desired consistency. Where aggregate is added, mix as specified in 9.1.2.2.

9.1.3 *Mechanical Mixing:*

9.1.3.1 *Gypsum Neat Plaster*—Add aggregate at the job. Follow the following cycle of operations while the mixer is in continuous operation:

(1) Put in the approximate (less than total required) amount of water.
(2) If sand is used, add approximately half the sand. If vermiculite or perlite is used, add all of it.
(3) Add all of the plaster.
(4) Add the remainder of the sand.
(5) Adding water as necessary, mix to the proper consistency. Do not overmix.
(6) Dump the entire batch and use. Clean the mixer for the next batch.

9.1.3.2 *Gypsum Ready-Mixed Plaster*—Aggregates are added at the manufacturer's plant. Follow the following cycle of operation while the mixer is in continuous operation.

(1) Put in the approximate (slightly less than total required) amount of water.

(2) Add the ready-mixed plaster.

(3) Adding water or plaster as necessary, mix to the desired consistency.

(4) Dump the entire batch and use. Clean the mixer for the next batch.

9.1.3.3 *Gypsum Wood-Fibered Plaster*—Where used with aggregate added, the loading and mixing cycle shall be as prescribed in 9.1.3.1. Where used without aggregate, the loading and mixing cycle shall be as prescribed in 9.1.3.2.

9.2 *Base Coat Proportions:*

9.2.1 For *Job-Mixed Plaster*, proportions (Note 11) of sand, perlite, or vermiculite aggregate to 100 lb (45.36 kg) of gypsum neat plaster shall not exceed those specified in Table 4. (See Annex A1.4 for equivalent measure for aggregates.)

9.2.2 *Gypsum Ready-Mixed Plaster* may be used instead of job-prepared mixes of gypsum neat plaster and aggregate provided the proportion of aggregate to plaster does not exceed that specified in Table 4. Gypsum ready-mixed plaster shall be used with the addition of water only, except as otherwise specified herein.

9.2.3 *Gypsum Wood-Fibered Plaster*—The addition of damp, loose sand or perlite or vermiculite shall not exceed 1 ft³ (0.028 m³) to 100 lb (45.36 kg) of gypsum wood-fibered plaster.

NOTE 11—Gypsum wood-fibered plaster is normally applied without the addition of aggregate.

9.3 *Preparation of Lime Putty*—Prepare lime putty from Type N or Type S hydrated lime or pulverized quicklime, in accordance with the printed directions of the manufacturer.

9.3.1 *Hydrated Lime*, hydrated at atmospheric pressure (Type N), shall be used after the soaking period recommended by the manufacturer.

9.3.2 *Hydrated Lime*, hydrated at elevated pressure (Type S), shall be used after the soaking period recommended by the manufacturer.

9.3.3 *Pulverized Quicklime* shall be slaked by sifting it into the amount of water called for in the printed directions of the manufacturer and then allowing it to cool before using.

9.4 *Finish Coat Proportions*—Finish coats may be ready-mixed, requiring the addition of water only or be job mixed in accordance with the applicable requirements of Table 5.

9.4.1 *Troweled Finishes of Lime Putty* gauged with gypsum gauging plaster shall be proportioned in accordance with Table 5. Where such finish coats are to be applied over base coats containing perlite or vermiculite, the addition of a minimum of ½ ft³ (0.14 m³) or a maximum of 1 ft³ (0.028 m³) of fine aggregate, meeting the sieve analysis of Table 1 shall be added to the mix.

9.4.1.1 The nominal proportions of fine aggregate in 9.4.1 are equivalent to:

Fine Aggregate[A,B]

Not Less Than	Not More Than	Per
½ ft³ (0.14 m³)	1 ft³ (0.028 m³)	100 lb (45.36 kg) gypsum gauging
or ⅛ ft³ (0.0035 m³)	¼ ft³ (0.007 m³)	50 lb (22.68 kg) dry hydrated lime
or 1 U.S. gal (0.0038 m³)	2 U.S. gal (0.0076 m³)	50 lb (22.68 kg) dry hydrated lime
or 1 U.S. gal (0.0038 m³)	2 U.S. gal (0.0076 m³)	1 ft³ (0.028 m³) lime putty
or 1 pt (0.0047 m³)	1 qt (0.00094 m³)	1 U.S. gal (0.0038 m³) lime putty

[A] For any of the above proportions, up to ½ ft³ (0.014 m³) of fine aggregate complying with Table 1 may be added to the mix per each 100 lb (45.36 kg) of gypsum Keene's cement.

[B] The largest particle size of aggregate and its proportion ratio will determine the degree of coarseness of the sand finish. The specifier should indicate the maximum size desired.

9.4.2 Troweled finishes of lime putty, gauged with gypsum Keene's cement shall be specified as medium or hard and shall be proportioned in accordance with Table 5.

9.4.2.1 When mechanically mixing, the water shall be put in the mixer first, then the lime, the fine aggregate (if used), and finally the gypsum Keene's cement.

9.4.3 Troweled finishes of prepared gypsum (ready-mixed) finish shall be mixed in accordance with Table 5.

9.4.4 Troweled finishes of vermiculite finish (for use over vermiculite base coats only) shall be proportioned in accordance with Table 5 and shall be mixed by placing the gypsum in water and mixing. Then add the vermiculite finish aggregate and thoroughly mix all materials (3 to 5 min). Additional water shall be added, as necessary, to obtain workable consistency. The mixed plaster shall set in not more than 3 h.

9.4.5 Floated finishes of lime putty, gauged with gypsum gauging plaster shall be proportioned in accordance with Table 5. Sand aggregate, where used, shall be graded within the limits shown for basecoats in Specification C 35 except that all of the sand pass the No. 8 (2.36 mm) sieve.

9.4.6 Floated finishes of lime putty, gauged with gypsum Keene's cement shall be proportioned in accordance with Table 5.

9.4.6.1 When mechanically mixing wet materials follow procedures in 9.4.2.1. For mechanically mixing dry materials, add the lime, gypsum Keene's cement, and sand in that order, mix dry to a uniform color, then add water to produce desired consistency.

9.4.7 Floated finishes of gypsum-sand (job-mixed) shall be proportioned in accordance with Table 5.

9.4.8 Floated finishes of prepared gypsum-sand float (ready-mixed) shall be mixed with water only in accordance with Table 5.

9.4.9 Floated finishes of gypsum-vermiculite (job-mixed) shall be proportioned in accordance with Table 5 and shall be mixed as specified in 9.4.4.

9.4.10 Floated or textured colored finishes, where ready-mixed, shall be mixed in accordance with the printed directions of the manufacturer. Where job mixed, the proportions

shall be in accordance with the applicable provisions of Table 5.

9.4.11 Special finishes shall meet the applicable provisions for proportioning of Table 5. Where thickness of the finish coat will exceed ⅛ in. (3.2 mm) the proportion of the cementitious (setting) material shall be increased, as required to produce a finish free of shrinkage type cracking.

10. Application

10.1 *General*—Apply the plaster by hand or machine as specified herein.

10.2 *Thickness of Plaster:*

10.2.1 Apply gypsum plaster to the thickness specified in Table 6. Plaster thickness shall be measured from the face plane of all plaster bases.

10.2.2 Plaster screeds shall be installed by the plastering contractor.

NOTE 12—Installation of wood or metal grounds or plaster screeds will assist in achieving specified thicknesses.

10.2.3 Metal grounds, beads, or screeds shall be furnished and installed by the plastering or lath contractor.

10.2.4 Wood grounds shall be installed by others.

10.3 *Application of Base Coats:*

10.3.1 *Two-Coat Work* (see 3.2.4)—Apply the scratch (first) coat with sufficient material and pressure to form a good bond to the solid plaster base, and cover completely. Before the scratch coat has set, and without scoring its surface, double back material of the same proportions to bring the plaster out to the grounds or specified thickness. Straighten to a true plane without application of water, and leave the surface porous and sufficiently rough to provide a mechanical bond for the finish (second) coat.

10.3.2 *Three-Coat Work* (see 3.2.5)—Apply the scratch (first) coat with sufficient material and pressure to form tight contact with, and good bond to solid plaster bases and to form full keys through and to embed metal reinforcement and with sufficient depth of material over metal and solid bases to provide mechanical key for the brown (second) coat.

10.3.2.1 After the scratch (first) coat has set firm and hard, apply the brown (second) coat with sufficient material and pressure to ensure tight contact with the scratch coat and to bring the combined thickness of the first and second coats out to the grounds or specified thickness (Note 13), straighten to a true plane without application of water, and leave the surface porous and sufficiently rough to provide mechanical bond for the finish (third) coat.

NOTE 13—To ensure full plaster thickness, plaster screeds should be applied over the scratch coat prior to application of the brown coat.

10.3.2.2 Use three-coat work on the following types of construction:

(1) Over gypsum lath attached to ceiling supports spaced more than 16 in. (406 mm) on center.

(2) Over gypsum lath attached to ceiling supports by clips.

(3) Over ⅜-in. (9.5-mm) thick gypsum lath

where a vapor barrier is used adjacent to the back of the lath.

(4) Over metal plaster bases.

10.3.2.3 Metal base and metal frames for hollow partitions shall be plaster grouted prior to plastering, or shall be filled solid between base or frame and plaster base at the time of plastering (see 8.3.1 and 8.3.2).

10.3.3 *Plastering on Monolithic Concrete* — Observe the applicable requirements of 8.1, 8.1.1, 8.2.3 through 8.2.3.3 and Table 6 footnotes.

10.3.3.1 Walls and columns shall have a coat of dash-bond, bonding compound, or metal reinforcement as provided in 8.2.3 or shall have a scratch coat of plaster followed by a brown coat, the proportions of which are 100 lb (45.36 kg) of gypsum neat plaster to not more than 3 ft³ (0.085 m³) of sand, perlite, or vermiculite applied by the scratch and double-back method. Bring out the brown coat to grounds or specified thickness, using a double-back application if required, straighten to a true plane without application of water, and leave porous and sufficiently rough to provide mechanical bond for the finish coat.

10.3.3.2 Ceilings shall provide bond as prescribed in 8.2.3 and shall be scratched in thoroughly, double back, filled out to a true even plane, without the application of water, and be scored or left porous and sufficiently rough to provide bond for the finish coat.

10.3.4 *Solid Plaster Partitions with Steel Studs (Metal Lath and Plaster)* — These partitions shall not be less than 2 in. (50.8 mm) in thickness and shall have scratch, back-up, and finish coats in accordance with 10.3.4.1 through 10.3.4.4.

10.3.4.1 Where temporary bracing of studs are installed, maintain the bracing until the scratch coat on the lath side has set. Apply the scratch coat on the lath side with sufficient material and pressure to form full keys and embed the metal reinforcement and with sufficient depth of material to be scored to a rough surface immediately following its application.

10.3.4.2 Apply the back-up coat on the channel (stud) side in not less than two applications after the scratch coat on the lath side has set firm and hard. The first application shall completely cover the keys of the scratch coat. The second coat or coats shall bring the plaster out to the grounds. Straighten to a true plane without application of water, and leave porous and sufficiently rough to provide mechanical bond for the finish coat.

10.3.4.3 Apply the brown (second) coat on the lath side over the scratch coat, after the back-up coat on the channel (stud) side has set, and leave as specified for the back-up coat.

10.3.4.4 Plaster shall extend to the floor, except that where plaster grouted combination metal bases and screeds are used, the plaster shall extend to the grout below the top of the base. Fill all spaces between grounds. Fill hollow metal door frames solid.

10.3.5 *Studless Solid Partitions (Metal Lath and Plaster)* — These partitions shall be not less than 2 in. (50.8 mm) in thickness, and plaster shall be applied in the same number of coats as for solid partitions with steel studs (see 10.3.4).

10.3.5.1 Where rib metal lath is used, apply the scratch coat on the flat side and the temporary bracing on the rib side. Where diamond mesh (flat expanded) metal lath is used, the temporary bracing may be attached to either side, and the scratch applied on the opposite side.

10.3.5.2 Sequence of application of brown coat and back-up coats shall be in accordance with one of these alternate methods:

(1) Apply the brown coat over the face of the scratch coat; then apply the back-up coat on the reverse side of the partition, or

(2) Apply the back-up coat on the rib side of the metal lath, or on the back side of the scratch coat, and follow by the brown coat applied over the face of the scratch coat.

10.3.6 *Studless Solid Partitions (Gypsum Lath and Plaster)* — These partitions shall be not less than 2 in. (50.8 mm) thick overall and shall be three-coat work having scratch, brown, and finish on both sides.

10.3.6.1 Apply the scratch coat approximately ³/₈ in. (9.5 mm) in thickness with a minimum thickness of ³/₁₆ in. (4.8 mm) at any point first to the side opposite the temporary bracing, followed by a scratch coat of approximately ³/₈ in. thick applied to the side on which the bracing occurs without removing the bracing.

10.3.6.2 Apply the brown coat to the side opposite the bracing, after the scratch on that side has set firm and hard and is partially dry. When this brown coat has set firm and hard and is partially dry, remove the temporary bracing and apply the brown coat to the bracing side. Leave the brown coat as specified in 10.3.4.2, allowing ¹/₁₆ to ¹/₈ in. (1.6 to 3.2 mm) on each side for the finish coat to bring the overall partition thickness to not less than 2 in. (50.8 mm).

10.3.7 *Studless Solid Partitions (Multiple-Thickness Gypsum Lath and Plaster)* — These partitions shall be not less than 2 in. (50.8 mm) thick overall, and the total thickness of the scratch, brown, and finish coats on each side shall be not less than ¹/₂ in. (12.7 mm). The thickness of the separate coats and the sequence of plastering on each side depending on whether bracing is used, shall be in accordance with the proprietary specifications for the systems employed.

10.3.7.1 Apply each scratch coat with sufficient pressure to form a good bond and leave sufficiently rough to provide a mechanical bond for the brown coat.

10.3.7.2 Apply each brown coat after the scratch coat has set firm and hard and is partially dry. Allow ¹/₁₆ to ¹/₈ in. (1.6 to 3.2 mm) on each side for the finish coat to bring the overall partition thickness to not less than 2 in. (50.8 mm).

10.3.8 *Gypsum Lath Ceilings Attached by Clips* — Plaster for ceilings with clip attachments of gypsum lath shall be three-coat work (scratch, brown, and finish). Each coat shall be set before the next coat is applied. Leave the scratch and brown coats sufficiently rough to provide mechanical bond for the following coat.

10.3.8.1 The scratch shall be from ³/₁₆ to ¹/₄ in. (4.8 to 6.4 mm) thick over the face of the plaster base.

10.3.8.2 The total thickness of scratch and brown coats shall be not less than ⁷/₁₆ in. (11.1 mm).

10.4 *Application of Finishes:*

10.4.1 Apply finish coat(s) to a partially dry base coat or to a thoroughly dry base coat that has been evenly wetted by brushing or spraying. Do not apply finish coat(s) to base coats having free water standing on the base coat surface. Avoid the use of excessive water (Note 14) in the application of all types of finish coat plastering.

NOTE 14—It is especially important to limit the amount of water used in the finishing of colored floated or textured finishes. Color is not recommended for smooth-trowel finishes.

10.4.2 Apply trowel finishes over the base coat, scratch in tightly, covering the surface of the base coat completely, then double back and fill out to a true even surface. The nominal thickness shall be ¹/₁₆ in. to ¹/₈ in. (1.6 to 3.2 mm). Allow the applied finish to dry (begin to lose its moisture to the base coat and to the ambient air) for a few minutes and then trowel it well (firmly compacting the material) with water, free from catfaces and other blemishes or irregularities. Do the final troweling after the finish has begun to set.

10.4.2.1 Apply lime putty-gypsum gauged finish in accordance with 10.4.2.

10.4.2.2 Apply lime putty-gypsum Keene's cement finish in accordance with 10.4.2 except continue the final troweling until the finish has set. The thickness may be increased for decorative work as required.

10.4.2.3 Apply ready-mixed gypsum trowel finish in accordance with the printed directions of the manufacturer and finish in accordance with 10.4.2.

10.4.2.4 Apply vermiculite-gypsum trowel finish and finish in accordance with 10.4.2.

10.4.3 Apply float finishes over the base coat, scratch in tightly, covering the base coat surface completely, then double back and fill out to a true even surface. The nominal thickness shall be ¹/₁₆ to ¹/₈ in. (1.6 to 3.2 mm). Float the trowel-applied finish to a true, even surface free from slick spots, catfaces, and other blemishes or irregularities. The texture specified or desired will determine the type of float surface to be used (wood, carpet, cork, rubber, or other) and the length of time between trowel application and floating (take-up).

10.4.3.1 Apply the mixes in 10.4.2.1 through 10.4.2.3 where used for a float finish, and finish in accordance with 10.4.3.

10.4.4 Apply textured or special finishes where ready-mixed in accordance with the printed directions of the manufacturer and finish as specified. Where job mixed, apply them and finish in accordance with the application procedures of 10.4.2 or 10.4.3 for scratch coat. The texture coat or special finish shall not reduce the total thickness of the finish below $1/16$ in. (1.6 mm) and shall be finished as specified or closely match an approved sample.

10.4.4.1 Special finishes shall not reduce the combined thickness of the base coat and finish coat to less than that specified in Table 6. Inorganic coloring material added to the job mix shall be in accordance with Annex A1.7.

10.4.4.2 Apply acoustical plaster and finish in accordance with the printed directions of the manufacturer.

TABLE 1   Aggregate for Finish Coat Plasters, Percentage Retained on Each Sieve, Cumulative

Sieve Size	Perlite, Natural and Manufactured Sand			
	Volume %		Weight %	
	max	min	max	min
No. 20 (850 μm)	0	...	0	...
No. 30 (600 μm)	10	...	0.5	...
No. 100 (150 μm)	100	40	100	40
No. 200 (75 μm)	100	70	100	70

TABLE 2   Sand for Job-Mixed Lime Putty-Gypsum Gauged Sand Float Finish, Percent Retained on Each Sieve by Weight, Cumulative

Sieve Size	max	min
No. 16 (1.18 mm)	0	0
No. 30 (600 μm)	50	20
No. 50 (300 μm)	70	50
No. 100 (150 μm)	100	80

TABLE 3   Vermiculite for Job-Mixed Finish Coat Plaster, Percent Retained on Each Sieve by Weight, Cumulative

Sieve Size	max	min
No. 8 (2.36 mm)	0	0
No. 16 (1.18 mm)	5	0
No. 30 (600 μm)	65	15
No. 50 (300 μm)	98	60
No. 100 (150 μm)	100	90

TABLE 4   Base Coat Proportions[A]

Plaster Base	Aggregates[B]		
	Sand		Perlite or Vermiculite[C]
	By Volume, ft³ (m³), Damp and Loose	By Weight, lb (kg), Damp and Loose	By Volume, ft³ (m³)
Over Gypsum Lath			
Two-coat work:			
Base coat	2½ (0.071)	250 (91.5)	2 (0.05)
Three-coat work:			
Scratch coat	2 (0.056)	200 (90.7)	2 (0.05)
Brown coat	3 (0.085)	300 (136)	2 (0.05)
or			
Scratch and brown coats	2½ (0.071)	250 (91.5)	...
Over Metal Lath			
Three-coat work:			
Scratch coat	2 (0.056)	200 (90.7)	2 (0.05)
Brown coat	3 (0.085)	300 (136)	2 (0.05)
or			
Scratch and brown coats	2½ (0.071)	250 (91.5)	...
Over Unit Masonry (Note 6)			
Two-coat work:			
Base coat	3 (0.085)	300 (136)	3 (0.085)
Three-coat work:			
Scratch coat	3 (0.085)	300 (136)	3 (0.085)
Brown coat	3 (0.085)	300 (136)	3 (0.085)

Over Monolithic Concrete[D]
For base coat proportions applicable to monolithic concrete, see 10.2.1.

[A] The proportions in Table 4 are applicable for both hand and machine application of plaster. See plaster manufacturer's instructions for application of machine-applied plaster.

[B] Use of an accurate device to measure quantities, such as a measuring box or container of known capacity, is highly encouraged. Where such a device is not available, six No. 2, square-edge (not scoop) shovels, with a blade approximately 8½ in. (216 mm) wide and 11 in. (279 mm) long, with the maximum depth of sides not more than 1½ in. (38 mm) higher than the face of the blade, and filled to an average depth of 4 in. (102 mm) of damp, loose sand, may be considered as the approximate equivalent to 1 ft³ (0.028 m³).

[C] Where the plaster is 1 in. (25 mm) or more in total thickness, or where the finish coat is sand float or acoustical, the proportions for the brown coat may be increased to 3 ft³ (0.085 m³).

[D] For use of bonding compounds for plastering on monolithic concrete, see 8.2.3, 8.2.3.1, 8.2.3.2, 8.2.3.3., 10.2.1 and footnotes in Table 6.

TABLE 5   Proportion of Gypsum To Not More Than Lime/Aggregate, with Dry and Wet Equivalents

	Dry						Lime Putty Wet Equivalent		
	Weight, lb (kg)			Volume, ft³ (m³)			ft³ (m³)	U.S. gal (litres)	lb (kg)
	Gypsum	Lime	Aggregate	Gypsum	Lime	Aggregate			
Troweled Finishes:[A]									
Lime putty with:[B]									
Gypsum gauging	100	225	0[A]	1	4½	0	6.75	52.5	450
Gypsum Keene's cement:									
Medium	100	50	0[A]	1	1	0	1⅛	8¾	100
Hard	100	25	0[A]	1	½	0	⅝	4½	50
Ready-mixed gypsum	100	0	0	1	0	0	0	0	0
Gypsum vermiculite	100	0	7 to 15	1	0	1	0	0	0
Floated Finishes:									
Lime putty with:									
Gypsum gauging	100	225	200	1	4½	2	6.75	52.5	450
Gypsum Keene's cement:									
Medium	150	100	450	1½	2	4½	2¼	17½	200
Ready-mixed gypsum[C]	100	0	0	1	0	0	0	0	0
Gypsum-vermiculite	100	0	7 to 15	1	0	1	0	0	0
Gypsum-sand (job-mixed)[D]	100	0	200	1	0	2	0	0	0

[A] See 9.4.1, 9.4.1.1, and 9.4.2.

[B] If additional hardness of finish coat is desired, increased amounts of gypsum may be used; however, hard finishes should not be used over lightweight aggregate base coats.

[C] Mixed with water only, in accordance with manufacturers' printed directions.

[D] Gypsum shall be neat, unfibered plaster.

TABLE 6 Thickness of Plaster

Plaster Base	Thickness of Plaster Including Finish Coat, in. (mm)
Metal plaster base	⁵/₈ (15.87) min
All other types of plaster base	¹/₂ (12.70) min
Unit masonry	⁵/₈ (15.87) min
Monolithic concrete surfaces:[A]	
Vertical[B, C]	⁵/₈ (15.87) min
Horizontal[C]	¹/₈ (3.17) to ³/₈ (9.52)

[A] Base coat plastering of the same proportions as specified herein for unit masonry may be applied over plain or reinforced monolithic concrete, provided the surface is first covered with a metal plaster base or first coated with a bonding compound.

[B] Finish coat plaster applied direct to a bonding compound over vertical monolithic concrete shall not exceed ³/₁₆ in. (4.76 mm) in thickness. Where more than ³/₁₆ in. of finish coat is required to bring such vertical surface to a true plane, a base coat of plaster shall first be applied to the bonding compound.

[C] Where horizontal or vertical monolithic concrete surfaces require more than ³/₈ in. (9.52 mm) or ⁵/₈ in. (15.87 mm) of plaster, respectively, to produce required lines or surfaces, metal plaster base shall be attached to the concrete before application of plaster. Where concrete surface requires the application of more than 1 in. (25.4 mm) of plaster to produce required lines or surfaces, lath shall be applied over furring secured to the concrete.

ANNEX

(Mandatory Information)

A1. TECHNICAL INFORMATION

A1.1 Details of construction to achieve required fire resistance may be obtained from official reports of fire tests conducted at recognized fire testing laboratories in accordance with Methods E 119.

A1.2 Details of construction to achieve required sound control may be obtained from official reports of sound tests conducted at recognized sound testing laboratories in accordance with the applicable sound tests, Test Method C 423, Method E 90, or Method E 492.

A1.3 Ventilation above gypsum plaster ceilings, attics, or similar unheated spaces above lath and plaster ceilings shall be adequately ventilated by providing effective cross ventilation for all spaces between roof and top floor ceilings by screened louvers or other approved or acceptable means.

A1.3.1 The ratio of total net free ventilating area to ceiling area shall be not less than $L/150$, except that the ratio may be $L/300$ provided:

(1) A vapor barrier having a transmission rate not exceeding 1 perm is installed on the warm side of the ceiling, or

(2) At least 50 % of the required ventilation area is provided by ventilators located in the upper portion of the space to be ventilated (at least 3 ft (914.4 mm) above eave or cornice vents) with the balance of the required ventilation area provided by eave or cornice vents.

A1.3.2 Attic space that is accessible and suitable for future habitable rooms or walled-off storage space shall have at least 50 % of the required ventilating area located in the upper part of the ventilated space as near the high point of the roof as practical and above the probable level of any future ceilings.

A1.4 *Equivalent Measure for Aggregates* — Six No. 2 shovels of damp loose sand are approximately equivalent to 1 ft³ (0.283 m³) and weigh approximately 100 lb (45.36 kg). Perlite and vermiculite are normally packed in bags marked with the cubic foot (or cubic metre) content.

A1.5 *Job Addition of Fiber*—Job added fiber where specified, shall be nonstaining natural or synthetic fiber, well-shredded and free from grease, oil, dirt, or other materials that may adversely affect the strength bond or setting time of the plaster. Uniform quantities of the fiber, as specified, shall be added to the mixer, batch by batch, after mixing water, aggregate and plaster have been added. The type and amount of fiber to be used is dependent on job specifications or on that quantity of the selected fiber that provides the desired application characteristics. Where the type and quantity of fiber is not specified by contractural agreement, gradually increased or decreased quantities shall be used until the desired application characteristics are obtained, and that quantity used thereafter.

A1.6 *Ready-Mixed Colored Finish Plasters*, shall be mixed and applied in strict accordance with the directions of the manufacturer. Where specified by contractural agreement, sample panels prepared in accordance with the job specifications shall be submitted to the architect or builder for approval of final color prior to the start of any job application.

A1.7 *Job-Mixed Colored Finish Plasters*—Job-mixed floated or textured colored finish plasters shall be prepared with inorganic or organic water-disperseable, nonbleeding, nonfading, lime-proof coloring materials, dry powdered, paste or liquid. Coloring materials shall be proportioned and thoroughly dispersed into the finish plaster in strict accordance with the directions of the manufacturer of the coloring material. In the absence of specific directions by the manufacturer of the coloring material, uniform quantities, mixer batch by mixer batch, shall be thoroughly dispersed into the mixing water. Uniform, identical proportioning by weight of coloring material, mixing water and dry finish plaster components, and uniform, identical mixing procedure, mixing time and application. Each unbroken area shall be completed in one continuous operation from angle to angle or natural interruption for best uniformity of color. Where specified by contractural agreement, sample panels prepared in accordance with the job specifications, including floating or texturing procedures, shall be submitted to the architect or builder for final color approval prior to the start of any job application. Floating or texturing procedure used for sample preparation shall be strictly followed for job application. The minimum quantity of coloring material necessary to obtain the specified or approved color shall be used, but in no case shall the proportion of coloring material exceed 8 weight % of the cementitious materials used.

SPECTEXT SPECIFICATION FOR
GYPSUM PLASTER
SECTION 09210

This section includes wet applied gypsum plaster, in either a two coat application over masonry, clay tile, concrete, gypsum lath or other solid surfaces, or a three coat application over metal lath surfaces; with smooth trowelled and decorative surface finish. This section can be edited with furring and lathing referenced to Section 09206 or furring and lathing can be specified in this section. This section includes proprietary and descriptive type specifications. Use only one specifying type to avoid any conflicting requirements. Edit accordingly.

PART 1   GENERAL

1.01   WORK INCLUDED

A.   [Hardwall] [Vermiculite] [_____] plaster system.

B.   [Metal furring and lath.]

C.   [Gypsum lath.]

D.   Plaster fireproofing for building structural members.

E.   [Smooth] [_____] surface finish.

1.02   WORK INSTALLED BUT FURNISHED UNDER OTHER SECTIONS

A.   Section [_____-_____]: Metal frames for [recessed light fixtures.] [_____.]

1.02 List sections which furnish products for installation under this section. When a product will be furnished "by Others," outside the Contract, it is considered to be furnished "by Owner."

1.03   RELATED WORK

A.   Section [_____-_____]: Wall substrate surface.

B.   Section 05400 - Cold Formed Metal Framing: Metal studding and framing behind plaster finish.

C.   Section 07900 - Joint Sealers.

D.   Section 09111 - Metal Stud Framing System: Metal studding and framing behind plaster finish.

E.   [Section 09206 - Metal Furring and Lathing: [Metal furring] [and] [lathing] for plaster finish.]

F.   [Section [_____-_____]: Gypsum Lath.]

G.   Section 09220 - Portland Cement Plaster.

H.   Section 09260 - Gypsum Board Systems.

I.   Section [_____-_____]: Surface finish of [_____.]

1.04   REFERENCES

A.   ANSI/ASTM C28 - Gypsum Plaster.

B.   ANSI/ASTM C35 - Inorganic Aggregates for use in Gypsum Plaster.

C.   ANSI/ASTM C37 - Gypsum Lath.

D.   ANSI/ASTM C61 - Gypsum Keene's Cement.

1.04 Include only reference standards that are to be indicated within the text of this section. Refer to TAS Sections 09210 and 09212. Edit the following, adding and deleting as required for project and product selection.

E. ANSI/ASTM C206 - Finishing Hydrated Lime.

F. ANSI/ASTM C631 - Bonding Compounds for Interior Plastering.

G. ANSI/ASTM C842 - Application of Interior Gypsum Plaster.

H. ANSI/ASTM E90 - Laboratory Measurement of Sound Transmission Loss of Building Partitions.

I. FS HH-I-521 - Insulation Blankets, Thermal (Mineral fiber, for Ambient Temperatures).

J. GA(Gypsum Association) - 201 - Gypsum Board for Walls and Ceilings.

1.05 SYSTEM DESCRIPTION

A. Acoustic Attenuation for [Identified] Interior Partitions: [_____] STC in accordance with ANSI/ASTM E90.

1.06 QUALITY ASSURANCE

A. Applicator: Company specializing in application of gypsum plaster work [with [_____] years [documented] experience.] [approved by manufacturer.]

1.07 REGULATORY REQUIREMENTS

A. Conform to [applicable] [_____] code for fire rated assemblies in conjunction with Section [09206] [_____] as follows:
   1. Fire Rated Partitions: Listed assembly by [UL] [FM] No. [_____.]
   2. Fire Rated Ceiling [and Soffits]: Listed assembly by [UL] [FM] No. [_____.]
   3. Fire Rated Structural Column Framing: Listed assembly by [UL] [FM] No. [_____.]
   4. Fire Rated Structural Beam Framing: Listed assembly by [UL] [FM] No. [_____.]

1.08 SUBMITTALS

A. Submit product data under provisions of Section [01300.] [01340.]

B. Provide product data on plaster materials, characteristics and limitations of products specified.

C. Submit manufacturer's installation instructions under provisions of Section [01300.] [01340.]

D. Submit manufacturer's certificate under provisions of Section [01400] [01405] that [products for fire resistance ratings] [_____] meet or exceed [specified requirements.] [_____.]

1.09 FIELD SAMPLES

A. Provide sample panel under provisions of Section [01300.] [01340.]

B. [Provide] [Construct] [_____] field sample panel, [_____] inch ([_____] mm) long by [_____] inch ([_____] mm) wide, illustrating [surface finish.] [_____.]

C. Locate [where directed.] [_____.]

D. Accepted sample may remain as part of the work.

1.10 ENVIRONMENTAL REQUIREMENTS

A. Do not apply plaster when substrate or ambient air temperature is less than 50 degrees F (10 degrees C) nor more than 80 degrees F (27 degrees C).

B. Maintain minimum ambient temperature of 50 degrees F (10 degrees C) during and after installation of plaster.

1.05 Use this Article carefully; restrict descriptions to total system parameters, performance or design requirements. Do not repeat statements made in "Work Included" Article.

1.08C When manufacturers' instructions for specific installation requirements are utilized, carefully edit Part 3 Execution requirements to avoid conflict with those instructions.

1.09 Use this Article for field applied finish samples.

PART 2   PRODUCTS

2.01   ACCEPTABLE MANUFACTURERS

    A.   [_____.]

    B.   [_____.]

    C.   [_____.]

    D.   Substitutions: Under provisions of Section [01600.]
          [01630.]

2.02   PLASTER BASE MATERIALS

    A.   Plaster: [ANSI/ASTM C28;] gypsum neat hardwall type,
          [fibrated] [unfibrated]; [[_____] manufactured by
          [_____].]

                    ****** [OR] ******

    B.   Plaster: [ANSI/ASTM C28;] gypsum mill aggregated type;
          [[_____] manufactured by [_____].]

                    ****** [OR] ******

    C.   Plaster: [ANSI/ASTM C28;] gypsum wood fiber type;
          [[_____] manufactured by [_____].]

                    ****** [OR] ******

    D.   Plaster: [ANSI/ASTM C28;] gypsum bonding type;
          [[_____] manufactured by [_____].]

    E.   Aggregate: [ANSI/ASTM C35;] [sand] [vermiculite] [perlite]
          type; [[_____] manufactured by [_____].]

    F.   Water: Clean, fresh, potable and free of mineral or
          organic matter which can affect plaster.

    G.   Bonding Agent: [ANSI/ASTM C631;] type recommended for
          bonding plaster to [concrete] [and] [concrete block]
          surfaces; [[_____] manufactured by [_____].]

2.03   FINISHING PLASTER

    A.   Gypsum/Lime Putty Type: [ANSI/ASTM C28;] mixture of gaging
          plaster and lime; [[_____] manufactured by
          [_____].]

                    ****** [OR] ******

    B.   Keene's Cement/Lime Putty Type: [ANSI/ASTM C61 and C206;]
          mixture of Keene's cement and lime; [[_____]
          manufactured by [_____].]

2.01   If only one manufacturer is acceptable, list in Article 2.02 and delete this Article; if more than one, list in this Article. If product substitution procedure is used, include Paragraph D. Refer to TAS Sections 09210 and 09212.

2.02   Edit the following descriptive specifications for any conflicts with manufacturer's products specified above.

2.02   There are a variety of plasters that are used to provide base coats. Select the type required, based on intended end result. The plaster indicated in the following paragraph is required to be mixed with an aggregate and water to provide a base coat.

2.02A   The plaster indicated in the following paragraph only requires to be mixed with water to provide a base coat.

2.02B   The plaster indicated in the following paragraph can be used without the addition of an aggregate to provide a base coat, depending upon project requirements.

2.02C   The plaster indicated in the following paragraph is intended for use on monolithic concrete or concrete block surfaces and only requires mixing with water to provide a base coat.

2.02E   Include for mixing aggregates only when required, to provide base coat or coats.

2.02G   Include the following paragraph only for situations when gypsum plaster is to be applied directly on masonry or concrete.

2.03   There are a variety of materials used to provide plaster finish coats. Select the system required, based on intended end result.

2.03C   A sand float finish can be achieved by using a prepared gypsum plaster or a Keene's cement/lime putty plaster with the addition of sand.

C. Sand Float Type: |ANSI/ASTM C28 and C35;| prepared mixture of gypsum plaster and sand; || | manufactured by | |.|

****** |OR| ******

D. Sand Float Type: |ANSI/ASTM C61 and C35;| prepared mixture of Keene's cement/lime putty and sand; || | manufactured by | |.|

E. Water: Clean, fresh, potable and free of mineral and organic matter which can affect plaster.

2.04 GYPSUM LATH

A. Gypsum Lath: [ANSI/ASTM C37,] [standard] [perforated] [fire rated] [insulating] [_____] type; [[3/8] [1/2] inch ([10] [13] mm) thick;] [thickness indicated on Drawings;] [[_____] manufactured by [_____].]

2.05 METAL LATH

A. Metal Lath: 2.5 lb/sq yd (1.12 kg/sq m) expanded metal [prime painted] [galvanized] finish, [self-furring type.]

B. Corner Mesh: Formed steel, minimum [26] [_____] gage ([0.5] [_____] mm) thick; [perforated] [expanded] flanges shaped to permit complete embedding in plaster; minimum [2] [_____] inches ([50] [_____] mm) wide; [galvanized] [rust inhibitive prime paint] finish.

2.06 ACCESSORIES

A. Corner Beads: Formed steel, minimum [26] [_____] gage ([0.5] [_____] mm) thick; [beaded] [bullnosed] edge, of longest possible length, sized and profiled to suit application; [galvanized] [rust inhibitive prime paint] finish.

B. Base Screeds: Formed steel, minimum [26] [_____] gage ([0.5] [_____] mm) thick; [square] [_____] edge, of longest possible length; sized and profiled to suit application; [galvanized] [rust inhibitive prime paint] finish.

C. Casing Bead: Formed steel, minimum [26] [_____] gage (0.5] [_____] mm) thick; thickness governed by plaster thickness; maximum possible lengths; [expanded metal] [solid] flanges, with [square] [bullnosed] [bevelled] [_____] edges; [galvanized] [rust inhibitive prime paint] finish.

D. Control [and Expansion] Joint Accessories: Formed steel; minimum [25] [_____] gage ([0.5] [_____] mm) thick; [accordian] [_____] profile, 2 inch (50 mm) [expanded metal] [solid] flanges each side; [galvanized] [rust inhibitive prime paint] finish.

E. Anchorages: Nails, staples, or other approved metal supports, of type and size to suit application, galvanized, to rigidly secure lath and associated metal accessories in place.

2.07 ACOUSTIC ACCESSORIES

A. Resilient Furring Channels: [_____]; [_____] manufactured by [_____].

B. Acoustic Insulation: FS HH-I-521, friction fit type, unfaced; [_____] inch ([_____] mm) thick; [[_____] manufactured by [_____]].

C. Acoustic Sealant: Non-hardening, non-skinning type, for use in conjunction with gypsum plaster system; [_____] manufactured by [_____].

2.04 This Article is intended for use when gypsum lath is to be part of the work of this section. If specified elsewhere, coordinate the statement in this Article to avoid conflicting statements.

2.05 This Article is intended for use when a metal furring and lathing section, such as Section 09206, is not used. If section on metal lath is used, coordinate the paragraphs in this Article for conflicting statements.

2.06 The following accessory listings describe components manufactured in steel. These items are also available in extruded plastic and zinc for special applications. If lathing accessories are specified in another section, coordinate the statements in this Article to avoid conflicting statements.

2.06C Control and expansion joint accessories are available in a variety of designs, sizes and shapes. Beware of joint components that are relatively inflexible; they can cause adjacent plaster work to crack just beyond the joint.

2.07 Delete the following requirements for acoustic accessories if specified in another section or if not required.

2.08   PLASTER MIX

A.   Mix and proportion plaster in accordance with [ANSI/ASTM
     C842.] [Lath and Plaster Institute of [_____].]
     [manufacturer's instructions.] [_____.]

PART 3   EXECUTION

3.01   INSPECTION

A.   Verify that surfaces and site conditions are ready to
     receive work.

B.   Masonry: Verify joints are cut flush and surface is ready
     to receive work of this Section. Verify no bituminous or
     water repellent coatings exist on masonry surface.

C.   Concrete: Verify surfaces are flat, honeycomb is filled
     flush, and surface is ready to receive work of this
     Section. Verify no bituminous, water repellent, or form
     release agents exist on concrete surface that are
     detrimental to plaster.

D.   Grounds and Blocking: Verify items within walls for other
     Sections of work have been installed.

E.   Gypsum Lath and Accessories: Verify substrate is flat,
     [joints are taped and sanded,] and surface is ready to
     receive work of this Section. Verify joint and surface
     perimeter accessories are in place.

                    ****** [OR] ******

F.   Metal Lath and Accessories: Verify lath is flat, secured
     to substrate, and joint and surface perimeter accessories
     are in place.

G.   Mechanical and Electrical: Verify services within walls
     have been tested and approved.

H.   Beginning of installation means acceptance of existing
     conditions.

3.02   PREPARATION

A.   Protect elements surrounding the work of this Section from
     damage or disfiguration.

B.   Dampen masonry surfaces to reduce excessive suction.

C.   Clean concrete surfaces of foreign matter. Thoroughly
     dampen surfaces before using acid solutions, solvent, or
     detergents to perform cleaning. Wash surface with clean
     water.

D.   Roughen smooth concrete surfaces [and smooth faced
     masonry.]

E.   [Apply bonding agent in accordance with manufacturer's
     instructions.]

3.03   INSTALLATION - LATH MATERIALS

A.   Install gypsum lath in accordance with GA 201.

B.   Install gypsum lath perpendicular to framing members, with
     lath face exposed. Stagger end joint of alternate
     courses. Butt joints tight. Maximum gap allowed: [1/8]
     [_____] inch ([3] [_____] mm).

C.   Place corner reinforcement diagonally over gypsum lath and
     across corner immediately above and below openings.
     Secure to gypsum lath only.

2.08 Base plaster mixes vary greatly for application to different types
of substrates. Finish plaster mixes
will vary depending on the type of
smooth or textured finish desired. If
a variety of mixes are required, consider adding the list to the schedule
at the end of this section.

3.01A Coordinate the following
requirements with masonry or concrete sections. Edit those sections
accordingly. Masonry of lightweight
or other absorptive materials can
cause uneven suction of moisture
from applied plaster, resulting in
"telegraphing" of base. Consider applying a bond coat to such surfaces.

3.01D Include one of the following paragraphs only when lath, accessories and supporting framing
are included under another section.

3.03 Include this Article when
gypsum or metal lathing is included
in this section.

D. Apply metal lath taut, with long dimension perpendicular to supports.

E. Lap ends minimum [1] [_____] ([25] [_____] mm). Secure end laps with tie wire where they occur between supports.

F. [Lap sides of diamond mesh lath minimum [1-1/2] [_____] inches ([38] [_____] mm). Nest outside ribs of rib lath together.]

G. Attach metal lath to wood supports using nails at maximum [_____] inches ([_____] mm) on center.

****** [OR] ******

H. Attach metal lath to metal supports using [tie wire] [_____] at maximum [6] [_____] inches (150] [_____] mm) on center.

****** [OR] ******

I. Attach metal lath to [concrete] [concrete masonry] using wire [hair pins,] [hooks,] [or] [loops]. Ensure that anchors are securely attached to concrete and spaced at maximum [24] [_____] inches ([600] [_____] mm) on center.

3.04   INSTALLATION-ACCESSORIES

A. Continuously reinforce internal angles with corner mesh, except where the metal lath returns 3 inches (75 mm) from corner to form the angle reinforcement; fasten at perimeter edges only.

B. Place [beaded] [bullnosed] external angle with mesh at corners; fasten at outer edges only.

C. Place strip mesh diagonally at corners of lathed openings. Secure rigidly in place.

D. Place [4] [_____] inch ([100] [_____] mm) wide strips of metal lath centered over junctions of dissimilar backing materials. Secure rigidly in place.

E. Place casing beads at terminations of plaster finish. Butt and align ends. Secure rigidly in place.

3.05   INSTALLATION - ACOUSTIC ACCESSORIES

A. Install resilient furring channels [24] [_____] inches ([600] [_____] mm) on center at right angles to framing members. Place end joints over framing members. Terminate channels [_____] inches ([_____] mm) short of door frames and partition perimeter.

B. Fit acoustic insulation tight between partition framing members. Pack insulation around mechanical, electrical, or other components in partition.

C. Place acoustic sealant at gypsum [backing board] [lath] partition perimeter in accordance with manufacturer's instructions. Seal penetrations of conduit, pipe, ductwork, rough-in boxes, and other components.

3.06   CONTROL [AND EXPANSION] JOINTS

3.05 Delete the following requirements for acoustic accessories if specified in another section or if not required.

3.06 At interior stable ambient temperatures, control joints should be located every 20 to 30 feet (6 to 9 m), at intersections of natural breaks in walls, and above door frame jambs. Expansion joints can be positioned on the building expansion joint lines in conjunction with joints of other materials. Review industry technical documents for detailed recommendations in this regard. Coordinate the following text with other affected sections to avoid repeating or creating conflicting requirements and coordinate with the Drawings.

A. Locate control [and expansion] joints [every [20] [_____] feet ([6] [_____] m).] [as indicated.]

B. Use [double casing bead [butted tight] [spaced [1/4] [_____] inch ([6] [_____] mm) apart]] [accordian profile accessory] to form joint.

C. Coordinate joint placement with [other related work.] [Section 09206.] [_____.]

.07 PLASTERING

A. Apply gypsum plaster in accordance with [ANSI/ASTM C842.] [manufacturer's instructions.] [_____.]

B. Apply brown and finish coats over [gypsum lath] [masonry] [concrete] [clay tile] [and] [_____] surfaces. [Apply brown coat to a nominal thickness of 3/8 inch (9 mm).]

C. Apply scratch, brown, and finish coats over [metal lath] [and] [_____] surfaces. [Apply scratch and brown coats to a nominal thickness of 3/8 inch (9 mm) each.]

D. [Apply color tinted bond coat to prepared surfaces within [_____] hours of plaster application. Apply in accordance with manufacturer's instructions.]

E. Apply finish coat to minimum [1/8] [_____] inch ([3] [_____] mm) thickness.

F. Work the finish coat [flat and smooth, with steel trowel.] [_____.]

.08 TOLERANCES

A. Maximum Variation from True Flatness: [1/8] [_____] inch in 10 feet ([3] [_____] mm in 3 m).

.09 SCHEDULE

3.07A This Article allows for applying materials over other building materials in renovation work. If existing surfaces have been painted, consider a three coat application with metal lath. Edit accordingly.

3.07D The following two paragraphs describe a smooth plaster finish. If a textured finish is scheduled, the finish thickness will vary. Edit accordingly.

3.09 Provide a schedule to identify fire rating assembly classifications as applicable to this section, varying finish requirements, and any other special conditions.

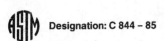

Designation: C 844 – 85

Standard Specification for

APPLICATION OF GYPSUM BASE TO RECEIVE GYPSUM VENEER PLASTER[1]

This standard is issued under the fixed designation C 844; the number immediately following the designation indicates the year of original adoption or, in the case of revision, the year of last revision. A number in parentheses indicates the year of last reapproval. A superscript epsilon (ε) indicates an editorial change since the last revision or reapproval.

This specification has been approved for use by agencies of the Department of Defense and for listing in the DoD Index of Specifications and Standards.

1. Scope

1.1 This specification covers the minimum requirements for, and methods of, application of gypsum veneer base for gypsum veneer plasters.

NOTE 1—Gypsum veneer base shall not be used as a base for direct adhesive application of ceramic, metal, or plastic tile in areas exposed to abnormal moisture or humidity or temperature.

1.2 Where a specific degree of fire resistance is required for gypsum base and veneer plaster systems, a special resistant base may be required. Details of constructions to achieve the required fire resistance may be obtained from official reports of fire tests conducted at recognized fire testing laboratories in accordance with Methods E 119.

1.3 Where a specific degree of sound control is required for veneer plaster assemblies and constructions, details of construction shall be in accordance with official reports of tests conducted in recognized sound testing laboratories in accordance with the applicable sound tests, Test Method C 423, Method E 90, or Method E 492.

1.4 Where this specification is more stringent (size or thickness of framing; spacing of fasteners) than the fire-rated construction, this specification shall govern; otherwise, the construction described in the fire test report shall govern.

1.5 Unheated spaces above gypsum veneer base ceilings shall be properly ventilated (see Appendix X3).

1.6 *General Information*—The Appendixes give general information and also suggestions for inclusions to be made elsewhere by the specifier and are not a part of this specification. The work shall be properly coordinated with the work of other trades.

2. Applicable Documents

2.1 *ASTM Standards:*

C 11 Definitions of Terms Relating to Gypsum and Related Building Materials and Systems[2]

C 423 Test Method for Sound Absorption and Sound Absorption Coefficients by the Reverberation Room Method[3]

C 442 Specification for Gypsum Backing Board and Coreboard[2]

C 473 Test Methods for Physical Testing of Gypsum Board Products and Gypsum Lath[2]

C 514 Specification for Nails for the Application of Gypsum Wallboard[2]

C 557 Specification for Adhesives for Fastening Gypsum Wallboard to Wood Framing[2]

C 587 Specification for Gypsum Veneer Plaster[2]

C 588 Specification for Gypsum Base for Veneer Plasters[2]

C 645 Specification for Non-Load (Axial) Bearing Steel Studs, Runners (Track), and

Rigid Furring Channels for Screw Application of Gypsum Board[2]

C 754 Specification for Installation of Steel Framing Members to Receive Screw-Attached Gypsum Wallboard, Backing Board, or Water-Resistant Backing Board[2]

C 843 Specification for Application of Gypsum Veneer Plaster[2]

C 955 Specification for Load-Bearing (Transverse and Axial) Steel Studs, Runners (Track), and Bracing or Bridging, for Screw Application of Gypsum Board and Metal Plaster Bases[2]

C 1002 Specification for Steel Drill Screws for the Application of Gypsum Board[2]

C 1007 Specification for the Installation of Load-Bearing (Transverse and Axial) Steel Studs and Related Accessories[2]

E 84 Test Method for Surface Burning Characteristics of Building Materials[4]

E 90 Method for Laboratory Measurement of Airborne Sound Transmission Loss of Building Partitions[3]

E 96 Test Methods for Water Vapor Transmission of Materials[3]

E 119 Methods of Fire Tests of Building Construction and Materials[4]

E 492 Method of Laboratory Measurement of Impact Sound Transmission Through Floor-Ceiling Assemblies Using the Tapping Machine[3]

2.2 *American Lumber Softwood Standards:* PS–20

3. Definitions

3.1 *base ply*—first ply of gypsum base in multiply application.

3.2 *control joint*—a predetermined opening between adjoining panels or surfaces at regular intervals to relieve the stresses of expansion and contraction transverse to the joint in large wall and ceiling areas.

3.3 *fastener*—nails, screws, or staples used for the application of the gypsum base or backing board.

3.4 *framing member*—that portion of the framing, furring, blocking, etc., to which the gypsum base is attached. Unless otherwise specified, the surface to which abutting edges or ends are attached shall be not less than 1½ in. (38 mm) wide for wood members, not less than 1¼ in. (32 mm) wide for steel members, and not less than 6 in. (152 mm) wide for gypsum studs. For internal corners or angles, the bearing surface shall be not less than ¾ in. (19 mm).

3.5 *parallel application*—gypsum base application where the paperbound edges are applied parallel to the framing members.

3.6 *perpendicular application*—gypsum base application where the paperbound edges are applied at right angles to the framing members.

3.7 Other definitions relating to gypsum shall be as defined in Definitions C 11.

4. Materials

4.1 Materials shall conform to the respective specifications and standards and to the requirements specified herein.

4.2 *Gypsum Base for Veneer Plasters* (hereinafter referred to as "gypsum base")—Specification C 588.

4.3 *Gypsum Backing Board or Coreboard*—Specification C 442.

4.4 *Special Fire-Retardant Gypsum Base*—Specification C 588, Type X.

4.5 *Foil-Backed Gypsum Base*—Specification C 588.

4.6 *Gypsum Veneer Plaster*—Specification C 587.

4.7 *Nails*—Specification C 514 or as recommended by the manufacturer of the gypsum base.

4.8 *Screws*—Specification C 1002 or as recommended by the manufacturer of the gypsum board.

4.9 *Backing Board Staples*—No. 16 USS gage (1.6-mm) flattened galvanized wire with ⁷⁄₁₆-in. (11.1-mm) wide-crown minimum outside measure and divergent points. Staples shall be used only for the base ply of two-ply application over wood framing. Leg lengths shall be as follows:

Backing Board Thickness, in. (mm)	Leg Length, in. (mm)
⅜ (9.5)	1 (25.4)
½ (12.7)	1⅛ (28.6)
⅝ (15.9)	1¼ (31.8)

4.10 *Reinforcing Staples*—Flattened galvanized wire with legs of sufficient length to hold the reinforcing in place.

4.11 *Steel Framing* shall meet or exceed the requirements of Specification C 645 for the specified design criteria.

4.12 *Accessories* (cornerbead, edge trim, control joints, etc.) shall be made from galvanized steel, plastic, or other corrosion-resistant materials of suitable thickness and design for the intended use. Flanges shall be free of dirt or grease, or other materials that may affect the bond of veneer plaster and shall be installed in a manner to ensure full coverage of the flanges.

4.12.1 Designations used to identify commonly specified types of steel cornerbead and trim are shown in Fig. 1.

4.13 *Adhesive:*

---

[1] This specification is under the jurisdiction of ASTM Committee C-11 on Gypsum and Related Building Material and Systems, and is the direct responsibility of Subcommittee C11.03 on Specifications for Application of Gypsum and Other Products in Assemblies.

Current edition approved April 30, 1985. Published June 1985. Originally published as C 844 – 78. Last previous edition C 844 – 79.

[2] *Annual Book of ASTM Standards*, Vol 04.01.

[3] *Annual Book of ASTM Standards*, Vol 04.06.

[4] *Annual Book of ASTM Standards*, Vol 04.07.

4.13.1 For application of gypsum base to wood framing see Specification C 557.

4.13.2 For application of gypsum base to steel framing, the adhesive shall be as recommended by the manufacturer of the gypsum base.

5. Delivery of Materials

5.1 All materials shall be delivered in the original packages, containers, or bundles bearing the brand name and manufacturer (or supplier) identification.

6. Protection of Materials

6.1 All materials shall be kept dry, preferably by being stored under roof. When necessary to store gypsum base outside, it shall be stacked off the ground, supported on a level platform, and protected from the weather, direct sunlight, and surface contamination.

6.2 Gypsum base shall be neatly stacked flat with care taken to avoid sagging or damage to edges, ends, or surfaces.

7. Application of Gypsum Base

7.1 *General Requirements:*

NOTE 2—For general wood framing requirements, see Appendix X2. Installation of steel framing shall be in accordance with Specification C 754 except for spacing as specified in Table 1.

7.1.1 *Method of Cutting and Installation*—Cut the gypsum base by scoring and breaking or by sawing, working from the face side. When cutting by scoring, cut the face paper with a knife or other suitable tool. Then snap the gypsum base back away from the cut face. The back paper may be broken by snapping the gypsum base in the reverse direction, or the back paper can be cut. Smooth all cut edges and ends of the gypsum base to obtain neat jointing. Score cut-outs for pipes, fixtures, or other small openings on the face and back in outline before removing or cut out with a saw or other suitable tool. Where it meets projecting surfaces, scribe the gypsum base and neatly cut and fit. Bring base panels tightly into contact with adjacent panels but do not force into place. Fit abutting ends and edges neatly.

7.1.2 Sheathing or other weather protection shall be in place before application of the gypsum base. In general, the gypsum base should be applied first to the ceiling and then to the walls.

7.2 *Fastening, Mechanical:*

7.2.1 Hold the gypsum base in firm contact with the underlying support while driving any fastener.

7.2.2 Fastener application shall proceed from the center of the field of the gypsum base to the ends and edges.

7.2.3 Except where required for fire rating, structural performance, or other special considerations, fastening to top or bottom plates is not required.

7.2.4 Space fasteners from the edges and ends of the gypsum base not less than ⅜ in. (9.5 mm) nor more than ½ in. (12.7 mm) (see also 8.4).

7.2.5 Drive nails and screws to a point flush with the gypsum base surface without breaking the surface paper or damaging the surrounding gypsum core and without stripping the framing member around the screws.

7.2.6 Drive staples in such a manner that both legs penetrate the support member and the crown

bears tightly against the backing board, but does not cut into the face paper. The staple crown shall be at right angles to the paperbound edges of the backing board, except where the paperbound edges fall on and are parallel to the framing members; then drive the staples parallel to the edges.

NOTE 3—If used, staples shall be used only for the first ply in two-ply gypsum base systems.

7.2.7 Minimum penetration of mechanical fasteners into supporting framing members shall be not less than:

	Nails	Screws	Staples
Wood:			
Single Ply	⅞ in.	⅝ in.	⅝ in., each leg
Steel:			
Single Ply	...	⅜ in.	...

7.3 *Fasteners (Single-Ply Application):*

7.3.1 *Nail Spacing*—Space nails a maximum of 7 in. (178 mm) on centers on the ceiling and a maximum of 8 in. (203 mm) on centers on the walls.

7.3.2 *Screw Spacing*—On wood framing, space screws a maximum of 12 in. (305 mm) on centers for ceilings. On walls, space screws a maximum of 16 in. (406 mm) on centers where the framing members are 16 in. on centers and 12 in. on centers where the framing members are 24 in. (610 mm) on centers. On light gage steel framing, space screws a maximum of 12 in. on centers.

7.3.3 Nails for single-ply application over wood members shall be as follows:

Gypsum Base Thickness, in. (mm)	Nails, in. (mm)
⅜ (9.5)	1¼ (32)
½ (12.7)	1⅜ (35)
⅝ (15.9)	1½ (38)

NOTE 4—Where a specific degree of fire resistance is required for gypsum board assemblies and constructions, nails of same or larger length, shank diameter, and head bearing area, as those described in the applicable fire test report may be used.

7.4 *Control Joints*—Install control joints in ceilings exceeding 2500 ft² (232 m²) in area. The distance between ceiling joints shall not exceed 50 ft (15 m) in either direction. Install a control joint where ceiling framing or furring changes direction. Install control joints where expansion joints occur in the base exterior wall. In partitions, walls, or wall furring, the distance between control joints shall not exceed 30 ft (9 m). Wall or partition height door frames may be considered as control joints.

8. Application of Single-Ply Gypsum Base

8.1 The maximum spacing of framing members for single-ply gypsum base construction shall not exceed those shown in Table 1.

8.2 In single-ply installation, all ends and edges of gypsum base shall occur over framing members or other solid backing except where treated joints occur at right angles to framing or furring members.

8.3 Stagger end joints and arrange joints on opposite sides of a partition to occur on different framing members.

8.4 *Floating Interior Angles*—To help minimize the possibility of cracking in areas adjacent to a wall and ceiling intersection, the floating angle method of application may be used. This method is applicable where single

nailing, double nailing, or screw attachment is used. Apply the base to ceilings first.

8.4.1 *Ceilings*—Locate the first attachment into each ceiling framing member framed perpendicular to the intersection 7 in. (178 mm) out from the wall intersection for single nailing, and 11 to 12 in. (279 to 305 mm) for double nailing or screw application.

8.4.2 *Sidewalls*—Apply the gypsum base on sidewalls so as to provide a firm support for the floated edges of the ceiling gypsum base. Locate the top attachment into each stud 8 in. (203 mm) down from the ceiling intersection for single nailing, and 11 to 12 in. (279 to 305 mm) for double nailing or screw application (see Figs. 4 and 5). At sidewall vertical angles (Fig. 6), fasten only the overlapping base so as to bring the back of the underlying board into firm contact with the face of the framing member behind it.

9. Adhesive-Mechanical Fastener Application

9.1 Except as modified herein, application shall be in accordance with Section 8.

9.2 Adhesive shall comply with the requirements of Specification C 557.

9.3 Surfaces of gypsum base and framing to be adhered by the adhesive shall be free of dust, dirt, grease or any other foreign matter that could impair the bond.

9.4 Apply a bead of adhesive ⅜ in. (9.5 mm) in diameter to the face of all wood framing members, except plates, that support the gypsum base. The adhesive shall spread to an average width of ¾ in. (19 mm) approximately ⅛ in. (3.2 mm) thick. Application patterns are shown in Fig. 7.

9.4.1 Where a joining of two adjacent pieces of gypsum base occurs (ends or edges) on a framing member, apply parallel beads of adhesive.

9.5 The adhesive container shall be clearly marked with the "open time" of the adhesive. Apply adhesive to no greater area than can be covered with gypsum base within the "open time."

9.6 *Fastener Spacing.*

9.6.1 If the properties of the adhesive are such as to ensure bridging between the gypsum base and the wood framing, no fastening is required in the field of the base for walls. In such cases, perimeter fastening, 16 in. (406 mm) on centers, is required.

9.6.2 If the properties of the adhesive are such that there is no positive bridging between the board and the wood framing, either temporary field nailing or temporary bracing is required to ensure contact between the base, the adhesive, and the stud face, until the adhesive develops adequate bond strength. Unless specifically recommended otherwise by the adhesive or mastic manufacturer, space fasteners in accordance with Table 2.

10. Application over Existing Surfaces

10.1 Gypsum base used over uneven or broken surfaces, and on the interior of all exterior masonry or concrete walls or columns, shall be applied over furring strips or furring channels in accordance with Section 8 of this specification and Appendix X2.

10.2 Over reasonably smooth, sound, existing surfaces, such as gypsum wallboard or plaster in good condition, applied to wood framing, apply the gypsum base by using fasteners long enough to penetrate not less than ⅞ in. (22 mm) into the underlying framing members. Space fasteners in accordance with 7.2.1 or 7.2.2. Alternatively, the gypsum base may be applied by fastener and adhesive attachment with fastener spacing in accordance with Section 9 of this specification and adhesive application in strict accordance with the gypsum base manufacturer's recommendations.

10.3 Over reasonably smooth, solid interior surfaces, such as concrete or unit masonry, apply the gypsum base by adhesive application in strict accordance with the gypsum base manufacturer's recommendations over furring strips or furring channels in accordance with 10.1.

11. Foil-Backed Gypsum Base

11.1 The application of foil-backed gypsum base shall conform to all of the foregoing specifications for the application of gypsum base except that it shall not be used in the following areas:

11.1.1 For the second ply on a two-ply laminating system.

11.1.2 For laminating directly to masonry.

11.1.3 For remodeling directly over plaster or other existing surfaces.

11.2 Place the reflective surface of the foil-backed gypsum base against the face of the framing members.

12. Application of Two-Ply Gypsum Base to Framing Members

NOTE 5—The base ply in two-ply construction may be gypsum board other than gypsum veneer base (see Section 4).

12.1 The maximum spacing for framing members and the application direction for both plies of two-ply gypsum base construction with no adhesive between plies shall comply with Table 1.

12.2 The maximum spacing for framing members for two-ply gypsum base construction with adhesive between plies shall comply with Table 3.

12.3 Apply the first ply of gypsum base with nails, screws, or staples of the size and type required in Section 4 and 7.3.3, spaced in accordance with Table 4.

12.4 Where no adhesive is used between plies, the two plies of gypsum base may be applied in opposite directions or in parallel directions. Apply the face layer with the number of nails or screws required for normal single-ply application. Face layer joints that are parallel to the framing shall fall over the framing when two plies of gypsum base are parallel.

12.5 Where a laminating adhesive is used between plies, the adhesive shall be an adhesive specifically recommended by the gypsum base manufacturer for two-ply gypsum base application. Joints in the face ply need not occur over framing members. Stagger parallel joints of the two plies (offset) at least 8 in. (203 mm). Laminating procedures shall be as specified by the gypsum base manufacturer.

12.6 Where gypsum base is adhesively installed, hold the face ply firmly in place with temporary fasteners or shoring (supports, props, headers, etc.) in sufficient quantity to ensure pressure for bonding. When a bond has developed, remove the supports. Where temporary nails are used, either remove the temporary nails, or drive them flush with the surface.

13. Solid Gypsum Base Partitions

13.1 *Face Panels*—½ in. (12.7 mm), ⅝ in. (15.9 mm), or multiple laminations of regular or Type X gypsum base of 4-ft (1.2-m) width.

13.2 *Core*—1-in. thick gypsum board, either single or multiple layers in 2-ft (0.6-m) or 4-ft (1.2-m) widths.

13.3 *Runners*—A minimum of 26-gage metal or construction grade wood.

13.4 *Laminating Adhesive*—Laminating adhesive shall be as recommended by the gypsum base manufacturer.

13.5 Install floor and ceiling runners in accordance with the partition layout and secure not less frequently than 24 in. (610 mm) on centers. Install vertical runners at required locations, such as exterior walls, exterior corners, and door frames. Secure partitions at the floor and ceiling in accordance with the gypsum base manufacturer's details or as otherwise required.

13.6 Application of laminating adhesive and erection of the panels shall be in accordance with the gypsum base manufacturer's recommendations. Laminate the gypsum base to the backing board with pressure to ensure a secure, firm bond. Stagger joints of gypsum base and backing board. Use fasteners to ensure an intimate and continuous bond between the gypsum base and the backing board.

13.7 Do not apply the veneer plaster until the veneer base is firmly bonded.

14. Adhesive Application of Gypsum Base to Interior Masonry or Concrete Walls

14.1 Only interior masonry or concrete walls above grade, that are not exposed to moisture on either side, are recommended for direct adhesive application.

14.2 Foil-backed gypsum base is not recommended for direct adhesive application.

14.3 Adhesive shall be as recommended by the gypsum base manufacturer. Gypsum base and masonry surfaces to be adhered shall be free of any foreign matter that could impair the bond. Remove projections or elevations in the masonry by chipping or wire brushing. Back-plaster large hollows or depressions with portland cement and allow to dry.

14.4 For application to monolithic concrete, porous brick, or concrete block, apply adhesive to the back of the gypsum base or on the wall in continuous beads spaced not to exceed 12 in. (305 mm) on center. Beads shall be of sufficient size to provide a continous bond between the gypsum base and the wall surface along the length of the gypsum base.

14.4.1 **Caution**—Be sure that the adhesive will bond to monolithic concrete. Form release compounds can cause bond failure.

14.4.2 *Gypsum Base Installation*—Position the gypsum base to provide a tight fit at abutting edges. Do not slide the base. Use moderate pressure to develop a full adhesive bond.

14.4.3 Do not apply the veneer plaster until the gypsum base is firmly bonded.

14.5 *Gypsum Base Application over Rigid Plastic Foam*—Gypsum base may be applied over rigid plastic foam insulation panels that have been applied to masonry or concrete walls.

14.5.1 Apply the foam insulation to the masonry or concrete in accordance with the foam manufacturer's specifications.

14.5.2 Attach furring strips or special metal furring members by mechanical means to the masonry or concrete wall surface. This may be done either before or after application of insulation, depending on the system used.

14.5.2.1 Install furring members in accordance with Table 1 of this specification and at the floor-wall angle, wall-ceiling angle (or at the termination of the gypsum base above suspended ceilings), around door, window, and other openings, and where required for cabinet or fixture attachment.

14.5.3 Apply the gypsum base to the furring as described in 7.3.1 or 7.3.2. Mechanical fasteners shall be of such length that they shall not penetrate completely to the masonry or concrete.

NOTE 6—Gypsum base applied over rigid plastic foam insulation in the manner described in 14.5 may not necessarily provide finish ratings required by local building codes.

15. Gypsum Base Application over Steel Framing and Furring

15.1 Steel framing and furring shall comply with the requirements of Specification C 754 except for spacing of framing.

15.2 *Screw Application*—Apply screws as specified in 7.2.5.

15.3 *Framing Spacing*—Maximum spacing of steel framing and furring for screw application shall be as specified in Table 1 for single-ply gypsum base and as specified in Table 3 for two-ply gypsum base.

15.4 *Screw Spacing:*

15.4.1 Maximum screw spacing for single-ply gypsum base and face ply of two-ply gypsum base with no adhesive shall be as specified in 7.3.2.

15.4.2 Maximum screw spacing for parallel applied first-ply of two-ply gypsum base over steel framing on side walls with no adhesive between the plies shall be 12 in. (305 mm) on centers along the edges of the gypsum base and at third points into each stud or furring channel in the field of the gypsum base.

15.4.3 Maximum screw spacing for a perpendicular applied first-ply of two-ply gypsum base over steel framing on side walls with no adhesive between the plies shall be one screw at each paper bound edge at each stud or furring channel intersection and one screw midway between paperbound edges at each stud or furring channel.

15.4.4 Maximum screw spacing for perpendicular or parallel applied first-ply of two-ply gypsum base over steel framing with adhesive between plies shall be as specified for single-ply gypsum base in 7.3.2.

15.4.5 Maximum screw spacing for the first ply of two-ply gypsum base over steel framing with adhesive between plies on ceilings shall be the same as specified for the face-ply of two-ply gypsum base in 12.4. On wall surfaces, use only a sufficient number of screws to hold the gypsum base in place until the adhesive, recommended and applied as specified by the gypsum base manufacturer, develops a bond.

16. Application to Arches

16.1 Carefully bend into place the gypsum base to be applied to the soffit of arches (see Table 5). If necessary, dampen or score it first approximately 1 in. (25.4 mm) on centers on the back side. In the latter case, after the core

has been broken at each cut, apply the gypsum base to the curved contour and anchor in place. If the base is dampened, allow it to dry before plastering.

NOTE 7—To apply the base, place a stop at one end of the curve and then gently and gradually push on the other end of the base, forcing the center against the framing until the curve is complete.

NOTE 8—By moistening the face and back paper thoroughly and allowing the water to soak well into the core, the base may be bent to still shorter radii. When it dries thoroughly, the base will regain its original hardness.

APPENDIXES

(Nonmandatory Information)

These Appendixes provide general information and also suggestions for inclusions to be made elsewhere by the specifier, and are not a part of this specification.

X1. GENERAL INFORMATION

X1.1 Scaffolding shall be constructed and maintained in strict conformity with applicable laws and ordinances and shall not interfere with or obstruct the work of others. The work shall be properly coordinated with the work of other trades.

X1.2 The bond of veneer plasters to gypsum base is impaired if the gypsum base is exposed to direct sunlight for extended periods. Therefore, the surfaces of the gypsum base should be protected from direct sunlight in storage and should be plastered as soon as possible after application.

X2. WOOD FRAMING REQUIREMENTS

X2.1 Requirements covering framing, furring, spacing, etc., are essential to provide a proper surface to receive the gypsum base. The following requirements should be included in the project specifications for framing and furring:

X2.1.1 All framing members to which gypsum base will be fastened shall be straight and true and spaced not to exceed the maximum spacings shown in Table 1. Wood framing, bridging, and furring members shall be the proper grade for the intended use and members 2 by 4 in. (50 by 100 mm) nominal size or larger shall bear the grade mark of a recognized inspection agency. Framing, bridging, and furring shall be adequate to carry the design or code loading, or both. In place of an applicable local code, the framing shall meet the minimum requirements of FHA for single or multifamily dwellings. The deflection of members supporting gypsum base shall not exceed 1/360 of the span at full design load. Headers shall be provided as necessary for the support of fixtures.

X2.1.2 When the gypsum base is nailed to wood cross furring on ceilings, these members shall have a minimum cross section of 1½ by 1½ in. (38 by 38 mm) actual and shall be spaced in accordance with the requirements of 5.1 and 10.2. Where screw application is used, the furring member may be ¾ by 2½ in. (19 by 64 mm) actual size. Lumber shall conform to American Softwood Lumber Standard PS-20.

X2.1.3 All vertical solid surfaces that are furred to receive gypsum base shall be furred with wood framing members not less than ¾ in. (19 mm) thick and 1½ in. (38 mm) wide. (Fastener penetration into the framing member shall be ⅝ in. (15.9 mm) minimum.)

X2.1.4 Insulating blankets or flanges of blankets shall not be applied over framing members that are to receive gypsum base. Foil-backed gypsum base may be used as a vapor retarder where required.

X3.1.1 The ratio of total net free ventilating area to area of ceiling shall be not less than 1/150, except that the ratio may be 1/300 provided:

X3.1.1.1 A vapor retarder having a transmission rate not exceeding 1 perm dry cup is installed on the warm side of the ceiling, or

X3.1.1.2 At least 50 % of the required ventilating area is provided by ventilators located in the upper portion of the space to be ventilated (at least 36 in.

(915 mm) above eave or cornice vents) with the balance of the required ventilation area provided by eave or cornice vents.

X3.1.2 Attic space that is accessible and suitable for future habitable rooms or walled-off storage space shall have at least 50 % of the required ventilating area located in the upper part of the ventilated space as near the high point of the roof as practicable and above the probable level of any future ceilings.

TABLE 1 Maximum Framing Spacing

Single-ply Base (Thickness), in. (mm)	Application to Framing	Maximum On Center Spacing of Framing, in. (mm)
Ceilings:		
½ (12.7)	parallel	16 (406)
½ (12.7)[A]	perpendicular	24 (610)
⅝ (15.9)[B]	perpendicular	24 (610)
Sidewalls:		
⅜ (9.5)[C]	perpendicular	16 (406)
½ (12.7)	parallel	16 (406)
½ (12.7)[A]	perpendicular	24 (610)
⅝ (15.9)[B]	parallel or perpendicular	24 (610)

[A] ½-in. (12.7-mm) gypsum base applied perpendicular on 24-in. (610-mm) on centers framing is minimum construction and may be subject to ridging and cracking at joints and to sagging or bowing.
[B] For ⅝-in. (15.9-mm) base, perpendicular on ceilings and either perpendicular or parallel on sidewalls at 24-in. (610-mm) on centers spacing, only two-component veneer plaster or special joint reinforcement, as recommended by the gypsum veneer plaster manufacturer, shall be used.
[C] ⅜-in. (9.5-mm) gypsum base shall be used over wood framing with two-component veneer plaster systems only.

TABLE 2 Fastener Spacing with Adhesive or Mastic Application and Supplemental Fastening, in. (mm) On Center

Framing Member Spacing	Ceilings		Partitions Load Bearing		Partitions Nonload-Bearing	
	Nail	Screw	Nail	Screw	Nail	Screw
16 (406)	16 (406)	16 (406)	16 (406)	24 (610)	24 (610)	24 (610)
24 (610)	12 (305)	16 (406)	12 (305)	16 (406)	16 (406)	24 (610)

TABLE 3 Maximum Framing Spacing with Adhesive Between Plies

Gypsum Base Thickness, in. (mm)		Application Direction		Maximum On Center Spacing of Framing. in. (mm)
Ceilings				
Base	Face	Base	Face	
⅜ (9.5)	½ (12.7)	perpendicular	perpendicular or parallel	16 (406)
½ (12.7)	½ (12.7)	parallel	perpendicular or parallel	16 (406)
		perpendicular	perpendicular or parallel	24 (610)
⅝ (15.9)	½ (12.7)	perpendicular or parallel	perpendicular or parallel	24 (610)
⅝ (15.9)	⅝ (15.9)	perpendicular	perpendicular or parallel	24 (610)
Sidewalls				
⅜ (9.5)	½ (12.7)	perpendicular or parallel	perpendicular or parallel	16 (406)
½ (12.7)	½ (12.7)	perpendicular or parallel	perpendicular or parallel	24 (610)
⅝ (15.9)	½ or ⅝ (12.7 or 15.9)	perpendicular or parallel	perpendicular or parallel	24 (610)

TABLE 4 Attachment Spacing for Base Ply over Wood or Steel Framing

Fastener	Nails	Screws	Staples
Framing	wood	wood or metal	wood
No adhesive between plies	24 in. (610 mm) on centers	24 in. (610 mm) on centers	16 in. (406 mm) on centers
Adhesive between plies	Section 7.2.1	Section 7.2.2	7 in. (178 mm)

TABLE 5 Bending Radii for Application to Arches

Gypsum Base Thickness, in. (mm)	Bent Lengthwise, ft (m)	Bent Widthwise, ft (m)
½ (12.7)	10 (3.05)[A]	—
⅜ (9.5)	7½ (2.286)	25 (7.62)

[A] Bending two ¼-in. (6.4-mm) pieces successively permits radii shown for ¼ in.

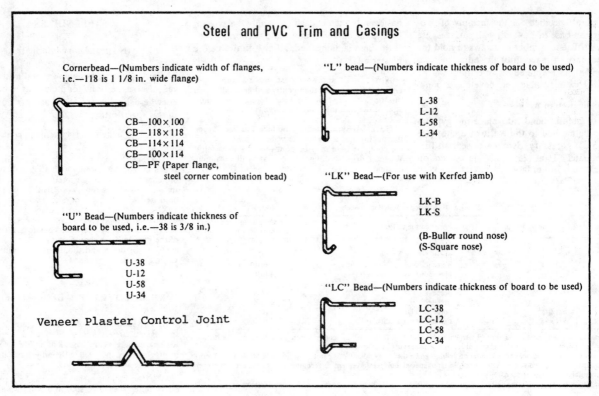

Steel and PVC Trim and Casings

Cornerbead—(Numbers indicate width of flanges, i.e.—118 is 1 1/8 in. wide flange)

CB—100 × 100
CB—118 × 118
CB—114 × 114
CB—100 × 114
CB—PF (Paper flange, steel corner combination bead)

"U" Bead—(Numbers indicate thickness of board to be used, i.e.—38 is 3/8 in.)

U-38
U-12
U-58
U-34

Veneer Plaster Control Joint

"L" bead—(Numbers indicate thickness of board to be used)

L-38
L-12
L-58
L-34

"LK" Bead—(For use with Kerfed jamb)

LK-B
LK-S

(B-Bullor round nose)
(S-Square nose)

"LC" Bead—(Numbers indicate thickness of board to be used)

LC-38
LC-12
LC-58
LC-34

NOTE—All dimensions are in U.S. Customary units
FIG. 1 Accessories

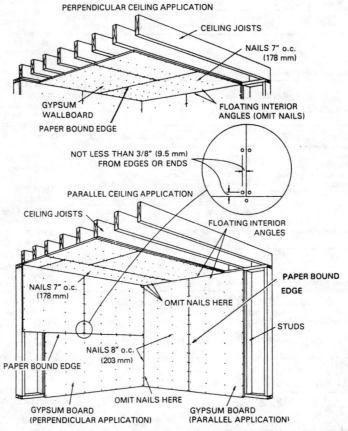

PERPENDICULAR CEILING APPLICATION
CEILING JOISTS
NAILS 7" o.c. (178 mm)
GYPSUM WALLBOARD
PAPER BOUND EDGE
FLOATING INTERIOR ANGLES (OMIT NAILS)
NOT LESS THAN 3/8" (9.5 mm) FROM EDGES OR ENDS
PARALLEL CEILING APPLICATION
CEILING JOISTS
FLOATING INTERIOR ANGLES
PAPER BOUND EDGE
NAILS 7" o.c. (178 mm)
OMIT NAILS HERE
STUDS
NAILS 8" o.c. (203 mm)
PAPER BOUND EDGE
OMIT NAILS HERE
GYPSUM BOARD (PERPENDICULAR APPLICATION)
GYPSUM BOARD (PARALLEL APPLICATION)

FIG. 2 Single Nailing

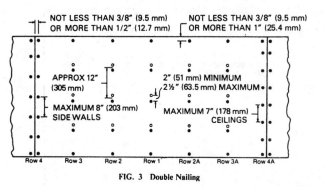

FIG. 3 Double Nailing

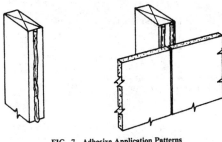

FIG. 7 Adhesive Application Patterns

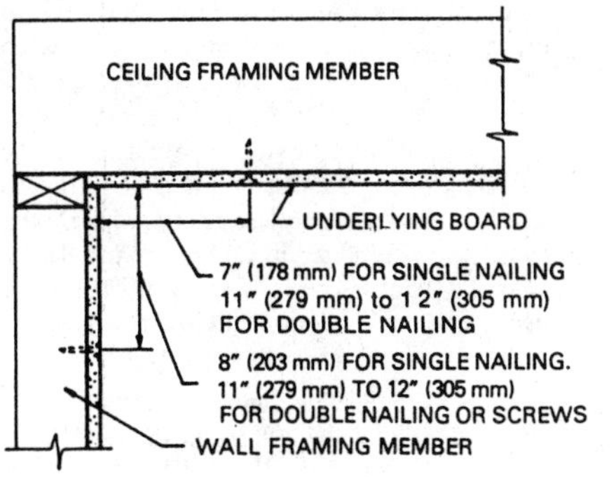

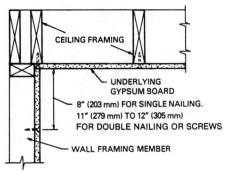

FIG. 4 Vertical Section, Ceiling Framing Perpendicular To Wall

X3. VENTILATING ABOVE GYPSUM BASE CEILINGS

X3.1 Attics or similar unheated spaces above gypsum base ceilings shall be adequately ventilated as follows: Provide effective cross ventilation for all spaces between the roof and top floor ceilings by screened louvers or other approved and acceptable means.

X4. CABLE ELECTRIC RADIANT HEATING SYSTEMS FOR CEILINGS

X4.1 The application of gypsum base under this specification shall be performed in accordance with the requirements of this specification except as follows:

X4.1.1 The gypsum base shall be a minimum of ½ in. (12.7 mm) in thickness. The gypsum base shall be applied perpendicular to framing members. Attachment shall be as required in Section 8.

X4.1.2 Electric heating cables shall be securely attached to the gypsum base in accordance with the recommendations of the cable manufacturer or supplier.

X4.1.3 Cables shall run at right angles to the gypsum base paperbound edges. Do not position cables directly over joints that run parallel to the direction of cable run.

X4.1.4 Position cables no closer than 6 in. (152 mm) from all wall-ceiling intersections and openings.

X4.1.5 All inspections and testing of the heating system shall be completed before application of the veneer plaster.

X4.1.6 Under no operating conditions shall the heating cable exceed a temperature of 125°F (52°C).

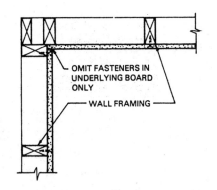

FIG. 5 Vertical Section, Ceiling Framing Parallel to Wall

FIG. 6 Horizontal Section Through Interior Vertical Angle

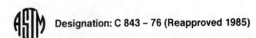

Designation: C 843 – 76 (Reapproved 1985)

Standard Specification for

APPLICATION OF GYPSUM VENEER PLASTER[1]

This standard is issued under the fixed designation C 843; the number immediately following the designation indicates the year of original adoption or, in the case of revision, the year of last revision. A number in parentheses indicates the year of last reapproval. A superscript epsilon (ε) indicates an editorial change since the last revision or reapproval.

This specification has been approved for use by agencies of the Department of Defense and for listing in the DoD Index of Specifications and Standards.

1. Scope

1.1 This specification covers the minimum requirements for and methods of applying gypsum veneer plaster.

1.2 Where a specific degree of fire resistance is required for veneer plaster systems, applicable building code regulations shall be followed. Details of construction that have achieved required fire resistance endurance ratings may be obtained from official reports of fire tests conducted at recognized fire testing laboratories in accordance with Methods E 119.

1.3 Where a specific degree of sound control is required for veneer plaster assemblies and constructions, details of construction shall be in compliance with official reports of tests conducted in recognized sound testing laboratories in accordance with the applicable sound tests of Method E 90, Test Method C 423, or Method E 492.

NOTE 1—To ensure desirable results, this specification should be coordinated with the provisions of Specification C 844.

NOTE 2—General information regarding matters of a contractual nature concerning veneer plaster work will be found in Annex A1. Additional technical information related to veneer plastering is provided in Annex A2.

1.4 The values stated in inch-pound units are to be regarded as the standard.

2. Applicable Documents

2.1 *ASTM Standards:*

C 11 Definitions of Terms Relating to Gypsum and Related Building Materials and Systems[2]

C 423 Test Method for Sound Absorption and Sound Absorption Coefficients by the Reverberation Room Method[4]

C 472 Methods for Physical Testing of Gypsum Plasters and Gypsum Concrete[2]

C 587 Specification for Gypsum Veneer Plaster[2]

C 588 Specification for Gypsum Base for Veneer Plasters[2]

C 631 Specification for Bonding Compounds for Interior Plastering[2]

C 844 Specification for Application of Gypsum Base to Receive Gypsum Veneer Plaster[2]

E 84 Test Method for Surface Burning Characteristics of Building Materials[3]

E 90 Method for Laboratory Measurement of Airborne Sound Transmission Loss of Building Partitions[4]

E 119 Methods of Fire Tests of Building Construction and Materials[3]

E 492 Method of Laboratory Measurement of Impact Sound Transmission Through Floor-Ceiling Assemblies Using the Tapping Machine[4]

3. Definitions

3.1 *veneer plaster system* — gypsum veneer plaster applied to (*1*) a gypsum base in accordance with 10.1 or (*2*) properly prepared masonry or monolithic concrete surfaces, neither side of which is exposed to moisture. The plaster may be applied with one or more components normally not exceeding ¹/₄ in. (6.4 mm) in total thickness.

3.2 *two-component system* — a veneer plaster system involving two separate materials mixed and applied separately for base coat and finish coat.

3.3 *one-component system* — a veneer plaster system designed for application directly over approved bases in a single plaster mix applied in a single coat or double-back operation with the same material.

3.4 *base coat* — veneer plaster trowel or machine applied as the first coat of a two-component system.

3.5 *finish coat* — veneer finish plaster trowel or machine applied as the second coat over the base coat plaster in a two-component system.

3.5.1 *smooth-trowel finish* — a finish resulting from steel troweling with a minimum of water after the plaster has become firm. A smooth finish free of trowel marks, blemishes, or other imperfections.

3.5.2 *texture finish* — a finish resulting from (*1*) trowel application followed by floating or texturing of the surface with any of a variety of tools using a minimum of water or (*2*) machine application which may or may not be hand textured. The surface should be textured uniformly, free of slick spots or other blemishes.

3.6 *reinforced joint* — a joint between gypsum base boards that is reinforced with strip material embedded in a cementitious material.

3.7 *joint-reinforcing embedment* — the cementitious material used to embed the strip material that is compatible with the veneer plaster used.

3.8 Other terms relating to gypsum shall be as defined in Definitions C 11.

4. Delivery of Materials

4.1 All manufactured materials shall be delivered in the original packages, containers, or cartons bearing the brand name and manufacturer identification.

5. Protection of Materials

5.1 Plasters and other cementitious materials shall be kept dry until ready to be used; they shall be kept off the ground, under cover, and away from damp walls and surfaces.

6. Environmental Conditions

6.1 *Temperature* — When the ambient outdoor temperature at the building site is less than 50°F (10°C) maintain a continuous uniform temperature at the minimum comfortable working level, but not less than 50°F and not more than 80°F (27°C) in the building for a period of not less than 1 week prior to the application of veneer plaster (Note 3), while the plastering is being done, and for a period of at least 1 week after the plaster is set. Distribute the heat evenly in all areas, and use deflective or protective screens to prevent concentrated or uneven heat on the veneer plaster near the heat source.

6.2 *Ventilation* — Provide ventilation to remove the water in excess of that required for hydration of the veneer plaster, immediately subsequent to its application and set (Note 4). See Annex A2.2 for provisions for ventilating portions of underside of roofs enclosed with veneer plaster ceilings.

NOTE 3—The building should be permitted to become adjusted to the uniform temporary heating conditions for at least 1 week prior to the start of veneer plastering to minimize the possibility of plaster cracking due to structural movements caused by thermal changes. Blasts of cold air through openings should be avoided.

NOTE 4—In glazed buildings ventilation shall be accomplished by keeping windows open approximately 2 in. (50 mm) top and bottom (or side-pivoted windows approximately 4 in. (100 mm) to provide air circulation. Avoid rapid drying. During periods of low indoor humidity, minimum air circulation is desirable following plastering and until the plaster is dry. For enclosed areas of buildings that lack natural ventilation, provisions shall be made to exhaust the moist air to the outside by mechanical means such as temporary circulators, exhaust fans, or the air conditioning system. A portion of the water used in mixing veneer plaster is necessary for the chemical reaction that rehydrates and hardens the plaster. The excess amount required for mixing and application must be exhausted from the plaster after set has taken place. Minimum circulation of air is desirable to prevent dryouts (see 10.4.1).

7. Materials

7.1 *Gypsum Base for Veneer Plasters* (hereinafter referred to as "gypsum base") Specification C 588.

7.2 *Veneer Plasters* — Specification C 587.

7.3 *Reinforcing* — Noncorrosive strip providing the joint strength requirements of Specification C 587.

7.4 *Liquid Bonding Compounds* — Bonding compounds may be used at the discretion of the design authority and shall conform to the requirements of Specification C 631. See Annex A2.

---

[1] This specification is under the jurisdiction of ASTM Committee C-11 on Gypsum and Related Building Materials and Systems and is the direct responsibility of Subcommittee C11.03 on Specifications for Application of Gypsum and Other Products on Assemblies.

Current edition approved July 30, 1976. Published September 1976.

[2] *Annual Book of ASTM Standards*, Vol 04.01.

[3] *Annual Book of ASTM Standards*, Vol 04.07.

[4] *Annual Book of ASTM Standards*, Vol 04.06.

7.5 *Water* — Water shall be clean, fresh, and of potable water quality.

8. Surface Preparation

8.1 *Examination* — Carefully examine all surfaces, including but not limited to unit masonry, monolithic concrete, gypsum bases of all types, and accessories to receive veneer plaster, before the veneer plaster is applied. Notify the proper authorities promptly of all unsatisfactory conditions. Gypsum base exposed to excessive sunlight may cause bond failure; therefore, the examination shall determine if this condition exists. Do not apply veneer plaster until after such unsatisfactory conditions are rectified to the satisfaction of the plastering contractor.

8.1.1 Do not apply veneer plaster over any surfaces of unit masonry or concrete that have been coated with any bituminous compound or other detrimental waterproofing or dampproofing or form release agent.

8.2 *Conditioning of Surfaces:*

8.2.1 Immediately before the plaster is applied wet down the masonry surfaces on which suction must be reduced. Visible water shall not remain on the surface.

8.2.2 Carefully examine accessories, such as corner beads, control joints, casing beads, etc., to assure that they are straight, curved, plumb, level, square, or true to the required angles and have been applied in a manner to ensure full coverage of flanges before the plaster is applied.

8.3 *Monolithic Concrete Surfaces* — Clean monolithic concrete surfaces of all dust, loose particles, and other foreign matter. Completely remove all grease, oil, noncompatible curing compounds, and form releasing agents. Remove all ridges and protrusions greater than $1/8$ in. (3.2 mm) and fill all depressions greater than $1/4$ in. (6.4 mm) level with portland cement mortar and allow to set and dry. Any further preparation shall be as recommended by the veneer plaster manufacturer.

8.4 *Gypsum Base Joint Reinforcement* — Reinforce all interior angles and flat joints prior to application of the veneer plaster over the base.

8.4.1 *Interior Angles* — Position and secure reinforcement with staples (on 12-in. (305-mm) centers, one side only), veneer plaster, or other cementitious material compatible with the veneer plaster or by using self-adhering strip reinforcement. When stapling, staple along the ceiling edge only for wall-to-ceiling angles and along one edge for wall-to-wall angles. When securing reinforcement with veneer plaster, reinforcement must be thoroughly embedded so that embedment material is both under and covering reinforcement.

8.4.2 *Flat Joints* — Center reinforcement over the joint line of the gypsum base and secure with staples or veneer plaster if not using self-adhering reinforcement. Keep reinforcement tight and flat against the gypsum base and, when stapling, position the staples no farther than 24 in. (610 mm) apart, staggered along each edge. Embed reinforcement and continue application of the veneer plaster over the field of the base. **Caution** — This method provides minimum reinforcement of the joints. To minimize incidence and severity of joint ridging or cracking, or both, one of the following methods (8.4.2.1, 8.4.2.2, or 8.4.2.3) shall be used in the absence of any specific method recommended by the veneer plaster manufacturer. Where paper tape and setting-type joint compounds are used, procedure 8.4.2.3 shall be emphasized.

8.4.2.1 *Reinforcement over Set Veneer Plaster* — Tightly trowel veneer plaster over the joint line leaving the plaster feathered out to a width of about 6 in. (152 mm) flush with the face of the gypsum base. Allow the plaster to set; then secure reinforcement over the joint line as described under 8.4.2.

8.4.2.2 *Reinforcement Secured and Embedded with Veneer Plaster* — Apply reinforcement over the joint line as described under 8.4.2. Tightly trowel the plaster over the reinforcement along the joint line to provide thorough embedment of the reinforcement. Allow the joint embedment to set before proceeding with general plastering.

8.4.2.3 *Reinforcement Embedded (no staples)* — Tightly trowel the embedment material to a depth of about $1/32$ in. (0.8 mm), working the trowel in both directions along the joint line. Center the reinforcement over the joint line. Firmly and evenly press the reinforcement into the soft embedment material using a little soft material on the trowel to bury it completely, and leave the embedment material feathered out to a width of about 6 in. (152 mm) flush with the face of the gypsum base. Allow the embedment material to set before proceeding with general plastering.

9. Mix Design

9.1 Do not use frozen, caked, or lumpy material. Do not retemper or use material that has partially set. Mix each batch separately. Clean mixers thoroughly after each batch so as not to accelerate the following batches. This can be done by spray hosing the mixer paddle and containers immediately after each batch. Water ratios and other techniques used for mixing shall conform to the manufacturer's recommendation for specific veneer plaster products. In the absence of explicit directions, the following practices and recommendations shall apply:

9.1.1 *Equipment* — A paddle-type agitator fitted to a $1/2$-hp (373-W) heavy-duty, electric drill, rated from 900 to 1000 rpm (no load), and a clean drum of convenient size are recommended for rapid, efficient mixing of veneer plasters.

9.2 *Mixing Procedure:*

9.2.1 Put all but 1 or 2 qt. (0.91 or 1.91 litres) of the proper amount of water in the mixer for each bag of veneer plaster to be mixed.

9.2.2 Add the veneer plaster to the water and immediately mix until uniformly wetted. Veneer plaster may be added while the agitator is turning.

9.2.3 Continue mixing, adding water to obtain the desired lump-free mortar consistency.

9.2.4 When mixing plaster for spray application, mix to a consistency so that 75 to 90 % of the batched mortar will pass a No. 8 (2.36-mm) sieve without shaking. Make sure the plaster is specifically designed for spray application to avoid quick set problems.

9.2.5 Do not overmix. Three to four minutes is usually sufficient to attain fluidity.

9.3 *Setting Time* — Setting time is carefully controlled by the manufacturer and no job adjustment should be made. Should problems in veneer plaster setting time develop under unusual job situations, consult the manufacturer of the veneer plaster for recommendations.

9.3.1 Do not use gauging, moulding, or casting plasters, lime, gypsum, Keene's cement, portland cement, etc., to adjust the veneer plaster setting time.

9.4 Occasionally special textures are desired that require additional sand. Approval of the veneer plaster manufacturer must be obtained prior to any such additions.

10. Application

10.1 *General* — Veneer plasters have widely differing working properties, physical characteristics, and limitations regarding their compatibility with one another and with various bases. To secure the desired results, only those particular application methods or techniques, or both, limitations, and procedures set forth by individual manufacturers for their products shall be followed. In the absence of manufacturer's directions, the following practices and recommendations shall apply.

10.2 *Procedures:*

10.2.1 *One-Component Plasters* — Apply the plaster with sufficient material and pressure to provide a good bond on the gypsum bases or other bases as approved by the veneer plaster manufacturer. Double back immediately with the same mixer batch of plaster to the desired thickness. Straighten to a true surface without application of water (water will cause blistering when applied at this time). Allow the material to "take-up" and texture with a float or sponge as desired. If a smooth finish is required, allow the plaster to become firm (so that water can be used without blistering) and trowel, using a minimum of water, to achieve a smooth finish, free of catfaces, trowel marks, blemishes, or other imperfections. Complete finishing before the veneer plaster sets.

10.2.2 *Two-Component Plasters* — The finish component coat may be applied over set but still "green", partially dry, or dry base coat.

NOTE 5 — Application of a second component over a base coat that is damp on joints and dry in the field of the base may result in "photographing" of joints and should be avoided.

10.2.3 *Base Coat-Hand Application* — Apply the base coat with sufficient material and pressure to provide a good bond on the gypsum bases or other bases as approved by the veneer plaster manufacturer and straighten to a true surface without application of water. Leave the surface sufficiently rough to provide a mechanical key for the finish coat.

10.2.4 *Base Coat-Machine Application* – First spray flat joints, beads, and interior angles, and work the material with hand tools to cover the reinforcement, fill the beads, straighten the angles, and feather out the plaster over the joints. Proceed with full areas by spraying with a broad sweeping motion, holding the nozzle close enough to the surface to avoid excessive overspray and buildup in the angles. First apply a light coating while moving in one direction; then immediately double back over the same area, sweeping the nozzle in the opposite direction to obtain a level coat that requires no further working, and that is suitable to receive the finish component coat on setting.

10.3 *Plaster Thickness* – Measure the plaster thickness from the face of the base to which it is applied, exclusive of joint treatment. The minimum thickness shall be as recommended by the manufacturer of the plaster used but in no case less than shown as follows:

Base coats (trowel applied)	⅟₁₆ in. (1.6 mm) min
Base coats (spray applied)	⅟₁₆ in. (1.6 mm) min
Finishes (two-component systems)	⅟₃₂ in. (0.8 mm) to ⅟₁₆ in. (1.6 mm) min
One-component systems	⅟₁₆ in. (1.6 mm) min

10.4 *Curing:*

10.4.1 Screen openings with cheesecloth or similar materials if glazed sashes are not in place and the building is subjected to hot, dry winds.

NOTE 6—Free circulation of hot, dry winds over the face of freshly applied plaster should be prevented. It causes "dry-outs" by removing the water from the plaster before setting is complete.

10.5 *Finishes:*

10.5.1 *Smooth-Trowel Finish* – Apply the plaster by troweling with firm pressure, then doubling back and filling out to a true, even surface. After the plaster has become firm, trowel it well with a minimum amount of water to a smooth finish, free of catfaces, blisters, trowel marks, blemishes, or other imperfections.

10.5.2 *Texture Finishes* – Apply the plaster over the base coat by troweling thoroughly, building up to an even surface, and then floating with a variety of tools using a minimum of water. It may also be sprayed (depending on the type of textured desired) to a true, even surface free of slick spots or other blemishes.

ANNEXES

A1. GENERAL INFORMATION

A1.1 Scaffolding shall be constructed and maintained in strict conformity with applicable laws and ordinances, and so as not to interfere with or obstruct the work of others.

A1.2 The work shall be properly coordinated with the work of other trades.

A2. TECHNICAL INFORMATION

A2.1 *Bonding Compounds* – Bonding compounds meeting Specification C 631 may be used to bond plaster to portland cement concrete and other sound surfaces. Such bonding agents may be used at the discretion of the specifier who may require performance records from the bonding agent manufacturer. The bonding agent shall be applied in accordance with the bonding agent manufacturer's directions.

A2.2 *Heating and Ventilation Requirements* – Proper ventilation above veneer plaster ceilings shall be provided by cross ventilating the space between the roof and the top floor ceilings by screened louvers or other acceptable means.

A2.2.1 The ratio of total net free ventilating area to the ceiling area shall be not less than ¹/₁₅₀, except that the ratio may be ¹/₃₀₀ provided: (*1*) insulating (foil back) gypsum base is installed or an approved vapor barrier is installed on the warm side of the ceiling and above bases or (*2*) at least 50 % of the required ventilating area is provided by ventilators at least 3 ft (0.91 m) above the eave and cornice vents which provide the balance of the required ventilation.

A2.2.2 Attic space accessible and suitable for future rooms or walled-off storage shall have at least 50 % of the required ventilating area located as near the high point of the ventilated space as practicable and above the level of any future ceiling.

A2.2.3 Ventilation should be provided to remove excess moisture prior to, during, and after application of the gypsum base and plastering.

A2.3 *Application of Veneer Plaster over Electric Radiant Heat Cable* – Application of veneer plaster over electric radiant heating cable shall be performed in accordance with the requirement of this specification except as follows:

A2.3.1 Treatment of flat joints in accordance with 8.4.2 shall be done before installation of the electric heating cable.

A2.3.2 All inspections and testing of the heating system shall be completed before application of the veneer plaster.

A2.3.3 Veneer plaster shall be applied in a tight scratch coat working the trowel parallel to the heating cable. Plaster thickness shall be flush with the cables. Immediately double back before the scratch application has set and build up to a thickness to cover the cables with a minimum of ¹/₁₆ in. (1.6 mm) and a maximum of ¹/₈ in. (3.2 mm) of veneer plaster.

A2.3.4 Apply the finish as recommended by the veneer plaster manufacturer or as described in 10.5.

A2.3.5 The heating system shall not be turned on until the veneer plaster has dried. The heating system should be deactivated and supplemental heat should be provided, as required, while paint is being applied. If the room temperature is more than 15°F (8°C) below the desired temperature after decoration is completed and dried, the heat should be turned on in increments to raise the temperature 5°F (3°C) in a 24-h period.

A2.3.6 Veneer plaster shall not be applied over heating systems that will exceed 125°F (52°C) wire temperature at any time.

SPECTEXT SPECIFICATION FOR
VENEER PLASTER
SECTION 09215

This section is intended to include interior veneer plaster in one or two coats applied to sheet gypsum base, masonry, or concrete substrate surfaces. Acoustic accessories are included in this section; edit accordingly.

PART 1   GENERAL

.01   WORK INCLUDED

A.   Acoustic accessories attached to [metal] [wood] framed substrate.

B.   Substrate surface of gypsum backing board, [single] [double] layer.

C.   Veneer plaster, [one] [two] coat application over [sheet gypsum] [masonry] [concrete] surface.

.02   RELATED WORK

A.   Section [_____ - _____:] Masonry partition substrate.

B.   Section [03001 - Concrete:] [03300 - Cast-in-Place Concrete.]

C.   Section 05400 - Cold-Formed Metal Framing: Metal framing system.

D.   Section [06001 - Carpentry Work:] [06112 - Framing and Sheathing:] Wood framing system.

E.   Section 07213 - Batt and Blanket Insulation: Thermal insulation and vapor barrier in exterior walls.

F.   Section 09111 - Metal Stud Framing System: Metal framing system.

G.   Section 09260 - Gypsum Board Systems: Metal framing and gypsum backing board substrate[.] [and acoustic accessories in interior partitions.]

H.   Section 08305 - Access Doors.

1.03   REFERENCES

1.03   Include only reference standards that are to be indicated within the text of this section. Refer to TAS Sections 09215, 09250, 09260 and 09280. Edit the following, adding and deleting as required for project and product selection.

A.   ANSI/ASTM C442 - Gypsum Backing Board.

B.   ANSI/ASTM C587 - Gpysum Veneer Plaster.

C.   ANSI/ASTM C588 - Gypsum Base for Veneer Plasters.

D.   ANSI/ASTM C631 - Bonding Compounds for Interior Plastering.

E.   ANSI/ASTM C843 - Application of Gypsum Veneer Plaster.

F.   ANSI/ASTM C844 - Application of Gypsum Base to Receive Gypsum Veneer Plaster.

G. ANSI/ASTM E90 - Laboratory Measurement of Sound
Transmission Loss of Building Partitions.

H. FS HH-I-521 - Insulation Blankets, Thermal (Mineral Fiber,
for Ambient Temperatures).

I. GA 216 - Recommended Specifications for the Application
and Finishing of Gypsum Board.

1.04   SYSTEM DESCRIPTION

A. Acoustic Attenuation for [Identified] Interior Partitions:
[_____] STC in accordance with ANSI/ASTM E90.

1.05   QUALITY ASSURANCE

A. Apply gypsum backing board in accordance with ANSI/ASTM
C844 and GA 216.

B. Apply gypsum veneer plaster in accordance with ANSI/ASTM
C843.

1.06   REGULATORY REQUIREMENTS

A. Fire Rated Partitions: Listed assembly by [UL] [FM] No.
[_____.]

B. Fire Rated Ceiling: Listed Assembly by [UL] [FM] No.
[_____.]

1.07   SUBMITTALS

A. Submit product data under provisions of Section [01300.]
[01340.]

B. Submit manufacturer's installation instructions under
provisions of Section [01300.] [01340.]

1.08   ENVIRONMENTAL REQUIREMENTS

A. Do not apply veneer plaster when substrate or ambient air
temperature is less than 50 degrees F (10 degrees C) nor
more than 80 degrees F (27 degrees C).

PART 2   PRODUCTS

2.01   ACCEPTABLE MANUFACTURERS

A. [_____ .]

B. [_____ .]

C. [_____ .]

D. Substitutions: Under provisions of Section [01600.]
[01630.]

2.01 If only one manufacturer is
acceptable, list in Paragraph A of
product description; if more than
one, list in this Article. If product
substitution procedure is used, include Paragraph D. Refer to TAS
Section 09215.

2.01D Edit the following descriptive specifications for any conflict
with manufacturers' products specified above.

2.02   MATERIALS

A. Gypsum Backing Board: ANSI/ASTM C588, [standard type;]
[fire rated Type X;] aluminum foil backed; [3/8] [1/2]
[5/8] inch ([9.5] [12.7] [15.9] mm) thick, [[_____ x

_____] inch ([_____ x _____] mm) sheet size;] [length to
match partition height;] [square] [tapered] [rounded]
edges, ends square.

B. Gypsum Board Metal Accessories: GA 216.

C. Reinforcing Tape, Joint Compound, Adhesive, Water,
Fasteners: GA 216.

D. Gypsum Veneer Plaster: ANSI/ASTM C587.

E. Bond Coat: ANSI/ASTM C631.

2.03 ACOUSTIC ACCESSORIES

A. Resilient Furring Channels: [_____ _____]
manufactured by [_____ _____ _____] or
[_____ _____ _____.]

B. Acoustic Insulation: FS HH-I-521, friction fit type,
unfaced; [_____] inch ([_____] mm) thick.

C. Acoustic Sealant: Type recommended for use in conjunction
with gypsum board; [_____ _____] manufactured by
[_____ _____ _____] or [_____
_____ _____.]

2.04 MIX DESIGN

A. Develop plaster mix in accordance with manufacturer's
instructions.

PART 3 EXECUTION

3.01 INSPECTION

A. Verify masonry mortar joints are cut flush and surface is
ready to receive work of this Section. Verify no
bituminous or water repellent coatings exist on masonry
surface.

B. Verify concrete surfaces are flat, honeycomb is filled
flush, and surface is ready to receive work of this
Section. Verify no bituminous, water repellent, or form
release agents exist on concrete surface.

C. Verify gypsum board substrate is flat, joints are taped
and sanded, and surface is ready to receive work of this
Section. Verify joint and surface perimeter accessories
are in place.

D. Beginning of installation means acceptance of substrate.

3.02 PREPARATION

A. Clean surfaces of dust or loose matter.

B. Remove projections greater than 1/8 inch (3 mm) and fill
depressions greater than 1/4 inch (6 mm) with [Portland
cement mortar.] [latex filler.] [_____ _____.]

C. Apply color tinted bond coat to prepared masonry surfaces
within [_____] hours of plaster application. Apply in
accordance with manufacturer's instructions.

3.03 INSTALLATION - ACOUSTIC ACCESSORIES

A. Install resilient furring chanels at [24] [_____] inches
([600] [_____] mm) on center at right angles to framing
members. Make end joints occur over framing members.
Terminate channels [_____] inches ([_____] mm) short of
door frames and partition perimeter.

B. Fit acoustic insulation tight between partition framing
members. Pack insulation around mechanical, electrical,
or other components in partition.

C. Place acoustic sealant at gypsum backing board partition
perimeter in accordance with manufacturer's instructions.
Seal penetrations of backing board by conduit, pipe,
ductwork, rough-in boxes, and other components in
partition.

3.04 INSTALLATION - GYPSUM BASE

2.03 Delete the following requirements for acoustic accessories if specified in another section or if not required.

3.01 Coordinate the following preparation requirements with masonry or concrete sections. Edit those sections accordingly.

3.01B Included the following paragraph when gypsum backing board and supporting framing is included under another section.

3.02B Masonry units of lightweight or other absorptive materials can cause uneven suction of moisture from applied veneer plaster, resulting in "telegraphing" or "photographing" of base. Consider applying a bond coat to such surfaces.

3.03 Delete the following requirements for acoustic accessories if specified in another section or if not required.

3.04 Include the following paragraphs if gypsum base wallboard is included in this section.

A. Install gypsum board in accordance with GA 216.

B. Use screws to fasten gypsum board to metal framing. [Use [nails] [screws] to fasten gypsum board to wood framing.]

C. Erect single layer gypsum board [vertical,] [horizontal,] [in direction most practical and economical,] with ends [and edges] occurring over firm bearing.

****** [OR] ******

D. Erect single layer fire rated gypsum board vertically, with edges and ends over firm bearing.

E. At resilient furred partition faces, place [4] [_____] inch ([100] [_____] mm) wide strip of gypsum board, same thickness as furring channel, at perimeter of wall openings and partition. Erect single layer gypsum board vertically, with edges and ends over firm bearing.

F. For double layer application, use gypsum backing board for first layer, placed [perpendicular] [parallel] to framing or furring members.

G. Place second layer [perpendicular] [parallel] to first layer. Ensure joints of second layer do not occur over joints of first layer.

H. Secure second layer with [adhesive and sufficient support to hold in place.] [fasteners.] [Apply adhesive in accordance with manufacturer's instructions.]

I. Tape, fill, and sand filled joints, edges, corners, openings, and fixings to produce surface ready to receive veneer finish. Feather coats onto adjoining surfaces so that camber is maximum [1/32] [_____] inch ([0.8] [_____] mm).

3.05 APPLICATION - VENEER PLASTER

A. Apply gypsum veneer plaster in accordance with ANSI/ASTM C843.

B. Install angle, corner, and joint reinforcement.

C. Dampen [masonry] [and] [concrete] surfaces without visible water on surface to minimize suction from plaster materials.

D. Apply single coat work [immediately after dampening substrate] [over substrate] to a thickness of [3/16] [_____] inch ([4.8] [_____] mm), plus or minus 1/64 inch (0.4 mm).

E. [Immediately after dampening,] apply base coat to a thickness of [1/8] [_____] inches ([3.2] [_____] mm) plus or minus 1/64 inch (0.4 mm).

F. Apply final coat over slightly green, almost dry base coat, to a thickness of [1/16] [_____] inch ([1.6] [_____] mm), plus or minus 1/64 inch (0.4 mm).

G. Total Thickness: [3/16] [_____] inch ([4.8] [_____] mm), plus or minus 1/64 inch (0.4 mm).

H. Finish surface to [flat, smooth, hard trowel finish.] [_____ _____ _____.]

3.04B Use Paragraphs C, D or E for installation of single layer wallboard base. Paragraphs F, G and H describe double layer base.

3.05 Edit the following paragraphs as appropriate to applying base coat to substrate of masonry, concrete or one or two coat work on gypsum wallboard base. Use Paragraph D for one coat work. Use Paragraphs E, F and G for two coat work.

Designation: C 840 – 87

APPLICATION AND FINISHING OF GYPSUM BOARD[1]

This standard is issued under the fixed designation C 840; the number immediately following the designation indicates the year of original adoption or, in the case of revision, the year of last revision. A number in parentheses indicates the year of last reapproval. A superscript epsilon (ε) indicates an editorial change since the last revision or reapproval.

This specification has been approved for use by agencies of the Department of Defense and for listing in the DoD Index of Specifications and Standards.

1. Scope*

1.1 This specification describes the minimum requirements for, and the methods of, application and finishing of gypsum board, including related items and accessories.

1.2 Where a fire resistance rating is required for a gypsum board assembly, details of construction shall be in accordance with reports of fire tests of assemblies that have met the requirements of the fire rating imposed.

1.3 Where sound control is required for a gypsum board assembly, details of construction shall be in accordance with reports of acoustical tests of assemblies that have met the required acoustical values.

1.4 Where this specification is more stringent (size or thickness of framing; size and spacing of fasteners) than the fire-rated construction, this specification shall govern; otherwise, the construction described in the fire test report shall govern.

1.5 Unheated spaces above gypsum board ceilings shall be properly ventilated (See Appendix X5).

1.6 The values stated in inch-pound units are to be regarded as the standard.

1.7 The following precautionary caveat pertains only to Sections 8, 9, and 10: *This standard may involve hazardous materials, operations, and equipment. This standard does not purport to address all of the safety problems associated with its use. It is the responsibility of the user of this standard to establish appropriate safety and health practices and determine the applicability of regulatory limitations prior to use. For a specific precautionary statement, see 10.5.9.*

2. Referenced Documents

2.1 *ASTM Standards:*

C 11 Definitions of Terms Relating To Gypsum and Related Building Materials and Systems[2]

C 36 Specification for Gypsum Wallboard[2]

C 442 Specification for Gypsum Backing Board and Coreboard[2]

C 475 Specification for Joint Treatment Materials for Gypsum Wallboard Construction[2]

C 514 Specification for Nails for the Application of Gypsum Wallboard[2]

C 557 Specification for Adhesives for Fastening Gypsum Wallboard to Wood Framing[2]

C 630 Specification for Water-Resistant Gypsum Backing Board[2]

C 645 Specification for Non-Load (Axial) Bearing Steel Studs, Runners (Track), and Rigid Furring Channels for Screw Application of Gypsum Board[2]

C 754 Specification for Installation of Steel Framing Members to Receive Screw-Attached Gypsum Wallboard, Backing Board,

or Water-Resistant Backing Board[2]

C 1002 Specification for Steel Drill Screws for the Application of Gypsum Board[2]

E 84 Test Method for Surface Burning Characteristics of Building Materials[3]

E 90 Method for Laboratory Measurement of Airborne Sound Transmission Loss of Building Partitions[4]

E 119 Methods for Fire Tests of Building Construction and Materials[3]

E 336 Test Method for Measurement of Airborne Sound Insulation in Buildings[4]

E 492 Method of Laboratory Measurement of Impact Sound Transmission Through Floor-Ceiling Assemblies Using the Tapping Machine[4]

2.2 *ANSI Standards:*

A108.4 Standard for Installation of Ceramic Tile with Water-Resistant Organic Adhesives[5]

A136.1 Standard for Organic Adhesives for Installation of Ceramic Tile, Type I[5]

2.3 *HUD Minimum Property Standard:*

U.S. Department of Commerce (DOC) Publication PS 20 American Softwood Lumber Standard

3. Definitions

3.1 *General*—Definitions shall be in accordance with Definitions C 11 except as otherwise specified.

3.1.1 *face panel*—a gypsum board which is applied to gypsum studs or backer board.

3.1.2 *fasteners*—nails, screws, or staples used for the mechanical attachment of the gypsum board.

3.1.3 *finishing*—the taping of joints, the concealment with joint treatment compound of such joints, heads of fasteners and flanges of corner protective devices and sanding to prepare them to receive the field application of priming, painting, coating, decorative coating and coverings such as wallpaper and vinyl materials. (See Section 10, X3, and X4.)

3.1.4 *framing member*—that portion of the framing, furring and blocking to which the gypsum board is attached. Unless otherwise specified herein, the surface to which abutting edges or ends are attached shall be not less than 1½ in. (38.1 mm) wide for wood members and not less than 1¼ in. (31.8 mm) wide for metal members and not less than 6 in. (152.4 mm) wide for gypsum studs. For internal corners or angles, the bearing surface shall be not less than ¾ in. (19.1 mm).

3.1.5 *laminating compound (adhesive)*—joint-treatment compound of the type used for embedding tape, or an adhesive, or laminating compound recommended by the manufacturer of the gypsum board (see Specification C 475).

3.1.6 *parallel or vertical application*—gypsum board applied with the edges parallel to the member to which it is attached.

3.1.7 *perpendicular or horizontal application*—gypsum board application with the edges applied at right angles to the member to which it is attached.

3.1.8 *treated joint*—a joint between gypsum boards that is reinforced and concealed with tape and joint treatment compound, or covered by strip-moldings.

3.1.9 *untreated joint*—a joint that is left exposed.

4. Delivery, Identification, Handling, and Storage

4.1 Deliver all materials in the original packages, containers, or bundles bearing the brand name, applicable standard designation, and the name of the manufacturer, or the supplier for whom the product is manufactured.

4.2 Keep all materials dry, preferably by being stored inside the building under roof. Where necessary to store gypsum board outside, off the ground, properly supported on a level platform and fully protected from the weather or direct sunlight exposure. Provide adequate ventilation to prevent condensation.

4.3 Neatly stack gypsum board flat with care taken to prevent sagging or damage to the edges, ends, and surfaces.

5. Environmental Conditions

5.1 *Application of Gypsum Board, Joint Treatment Materials, and Adhesives*—Maintain a room temperature of not less than 40°F (4°C) during application of gypsum board except, when adhesive is used for the attachment of gypsum board. For the bonding of adhesive, joint treatment, texturing and decoration, the room temperature shall be maintained at least at 50°F (10°C) for 48 h prior to application and continuously thereafter until completely dry.

NOTE 1—Precaution: When a temporary heat source is used, the temperature shall not exceed 95°F (35°C) in any given room or area.

NOTE 2—Precaution: Maintain adequate ventilation in the working area during installation and curing periods.

5.2 Protect gypsum board products from direct exposure to rain, snow, sunlight, or other excessive weather conditions.

NOTE 3—Where manufacturers recommendations differ from the above, follow their recommendation.

[1] This specification is under the jurisdiction of ASTM Committee C-11 on Gypsum and Related Building Materials and Systems and is the direct responsibility of Subcommittee C11.03 on Specifications for Application of Gypsum and Other Products in Assemblies.
Current edition approved Feb. 27, 1987. Published April 1987. Originally published as C 840 – 79. Last previous edition C 840 – 84a.
[2] *Annual Book of ASTM Standards*, Vol 04.01.
[3] *Annual Book of ASTM Standards*, Vol 04.07.
[4] *Annual Book of ASTM Standards*, Vol 04.06.
[5] Available from American National Standards Institute, 1430 Broadway, New York, NY 10018.

* A Summary of Changes section appears at the end of this specification.

6. Materials

6.1 *Gypsum Wallboard*—Specification C 36, Type X, regular, and foil-backed.

6.1.1 *Type X (Special Fire-Resistant) Gypsum Wallboard, Gypsum Backing Board, or Water-Resistant Gypsum Backing Board*—Gypsum board that provides a greater degree of fire resistance than regular gypsum board as defined in Specifications C 36, C 442, and C 630.

6.1.2 *Foil-Backed Gypsum Board*, either regular or Type X gypsum wallboard or gypsum backing board with foil laminated to the back surface. The foil is a vapor retarder.

6.1.3 *Predecorated Gypsum Board*, gypsum board with a decorative wall covering or coating applied in-plant by the gypsum board manufacturer.

6.1.4 *Gypsum Backing Board*—Specification C 442, Type X, regular, and foil-backed.

6.1.5 *Gypsum Board*—A family of gypsum sheet products, as defined in Definitions C 11.

6.2 *Joint Treatment Materials*—Specification C 475.

6.2.1 *Taping or Embedding Compound*, specifically formulated and manufactured for use in embedding of tape at gypsum wallboard joints and shall be completely compatible with the tape and the substrate.

6.2.2 *Finishing or Topping Compound*, specifically formulated and manufactured for use as a finishing compound.

6.2.3 *All-Purpose Compound*, specifically formulated and manufactured to serve as both a taping and finishing compound and compatible with the tape and the substrate.

6.2.4 *Joint Tape*—Specification C 475.

6.3 *Water-Resistant Gypsum Backing Board*—Specification C 630, Type X, and regular.

6.4 *Nails*—Specification C 514. Special nails for the predecorated gypsum board shall be as recommended by the predecorated gypsum board manufacturer.

6.5 *Screws*—Specification C 1002. Type S steel self-drilling and self-tapping screws. Specially designed steel screws as recommended by the manufacturer of the gypsum board shall be used for the screw application of gypsum board to wood or steel framing.

6.6 *Screws*—Specification C 1002. Type G steel self-drilling and self-tapping screws, specifically designed for the screw application of gypsum board to gypsum board.

6.7 *Screws*—Specification C 1002. Type W steel self-drilling and self-tapping screws, specifically designed for screw application of gypsum board to wood framing.

6.8 *Staples*—No. 16 USS gage flattened galvanized wire staples with 7/16 in. (11.1 mm) wide crown outside measurement and divergent point for the base ply only of two-ply gypsum board application:

Length of Legs, in. (mm)	Thickness of Gypsum Board, in. (mm)
1 (25.4)	3/8 (9.5)
1⅛ (28.6)	½ (12.7)
1¼ (31.8)	⅝ (15.9)

6.9 *Adhesives*—Specification C 557:

6.9.1 For fastening gypsum board to metal framing the adhesive shall be as recommended by the gypsum board manufacturer.

6.9.2 *Ceramic Tile Adhesive*—ANSI A136.1, Type 1.

6.10 *Framing Members:*

6.10.1 Wood framing members shall conform to PS 20, American Softwood Lumber Standard.

6.10.2 Light-gage, nonload bearing steel framing shall comply with Specification C 645.

6.10.3 Gypsum stud shall be 1 in. (25.4 mm) minimum thickness and 6 in. (152.4 mm) minimum width. They may be of 1 in. (25.4 mm) gypsum board material or multilayers laminated to the required thickness. Material shall conform to Specifications C 36 or C 442.

6.10.4 *Steel Studs, Furring Channels, and Runners*—Specification C 645.

6.10.5 *Cornerbead and Edge Trim*, made from protective coated steel or plastic designed for its intended use. Flanges shall be free from dirt, grease, or other materials that may adversely effect the bond of joint treatment materials or decoration. See Fig. 1.

6.11 Other types of corner, edge trim, or decorative dividers between gypsum wallboard panels may be used.

6.12 Water shall be clean, fresh, and potable (suitable for domestic consumption).

6.13 *Face Panels*, shall be ½ in. (12.7 mm), ⅝ in. (15.9 mm), or multiple laminations of regular or Type X gypsum board.

6.14 *Core Board*, shall be ¾ in. (19.1 mm) or 1 in. (25.7 mm) either single thickness or multiple layers to the required thickness.

7. System Classification

7.1 The various application systems are classified as follows:

I	Application of single-ply gypsum board to wood framing members
II	Application of two-ply gypsum board wood framing members
III	Application of gypsum board adhesive/nail-on to wood framing members
IV	Semi-solid gypsum board partitions
V	Solid gypsum board partitions
VI	Application of gypsum board with adhesive to interior masonry or concrete walls
VII	Application of gypsum board to rigid foam insulation
VIII	Application of gypsum board to steel framing and furring
IX	Arches and bending radii
X	Gypsum board used as a substrate to receive ceramic tile or plastic wall panels
XI	Exterior application of gypsum board
XII	Floating interior angles
XIII	Control joints
XIV	Foil-backed gypsum board

8. Substrate, Surface Preparation

8.1 Wood framing shall be of the proper grade and size as set by the American Softwood Lumber Standard PS 20 for the use intended. Wood framing shall be as straight and true as possible. Wood framing shall be securely attached following acceptable engineering practice and as required by the plans and specifications for the intended design.

NOTE 4—For installation of wood framing, Appendix X2 shall apply.

8.2 Metal framing members shall be of the proper size and design for their intended use and, when applicable, shall comply with Specification C 645 and shall be installed in accordance with Specification C 754.

8.3 Masonry or concrete walls shall be dry, free of dust, oil, or other form release agents, protrusions or voids, or foreign matter that would not permit proper bond for adhesively applied gypsum board.

8.4 Install all framing members and substrate so that after the gypsum board has been applied, the finished surface will be in an even plane.

8.5 Gypsum board shall be kept free of any dirt, oil, or other foreign matter that could cause a lack of bond. All dents or gouges shall be brought up to a smooth level plane of the surface of the board. Mechanical fasteners shall be set below the plane of the surface of the board. All joints shall be true and even. The board shall be tight against the framing member or substrate.

9. Application of Gypsum Board

9.1 *General:*

9.1.1 *Method of Cutting and Installation*—Cut the gypsum board by scoring and breaking or by sawing, working from the face side. When cutting by scoring, cut the face paper with a sharp knife or other suitable tool. Break the gypsum board by snapping the gypsum board in the reverse direction, or the back paper may be cut.

9.1.2 Smooth cut edges and ends of the gypsum board where necessary to obtain neat jointing when installed. Score holes for pipes, fixtures, or other small openings on the back and the face in outline before removal or cut out with a saw or special tool designed for this purpose. Where gypsum board meets projecting surfaces, scribe and cut neatly.

9.1.3 When gypsum board is to be applied to both ceiling and walls, apply the gypsum board first to the ceiling and then to the walls.

9.1.4 Space the fasteners, when used at edges or ends, not more than 1 in. (25.4 mm) from edges and not less than ⅜ in. (9.5 mm) from edges and ends of gypsum board (except where floating angles are used). Perimeter fastening into partition plate or sole at the top and bottom is not required or recommended except where the fire rating, structural performance or other special conditions require such fastening. While driving the fasteners, hold the gypsum board in firm contact with the underlying support. Application of fasteners shall proceed from the center or field of the gypsum board to the ends and edges.

9.1.5 Drive the nails with the heads slightly below the surface of the gypsum board. Avoid damage to the face and core of the board, such as breaking the paper or fracturing the core.

9.1.6 Drive the screws to provide screwhead penetration just below the gypsum board surface without breaking the surface paper of the gypsum board or stripping the framing member around the screw shank.

9.1.7 Drive the staples with the crown parallel to the framing members. Drive the staples in such a manner that the crown bears tightly against the gypsum board but does not cut into the face paper.

NOTE 5—Staple attachment is restricted to the base ply only of gypsum board in a two-ply system.

9.1.8 Keep the board tight against the framing.

9.1.9 Protect the external corners with a metal

bead or other suitable type of corner protection that generally are attached to supporting construction with fasteners spaced nominally 6 in. (152.4 mm) on centers. (See Section 3 and Fig. 1.) Corner beads may also be attached with a crimping tool.

9.2 *Application of Single-Ply Gypsum Board to Framing Members, System I:*

9.2.1 The maximum spacing for framing members for single-ply gypsum board assembly shall not exceed those shown in Table 1.

9.2.2 In single-ply installation, all ends and edges of gypsum board shall occur over framing members or other solid backing except where treated joints occur at right angles to framing or furring members.

9.2.3 End joints shall be staggered and joints on opposite sides of a partition shall be arranged to occur on alternate framing members.

9.2.4 *Nailing for Single-Ply Application of Gypsum Board Over Wood Members*—Length of nails for single-ply application shall be as shown in Table 2.

Note 6—Where a specific degree of fire resistance is required for gypsum board assemblies, nails of the same length, shank diameter, and head bearing area, as those described in the fire test report, shall be used.

9.2.4.1 *Single Nailing*—Nails shall be spaced a maximum of 7 in. (177.8 mm) on centers on ceilings, and a maximum of 8 in. (203.0 mm) on centers on walls (see Fig. 2).

9.2.4.2 *Double Nailing*—Nails shall be spaced and driven as shown in Fig. 3.

(1) Starting at center of the board, apply nails shown by solid dots in row 1, then rows 2 and 2A, 3 and 3A, 4 and 4A, always nailing from center to edges of sheet. Keep the board tight against the framing.

(2) Apply second nails shown by circles in the same manner as the first nails, also starting at row 1.

(3) As an alternative procedure, the second nail may be applied immediately after all nails in each row are driven in accordance with *(2)*.

(4) Use single nails on the perimeter of the board, unless otherwise specified.

Note 7—It may be necessary to reset the first nails in each row after the second nails have been set.

9.2.5 *Spacing of Screws*—Space the screws a maximum of 12 in. (304.8 mm) on centers for ceilings and 16 in. (406.4 mm) on centers for walls where the framing members are 16 in. on centers. Space the screws a maximum of 12 in. on centers for ceilings and walls where the framing members are 24 in. (609.6 mm) on centers.

9.3 *Application of Two-Ply Gypsum Board to Framing Members, System II:*

9.3.1 The maximum spacings for framing members for two-ply gypsum board assemblies shall not exceed those shown in Tables 3 and 4.

Note 8—See 9.4 for adhesive application method.

9.3.2 The base ply of gypsum board shall be applied with nails, screws or staples of the size and type required in Sections 6 and 9.

9.3.3 When adhesive is not used between the plies, apply the two plies of gypsum board as indicated in Tables 3 and 5. Apply the face ply with the number of nails or screws required for normal single-ply application. Nails used for the face ply shall be long enough to penetrate at least 7/8 in. (22.3 mm) into wood framing members. Screw length shall be long enough to penetrate at least 5/8 in. (15.9 mm) into wood. Face ply joints that are parallel to framing shall fall over framing members and be offset from the base ply joints when two plies of gypsum board are parallel.

9.3.4 When an adhesive is used between the plies, apply the two plies as indicated in Table 4. If the two plies are applied in parallel direction, the joints of the face ply shall be offset from the base ply.

9.3.4.1 Joints in the face ply need not occur over framing members. The adhesive to be used between the two plies of gypsum board shall be uniformly applied over the back surface of the face ply of the gypsum board before it is erected or to the face surface of the base ply. Place the face ply of gypsum board in position and fasten with a sufficient number of nails or screws to hold gypsum board in place until the adhesive develops adequate bond.

9.3.4.2 Use permanent fasteners around the perimeter 12 in. (304.8 mm) on centers and 16 in. (406.4 mm) on centers in the field of the face ply of gypsum board applied on ceilings.

9.3.4.3 In place of nails or screws, the face ply of gypsum board applied on sidewalls may be held in position by shoring with props and headers, or other temporary supports to ensure adequate pressure for bonding. Permanent fasteners shall be used on top and bottom of wall 16 in. (406.4 mm) on centers maximum. Nails or screws used to hold the gypsum board face ply shall be left in place and finished in the same manner as for single-ply gypsum board application (see 9.4).

9.4 *Application of Adhesive Nail-On to Wood Framing Members, System III:*

Note 9—Except as herein modified, application shall be in accordance with 9.2.

9.4.1 Surfaces of gypsum board and framing to be adhered by the adhesive shall be free of dust, dirt, grease, or any other foreign matter that could impair bond.

9.4.2 Apply a bead of adhesive 3/8 in. (9.5 mm) in diameter to the face of all wood framing members, except plates, that support the gypsum board. The adhesive should spread to an average width of 3/4 in. (19.1 mm) approximately 1/16 in. (1.6 mm) thick. Application patterns are shown in Fig. 7.

9.4.2.1 Where a joining of two adjacent pieces of gypsum board occurs on a framing member, two parallel beads of adhesive shall be applied, one near each edge of the framing member.

9.4.3 Adhesive shall be applied to no greater area than can be covered with gypsum board within the "open time".

Note 10—"Open time" is the time period available to work with certain adhesives before they set up in accordance with the adhesive manufacturer's specifications.

9.4.4 *Fastener Spacing:*

9.4.4.1 If the properties of the adhesive are such as to assure bridging between the gypsum board and the wood framing, nailing is not required in the field of the board for walls. In such cases, perimeter nailing, 16 in. (406.4 mm) on centers, is required.

9.4.4.2 If the properties of the adhesive are such that there is no positive bridging between the board and the wood framing, either temporary field nailing or temporary bracing is required to ensure contact between the board, the adhesive, and the stud face, until the adhesive develops adequate bond strength.

9.4.4.3 Unless specifically recommended otherwise by the adhesive manufacturer, space the fasteners in accordance with Table 6.

9.5 *Semi-Solid Gypsum Board Partitions, System IV:*

9.5.1 *Installation:*

9.5.1.1 Erect vertical runners or studs where required to provide proper support at locations; such as exterior walls, partition junctions, terminals, external corners, and door frames.

9.5.1.2 Position vertically the gypsum board face panels and studs.

9.5.1.3 Laminate the gypsum studs to face panels not over 24 in. (609.6 mm) on centers and locate at face panel vertical joints and at vertical center line of panel.

9.5.1.4 Gypsum studs may be laminated to face panels prior to erection or as erection of partition proceeds. Erect a starter face panel vertically at an intersecting wall. The starter panel shall be plumb and secured to the floor, the ceiling and the vertical runners.

9.5.1.5 Erect the next face panel adjacent to the starter panel, butting its edge and end firmly to the starter panel and the ceiling. Continue erection of the face panels, laminating exposed faces of gypsum studs as work progresses.

9.5.1.6 Carefully and accurately mark and cut the openings in partitions for doors and electrical outlets.

9.5.1.7 Laminating compounds or adhesives should be of a consistency and volume that will cover approximately three-fourths of the stud surface after lamination.

9.5.1.8 Use Type G screws as required to ensure a continuous bond between face panels and studs, and place a maximum of 36 in. (914.4 mm) on centers.

9.5.1.9 Reinforce the openings or changes in direction of partition with additional studs laminated in place at the following locations:

(1) External Corners—Between the face panels in the corner opposite the vertical runners.

(2) Abutting Walls—Between the face panels of a partition to reinforce junction of an abutting wall.

(3) Door Openings—Locate a vertical stud within 3 in. (76.2 mm) of door frame for reinforcement, and place a stud horizontally over the door header.

9.6 *Solid Gypsum Board Partitions, System V:*

9.6.1 Non-load bearing solid partition consisting of gypsum core-board faced on each side with gypsum board panels.

9.6.2 *Installation:*

9.6.2.1 Install floor and ceiling runners according to the layout and secure not more than 24 in. (609.6 mm) on centers. Install vertical runners at required locations, such as exterior corners and door frames.

9.6.2.2 For partitions located parallel to and between ceiling members, steel or wood blocking not more than 24 in. (609.6 mm) on centers must be provided to fasten ceiling runners prior to erection of ceiling.

9.6.2.3 Face panels shall be attached to runners at 24 in. (609.6 mm) on centers.

9.6.2.4 Core board may be installed prior to the installation of the face panel. Core board shall be attached not more than 24 in. (609.6 mm) on centers to the steel angles when steel angles are used as runners. When steel channels are used as runners to secure core board, attachment is not required.

NOTE 11—Combinations of wood and steel channels or angles can vary the installation procedure.

9.6.2.5 Apply the joint compound or laminating adhesive to the back surface of the face panels or the face of the core board as described for two-ply gypsum board construction (see 9.3).

9.6.2.6 Laminate the face panels to the core with sufficient pressure to ensure bond. Stagger the joints of face panels and core. Fasteners may be used to ensure bond between face panels and core.

9.7 *Application of Gypsum Board with Adhesive to Interior Masonry or Concrete Walls, System VI:*

9.7.1 When applying gypsum board to monolithic concrete, brick, or concrete block, apply the adhesive directly to the back of the gypsum board or on the wall in continuous beads not more than 12 in. (304.8 mm) on centers or daubs spaced not exceeding 12 in. on centers each way.

9.7.1.1 Beads shall not be less than $3/8$ in. (9.6 mm) in diameter to provide continuous bond between the gypsum board and the wall surface.

9.7.1.2 Daubs shall be 2 to 3 in. (50.8 to 76.2 mm) in diameter.

9.7.2 Position the gypsum board $1/8$ in. (3.2 mm) from the floor and provide a tight fit at abutting edges or ends. Do not slide the board. Use mechanical fasteners, or temporary bracing, as required, to support gypsum board until adhesive sets.

9.7.3 Delay the joint treatment until the gypsum board is firmly bonded.

NOTE 12—Foil-backed gypsum board is not recommended for direct adhesive application. Recommended for direct adhesive application are only interior masonry or concrete walls above grade, or the inside of exterior masonry cavity walls with 1-in. (25.4-mm) minimum width cavity between the inside and outside masonry for the full height of the above grade surface to receive gypsum board. Surfaces to which gypsum board is to be adhered shall be free of any foreign matter, projections, or depressions that will impair the bond.

9.8 *Application of Gypsum Board to Rigid Plastic Foam Insulation, System VII:*

9.8.1 *Application of Furring and Plastic Foam Insulation to Masonry and Concrete Walls:*

9.8.1.1 Apply the foam insulation to the masonry or concrete in accordance with the foam manufacturer's specifications.

9.8.1.2 Attach furring strips or special metal furring members by mechanical means to the masonry or concrete wall surface either before or after application of the insulation, depending on the system used. Install the furring members in accordance with Table 1 and at gypsum board termination above suspended ceilings, around door, window, and other openings, and where required for cabinet and fixture attachment.

9.8.2 Apply the gypsum board to furring as described in 9.2.4.1 or 9.2.5. The mechanical fasteners length shall not completely penetrate to the masonry or concrete.

NOTE 13—Gypsum board applied over rigid plastic foam insulation in the manner described in this section may not necessarily provide finish ratings required by local building codes.

9.9 *Gypsum Board Application to Steel Framing and Furring, System VIII:*

NOTE 12—Installation of steel framing shall be in accordance with Specification C 754.

9.9.1 *Screw Application* shall be applied in accordance with 9.1.6.

9.9.2 *Framing Spacing*—Maximum spacing of steel framing and furring for screw application shall be as specified in Table 1 for single-ply gypsum board and as specified in Tables 3 and 4 for two-ply gypsum board.

9.9.3 *Screw Spacing:*

9.9.3.1 Maximum screw spacing for single-ply gypsum board and face ply of two-ply gypsum board with no adhesive shall be in accordance with 9.2.5.

9.9.3.2 Maximum screw spacing for parallel applied base-ply of two-ply gypsum board over steel framing with no adhesive between the plies shall be 12 in. (304.8 mm) on centers along the edges of the gypsum board and 24 in. (609.6 mm) on centers into the stud or furring channel in the field of the gypsum board.

9.9.3.3 Maximum screw spacing for a perpendicularly applied base-ply of two-ply gypsum board over steel framing with no adhesive between the plies shall be one screw at each edge of the gypsum board at each framing member and one screw midway between the edges of each framing member.

9.9.3.4 Maximum screw spacing for perpendicular or parallel applied base-ply of two-ply gypsum board over steel framing with adhesive between plies shall be as specified for single-ply gypsum board in 9.2.5.

9.9.3.5 Maximum screw spacing on ceilings for the face-ply of two-ply gypsum board over steel framing with adhesive between plies shall be the same as specified for the base-ply of two-ply gypsum board in 9.9.3.2 and 9.9.3.3.

9.9.3.6 On wall surfaces with adhesive between the plies, the face-ply shall have only a sufficient number of screws to hold gypsum board in place.

9.10 *Arches and Bending Radii, System IX:*

9.10.1 Where gypsum board is to be applied to the soffit of arches, it shall be carefully bent into place (see Table 7). If necessary, it first shall either be dampened or cut approximately 1 in. (25.4 mm) on centers on the back side. In the latter case, after the core has been broken at each cut, the gypsum board shall be applied to the curved framing member and fastened in place. At the arrises of the arch (exterior or interior "corners" formed at the meeting of the adjoining angle surfaces), joint compound and reinforcing tape or corner bead shall be applied. Snip the tape or corner bead at intervals along one side to permit it to conform to the curved contour.

NOTE 15—To apply the board, place a stop at one end of the curve, then gently and gradually push on the other end of the board, forcing the center against the framing until the curve is complete.

NOTE 16—By thoroughly moistening the face and back paper and allowing the water to soak well into the core, the board may be bent to still shorter radii. When the board thoroughly dries, it will regain its original hardness. Any subsequent joint treatment or decoration shall not be started until gypsum board is thoroughly dry.

9.11 *Application of Gypsum Board to Receive Tile by Adhesive Application, System X:*

9.11.1 Framing around tub enclosures and shower stalls shall allow sufficient room so that the inside lip of the tub, prefabricated receptor, shower pan, or membrane can be applied as shown in Figs. 8, 9, and 10.

NOTE 17—This may necessitate furring out from the framing members the thickness of the gypsum board to be used ($1/2$ or $5/8$ in.) (12.7 or 15.9 mm) less the thickness of the lip, on each wall abutting a tub receptor or sub-pan.

9.11.2 *Blocking or Backing*—Framing shall not be spaced more than 16 in. (406.4 mm) on centers. Suitable blocking or backing shall be located approximately 1 in. (25.4 mm) above the top of the tub or receptor and at gypsum board horizontal joints in area to receive tile. Reinforce interior angles with framing members to provide rigid corners.

NOTE 18—Provide appropriate blocking, headers, or supports to support the tub and other plumbing fixtures and to receive soap dishes, grab bars, towel racks, and similar items as may be required.

9.11.3 *General:*

9.11.3.1 Use water-resistant gypsum board as a base for adhesive application of ceramic or plastic wall tile in wet areas such as tub and shower enclosures. Regular gypsum board may be used as a base for tile in any other areas except do not use water-resistant gypsum board nor regular board in extremely critical areas such as saunas, steam rooms, or gang shower rooms. Do not use foil-backed gypsum board as a base for tile or wall panels in the tub and shower enclosures and do not apply directly over a vapor barrier. Do not use water-resistant gypsum board on ceilings.

NOTE 19—Asphalt-impregnated felt is not considered a vapor barrier.

9.11.3.2 Install all multiple plies of gypsum board in wet areas without adhesive.

9.11.3.3 Waterproof receptors, pans, or sub-pans shall have an upstanding lip or flange that shall be a minimum of 1-in. (25.4-mm) higher than the water dam or threshold contained in the entry way to the shower.

9.11.4 *Installation:*

9.11.4.1 Apply the water-resistant gypsum backing board with the factory edge spaced a minimum of $1/4$ in. (6.4 mm) above the lip of the receptor, tub, or sub-pan. Install shower pans, receptor, or tubs prior to the erection of the water-resistant gypsum backing board.

9.11.4.2 Attach the water-resistant gypsum backing board by nails or screws spaced not more than 8 in. (203.2 mm) on centers. When ceramic tile over $3/8$ in. (9.5 mm) thick is to be applied, the nail or screw spacing shall not exceed 4 in. (101.6 mm) on centers.

9.11.4.3 If treatment of water-resistant gypsum backing board joints under tile in wet areas (tub and shower enclosures) is specified by the tile manufacturer or the architect, the joint compound and application thereof, shall be recommended by the joint compound manufacturer for this specific use, or shall be protected from penetration of moisture or water (see Note 19).

9.11.4.4 Fire or sound-rated partitions may require the installation of an additional layer of water-resistant gypsum backing board (see Fig. 10). Maintain the required ¼-in. (6.4-mm) clearance between the wallboard and tub ring, or shower base, as shown in Figs. 8 and 9.

NOTE 20—Caulk all cut edges and openings around pipes and fixtures flush with waterproof, nonhardening caulking compound or Type I adhesive complying with ANSI A136.1.

NOTE 21—Wall tile shall be used in combination with adhesive to protect the gypsum board and any water sensitive materials if present (such as joint compound) from penetration of moisture or water. Responsibility for performance of completed installation shall rest with the surface materials manufacturer or the applicator. Ceramic wall tile shall comply with ANSI A108.4. The adhesive for this tile shall comply with ANSI A136.1, Type I.

NOTE 22—For application of ceramic, plastic or metal wall tile, or plastic finished rigid wall panels, or other types of surfacing materials over gypsum board in wet or dry areas, the recommendations of the manufacturer of tile, wall panel, or other surfacing materials shall be followed.

NOTE 23—Apply the surfacing material down to the top surface or edge of the finished shower floor, return, or tub, and install so as to overlap the top lip of receptor, sub-pan, or tub. It shall completely cover the following areas:

(1) Over tubs without showerheads—6 in. (152.4 mm) above the rim of the tub.
(2) Over tubs with showerheads—a minimum of 5 ft (1.52 m) above the rim or 6 in. (152.4 mm) above the height of the showerhead, whichever is higher.
(3) Shower stalls—a minimum of 6 ft (1.83 m) above the shower dam or 6 in. (152.4 mm) above the showerhead, whichever is higher.
(4) All gypsum board window sills and jambs in shower or tub enclosures shall be covered to a like height.
(5) The surfacing material shall be applied to the full specified height for a distance of at least 4 in. (101.6 mm) beyond the external face of the tub or receptor. Areas beyond an exterior corner are included.

NOTE 24—Where plastic finished rigid wall panels are used as a surfacing material, the following precautions shall be taken:

(1) The type and shape of moldings recommended by the manufacturer of the surfacing material shall be used. Recommended tub moldings shall be used at the base where the surfacing materials abut the tub, shower, floor, or curb. Such moldings shall be set in waterproof, nonhardening caulking compound.
(2) Joints shall be filled in such a manner as to leave no voids for water penetration.
(3) A bead of adhesive shall be applied as a dam between the back surface of the finishing material and the tub or receptor to prevent any leakage of water at the joint.

9.12 *Exterior Application of Gypsum Wallboard, System XI:*

9.12.1 Gypsum wallboard may be used for ceiling of carports, open walkways, porches, and the soffits of eaves that are horizontal or inclined downward away from the building. Gypsum wallboard shall be either ½ or ⅝ in. (12.7 or 15.9 mm) in thickness. Framing shall be no more than 16 in. (406.4 mm) on centers for ½ in. thick gypsum wallboard nor more than 24 in. (609.6 mm) on centers for ⅝ in. thick gypsum wallboard and shall be installed perpendicularly in accordance with the foregoing specifications except as herein modified.

NOTE 25—Water-resistant gypsum backing wallboard shall not be used in an exterior location.

9.12.2 Provide suitable facia and molding around the perimeter to protect the gypsum wallboard from direct exposure to water. Unless protected by metal or other water stops, place the edges of the wallboard not less than ¼ in. (6.4 mm) away from abutting vertical surfaces. (See Figs. 11, 12, and 13.) Treat joints and fastener heads as specified in Section 10.

9.12.3 Treat exposed surfaces of gypsum wallboard as specified in X3.6.

9.12.4 Provide adequate ventilation for the space immediately above installations as specified in Appendix X5.

9.13 *Floating Interior Angles, System XII*—To minimize the possibility of the fastener popping in areas adjacent to the wall and ceiling intersection, the floating angle method of application may be used for single-ply or two-ply application of gypsum board to wood framing. This method is applicable where single nailing, double nailing, or screw attachment is used. Gypsum board shall be applied to ceiling first. (See Figs. 4, 5, and 6.)

9.13.1 *Ceilings*—Locate the first attachment into each ceiling framing member framed perpendicular to the intersection 7 in. (177.8 mm) out from the wall intersection for single nailing, and 11 to 12 in. (279.4 to 304.8 mm) for double nailing, or screw application.

9.13.2 *Sidewalls*—Apply gypsum board on sidewalls so as to provide a firm support for the floated edges of the ceiling gypsum board. Locate the top attachment into each vertical framing member 8 in. (203.2 mm) down from the ceiling intersection for single nailing, and 11 to 12 in. (279.4 to 304.8 mm) for double nailing, or screw application. (See Figs. 4 and 5.) At sidewall vertical angles, Fig. 6, apply only the over-lapping board so as to bring the back of the underlying board into firm contact with the face of the framing member behind it. Omit fasteners from the underlying board at the intersection (See Fig. 6.)

9.14 *Control Joints, System XIII*—Install expansion and contraction (control) joints in ceilings exceeding 2500 ft² (232.2 m²) in area and in partition, wall and wall furring runs exceeding 30 ft (9.1 m). Do not exceed a distance of 50 ft (15.2 m), in either direction, between ceiling control joints and install a control joint where ceiling framing or furring changes direction. Do not exceed a distance of 30 ft between control joints in walls or wall furring, and install a control joint where an expansion joint occurs in the base exterior wall. Wall or partition height door frames can be considered a control joint.

9.15 *Foil-Backed Gypsum Board, System XIV*—The application of foil-backed gypsum board shall conform to the foregoing specifications for the application of gypsum board. Place the reflective surface against the face of the framing members. Foil-backed gypsum board shall *not* be used in the following areas:

9.15.1 As a backing material for tile.

9.15.2 For the second ply on a two-ply laminating system.

9.15.3 For laminating direct to masonry.

9.15.4 In conjunction with electric heating cables.

10. Finishing of Gypsum Wallboard

10.1 Compounds for taping and finishing may be either a drying or setting type. Do not mix drying and setting formulations unless recommended by the joint compound manufacturer.

10.2 When applied, the compounds must be of a chemical composition compatible with previous and successive coats.

10.3 No finishing operation shall be done until the interior temperature has been maintained at a minimum of 50°F (10°C) for a period of at least 48 h and thereafter until the compounds have completely dried. When two-ply application is used with adhesives between plies, precautions shall be taken to ensure that the adhesive is thoroughly dried before any decorative finish is applied.

10.4 Adequate and continuous ventilation shall be provided to ensure proper drying, setting, or curing of the taping and finishing compounds.

10.5 *Taping and Finishing:*

10.5.1 Keep the gypsum wallboard free of any dirt, oil, or other foreign matter that could cause a lack of bond. All dents or gouges shall be filled. Set the mechanical fasteners below the plane of the board. All joints shall be even and true. Keep the board tight against the framing members.

10.5.2 Tape and finish using proper hand tools, such as broad knives or trowels with straight and true edges or mechanical tools designed for this purpose.

10.5.3 Apply the tape properly either by applying compound to the joint (buttering), pressing in the tape and wiping off the excess compound, or by mechanical tools designed for this process, taking care to leave sufficient compound under the tape to provide a proper bond of the tape to the board.

10.5.4 Apply the second coat with tools of sufficient width to extend beyond the joint center approximately 3 in. (76.2 mm). Draw the compound down to a smooth even plane. After drying or setting, sand the treated surface as needed (see 10.5.9) to eliminate any high spots or excessive compound. Non-setting compounds must be allowed to dry thoroughly between successive coats before sanding (see 10.5.9) and the application of the next coat.

10.5.4.1 Setting type compounds can receive additional coats as soon as the material has set and before it dries completely.

10.5.4.2 If a texture finish is to be used, a third coat may not be necessary unless required by a fire test or other performance test for the particular assembly.

10.5.5 Apply the third coat with tools that will permit the feathering of joint treatment edges approximately 6 in. (15.6 mm) from the center of the joint. After drying, lightly sand this final coat (see 10.5.9) to leave a smooth surface covering the joint. Do not raise the nap of the paper when sanding.

10.5.6 Cover the fastener heads with three successive coats each applied in a different direction except as noted in 10.5.2.

10.5.6.1 Back fill all cut-outs with compound used for taping or finishing so that there is no opening larger than ¼ in. (6.4 mm) between the gypsum board and a fixture or receptor.

10.5.7 Finish the corner bead with successive coats of joint treatment compound as directed by the compound manufacturer, so as to make the corner bead flush with the board surface.

10.5.8 Take care to ensure that all tools and containers are kept clean and free from foreign materials. Use only potable water for mixing powder compounds or to thin premixed materials. Compounds shall not be allowed to freeze.

10.5.9 **Caution:** *Wear approved protective respirators when mixing powder or sanding. Mix compounds in accordance with the manufacturers directions. Mixer speeds shall not exceed those recommended by the manufacturer of the compounds.*

TABLE 1   Maximum Framing Spacing for Single Ply Construction[A]

Single-Ply Gypsum Board Thickness, in. (mm)	Application[B]	Maximum Framing Members On Centers Spacing, in. (mm)
Ceilings:		
⅜ (9.5)[C]	perpendicular	16 (406.4)
½ (12.7)	parallel	16 (406.4)
⅝ (15.9)	parallel	16 (406.4)
½ (12.7)	perpendicular	24 (609.6)
⅝ (15.9)	perpendicular	24 (609.6)
Sidewalls:		
⅜ (9.5)	perpendicular or parallel	16 (406.4)
½ (12.7) or ⅝ (15.9)	perpendicular or parallel	24 (609.6)

[A] This table does not apply where water-based spray textures are used. Refer to Appendix X4.

[B] Nails for gypsum board applied over existing surfaces shall have a flat head and diamond point, and shall penetrate not less than ⅞ in. (22.2 mm), nor more than 1¼ in. (31.8 mm) into the framing member.

[C] ⅜-in. (9.5-mm) single-ply gypsum board shall not be applied to ceilings where the gypsum board supports insulation.

TABLE 2   Nail Lengths for Single-Ply Gypsum Board Assembly

Gypsum Board Thickness, in. (mm)	Minimum Nail Length, in. (mm)[A]
Application of gypsum board over existing surfaces:	Flat or concave head, diamond or needle point of such length as to provide not less than ⅞ in. (22.2 mm) nor more than 1¼ in. (31.8 mm) penetration into the following framing members:
⅜ (9.5)	1¼ (31.8)
½ (12.7)	1⅜ (34.9)
⅝ (15.9)	1½ (38.1)

[A] Nails that meet the performance criteria in Specification C 514 may be used.

TABLE 3   Maximum Framing Spacing for Two-Ply Assemblies, Fasteners Only, No Adhesive Between Plies[A]

Gypsum Board Thickness, in. (mm)		Application Direction		Maximum On Centers Spacing of Framing, in. (mm)
Base Ply	Face Ply	Base Ply	Face Ply	
Ceilings:				
⅜ (9.5)	⅜ (9.5)	perpendicular	perpendicular	16 (406.4)
½ (12.7)	⅜ (9.5) or ½ (12.7)	parallel	perpendicular	16 (406.4)
½ (12.7)	½ (12.7)	perpendicular	perpendicular	24 (609.6)
⅝ (15.9)[B]	½ (12.7) or ⅝ (15.9)	perpendicular	perpendicular	24 (609.6)
Sidewalls:				
⅜ (9.5)	⅜ (9.5)	perpendicular or parallel	perpendicular or parallel	16 (406.4)
½ (12.7)	⅜ (9.5) or ½ (12.7)	perpendicular or parallel	perpendicular or parallel	24 (609.6)
⅝ (15.9)	½ (12.7) or ⅝ (15.9)	perpendicular or parallel	perpendicular or parallel	24 (609.6)

[A] This table does not apply where water based spray textures are used. Refer to Appendix X4.

[B] ⅝-in. (15.9-mm) board may be applied perpendicularly at 16 in. (406.4 mm) spacing.

TABLE 4   Maximum Framing Spacing for Two-Ply Assembly Fasteners with Adhesive Between Plies[A]

Gypsum Board Thickness, in. (mm)		Application Direction		Maximum On Centers Spacing of Framing, in. (mm)
Base Ply	Face Ply	Base Ply	Face Ply	
Ceilings:				
⅜ (9.5)	⅜ (9.5)	perpendicular	perpendicular or parallel	16 (406.4)
½ (12.7)	⅜ (9.5) or ½ (12.7)	parallel	perpendicular or parallel	16 (406.4)
⅝ (15.9)	½ (12.7)	perpendicular or parallel	perpendicular or parallel	16 (406.4)
⅝ (15.9)	⅝ (15.9)	perpendicular	perpendicular or parallel	24 (609.6)
Sidewalls:				
⅜ (9.5)	⅜ (9.5)	perpendicular or parallel	perpendicular or parallel	24 (609.6)
½ (12.7)	⅜ (9.5) or ½ (12.7)	perpendicular or parallel	perpendicular or parallel	24 (609.6)
⅝ (15.9)	½ (12.7) or ⅝ (15.9)	perpendicular or parallel	perpendicular or parallel	24 (609.6)

[A] Adhesive between plies shall be dried or cured prior to joint treatment application.

TABLE 5   Fastener Spacing, Base-Ply of Two-Ply System, in. (mm) On Centers

	Nails	Screws	Staples
No adhesive between plies	24 (609.6)	24 (609.6)	16 (406.4)
Adhesive between plies	See 9.3	See 9.3	7 (177.8)

TABLE 6   Fastener Spacing with Adhesive or Mastic Application and Supplemental Fastening

Framing Member Spacing, in. (mm) On Centers	Ceilings, in.		Partitions Load Bearing, in.		Partitions Nonload Bearing, in.	
	Nail	Screw	Nail	Screw	Nail	Screw
16 (406.4)	16	16	16	24	24	24
24 (609.6)	12	16	12	16	16	24

TABLE 7   Bending Radii

Gypsum Board Thickness, in. (mm)	Bent Lengthwise, ft (m)	Bent Widthwise, ft (m)
½ (12.7)	10 (3.05)[A]	—
⅜ (9.5)	7½ (2.29)	25 (7.62)
¼ (6.4)	5 (1.52)	15 (4.57)

[A] Bending two ¼-in. (6.4-mm) pieces successively permits radii shown for ¼ in. (6.4 mm).

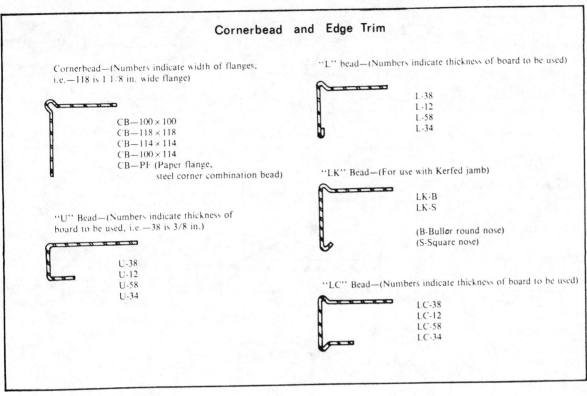

FIG. 1   Accessories

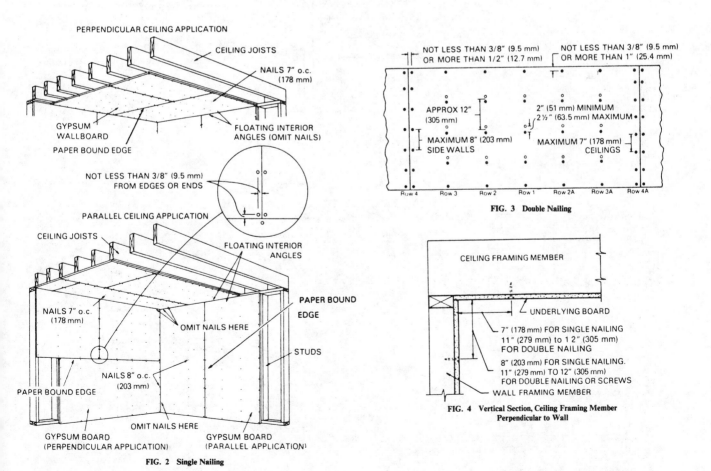

FIG. 2   Single Nailing

FIG. 3   Double Nailing

FIG. 4   Vertical Section, Ceiling Framing Member
Perpendicular to Wall

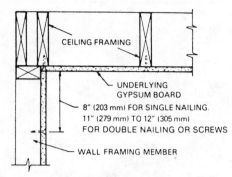

FIG. 5   Vertical Section, Ceiling Framing Parallel to Wall

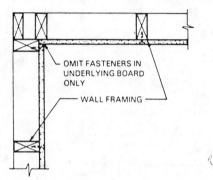

FIG. 6   Horizontal Section Through Interior Vertical Angle

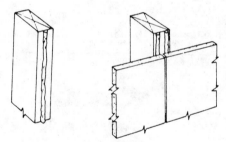

FIG. 7   Adhesive Application Patterns

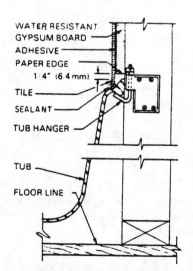

FIG. 8   Application of Gypsum Board at Bathtub Where WR Board Will Receive Ceramic Tile or Other Protective Covering

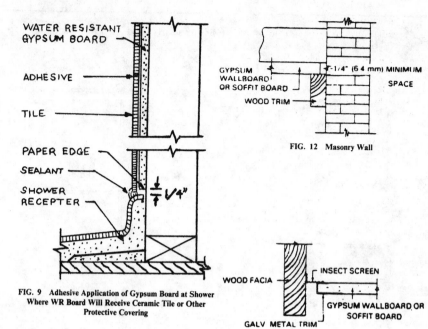

FIG. 9   Adhesive Application of Gypsum Board at Shower Where WR Board Will Receive Ceramic Tile or Other Protective Covering

FIG. 12   Masonry Wall

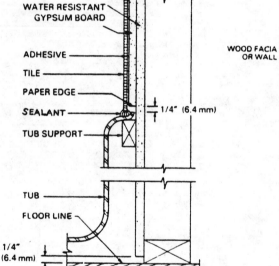

FIG. 10   Adhesive Application of Tile Over Gypsum Board

FIG. 13   Alternate Facia Details

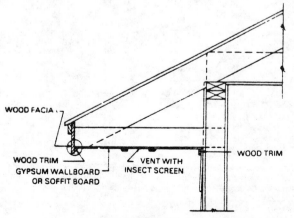

FIG. 11   Frame Wall

APPENDIXES
(Nonmandatory Information)

These Appendixes give general information and also suggestions for inclusions to be made elsewhere by the specifier. They are not a part of this specification.

X1. GENERAL INFORMATION

X1.1 Scaffolding shall be constructed and maintained in strict conformity with applicable laws and ordinances.

X1.2 This work shall be properly coordinated with the work of other trades.

X1.3 Gypsum board shall be protected from the elements of weather before, during, and after application.

X2. WOOD FRAMING REQUIREMENTS

X2.1 The following requirements should be included in the project specifications for framing and furring and are essential to provide a proper base to receive the gypsum board.

X2.2 All framing members to which gypsum board will be fastened shall be straight and true. Framing shall be in alignment and spaced not to exceed the maximum spacings shown in Table 1. Framing, bridging, and furring members shall be the proper grade for the intended use and members 2 by 4 in. nominal size or larger shall bear the grade mark of a recognized inspection agency. Framing, bridging, and furring shall be adequate to meet the design, or code loading, or both. Where there is an applicable local code, the framing shall be in accordance with the Manual for House Framing by the National Lumber Manufacturers. Supports shall be provided as necessary for the support of fixtures.

X2.3 When gypsum board is nailed to wood cross furring on ceilings, these furring members shall have a minimum cross section of 1½ by 1½ in. (38.1 by 38.1 mm) actual size and be spaced in accordance with 9.2 and 9.3. Where screw application is used, the furring member may be ¾ in. by 2½ in. (19.1 by 63.5 mm) actual size.

X2.4 Where wood furring is used over masonry or concrete, fasteners should be of a length that does not come into contact with the masonry surface.

X2.5 Insulating blankets or flanges of blankets shall not be applied over framing members that are to receive gypsum board.

X2.6 Foil-backed gypsum board may be used where a vapor retarder is required.

X3. JOB APPLIED DECORATION

X3.1 Prolonged exposure of gypsum board to sunlight may cause problems in decoration.

X3.2 With the joints and fastener head depressions treated as specified in Section 10, interior walls of gypsum board may be decorated in any of the popular variety of finishes, such as texture or stipple, flat paint or flat enamel paint, wall paper, or vinyl wall coverings.

X3.3 Because the porosity and texture of the gypsum board differs from that of the joint treatment, the surface shall be primed and sealed as may be required for the subsequent finish coats.

X3.4 In rooms where high humidity may be encountered, such as the kitchen, bath or utility room, a flat or semigloss enamel finish is recommended.

X3.5 Care should be exercised in the selection of primer and sealer paints to make sure they will perform satisfactorily, and fulfill the following functions:

X3.5.1 Equalize variations of suction over the entire surface.

X3.5.2 Provide a bonding surface or "tooth" for the paint to be applied.

X3.5.3 Avoid nap raising.

X3.6 Before applying the sealer, remove all loose dirt and dust by brushing with a soft brush or by rubbing with a dry cloth. Be sure the joint treatment is thoroughly dry before any application of sealer or paint.

X3.7 In applying primers or sealers, apply sufficient quantity to assure that the surface is completely covered. Follow the manufacturer's printed directions and do not over thin. It is good practice to tint the sealer to approximately the shade of the finish coat. This will lead to better results in the finished job.

X3.8 In all cases where deep tones are to be used in the finish paint, best results will be achieved if the surface is first sealed. More than one coat of sealer may be necessary. Each coat must be thoroughly dry before applying another.

X3.9 Under normal atmospheric conditions, a waiting period of 12 to 18 h after application of primer-sealer should be observed before decoration is applied. In rainy, humid, and cold weather, a longer waiting period; sometimes as long as 36 to 48 h; may be necessary to make certain the sealer coat is absolutely dry.

X3.10 Exposed surfaces of gypsum board, as specified in Section 10 shall be painted with not less than two coats of exterior paint.

X3.11 When semi-gloss or high-gloss paints are to be used, a skim coat of joint compound shall be applied to the entire surface prior to the application of the sealer.

X4. PRECAUTIONS TO MINIMIZE POTENTIAL OF SAGGING

X4.1 Ensure framing spacing is adequate for thickness of board to be used. Ensure board is applied perpendicular to framing.

X4.2 Determine that excessive weight of insulation will not be added.

X4.3 Control the relative humidity within the structure by providing adequate ventilation before, during, and after board application. Watch for pouring of basement floors after board application.

X4.4 In cold weather, maintain inside temperature between 50°F (10°C) and 70°F (20°C). Where portable heaters are used, make sure to remove the extra humidity they produce.

X4.5 Ensure the gypsum board is thoroughly dry and at ambient temperature before application.

X4.6 Ensure joint treatment is thoroughly dry before applying any decoration.

X4.7 Ensure that primer and paint coats are dry before the application of successive coats.

X4.8 Where gypsum board supports ceiling insulation and is finished with a water-based texture material, either sprayed or hand applied, use only ⅝-in. (15.9-mm) gypsum board applied perpendicular to the framing members. Where severe adverse conditions are anticipated, reduce framing spacing to 16 in. (400 mm) on centers.

X5. VENTILATION ABOVE GYPSUM BOARD CEILINGS

X5.1 Attics or similar unheated spaces above gypsum board ceilings shall be adequately ventilated by providing effective cross ventilation for all spaces between roof and top floor ceilings by screened louvers or other approved and acceptable means.

X5.2 The ratio of total net free ventilating area to ceiling area shall be not less than 1:150, except that the ratio may be 1:300 provided:

X5.2.1 A vapor barrier having a transmission rate not exceeding 1 perm is installed on the warm side of the ceiling, or

X5.2.2 At least 50 % of the required ventilating area is provided by ventilators located in the upper portion of the space to be ventilated (at least 3 ft (914.4 mm) above eave or cornice vents) with the balance of the required ventilation area provided by eave or cornice vents.

X5.3 Attic space that is accessible and suitable for future habitable rooms or walled-off storage space shall have at least 50 % of the required ventilating area located in the upper part of the ventilated space as near the high point of the roof as practicable and above the probable level of any future ceilings.

X6. ELECTRIC RADIANT HEATING SYSTEMS FOR GYPSUM BOARD CEILINGS

X6.1 The application of gypsum board under this specification shall be performed in accordance with Specification C 840, except as herein modified.

X6.2 Gypsum boards shall have no less thickness than that specified for perpendicular application for the ceiling framing spacing. Both layers shall be applied perpendicularly to framing members. Attachment shall be as required in Section 9.

X6.3 Electric heating cables shall be securely attached to gypsum board in accordance with the recommendations of the cable manufacturer and the National Electrical Code by other trades.

X6.4 Cables shall be parallel to and between framing members with at least 1¼-in. (31.8-mm) clearance from center of framing member on each side so that at least 2½-in. (63.5-mm) wide unobstructed strip is provided under each framing member.

X6.5 Cables shall cross framing members only at ends of ceiling 4 to 6 in. (101.6 to 152.4 mm) from the wall. There shall be at least a 4 in. (101.6 mm) space completely around the perimeter of each ceiling that is clear of cables. Cables shall be kept at least 8 in. (203.2 mm) clear of all openings.

X6.6 All inspections and testing of the heating system shall be completed before application of the non-insulating filler.

X6.7 Under no operating conditions shall the gypsum board core be exposed to a temperature exceeding 125°F (51.7°C) unless otherwise specifically recommended by the gypsum board manufacturer.

X6.8 Complete embedding of the heating cable is necessary. The area between the heating cables shall be filled with a nonshrinking, noninsulating filler that shall be leveled.

X6.9 If a face layer of gypsum board is to be applied then the nails or screws around the perimeter of the ceilings shall be spaced 8 to 10 in. (203.2 to 254.0 mm) away from the wall to prevent striking the heating elements where they cross the framing members.

X6.9.1 All joints and heads of fasteners shall be finished in accordance with the recommendations of the gypsum board manufacturer. Allow a minimum of one week with good drying conditions and two weeks in cold conditions before operating the heating system.

X6.10 If gypsum board is not used as a finish, a special spray texture for radiant heat ceilings may be applied.

X6.11 *Design Data:*

X6.11.1 It is the responsibility of the architect to prepare and correlate complete plans and specifications.

X6.11.2 It is the responsibility of the heating engineer to properly design the radiant heating system to work effectively with these materials.

X6.11.3 The data in Table X6.1 are recommended for use in calculating the heat transmission coefficient (U) for gypsum board constructions.

X6.12 *Decoration:*

X6.12.1 Decoration of the finished work, including sizing or sealing shall not proceed until the heating system has been test operated for a minimum of 24 h.

X6.12.2 The heating system should then be turned off and allowed to cool before decorating.

X6.12.3 Where texture finishes are used, such finishes must be allowed to thoroughly set or dry prior to the operation of the heating system.

TABLE X6.1 Thermal Conductivity Values

Thickness, in. (mm)	Conductance C		Resistance R	I/C
	Btu/h·ft²·°F	W/m²k	°F·h·ft²/Btu	k·m²/w
⅜ (9.5)	3.10	17.6	0.32	0.057
½ (12.7)	2.22	12.6	0.45	0.079
⅝ (15.9)	1.78	10.1	0.56	0.099
1 (25.4)	1.20	6.8	0.83	0.147

SUMMARY OF CHANGES

This section identifies the location of changes to this specification that have been incorporated since the last issue. Committee C-11 has highlighted those changes that affect the technical interpretation or use of this specification.

(1) Section 5 was revised.

(2) Notes 1 and 2 were added to 5.1.

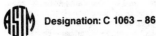

Standard Specification for

INSTALLATION OF LATHING AND FURRING FOR PORTLAND CEMENT-BASED PLASTER[1]

This standard is issued under the fixed designation C 1063; the number immediately following the designation indicates the year of original adoption or, in the case of revision, the year of last revision. A number in parentheses indicates the year of last reapproval. A superscript epsilon (ε) indicates an editorial change since the last revision or reapproval.

1. Scope

1.1 This specification covers the minimum requirements for exterior lathing and furring for portland cement-based plastering as specified in Specification C 926.

1.2 Where a fire resistance rating is required for plastered assemblies and constructions, details of construction shall be in accordance with reports of fire tests of assemblies that have met the requirements of the fire rating imposed.

1.3 Where a specific degree of sound control is required for plastered assemblies and constructions, details of construction shall be in accordance with official reports of tests conducted in recognized testing laboratories in accordance with the applicable requirements of Method E 90.

2. Referenced Documents

2.1 *ASTM Standards:*

A 526 Specification for Steel Sheet, Zinc-Coated (Galvanized) by the Hot-Dip Process, Commercial Quality[2]

A 641 Specification for Zinc-Coated (Galvanized) Carbon Steel Wire[2]

B 69 Specification for Rolled Zinc[3]

C 11 Definitions of Terms Relating to Ceilings and Walls[4]

C 841 Specification for Installation of Interior Lathing and Furring[4]

C 847 Specification for Metal Lath[4]

C 926 Specification for the Application of Portland Cement Based Plaster[4]

C 933 Specification for Welded Wire Lath[4]

C 1032 Specification for Woven Wire Plaster Base[4]

D 1784 Specification for Rigid Poly (Vinyl Chloride) (PVC) Compounds and Chlorinated Poly (Vinyl) Chloride (CPVC) Compounds[5]

E 90 Method for Laboratory Measurement of Airborne-Sound Transmission Loss of Building Partitions[6]

3. Terminology

3.1 *Definitions*—For definitions relating to Ceilings and Walls, see Definitions C 11.

3.2 *Descriptions of Terms Specific to This Standard:*

3.2.1 *hangers*—wires or steel rods or straps used to support main runners for suspended ceilings beneath floor or roof constructions.

3.2.2 *inserts*—devices embedded in concrete structural members to provide a loop or opening for attachment of hangers.

3.2.3 *saddle tie*—see Figs. 1 and 2.

3.2.4 *self-furring*—a metal plaster base manufactured with evenly-spaced indentations that hold the body of the lath approximately ¼ in. (6.4 mm) away from solid surfaces to which it is applied.

4. Delivery of Materials

4.1 All materials shall be delivered in the orig-inal packages, containers, or bundles bearing the brand-name and manufacturer's (or supplier's) identification.

5. Storage of Materials

5.1 All materials shall be kept dry. Materials shall be stacked off the ground, supported on a level platform, and protected from the weather and surface contamination.

5.2 Materials shall be neatly stacked with care taken to avoid damage to edges, ends, or surfaces.

6. Materials

6.1 *Expanded Metal Lath*—As specified in Specification C 847, galvanized.

6.2 *Wire Laths:*

6.2.1 *Welded Wire Lath*—As specified in Specification C 933.

6.2.2 *Woven Wire Lath*—As specified in Specification C 1032.

6.3 *Accessories:*

6.3.1 *General*—All accessories shall have perforated or expanded flanges or clips shaped so as to permit complete embedment in the plaster, to provide means for accurate alignment, and to secure attachment of the accessory to the underlying surface. Accessories shall be designed to receive or to permit application of the specified plaster thickness.

6.3.2 Accessories shall be fabricated from zinc-coated (galvanized) steel, zinc alloy, or rigid PVC plastic.

6.3.2.1 *Steel*—As specified in Specification A 526 and shall have a G60 coating.

6.3.2.2 *PVC Plastic*—As specified in Specification D 1784, cell classification 13244C.

6.3.2.3 *Zinc Alloy*—As specified in Specification B 69.

6.3.3 Thickness shall be as shown in Table 3.

6.3.4 *Cornerite*—1.75 lb/yd² (0.059 kg/m²), galvanized expanded metal lath, galvanized, 1.7 lb/yd² (0.057 kg/m²) galvanized woven or welded wire fabric of 0.0410 in. (1.04 mm) wire. When shaped for angle reinforcing, it shall have outstanding flanges (legs) of not less than 2 in. (51 mm).

6.4 *Steel Channels*—Hot-rolled or cold-rolled steel, galvanized or coated with rust inhibitive paint, of the following weights per thousand linear feet:

Sizes, in. (mm)	Hot-Rolled, lb (kg/m)	Cold-Rolled, lb (kg/m)
¾ (19.1)	300 (0.446)	300 (0.446)
1 (25)	410 (0.610)	...
1½ (38.1)	1120 (1.660)	475 (0.707)
2 (51)	1260 (1.870)	590 (0.878)

NOTE 1—Channels used in areas subject to corrosive action of salt air shall be hot-dipped galvanized, G60 coating.

6.5 *Wire*—As specified in Specification A 641 with a Class I coating (galvanized), soft temper.

6.6 *Rod and Strap Hangers*—Mild steel, zinc or cadmium plated, or protected with a rust-inhibiting paint.

6.7 *Clips*—Form of zinc-coated (galvanized) steel wire as specified in Specification A 641 or steel sheet as specified in Specification A 526, depending on use and manufacturer's requirements.

6.8 *Nails*—For attaching metal plaster bases, 0.1205-in. (3.06-mm) diameter, 7/16-in. (11.1-mm) head, barbed, galvanized roofing nails or galvanized common nails.

7. Installation

7.1 *Workmanship*—Erect metal furring and lathing so that the finished plaster surfaces are true to line, level, plumb, square, or curved as required to receive the specified plaster thickness.

7.2 *Hangers and Inserts:*

7.2.1 Hangers shall be of ample length and shall conform to the requirements of Table 3 both as to size and maximum area to be supported, except as modified in this section.

7.2.2 Where 1 by 3/16-in. (25 by 4.8-mm) flat inserts and hangers are used, 7/16-in. (11.1-mm) provide diameter holes on the center line at the lower end of the insert and upper end of the hanger to permit the attachment of the hanger to the insert. The edge of the holes in both the inserts and the hangers shall not be closer than 3/8 in. (9.5 mm) from the ends.

7.2.3 In concrete, attach hangers to inserts embedded in the concrete or to other attachment devices designed for this purpose and able to develop full strength of the hanger.

7.2.4 Bolt flat, steel hangers to 1 by 3/16-in. (25 by 4.8-mm) inserts with 3/8-in. (9.5-mm) diameter round-head stove bolts.

7.2.5 Draw the nuts of the bolts up tight.

NOTE 2—Where upward wind pressures may be exerted, hangers shall be of a type to resist compression. Struts of channels may be used.

7.3 *Installation of Hangers for Suspended Ceilings Under Wood Constructions*—Attach hangers to supports by one of the following methods:

7.3.1 Drill a hole through the wood member a minimum of 3 in. (76 mm) above the bottom with the upper end of the wire hanger passed through the hole and twisted three times around itself. See Fig. 3.

7.3.2 Drive three 12d nails, on a downward slant, into the sides of the wood member with not less than 1¼ in. (31.8 mm) penetration and at least 5 in. (127 mm) from the bottom edges, and not over 36 in. (914 mm) on the center with

[1] This specification is under the jurisdiction of ASTM Committee C-11 on Gypsum and Related Building Materials and Systems and is the direct responsibility of Subcommittee C11.0 on Specifications for Application of Gypsum and Other Products in Assemblies.

Current edition approved March 27, 1986. Published May 1986.

[2] *Annual Book of ASTM Standards,* Vol 01.06.
[3] *Annual Book of ASTM Standards,* Vol 02.04.
[4] *Annual Book of ASTM Standards,* Vol 04.01.
[5] *Annual Book of ASTM Standards,* Vols 08.02 and 08.04.
[6] *Annual Book of ASTM Standards,* Vol 04.06.

e upper end of the wire hanger wrapped around
e nails and twisted three times around itself.
ee Fig. 4.

7.3.3 Form a loop in the upper end of the wire
anger and secure it to the wood member by four
½ in. (38.1 mm), 0.1483-in. (3.77-mm) diame-
r wire staples driven horizontally or on a down-
ard slant into the sides of the wood members,
ree near the upper end of the loop and the
urth to fasten the loose end. See Fig. 5.

7.3.4 Where supports for flooring thicker than
½ in. (38.1 mm) are spaced more than 4 ft. (1.2
) on center, 1½-in. (38.1-mm) No. 0 screw eyes,
equivalent, spaced not more than 3 ft (0.9 m)
centers may be screwed into the flooring with
e upper end of the wire hanger inserted through
e screw eyes and twisted three times around
self.

7.3.5 Drill two holes in the upper end of the
at hangers and nail to the sides of the wood
embers with 12d nails driven through the holes
d clinched. Nails shall be at least 3 in. (76 mm)
ove the bottom of member. See Fig. 6.

7.4 *Attachment of Hangers to Main Runners:*
7.4.1 Saddle-tie wire hangers to the runners.
ee Fig. 1.

7.4.2 Fasten smooth or threaded rod hangers
the runners with special attachments appro-
riate to the design.

7.4.3 Bolt the lower ends of flat hangers to the
ain runners, or bend tightly around the runners
d carry up and above the runners and bolt to
e main part of the hanger. Bolts shall be ⅜-in.
.5-mm) diameter, round-head stove bolts. See
g. 7.

7.5 *Installation of Main Runners:*
7.5.1 Minimum sizes and maximum spans
d spacings of main runners for the various
ans between hangers or other supports shall
onform to the requirements of Table 1.

7.5.2 Maintain a clearance of not less than 1
. (25 mm) between the ends of the main run-
ers and the abutting masonry or the concrete
alls, partitions, and columns. Where special
onditions require that main runners let into
butting masonry or concrete construction,
ithin such constructions maintain a clearance
f not less than 1 in. (25 mm) from the ends and
ot less than ⅛ in. (3.2 mm) from the tops and
des of the runners.

7.5.3 A main runner shall be located within 6
. (152 mm) of the paralleling walls to support
e ends of the cross furring. The ends of main
nners shall be supported by hangers located
ot more than 6 in. (152 mm) from the ends.

7.5.4 Where main runners are spliced, overlap
he ends not less than 12 in. (305 mm) with
anges of channels interlocked and securely tied
ear each end of the splice with double loops of
.0625 in. (1.59 mm) or double loops of twin
rands of 0.0475-in. (1.21-mm) galvanized wire.

7.6 *Installation of Cross Furring:*
7.6.1 Minimum size and maximum spans and
pacings of various types of cross furring for
arious spans between main runners and sup-
orts shall conform to the requirements of Table

7.6.2 Securely saddle-tie cross furring to main
nners with 0.0625-in. (1.59-mm) galvanized
ire, or a double strand of 0.0475-in. (1.21-mm)
alvanized wire or with special galvanized clips,
r equivalent attachments.

TABLE 1 Suspended and Furred Ceilings Minimum Sizes for Wire, Rod, and Rigid Hangers; Minimum Sizes and Maximum Spans and Spacings for Main Runners; and Minimum Sizes and Maximum Spans and Spacings for Cross Furring

NOTE—1 in. = 25.4 mm; 1 ft² = 0.093 m²

	Hangers	
	Maximum Ceiling Area Supported ft²	Minimum Size of Hangers, in.
Hangers for Suspended Ceilings	12.5	0.1483 wire
	16	0.1620 wire
	18	3/16 in. diameter, mild steel rod[A]
	20	7/32 in. diameter, mild steel rod[A]
	22.5	¼ in. diameter, mild steel rod[A]
	25.0	1 by 3/16 in. mild steel strap[B]
Attachments for Tying Runners and Furring Directly to Beams and Joists: For Supporting Runners: Single Hangers Between Beams[C]	8	0.1055 wire
	12	0.1350 wire
	16	0.1620 wire
Double Wire Loops at Beams or Joists[C]	8	0.0800 wire
	12	0.1055 wire
	16	0.1205 wire
For Supporting Furring Without Runners[C] (Wire Loops at Supports): Types of Support: Concrete	8	0.0800 wire
Steel	8	0.0625 (2 loops)[D]
Wood	8	0.0325 (2 loops)[D]

Spans and Spacings of Main Runners[E,F]		
Minimum Size and Type	Maximum Span Between Hangers or Support, in.	Maximum Center to Center Spacing of Runners, in.
¾ in.—0.3 lb/ft, cold or hot-rolled channel	24	36
1-½ in.—0.475 lb/ft, cold-rolled channel	36	48
1-½ in.—0.475 lb/ft, cold-rolled channel	42	42
1-½ in.—0.475 lb/ft, cold-rolled channel	48	36
1-½ in.—1.12 lb/ft, hot-rolled channel	48	54
2 in.—0.59 lb/ft, cold-rolled channel	60	48
2 in.—1.26 lb/ft, hot-rolled channel	60	60
1-½—1-½ by 3/16 in. angle	60	42

Spans and Spacings of Cross Furring[E,F]		
Minimum Size and Type	Maximum Span Between Runners or Supports, in.	Maximum Center to Center Spacing of Cross Furring Members, in.
¼ in. diameter pencil rods	24	12
⅜ in. diameter pencil rods	24	19
⅜ in. diameter pencil rods	30	12
¾ in.—0.3 lb/ft, cold or hot-rolled channel	36	24
	42	19
	48	16
	48	24
1 in.—0.410 lb/ft, hot-rolled channel	54	19
	60	12

[A] It is highly recommended that all rod hangers be protected with a zinc or cadmium coating.
[B] It is highly recommended that all flat hangers be protected with a zinc or cadmium coating or with a rust-protective paint.
[C] Inserts, special clips, or other devices of equal strength may be substituted for those specified.
[D] Two loops of 0.0475-in. wire may be substituted for each loop of 0.0625-in. wire for attaching steel furring to steel or wood joists.
[E] These spans are based on webs of channels being erected and maintained in a vertical position.
[F] Other sections of hot- or cold-rolled members of equivalent beam strength may be substituted for these specified.

7.6.3 Where cross furring members are
spliced, overlap the ends not less than 8 in. (203
mm), with flanges of channels interlocked, and
securely tied near each end of the splice with
double loops of 0.0625-in. (1.59-mm) galvanized
wire or twin strands of 0.0475-in. (1.21-mm)
galvanized wire.

7.6.4 Cross furring shall not come into contact
with abutting masonry or reinforced concrete
walls or partitions except, where special conditions require that cross furring be let into abutting
masonry or concrete construction, the applicable
provisions of 7.5.2 shall apply.

7.6.5 Main runners and cross furring shall be
interrupted at control (expansion) joints.

7.7 *Metal Furring for Walls:*
7.7.1 Attachments for furring shall be concrete nails driven securely into concrete or into
masonry joints, short pieces of ¾-in. (19.1-mm)
channels used as anchors driven into masonry
joints, or other devices specifically designed as
spacer elements, spaced horizontally not to exceed 2 ft (0.6 m) on centers. They shall be spaced
vertically in accordance with horizontal stiffener
spacing so that they project from the face of the
wall to permit ties to be made.

7.7.2 Horizontal stiffeners shall be not less
than ¾-in. (19.1-mm) hot-rolled or cold-rolled
channels, spaced not to exceed 54 in. (1372 mm)
on centers vertically, with the lower and upper
channels not more than 6 in. (152 mm) from the
ends of vertical members and not less than ¼ in.
(6.4 mm) clear from the wall face, securely tied
to attachments with three loops of galvanized,

TABLE 2 Types and Weights of Metal Plaster Bases and Corresponding Maximum Permissible Spacing of Supports

Type of Metal Plaster Base	Minimum Weight of Metal Plaster Base, lb/yd² (kg/m²)	Maximum Permissible Spacing of Supports Center to Center, in. (mm)				
		Walls (Partitions)			Ceilings	
		Wood Studs or Furring	Solid Partitions[A]	Steel Studs or Furring	Wood or Concrete	Metal
U.S. Nominal Weights:						
Diamond Mesh[D]	2.5 (1.4)	16 (406)[C]	16 (406)	16 (406)[C]	12 (305)	12 (305)
	3.4 (108)	16 (406)[C]	16 (406)	16 (406)[C]	16 (406)	16 (406)
Flat Rib	2.75 (1.5)	16 (406)	16 (406)	16 (406)	16 (406)	16 (406)
	3.4 (1.8)	19 (402)	24 (610)	19 (482)	19 (482)	19 (482)
¾ in. Rib	3.4 (1.8)	24 (610)	N/A[B]	24 (619)	24 (610)	24 (610)
	4.0 (2.1)	24 (610)	N/A	24 (610)	24 (610)	24 (610)
¾ in. Rib	5.4 (2.9)	24 (610)	N/A	24 (610)	36 (914)	36 (914)
Sheet	4.5 (2.4)	24 (610)	N/A	24 (610)	24 (610)	24 (610)
Welded Wire[D]	1.4 (0.8)	16 (406)	16 (406)	16 (406)	16 (406)	16 (406)
	1.95 (1.1)	24 (610)	24 (610)	24 (610)	24 (610)	24 (610)
Woven Wire[D]	1.1 (0.6)	24 (610)	16 (406)	16 (406)	16 (406)	24 (610)
	1.4 (0.6)	24 (610)	16 (406)	16 (406)	24 (610)	16 (406)
Canadian Nominal Weights:						
Diamond Mesh[D]	2.5 (1.4)	16 (406)	12 (305)	12 (305)	12 (305)	12 (305)
	3.0 (1.6)	16 (406)	12 (305)	12 (305)	12 (305)	12 (305)
	3.4 (1.8)	16 (406)	16 (406)	16 (406)	16 (406)	16 (406)
Flat Rib	2.5 (1.4)	16 (406)	12 (305)	12 (305)	12 (305)	12 (305)
	3.0 (1.6)	16 (406)	16 (406)	16 (406)	16 (406)	13-½
⅜ in. Rib	3.0 (1.6)	19 (482)	N/A	16 (406)	16 (406)	16 (406)
	3.5 (1.9)	24 (610)	N/A	19 (482)	19 (482)	19 (482)
	4.0 (2.1)	24 (610)	N/A	24 (610)	24 (610)	24 (610)

[A] Where lath with backing is used, backing shall be absorbent and slotted paper.
[B] Not applicable.
[C] These spacings are based on unsheathed walls. Where self-furring lath is placed over sheathing or a solid surface, the permissible spacing of supports may be increased to 24 in.
[D] Metal plaster bases shall be furred away from vertical supports or solid surfaces at least ¼ in. Self-furring lath meets furring requirements; except, furring of expanded metal lath is not required on supports having a bearing surface of 1⅝ in. or less.

soft-annealed wire, or equivalent devices.

7.7.3 Approved furring devices, securely attached to concrete or masonry, may be used instead of horizontal stiffeners for support of vertical members.

7.7.4 Vertical members shall be not less than ¾-in. (19.1-mm) hot-rolled or cold-rolled channels in accordance with the requirements of Table 2. Saddle-tie vertical members to horizontal stiffeners with three loops of 0.0475-in. (1.21-mm) galvanized soft-annealed wire, or equivalent devices, at each crossing, and securely anchor to the floor and ceiling constructions. Where furring is not in contact with the wall, install channel braces between horizontal stiffeners and the wall, spaced horizontally not to exceed 2 ft (600 mm) on centers.

7.7.5 Where dampproofing has been damaged during installation of attachments, repair the dampproofing with the same material before proceeding with the installation of the furring.

7.8 *Lapping of Metal Plaster Bases:*

7.8.1 Secure side laps of metal plaster bases to supports. They shall be tied between supports with 0.0475-in. (1.21-mm) wire at intervals not to exceed 9 in. (229 mm).

7.8.2 Lap metal lath ½ in. (12.7 mm) at the sides, or nest the edge ribs. Wire lath shall be lapped one mesh at the sides and the ends. Lap metal lath 1 in. (25 mm) at ends. When end laps occur between the supports, lace or wire-tie the ends of the sheets of all metal plaster bases with 0.0475-in. (1.21-mm) galvanized, annealed steel wire.

7.8.3 Where metal plaster base with backing is used, the vertical and horizontal lap joints shall be backing on backing and metal on metal.

7.8.3.1 Lap backing not less than 1 in. (25 mm). On walls, lap the backing so water will flow to the exterior.

7.9 *Spacing of Attachments for Metal Plaster Bases*—Space attachments for securing metal plaster bases to supports not more than 7 in. (178 mm) apart for diamond mesh and flat rib laths and at each rib for ⅜-in. (9.5-mm) rib lath.

TABLE 3 Minimum Thickness of Accessories

Accessory	Base Material, in. (mm)		
	Steel	Zinc Alloy	P.V.C.
Corner Beads	0.0172 (0.44)	0.0207 (0.53)	0.035 (0.89)
Casing Beads			
Drip Screeds	0.0225 (0.57)	0.024 (0.61)	0.050 (1.27)
Control Joints	0.0172 (0.44)	0.018 (0.46)	0.050 (1.27)

7.10 *Application of Metal Plaster Bases to Supports:*

7.10.1 *General:*

7.10.1.1 The spacing of supports for the type and weight of metal plaster base used shall conform to the requirements of Table 2. Attach all metal plaster bases to supports at no more than 7 in. (178 mm) along supports except for ⅜-in. (9.5-mm) rib metal lath that shall be attached at each rib.

7.10.1.2 Apply all lath with the long dimension at right angles to the supports, unless otherwise specified.

7.10.1.3 Stagger ends of adjoining plaster bases.

7.10.1.4 Lath shall not be continuous through control joints but shall be stopped and tied at each side.

7.10.1.5 Clips or other methods of attachment that are shown to have carrying strength equal to those specified may be used.

7.10.1.6 Where furred or suspended ceilings butt into or are penetrated by columns, walls, beams, or other elements, terminate the sides and ends of the ceiling lath at the horizontal internal angles with a casing bead, control joint, or similar device designed to keep the sides and ends of the ceiling lath and plaster free of the adjoining vertically oriented elements. Do not use Cornerite at these locations.

7.10.1.7 Where load bearing walls or partitions butt into structural walls, columns, or floor or roof slabs, terminate the sides or ends of the wall or partition lath at the internal angles with a casing bead, control joint, or similar devic designed to keep the sides and ends of the wal or partition lath free of the adjoining elements Do not use Cornerite at these internal angles.

7.10.2 *Attachments for Metal Plaster Bases t Wood Supports:*

7.10.2.1 Attach lath to supports with attachments spaced no more than 7 in. (178 mm) o centers along supports.

7.10.2.2 Attach diamond-mesh expande metal lath, flat-rib expanded metal lath, and wir lath to horizontal wood supports with 1½-in (38.1-mm) roofing nails driven flush with th plaster base and attached to vertical wood supports with 6d common nails, or 1-in. (25-mm roofing nails driven to a penetration of at least ¾ in. (19.1 mm), or 1-in. (25-mm) wire staple driven flush with the plaster base.

7.10.2.3 Attach rib-expanded metal lath an sheet lath of ⅜ in. (9.5 mm) to horizontal an vertical wood supports with nails or staples t provide at least 1¾-in. (34.9-mm) penetratio into horizontal wood supports, and ¾-in. (19.1 mm) penetration into vertical wood supports.

7.10.2.4 On vertical wood supports, bend ove common nails to engage at least three strands c lath, or bend over a rib when installing rib lath.

7.10.3 *Attachments for Metal Plaster Bases t Metal Supports:*

7.10.3.1 Except as described in 7.10.3.2, securely attach all metal plaster bases to meta supports with 0.0475-in. (1.21-mm) wire ties o by other means of attachment which afford carrying strength and resistance to corrosion equal t or superior to that of the wire.

7.10.3.2 Attach rib metal lath to open-wet steel joists by single ties of galvanized, annealec steel wire, not lighter than 0.0475 in. (1.21 mm) with the ends of each tie twisted together 1½ times.

7.10.4 *Attachments for Metal Plaster Bases tc Concrete Joists*—Attach rib metal lath to concrete joists by loops of 0.0800-in. (2.03-mm galvanized, annealed steel wire, with the ends o each loop twisted together.

7.10.5 *Attachments for Metal Plaster Bases tc Masonry*—Attach metal plaster bases to masonr 16 in. (406 mm) on centers along the sheet, usin 5 nails across the sheet. Securely wire tie side lap or lace between the cross rows.

7.11 *Application of Accessories:*

7.11.1 *General*—Install all metal accessorie in such a manner that flanges and clips provide for their attachment are completely embedded i the plaster.

7.11.1.1 Attach accessories to substrate i such a manner as to ensure proper alignmen during application of plaster.

7.11.2 *Corner Beads*—Install corner beads t protect all external corners and to establisl grounds.

7.11.3 *Casing Beads*—Isolate nonload-bearing members from load-bearing members witl casing beads or other suitable means, to avoi transfer of structural loads.

7.11.4 *Control Joints-General*—Form contro joints by using a single prefabricated member o fabricate by installing casing beads back to bacl with a flexible barrier membrane behind the casing beads. The separation spacing shall be no less than ⅛ in. (3.2 mm) or as required by the anticipated thermal exposure range.

7.11.4.1 *Control Joints*—Install control (expansion and contraction) joints in ceilings or walls exceeding 144 ft² (13.4 m²) in area.

7.11.4.2 The distance between control joints shall not exceed 18 ft (5.5 m) in either direction or a length-to-width ratio of 2½ to 1. Install a control joint where the ceiling framing or furring changes direction.

7.11.4.3 Install a control joint where an expansion joint occurs in the base exterior wall.

7.11.4.4 Wall or partition height door frames may be considered as control joints.

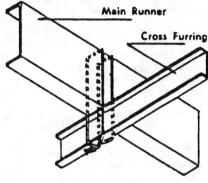

FIG. 2 Saddle Tie

FIG. 5 Hanger Attached to Support Using Staples

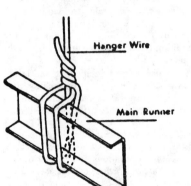

FIG. 1 Saddle Tie

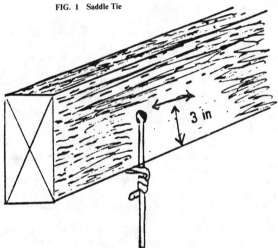

FIG. 3 Hanger Attached to Support Through a Drilled Hole

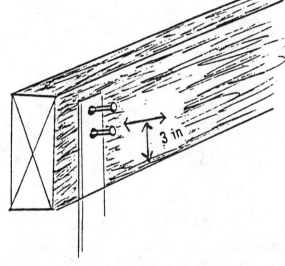

FIG. 6 Flat Hanger Attached to Support Using Nails

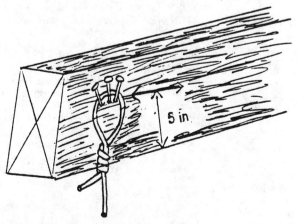

FIG. 4 Hanger Attached to Support Using Nails

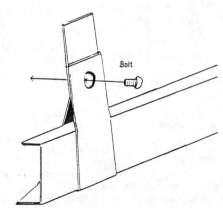

FIG. 7 Flat Hanger Attached to Support Using Round-Head Stove Bolt

 **Designation: C 926 – 86**

Standard Specification for
APPLICATION OF PORTLAND CEMENT-BASED PLASTER[1]

This standard is issued under the fixed designation C 926; the number immediately following the designation indicates the year of original adoption or, in the case of revision, the year of last revision. A number in parentheses indicates the year of last reapproval. A superscript epsilon (ε) indicates an editorial change since the last revision or reapproval.

1. Scope

1.1 This specification covers the minimum requirements for the application of portland cement-based plaster for exterior (stucco) and interior work.

1.2 This specification sets forth tables necessary for proportioning of various plaster mixes and plaster thickness.

NOTE 1—General information will be found in Annex A1. Design considerations will be found in Annex A2.

2. Referenced Documents

2.1 *ASTM Standards:*

C 11 Definitions of Terms Relating to Gypsum and Related Building Materials and Systems[2]

C 25 Methods of Chemical Analysis of Limestone, Quicklime, and Hydrated Lime[2]

C 91 Specification for Masonry Cement[2]

C 150 Specification for Portland Cement[2]

C 206 Specification for Finishing Hydrated Lime[2]

C 207 Specification for Hydrated Lime for Masonry Purposes[2]

C 219 Terminology Relating to Hydraulic Cement[2]

C 260 Specification for Air-Entraining Admixtures for Concrete[3]

C 595 Specification for Blended Hydraulic Cements[2,3]

C 897 Specification for Aggregate for Job-Mixed Portland Cement-Based Plasters[2]

E 90 Method for Laboratory Measurement of Airborne-Sound Transmission Loss of Building Partitions[4]

E 119 Methods of Fire Tests of Building Construction and Materials[5]

E 492 Method of Laboratory Measurement of Impact Sound Transmission Through Floor-Ceiling Assemblies Using the Tapping Machine[4]

2.2 *ANSI Standards:*

A42.3 Lathing and Furring for Portland Cement, Portland Cement-Lime Plastering, Exterior (Stucco) and Interior (out of print, not listed in ANSI catalogs, only available by photocopy from ANSI microfilm records)[6]

A108.1 Installation of Ceramic Tile[6]

2.3 *Military Specification:*

MIL-B-199235 Bonding Compounds, Concrete[7]

3. Descriptions of Terms Specific to This Standard

3.1 Terms shall be defined as in Definitions C 11 and Terminology C 219, except as modified herein.

3.2 *accelerator*—an admixture that will shorten the setting time of plaster.

3.3 *admixture*—a material other than water, aggregate, or basic cementitious material added to the batch before or during job mixing.

3.4 *acid etching*—the cleansing and controlled erosion of a solid surface, using an acid wash.

3.5 *air entrainment*—the use of an air-entraining admixture of air-entraining cementitious material in the plaster mix to yield a controlled quantity of minute (typically between 10 and 1000 μm in diameter) disconnected air bubbles in the plaster (see *entrapped air*).

3.6 *backplaster*—plaster applied to the face of metal lath opposite a previously applied plaster.

3.7 *bond*—the state of adhesion between plaster coats or between plaster and plaster base.

3.8 *bonding compound or agent*—compounds surface applied or integrally mixed with plaster to improve the quality of bond between plaster and plaster base or between plaster coats.

3.9 *cementitious material*—a material that, when mixed with water and with or without aggregate, provides the plasticity and the cohesive and adhesive properties necessary for placement and the formation of a rigid mass.

3.10 *coat*—a thickness of plaster applied in a single operation.

3.10.1 *scratch coat*—the first coat of plaster applied to a plaster base in multiple-coat work.

3.10.2 *brown coat*—in multiple-coat work, the second coat, applied over the scratch coat. In two-coat work, brown coat refers to the double-up basecoat. In either use, the brown coat is the coat directly beneath the finish coat.

3.10.3 *basecoat*—any plaster coat applied before the application of the finish coat.

3.10.4 *bedding coat*—a plaster coat that receives aggregate or other decorative material impinged into its surface before it sets.

3.10.5 *dash-bond coat*—a thick wet mixture of portland cement and water, with or without aggregate, dashed onto the surface of a plaster base such as smooth monolithic concrete or concrete block surfaces to improve the mechanical key for subsequent plaster coats.

3.10.6 *double-up coat*—the brown-coat plaster applied to the scratch coat plaster before the scratch-coat plaster has set.

3.10.7 *finish coat*—the final layer of plaster applied over basecoat plaster.

3.10.8 *fog coat*—a light coat of cement and water, with or without aggregate or color pigment, applied by machine spray to improve color consistency.

3.10.9 *skim coat*—a thin finish coat applied to an existing plaster surface or other substrate to improve appearance.

3.10.10 *three-coat work*—application of plaster in three successive coats with time between coats for setting or drying, or both.

3.11 *cold joint ("joining" or "jointing")*—the juncture of fresh plaster application adjacent to set plaster.

3.12 *curing*—the act or processes of producing a moisture environment favorable to cement hydration or lime carbonation, resulting in the setting or hardening of the plaster.

3.13 *entrapped air*—unintentional air voids in the plaster generally larger than 1 mm.

3.14 *fiber, natural or synthetic*—an elongated fiber or strand admixture added to plaster mix to improve cohesiveness or pumpability, or both.

3.15 *factory prepared ("mill-mixed" or "ready mixed")*—pertaining to material combinations that have been formulated and dry-blended by the manufacturer, requiring only the addition of and mixing with water to produce plaster.

3.16 *floating*—act of compacting and leveling brown-coat plaster to a reasonably true surface plane using a float and the act of bringing the aggregate to the surface of finish-coat plaster.

3.17 *key (also mechanical key)*—plaster that physically surrounds, penetrates, or deforms to lock onto the perforations or irregularities of the plaster base or previous coat of plaster.

3.18 *metal plaster base*—expanded metal lath, or welded or woven wire lath.

3.19 *plaster*—portland cement-based cementitious mixture (see *stucco*).

3.20 *rustication (also "break")*—an interruption or change in plane of a plastered surface.

3.21 *required*—pertaining to a mandatory obligation imposed by a force outside of this specification, such as a building code, project specification, contract, or purchase order.

3.22 *scoring (also known as "scratching")*—the grooving of the surface of an unset plaster coat to provide a key for a subsequent coat.

3.23 *set*—the chemical and physical change in plaster as it goes from a plastic, workable state to a solid rigid state.

3.24 *specified*—pertaining to a mandatory requirement of this specification (see *required*).

3.25 *stucco*—portland cement plaster used on exposed exterior locations.

3.26 *stucco finish*—a factory-prepared, dry blend of materials for finish coat applications, that optionally may be integrally colored.

3.27 *temper, v.*—to mix or restore unset plaster with water to a workable consistency.

3.28 *texture*—any surface appearance as contrasted to a smooth surface.

4. Identification and Marking

4.1 Packaged materials shall be clearly marked or labeled to indicate product, brand name, the manufacturer, and the weight of the

---

[1] This specification is under the jurisdiction of ASTM Committee C-11 on Gypsum and Related Building Materials and Systems and is the direct responsibility of Subcommittee C11.03 on Specifications for Application of Gypsum and Other Products on Assemblies.

Current edition approved Aug. 29, 1986. Published October 1986. Originally published as C 926 – 81. Last previous edition C 926 – 81.

[2] *Annual Book of ASTM Standards*, Vol 04.01.

[3] *Annual Book of ASTM Standards*, Vol 04.02.

[4] *Annual Book of ASTM Standards*, Vol 04.06.

[5] *Annual Book of ASTM Standards*, Vol 04.07.

[6] Available from American National Standards Institute, 1430 Broadway, New York, NY 10018.

[7] Available from Naval Publications and Forms Center, 5801 Tabor Ave., Philadelphia, PA 19120.

material contained therein. Provide similar information in the shipping advices accompanying the shipment of bulk materials.

5. Delivery of Materials

5.1 Deliver packaged materials in factory-sealed, unopened, and unbroken packages, containers, or bundles.

5.2 Deliver bulk materials in clean transport vessels, free of contaminates.

6. Protection of Materials

6.1 Keep weather-sensitive materials in a dry condition until ready for use. (See Annex A2.4.)

6.2 Store bulk materials to prevent subsequent contamination and segregation.

7. Environmental Conditions

7.1 Do not apply portland cement-based plaster to frozen base or to a base containing frost. Plaster mixes shall not contain frozen ingredients. Protect plaster coats from freezing for a period of not less than 24 h after set has occurred.

7.2 Protect portland cement plaster from uneven and excessive evaporation during dry weather and from strong blasts of dry air.

8. Materials

8.1 Materials shall conform to the requirements of the referenced specifications and standards and to the requirements specified herein.

8.2 *Cement:*

8.2.1 *Portland Cement*—Specification C 150, Type I, II, III, as required. White where required.

8.2.2 *Air-Entraining Portland Cement*—Specification C 150, type as required. White where required.

8.2.3 *Masonry Cement*—Specification C 91. White where required.

8.2.4 *Blended Hydraulic Cement*—Specification C 595, type as required.

8.2.5 *Plastic Cement*—Plastic cement shall meet the requirements of Specification C 150, except in respect to the limitations on insoluble residue, air-entrainment, and additions subsequent to calcination. Plasticizing agents may be added to portland cement, Types I and II, in the manufacturing process, but not in excess of 12 % of the total volume. Plastic cements are not available nationally.

8.3 *Lime*—Lime shall contain not more than 8 weight % of unhydrated oxide when tested in accordance with Method C 25.

8.3.1 *Special Finishing Hydrated Lime*—Specification C 206.

8.3.2 *Air-Entraining Special Finishing Hydrated Lime*—Specification C 206 or C 207.

8.3.3 *Hydrated Lime for Masonry Purposes*—Specification C 207.

8.4 *Aggregates:*

8.4.1 *Sand for Base Coats*—Specification C 897.

8.4.2 *Perlite*—Specification C 897.

8.4.3 *Sand for Job-Mixed Finish Coats*—Specification C 897. Exceptions to sand gradation for finish coat shall be stipulated where textured surfaces are specified and aggregate gradation has a significant role. (See Annex A2.6.)

8.5 *Water*—Water used in mixing, application, and finishing of plaster shall be clean, fresh, suitable for domestic water consumption, and free of such amounts of mineral or organic substances as would affect the set, the plaster, or any metal in the system.

8.6 *Admixtures*—See 3.3 and Annex A2.5.

9. Requirements for Bases to Receive Portland Cement-Based Plaster

9.1 Metal bases and accessories used to receive plaster shall be installed in conformance with ANSI A42.3, except as otherwise required.

9.1.1 These members shall be installed straight and true and properly attached to receive the specified amount of plaster.

9.1.2 All metal members shall be free of deleterious amounts of rust, oil, or other foreign matter, which could cause bond failure or unsightly discoloration.

9.2 Surfaces of solid bases to receive plaster, such as masonry, stone, cast-in-place or precast concrete shall be straight and true within ¼ in. in 10 ft (2.1 mm/m) and shall be free of form oil or other elements, which would interfere with proper bonding. Form ties or other metal obstructions and projecting joint mortar shall be removed or trimmed back even with the surface of the solid base by others.

9.2.1 Solid surfaces shall have the suction (ability to absorb water) or surface roughness, or both, to provide the bond required for the plaster.

9.2.2 Prepare smooth or nonabsorbent solid surfaces, such as cast-in-place or precast concrete to receive portland cement plaster by one of the following methods:

9.2.2.1 Sandblasting, wire brushing, acid etching, or chipping or a combination of these.

9.2.2.2 Application of a dash-bond coat applied forcefully against the surface, left untroweled, undisturbed, and moist cured for at least 24 h.

9.2.2.3 Application of a bonding compound suitable for exterior or interior exposure solid surfaces in accordance with the manufacturer's written directions.

9.2.3 Where bond cannot be obtained over the entire surface to receive plaster by one or more of the methods in 9.2.2, or where total plaster thickness will exceed the maximum thickness specified in Table 4 for types of solid bases, install furred or self-furring metal plaster base in accordance with ANSI A42.3.

10. Plaster Proportions and Mixing

10.1 *Proportions:*

10.1.1 Proportion and mix all portland cement plasters in accordance with the following tables and accompanying requirements, using measuring devices of known volume with successive batches proportioned alike.

10.1.2 Plaster mix used shall be as designated and referenced to Table 1.

10.1.3 Base-coat proportions shall be as shown in Table 2 for the mix specified from Table 1.

10.1.4 Finish-coat proportions for job-mixed finish coats shall be as specified in Table 3.

10.1.5 Factory-prepared finish coats shall require the addition of water only.

10.1.6 Dash-bond coat proportions shall be 1 volume part portland cement to 0 to 2 volume parts of aggregate mixed to a consistency that will permit application as specified in 3.9.5.

10.1.7 Admixtures shall be proportioned, mixed, and applied in accordance with the printed directions of the manufacturer. (See Annex A2.5.)

10.2 *Mixing:*

10.2.1 Prepare all plaster in a mechanical mixer, using sufficient water to produce of a workable consistency and uniform color. (See Annex A2.8.)

10.2.2 Plaster mixes for either base coat that has stiffened because of evaporation of water, may be re-tempered one time by remixing with additional water to restore the required consistency. Discard plaster not used within 2½ h from start of initial mixing.

NOTE 2—Severe hot, dry climatic conditions accelerate the stiffening of plaster. Such severe conditions may require this limit to be reduced.

10.2.3 Finish-coat plaster shall not be tempered.

11. Application

11.1 *General:*

11.1.1 Apply portland cement plaster, by hand or machine, to the thickness specified in Table 4.

11.1.2 Measure plaster coat thickness from the back plane of the metal plaster base, exclusive of ribs or dimples, or from the face of the solid backing, with or without metal plaster base, to the outer surface exclusive of texture variations.

11.1.3 Do not apply portland cement-based plaster directly to the surface of solid backing consisting of gypsum board, gypsum plaster, wood, or rigid foam board-type products without the application of metal plaster base.

NOTE 3—On horizontal ceiling supports or roof soffits protected by a drip edge, gypsum board products may be used as backing for metal base to receive portland cement plaster.

11.1.4 Provide separation where plaster abuts, either of dissimilar construction materials or openings. (See Annex A2.1.5.)

11.1.5 Apply each plaster coat to an entire wall or ceiling panel without interruption to avoid cold joints and abrupt changes in the uniform appearance of succeeding coats. Wet plaster shall abut set plaster at naturally occurring interruptions in the plane of the plaster, such as corner angles, rustications, openings, and control joints where this is possible. Cut joinings, where necessary, square and straight and at least 6 in. (152 mm) away from a joining in the preceding coat.

11.1.6 Use three-coat work over all metal plaster base, with or without solid backing. The combined total thickness shall be as shown in Table 4. A dash-bond coat shall not replace one of the specified number of coats.

11.1.7 Two-coat work may be used only over solid bases meeting the requirements of 9.2. The combined total thickness shall be as shown in Table 4. A dash-bond coat shall not replace one of the specified number of coats.

11.1.8 Apply a backplaster coat, where required, only after the coat on the opposite side has set sufficiently to resist breaking or cracking the plaster keys.

11.1.9 Permit each coat to cure before the next coat is applied. (See Annex A2.7 and 3.12.)

11.1.10 Dampen plaster coats that have become dry evenly with water prior to applying subsequent coats to obtain uniform suction. There shall be no visible water on the surface when plaster is applied.

11.2 *Plaster Application on Metal Plaster Bases:*

11.2.1 Apply the first coat with sufficient material and pressure to form full keys through, and to embed the metal base, and with sufficient thickness of material over the metal to allow for scoring the surface.

11.2.1.1 As soon as the first coat becomes firm, score its entire surface in one direction only. Score vertical surfaces horizontally.

11.2.1.2 The first coat shall become sufficiently rigid to support the application of the second coat without damage to the monolithic continuity of the first coat or its key.

11.2.2 Apply the second coat with sufficient material and pressure to ensure tight contact with the first coat and to bring the combined thickness of the basecoat to the thickness shown in Table 4.

11.2.2.1 Bring the surface of the second coat to a true, even plane with a rod or straightedge, filling surface defects in plane with plaster. Dry rodding the surface of the brown coat shall be permitted.

11.2.2.2 Float the surface uniformly to promote densification of the coat and to provide a surface receptive to bonding of the finish coat.

11.2.3 Apply the third (finish) coat with sufficient material and pressure to ensure tight contact with, and complete coverage of the basecoat and to the thickness shown in Table 4 and footnote g.

11.3 *Plaster Application on Solid Plaster Bases:*

11.3.1 Dampen high-suction bases with clean water prior to the application of plaster. Do not dampen low-suction solid bases, such as dense concrete or smooth brick.

11.3.2 *Three-Coat Application on Solid Bases:*

11.3.2.1 Apply the first coat with sufficient material and pressure to ensure tight contact and complete coverage of the solid base, to the thickness shown in Table 4. Score this coat immediately. Score vertical surfaces horizontally.

11.3.2.2 Apply the second coat using the same procedures specified in 11.2.2 and 11.2.2.1. Bring the surface to a true, even plane with a rod or straightedge, filling any defects in plane with plaster and darbying. Float the surface uniformly to provide a surface receptive to the application of the finish coat.

11.3.2.3 Apply the third (finish) coat as specified in 11.2.3.

11.3.3 *Two-Coat Application on Solid Plaster Bases:*

11.3.3.1 Apply the first coat as specified in 11.3.2.1.

11.3.3.2 Apply the finish coat as specified in 11.2.3.

11.4 *Finish-Coat Application:*

11.4.1 Apply job-mixed or factory-prepared finish coats, by machine or by hand, as specified in 11.2.3.

11.4.2 Avoid the use of excessive water during the application and finishing of finish-coat plaster.

11.4.3 Apply special finishes requiring thicknesses greater than shown in Table 4 or aggregates other than specified in 8.4.3, or other modifications, and finish as required in the contract specifications. (See Annex A2.6.)

11.5 *Fog-Coat Application*—Apply job-mixed or factory-prepared fog coats in accordance with the directions of the manufacturer.

12. Curing and Time Between Coats

12.1 Provide sufficient moisture in the plaster mix or by curing to permit continuous hydration of the cementitious materials. The most effective procedure for curing and time between coats, shall be based on climatic and job conditions. (See Annex A2.9.)

12.2 Allow sufficient time between coats to permit each coat to cure or develop enough rigidity to resist cracking or other physical damage when the next coat is applied. (See Annex A2.9.)

TABLE 1 Plaster Bases—Permissible Mixes

NOTE—See Table 2 for plaster mix symbols.[A]

| | Mixes for Plaster Coats | |
	First (Scratch)	Second (Brown)
Low absorption, such as dense, smooth clay tile or brick	C	C or L or M or CM
	CM	CM or M
	CP	CP or P
	P	P
High absorption, such as concrete masonry, porous clay brick, or tile	L	L
	M	M
	P	P
Metal plaster base[B]	C	C or L or M or CM
	L	L
	CM	CM or M
	M	M
	CP	CP or P

[A] The letter designations C, CM, L, M, CP, and P as defined in Table 2 are convenience symbols for Tables 1 and 2. They should not be used in contract specifications.

[B] Metal plaster base with paperbacking, under extremely severe drying conditions, may require dampening of the paper, prior to the application of plaster.

TABLE 3 Job-Mixed Finish Coat Proportion Parts by Volume

| Plaster Mix Symbols[A] | Cementitious Materials | | | | Volume of Aggregate per Sum of Cementitious Material | |
	Portland Cement	Lime	Masonry Cement	Plastic Cement	Sand	Perlite[C]
F[B]	1	¾–1½			3	2
FL	1	1½–2			3	2½
FP				1	1½	1
FM			1		1½	1½
FPM	1		1		3	2

[A] The letter designations F, FL, FP, FM, and FPM are convenience symbols for Table 3. They should not be used in contract specifications.

[B] For finish surfaces subject to abrasive action, specify plaster mix F, FP, or FPM.

[C] Recommended only over basecoat plaster containing perlite aggregate and areas not subject to impact.

TABLE 2 Base-Coat Proportions,[A] Parts by Volume[B]

| Plaster Mix Symbols | Cementitious Materials | | | | Volume of Aggregate per Sum of Cementitious Materials | | |
	Portland Cement	Lime	Masonry Cement	Plastic Cement	1st Coat	2nd[C] Coat	1st and 2nd Coats
C	1	0–¾			2½–4	3–5	3–4
CM	1		1–2		2½–4	3–5	1½–2
L	1	¾–1½			2½–4	3–5	2–3
M			1		2½–4	3–5	
CP	1			1	2½–4	3–5	2–3
P				1	2½–4	3 to 5	3–4

[A] The mix proportions for plaster scratch and brown coats to receive ceramic tile shall be in accordance with the applicable requirements of ANSI A108.1.

[B] Variations in lime, sand, and perlite contents are allowed due to variation in local sands and insulation and weight requirements. A higher lime content will generally support a higher aggregate content without loss of workability. The workability of the plaster mix will govern the amounts of lime, sand, or perlite.

[C] The same or a greater sand proportion shall be used in the second coat, than is used in the first coat, within the limits shown.

ANNEXES

(Mandatory Information)

A1. GENERAL INFORMATION

A1.1 The work includes all labor, materials, services, equipment, and scaffolding required to complete the plastering of the project in accordance with the drawings and specifications, except heat, electric power, and potable water.

A1.2 Where a specific degree of fire resistance is required for plastered assemblies and constructions, details of construction should be in accordance with official reports of fire tests conducted by recognized testing laboratories, in accordance with Method E 119.

A1.3 Where a specific degree of sound control is required for plastered assemblies and constructions, details of construction should be in accordance with official reports of tests conducted by recognized testing laboratories, in accordance with applicable sound tests Method E 90 or E 492.

A1.4 Scaffolding should be constructed and maintained in strict conformity with applicable laws and ordinances.

A1.5 The work should be coordinated with the work of other trades.

A1.6 Surfaces and accessories to receive plaster should be examined before plastering is applied thereto. The proper authorities should be notified and unsatisfactory conditions should be corrected prior to the application of plaster. Unsatisfactory conditions should be corrected by the party responsible for such conditions.

A1.6.1 Examine metal plaster bases, backing, attachment, and accessories to receive plaster and determine if the applicable requirements of ANSI A42.3 have been met unless otherwise required by the contract specifications.

TABLE 4 Nominal[A] Plaster Thickness for Three- and Two-Coat Work[B], in. (mm)

NOTE—See 11.1.2 for definition of plaster thickness.

Base	Vertical[C]				Horizontal			
	1st Coat	2nd Coat	3rd Coat[D]	Total	1st Coat	2nd Coat	3rd Coat[D]	Total
Interior								
Three-coat Work:[E]								
Metal plaster base	⅜ (9.5)	⅜ (9.5)	⅛ (3)	⅞ (22)	¼ (6)	¼ (6)	⅛ (3)	⅝ (16)
Solid plaster base:[F]								
Unit masonry	¼ (6)	¼ (6)	⅛ (3)	⅝ (16)	Use two-coat work.			⅜ (9.5) max, G, H
Cast-in-place or pre-cast concrete	¼ (6)	¼ (6)	⅛ (3)	⅝ (16)				
Metal plaster base over solid base	½ (12.5)	¼ (6)	⅛ (3)	⅞ (22)	½ (12.5)	¼ (6)	⅛ (3)	⅞ (22)
Two-coat work:[I]								
Solid plaster base:[F]								
Unit masonry	⅜ (9.5)	⅛ (3)		½ (12.5)				⅜ (9.5)[G, H]
Cast-in-place or pre-cast concrete	¼ (6)	⅛ (3)[G, H]		⅜ (9.5)				⅜ (9.5)
Exterior								
Three-coat work:[E]								
Metal plaster base	⅜ (9.5)	⅜ (9.5)	⅛ (3)	⅞ (22)	¼ (6)	¼ (6)	⅛ (3)	⅝ (16)
Solid plaster base:[F]								
Unit masonry	¼ (6)	¼ (6)	⅛ (3)	⅝ (16)	Use two-coat work.			⅜ (9.5) max
Cast-in-place or pre-cast concrete	¼ (6)	¼ (6)	⅛ (3)[G, H]	⅝ (16)				
Metal plaster base over solid base	½ (12.5)	¼ (6)	⅛ (3)	⅞ (22)	½ (12.5)	¼ (6)	¼ (6)	⅞ (22)
Two-coat work:[I]								
Solid plaster base:[F]								
Unit masonry	⅜ (9.5)	⅛ (3)		½ (12.5)[J]				⅜ (9.5) max
Cast-in-place pre-cast	¼ (6)	⅛ (3)		⅜ (9.5)				⅜ (9.5) max

[A] Approximate minimum thickness: 1st coat-⅜ in. (9.5 mm); 1st and 2nd coats-total ¾ in. (9.0 mm); 1st, 2nd, and finish coats-⅞ in. (22.2 mm).

[B] Where a fire rating is required, plaster thickness should conform to the applicable building code or to an approved test assembly.

[C] For solid plaster partitions, additional coats should be applied to meet the finished thickness specified.

[D] For exposed aggregate finishes, the second (brown) coat may become the "bedding" coat and should be of sufficient thickness to receive and hold the aggregate, but the total thickness shown in Table 4 should be achieved.

[E] Where three-coat work is required, a dash-bond or brush coat of plaster materials should not be accepted as a required coat.

[F] Where masonry and concrete surfaces vary in plane, plaster thickness required to produce level surfaces cannot be uniform.

[G] On horizontal solid base surfaces, such as ceilings or soffits, requiring more than ⅜ in. (9.5 mm) plaster thickness to obtain a level plane, metal plaster base should be attached to the concrete, and the thickness specified for three-coat metal plaster base over solid base should apply.

[H] Where horizontal solid base surfaces, such as ceilings or soffits, require ⅜ in. (9.5 mm) or less plaster thickness to level and decorate, and have no other requirements, a liquid bonding agent or dash-bond coat may be used.

[I] Table 4 shows only the first and finish coats for vertical surfaces and only the total thickness on horizontal surfaces for two-coat work.

[J] Exclusive of texture.

A1.6.2 The construction specifier should describe, in the proper section of the contract specifications, the physical characteristics of solid surface bases to receive plaster. The plane tolerence should not exceed ¼ in. in 10 ft (3.1 mm/m). The mortar joints should be flush and not struck. Dissimilar materials such as ties, reinforcing steel, etc., should be cut back ⅛ in. (3 mm) below the surface and treated with a corrosion-resistant coating. Masonry should be solid at corners and where masonry changes thickness in a continuous construction. Form release compounds must be compatible with plaster or be completely removed from surfaces to receive plaster. The plastering contractor should use this portion of the construction specifications for acceptance or rejection of such surfaces.

A2. DESIGN CONSIDERATIONS

A2.1 Exterior plaster (stucco) is applied to outside surfaces of all types of structures to provide a durable covering. Interior plaster is applied to inside surfaces that will be subjected to various exposures, such as abrasion, vibration, or to continuous or frequent moisture and wetting, or to freezing or thawing.

A2.1.1 Provide sufficient slope on faces of plastered surfaces to prevent water, snow, or ice from accumulating or standing. Air-entrained plaster properly proportioned, mixed, applied, and cured will exhibit good durability and resistance to its natural environment. Resistance to rain penetration is improved where plaster has been adequately densified during application and properly cured. Plaster should not, however, be considered to be "waterproof."

A2.1.2 Provide flashings made only from corrosion-resistant materials to prevent water from getting behind the plaster at all openings and perimeters. Cover detailed requirements for furnishing and the application of flashing in the appropriate section of

the project specifications. Clearly show flashings in large-scale detail. Aluminum flashing should not be used.

A2.1.3 Provide for sealing or caulking of V-grooves, exposed ends, and edges of plaster panels exterior work to prevent entry of water.

A2.1.4 To reduce spalling where plaster abuts openings, such as wood or metal door or window frames, or fascia boards, the edge of three-coat plaster shall be tooled through the second and finish coats to produce a continuous small V-joint of uniform depth and width. On two-coat work, tool the V-joint through the finish coat only.

A2.1.5 Provide in the appropriate project specification section that solid bases to receive plaster should not be treated with bond breakers, parting compounds, form oil, or other material that will prevent or inhibit the bond of the plaster to the base.

A2.2 *Provisions for Drainage Behind Exterior Plaster:*

A2.2.1 In multistory construction where lath and portland cement plaster exterior walls are continuous past a floor slab, tracks or plates and studs should be offset to provide a space (⅜ in. (9.5 mm) min) between the inner face of the exterior plaster and the edge of the floor slab. Fire stopping should be installed where specified.

A2.2.2 At the bottom of exterior walls where the wall is supported by a floor or foundation, a drip screed and throughwall flashing or weep holes or other effective means to drain away any water that may get behind the plaster should be provided.

A2.2.3 Where vertical and horizontal exterior plaster surfaces meet, terminate both surfaces with casing beads with the vertical surface so it protrudes at least ¼ in. (6 mm) below the intersecting horizontal plastered surface, thus providing a drip edge. Terminate the casing bead for the horizontal surface at least ¼ in. from the back of the vertical surface to provide drainage.

A2.3 *Relief from Stresses:*

A2.3.1 For information on control joints and perimeter relief see ANSI A42.3 Sections 2.7, 3.6, 5.7.3, 5.7.4, 5.7.5, 5.7.6, and 8.7.2.

A2.3.1.1 Control joints should be cleaned and clear of plaster within the control area after plaster application and before final plaster set.

A2.3.1.2 Prefabricated control joint members must be installed prior to the application of plaster; therefore the decision to use them, the type selected, their location and method of installation must be determined and specified in project specification sections other than the section on plastering.

A2.3.1.3 The creation of a groove or cut in plaster only, should be specified in the plastering section.

A2.3.2 Where plaster and metal plaster base continues across the face of a concrete column, or other structural member, place water-resistive building paper or felt between the metal plaster base and the structural member (or use paper or plastic-backed metal plasterbase). Where the width of the structural member exceeds the approved span capability of the metal plaster base, use self-furring metal plaster base and sparingly scatter nail to bring paper and metal base to general plane.

A2.3.3 Where dissimilar base materials abut and are to receive a continuous coat of plaster: (1) install a suitable metal accessory, such as casing beads back to back or a control joint member at the juncture of such bases; or (2) cover the juncture with an 8-in. (203-mm) wide strip of galvanized, self-furring metal plaster base extending 4 in. (102-mm) on either side of the juncture; or (3) where one of the bases is metal plaster base, extend self-furring metal plaster base 4 in. onto the abutting base.

A2.4 *Weather-Sensitive Materials*—Water-sensitive materials must be stored off the ground or floor and under cover, avoiding contact with damp floor or wall surfaces. Temperature-sensitive materials must be protected from freezing. Store bulk materials in area of intended use and exercise caution to prevent subsequent contamination and segregation of bulk materials prior to use.

A2.5 *Admixtures*—Admixtures must be proportioned and mixed in accordance with the published directions of the admixture manufacturer.

A2.5.1 The quantity of admixtures required to impart the desired performance is generally very small in relation to the quantities of the other mix ingredients. Accurate measurement of batch-to-batch quantities are extremely important.

A2.5.2 Air-entraining agents cause air to be incorporated in the plaster in the form of minute bubbles, usually to improve frost or freeze-thaw resistance. It may also improve the workability of the plaster during application. Air-entraining agents for portland cement based plaster should meet the requirements of Specification C 260.

A2.5.3 Bonding compounds or agents may be pre-applied to a surface to receive plaster. In this usage it is not considered an admixture. Bonding compounds that are integrally mixed with plaster prior to its application are considered admixtures. Where exterior exposure and cyclic wetting are anticipated, the re-emulsification capability of the bonding material must be considered. Bonding agents are only as good as the material surface to which they are applied; therefore form release materials must be removed from concrete or be compatible with the bonding material used. Bonding agents in plaster mixes may increase the cohesive properties of the plaster.

A2.5.4 Damproofing or waterproofing is possible by use of a suitable admixture to improve the plaster's resistance to moisture movement, but use of the term "proof" is misleading and should be discouraged.

A2.5.5 Natural or synthetic fibers, 1½ to 2 in. (38.1 to 50.8 mm) in length and free of contaminates may be specified to improve resistance to cracking or to impart improved pumpability characteristics. The quantities per batch shall be in accordance with the published directions of the fiber manufacturer. No more than 2 lb (0.90 kg) of fiber should be used per cubic foot of cementitious material. Asbestos fibers should not be used. Alkaline-resistant glass fibers are recommended where glass fiber is used.

A2.5.6 Plasticizers increase the workability of a portland cement plaster and may include hydrated lime putty, masonry cement, air-entraining agents, or approved fatteners. Plaster consistency and workability are affected by plasticizers that are beneficial in proper quantities from an economic standpoint, but in excess can be detrimental to the long-term performance of the plaster in place.

A2.5.7 Color material for integral mixing with plaster should not significantly alter setting, strength development, or durability characteristics of the plaster. Natural or mineral pigments that are produced by physical processing of materials mined directly from the earth appear to offer the best long-term performance with respect to resistance to fading. Plaster color is determined by the natural color of the cementitious materials, aggregate, and any color pigment, and their proportions to each other. The use of white cement with the desired mineral oxide pigment color material may result in truer color.

A2.5.7.1 The uniformity of color cannot be guaranteed by the materials manufacturer of the component materials or by the applicating contractor. Color uniformity is affected by the uniformity of proportioning, thoroughness of mixing, cleanliness of equipment, application technique, and curing conditions and procedure, which are generally under the control of the applicator. Color uniformity is affected to an even greater degree by variations in thickness and differences in the suction of the base coat from one area or location to another, the type of finish selected, the migration of color pigments with moisture, and with job site climatic and environmental conditions. These factors are rarely under the control of the applicator.

A2.6 *Finish Coat Categories* (applicable to both natural and colored finishes):

A2.6.1 Texture, as a description of surface appearance, is identified generally with the method and tools used to achieve the finish. Texture can be varied by the size and shape of the aggregate used, the equipment or tools employed, the consistency of the finish coat mix, the condition of the base to which it is applied, and by subsequent decorative or protective treatment.

A2.6.2 More often than not, functional or economic factors are outweighed by subjective preferences on the part of the architecture or client or by the colloquial limitations on the ability of the contractor-applicator. No amount of theoretical knowledge can compensate for the failure to understand the overriding importance of the conditions under which on-site finishes are accomplished. Practical, workable instructions that allow contractor and applicator to suit their methodology and experience to the requirements of the desired finish are more likely to be successful than those obtained with over-complicated and restrictive specifications which make demands altogether irreconcilable with site conditions.

A2.6.3 With the almost limitless variations possible for finish appearance or texture, the same term may not have the same meaning to the specifier, the contractor, and the actual applicator. The specifier is cautioned to use an approved range of sample panels. To provide some guidance, the following categories are generally understood and recognized to imply a particular method of application technique or resulting finished appearance:

A2.6.3.1 *Smooth Trowel*—Hand- or machine-applied plaster floated as smooth as possible and then steel-troweled. Steel troweling should be delayed as long as possible and used only to eliminate uneven points and to force aggregate particles into the plaster surface. Excessive troweling should be avoided.

A2.6.3.2 *Float*—A plaster devoid of coarse aggregate applied in a thin coat completely covering the basecoat, followed by a second coat which is floated to a true plane surface yeilding a relatively smooth to fine-textured finish, depending on size of aggregate and technique used. Also known as *sand finish*.

A2.6.3.3 *Trowel-Textured (such as Spanish Fan, Trowel Sweep, English Cottage)*—A freshly applied plaster coat is given various textures, designs, or stippled effects by hand troweling. The effects achieved may be individualized and may be difficult to duplicate by different applicators.

A2.6.3.4 *Rough-Textured (such as Rough Cast, Wet Dash, Scottish Harl)*—Coarse aggregate is mixed intimately with the plaster and is then propelled against the basecoat by trowel or by hand tool. The aggregate is largely unexposed and deep textured.

A2.6.3.5 *Exposed Aggregate (also Known as Marblecrete)*—Varying sizes of natural or manufactured stone, gravel, shell, or ceramic aggregates are embedded by hand or machine propulsion into a freshly applied finish "bedding" coat. The size of the aggregate determines the thickness of the "bedding" coat. It is generally thicker than a conventional finish coat.

A2.6.3.6 *Spray-Textured*—A machine-applied plaster coat directed over a previously applied thin smooth coat of the same mix. The texture achieved depends on the consistency of the sprayed mixture, moisture content of the base to which it is applied, the angle and distance of the nozzle to the surface, and the pressure of the machine.

A2.6.3.7 *Brush-Finish*—A method of surfacing or resurfacing new or existing plaster. The plaster is applied with a brush to a thickness of not less than ¹⁄₁₆ in. (1.6 mm). For an existing plaster surface the bond capability must be determined by test application or a bonding compound must be applied prior to the brush application.

A2.6.3.8 *Miscellaneous Types*—This finish coat category is somewhat similar to trowel-textured finishes, except that the freshly applied plaster is textured with a variety of instruments other than the trowel, such as swept with a boom or brush, corrugated by raking or combing, punched with pointed or blunt instrument, scored by aid of a straightedge into designs of simulated brick, block, stone, etc. A variation of texturing a finish coat involves waiting until it has partially set and then flattening by light troweling of the unevenly applied plaster or by simulating architectural terra cotta.

A2.6.3.9 *Scraffitto*—A method of applying two or more successive coats of different colored plaster and then removing parts of the overlaid coats to reveal the underlaying coats, usually following a design or pattern. Not generally considered a finish coat operation because of the number of thickness of coats.

A2.7 *Environmental Conditions for Plaster Application*—When artificial heat is required, locate heaters to prevent a concentration of heat on uncured plaster. Vent heaters to the outside to prevent toxic fumes and other products of combustion from adhering to or penetrating plaster bases and plaster. Maintain adequate ventilation in all areas, particularly in interior areas with little or no natural air movement.

A2.7.1 For interiors to be plastered, maintain a temperature of more than 40°F (4.4°C) for 48 h prior to application, while application is in progress, and during curing for at least 48 h.

A2.7.2 For exteriors, do not apply plaster when the ambient temperature is less than 40°F (4.4°C), unless the work area is enclosed and heat is provided as described in A2.7.

A2.7.3 To minimize the possibility of plaster cracking due to structural movement caused by thermal change in the building or by damage from frost, heat within the building should be maintained until normal occupancy conditions are established.

A2.8 *Mixing*:

A2.8.1 Hand mixing should not be permitted, except as approved by the contract specifier.

A2.8.1.1 After all ingredients are in the mixer, mix the plaster for 3 to 5 min or until the requirements of 10.2.1 are met.

A2.8.2 The amount of water used in the plaster mix should be determined by the plasterer. Factors such as the suction of the base, or of the previous coat, water content of the aggregate, drying conditions, and finishing operations should be considered in determining water usage. Use of excessive water may result in dropouts, fall or slide off, excessive shrinkage, high porosity, and lower strength.

A2.9 *Time Between Coats and Curing for Portland Cement Based Plaster.*

A2.9.1 The timing between coats will vary with climatic conditions and types of plaster base. Temperature and relative humidity extend or reduce the time between consecutive operations. Cold or wet weather lengthens and hot or dry weather shortens the time period. Moderate changes in temperature and relative humidity can be overcome by providing additional heating materials during cold weather and by reducing the absorption of the base by pre-wetting during hot or dry weather.

A2.9.1.1 To provide more intimate contact and bond between coats and to reduce rapid water loss, as soon as the first coat is sufficiently rigid to resist cracking, the pressures of the second coat application, and leveling process, the second coat should be applied.

A2.9.2 The amount of water and the timing for curing portland cement plaster will vary with the climatic conditions, the type of base, and use or non-use of water-retentive admixtures.

A2.9.2.1 Some moisture must be retained in or added back to freshly applied plaster. If the relative humidity is relatively high (above 75 % relative humidity), the frequency for rewetting a surface may be reduced. If it is hot, dry, and windy, the frequency of rewetting must be increased.

A2.9.2.2 The method of curing selected should consider the physical characteristics of the structure as well as the previously mentioned conditions. The methods can be one or a combination of the following:

(1) Moist curing is accomplished by applying a fine fog spray of water as frequently as required, generally twice daily in the morning and evening. Care must be exercised to avoid erosion damage to plaster surfaces. Except for severe drying conditions, the wetting of finish coat should be avoided, that is, the basecoat prior to application of the finish coat.

(2) Plastic film, when taped or weighted down around the perimeter of the plastered area, can provide a vapor barrier to retain the moisture between the membrane and plaster. Care must be exercised in placing the film; if too soon, the film may damage surface texture; if too late, the moisture may have already escaped.

(3) Canvas, cloth, or sheet material barriers can be erected to deflect sunlight and wind, both of which will reduce the rate of evaporation. If the humidity is very low, this option may not provide adequate protection.

A2.10 *Staining of Plaster*—Staining and discoloration of plaster caused by free water draining from one plane of plaster to another or from a dissimilar material onto a plaster surface, can be minimized by providing sufficient depth and angle for drip caps and the use of water-resistive surface coatings.

A2.10.1 Staining of plaster due to entrapment of moisture behind the plaster, where insulation with or without vapor barrier, or other material containing asphaltic or coal tar derivatives, fireproofing salts, etc., can migrate with moisture movement to the finished plaster surface, can be avoided or minimized by providing an air space for ventilation between the back of the plaster and adjacent material.

SPECTEXT SPECIFICATION FOR

PORTLAND CEMENT PLASTER
SECTION 09220

This section includes wet applied portland cement plaster, in either a two coat application over masonry, clay tile, concrete or other solid surface, or a three coat application over metal lath surfaces; with smooth trowelled, decorative and aggregate surface finish. Some regions refer to the aggregate surfaced product as "stucco." This section can be edited with furring and lathing referenced to Section 09206, or furring and lathing can be specified in this section. This section includes proprietary and descriptive type specifications. Use only one specifying type to avoid any conflicting requirements. Edit accordingly.

PART 1   GENERAL

1.01   WORK INCLUDED

    A.   [Metal furring and lathing.]

    B.   Portland cement plaster system.

    C.   [Smooth] [Glass] [Aggregate] [Special rendered] surface finish.

1.02   WORK INSTALLED BUT FURNISHED UNDER OTHER SECTIONS

    A.   Section[_____-_____]: Metal frames for [recessed light fixtures.] [_____.]

1.02   List sections which furnish products for installation under this section. When a product will be furnished "by Others," outside the Contract, it is considered to be furnished "by Owner."

1.03   RELATED WORK

    A.   Section [_____-_____]: Wall substrate surface.

    B.   Section 05400 - Cold Formed Metal Framing: Structural metal studding and framing behind plaster base.

    C.   Section 07900 - Joint Sealers.

    D.   Section 09111 - Metal Stud Framing System: Metal studding and framing behind plaster base.

    E.   [Section 09206 - Metal Furring and Lathing: [Metal furring] [and] [lathing] for plaster base.]

    F.   Section 09210 - Gypsum Plaster.

    G.   Section 09260 - Gypsum Board Systems.

    H.   Section [_____-_____]: Surface finish of [_____.]

1.04   REFERENCES

    A.   ANSI/ASTM C91 - Masonry Cement.

    B.   ASTM C150 - Portland Cement.

    C.   ANSI/ASTM C206 - Finishing Hydrated Lime.

    D.   ANSI/ASTM C207 - Hydrated Lime for Masonry Purposes.

    E.   ANSI/ASTM C631 - Bonding Compounds for Interior Plastering.

    F.   ANSI/ASTM C897 - Aggregate for Job-Mixed Portland Cement-Based Plasters.

1.04   Include only reference standards that are to be indicated within the text of this section. Refer to TAS Section 09220. Edit the following, adding and deleting as required for project and product selection.

G. ANSI/ASTM C926 - Application of Portland Cement-Based Plaster.

H. FS HH-I-521 - Insulation Blankets, Thermal (Mineral Fiber, for Ambient Temperatures).

I. PCA (Portland Cement Association) - Plaster (Stucco) Manual.

1.05   SYSTEM DESCRIPTION

A. Fabricate vertical elements to limit finish surface to [1/180] [_____] deflection under lateral point load of [100] [_____] lbs ([445] [_____] N).

B. Fabricate horizontal elements to limit finish surface to [1/360] [_____] deflection under superimposed dead loads and wind uplift loads.

1.06   QUALITY ASSURANCE

A. Applicator: Company specializing in cement plaster work [with [_____] years [documented] experience.] [approved by manufacturer.]

B. Apply cement plaster in accordance with [ASTM C926.] [PCA Plaster (Stucco) Manual.] [_____.]

1.07   REGULATORY REQUIREMENTS

A. Conform to [applicable] [_____] code for fire rated assemblies on conjunction with Section [09206] [_____] as follows:
   1. Fire Rated Partitions: Listed assembly by [UL] [FM] No. [_____.]
   2. Fire Rated [Ceiling] [and Soffits]: Listed assembly by [UL] [FM] No. [_____.]
   3. Fire Rated Structural Column Framing: Listed assembly by [UL] [FM] No. [_____.]
   4. Fire Rated Structural Beam Framing: Listed assembly by [UL] [FM] No. [_____.]

1.08   SUBMITTALS

A. Submit product data under provisions of Section [01300.] [01340.]

B. Provide product data on plaster materials, characteristics and limitations of products specified.

C. Submit manufacturer's installation instructions under provisions of Section [01300.] [01340.]

D. Submit manufacturer's certificate under provisions of Section [01400] [01405] that [products for fire resistance ratings] [_____] meet or exceed [specified requirements.] [_____.]

1.09   FIELD SAMPLES

A. Provide sample panel under provisions of Section [01300.] [01340.]

B. [Provide] [Construct] [_____] field sample panel, [_____] inch ([_____] mm) long by [_____] inch ([_____] mm) wide, illustrating [surface finish.] [color of finish coat.] [_____.]

C. Locate [where directed.] [_____.]

D. Accepted sample may remain as part of the work.

1.10   ENVIRONMENTAL REQUIREMENTS

A. Do not apply plaster when substrate or ambient air temperature is less than 50 degrees F (10 degrees C) nor more than 80 degrees F (27 degrees C).

B. Maintain minimum ambient temperature of 50 degrees F (10 degrees C) during and after installation of plaster.

1.05   Use this Article carefully; strict descriptions to total syst parameters, performance or des requirements. Do not repeat sta ments made in "Work Included" ticle.

1.08B   When manufacturers'a structions for specific installation quirements are utilized, carefu edit Part 3 Execution requireme to avoid conflict with those instr tions.

1.09   Use this Article for field plied finish samples.

PART 2   PRODUCTS

.01   ACCEPTABLE MANUFACTURERS

A.   [_____.]

B.   [_____.]

C.   [_____.]

D.   Substitutions: Under provisions of Section [01600.] [01630.]

.02   PLASTER BASE COAT MATERIALS

A.   Cement: [ASTM C150, [Normal - Type I] [_____] Portland.] [ANSI/ASTM C91, [grey] [white] color.]

B.   Lime: ANSI/ASTM [C206, Type S.] [C207, Type S.]

C.   Aggregate: In accordance with [ANSI/ASTM C897.] [PCA Plaster (Stucco) Manual.]

D.   Aggregate: [Natural] [Manufactured] sand, within the following limits:

Sieve_Size	Percent_Retained
No. [4] [___] ([4.75] [___] mm)	[0] [___]
No. [8] [___] ([2.36] [___] mm)	[0 to 5] [___to___]
No. [16] [___] ([1.18] [___] mm)	[5 to 30] [___to___]
No. [30] [___] ([0.60] [___] mm)	[30 to 65] [___to___]
No. [50] [___] ([0.30] [___] mm)	[65 to 95] [___to___]
No. [100] [___] ([0.15] [___] mm)	[90 to 100] [___to___]

E.   Water: Clean, fresh, potable and free of mineral or organic matter which can affect plaster.

F.   Bonding Agent: [ANSI/ASTM C631;] type recommended for bonding plaster to [concrete] [and] [concrete masonry] surfaces; [[_____] manufactured by [_____].]

G.   Admixtures: [Air entrainment] [_____] type; [_____] manufactured by [_____.]

H.   [Plaster Mix Reinforcement: Glass fibers, 1/2 inch (13 mm) nominal length, alkali resistant.]

.03   PLASTER FINISH COAT MATERIALS

A.   Premixed Finishing Coat: [_____] type; [_____] manufactured by [_____.]

****** [OR] ******

B.   Cement: As specified for plaster base coat, [white] [grey] color.

C.   Lime: As specified for plaster base coat.

D.   Color Pigment: [_____] type, [_____] color; [_____] manufactured by [_____.]

E.   Water: Clean, fresh, potable and free of mineral or organic matter which can affect plaster.

2.01   If only one manufacturer is acceptable, list in Article 2.02 and delete this Article; if more than one, list in this Article. If product substitution procedure is used, include Paragraph D. Refer to TAS Section 09220.

2.01D   Edit the following descriptive specifications for any conflicts with manufacturers' products specified above.

2.02B   Adjust grading of ingredients limits to suit project. Grading tables are not required when specifying to ANSI/ASTM C 897 or Portland Cement Association's Plaster (Stucco) Manual.

2.02E   Include the following paragraph for situations when cement plaster is to be applied directly on masonry or concrete.

2.02G   Consider adding reinforcement to plaster materials for exterior applications.

2.03   There are a variety of materials used to provide cement plaster finishes. Many premix finish coat cement plasters are available for application over base coats. They are available in different colors and require strict quality control on project. Most premix plasters are designed for float or textured finish and are not recommended for smooth trowelling. Select the system required, based on intended end result.

2.03C   Fading characteristics of pigments or coloring agents and their effects upon cement plaster strength should be evaluated before making final selection. Determine the potential uniformity of the pigment when receiving a sample. Consider the use of naturally colored aggregate in lieu of pigments.

2.04   FINISH AGGREGATE

   A.   Mineral Aggregate: [_____] colored [limestone]
        [marble] [granite] [_____]; graded to No. [0] [1] [2]
        [3] [4] size chip.

   B.   Glass Aggregate: Crushed glass, color range of
        [_____], graded to No. [0] [1] [2] size chip.

2.05   FURRING AND LATHING

   A.   Wire Mesh Reinforcement: [2 x 2] [1-1/2 x 1-1/2] [_____]
        inch ([50 x 50] [38 x 38] [_____] mm) galvanized steel
        [24] [_____] gage ([0.6] [_____] mm) wire, woven mesh,
        [self-furring type.]

   B.   Metal Lath: 2.5 lb/sq yd (1.12 kg/sq m) expanded metal,
        [self-furring type;] [prime painted] [galvanized] finish.

   C.   Underlayment: [Asphalt saturated No. 15 felt.]
        [_____.]

   D.   Corner Mesh: Formed steel, minimum [26] [_____] gage
        ([0.5] [_____] mm) thick; [perforated] [expanded] flanges
        shaped to permit complete embedding in plaster; minimum
        [2] [_____] inches ([50] [_____] mm) wide; [galvanized]
        [rust inhibitive prime paint] finish.

   E.   Corner Beads: Formed steel, minimum [26] [_____] gage
        ([0.5] [_____] mm) thick; [beaded] [bullnosed] edge, of
        longest possible length; sized and profiled to suit
        application; [galvanized] [rust inhibitive prime paint]
        finish.

   F.   Base Screeds: Formed steel, minimum [26] [_____] gage
        ([0.5] [_____] mm) thick; [square] [_____] edge, of
        longest possible length; sized and profiled to suit
        application; [galvanized] [rust inhibitive prime paint]
        finish.

   G.   Casing Bead: Formed steel; minimum [26] [_____] gage
        ([0.5] [_____] mm) thick; thickness governed by plaster
        thickness; maximum possible lengths; [expanded metal]
        [solid] flanges, with [square] [bullnosed] [bevelled]
        [_____] edges; [galvanized] [rust inhibitive prime
        paint] finish.

   H.   Control [and Expansion] Joint Accessories: Formed steel;
        minimum [26] [_____] gage ([0.5] [_____] mm) thick;
        [accordian] [_____] profile, 2 inch (50 mm) [expanded
        metal] [solid] flanges each side; [galvanized] [rust
        inhibitive prime paint] finish.

   I.   Anchorages: Nails, staples, or other approved metal
        supports, of type and size to suit application, galvanized
        to rigidly secure lath and associated metal accessories in
        place.

   J.   Polyethylene Sheet: Clear, 6 mil (0.15 mm) thick.

2.06   ACOUSTIC ACCESSORIES

   A.   Resilient Furring Channels: [_____;] [_____]
        manufactured by [_____.]

   B.   Acoustic Insulation: FS HH-I-521, friction fit type,
        unfaced; [_____] inch ([_____] mm) thick; [_____]
        manufactured by [_____.]

   C.   Acoustic Sealant: Non-hardening, non-skinning type, for
        use in conjunction with cement plaster system; [_____]
        manufactured by [_____.]

2.07   CEMENT PLASTER MIXES

   A.   Mix and proportion cement plaster [in accordance with
        ANSI/ASTM C926.] [in accordance with PCA Plaster (Stucco)
        Manual.] [in accordance with manufacturer's instructions.]
        [as indicated.]

2.04   Aggregate sizes and grading can be established by the same method used to grade terrazzo aggregates. Refer to NTMA documents.

2.05   This Article is intended for use when a furring and lathing section, such as Section 09206, is not used. If a furring and lathing section is used, carefully coordinate the statements in this Article to avoid conflicting statements. Specialized cement plaster lath mesh interwoven with treated papers are available; edit Paragraphs A or B to suit.

2.05C   The following accessory listings describe components manufactured in steel. These items are also available in extruded plastic and zinc for special applications.

2.05G   Control and expansion joint accessories are available in a variety of designs, sizes and shapes. Beware of joint components that are relatively inflexible; they can cause adjacent plaster work to crack just beyond the joint.

2.06   Delete the following requirements for acoustic accessories if specified in another section or if not required.

2.07A   Adjust cement plaster mix or mixes to suit project requirements. If basecoat and finish coat plasters are to be of same material and mix, edit the following paragraph accordingly.

B. Basecoat [and Brown Coat]: One part cement, minimum [3-1/2] [_____] and maximum [4] [_____] parts aggregate, and minimum [15] [_____] percent and maximum [25] [_____] percent hydrated lime [,] [and glass fibers at a rate of [1-1/2] [_____] lbs ([0.7] [_____] Kg) per sack of cement.]

C. Finish Coat: One part cement, minimum [_____] and maximum [_____] parts aggregate, [minimum [_____] percent and maximum [_____] percent pigment,] and minimum [_____] percent and maximum [_____] percent lime.

D. Mix only as much plaster as can be used in [one] [_____] hour.

E. [Add color pigments in accordance with manufacturer's instructions. Ensure uniformity of mix and coloration.]

F. Mix materials dry, to uniform color and consistency, before adding water.

G. Add [air entrainment admixtures to provide [5-7] [_____] percent entrainment] [_____] in [all coats.] [[_____] coat.]

H. Protect mixtures from frost, contamination, and evaporation.

I. Do not retemper mixes after initial set has occurred.

PART 3   EXECUTION

.01   INSPECTION

A. Verify that surfaces and site conditions are ready to receive work.

B. Masonry: Verify joints are cut flush and surface is ready to receive work of this Section. Verify no bituminous or water repellent coatings exist on masonry surface.

C. Concrete: Verify surfaces are flat, honeycomb is filled flush, and surface is ready to receive work of this Section. Verify no bituminous, water repellent, or form release agents exist on concrete surface that are detrimental to plaster.

D. Grounds and Blocking: Verify items within walls for other Sections of work have been installed.

E. Mechanical and Electrical: Verify services within walls have been tested and approved.

F. Beginning of installation means acceptance of existing conditions.

.02   PREPARATION

A. Protect surfaces near the work of this Section from damage or disfiguration.

B. Dampen masonry surfaces to reduce excessive suction.

C. Clean concrete surfaces of foreign matter. Clean surfaces using acid solutions, solvents, or detergents. Wash surfaces with clean water.

D. Roughen smooth concrete surfaces and apply bonding agent. Apply in accordance with manufacturer's instructions.

.03   INSTALLATION - LATHING MATERIALS

A. Apply one ply of felt underlayment over substrate; weatherlap edges 4 inches (100 mm) minimum. Fasten in place.

B. Apply self-furring reinforcement with self-furring ribs perpendicular to supports.

C. Apply metal lath taut, with long dimension perpendicular to supports.

2.07B Quantities of special aggregates must be controlled within limits which will not impair strength of plaster.

3.01A coordinate the following requirements with masonry or concrete sections. Edit those sections accordingly. Masonry of lightweight or other absorptive materials can cause uneven suction of moisture from applied plaster, resulting in "telegraphing" of base. Consider applying a bond coat to such surfaces.

3.03 Utilize underlayment on vertical exterior cement plaster work.

3.03A Include the following when metal furring and lathing is included in this section. If self-furring reinforcement is used in lieu of metal lath, include Paragraph B and edit the subsequent paragraphs accordingly.

197

D. Lap ends minimum [one] [_____] inch ([25] [_____] mm). Secure end laps with tie wire where they occur between supports.

E. [Lap sides of diamond mesh lath minimum [1-1/2] [_____] inches ([38] [_____] mm). Nest outside ribs of rib lath together.]

F. Attach metal lath to wood supports using nails at maximum [_____] inches ([_____] mm) on center.

G. Attach metal lath to metal supports using [tie wire] [_____] at maximum [6] [_____] inches ([150] [_____] mm) on center.

H. Attach metal lath to [concrete] [concrete masonry] using wire [hair pins,] [hooks,] [or] [loops]. Ensure that anchors are securely attached to concrete and spaced at maximum [24] [_____] inches ([600] [_____] mm) on center.

I. Continuously reinforce internal angles with corner mesh, except where the metal lath returns 3 inches (75 mm) from corner to form the angle reinforcement. Fasten at perimeter edges only.

J. Place [beaded] [bullnosed] external angle with mesh at corners. Fasten at outer edges only.

K. Place strip mesh diagonally at corners of lathed openings. Secure rigidly in place.

L. Place [4] [_____] inch ([100] [_____] mm) wide strips of metal lath centered over junctions of dissimilar backing materials. Secure rigidly in place.

M. Place casing beads at terminations of plaster finish. Butt and align ends. Secure rigidly in place.

N. Install accessories to lines and levels:

3.04 CONTROL [AND EXPANSION] JOINTS

A. Locate interior control [and expansion] joints [every [20] [_____] feet ([6] [_____] m).] [as indicated.]

B. After initial set, scribe contraction joints in exterior work [every [3] [_____] feet ([one] [_____] m) in each direction] [as indicated on reflected ceiling plan] by cutting through 2/3 of the cement plaster depth, neatly, in straight lines.

C. Locate exterior control [and expansion] joints [every [12] [_____] feet ([3.5] [_____] m) in each direction.] [as indicated on reflected ceiling plan.]

3.04 At exterior stable ambient temperatures, control joints should be located every 20 to 30 feet (6 to 9 m), at intersections of natural breaks in wall, and above door frame jambs. Expansion joints can be positioned on the building expansion joint lines in conjunction with joints of other materials. Review industry technical documents for detailed recommendations in this regard. Interior elevations and reflected ceiling plans should indicate locations of joints. Coordinate the following text with other affected sections to avoid repeating or creating conflicting requirements.

3.04A Exterior soffits require careful attention due to thermal and structural movement and shrinkage caused by curing of plaster. Jointing of cement plaster work should not exceed 3 to 4 feet (1 to 1.2 m) without scribed contraction joints. Control joints in this work should not exceed 12 feet (3.5 m). Joints constructed of back to back casing beads need to be filled with a sealant capable of flexible joint movement. Expansion joints are usually associated with building expansion joint lines and require special consideration. Edit the following paragraphs accordingly.

3.04C Include either of the next two paragraphs as the method for achieving interior or exterior joints

D. Establish control [and expansion] joints with [double casing beads butted tight.] [back to back casing beads, set [1/4] [_____] inch ([6] [_____] mm) apart.] Set both beads over [6] [_____] inch ([150] [_____] mm) wide strip of polyethylene sheet for air seal continuity.

E. Establish control [and expansion] joints with specified joint device.

F. Coordinate joint placement with [other related work.] [Section 09206.] [_____.]

05   PLASTERING

A. Apply plaster in accordance with [ASTM C926.] [manufacturer's instructions.] [PCA Plaster (Stucco) Manual.] [_____.]

B. Apply brown coat to a nominal thickness of [3/8] [_____] inch ([9] [_____] mm) and a finish coat to a nominal thickness of [1/8] [_____] inch ([3] [_____] mm) over [masonry] [concrete] [clay tile] [and] [_____] surfaces.

C. Apply scratch coat to a nominal thickness of [3/8] [_____] inch ([9] [_____] mm), brown coat to a nominal thickness of [3/8] [_____] inch ([9] [_____] mm), and a finish coat to a nominal thickness of [1/8] [_____] inch ([3] [_____] mm) over [metal lathed] [self-furring reinforcement] [and] [_____] surfaces.

D. Moist cure [scratch and] brown coats. [Apply brown coat immediately following initial set of scratch coat].

E. After curing, dampen base coat prior to applying finish coat.

F. Apply finish coat and [wood float] [steel trowel] to a smooth and consistent finish.

G. Avoid excessive working of surface. Delay trowelling as long as possible to avoid drawing excess fines to surface.

H. [[Hand] [Machine] apply aggregate surfacing to full surface coverage.]

I. Moist cure finish coat for minimum period of [48] [_____] hours.

06   TOLERANCES

A. Maximum Variation from True Flatness: [1/8] [_____] inch in 10 feet ([3] [_____] mm in 3 m).

07   SCHEDULE

3.05A   Paragraph B describes a two coat application; Paragraph C describes a three coat application. Edit accordingly.

3.07   Provide a schedule to identify fire rating assembly classifications as applicable to this section, varying finish requirements, and any other special conditions.

SPECTEXT SPECIFICATION FOR

GYPSUM BOARD SYSTEMS
SECTION 09260

This section includes metal channel and stud framing with paper or vinyl faced gypsum board for walls, ceiling and furred space enclosures; including taped joint or batten joint closures. This section also includes shaft wall systems, a subject which can alternately be specified as a separate section. This section includes proprietary and descriptive type specifications. Use only one specifying type to avoid any conflicting requirements. Edit accordingly.

PART 1   GENERAL

1.01   WORK INCLUDED

    A.   Metal stud wall framing.

    B.   Metal channel ceiling framing.

    C.   Acoustic insulation.

    D.   Gypsum board.

    E.   [Taped and sanded] [Batten] joint treatment.

1.02   WORK INSTALLED BUT FURNISHED UNDER OTHER SECTIONS

    A.   Section [_____-_____]: Frames for recessed [washroom accessories.] [access panels.] [_____.]

1.02 List sections which furnish products for installation under this section. When a product will be furnished "by Others," outside the Contract, it is considered to be furnished "by Owner."

1.03   RELATED WORK

    A.   Section 05400 - Cold Formed Metal Framing.

    B.   Section 06114 - Wood Blocking and Curbing: Wood blocking for support of [_____].

    C.   Section [_____-_____]: Thermal insulation.

    D.   Section 08111 - Standard Steel Doors and Frames.

    E.   Section 08305 - Access Doors: Metal access panels.

    F.   Section 09111 - Metal Stud Framing System.

    G.   Section [09900 - Painting] [_____-_____]: Surface finish.

    H.   Section 10616 - Demountable Gypsum Board Partitions.

1.04   REFERENCES

    A.   ANSI/ASTM C36 - Gypsum Wallboard.

    B.   ANSI/ASTM C79 - Gypsum Sheathing Board.

    C.   ANSI/ASTM C442 - Gypsum Backing Board.

    D.   ANSI/ASTM C475 - Joint Treatment Materials for Gypsum Wallboard Construction.

    E.   ANSI/ASTM C514 - Nails for the Application of Gypsum Wallboard.

    F.   ANSI/ASTM C557 - Adesive for Fastening Gypsum Wallboard to Wood Framing.

1.04 Include only reference standards that are to be indicated within the text of this section. Refer to TAS Sections 09250, 09280 and 09285. Edit the following, adding and deleting as required for project and product selection.

G. ANSI/ASTM C630 - Water Resistant Gypsum Backing Board.

H. ANSI/ASTM C645 - Non-Load (Axial) Bearing Steel Studs, Runners (Track), and Rigid Furring Channels for Screw Application of Gypsum Board.

I. ANSI/ASTM C646 - Steel Drill Screws for the Application of Gypsum Sheet Material to Light Gage Steel Studs.

J. ANSI/ASTM C754 - Installation of Framing Members to Receive Screw Attached Gypsum Wallboard, Backing Board, or Water Resistant Backing Board.

K. ANSI/ASTM E90 - Method for Laboratory Measurement of Airborne Sound Transmission Loss of Building Partitions.

L. ANSI/ASTM E119 - Fire Tests of Building Construction and Materials.

M. FS HH-I-521 - Insulation Blankets, Thermal (Mineral Fiber, for Ambient Temperatures).

N. GA-201 - Gypsum Board for Walls and Ceilings.

O. GA-216 - Recommended Specifications for the Application and Finishing of Gypsum Board.

1.05   SYSTEM DESCRIPTION

A. Acoustic Attentuation for [Identified] Interior Partitions: [_____] STC in accordance with ANSI/ASTM E90.

B. Shaft Wall: Perform to the following:
1. Air Pressure Within Shaft: [_____] psf ([_____] KPa) with maximum mid-span deflection of [_____] inches ([_____] mm).
2. Fire Rating Requirements: [_____] hour in accordance with [UL] [_____] listed assembly No. [_____].
3. Acoustic Attenuation: [_____] STC in accordance with ANSI/ASTM E90.

1.06   QUALITY ASSURANCE

A. Applicator: Company specializing in gypsum board systems work [with [_____] years [documented] experience.] [approved by manufacturer].

1.07   REGULATORY REQUIREMENTS

A. Conform to [applicable] [_____] code for fire rated assemblies in conjunction with Section [05400] [09111] [_____] as follows:

1. Fire Rated Partitions: Listed assembly by [UL] [FM] No. [_____.]
2. Fire Rated Ceiling [and Soffits]: Listed assembly by [UL] [FM] No. [_____.]
3. Fire Rated Structural Column Framing: Listed assembly by [UL] [FM] No. [_____.]
4. Fire Rated Structural Beam Framing: Listed assembly by [UL] [FM] No. [_____.]

1.08   SUBMITTALS

A. Submit [shop drawings] and product data under provisions of Section [01300.] [01340.]

B. [Indicate on shop drawings, special details associated with fireproofing, acoustic seals, and [_____].]

C. Provide product data on metal framing, gypsum board, joint [tape,] [batten,] [decorative finish,] and [_____.]

D. Submit samples under provisions of Section [01300.] [01340.]

E. Submit [two] [_____] samples of predecorated gypsum board [___x___] inch ([___x___] mm) in size illustrating decorative finish.

F. Submit manufacturer's installation instructions under provisions of Section [01300.] [01340.]

1.05   Use this Article carefully; restrict descriptions to total system parameters, performance or design requirements. Do not repeat statements made in "Work Included" Article.

1.08   Shop drawings are a costly element of the project; minimize their use for standard products such as gypsum board systems. Request shop drawings when special out-of-the-ordinary conditions exist.

1.08C   Use the following two paragraphs for submission of physical samples for selection of finish, color, texture, etc.

1.08E   When manufacturers' instructions for specific installation requirements are utilized, carefully edit Part 3 Execution requirements to avoid conflict with those instructions.

PART 2   PRODUCTS

2.01   ACCEPTABLE MANUFACTURERS - GYPSUM BOARD SYSTEM

    A.   [_____.]

    B.   Other acceptable manufacturers offering equivalent
        products:
        1.   [_____.]
        2.   [_____.]
        3.   [_____.]

    C.   Substitutions: Under provisions of Section [01600.]
        [01630.]

2.02   FRAMING MATERIALS

    A.   Studs and Tracks: [ANSI/ASTM C645;] [GA 201 and GA 216;]
        galvanized sheet steel, [26] [_____] gage (([0.5] [_____]
        mm) thick, ['C'] [_____] shape, [with serrated
        faces.]

    B.   Shaft Wall Studs: [_____.]

    C.   Furring, Framing and Accessories: [ANSI/ASTM C645.] [GA
        201 and GA 216.] [_____.]

    D.   Fasteners: [ANSI/ASTM [C514.] [C646.]] [GA 201 and GA
        216.] [_____.]

    E.   Adhesive: [ANSI/ASTM C557.] [GA201 and GA216.]
        [_____.]

2.03   GYPSUM BOARD MATERIALS

    A.   Standard Gypsum Board: ANSI/ASTM C36; [3/8] [1/2] [5/8]
        inch (([10] [13] [16] mm) thick, maximum permissible
        length; ends square cut, [tapered] [tapered and beveled]
        [square] [round] edges.
    B.   Fire Rated Gypsum Board: ANSI/ASTM C36; [fire resistive]
        [moisture resistant] type, UL rated; [1/2] [5/8] inch
        (([13] [16] mm) thick, maximum permissible length; ends
        square cut, [tapered] [tapered and beveled] [square]
        [round] edges.
    C.   Moisture Resistant Gypsum Board: ANSI/ASTM C630; [1/2]
        [5/8] inch (([13] [16] mm) thick, maximum permissible
        length; ends square cut, [tapered] [tapered and beveled]
        [square] [round] edges.
    D.   Insulating Gypsum Board: ANSI/ASTM C36; [3/8] [1/2] [5/8]
        inch (([10] [13] [16] mm) thick, maximum permissible
        length; back surface laminated with aluminum foil; ends
        square cut, [tapered] [tapered and beveled] [round] edges.
    E.   Gypsum Backing Board: ANSI/ASTM C442; [standard] [fire
        rated] [insulating] type; [3/8] [1/2] [5/8] inch (([10]
        [13] [16] mm) thick; [square] [round] [V-grooved] [book
        tongue and grooved] edges, ends square cut, maximum
        permissible length.
    F.   Gypsum Sheathing Board: ANSI/ASTM C79; moisture resistant
        [and fire resistant] type; [3/8] [1/2] [5/8] inch (([10]
        [13] [16] mm) thick, maximum permissible length; ends
        square cut, [square] [book tongue and grooved] edges;
        water repellent paper faces.
    G.   Gypsum Core Board: One inch (25 mm) thick, maximum
        permissible length; [square] [tongue and groove] edges,
        ends square cut.

    H.   Exterior Gypsum Ceiling Board: [Standard] [Fire rated]
        type, 1/2 inch (13 mm) thick, maximum permissible length;
        ends square cut, tapered and beveled edges.

2.04   ACCEPTABLE MANUFACTURERS - PREDECORATED GYPSUM BOARD

        A.   [_____.]

        B.   [_____.]

        C.   [_____.]

(Right margin annotations)

2.01   In this Article, list the manufacturers acceptable for this project. In Paragraph A, identify the one manufacturer (and the model name or number) of the product upon which this section is based. In Paragraph B, list other acceptable manufacturers. If product substitution is allowed, include Paragraph C. Refer to TAS Sections 09250, 09260 and 09280.

2.01C   Edit the following descriptive specifications for any conflict with manufacturers' products specified above.

2.02A   In the following paragraph specify the shaft wall products by proprietary system incorporating the products of this section, or special products not listed.

2.03E   Include the following paragraph if gypsum sheathing is not specified in the relevant wood framing sections of Division 6 or the roof deck preparation part of relevant sections in Division 7.

2.04   If only one manufacturer is acceptable, list in Article 2.05 and delete this Article; if more than one, list in this Article. If product substitution procedure is used, include Paragraph D. Refer to TAS Section 09250.

D. Substitutions: Under provisions of Section [01600.] [01630.]

05 PREDECORATED GYPSUM BOARD MATERIALS

A. Gypsum Board: ANSI/ASTM C36; [standard] [fire rated] type; [3/8] [1/2] [5/8] inches ([10] [13] [16] mm) thick, maximum permissible length; paperbound, edges and ends square cut.

B. Predecorated Facing: Sheet vinyl, [fabric reinforced,] flame/fuel/smoke rated for [_____/_____/_____] in accordance with ANSI/ATSM E119; [_____] color [as selected.]

C. Batten Joints: [_____.]

06 ACCESSORIES

A. Acoustical Insulation: FS-HH-I-521; preformed mineral wool, friction fit type without integral vapor barrier membrane, [[_____] inch ([_____] mm) thick.]

B. Acoustical Sealant: Non-hardening, non-skinning, for use in conjunction with gypsum board; [_____] manufactured by [_____.]

C. Corner Beads: [Metal.] [Metal and paper combination.]

D. Edge Trim: GA 201 and GA 216; Type [LC] [L] [LK] [U exposed reveal] bead.

E. Joint Materials: [ANSI/ASTM C475;] [GA 201 and GA 216;] reinforcing tape, joint compound, adhesive, water, and fasteners.

PART 3 EXECUTION

01 INSPECTION

A. Verify that site conditions are ready to receive work and opening dimensions are as [indicated on shop drawings] [instructed by the manufacturer.]

B. Beginning of installation means acceptance of [existing surfaces] [substrate.]

02 METAL STUD INSTALLATION

A. Install studding in accordance with [ANSI/ASTM C754.] [GA 201 and GA 216.] [manufacturer's instructions.]

B. Metal Stud Spacing: [16] [_____] inches ([400] [_____] mm) on center.

C. Partition Heights: [To suspended ceilings.] [To [_____] inches ([_____] mm) above suspended ceilings.] [Full height to floor or roof construction above.] [Install additional bracing for partitions extending above ceiling.]

D. Door Opening Framing: Install double studs at door frame jambs. [Install stud tracks on each side of opening, at frame head height, and between studs and adjacent studs].

E. Blocking: [Nail wood blocking to studs.] [Bolt or screw steel channels to studs.] Install blocking for support of [plumbing fixtures,] [toilet partitions,] [wall cabinets,] [toilet accessories,] [hardware,] and [_____.]

F. Coordinate installation of bucks, anchors, blocking, electrical and mechanical work placed in or behind partition framing.

03 WALL FURRING INSTALLATION

A. Erect wall furring for direct attachment to [concrete block] [_____] [and] [concrete] walls.

B. Erect furring channels [horizontally.] [vertically.]. Secure in place on alternate channel flanges at maximum [24] [_____] inches ([600] [_____] mm) on center.

2.04D Edit the following descriptive specifications for any conflicts with manufacturer's products specified above.

2.06 Delete the following requirements for acoustical batt insulation of specified in another section. Delete reference to thickness if performance specifying acoustical STC rating.

3.02C Use the following paragraph for more rigid construction at door frames.

3.03 Do not use this Article if furring is specified in Section 09206. Use this Article as appropriate to describe conditions that are not specified elsewhere.

C. Space furring channels maximum [16] [24] inches ([400] [_____] mm) on center, not more than [4] [_____] inches ([100] [_____] mm) from [floor and ceiling lines.] [abutting walls.]

D. Install thermal insulation [and Z-furring channels] directly attached to [concrete masonry] [_____] [and] [concrete] walls in accordance with manufacturer's instructions.

E. Install thermal insulation vertically and hold in place with Z-furring channels spaced maximum 24 inches (600 mm) on center, not more than 3 inches (75 mm) at external corners and 12 inches (300 mm) at internal corners.

F. Secure Z-furring channels at maximum 24 inches (600 mm) on center. Secure thermal insulation by [_____.]

G. Erect free-standing metal stud framing tight to [concrete] [concrete masonry] [_____] walls, attached by adjustable furring brackets in accordance with manufacturer's instructions.

3.04 FURRING FOR FIRE RATINGS

A. Install furring as required for fire resistance ratings indicated.

3.05 SHAFT WALL INSTALLATION

A. Shaftwall Framing: [In accordance with manufacturer's installation instructions.] [_____.]

3.06 CEILING FRAMING INSTALLATION

A. Install in accordance with [ANSI/ASTM C754.] [GA 201 and GA 216.] [manufacturer's instructions.]

B. Coordinate location of hangers with other work.

C. Install ceiling framing independent of walls, columns, and above-ceiling work.

D. Reinforce openings in ceiling suspension system which interrupt main carrying channels or furring channels, with lateral channel bracing. Extend bracing minimum [24] [_____] inches ([600] [_____] mm) past each end of openings.

E. Laterally brace entire suspension system.

3.07 ACOUSTICAL ACCESSORIES INSTALLATION

A. Install resilient channels at maximum [24] [_____] inches ([600] [_____] mm) on center. [Locate joints over framing members.]

B. Place acoustical insulation in partitions tight within spaces, around cut openings, behind and around electrical and mechanical items within or behind partitions, and tight to items passing through partitions.

C. Install acoustical sealant within partitions in accordance with manufacturer's instructions.

D. Install acoustical sealant at gypsum board perimeter at:
1. Metal Framing: [One] [Two] beads.
2. [Base Layer.]
3. [Face Layer.]
4. Calk all penetrations of partitions by conduit, pipe, ductwork, rough-in boxes, and [_____.]

3.08 GYPSUM BOARD INSTALLATION

A. Install gypsum board in accordance with [GA 201 and GA 216.] [manufacturer's instructions.]

B. Erect single layer [standard gypsum] [_____] board in most economical direction, with ends and edges occurring over firm bearing.

C. Erect single layer fire rated gypsum board vertically, with edges and ends occurring over firm bearing.

3.05 If proprietary specifying the shaft wall framing, include in the following paragraph special installation statements. If prescriptive specifying, expand installation statements as appropriate.

3.07A Include the following paragraph only if acoustical insulation to be installed in this section.

3.08C Include the following paragraph only if gypsum sheathing not specified in other sections.

D. [Erect exterior gypsum sheathing horizontally, with edges butted tight and ends occurring over firm bearing.]

E. Use screws when fastening gypsum board to metal furring or framing.

F. Use [nails] [or] [screws] when fastening gypsum board to wood furring or framing. [Staples may only be used when securing the first layer of double layer applications.]

G. Double Layer Applications: Use gypsum backing board for first layer, placed [perpendicular] [parallel] to framing or furring members. [Use fire rated gypsum backing board for fire rated partitions.] Place second layer [perpendicular] [parallel] to first layer. Offset joints of second layer from joints of first layer.

H. Double Layer Applications: Secure second layer to first with [fasteners.] [adhesive and sufficient support to hold in place.] [Apply adhesive in accordance with manufacturer's instructions.]

I. Erect exterior gypsum soffit board perpendicular to supports, with staggered end joints over supports.

J. Treat cut edges and holes in [moisture resistant gypsum board] [and] [exterior gypsum ceiling board] with sealant.

K. Place control joints consistent with lines of building spaces [as indicated.] [as directed.]

L. Place corner beads at external corners [as indicated.] Use longest practical length. Place edge trim where gypsum board abutts dissimilar materials [as indicated.]

.09   JOINT TREATMENT

3.09 For taped and filled joints, use the following three paragraphs.

A. Tape, fill, and sand exposed joints, edges, and corners to produce smooth surface ready to receive finishes.

B. Feather coats onto adjoining surfaces so that camber is maximum [1/32] [1/16] inch ([0.8] [1.6] mm).

C. Taping, filling, and sanding is not required at surfaces behind [adhesive applied ceramic tile.] [_____.]

3.09C For battern jointing method, edit the following two paragraphs as required.

D. Erect pre-decorated gypsum board vertically, with exposed batten fastening system.

E. Erect in accordance with manufacturer's instructions.

.10   TOLERANCES

A. Maximum Variation from True Flatness: [1/8] [_____] inch in 10 feet ([3] [_____] mm in 3 m) in any direction.

.11   SCHEDULE

3.11 Provide a schedule for identifying different types of gypsum wallboard (moisture resistant, insulating, etc.), partition types, heights and special conditions.

CHAPTER THREE-A

SPECIFICATIONS FOR GYPSUM BOARD (DRYWALL) CONSTRUCTION

GYPSUM BOARD CONSTRUCTION

The Gypsum Association, head-quartered in Evanston, Illinois, is the nationally recognized trade association of the manufacturers of gypsum products dedicated to upgrading and promoting the use of gypsum products through research and development, education, building codes and the dissemination of information in general through printed materials.

One of its most valuable publications is its "GA-201-85," *"Using Gypsum Board For Walls And Ceilings".* With permission of the Gypsum Association GA-201-85 is reproduced in a slightly abridged version in the section of this book that follows.

A second very informative Gypsum Association document "GA-216-85," is reproduced immediately following "GA-201-85." This document, the Gypsum Association's recommended specification for gypsum wallboard installations, is primarily for the benefit of specification writers and is an alternate choice for the wallboard application specifications appearing as Section 12 of Chapter Two of this manual.

It will be noted that there is a duplication of some of the illustrations and material in the two Gypsum Association documents which follow. These have been left in place for the convenience of the reader.

Gypsum board is the generic name for a family of panel products consisting of a non-combustible core, primarily of gypsum, with a paper surfacing on the face, back and long edges. A typical board application is shown in figure 1.

Gypsum board is often called drywall, wallboard, or plasterboard and differs from products such as plywood, hardboard and fiber board because of its noncombustible core. It is designed to provide a monolithic surface when installed with joint treatment compound.

Gypsum is a mineral found in sedimentary rock formations in a crystalline form known as calcium sulfate dihydrate $CaSO_4 \bullet 2H_2O$. One hundred pounds of gypsum rock contains approximately 21 pounds (or 10 quarts) of chemically combined water. Gypsum rock is mined or quarried and then crushed. The crushed rock is heated to about 350 F, driving off three fourths of the chemically combined water in a process called calcining. The calcined gypsum (or hemihydrate) $CaSO_4 \bullet \frac{1}{2}H_2O$, then is ground into a fine powder used in producing the base for gypsum plaster, wallboard and other gypsum products.

To produce gypsum board, the calcined gypsum is mixed with water and additives to form a slurry which is fed between continuous layers of paper on a board machine. As the board automatically moves down a conveyer line, the calcium sulfate recrystalizes or rehydrates, reverting to its original rock state. The paper becomes chemically and mechanically bonded to the core. The board is then cut to length and conveyed through dryers to remove any free moisture.

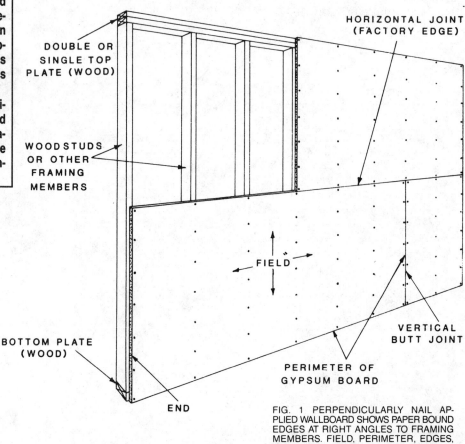

DOUBLE OR SINGLE TOP PLATE (WOOD)

WOOD STUDS OR OTHER FRAMING MEMBERS

BOTTOM PLATE (WOOD)

END

HORIZONTAL JOINT (FACTORY EDGE)

FIELD

VERTICAL BUTT JOINT

PERIMETER OF GYPSUM BOARD

FIG. 1 PERPENDICULARLY NAIL APPLIED WALLBOARD SHOWS PAPER BOUND EDGES AT RIGHT ANGLES TO FRAMING MEMBERS. FIELD, PERIMETER, EDGES, ENDS, AND JOINTS ARE INDICATED.

TYPES OF GYPSUM BOARD PRODUCTS

Many types of gypsum board are available for a variety of building needs. Developed through modern technology as a result of specific requirements, gypsum board panels are mainly used as the surface layer of interior walls and ceilings; as a base for ceramic, plastic and metal tile; for exterior soffits; for elevator and other shaft enclosures; and to provide fire protection to structural elements. Most gypsum board products are available with aluminum foil backing which provides an effective vapor retarder for exterior walls when applied with the foil surface against the framing. Foil backed gypsum board should not be used as a backing material for tile, a second face ply on a two-ply system, in conjunction with heating cables, or when laminating directly to masonry, ceiling and roof assemblies.

The various thicknesses of gypsum wallboard available in regular, type X and predecorated board are as follows:

¼-in. — a low cost board used as a base in a multi-layer application for improving sound control, or to cover existing walls and ceilings in remodeling.

⁵⁄₁₆-in. — a gypsum board used in manufactured housing.

⅜-in. — a board principally applied in a double-layer system over wood framing and as a face layer in repair or remodeling.

½-in. — generally used as a single-layer wall and ceiling material in residential work and in double-layer systems for greater sound and fire ratings.

⅝-in. — used in quality single-layer and double-layer wall systems. The greater thickness provides additional fire resistance, higher rigidity, and better impact resistance.

¾ & 1-in. — Used in interior partitions, shaft walls, stairwells, chaseways, area separation walls and corridor ceilings. Special edged panels are used in some interior partitions.

Standard size gypsum boards are 4 ft. wide and 8, 10, 12 or 14 ft. long. The width is compatible with the standard framing of studs or joists spaced 16 in. and 24 in. o.c. (Other lengths and widths are available from the manufacturer on special order).

Edges available are tapered, square edge, beveled, rounded and tongue and grooved. (Fig. 3)

Eased taper edge gypsum board has a tapered and slightly rounded or beveled factory edge. It may be used as an aid in custom finishing of joints.

Regular gypsum board is used as a surface layer on walls and ceilings.

Type X gypsum board is available in ½ in. and ⅝ in. thicknesses and has an improved fire resistance made possible through the use of special core additives. It is also available with a predecorated finish. Type X gypsum board is used in most fire rated assemblies. (Fig. 4)

Pre-decorated gypsum board has a decorative surface which does not require further treatment. (Fig. 5) The surfaces may be coated, printed, or have a vinyl film. Other predecorated finishes include factory painted and various textured patterns.

Water-resistant gypsum board has a water resistant gypsum core and a water repellent paper. It serves as a base for application of ceramic or plastic wall tile or plastic finish panels in bath, shower, kitchen and laundry areas. It is available with a regular or Type X core and in ½ in. and ⅝ in. thicknesses. This product should not be used on ceilings or soffits.

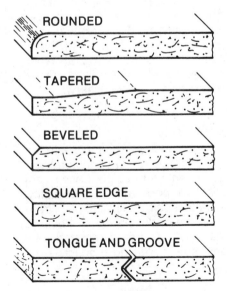

ROUNDED

TAPERED

BEVELED

SQUARE EDGE

TONGUE AND GROOVE

EDGES AVAILABLE
Fig. 3

Fig. 4 Regular gypsum board is shown with board surface paper and both tapered and square edges. (Type X board has the same appearance, but is labeled Type X on the tapered edge or on the back.)

Fig. 5 Predecorated gypsum board is available with a variety of surface finishes.

Gypsum backing board is designed to be used as a base layer or backing material in multi-layer systems. It is available with aluminum foil backing and with regular and Type X cores.

Gypsum coreboard is available as a 1 in. thick solid coreboard or as a factory laminated board composed of two ½ in. boards. It is used in shaft walls and laminated gypsum partitions with additional layers of gypsum board applied to the coreboard to complete the wall assembly. It is available in 24 in. widths with a variety of edges.

Gypsum liner board is used as a liner panel in shaft walls, stairwells, chaseways, area separation walls and corridor ceilings. It has a special fire-resistant core encased in moisture-resistant paper. Liner board is available ¾ or 1 in. thick, in widths of 24 or 48 in., and with square edges (sometimes eased square edges).

Exterior gypsum soffit board is board designed for use on the underside of eaves, canopies, carports, and other commercial and residential exterior applications with indirect exposure to the weather. Available in ½ or ⅝ in. thicknesses with both regular and Type X core.

Gypsum sheathing is used as a protective, fire resistive membrane under exterior wall surfacing materials such as wood siding, masonry veneer, stucco and shingles. It also provides protection against the passage of water and wind and adds structural rigidity to the framing system. The noncombustible core is surfaced with firmly bonded water repellent paper; in addition may also have a water resistant core. It is available in 2 ft. and 4 ft. widths, ½ in. and ⅝ in. thick. It is also available with Type X core.

Gypsum board substrate for floor or roof assemblies has a Type X core ½ or ⅝ in. thick and is available in 24 in. or 48 in. widths.

It is used under combustible roof coverings to protect the structure from fires originating on the roof. It can also serve as an underlayment when applied to the top surfaces of floor joists and under sub-flooring. It may also be used as a base for built-up roofing applied over steel decks.

Gypsum base for veneer plaster is used as a base for thin coats of hard, high strength gypsum veneer plaster.

Gypsum lath is a board product used as a base to receive hand or machine applied plaster. It is available in a ⅜ in. or ½ in. thickness, 16 in. or 24 in. nominal widths, and normally 48 in. lengths. Other lengths are available on special order.

ADVANTAGES OF GYPSUM BOARD CONSTRUCTION

Gypsum board walls and ceilings have a number of outstanding advantages:
- Fire resistance
- Sound isolation
- Durability
- Economy
- Versatility

Fire resistance

Gypsum board is an excellent fire resistive material. It is the most commonly used interior finish where fire resistance classifications are required. Its noncombustible core contains chemically combined water which, under high heat, is slowly released as steam, effectively retarding heat transfer. Even after complete calcination, when all the water has been released, it continues to act as a heat insulating barrier. In addition, tests conducted in accordance with ASTM Method E84 show that it has a low flame spread index and smoke density index. When installed in combination with other materials it serves to effectively protect building elements from fire for prescribed time periods.

Sound isolation

Control of unwanted sound that might be transmitted to adjoining rooms is a key consideration in the design stage of a building, taking into account the environment desired for the particular activity of the occupants. It has been determined that low density paneling transmits annoying amounts of noise and that sound absorbing acoustical surfacing materials, while they reduce the reflection of sound within a room, do not greatly reduce transmission of sound into adjoining rooms. Gypsum board wall and ceiling systems do effectively help control sound transmission. Suggested construction techniques for sound isolation are described and illustrated in Chapter VII along with recommended procedures and materials necessary to obtain adequate sound control.

Durability

Gypsum board is used to construct strong high quality walls and ceilings with excellent dimensional stability and durability. The surfaces are easily decorated and refinished.

Economy

Gypsum board products are readily available and easy to apply. They are inexpensive wall surfacing materials offering a fire-resistant interior finish. Both regular and predecorated wallboard may be installed at relatively low cost. When predecorated board is used, further decoration is unnecessary.

Versatility

Gypsum board products satisfy a wide range of architectural requirements for design. Ease of application, performance, availability, ease of repair, and its adaptability to all forms of decoration combine to make gypsum board unmatched by any other surfacing product.

APPLICATION OF GYPSUM BOARD

Gypsum board panels can be applied over wood or metal framing, or furring. They can be applied to masonry and concrete surfaces, either directly or to wood or metal furring strips. If the board is to be applied directly, any irregularities in the masonry or concrete surfaces must be smoothed or filled. Furring is a means to provide a flat surface for standards fastener application as well as provide separation to overcome dampness in exterior walls. Gypsum board must not come in direct contact with surfaces such as concrete or soil that can have high moisture content.

Most common in residential construction is the gypsum wallboard system with the joints between the panels and internal corners reinforced with tape and covered with joint treatment compound to prepare them for decoration. External corners are normally reinforced with corner bead which in turn is covered with joint compound. Exposed edges are covered with metal or plastic trim. The result is a smooth, unbroken surface ready for final decoration of paint, textures, wallpaper, tile, paneling or other materials. When predecorated board is used, no further finishing is necessary, but moldings or battens can be applied to cover the joints if desired.

Single and Multi-ply Application

In light commercial and in residential construction, single-ply gypsum board systems (Fig. 6-A — 6-B) are commonly used. Usually they are adequate to meet fire resistance and sound control requirements.[1]

Multi-ply systems (Fig. 7) have two or more layers of gypsum board and therefore can increase sound isolation and fire resistive performance. They also provide better surface quality because face layers are often laminated over base layers thereby reducing the number of fasteners. As a result, surface joints of the face layer are reinforced by the continuous base layers cf gypsum board. Nail popping and ridging problems are less frequent and imperfectly aligned supports have less effect on the finished surface.

Satisfactory results can be assured with either single-ply or multi-ply assemblies by requiring proper:

- framing details (straight, correctly spaced, properly cured lumber)
- job conditions (controlled temperature and adequate ventilation during application)
- application of the board (measuring, cutting, aligning, fastening)
- joint and fastener treatment
- special requirements for proper sound isolation, fire resistance, thermal properties, or moisture resistance.

Greater details for single and multi-ply attachment are given in Chapters III and IV.

Control Joints

Control joints should be installed in gypsum board systems wherever expansion or control joints occur in the base exterior wall and not more than 30 ft. o.c. in long wall furring runs. Wall or partition height door or window frames may be considered control joints. The gypsum board should be isolated from structural elements (columns, beams, load-bearing interior walls, etc.) and from dissimilar wall or ceiling finishes by control joints, metal trim or other means.

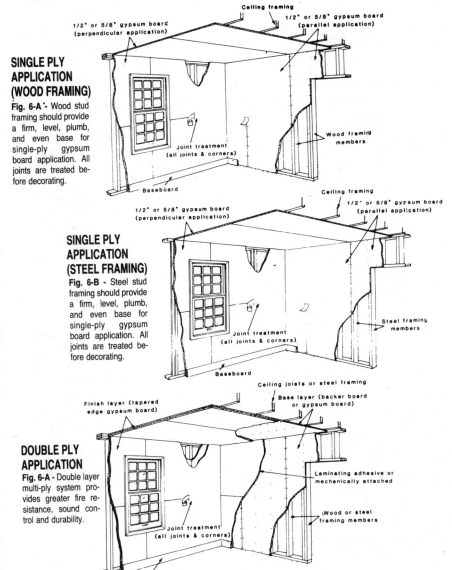

SINGLE PLY APPLICATION (WOOD FRAMING)
Fig. 6-A - Wood stud framing should provide a firm, level, plumb, and even base for single-ply gypsum board application. All joints are treated before decorating.

1/2" or 5/8" gypsum board (perpendicular application)
Ceiling framing
1/2" or 5/8" gypsum board (parallel application)
Wood framing members
Joint treatment (all joints & corners)
Baseboard

SINGLE PLY APPLICATION (STEEL FRAMING)
Fig. 6-B - Steel stud framing should provide a firm, level, plumb, and even base for single-ply gypsum board application. All joints are treated before decorating.

1/2" or 5/8" gypsum board (perpendicular application)
Ceiling framing
1/2" or 5/8" gypsum board (parallel application)
Steel framing members
Joint treatment (all joints & corners)
Baseboard

DOUBLE PLY APPLICATION
Fig. 6-A - Double layer multi-ply system provides greater fire resistance, sound control and durability.

Finish layer (tapered edge gypsum board)
Ceiling joists or steel framing
Base layer (backer board or gypsum board)
Laminating adhesive or mechanically attached
Wood or steel framing members
Joint treatment (all joints & corners)
Baseboard

Jobsite Preparations

Job conditions such as temperature and humidity can affect the performance of joint treatment materials and the appearance of the joint. These conditions can sometimes affect adhesive materials and their ability to develop adequate bond. During the cold season, interior finishes should not be installed unless the building has a controlled heat of not less than 50 F nor more than 80 F. These temperatures should be maintained at least 48 hours before, during, and 48 hours after, the installation. All materials should be protected from the weather.

If humidity is excessive ventilation should be provided. In glazed buildings windows should be kept open to provide air circulation. In enclosed areas without natural ventilation, fans should be used. When drying is slow, additional drying time between coats of joint compound should be allowed. During hot dry weather, drafts should be avoided so that joint compound will not dry too rapidly.

When ceilings are to receive water-based spray texture finishes, special attention must be given to the spacing of framing members, thickness of board, ventilation, vapor retarder, insulation and other items which can affect the performance of the system. Sagging may occur if recommendations for joint spacing, board application, vapor retarder and insulation are not followed. During cold weather insulate *before* hanging board.

Heavy water-based texturing materials may cause sag in gypsum board ceilings under the following adverse conditions: high humidity, improper ventilation, panels applied parallel to framing and panels having insufficient thickness to span the distance between framing. The following table gives max. framing spacing for panels that are to be covered with water-based texturing materials.

Frame Spacing —
Textured Gypsum Panel Ceilings

board thickness	application method (long edge relative to frame)	max. framing spacing o.c.
in.		in.
3/8	not recommended	—
1/2	perpendicular only	16
5/8	perpendicular only	24

Note: For adhesively laminated double-layer applications with 3/4" or more total thickness, 24" o.c. max.

Water-based texturing materials applied to ceilings should be completely dry before insulation and vapor retarder are installed.

Lumber must be kept dry during storage and installation at the job site. (*Moisture content of lumber should not exceed 15 percent at the time of gypsum board application.*) "Green lumber" should not be used for framing. Since lumber shrinks across the grain as it dries it tends to expose the shanks of nails driven into the edge of the framing members, (Fig. 8). If shrinkage is substantial or nails are longer than necessary, separation between the gypsum board and its framing lumber can result in protrusion of the nail head above the board surface (nail pops).

Delivery of gypsum board should coincide with the installation schedule. Boards should be stored flat and protected from the elements. Materials used as storage supports

Fig. 9 A boom truck delivers gypsum board economically to upper stories.

should be at least 4 inches in width and of uniform depth. As the units are tiered, supports should be carefully aligned from bottom to top so that each tier rests on solid bearing.

Stacking long lengths on short lengths should be avoided to prevent the longer boards from breaking. Leaning boards against the framing for prolonged periods with the long edges horizontal is not recommended. Leaning of boards should also be avoided during periods of high humidity as the boards may be subject to warping. All materials should remain stored in their original wrappings or containers until ready for use on the job site. When boards are moved on the job they should be carried, not dragged, so that the edges are not abused.

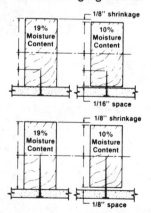

Fig. 8 Wood studs may shrink as they dry out, and separation can occur between the stud and the gypsum board in a wall. As studs shrink, separation may be as much as 1/16 in. when nails penetrate one in. or 1/8 in. with two in. penetration.

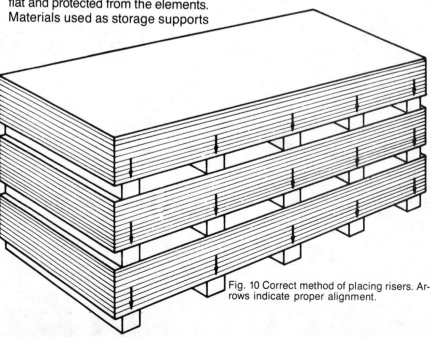

Fig. 10 Correct method of placing risers. Arrows indicate proper alignment.

Cutting and Fitting Procedures

Gypsum board installations should be carefully planned. Accurate measuring, cutting and fitting are very important. In residential buildings with less than 8'1" ceiling heights it is preferred that wallboard be installed at right angles (perpendicular) to framing members; thus creating fewer joints to finish. On long walls, boards of maximum practical length should be used to minimize the number of butt joints. Scored, scratched, broken or otherwise damaged board should not be used.

Measurements should be taken accurately at the correct ceiling or wall location for each edge or end of the board (Figure 11). Accurate measuring will usually reveal irregularity in framing and furring so that corrective allowances can be made in cutting. Poorly aligned framing should be corrected before applying gypsum board (see Chapter II, Supporting Construction).

Gypsum board should be cut by first scoring (cutting) through the paper to the core with a sharp knife, working from the face side (Figure 12). The board is then snapped back away from the cut face (Figure 13). The back paper should be broken by snapping the board in the reverse direction or by cutting the backpaper with scoring knife. Gypsum board may also be cut by sawing. All cut edges and ends of the gypsum board should be smoothed to form neat tight-fitting joints when installed. Ragged cut ends or broken edges can be smoothed with a rasp or sandpaper or trimmed with a sharp knife. If burrs on the cut ends are not removed they will form a visible ridge in the finished surface.

The following procedures should be followed to assure a successful application.

1. Install ceiling panels first, then the wall panels.
2. Cut panels should fit easily into place without force.
3. Match similar edges and ends, i.e.: tapered to tapered, square cut ends to square ends.
4. Plan to span the entire length of ceilings or walls with single boards if possible to reduce the

Fig.11

Fig. 12

Fig. 13

number of butt joints which are more difficult to finish. Stagger butt joints and locate them as far from the center of the wall and ceiling as possible so they will be inconspicuous.

5. In a single-ply application, the board ends and edges parallel to framing members should fall on these members to reinforce the joint. (Exception: In a two-ply assembly, with adhesive between the plies, the ends and edges of face layers need not fall on supporting members).

Mechanical and electrical equipment should be installed to allow for the gypsum board thickness when applying the trim components such as cover plates, registers and grilles. The depth of electrical boxes should not exceed the framing depth and boxes should not be placed back to back in the same stud cavity space. Electrical boxes, cabinets, and other devices preferably should not penetrate completely through walls. This can be detrimental to sound and fire resistance.

SUPPORTING CONSTRUCTION

Wood Framing

All wood framing and furring must be accurately aligned in the same plane so that the gypsum board fits flat against it at all points (Fig. 14). Framing member surfaces should be in alignment with the plane of the faces of adjacent supports.

Improper Framing

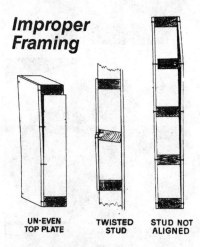

UN-EVEN TOP PLATE   **TWISTED STUD**   **STUD NOT ALIGNED**

Fig. 14 Framing that is not accurately aligned can cause loose board attachment and problems in finished surfaces.

Furthermore, the spacing of framing should not exceed the maximum recommended for the gypsum board thickness as shown in Table I for single-ply and in Tables II & III for multi-ply construction. Avoid placing and stapling paper flanges of insulation batts over framing faces, because the paper may cause poor nailing, which might cause eventual ridging and nail popping problems.

When selecting the wood for framing and furring, care should be taken to use only properly cured lumber. Excessive moisture in wood can cause warping as the wood dries. If warped or crooked studs and joists have been used, they should be replaced with straight lumber. Gypsum board cannot compensate for improper or misaligned framing.

When proper framing practices are followed, a firm, even structure for the application of gypsum board will result. Headers or lintels should be provided over openings to support structural loads and special construction should be provided

where required to support wall hung equipment and fixtures. The bath and shower areas are examples where special framing must be provided for grab bars and heavy fixtures. Provision should also be made for cabinets or wall hung appliances in the kitchen or utility areas.

Table I
Maximum Framing Spacing

Single ply Gypsum Board (Thickness)	Application to Framing	Maximum o.c. Spacing of Framing
Inches		Inches
Ceilings:		
†3/8	*Perpendicular	16
1/2	Perpendicular	16
1/2	*Parallel	16
5/8	Parallel	16
1/2	*Perpendicular	24
5/8	Perpendicular	24
Sidewalls:		
3/8	Perpendicular or Parallel	16
1/2	Perpendicular or Parallel	24
5/8	Perpendicular or Parallel	24

*On ceilings to receive a water base texture material, either hand or spray applied, install gypsum board perpendicular to framing and increase board thickness from 3/8 in. to 1/2 in. for 16 in. o.c. framing and from 1/2 in. to 5/8 in. for 24 in. o.c. framing.

†Should not support thermal insulation.

Table II
Two-Ply Application Without Adhesive Between Plies

Gypsum Board (Thickness)		Application to Framing		Maximum o.c. Spacing of Framing
Base	Face	Base	Face	
Inch	Inch			Inch
Ceilings:				
3/8	3/8	Perpend	Perpend*	16
1/2	3/8	Parallel	Perpend*	16
1/2	1/2	Parallel	Perpend	16
1/2	1/2	Perpend	Perpend*	24
5/8	1/2	Perpend	Perpend*	24
5/8	5/8	Perpend	Perpend*	24

Sidewalls:
For two-ply application with no adhesive between plies, 3/8 in., 1/2 in. or 5/8 in. thick gypsum board may be applied perpendicular or parallel on framing spaced a maximum of 24 in. o.c. Maximum spacing should be 16 in. o.c. when 3/8 in. thick board is used as face ply.

*On ceilings to receive a water base texture material, either hand or spray applied, install gypsum board perpendicular to framing and increase board thickness from 3/8 in. to 1/2 in. for 16 in. o.c. framing and from 1/2 in. to 5/8 in. for 24 in. o.c. framing.

Table III
Two-Ply Application With Adhesive Between Plies*

Gypsum Board (Thickness)		Application to Framing		Maximum o.c. Spacing of Framing
Base	Face	Base†	Face†	
Inch	Inch			Inch
Ceilings:				
3/8	3/8	Perp	Perp or Par	16
1/2	3/8	Perp or Par	Perp or Par	16
1/2	1/2	Perp or Par	Perp or Par	16
5/8	1/2	Par	Perp or Par	24
5/8	5/8	Perp or Par	Perp or Par	24

Sidewalls:
For two-ply application with adhesive between plies, 3/8 in., 1/2 in. or 5/8 in. thick gypsum board may be applied perpendicular or parallel on framing spaced a maximum of 24 in. o.c.

Adhesive between plies should be dried or cured prior to any decorative treatment. This is especially important when water based texture material (hand or spray applied) is to be used.

† Perp = perpendicular
  Par = parallel

Furring

Cross furring should be used to correct surface unevenness in the existing framing. The fastening surface of wood furring strips must be no less than 1½ in. actual dimension. In general, wood furring should not be less than nominal 2 × 2 to provide a rigid support during nailing when the wood furring strips are directly attached to the underlying framing. Where wood furring strips are attached to concrete or masonry walls and where screws are used to attach the gypsum board, wood furring may be nominal 1 × 3 (¾ in. minimum thickness.) The maximum spacing between furring strips should be the same as shown for framing members in Table I or Table II.

In wood framed construction where a higher degree of sound control is desired, gypsum board can be screw attached to resilient metal furring channels, laminated, or attached as specified in specific sound control assemblies.

Studs

Wood studs in load-bearing partitions usually are 2 × 4 or larger (can be 2 × 3 in double wall). In non-load-bearing single row stud or staggered stud partitions, 2 × 3 wood studs may be used.

Back-up framing or special clips should be provided at all interior corners for support or as a nailing base for the gypsum board.

Joists

Ceiling joists should be evenly spaced with faces aligned in a level plane. Excessively bowed or crooked joists should not be used. Joists with a slight crown may be used if they are installed with the crown up. Slightly crooked or bowed joists can sometimes be aligned by nailing bracing members (strong-back) across the joists approximately at mid span (Fig. 16).

Wood trusses to which a ceiling is to be attached might have irregularities in spacing and leveling. When wide variances are found, cross furring is recommended to provide a

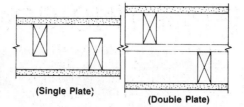

(Single Plate)   (Double Plate)

Fig. 15 Staggered stud partition.

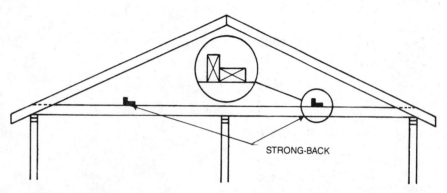

STRONG-BACK

Fig. 16 Slightly crooked or bowed joists or truss chords may be corrected by nailing stringers approximately at the mid spans. 2 in × 4 in. stringers, laid flat, regulate spacing, and other stringers, set on edge, accomplish some leveling.

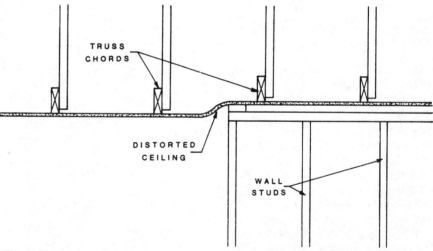

TRUSS CHORDS

DISTORTED CEILING

WALL STUDS

Fig. 17 Ceiling distortion can result under truss roofs where the weight of the roof forces the ceiling down while the interior partitions remain firmly in place. Non-load-bearing partitions should be installed only after truss roofs and ceilings are completed and the roof covering is finished.

level surface to support the gypsum board.

When trusses change direction in the middle of a room, proper blocking should be installed so that maximum spacing of framing does not exceed requirements indicated in Tables I, II & III with regard to board application.

With truss roof construction, the exterior walls and ceilings often are finished before interior partitions are

erected and finished. When this method is used, the roofing and all other construction elements which increase roof loads should be installed before interior partitions are erected. If substantial roof loads are introduced after partitions are installed, the ceiling may be forced down against the partitions and may be distorted as the roof trusses deflect (Fig. 17).

METAL STUD PARTITIONS AND CEILING SYSTEMS

The advantages of metal framing are noncombustibility, uniformity of dimension, lightness of weight, freedom from rot and moisture problems, and relative ease of erection. The components of metal frame systems are manufactured to fit together easily, generally are friction fit (no fasteners) except at doors, windows, etc.

There are a variety of metal framing systems, both load-bearing and non-load-bearing. Some non-load-bearing gypsum partition systems are designed to be demountable or movable and still meet requirements for sound isolation and fire resistance.[1]

Metal stud partition systems are commonly used in highrise and other buildings where noncombustible partition framing and furring are required (Fig. 18); use is also increasing for residential construction. Ceilings usually are applied to furring members secured to or suspended from open-web joists or light beams.[2]

The spacing for metal framing to receive gypsum board should not exceed the maximum spans recommended for single and multi-ply construction as shown in Tables I and II on page 8.

Fig. 18 Metal Studs go up quicky in this office building.

Studs

Typically, metal studs in non-load-bearing partition framing are "C" shaped, 25 gage steel (nominal), and have a protective coating to prevent corrosion in normal use. (Fig. 19). They are available in 1⅝, 2½, 3⅝, 4 or 6 inches deep and from 6 ft. to 16 ft. long.[3]

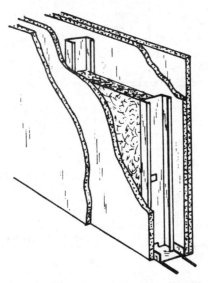

Fig. 19 "C" shaped steel studs placed in metal runners support this multi-ply, sound-insulating partition.

Joists

Open-web steel bar joists are not designed to receive wallboard directly, and they often are spaced more than 24 in. o.c. Hence, it is necessary to provide suitable furring or suspension systems with appropriate spacing for attaching the gypsum board ceiling.

Metal Furring or Framing Channels

There are four general types of metal furring channels: (Fig. 20) cold-rolled channels for suspension systems; furring channels used to space a gypsum board away from framing members; directly suspended interconnecting grid type channel systems; and resilient furring channels used to reduce sound attenuation through a horizontal or vertical assembly. Steel studs may also be used as furring channels where spans over 48 inches are desired. The minimum width for the fastening face of any metal furring channel must be 1¼ in. to accommodate abutting edges or ends of gypsum panels.

Cold-rolled channels may be used in furred partitions and in most types of suspended ceiling assemblies. Ordinarily, they are suspended by wire or rods with the furring channels tied or clipped to them. These channels are usually of 16 gage steel with either a galvanized or black asphaltum finish. They are available in sizes ranging from ¾ in. to 2 in. wide in lengths up to 20 ft.

---

[1] The Gypsum Association Fire Resistance Design Manual (GA-600) contains a wide variety of designs tested for fire resistance and sound control.

[2] Detailed specifications for application of gypsum board to metal framing are given in Gypsum Association Recommended Specifications for the Application and Finishing of Gypsum Board, GA-216 and in ASTM Standard C 754.

[3] Studs should be manufactured in conformance with ASTM Standard Specification for Non-Load (Axial) Bearing Steel Studs, Runners (track), and Rigid Furring Channels for Screw Application of Gypsum Board, C645.

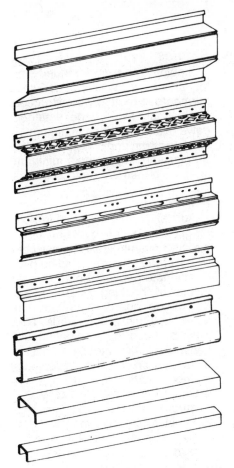

Fig. 20 Plain drywall channel (top) has a hat shaped cross section, while resilient drywall channels (center four) come in several designs. All are of galvanized steel. Cold-rolled channels (bottom two) are "C" shaped and generally of 16 gage black asphaltum painted steel.

Rigid furring channels are 25 gage (minimum) galvanized steel and are generally "hat" shaped. Rigid furring channels are designed for gypsum board attachment with gypsum drywall screws described on page 12. The furring channels should be installed at right angles to the furring supports. They are fastened to wood framing with 1¼ in. long screws or 1¾ in. 5d nails, and to metal framing with screws, wire ties or clips.

Rigid furring channels over masonry or concrete surfaces may be installed horizontally or vertically. Metal studs may be used in place of furring channels. The spacing should be as shown in Table I or Table II on page 8. (Fig. 20)

Suspended grid furring channels are grid type furring systems used below both metal and wood framing. They include directly suspended inverted T-shaped main runners (beams), typically 4' o.c., with interconnecting cross channels 16 in. or 24 in. o.c. (Figs. 20 & 21)

Resilient furring channels are used over both metal and wood framing to provide a sound absorbent spring mounting for gypsum board and should be attached as specified by the manufacturer. They not only improve sound isolation, but they also help to isolate the gypsum board from structural movement, minimizing the possibility of cracking. Resilient furring channels may also be used for the application of gypsum board over masonry or concrete walls. (Fig. 20)

FURRING CHANNEL DETAILS

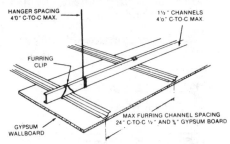

Fig. 21 In suspended gypsum board ceilings, drywall channels are clipped to suspended cold-rolled channels rather than to framing members. Various clips may be used.

ATTACHMENTS AND FASTENERS

Nails and screws are commonly used to attach gypsum board in both single and multi-ply installations; clips and staples are used only to attach the base layer in multi-ply construction.

Special drywall adhesives can be used to secure single-ply gypsum board to framing and furring, masonry and concrete, or to laminate a face ply to a base layer of gypsum board or other base material. Adhesives must be supplemented with mechanical fasteners.

Mechanical application of gypsum board requires special fasteners. To penetrate the board without damage, hold it tightly against framing and permit correct countersinking for proper fastener concealment.

Where fasteners are used at the board perimeter, they should be placed at least ⅜ in. from board edges and ends. Fastening should start in the middle of the board and proceed outward toward the perimeter. Fasteners should be driven as near to perpendicular as possible **while the board is held firmly against the supporting construction.** Nails should be driven with a crown-headed hammer which forms a uniform depression or "dimple" not more than ¹⁄₃₂ in. deep around the nail head (Fig. 22). Particular care should be taken not to break the face paper or crush the core with too heavy a blow.

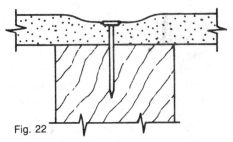

Fig. 22

Nails

Examples of acceptable nails for gypsum board application are show in Fig. 23. Preferably, the nails should have heads that are flat or concave and thin at the rim. The heads should be at least ¼ in. in diameter and not more than ⁵⁄₁₆ in. in diameter to provide adequate holding power without cutting the face paper when the nail is dimpled. The heads of casing nails and common nails are too small in relation to the shank; they easily cut into the face paper and should not be used. Nail heads that are too large are also likely to cut the paper surface if the nail is driven incorrectly at a slight angle.

Nails should be long enough to go through the wallboard layers and far enough into supporting construction to provide adequate holding power. Nail penetration into the framing member should be ⅞ in. for smooth shank nails, but only ¾ for annular ringed nails which provide more withdrawal resistance. For fire rated assemblies, greater penetration is required; generally 1 1/8 in. to 1 1/4 in. for one hour assemblies.

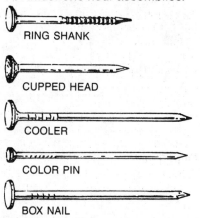

Fig. 23 Nails used in drywall construction have thin, 1/4 in. diameter heads (some cupped). Small-headed nails shown are colored for use with predecorated wallboard. These are used according to instructions of the manufacturers.

Screws

Drywall screws are used to attach gypsum board to wood or steel framing or to other gypsum boards. They have cupped Phillips head designed for use with a drywall power screwdriver. These screws pull the board tightly to the supports without damaging the board, and minimize fastener and surface defects due to loose boards. The specially contoured head when properly driven makes a uniform depression free of ragged edges.

The three basic types of drywall screws, Type W for wood, Type S for sheetmetal, and Type G for solid gypsum construction, are shown in Figure 24. Also shown are Type S-12 screws for attaching wallboard to heavier gage metal studs.

Type W gypsum drywall screws are designed for fastening gypsum board to wood framing or furring. Diamond-shaped points on Type W screws provide efficient drilling action through both gypsum and wood, and a specially designed thread gives quick penetration and increased holding power. Recommended minimum penetration into supporting construction is 5/8 in. However, in two-ply construction where the face layer is screw attached, additional holding power is developed in the base ply which permits reduced penetration into supports to 1/2 in. Type S *screws*

may be substituted for Type W in two-ply construction.

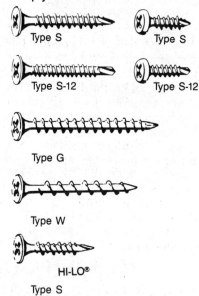

Fig. 24 Type S screws should penetrate at least 3/8 in. beyond gypsum board surface to hold tightly against metal framing. Type G screws should penetrate at least 3/8 in. into supporting layers of gypsum board to hold tightly, and Type W screws should penetrate at least 5/8 in. into wood framing under gypsum board.

Type S gypsum drywall screws are designed for fastening gypsum board to metal studs or furring. They are self-drilling, have a self-tapping thread and generally a mill-slot or hardened drill point designed to penetrate metal with little pressure. (Easy penetration is important because steel studs often are flexible and tend to bend away from the screws.)

Type G gypsum drywall screws are used for fastening gypsum board panels to gypsum backing boards. Type G are similar to Type W screws, but they have a deeper special thread design. They are generally 1 1/2 in. long, but other lengths are available. Gypsum drywall screws require penetration of at least 1/2 in. of the threaded portion into the supporting board.

Gypsum drywall screws should not be used to attach wallboard to 3/8 in. backing board because sufficient holding strength is not available. Nails or longer screws should be driven through both the surface layer and the 3/8 in. base ply to give the proper penentration in supporting wood or metal construction.

Staples

Staples are recommended only for attaching base ply to wood members in multi-ply construction. They should be 16 gage, flattened, galvanized wire with a minimum 7/16 in. wide crown and spreading points. Staples should provide a minimum of 5/8 in. penetration into supports.

Adhesives

Adhesives may be used to bond single layers of gypsum board directly to framing, furring, masonry, or concrete. They can also be used to laminate gypsum board to base layers of backing board, sound deadening board, rigid foam, or other rigid insulation boards. They must be used in combination with nails, staples (in mobile homes) or screws which provide supplemental support. Adhesives for applying wallboard finishes are classed:
1. stud adhesives
2. laminating adhesives:
   a. joint compound adhesives
   b. drywall contact and modified contact adhesives

Stud adhesives are specially prepared to adhere to single-ply wallboard to steel or wood supports and are used with adhesive nail-on application. Some permit even further reduction in mechanical fasteners, but these also require fastening at least at the board perimeters. They should be of caulking consistency so that they bridge framing irregularities. Also they should meet the requirements of the Standard Specification for Adhesives for Fastening Gypsum Wallboard to Wood Framing (ASTM Designation: C 557) for workability, consistency, open time, wetting characteristics, strength, bridging ability, aging and freeze-thaw resistance. They are applied with a gun (Fig. 25) in a continuous or semi-continuous bead. If the stud adhesive has a solvent base it should not be used near an open flame, in poorly ventilated areas, or for lamination of predecorated gypsum board. Special adhesives are available for predecorated gypsum board.

Joint compound adhesives are used to laminate gypsum boards to each other or to suitable masonry or concrete surfaces. They are not intended for adhesive bonding to wood framing or furring.

Mix only as much adhesive as can be used within the working time specified by the manufacturer. Water, if used, should be at room temperature and clean enough to drink. The adhesive may be applied over the entire board area with a suitable notched spreader, or it may be applied in spaced parallel ribbons or a pattern of spots as recommended by the manufacturer. All dry powder adhesives require permanent mechanical fasteners at the perimeter of the boards. If the board is applied vertically on sidewalls, fasteners are placed at the top and bottom. Face boards may require temporary support or supplemental fasteners until full bond strength is developed.

Fig. 25 Application to stud with an adhesive gun.

Drywall Contact and Modified Contact Adhesives: These adhesives require permanent mechanical fasteners at least at the perimeter of boards applied to walls, ceilings and soffits. If used to apply predecorated gypsum wallboard vertically on sidewalls, permanent mechanical fasteners are required only at the top and bottom of the boards where they will be concealed by base and ceiling moldings or other decorative trim. On ceilings, fasteners should be no further apart than 24 in. o.c. regardless of the type adhesive used. Contact adhesives may be used to laminate gypsum boards to each other or to metal studs. The adhesive is applied by roller, spray gun, or brush, in a thin, uniform coating to both surfaces to be bonded. For most contact adhesives, some dry-

ing time usually is required before surfaces can be joined and bond can be developed. To assure proper adhesion between mating surfaces, the face board should be impacted over its entire surface with a suitable tool, such as a rubber mallet. No temporary supports are needed while a contact adhesive sets and the bond forms. One disadvantage of contact adhesives is their inability to fill in irregularities between surfaces, which leaves areas without adhesive bond. Another disadvantage is that most of these adhesives do not permit moving the boards once contact has been made.

Extra care and judgment should be taken when contact adhesives

are used. Manufacturer's recommendations should be followed.

Modified contact adhesives provide a longer placement time. They have an open time of up to ½ hour during which the board can be repositioned if necessary. They combine good long-term strength with sufficient immediate bond to permit erection with a minimum of temporary fasteners. In addition, the adhesive has bridging ability.

Modified contact adhesive is intended for attaching wallboard to all kinds of supporting construction, including solid walls, other gypsum board, and various insulating boards, including rigid foam insulation.

MASONRY AND CONCRETE WALLS

Gypsum board panels can be laminated directly to above grade interior masonry and concrete wall surfaces if the surface is dry, smooth, clean and flat. Gypsum board can be laminated directly to exterior cavity walls if the cavities are properly insulated to prevent condensation and the inside face of the cavity is properly water proofed.

Predecorated gypsum board with a surface highly resistant to water vapor should not be laminated directly to concrete or masonry since moisture may become trapped within the gypsum core of the board.

The base surface must be made as level as possible. Rough or protruding edges and excess joint mortar must be removed and depressions filled with mortar to make the wall surface level.

Base surfaces should be cleaned of form oils, curing compounds, loose particles, dust, or grease to

assure adequate bond. Concrete should be allowed to cure for at least 28 days before gypsum board is laminated directly to it.

Exterior below-grade walls or surfaces should be furred and protected with the installation of a vapor barrier and insulation in order to provide a suitable base for attaching the gypsum board. This is also true for any surface which cannot be prepared readily for direct adhesive lamination.

Supplemental mechanical fasteners spaced 16 in. o.c. may be used to hold gypsum board in place while adhesive is developing bond.

A variety of clips, runners and adjustable brackets are available with furring systems to facilitate installation over irregular masonry walls. (Fig. 26) When special clips are used the manufacturer's instructions for their use should be followed.

ceiling attachments
rough or finished ceiling

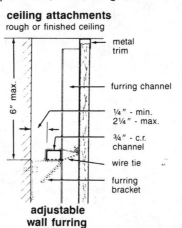

metal trim

furring channel

¼" - min.
2¼" - max.

¾" - c.r. channel

wire tie

furring bracket

adjustable wall furring

floor attachments

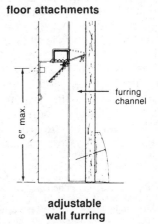

furring channel

6" max.

adjustable wall furring

Fig. 26 Adjustable brackets make installation of furring easy over irregular masonry walls.

SINGLE-PLY APPLICATION

The quality of single-ply surfaces is dependent on accurately aligned framing supports as described in Chapter II. Before a single-ply application of gypsum board is made, the framing or furring members should be checked for firmness and alignment (Fig. 14). Gypsum board is generally attached to framing through the use of nails, screws, and staples. Nails are most commonly used; however screws are often used in lieu of nails in wood construction because they are applied with automatic screw guns and have excellent holding qualities. Staples are used because they are economical and can be quickly applied with staple guns; however, their application is limited to the base-ply of two layers over wood framing.

Nail Attachment and Spacing

Gypsum board can be attached by either a single-nailing or double-nailing method. Double-nailing produces a tighter board-to-stud contact. Wherever fire resistive construction is required, nail spacing specified in the fire test should be followed.

Single nails should be spaced a maximum of 7 in. o.c. on ceilings and 8 in. o.c. on walls along framing supports (Fig. 27)[1]. Nails are first driven in the center or field of the board and then outward toward the edges and ends. In single-ply installation, all ends and edges of gypsum board are placed over framing members or other solid backing except where treated joints are at right angles to framing members.

In double-nailing, the spacing of the first set of nails is 12 in. o.c. with the second nailing 2 in. to 2½ in. from the first. (Fig. 28) The second set of nails is applied in the same sequence as the first set, but not on the perimeter of the board. The first nails driven should be reseated as necessary following application of the second set.

[1] For fire and sound rated construction, framing and fastener size and spacing must follow that of the system tested. Refer to Gypsum Association Fire Resistance Design Manual (GA-600) for details of fire resistive construction.

Attachment Procedures

1. Carefully measure and cut the board.
2. Prior to nailing, mark the gypsum board to indicate the location of the framing.
3. Hold board firmly against framing when nailing to avoid nail pops or protrusions.
4. Drive nails straight into the member. Nails that miss the supports should be removed, the nail hole dimpled, and then covered with joint compound.
5. Damage to the board caused by over driving nails is corrected by driving a new nail 2 in. away to provide firm attachment. Damage to the board should be repaired with joint treatment compound.

Examples of correctly and incorrectly driven nails are shown in Figures 29 and 30. Other common causes of face paper fractures are misaligned or twisted supporting framing members and projections of improperly installed blocking or bracing. (Fig. 14). These framing faults prevent solid contact between gypsum board and supporting members, and hammer impact causes the board to rebound and rupture the paper. Defective framing should be corrected prior to application of gypsum board. Protruding supporting members should be trimmed or reinstalled. The use of screws, adhesives or two-ply construction may minimize problems resulting from these defects.

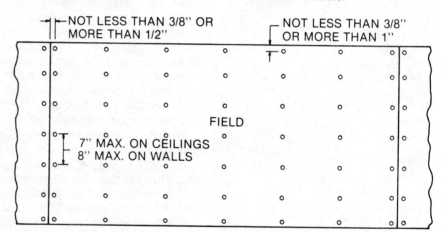

Fig. 27 In single nailed gypsum board, nails are spread 7 to 8 in. apart along supports and not less than ⅜ in. from edges and ends. Nails are first driven in the middle area of the board and then out toward the edges and ends.

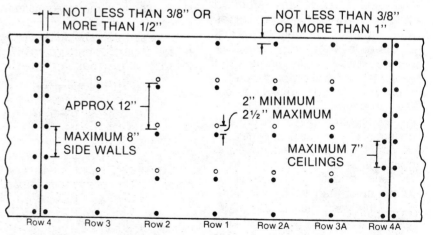

Fig. 28 In double nailing, first nails, represented by black dots, are driven starting with row 1, then row 2 and 2A, working toward the ends. Second nails, represented by the circles, are then driven in the same sequence.

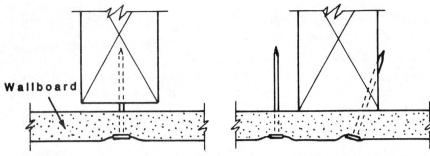

Fig. 29 Loose nails are caused by poor fit of gypsum board to framing surface (left) or by nails missing the framing members when driven (right).

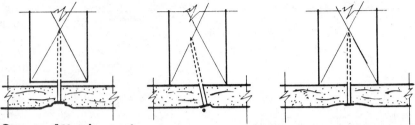

Screw Attachment

Because fewer fasteners are required when screws are used to attach gypsum board, the number to be treated is reduced and possible application defects minimized.

Screws should be spaced 12 in. o.c. on ceilings and 16 in. o.c. on walls where framing members are 16 in. o.c. Screws should be spaced a maximum of 12 in. o.c. on walls and ceilings where framing members are 24 in. o.c. The necessary penetration for screws is given in Figure 24.

Gypsum board should be attached to metal framing and furring with Type S screws spaced not more than 12 in. o.c along supports for both walls and ceilings. Type S-12 screws are required for metal framing 20 gage or heavier. These screw spacings are appropriate also when gypsum board is mounted on resilient furring channels over wood framing.[1]

Floating Interior Angle Construction

To minimize the possibility of fastener-popping in areas adjacent to the wall and ceiling intersection, and to minimize cracking due to structural stresses, the floating angle method may be used for either single or double layer application of gypsum board to wood framing. This method is applicable for single nailing, double nailing or screw attachment. The same nail-free clearances at corners should be maintained in double nailing. See fig. 31.

Fig. 30 Nails driven in loose board will probably fracture the face paper (top). Nails driven at an angle also usually fracture the face paper (center). Properly driven nails produce a tight attachment with a slight, uniform dimple (bottom).

In floating interior angle construction where the ceiling framing members are perpendicular to the wall/ceiling intersection, the ceiling fasteners should be located 7 in. from the intersection for single nailing and 11 to 12 in. for double nailing or screw application. On ceilings, where the joists are parallel to the intersection with a wall, nailing should start at the intersection. Gypsum board should be applied to ceilings first and then to walls.

Gypsum board on sidewalls should be applied to provide a firm, level support for the floating edges of the ceiling board. The top attachment into each stud should be located 8 in. down from the ceiling intersection for single nailing, and 11 to 12 in. for double nailing or screw applications. (Figs. 32a and 32b) At verticle angles, (Fig. 32c) apply the overlapping board firmly against the underlying board to bring the underlying board into firm contact with the face of the framing member behind it. The overlapping board should be nailed or screwed and fasteners omitted from the underlying board at the vertical intersection.

Adhesive Nail-On Attachment

An adhesive nail-on system has been developed for the application of gypsum board to wood framing.

The advantages of this system are that the number of nails can be reduced by 50 percent or more, a continuous bond between the gypsum board and the framing is provided, and a stronger assembly with fewer fasteners is the result. The adhesive also serves to bridge minor framing irregularities.

Stud Adhesives

Stud adhesives should be applied with a caulking gun in accordance with the manufacturer's recommendations. A straight bead, approximately $3/8$ in. diameter, is applied to the face of supports in the field of the panel (Fig. 33). Where two gypsum panels abut over a supporting member, two parallel beads of adhesive should be applied, one near each edge of the member. The adhesive should not squeeze out at the joints. Zig-zag or serpentine beads should be avoided under abutting boards because the adhesive can be forced out at the joint when boards are pressed tightly to the framing.

In the adhesive application of gypsum board to wall framing, supplemental fasteners are used in the perimeter of the board. The fasteners should be spaced 16 in. o.c. along edges or ends that fall on parallel supports, and at each point where edges or ends cross perpendicular supports. For ceiling application, supplemental fasteners are required in the perimeter of the gypsum board (the same as for walls) and in the field 24 in. o.c. (Fig. 34). Adhesive is not required at inside corners, top or bottom plates, bridging, bracing or fire stops.

Where fasteners at vertical joints are undesirable, gypsum panels may be prebowed and adhesively attached to the framing. Supplemental fasteners 16 in. o.c. are then used at the top and bottom plates. Prebow gypsum board by stacking face up with ends resting on 2 × 4 lumber or other blocks and with center of boards resting on floor. Allow to remain overnight or until boards have a 2 in. permanent bow. Predecorated panels can be installed in this manner but care should be taken to avoid adhesive contact to the decorated face. Position within the open time specified for the adhesive and use a rubber mallet to tap the gypsum board along the studs to assure continuous bond of the board to the framing. Follow the manufacturer's specifications for predecorated gypsum board.

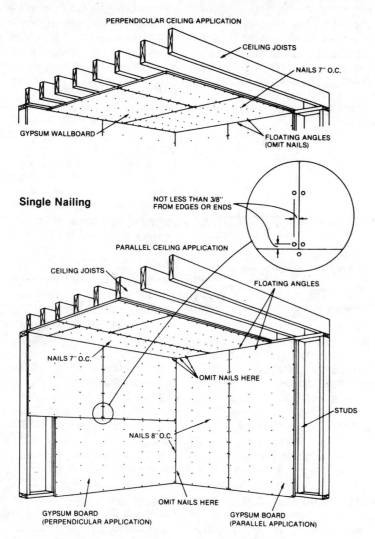

Fig. 31 Floating angle construction helps eliminate nail popping and corner cracking. Fasteners at the intersections of walls or ceilings are omitted.

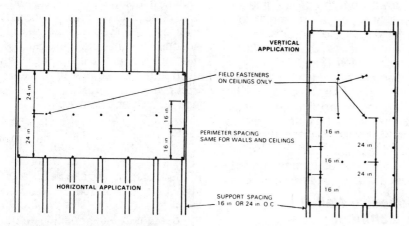

Fig. 34 Single-ply gypsum board systems attached with studs adhesives require supplemental fasteners on the perimeter. Spacing is the same whether board edges are perpendicular or parallel to supports. Ceiling installations require supplemental fasteners in the field as well as on the perimeter.

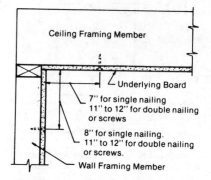

Fig. 32a Vertical section, ceiling framing member perpendicular to wall

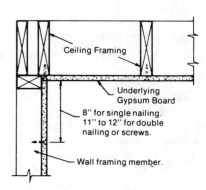

Fig. 32b Vertical section, ceiling framing parallel to wall.

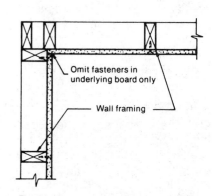

Fig. 32c Cross section through interior vertical angle.

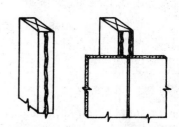

Fig. 33 Beads of stud adhesive are applied (left) straight under the field of a board, and (right) parallel under predecorated joints; or joints to be taped.

MULTI-PLY APPLICATION

Multi-ply construction has one or more layers of gypsum board applied over a base layer. This results in improved surface finish, greater strength and higher fire resistance and sound classifications. The base layer can be a gypsum backing board, with or without foil, regular gypsum board or other base material.

The maximum support spacing for multi-ply systems depends mainly upon the base-ply thickness and placement. Framing spacing for wood and metal framing or furring is given in Tables I and II on page 8. Table IV on this page shows fastener spacing for the base-ply attachment.

TABLE IV

BASE-PLY FASTENER SPACING[1] FOR MULTI-LAYER APPLICATION

	NAIL SPACING		SCREW SPACING		STAPLE SPACING	
	Face Ply Laminated	Face Ply[3] Nailed	Face Ply[2] Laminated	Face Ply[3] Screwed	Face Ply Laminated	Face Ply[3] Nailed or Screwed
Walls	8 in. o.c.	16 in. o.c.	16 in. o.c.	24 in. o.c.	7 in. o.c.	16 in. o.c.
Ceilings	7 in. o.c.	16 in. o.c.	16 in. o.c.	24 in. o.c.	7 in. o.c.	16 in. o.c.

[1] Fastener size and spacing for applying boards for fire and sound rated constructions may vary from these recommendations. The manufacturer's recommendations should be followed.

[2] 12 in. o.c for both ceilings and walls when supports are spaced 24 in. o.c.

[3] Fastener spacing for face ply shall be the same as for single layer application.

INSTALLATION OVER WOOD FRAMING AND FURRING

Base ply ceilings

When a multi-ply system with a laminated face ply is to be used over wood supports, the base ply should be fastened as recommended for single-ply construction. Double nailing is not needed because the fasteners used on a two-ply application will produce a firmly fastened system. The base ply should be applied with long edges perpendicular or parallel to framing members. End joints may occur on or between framing members. Face-ply joints, however, should occur over framing and be offset from base ply joints. If the base ply is foil backed board, apply the foil side against framing.

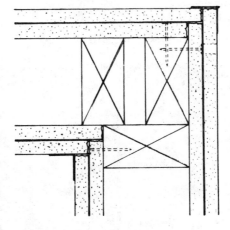

Fig. 35 Floating angle construction for multi-ply systems has overlapping side of base ply only nailed at interior corner.

Base ply walls

The base ply should be applied with the long edges parallel with the framing members unless tested assembly specifies otherwise. When predecorated face panels are to be used, the base ply should be applied perpendicular to the framing. At inside corners, it is recommended that only the overlapping base board end be nailed or screwed and that fasteners be omitted from the face ply. The floating corner treatment is better able to resist structural stresses. (Fig. 35)

Staples used to fasten base plies should be driven with the crown perpendicular to the edges of the board. Where edges fall over supports, staples should be driven parallel with the edges. (Fig. 36)

Crowns of driven staples should bear tightly against the board without breaking the face paper.

Base Ply Attachment

Base-ply gypsum board is normally attached to metal framing and furring with screws at least $5/16$ in. longer than the thickness of the board. Application may be either perpendicular or parallel except where fire ratings are involved. In that event board should be applied as tested by the manufacturer. In perpendicular application where no adhesive is used between plies, the base ply should be fastened with a single screw into each stud or furring channel at the board edges and with one screw at the middle of the board at each stud or channel.

In parallel application with no adhesive between plies, the base ply is fastened with screws 12 in. o.c. along the edges of the board and 24 in. o.c. in each stud in the field of the board.

When the base ply is to be attached either perpendicular or parallel to metal framing 16 in. o.c. and with adhesive between the plies, screw spacing should be 12 in. o.c. for ceilings and 16 in. o.c. for walls. Maximum screw spacing on metal framing at 24 in. o.c. is 12 in. o.c. for both walls and ceilings.

Face Ply Attachment

The joints in the face ply should be offset at least 10 inches from joints in the base ply. Gypsum board can be applied to framing either perpendicularly or parallel, whichever results in the least waste of materials. Perpendicular application is preferred on walls since it usually results in fewer joints. When the face ply is attached with mechanical fasteners and with no adhesive between plies, the maximum spacing and minimum penetration recommended for screws should be the same as for single ply application. These are given in Chapter III under "Screw Attachment." Predecorated board is normally installed parallel and does not require joint treatment. Some systems are designed to utilize decorative battens over the joints.

Adhesive Attachment

Typical multi-ply construction may employ sheet lamination, strip lamination, or spot lamination to at-

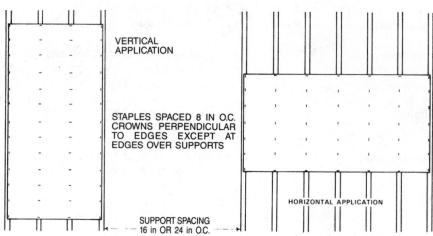

Fig. 36 Staples are driven with crowns perpendicular to gypsum board edges except where edges fall along supporting members.

METAL FRAMING OR FURRING

tach the face-ply to the base-ply.

Sheet lamination involves covering the entire back of the face-ply with laminating adhesive using a notched spreader, box spreader, or other suitable tool (Fig. 37). The size and spacing of the notches are determined by the type of adhesive being used. The gypsum board should be erected using moderate pressure, and any adhesive squeezed out at the joints should be promptly removed.

Strip and spot lamination are preferred to sheet lamination in sound rated partitions. In strip lamination, adhesive is applied in ribbons with a special spreader. The ribbons are normally spaced 16 in. to 24 in. o.c.

(Fig. 38). In spot lamination, spots of adhesive are brushed or daubed on in a regular pattern.

Supplemental fasteners

In order to assure satisfactory adhesive bond it is necessary to hold the face ply firmly against the base ply with supplemental fasteners, shoring or bracing while the adhesive is setting. Generally, mechanical fasteners are applied in the field of each laminated face ply in ceiling applications. On side walls these fasteners can be placed at the perimeter of the board where they will be concealed by joint treatment or trim. It should be remembered that in fire rated assemblies the

Fig. 37 Notched spreaders are used to apply adhesive for sheet lamination.

specific fastener spacing is given for the particular assembly tested and may not be related to whether or not adhesive is used between the plies. Fastener spacing details in fire rated assemblies are available from the sponsor of the test and, in many cases from the Gypsum Association. In sound-rated partitions where fire resistance is not a consideration, an acceptable practice is to attach the face ply vertically over sound insulation board or backing board with permanent mechanical fasteners in the board ends. Intermediate fasteners are omitted and panels temporarily braced until the adhesive has developed sufficient bond strength.

Fig. 38 Notched spreaders may also be used after base layer of board has been installed.

PREDECORATED COMBUSTIBLE PANELING OVER GYPSUM SUBSTRATE

The use of gypsum board as a substrate when applying wood paneling provides increased fire resistance and sound control. In new or existing construction a ⅜ in. or ½ in. gypsum board substrate is recommended before applying combustible paneling. (Fig. 39)

Data have been developed to show the increased fire resistance, sound control, and impact resistance achieved through the use of a gypsum substrate under combustible paneling. Examples are shown in Table V on page 22.

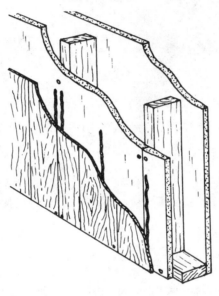

Fig. 39 Predecorated combustible paneling over gypsum substrate.

Application

The gypsum board substrate should be attached parallel to framing using 1⅜ in. drywall nails, 1 in. drywall screws or a drywall stud adhesive. The spacing of fasteners should be the same as given in Chapter II. The edges of the board should be centered on framing members. The joints need not be taped and finished.

The predecorated panels should be applied using a bead of panel adhesive over each stud and a bead midway between the studs. Joints in the wood paneling should be staggered from the joints in the base ply of gypsum board. Secure the paneling at the top and bottom with 4d finishing nails, 12 in. o.c. and with one nail at mid-height per stud.

TABLE V	¼" Paneling (No Gypsum Board)	¼" Paneling With ⅜ in. Gypsum Board	¼" Paneling With ½ in. Gypsum Board
Burn thru Time	8 min.	42 min.	73 min. (plus)
Better Sound Rating	28 STC	40 STC	40 STC
Better Impact Resistance	130 ft-lb	410 ft-lb	410 ft-lb

RESURFACING EXISTING CONSTRUCTION

Gypsum board may be used to provide a new finish on existing walls and ceilings of wood, plaster, masonry or wallboard. If the existing surface is structurally sound and provides a sufficiently smooth and solid backing without shimming, ¼ in. or thicker gypsum board can be applied with adhesives, nails or screws. Drywall nails shall be of sufficient length to penetrate framing ⅞ in. When power driven screws are to be used the threaded portion of the screws must penetrate the framing ⅝ in.

Existing surfaces which are too irregular to receive gypsum board directly should be furred and shimmed to provide a suitable fastening surface. Minimum gypsum board thickness for various support spacings and installation methods should be as recommended for new construction over furring.

(See Chapter II.)[1]

Surface trim for mechanical and electrical equipment such as switch plates, outlet covers, and ventilating grilles should be removed and saved for reinstallation. Electrical boxes should be reset prior to the installation of new gypsum board.

[1] See also Gypsum Association publication *Covering Existing Walls and Ceilings with Gypsum Board Products* (GA-660).

JOINT AND FASTENER TREATMENT

After the gypsum board is installed and secured with the proper fasteners, it is necessary to reinforce and conceal the joints, fasteners and corner beads. Joint compound is used for this purpose to achieve the appearance of a monolithic surface. Reinforcing tape and joint compound are the products recommended. They should conform to ASTM C 475 "Standard Specifications for Joint Treatment Materials for Gypsum Wallboard Construction." All compounds used should be compatible. Do not mix different formulations unless recommended by the joint compound manufacturer. During the taping and finishing operations, adequate and continuous ventilation should be provided to insure proper setting and drying of the taping and finishing compounds. Precreased tape is available and should be applied in the corners of wall and ceiling intersections. Exposed edges of the gypsum board may require some type of metal casing or appropriate trim.

Fig. 40 Gypsum board walls and ceilings usually have joints treated with tape and compound to give a smooth, monolithic appearance.

JOINT TREATMENT PRODUCTS AND PRECAUTIONS

Joint tape is designed for use with ready mixed or dry joint compounds to reinforce and finish the joints between adjacent gypsum boards. It may consist of paper, either perforated or unperforated, glass mesh, or other material. Paper tape is available with metal strips which reinforce exterior corners.

Joint reinforcing compounds are of three general types:
1. a taping or bedding compound used to adhere the tape to board
2. a finishing or topping compound used especially for finishing
3. an all-purpose compound to be used for both embedding and finishing.

Most joint compounds contain water-soluble dispersible organic adhesives or synthetic resins. These products gain their strength and adhesion through drying (drying compounds). The loss of water is accompanied by shrinkage which is overcome by several thin applications of the compound. Each application should be thoroughly dry before the next application is started. Synthetic resin compounds such as "vinyls" will keep longer than the organic water soluble types. Another family of joint compounds gain strength by setting. The set may occur within 30 minutes or take as long as several hours. The quick-set type, having a shorter working time, must be used within the prescribed time limit. (Additional coats are possible before complete drying takes place.) It is common for setting compounds to be used for embedding the tape and "non-setting" types for the finishing operation.

Joint compounds of the pre-mixed type are also available and in two consistencies: one for hand application and the other for machine application. These compounds should not be allowed to freeze. Taping and finishing operations should not be started until the interior temperature has been maintained at a minimum of 50 F for a period of at least 24 hours. The temperature should be maintained until the compounds have completely dried.

Care should be taken not to contaminate containers or tools used for different types of joint compounds when mixing or storing. Even a small quantity of one type of joint compound in the seam of a mixing pail or inside a pump or on a tool can adversely affect the adhesive properties of a full mixture of another type of compound. All equipment *must* be clean. Tools should be disassembled and cleaned after each operation.

JOINT TREATMENT PROCEDURE

The depression formed by the tapered edges where gypsum boards join should be filled with joint embedding compound. Topping or finishing compound should not be used for embedding tape. Wipe off excess compound that is applied beyond the groove. Center reinforcing tape and press it down into the joint compound with a 4 in. broad knife or smooth-edged trowel. There should be sufficient compound under tape for a proper bond but not more than $1/32$ in. under the feathered edge of the tape.

The initial tape embedding can also be done with a semi-automatic tool that applies the joint compound and tape simultaneously. Allow the compound to dry completely. Drying can take 24 hours or more, depending on temperature and humidity.

After the embedding coat is completely dry, apply a second coat feathered about two inches beyond edges of the first coat. Spot fastener heads and allow to dry. On the end, or field cut board joint where there are no tapered edges, an application of joint compound about 12 to 16 inches wide is required to conceal the tape.

After the second coat is dry, sand lightly or wipe with a damp sponge and apply a thin finishing coat to joints and fastener heads. Feather edges to at least 6 inches on each side of the joint. the final step prior to decoration is to lightly sand the joints in order to eliminate laps where joints intersect, and in general to smooth the surface where necessary. Care should be taken to avoid scuffing the paper surface of the gypsum board as the scuffed areas may be visible after the decoration. All cut-outs should be backfilled with compound used for taping or finishing so that there is no opening larger than $1/8$ in. between the gypsum board and a fixture or receptor.

Approved protective respirators should be worn when mixing powder or when sanding. Mixing should be done according to the manufacturer's directions.

Texturing

It is often desired that the finished surface have a textured appearance. When gypsum board surfaces are to be textured, they must first be coated with a high quality primer sealer. Textures with and without aggregate are available to impart surface effects ranging from light stipples to heavy swirls or simulated acoustical finishes. These are applied by roller, brush or spray machine.

Trim and Casings

Figure 47 illustrates some of the more common sections used around doors, windows and other openings. They are also used when gypsum board is butted against a different surfacing material.

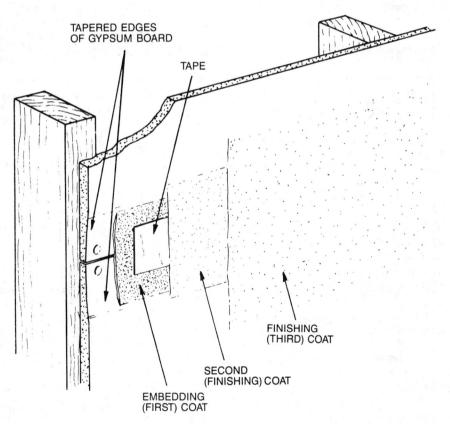

TAPERED EDGES OF GYPSUM BOARD

TAPE

FINISHING (THIRD) COAT

SECOND (FINISHING) COAT

EMBEDDING (FIRST) COAT

Fig. 41 Reinforcing joints with tape prevents cracks from appearing at filled gypsum board joints. The joint fill and first coat may be joint compound or all purpose compound. The second and third coats should be finishing compound or all-purpose compound.

Fig. 42 Tape bedding compound being applied to tapered edges of wallboard joint.

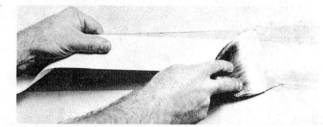

Fig. 43 Joint reinforcement tape being applied into bedding compound.

Fig. 44 Joint topping coats are applied

Fig. 45 Machine application of joint bedding compound and reinforcing tape.

Fig. 46 A variety of textured appearances can be obtained by using a spray machine.

Fig. 47

Gypsum Wallboard Corner Beads and Trim

Cornerbead — (Numbers indicate width of flanges, i.e. — 118 is 1⅛ in. wide flange)

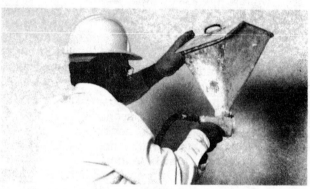

CB — 100 × 100
CB — 118 × 118
CB — 114 × 114
CB — 100 × 114
CB — PF (Paper flange, stzel corner
      combination bead)

"U" Bead — (Numbers indicate thickness of board to be used, i.e. — 38 is ⅜ in.)
U-38
U-12
U-58
U-34

Control Joint

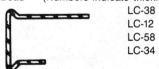

Removable Strip

Dimensions indicated are in U.S. Customary units.

"L" Bead — (Numbers indicate thickness of board to be used)
L-38
L-12
L-58
L-34

"LK" Bead — (For use with Kerfed jamb)
LK-B
LK-S
(B-Bull or round nose)
(S-Square nose)

"LC" Bead — (Numbers indicate thickness of board to be used.)
LC-38
LC-12
LC-58
LC-34

PROBLEMS AND REMEDIES

Nail Pops

Nail popping can be reduced by using the recommended drywall nails (Fig. 23 on page 12). Nails must be properly driven into correctly aligned framing free from excessive moisture. If nail popping does occur, the remedy is to drive another nail within 2 in. of the popped nail. The defective fastener should be reset using a nail set and the surface then repaired with joint treatment compound.

Cracking

Movement of the structure can impose severe stresses and cause cracks either at the joint or in the field of the board. Cracks are more likely to occur at an archway or over a door since this is usually the weakest point in the frame construction. In new construction, it is wise to wait until at least one heating season has passed before repairing or refinishing. The best repair method for cracks over ⅛ in. in width is to tape the crack and finish in the same manner as a conventional joint. It will be necessary to feather the joint out rather wide in order to hide the tape and produce a level surface. Minor cracks can be spackled, sanded and redecorated.

A source of cracking in nonbearing walls of highrise or commercial buildings, is the modern trend toward less rigid structures. Larger deflections in structural members and greater expansion and contraction of exterior columns can impose unexpected loads on nonbearing walls and lead to cracking. Detailed designs for perimeter relief of nonbearing partitions are available to improve this condition. (Fig. 49). The use of relief runners to attach nonbearing walls to ceiling and column members is effective in the prevention of cracking due to structural movement.

Shadowing

Joint and fastener head shadowing (show-through) can be due to any one of several causes. If the joint is crowned (higher than the plane of the board on each side) it can be seen in a low angle light. All joints should be finished in such a manner that the center of the joint is only slightly higher than the plane of the gypsum board surface and finished with a wide, flat, feathered edge. The remedy for a high or crowned joint is to widen the joint out to 14 in. to 16 in. or as necessary.

Another type of shadowing results from the difference in texture and porosity between gypsum board paper and joint compound materials. These problems can be significantly reduced by:
1. Making sure the joint compound is completely dry.
2. Ensuring that the entire wall surface has been sealed with a high quality primer/sealer.
3. Using a decoration that results in a light stipple or texture finish.

Beading or Ridging

Ridging, also called beading, is a uniform, fine line deformation at gypsum board joints. Factors that contribute to this condition include the use of wet lumber, prolonged high humidity and structural movement of framing members.

When this condition develops, it is advisable to wait through a complete heating cycle before repairing. This will permit the wallboard system to stabilize. Time repair during warm, dry weather.

Lightly sand the ridge down taking care not to damage the embedded joint reinforcing tape. Fill the surface over the joint with taping compound as wide as necessary to create an essentially plane surface. After 24 hours of drying, light sanding may be necessary to feather edges and remove trowel marks. If examination of the joint with strong sidelighting indicates the ridge is not concealed, additional feathering coats of joint compound may be required. The repaired joint should receive a coat of primer paint before any further decorative finish is applied.

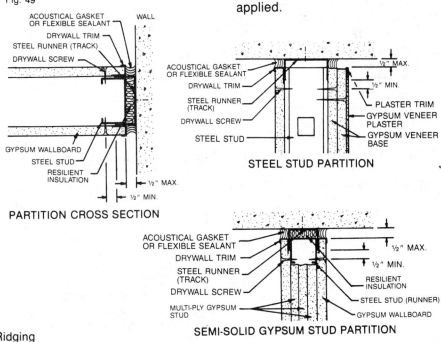

Fig. 49

PARTITION CROSS SECTION

STEEL STUD PARTITION

SEMI-SOLID GYPSUM STUD PARTITION

Fig. 48

Beading or Ridging

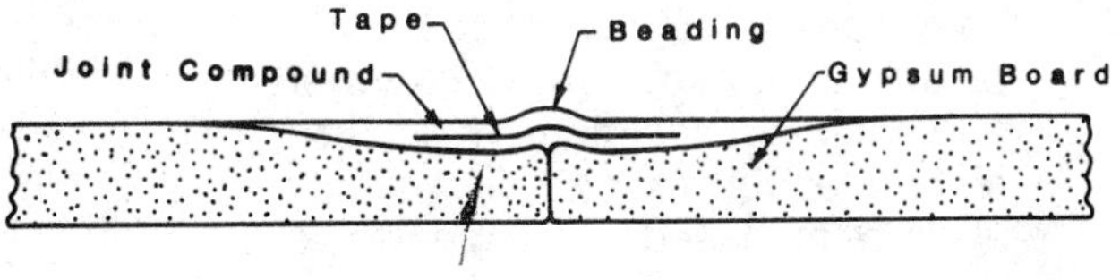

DECORATING

Predecorated Gypsum Board

Patterns or colors on some kinds of predecorated gypsum board may vary slightly from lot to lot. All the board used in a room should be from the same lot, but when this is not possible, the joint between two different lots should be in a corner.

Care should be taken to match the patterns and colors on adjacent pieces of predecorated gypsum board. One procedure is to simply stand the boards in the places they are to occupy. Boards that do not match well can be repositioned. If joints are to be covered with trim, plan the layout in advance so that a balanced pattern of joints is obtained. Avoid the appearance of two trimmed joints close together in a corner. If the stud spacing is not proper to allow an attractive arrangement of joints, furring strips may be necessary so that joint locations may be changed.

Cut edges that are to be covered with trim (moldings or battens) can be sawed or scored and snapped the same as regular gypsum board. If cut edge joints are to be exposed, however, the decorative facing should be cut approximately 1 in. wider than the gypsum core. The core should then be snapped and removed from the fabric — not the fabric from the core. The core should be trimmed straight and beveled on the face side under the flap. The overlapping face material is then pulled around the edges and fastened to the back with adhesive after removing any remaining paper in the flap.

Gypsum board finishes should be reinspected prior to decorating, particularly at joints and fasteners. Any imperfections should be sanded or repaired. Joint compounds must be thoroughly dry before the decoration is applied.

Gypsum board surfaces can be decorated with paint, paper, textures, fabric or vinyl wall coverings. Before decorating, always seal or prime the surface. Use a high quality alkyd or oil-based primer/sealer and allow to completely dry. Color or surface variations are thereby minimized and a more uniform texture for any surface covering is provided. The sealer allows the wall coverings to be removed more easily without marring the surface. Glue-size, shellac and varnish are not suitable as sealers or primers.

Paints may be spray, brush, or roller applied. The final coat can be textured if desired.

For Glossy Paints

After all irregularities have been eliminated and the joint treatment surfaces sanded or sponged where required, a thin skim coat of joint compound shall be applied to the entire surface of the board. This will minimize any suction, porosity or other variations between the joint compound and the face paper surfaces. Caution should be taken to eliminate laps or tool marks in the skim coating operation. The wall or ceiling surface should be lightly sanded or sponged where required to assure a smooth and even surface. A good quality, alkyd or oil-based primer/sealer should be applied prior to decoration.

For Severe Lighting Conditions

Where the wall or ceiling surfaces will be subjected to severe natural or artificial side lighting, it is recommended that a thin skim coat of joint compound be applied to the entire board surface to improve fastener and joint concealment. It is also recommended that a good quality alkyd or oil-based primer/sealer be applied prior to decoration.

NOTE: When using adhesives for the application of vinyl covered gypsum board, read and follow the gypsum board manufacturer's instructions. Improperly applied, or incompatible adhesives can damage vinyl finishes.

Most special constructions which are designed to improve sound, fire, moisture and thermal conditions depend on gypsum board assemblies built to specifications designed to accomplish the higher classifications. These include proper framing techniques, fasteners, caulking, taping and finishing of joints. In each case, the manufacturer's specific installation recommendations should be closely followed for individual systems.

SOUND ISOLATION CONSTRUCTION

The first essential for airborne sound isolation of any assembly is to close off air leaks and flanking paths. Since noise can travel over, under, or around walls, through windows and doors adjacent to them, through air ducts and through floors and crawl spaces below, these flanking paths must be correctly treated to reduce the transmission of sound. Hairline cracks and small holes will increase the transmission of sound. This can have a detrimental effect on the overall acoustical performance affecting the Sound Transmission Class (STC), particularly in higher rated assemblies. Where a very high STC performance is needed, air conditioning, heating and ventilating ducts should not be included in the assembly. Failure to observe special construction and design details can destroy the effectiveness of the best assembly.

Improved sound isolation is obtained by:

- Separate framing of the two sides of the wall
- Resilient channel mounting for the gypsum board
- Including sound absorbing materials in the wall cavity
- Using adhesively applied gypsum board of varying thicknesses in multi-layer construction
- Caulking the perimeter of gypsum board partitions, openings in walls and ceilings, partition-mullion intersections, outlet box openings, etc.
- Locating recessed wall fixtures in different stud cavities. (Medicine cabinets, electrical, telephone, television and intercom outlets,

plumbing, heating and air conditioning ducts should not be installed back to back.)

Any opening for such fixtures, piping and electrical outlets should be carefully cut to proper size and caulked. The entire perimeter of sound insulating partitions should be caulked around gypsum board edges to make it airtight as detailed in Figure 49 on pg. 30. The caulking should be nonhardening, nonshrinking, nonbleeding, nonstaining, resilient sealant. Sound control sealing must be covered in the specifications, understood by the workmen of all related trades, supervised by the foreman, and inspected carefully as construction progresses.

Separated Partitions

A staggered wood stud gypsum partition placed on separate plates will effectively decouple the system and provide an STC of 40-42. The addition of a sound absorbing material between the studs of one partition side can increase the STC as much as 8 points.

For the attachment of kitchen cabinets, lavatories, ceramic tile, medicine cabinets and other fixtures, a staggered stud wall or a metal stud chase wall rather than a resiliently mounted wall is recommended. The added weight and fastenings will acoustically "short circuit" a resiliently mounted wall.

Resilient Mounting

Resilient attachments, acting as

"shock absorbers," reduce passage of sound through the wall or ceiling and increase the STC rating. Further STC increases can result from more complex construction methods such as separated partitions, multiple layers of gypsum board, and sound absorbing materials.

Walls

Resilient furring channels are attached with the nailing flange down and at right angles to the wood studs (Fig. 50c). Drive 1¼ in. Type W or S screws or 6d coated nails through the pre-punched holes in channel flange. With extremely hard lumber, ⅞ in. or 1 in. Type S screws may be used. Locate channels 24 in. from the floor, within 6 in. of the ceiling line and no more than 24 in. o.c. Extend channels into all corners and fasten to corner framing. Attach a resilient channel or ½ in. by 3 in. wide continuous gypsum board filler strips to the bottom plate at floor line. Channels should be spliced directly over studs by overlapping ends and fastening both flanges to the support.

Apply gypsum board horizontally with long dimension parallel to channels using 1 in. Type S screws spaced 12 in. o.c. along channels. Abutting edges of board should be centered over channel flange and securely fastened.

Ceilings

Resilient furring channels are attached at right angles to wood joists

SOUND ISOLATION CONSTRUCTION — *(Cont.)*

in ceilings using 1¼ in. Type W or Type S screws or 6d coated nails 1⅞ in. long. Locate channels within 6 in. of the wall-ceiling intersection and no more than 24 in. o.c. for joists spaced 24 in. o.c. maximum. Extend the channel into all corners and fasten to corner framing. Channels should be spliced directly under joists by nesting channel and screwing through both flanges to support.

Apply the gypsum board with Type S screws spaced 12 in. o.c in the field and along abutting ends. Apply the long dimension of the board at right angles to the channels with end joints neatly fitted and staggered in alternate rows. Pieces of resilient channel should be used for back blocking butt joints not falling on furring members.

Sound Isolating Materials

- Mineral (including glass) fiber blankets and batts used in wood stud assemblies.
- Semi-rigid mineral or glass fiber blankets for use with metal studs.
- Mineral (including glass) fiber board.
- Gypsum core sound insulating board used behind gypsum board applied adhesively or mechanically fastened.
- Rigid plastic foam board used in exterior wall furring systems.
- Lead or other special shielding material.

Mineral wool or glass fiber insulating batts and blankets may be used in assembly cavities to absorb airborne sound within the cavity. They should be placed in the cavity and carefully fitted behind electrical outlets, around blocking and fixtures and around cutouts necessary for plumbing lines.

Insulating batts and blankets may be faced with paper or other vapor barrier and may have flanges or be unfaced friction fitted. They are installed by stapling or fitting to the inside surfaces of studs. Avoid fastening to the face of studs in order to assure good board-to-stud contact. In metal framed and in laminated all-gypsum board partitions, the blankets are attached to the back of the gypsum board. Batts and blankets without facings are installed by friction fitting within the stud space.

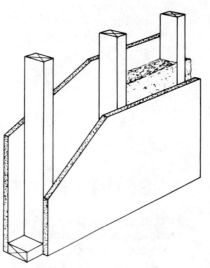

WITH INSULATION
STC 35-39
WITHOUT INSULATION
STC 30-34

Fig. 50a Adding sound absorbing insulation increases the STC four points in this wall.

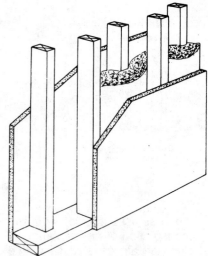

WITH INSULATION
STC 45-49
WITHOUT INSULATION
STC 40-44

Fig. 50b Staggered studs provide partially separated framing for wall surfaces and increase STC about eight points from wall in Fig. 50a. Addition of sound absorbing material can further raise STC to the 45-49 range.

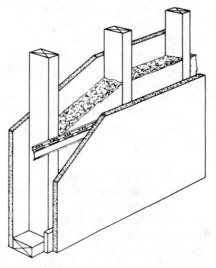

WITH INSULATION
STC 45-49

Fig. 50c Resilient mounting of gypsum board surfaces raises STC about 12 points above that of wall in Fig. 50a.

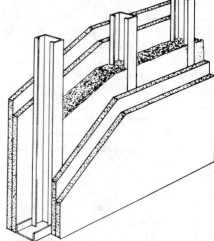

WITH INSULATION
STC 50-54

Fig. 50d Two-ply construction increases the weight of the wall surfaces and helps provide higher sound resistance of wall with STC of 50-54 obtainable.

SOUND ISOLATION CONSTRUCTION

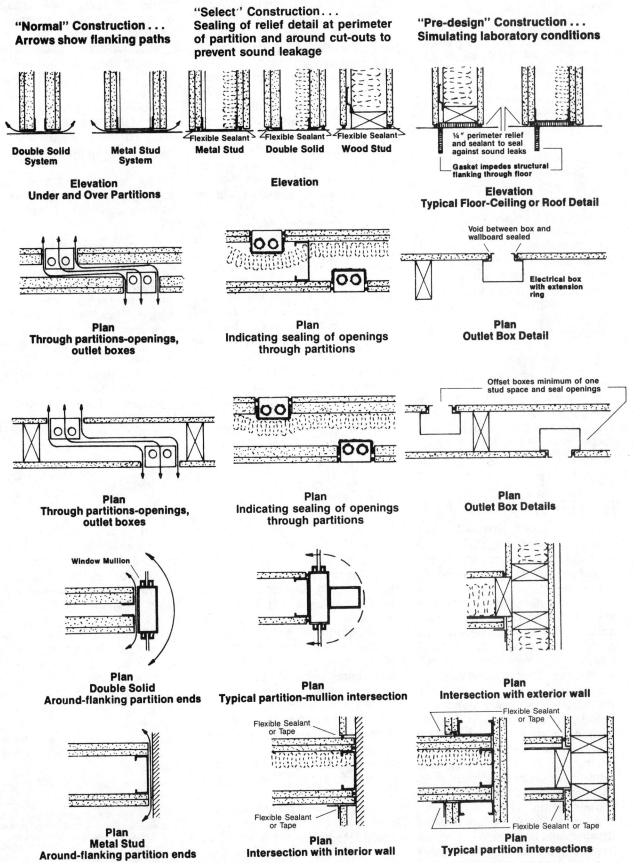

"Normal" Construction . . .
Arrows show flanking paths

Double Solid System | Metal Stud System

Elevation
Under and Over Partitions

"Select" Construction . . .
Sealing of relief detail at perimeter of partition and around cut-outs to prevent sound leakage

Flexible Sealant — Metal Stud | Flexible Sealant — Double Solid | Flexible Sealant — Wood Stud

Elevation

"Pre-design" Construction . . .
Simulating laboratory conditions

¼" perimeter relief and sealant to seal against sound leaks

Gasket impedes structural flanking through floor

Elevation
Typical Floor-Ceiling or Roof Detail

Plan
Through partitions-openings, outlet boxes

Plan
Indicating sealing of openings through partitions

Void between box and wallboard sealed

Electrical box with extension ring

Plan
Outlet Box Detail

Plan
Through partitions-openings, outlet boxes

Plan
Indicating sealing of openings through partitions

Offset boxes minimum of one stud space and seal openings

Plan
Outlet Box Details

Window Mullion

Plan
Double Solid
Around-flanking partition ends

Plan
Typical partition-mullion intersection

Plan
Intersection with exterior wall

Plan
Metal Stud
Around-flanking partition ends

Flexible Sealant or Tape

Flexible Sealant or Tape

Plan
Intersection with interior wall

Flexible Sealant or Tape

Flexible Sealant or Tape

Plan
Typical partition intersections

Fig. 49 Sound isolating techniques improve the sound transmission class (STC) and noise reduction of gypsum walls.

RIGID PLASTIC FOAM INSULATION

General Procedures

Gypsum board may be applied over rigid plastic foam insulation on the interior side of exterior masonry and concrete walls to provide a finished wall and protect the insulation from early exposure to fire originating within the building. These systems provide high insulation values needed for energy conservation.

Gypsum board applied over rigid plastic foam insulation in the manner described in this section may not necessarily provide finish ratings required by local building codes.

Many building codes require a minimum fire protection for rigid foam on interior surfaces equal to that provided by ½ inch gypsum board. Flammability characteristics of rigid foam insulation products vary widely, and the manufacturer's literature should be reviewed.

Mechanical Application

In applying gypsum board over rigid foam insulation, the entire insulated wall surface should be protected with the gypsum board, including the surface above ceilings and in closed, unoccupied spaces.

Single or double-ply, ½ in. or ⅝ in. thick, gypsum wallboard should be screw-attached to steel wall furring members attached to the masonry in accordance with the insulation manufacturer's directions, or with nails or screws directly into wood framing (See fig. 27 and 28.)

Furring members should be designed to minimize thermal transfer through the member and to provide a 1¼ in. minimum width face or flange for screw application of the gypsum board.

Furring members should be installed vertically, spaced 24 in. o.c. Blocking or other backing as required for attachment and support of fixtures and furnishings should be provided. Furring members should also be attached at floor-wall and wall-ceiling angles (or at the termination of gypsum board above suspended ceilings), and around door, window and other openings.

Single-ply gypsum boards should be applied vertically with the long edges of the board located over furring members. Gypsum boards should be placed so that end joints are avoided.

Fastener spacing should be as required for single ply application over framing or furring.

In double-ply application, the base ply should be applied vertically. The face ply may be applied either vertically or horizontally. Edge joints of vertically applied face ply and end joints of horizontally applied face ply should be offset at least one furring member space from base ply edge joints.

Fastener spacing should be as required for two-ply application over framing or furring (See Table IV page 19).

In all wallboard applications mechanical fasteners should be of such length that they do not penetrate completely to the masonry or concrete. In single layer application, all joints between gypsum boards should be reinforced with tape and finished with joint compound. In two-ply application the face layer joints may be concealed or left exposed. Vinyl faced gypsum wallboard face layers should not be adhesively applied over a rigid foam insulated wall, except as recommended by the gypsum board manufacturer.

BATH-SHOWER AREAS

Water resistant gypsum backing board complying with ASTM Standard C 630 is recommended for use on walls in bathrooms, laundries, kitchens, utility rooms and other interior areas where there is exposure to water. It is used as a base for the adhesive application of ceramic and plastic tile and plastic finished wall panels in shower and tub enclosures. Special adhesives are manufactured for this purpose. Water resistant gypsum board subjected to moisture should not be foil backed, or applied directly over a vapor barrier, and is not recommended for use in areas subject to extreme exposure to moisture, such as saunas, steam rooms and gang shower rooms.[1]

Water resistant gypsum backing board should be applied horizontally with the factory bound edge spaced a minimum of ¼ in. above the lip of the shower pan or tub. Shower pans, or tubs should be installed prior to the installation of the gypsum board. Shower pans should have an upstanding lip or flange a minimum of 1 in. higher than the threshhold in the entry way to the shower. It is recommended that the tub be supported at the walls on metal hangers or vertical blocking nailed to the studs. If necessary, the board should be furred away from the framing members so that the lip of the tub (Fig. 52a) or the upstanding leg of the pan (Fig. 52b) will be on the same plane with the face of the board.

An additional gypsum board extending the full height from floor to ceiling is required for a fire or sound rated construction. (Figure 53).

Suitable blocking should be provided approximately 1 in. above the top of tub or pan. Between-stud blocking should be placed behind the horizontal joint of the gypsum board above the tub or shower pan. Studs at least 3½ in. deep should be used 16 in. o.c. for a ceramic tile application. Appropriate blocking, headers or supports should be provided for tub, plumbing fixtures, and to receive soap dishes, grab bars, towel racks and similar items as required.

Interior angles should be reinforced with supports to provide rigid corners.

[1] Ceramic wall tile application to gypsum board should meet American National Standard Specifications for Installation of Ceramic Tile With Water-Resistant Organic Adhesives, ANSI A 108.4. The adhesives used should meet the specifications of the American National Standard for Organic Adhesives for Installation of Ceramic Tile, ANSI A 136.1.

Water-resistant gypsum backing board should be attached with nails or screws spaced not more than 8 in. o.c. When it is necessary that joints between adjoining pieces of gypsum board (including those at all angle intersections) and nail heads under areas to receive tile or wall panels be treated with joint compound and tape, seal treated joints and nail heads with a compatible sealer prior to tile installation, or use a water-resistant joint compound.

NOTE: The caulking compound or the sealer must be compatible for use with the adhesive to be used for application of the tile.

The cut edges and openings around pipes and fixtures should be caulked flush with waterproof, nonhardening caulking compound or adhesive complying with the American National Standard for Organic Adhesives for Installation of Ceramic Tile, Type I (ANSI A 136.1). Directions of the manufacturer of the tile, wall panel or other surfacing material should be followed.

The surfacing material should be applied down to the top surface or edge of the finished shower floor, return, or tub and installed to overlap the top lip of receptor, sub-pan or tub and should completely cover the following areas: (Fig. 54)

- over tubs without showerheads — 6 in. above the rim of the tub,
- over tubs with showerheads — a minimum of 5 ft. above the rim or 6 in. above the height of the showerhead, whichever is higher,
- shower stalls — a minimum of 6 ft. above the shower dam or 6 in. above the showerhead, whichever is higher,
- beyond the external face of the tub or receptor — a distance of at least 4 in. and to the full specified height,
- all gypsum board walls with window sills or jambs in shower or tub enclosures — to the full specified height.

For plastic finished wall panels, the recommendation of the manufacturer should be followed.

WATER RESISTANT GYPSUM BOARD
ADHESIVE
PAPER EDGE
¼"
TILE
FLEXIBLE SEALANT
TUB HANGER

TUB
FLOOR LINE

Fig. 52a

WATER RESISTANT GYPSUM BOARD
ADHESIVE
TILE
PAPER EDGE
FLEXIBLE SEALANT
TUB SUPPORT
1/4"

TUB
FLOOR LINE
¼"

Fig. 53 Details show installation of wall tile on gypsum board around tubs and showers. The gypsum board must be separated from the tub, receptor, or pan by at least ¼ in. gap to prevent moisture wicking.

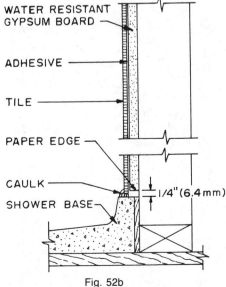

WATER RESISTANT GYPSUM BOARD
ADHESIVE
TILE
PAPER EDGE
CAULK
SHOWER BASE
1/4"(6.4mm)

Fig. 52b

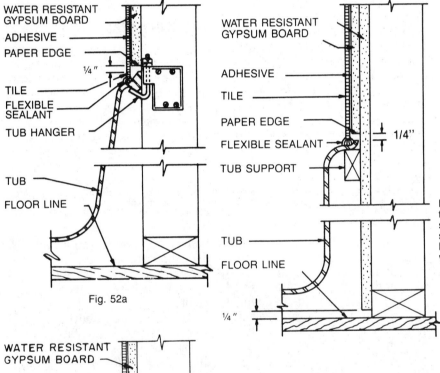

6" MIN
WITH SHOWERS 5' MIN
6" MIN
NO SHOWER 6" MIN
6" MIN
6'0"
TILE EXTENSION NOT REQUIRED AT EXTERIOR CORNER
EXTEND 4" MIN

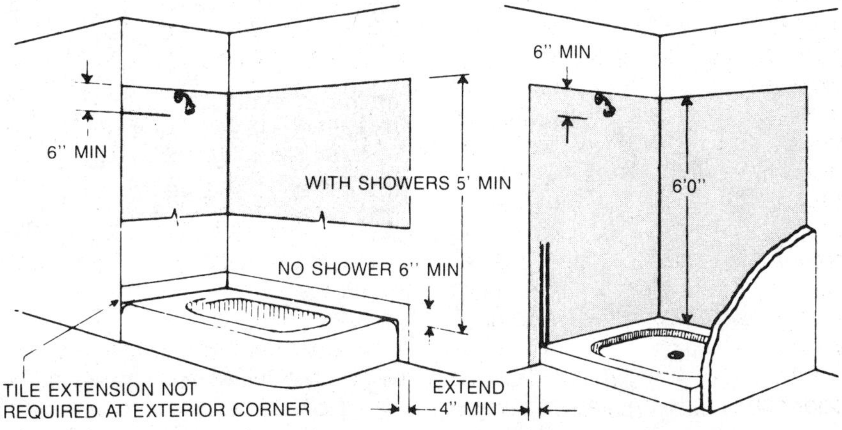

Fig. 54 Tile in baths and shower should be brought up to the levels shown.

EXTERIOR SOFFIT & CEILING CONSTRUCTION

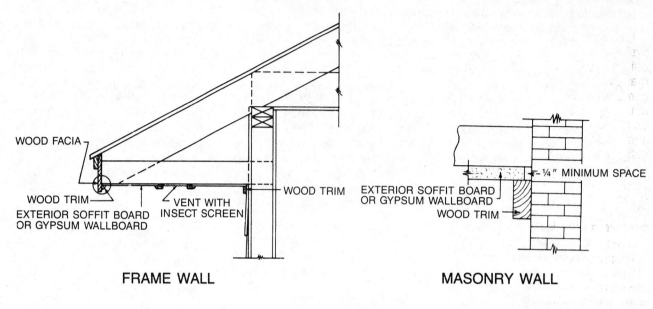

FRAME WALL   MASONRY WALL

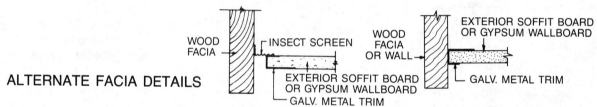

ALTERNATE FACIA DETAILS

Gypsum board should be protected from wetting in outdoor locations by facia boards with drip edges. Gypsum board edges should be separated from possible dampness by metal trim or by a ¼ in. space. Spaces enclosed by gypsum board must be adequately vented.

Gypsum wallboard and exterior soffit board are suitable for commercial use as ceilings for covered walkways and malls, large canopies and parking areas, and for residential use in open porches, breezeways, carports and exterior soffits. These areas must be horizontal or sloping downward away from the building. Framing should be no more than 16 in. o.c for ½ in. thick gypsum board, nor more than 24 in. o.c. for ⅝ in. board. Suitable facias and moldings should be provided around the perimeter to protect the board from direct exposure to water. Unless protected by metal or other water stops, the edges of the board should be placed not less than ¼ in. away from abutting vertical surfaces, (Fig. 51). The exposed surfaces of the gypsum board and metal trim should be sealed with two coats of exterior paint.

Where the area above opens to an attic space over habitable rooms, the attic space should be vented to the outside. Where the gypsum board is applied directly to joists or rafters, vents are required at each end of each joist space. Vents should be screened and be a minimum 2 in. by the full width between the joists and should be located within 6 in. of the outer edge of eaves. Gypsum board application should be identical to that for interior ceilings.

ELECTRIC RADIANT HEATING SYSTEMS

Packaged electric panel heating systems are commercially available utilizing gypsum board as the base material and can be installed as a complete system in new or existing structures. Also on-site systems can be fabricated in-place utilizing gypsum backing board, electric heating cable, a filler material and a face ply of gypsum board. Radiant panel heating systems should not be operated with wire temperature exceeding 125 F unless a higher temperature is specifically recommended by the gypsum board manufacturer.

Proper insulation adds to the efficiency of electric radiant panel heating systems. Unfaced, friction-fitted mineral wool or glass fiber insulating batts may be used in properly ventilated areas above the radiant heated ceiling. Blown-in insulation may also be utilized. Interior walls and foundation walls should also be efficiently insulated. Local electric utilities can offer specific recommendations.

It is recommended that a base layer of gypsum board at least ½ in. thick be attached perpendicular to ceiling supports with nails 7 in. o.c.

or with screws spaced 12 in. o.c. Taping of joints is not required. Electric heating cables should be securely attached to the backing board in accordance with the recommendations of the cable manufacturer and the requirements of the National Electrical Code. Cables should be parallel to and between framing members with at least 1¼ in. clearance from center of framing member on each side so that at least a 2½ in. wide unobstructed strip is provided under each framing member. Cables should cross framing members only at ends of ceiling 4 to 6 in. from the wall. There should be at least a 4 in. space clear of cables completely around the perimeter of each ceiling. Cables should be kept at least 8 in. clear of all openings such as light fixtures. (Fig. 55)

All inspections and testing of the heating system should be completed before the application of the filler material.

The heating cable must be completely embedded with a nonshrinking, noninsulating filler applied ¼ in.

thick, leveled and finished to a smooth surface. The filler material should be allowed to dry before the face board is installed.

To prevent striking heating cables where they cross framing members, attach the face layer of gypsum board (no thicker than ½ in.) with nails or screws 8 to 10 inches away from the wall around the perimeter of the ceiling. The spacing of fasteners should be the same as the base layer application, allowing ⅞ in. penetration into the framing members for nails and ⅝ in. for screws. The joints and fastener heads should be finished as detailed in Chapter V. A minimum of one week with carefully controlled drying conditions should be provided. Up to two weeks of drying time in cold conditions may be necessary before operating the heating system.

Decoration of the finished work, including sealing, should not proceed until the heating system has been tested a minimum of 24 hours. The heating system should then be turned off and allowed to cool before decorating.

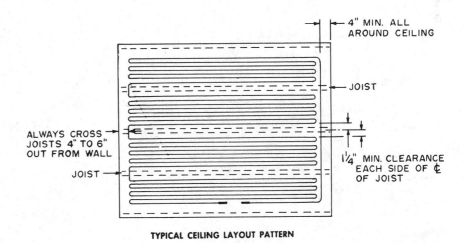

TYPICAL CEILING LAYOUT PATTERN

Fig. 55 Electrical heating panel is formed from cable fastened to gypsum board covered with a filling material.

GYPSUM BOARD AS A ROOFING UNDERLAYMENT

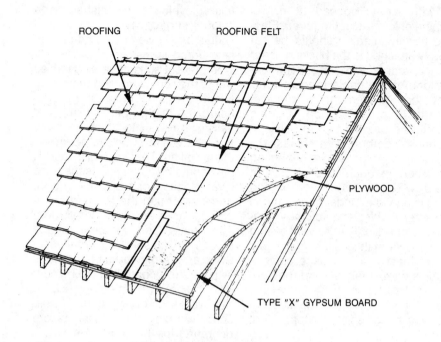

**PLYWOOD SHEATHED ROOF
OVER GYPSUM BOARD SUBSTRATE**

Purpose

Type X gypsum board may be added to the roof structure to help stop fire from burning down into the building. It also helps prevent the lateral spread of flame under the roof surface.

Application

Type X gypsum board should be delivered and stored so that it is protected from rain by a suitable covering. When installed, it can be elevated to the roof by the same method used to place plywood and shingles on the roof.

Apply Type X gypsum board with the long edge across rafters placed up to 24 in. o.c. Place ends on rafters. Nail with 4d common nails or 1⅜ in. gypsum wallboard nails, and use at least two nails at each rafter for every board. Butt end joints loosely, and fit edge joints snugly together without forcing. *Do not tape joints.*

Caution

Walk on rafter supported areas, not on unsupported gypsum board.

Install either plywood roof decking or spaced sheathing (1″ × 4″ stripping) over the Type X gypsum board. The plywood joints should be staggered with respect to joints in the gypsum board. Nail plywood or stripping through the Type X gypsum board and into the rafters in the regular pattern. *Use ½ in. longer nails than usual.*

Install roofing in the customary manner.

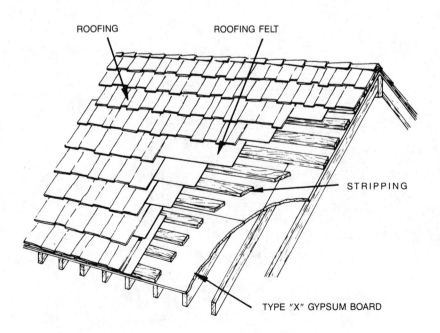

**STRIP SHEATHED ROOF
OVER GYPSUM BOARD SUBSTRATE**

Glossary

Adhesive, Contact.
Forms a strong bond after the two adhesive covered surfaces are brought together. Used to bond layers of gypsum board.

Adhesive, Dry Powder.
Used to laminate gypsum boards in multi-layer construction: gypsum board to gypsum studs or ribs and to sound insulating board. Not recommended for use on wood or wood fiber products.

Adhesive, Laminating.
Water based. Used in multi-layer construction or to bond gypsum board to suitable masonry or concrete surfaces or wood fiber board. The bond develops slowly as adhesive dries.

Adhesive, Solvent-Based.
Provides a more controlled vehicle evaporation in damp or cold weather. Can be used with foil-backed or predecorated gypsum board but must not come in contact with the vinyl surfaced gypsum board. Adequate ventilation is required and precautions must be taken to avoid sparks or fire near materials.

Adhesive, Stud.
Usually a solvent-base adhesive applied in ⅜ in. beads to metal or wood framing members for application of gypsum board. It will bridge minor framing irregularities.

Beading.
See **Ridging.**

Butt Joint.
Joints in which board ends (See "End") with the core exposed are butted together.

Corner Cracking.
Cracks occurring in the joint of inside corners between intersecting walls or at wall and ceiling intersections.

Corner-Floating.
See **Floating Angles.**

Crimping.
A method of fastening corner beads in place or fixing metal studs in runners by pinching and deforming the metal using a special tool.

Cross Furring.
Furring members installed across, or at right angles to, framing members or runners.

Crown Joint.
The height above the surface formed by treating compound that has been applied over the joint of two abutting square edged gypsum boards.

Edge (of gypsum board).
The paper bound edge as manufactured, the length dimension of the gypsum board.

End (of gypsum board).
The end perpendicular to the paper bound edge. The gypsum core is always exposed.

Fasteners.
Nails, screws or staples used for the mechanical attachment of the gypsum board.

Fastener Treatment.
Concealing fasteners which are set below the surface of the gypsum board by successive application of joint compound until a smooth surface is achieved.

Feather Edging. (Feathering.)
Tapering joint compound to a very thin edge to blend with the adjacent board surface creating a monolithic appearance.

Field.
The surface of the board exclusive of the perimeter. (Fig. 1)

Fire Resistance Classification.
A standard rating of fire resistive and protective characteristics of a building material or assembly when tested in conformance with requirements of ASTM Standard E 119.

Fire Taping.
The taping of gypsum board joints without subsequent finishing coats. A treatment method used in attic, plenum or mechanical areas where esthetics are not important.

Floating Angles.
Mechanical fasteners on one side of interior angles are omitted to relieve stresses that might otherwise cause cracking and nail popping.

Framing.
The supporting members such as studs, joists, and trusses, which may or may not bear part of the load of the building.

Furring.
Wood or metal members used to provide properly spaced, rigid and even support for gypsum board. They are fastened over, or attached to, underlying construction of framing, concrete, or masonry which is not in suitable condition for direct application of gypsum board.

Gypsum Board.
The name for a family of noncombustible sheet products consisting of a core primarily of gypsum between paper surfaces. The core may be regular, water resistant, special fire resistant (Type X) or a combination of these. The paper surface may be treated for water resistance, or where plaster is to be applied, treated for water absorption.

Joint Beading.
See **Ridging.**

Joint Compound.
A material used to conceal fasteners, joints, and surface indentations in gypsum board construction.

Joint Treatment.
The method of concealing gypsum board joints with tape and layers of joint compound.

Nail Popping.
A protrusion of nailhead and compound directly over the nailhead caused by outward movement of the nail relative to the gypsum board due to improper nailing or shrinkage of wood framing.

Nails, Drywall.
Nails designed for gypsum board application. See **Fasteners.**

Nail Spotting.
See **Fastener Treatment.**

Parallel Application.
Boards are applied with their edges parallel to framing members.

Perimeter.
Surface of the board at the edges and ends.

Perpendicular Application.
Boards are applied with edges at right angles to framing members.

Ridging.
A linear surface protrusion along with treated joints.

Screws, Drywall.
Screws designed for gypsum board application. See **Fasteners.**

Sheet Lamination.
Multi-ply construction with adhesives applied to the entire surface to be bonded. (See **Adhesive, Laminating.**)

Strip Lamination.
Multi-ply construction with adhesive applied in parallel strips usually spaced 16 in. or 24 in. apart.

Sound Transmission Class (STC).
A unit of measure used by engineers and architects to classify resistance of assemblies to airborne sound transmission.

Taping.
Embedding taping in joint compound in the process of joint treatment.

For a more detailed list see "Gypsum Board Products Glossary of Terminology (GA-505)."

GYPSUM BOARD (DRYWALL) APPLICATION SPECIFICATIONS

The following is a verbatim, authorized reproduction of "Recommended Specifications for the Application and Finishing of Gypsum Board," a publication (GA-216-85) of the Gypsum Association, national educational and promotional organization of the gypsum industry.

1. Scope and General Provisions

1.1 These are recommended specifications for application and finishing of gypsum board suitable to receive decoration, including related items and accessories.

1.2 Where a certain degree of fire resistance and sound control is required for gypsum board assemblies and constructions follow applicable building code regulations. Construction details are additionally described in the Gypsum Association's Fire Resistance Design Manual, and were extracted from official reports of fire and sound tests conducted by recognized testing laboratories in accordance with applicable Standards of the American Society for Testing and Materials (ASTM) including E 119, E 90, E 336 and E 492.

1.3 Requirements covering framing, furring, spacing, etc., are essential to provide a proper base to receive gypsum board. The following minimum requirements are recommended for wood framing and furring:

1.3.1 All framing members to which gypsum board will be fastened should be straight and true. Maximum spacing is shown in Table 1. Framing, bridging and furring members should be of proper grade for intended use and wood members 2 x 4 nominal size or larger should bear grade marks of recognized inspection agencies and conform to American Softwood Lumber Standard, NBS Voluntary Product Standard PS 20. Framing, bridging and furring members should be adequate to carry design or building code loading requirements. Deflections of horizontal members (ceiling) supporting gypsum board should not exceed L/240 of the span at full design load. Fastening surface of any framing or furring member should not vary more than 1/8 in. from plane of faces of adjacent framing, bridging or furring members. Additional supports should be provided as necessary for support of fixtures.

1.3.2 When gypsum board is nailed to wood cross furring on ceilings, these furring members should have a minimum cross section of 1-1/2 in. by 1-1/2 in. actual size and be spaced in accordance with sections 5.2 and 5.4. Where screw application is used, furring member may not be less than 3/4 in. x 2-1/2 in. actual size.

1.3.3 Where wood furring is used over solid surfaces, wood furring should be not less than 3/4 in. x 1-1/2 in. actual size.

1.3.4 Insulating blankets or flanges of blankets should not be applied to face of framing members that are to receive gypsum board.

1.3.5 During periods of cold or damp weather when a polyethylene vapor retarder is installed on ceilings behind the gypsum board, it is important to install the ceiling insulation BEFORE the board. Failure to follow this procedure can result in moisture condensation on the back side of the gypsum board, causing the board to sag.

1.3.6 Foil backed gypsum board may be used where a vapor retarder is required. (For exceptions to use of foil backed gypsum board see section 13.)

1.4 Attics or similar unheated spaces above gypsum board ceilings should be ventilated by providing effective cross ventilation for all spaces between roof and top ceiling by screened louvers or other approved and acceptable means.

1.4.1 The ratio of total net free ventilating area to ceiling area should be not less than 1/150, except the ratio may be 1/300 if:

(1) A vapor retarder having a transmission rate not exceeding one perm is installed on warm side of ceiling framing, or

(2) At least 50 percent of required ventilating area is provided by ventilators located in upper portion of space to be ventilated (at least 3 feet above eave or cornice vents) with balance of required ventilation area provided by eave or cornice vents.

1.4.2 Attic space which is accessible and suitable for future habitable rooms or walled-off storage space should have at least 50 percent of required ventilating area located in upper part of ventilated space as near high point of roof as practicable and above probable level of any future ceilings.

1.5 Where materials are being mixed and used for joint treatment or the laminating of one layer of board to another, the temperature in the building should be maintained at not less than 50°F for 48 hours before and continuously until applied materials are thoroughly dry.

1.6 Installation of gypsum board products and finishes should be done at temperatures at or above 50°F unless recommended otherwise by the manufacturer.

1.7 Use of gypsum board is not recommended where exposure to moisture is extreme or continuous.

1.8 Care should be taken to be sure the gypsum board will not be exposed to temperatures exceeding 125°F for extended periods of time, for example, when located adjacent to wood-burning stoves, electric heating appliances or other heating units, or hot air flues.

1.9 When gypsum board is used in air handling systems, the gypsum board surface temperature should be maintained above the air stream dew point temperature but not to exceed 125°F.

Note: Strict adherence to the National Electrical Code (NFPA 70A) is required for the proper installation of lighting and other heat producing electrical fixtures.

2. Definitions and Terms for Purpose of This Specification

2.1 Types of Gypsum Board.

2.1.1 Gypsum Wallboard — defined in Standard Specification for Gypsum Wallboard, ASTM C 36.

2.1.2 Gypsum Coreboard — defined in Standard Specification for Gypsum Backing Board and Coreboard, ASTM C 442. May be used as a coreboard in partition systems, typically one inch thick either single or multi-layer.

2.1.3 Gypsum Backing Board — defined in Standard Specification for Gypsum Backing Board and Coreboard, ASTM C 442. May be used as a base layer in multi-layer gypsum board systems.

2.1.4 Water Resistant Gypsum Backing Board — defined in Standard Specification for Water Resistant Gypsum Backing Board, ASTM C 630.

2.1.5 Type X (Special Fire Resistant) Gypsum Wallboard, Gypsum Backing Board or Water Resistant Gypsum Backing Board — gypsum board which provides greater fire resistance as defined in Standard Specification for Gypsum Wallboard, ASTM C 36, Standard Specification for Gypsum Backing Board and Coreboard, ASTM C 442, or Standard Specification for Water Resistant Gypsum Backing Board, ASTM C 630.

2.1.6 Foil Backed Gypsum Board — either regular or Type X gypsum wallboard or gypsum backing board with aluminum foil laminated to the back surface. The foil is a vapor retarder.

2.1.7 Gypsum Soffit Board — defined in Standard Specification for Exterior Gypsum Soffit Board, ASTM C 931.

2.2 Edge — paper bound edge, as manufactured.

2.3 End — mill-cut or field-cut end perpendicular to edge. At such cuts gypsum core is exposed.

2.4 Framing Member — that portion of framing, furring, blocking, etc., to which gypsum board is attached. Unless otherwise specified herein, the surface to which abutting edges or ends are attached should not be less than 1-1/2 in. wide for wood members and not less than 1-1/4 in. wide for metal members and not less than 6 in. wide for gypsum studs. For internal corners or angles, bearing surfaces should not be less than 3/4 in.

2.5 Fastener — nails, screws or staples used for mechanical application of gypsum board.

2.6 Perpendicular (Horizontal) Application — gypsum board application with edges applied at right angles to members to which it is attached.

2.7 Parallel (Vertical) Application — gypsum board applied with edges parallel to member to which it is attached.

2.8 Treated joint — a joint between gypsum boards which is reinforced and concealed with tape and joint treatment compound, or covered by strip moldings.

2.9 Untreated joint — a joint which is left exposed.

2.10 Control Joint — a predetermined separation installed or created between adjacent surfaces of gypsum board constructions to relieve movement stresses in large ceiling and wall areas.

2.11 Joint Treatment Compound — defined in Standard Specification for Joint Treatment Materials for Gypsum Wallboard Construction, ASTM C 475.

2.12 Finishing — the taping of joints, concealment with joint treatment compound of such joints, heads of fasteners, and edges of corner protective devices, and smoothing or leveling of such areas to prepare them to receive field application of priming, painting, coating, decorative coating, texturing and coverings such as wallpaper and vinyl materials. See Sections 18 and 19.

Note: Additional definitions are contained in GA 505, Gypsum Board Products Glossary of Terminology and in ASTM C 11, Terms Relating to Ceiling and Walls.

3. Materials

This specification is prepared on the basis that materials conform to the following ASTM specifications or requirements:

3.1 Gypsum Wallboard. Standard Specification for Gypsum Wallboard, ASTM C 36.

3.2 Gypsum Backing Board and Coreboard. Standard Specification for Gypsum Backing Board and Coreboard, ASTM C 442.

3.3 Water Resistant Gypsum Backing Board. Standard Specification for Water Resistant Gypsum Backing Board, ASTM C 630.

3.3.1 Gypsum Soffit Board — Standard Specification for Exterior Gypsum Soffit Board, ASTM C 931.

3.4 Joint Reinforcing Tape and Joint Compound. Standard Specification for Joint Treatment Materials for Gypsum Wallboard Construction, ASTM C 475.

3.5 Laminating Adhesive. Either joint compound, adhesive, or laminating compound, all as recommended by manufacturer of gypsum board.

3.6 Water. Should be clean, fresh and suitable for domestic consumption.

3.7 Nails. Standard Specification for Nails for Application of Gypsum Wallboard, ASTM C 514. Special nails for pre-decorated gypsum board should be as recommended by pre-decorated gypsum board manufacturer.

3.8 Screws. Standard Specification for Steel Drill Screws for the Application of Gypsum Board, ASTM C 1002 or C 954.

3.8.1 Type S screws are designed for gypsum board attachment to light gage steel framing and furring.

3.8.2 Type S-12 screws are designed for gypsum board attachment to heavier gage steel framing (not over 12 gage).

3.8.3 Type W screws are designed for gypsum board attachment to wood framing.

3.8.4 Type G screws are designed for attaching gypsum board to gypsum board.

3.9 Staples. Number 16 USS gage flattened galvanized wire staples with 7/16 in. wide crown outside measure and divergent point for first ply only of two-ply gypsum board application:

1 in. long legs for 3/8 in. thick gypsum board.

1-1/8 in. long legs for 1/2 in. thick gypsum board.

1-1/4 in. long legs for 5/8 in. thick gypsum board.

3.10 Adhesives.

3.10.1 Standard Specification for Adhesives for Fastening Gypsum Board to Wood Framing, ASTM C 557.

3.10.2 Adhesives for fastening gypsum board to metal framing should be as recommended by gypsum board manufacturer.

3.10.3 Adhesives for fastening gypsum board to gypsum board should be as recommended by gypsum board manufacturer.

3.11 Framing Members.

3.11.1 Wood framing members should conform to American Lumber Standards Specifications.

3.11.2 Light-gage, non-load bearing steel framing should comply with Standard Specification for Non-Load (Axial) Bearing Steel Studs, Runners (Track), and Rigid Furring Channels for Screw Application of Gypsum Board, ASTM C 645.

3.11.3 Gypsum studs should not be less than 6 in. wide and 1 in. thick.

3.12 Cornerbead and Edge Trim. Should be made from corrosive protected coated steel or plastic designed for its intended use. Flanges should be free from dirt, grease, or other materials that may adversely affect bond of joint treatment or decoration.

3.12.1 Designations used to identify commonly specified types of metal trim and casings are shown in Figure 1.

3.12.2 Other types of corner, edge trim, decorative dividers or control joints between gypsum wallboard panels may be used when they meet the general provisions of Section 3.12.

4. Delivery, Identification, Handling and Storage

4.1 All materials should be delivered in original packages, containers or bundles bearing brand name, applicable standard designation, and name of manufacturer or supplier for whom product is manufactured.

4.2 All materials should be kept dry, preferably by being stored inside building — under roof. Where necessary to store gypsum board products and accessories outside, they should be stacked above ground, properly supported on a level platform and fully protected from weather and direct sunlight exposure.

4.3 Gypsum board should be neatly stacked flat with care taken to prevent sagging or damage to edges, ends and surfaces.

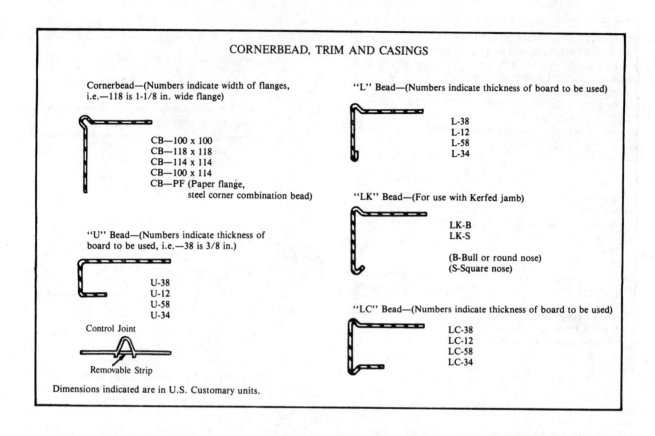

Figure 1

5. Application of Gypsum Board

NOTE: For wood framing, refer to Section 1.3. Installation of steel framing should be in accordance with ASTM Standard C 754.

5.1 Methods of Cutting and Installation. Gypsum boards may be cut by scoring and breaking or by sawing, working from face paper side. When scoring use a sharp knife to cut through the face paper into gypsum core. Gypsum board is then snapped back away from cut face. Back paper should be broken by snapping gypsum board in reverse direction, or preferably by cutting back paper. All cut edges and ends of gypsum boards should be smoothed where necessary to obtain neat fitting joints when gypsum board is installed. Holes for pipes, fixtures or other small openings should be scored on face and back before removal of cut-out with a saw or special tool designed for this use. Where gypsum board meets projecting surfaces, it should be scribed and neatly cut.

5.1.1 In general, gypsum board should be applied first to ceilings and then to walls.

5.1.2 Gypsum board joints at openings shall be located so that no joint will align with edges of opening unless control joints will be installed at these points. Joints shall be staggered, and joints on opposite sides of a partition shall not occur on the same stud.

5.1.3 Fasteners, when used at edges or ends, should be spaced not less than 3/8 in. from edges and ends of gypsum board. Fasteners used at edges or at board ends in horizontal application should not be more than 1 in. from edges or ends except when floating angle method (see 5.5) is used. Perimeter fastening into partition plate or sole at top and bottom is not required or recommended except where fire rating, structural performance or other special conditions require such fastening. **While driving fasteners, gypsum board should be held in firm contact with underlying support.** Application of fasteners proceeds from center or field of gypsum board toward ends and edges.

5.1.4 Nails should be driven so that heads are slightly below surface of gypsum board. Care should be taken to avoid damage to face and core of board, such as breaking the paper or fracturing core.

5.1.5 Screws should be driven so that screwheads are slightly below gypsum board surface without breaking surface paper or stripping framing member around screw shank.

5.1.6 Staples should be driven with crown parallel to framing members. Staples should be driven in such a manner that crown bears tightly against gypsum board but does not cut into face paper. *NOTE: Staple attachment is restricted to first ply only of gypsum board in a two ply system.*

5.2 Application of Single Ply Gypsum Board to Framing Members.

5.2.1 Maximum spacing of framing members for single ply gypsum board construction should not exceed those shown in Table 1.

Table 1
Maximum Framing Spacing

Single ply Gypsum Board (Thickness)	Application to Framing	Maximum o.c. Spacing of Framing
Inches		Inches
Ceilings:		
†3/8	*Perpendicular	16
1/2	Perpendicular	16
1/2	*Parallel	16
5/8	Parallel	16
1/2	*Perpendicular	24
5/8	Perpendicular	24
Sidewalls:		
3/8	Perpendicular or Parallel	16
1/2	Perpendicular or Parallel	24
5/8	Perpendicular or Parallel	24

*On ceilings to receive a water base texture material, either hand or spray applied, install gypsum board perpendicular to framing and increase board thickness from 3/8 in. to 1/2 in. for 16 in. o.c. framing and from 1/2 in. to 5/8 in. for 24 in. o.c. framing.

†Should not support thermal insulation.

5.2.2 In single ply installation, all ends and edges of gypsum board should occur over framing members or other solid backing except where treated joints occur at right angles to framing or furring members.

5.2.3 End joints should be staggered and joints on opposite sides of a partition assembly should be arranged to occur on alternate framing members.

5.3 Nailing for Single Ply Application Over Wood Members. Single ply application is shown in Table 2.

Table 2
Nailing Schedule for Single Ply Gypsum Board[1]

	Gypsum Board Thickness	Minimum Nail Length
	Inches	Inches
(2)	1/4	(2)
	3/8	1-1/4
	1/2	1-3/8
	5/8	1-1/2

(1) Where a certain degree of fire resistance is required for gypsum board assemblies and constructions, nails of same or larger length, shank diameter, and head bearing area, as those described in the fire test report must be used.

(2) For application over existing solid surfaces, nail should have a flat or concave head, a diamond point, and of such length as to provide not less than 7/8 in. nor more than 1-1/4 in. penetration into the nailing members.

5.3.1 Nails for single nailing should be spaced a maximum of 7 in. o.c. on ceilings, and a maximum of 8 in. o.c. on walls. (See Figure 2.)

5.3.2 Nails for double nailing should be spaced and driven as shown in Figure 3 and applied in the following sequence:

(1) Starting at center of board, nails shown by dot • are applied in row 1, then rows 2 and 2A, 3 and 3A, 4 and 4A, etc. always nailing from center to edges of sheet.

(2) Apply second nails shown by circle º in same manner as first nails, also starting at row 1.

(3) As an alternate procedure, second nail may be applied immediately after all nails in each row are driven according to step (2) above.

(4) Use single nails on perimeter of board.

NOTE: It may be necessary to reset first nails in each row after second nails have been set.

5.3.3 When screws are used in lieu of nails, they should penetrate framing not less than 5/8 in. and be spaced a maximum of 12 in. o.c. for ceilings and 16 in. o.c. for walls where the framing members are 16 in. o.c. Screws should be spaced a maximum of 12 in. o.c. for ceilings and walls where framing members are 24 in. o.c.

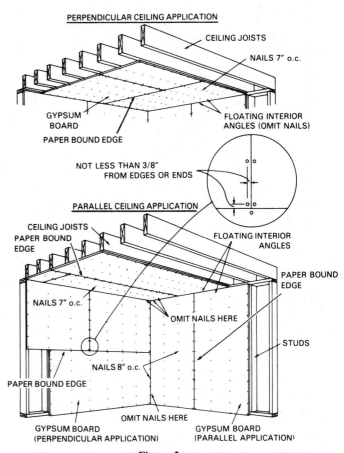

Figure 2
Single Nailing

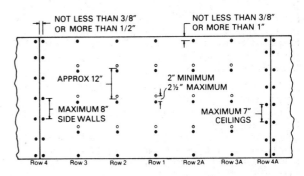

Figure 3
Double Nailing

5.4 Application of Two-Ply Gypsum Board to Framing Members.

5.4.1 Maximum spacings for framing members for two ply gypsum board constructions should not exceed those shown in Tables 3 and 4.

Table 3.
Two-Ply Application
Without Adhesive Between Plies

Gypsum Board (Thickness)		Application to Framing		Maximum o.c. Spacing of Framing
Base	Face	Base	Face	
Inch	Inch			Inch
Ceilings:				
3/8	3/8	Perpend	Perpend*	16
1/2	3/8	Parallel	Perpend*	16
1/2	1/2	Parallel	Perpend	16
1/2	1/2	Perpend	Perpend*	24
5/8	1/2	Perpend	Perpend*	24
5/8	5/8	Perpend	Perpend*	24

Sidewalls:
For two-ply application with no adhesive between plies, 3/8 in., 1/2 in. or 5/8 in. thick gypsum board may be applied perpendicular or parallel on framing spaced a maximum of 24 in. o.c. Maximum spacing should be 16 in. o.c. when 3/8 in. thick board is used as face ply.

*On ceilings to receive a water base texture material, either hand or spray applied, install gypsum board perpendicular to framing and increase board thickness from 3/8 in. to 1/2 in. for 16 in. o.c. framing and from 1/2 in. to 5/8 in. for 24 in. o.c. framing.

5.4.2 Base ply of gypsum board should be applied with nails, screws or staples of size and type described in Sections 3 and 5 and as spaced in Table 5.

5.4.3 When adhesive is not used between plies, the gypsum board should be applied as shown in Table 3. Face layer should be applied with spacing of nails or screws required for normal single ply application. Nails used for base and face layers should be long enough to penetrate at least 7/8 in. into wood framing members. Screw length should be long enough to penetrate at least 5/8 in. into wood. Face layer joints which are parallel to framing should fall over framing members and be offset from base layer joints.

5.4.4 When an adhesive is used between plies, two plies should be applied as shown in Table 4. If two plies are applied in parallel

Table 4.
Two-Ply Application
With Adhesive Between Plies[1]

Gypsum Board (Thickness)		Application to Framing		Maximum o.c. Spacing of Framing
Base	Face	Base[2]	Face[2]	
Inch	Inch			Inch
Ceilings:				
3/8	3/8	Perp	Perp or Par	16
1/2	3/8	Perp or Par	Perp or Par	16
1/2	1/2	Perp or Par	Perp or Par	16
5/8	1/2	Par	Perp or Par	24
5/8	5/8	Perp or Par	Perp or Par	24

Sidewalls:
For two-ply application with adhesive between plies, 3/8 in., 1/2 in. or 5/8 in. thick gypsum board may be applied perpendicular or parallel on framing spaced a maximum of 24 in. o.c.

1 Adhesive between plies should be dried or cured prior to any decorative treatment. This is especially important when water based texture material (hand or spray applied) is to be used.

2 Perp = perpendicular
Par = parallel

direction, joints of face ply should be offset from base layer joints. Joints in face ply need not occur over framing members. Adhesive to be used between two plies of gypsum board should be evenly applied over back surface of the face ply of gypsum board before it is erected or to face surface of base ply using type of adhesive and method of application recommended by the gypsum board manufacturer. Face ply of gypsum board should be placed in position and fastened with a sufficient number of nails or screws to hold gypsum boards in place until adhesive develops adequate bond strength. Unless otherwise recommended by the gypsum board manufacturer, permanent, rather than temporary, mechanical fasteners should be used on ceilings spaced 12 in. o.c. around perimeter and 16 in. o.c. in the field. In lieu of mechanical fastening, second ply of gypsum board applied on sidewalls may be held in position by shoring with props and headers or other temporary supports to insure adequate pressure for bonding. Unless otherwise recommended by the gypsum board manufacturer, permanent fasteners should be used on top and bottom of wall 16 in. o.c. maximum. Nails or screws used to hold gypsum board face ply should penetrate into framing and be finished in the same manner as for single ply gypsum board application. (See Section 5.3)

Table 5
Base Ply Fastener Spacing For Two-Ply Application[1]

Location	Nail Spacing		Screw Spacing		Staple Spacing	
	With Laminated Face Ply	With Nailed Face Ply[3]	With Laminated Face Ply[2]	With Screwed Face Ply[3]	With Laminated Face Ply	With Nailed or Screwed Face Ply[3]
Walls	8 in. o.c.	24 in. o.c.	16 in. o.c.	24 in. o.c.	7 in. o.c.	16 in. o.c.
Ceilings	7 in. o.c.	16 in. o.c.	12 in. o.c.	24 in. o.c.	7 in. o.c.	16 in. o.c.

1 Fastener size and spacing for applying sound deadening boards vary for different fire and sound rated constructions. The manufacturer's recommendations should be followed.

2 Use 12 in. o.c. for both ceilings and walls when supports are spaced 24 in. o.c.

3 Fastener spacing for face ply shall be the same as for single layer application.

5.5 Floating Interior Angles

To minimize possibility of fastener popping in areas adjacent to wall and ceiling intersection, floating angle method of application should be used for either single or double layer application of gypsum board to wood framing. This method is applicable where single nailing, double nailing or screw attachment is used. Gypsum board should be applied to ceiling first. (See Figures 4, 5 and 6.) Floating interior angles should not be used where fire ratings are required.

5.5.1 First attachment into each ceiling framing member framed either perpendicular or parallel to wall intersection should be located 7 in. out from intersection for single nailing, and 12 in. for double nailing or screw application.

5.5.2 Gypsum board on sidewalls should be applied to provide a firm support for floated edges of ceiling gypsum board. Top attachment into each stud should be located 8 in. down from ceiling intersection for single nailing, and 12 in. for double nailing or screw application. (See Figures 4 and 5.) At sidewall vertical angles, (Figure 6), over-lapping board should be applied so as to bring back of underlying board into firm contact with face of framing member behind it.

5.5.3 Special clips designed to provide anchorage at corners in lieu of the third corner stud may be used where approved by the authority having jurisdiction.

5.6 Control Joints

Control joints should be installed in ceilings exceeding 2500 sq. ft. in area and in partition, wall and wall furring runs exceeding 30 ft. Distance between ceiling control joints should not exceed 50 ft. in either direction, and a control joint should be installed where ceiling framing or furring changes direction. Distance between control joints in walls or wall furring should not exceed 30 ft. and a control joint should be installed where an expansion joint occurs in base exterior wall. Wall or partition height door frames may be considered a control joint. Whenever possible, control joints should coincide with any building control joints.

NOTE: Where a control joint occurs in an acoustically rated assembly, provision may be necessary to block the joint opening by using backing material such as gypsum board, mineral fiber or equivalent, or filling the void created with resilient insulating material.

6. Adhesive Nail-On Application to Wood Framing

Except as herein modified, application should be in accordance with Section 5.2, Single Ply Application.

6.1 Adhesive should comply with requirements of ASTM C 557.

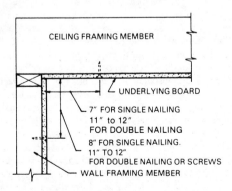

Figure 4
Vertical Section, Ceiling
Framing Member Perpendicular to Wall

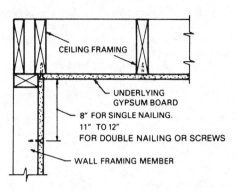

Figure 5
Vertical section, ceiling framing parallel to wall.

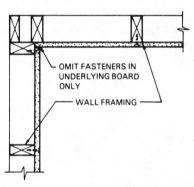

Figure 6
Horizontal section through interior vertical angle.

6.2 Surfaces of gypsum board and framing to be adhered by adhesive should be dry and free from dust, dirt, grease or any other foreign matter which could impair bond.

6.3 A bead of adhesive 3/8 in. in diameter should be applied to face of all wood framing members, except plates, which support gypsum board. When gypsum board is placed, adhesive should spread to an average width of 3/4 in. approximately 1/16 in. thick. Application patterns are shown in Figure 7.

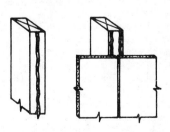

Figure 7
Adhesive Application Patterns

6.3.1 Where a joining of two adjacent pieces of gypsum board occurs (ends or edges) on a framing member, two parallel beads of adhesive 3/8 in. in diameter should be applied, one near each edge of the framing member.

6.4 The adhesive container should be clearly marked with "open time" of adhesive. Adhesive should be applied to no greater area than can be covered with gypsum board within that time.

6.5 If properties of adhesive are such as to assure bridging between gypsum board and wood framing, no nailing is required in field of board for walls. In such cases, perimeter nailing may be used, as required. If properties of adhesive are such that there is no positive bridging between board and wood framing, either temporary field nailing or temporary bracing may be required to insure contact between board, adhesive, and stud face, until adhesive develops adequate bond strength. Unless other recommendations are provided by the manufacturer of the gypsum board or the adhesive, fasteners should be spaced in accordance with Table 6.

Table 6
Fastener Spacing with Adhesive or Mastic Application and Supplemental Fastening

Framing Member Spacing	Ceilings		Partitions Load Bearing		Partitions Nonload-Bearing	
	Nail	Screw	Nail	Screw	Nail	Screw
16. in. o.c.	16 in.	16 in.	16 in.	24 in.	24 in.	24 in.
24 in. o.c.	12 in.	16 in.	12 in.	16 in.	16 in.	24 in.

7. Semi-Solid Gypsum Board Partitions

7.1 Materials

7.1.1 Face Panels, 1/2 in., 5/8 in., or multiple thicknesses of gypsum board of 4 ft. widths.

7.1.2 Gypsum studs should be 1 in. minimum thickness. They may be of 1 in. material or of multi-layer materials laminated to required thickness and a minimum of 6 in. wide and of lengths approximately 6 in. less than floor-ceiling height unless full lengths are required for fire stops or fire resistance.

7.1.3 Runners of wood or steel and of width equal to partition cavity should be used.

7.1.3.1 Floor and ceiling runners should be securely fastened with fasteners spaced not to exceed 24 in. o.c.

7.1.4 Laminating adhesive, joint compound for tape embedment, complying with ASTM C 475 or other laminating adhesive as recommended by gypsum board manufacturer should be used.

7.2 Installation

7.2.1 Erect vertical runners or gypsum studs where required to provide proper support at locations such as exterior walls, partition junctions, terminals, external corners and door frames.

7.2.2 Gypsum board face panels are positioned vertically. Gypsum studs are laminated to face panels not over 24 in. o.c. and located at face panel vertical joints and at vertical center line of panel.

7.2.3 Gypsum studs may be laminated to face panels prior to erection or as erection of partition proceeds. Erect a starter face panel vertically at an exterior wall. Starter panel should be plumb and secured to floor, ceiling and vertical runners.

7.2.4 Erect next face panel adjacent to starter panel, butting its edge and end firmly to starter panel and ceiling. Continue erection of face panels, laminating exposed faces of gypsum studs as work progresses.

7.2.5 Openings in partitions for doors, electrical outlets, etc., should be carefully and accurately marked and cut.

7.2.6 Laminating adhesives should be of a consistency and volume that will cover approximately three-fourths of the gypsum stud surface after lamination.

7.2.7 Type G screws should be used as required to insure a continuous bond between face panels and gypsum studs, and should be placed a maximum of 36 in. o.c.

7.2.8 Openings or changes in directions of partition should be reinforced with additional gypsum studs laminated in place at the following locations:

7.2.8.1 External Corners — between face panels in corner opposite vertical runners.

7.2.8.2 Abutting Walls — between face panels of a partition to reinforce junction of an abutting wall.

7.2.8.3 Door Openings — a vertical gypsum stud should be located within 3 in. of door frame for reinforcement, and a gypsum stud should be placed horizontally over door header.

8. Solid Gypsum Board Partitions

8.1 Materials

8.1.1 Face Panels — 1/2 in., 5/8 in., or multiple laminations of regular or type X gypsum board.

8.1.2 Core — 1 in. thick gypsum board, either single or multiple layers.

8.1.3 Runners — wood runners or steel channels or angles.

8.1.4 Laminating Adhesive — joint compound for tape embedment, complying with ASTM C 475, or other laminating adhesives recommended by gypsum board manufacturer should be used.

8.2 Installation

8.2.1 Floor and ceiling runners should be installed according to partition layout and secured not more than 24 in. o.c. Runners should be installed vertically at required locations such as exterior walls, exterior corners and door frames.

8.2.2 For partitions located parallel to and in between ceiling members, steel or wood headers not more than 24 in. o.c. should be provided to fasten ceiling runners prior to erection of ceiling.

8.2.3 Face panels should be attached to runners at 24 in. o.c.

8.2.3.1 Coreboard may be installed prior to installation of face panel. Coreboard should be attached using two fasteners evenly spaced per board and a maximum spacing of 24 in. o.c. to steel angles when steel angles are used as runners. When steel channels are used as runners to secure coreboard, attachment is not required. Combinations of wood and steel channels or angles can vary installation procedures.

8.2.4 Joint compound or laminating adhesive as recommended by the wallboard manufacturer should be applied to back surface of face panels or face of core as described for two ply gypsum board construction. (See Section 5.) Face panels should be laminated to core with sufficient pressure to insure bond. Joints of face panels and core should be staggered. Fasteners may be used to insure bond between face panels and core.

9. Adhesive Application of Gypsum Board to Interior Masonry or Concrete Walls

Foil backed gypsum board is not recommended for direct adhesive application. Only interior masonry or concrete walls above grade, or inside of exterior masonry cavity walls with 1 in. minimum width cavity between inside and outside masonry for full height of above grade surface to receive gypsum board, are recommended for direct adhesive application. Interior surfaces to which gypsum board is to be adhered should be free from any foreign matter, projections, or depressions that will impair bond, and exterior surfaces sealed against moisture penetration. In lieu of gypsum board manufacturer's specific recommendations refer to paragraph 9.1.

9.1 Apply adhesive directly to back of gypsum board or on wall in continuous beads spaced not to exceed 12 in. o.c. or daubs spaced not to exceed 16 in. when applying gypsum board to monolithic concrete, brick, or concrete block. Beads should not be less than 3/8 in. in diameter to provide continuous bond between gypsum board and wall surface and daubs should not be less than 2 in. diameter by 1/2 in. thick with a row centered at all vertical joint locations.

9.1.1 Position gypsum board to provide a tight fit at abutting edges or ends. Do not slide board. Use mechanical fasteners or temporary bracing as required to support gypsum board until adhesive sets or adequate bond strength is attained.

9.1.2 Joint treatment should be delayed until gypsum board is firmly bonded.

10. Application of Predecorated Combustible Panels to Gypsum Substrate

10.1 Gypsum board should be used as a substrate material to provide increased fire resistance and sound control in installations of rigid predecorated wall panels. (See Figure 8.)

10.2 Apply gypsum board to framing with the same framing and fastener recommendations as are specified in Section 5 for the base ply of two ply systems having adhesive between plies. Vertical joints should be centered on framing members and joint treatment is not required.

10.3 Apply rigid predecorated panels as recommended by the panel manufacturer. Stagger joints of rigid panels so they do not occur over joints of gypsum substrate.

11. Gypsum Board Application over Rigid Plastic Foam Insulation

NOTE: Gypsum board applied over foam insulation as recommended in this section may not necessarily meet Building Code finish rating requirements.

11.1 Gypsum board may be applied over rigid plastic foam insulation on interior side of exterior masonry and concrete walls to provide a wall finish and protect insulation from early exposure to fire originating within the building.

11.2 In applying gypsum board over rigid plastic foam insulation, the entire insulated wall surface should be protected with gypsum board, including walls above ceilings and in unoccupied spaces.

11.3 Single or double ply, 1/2 in. or 5/8 in. thick gypsum wallboard or gypsum veneer base should be fastened to steel or wood wall furring members attached to masonry in accordance with the furring manufacturer's directions. Single ply gypsum veneer base should be 5/8 in. thick.

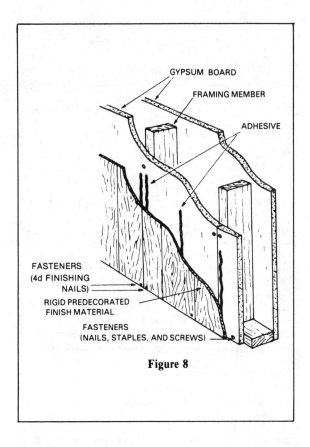

Figure 8

11.4 Furring members should be designed to minimize thermal transfer through the member and to provide 1-1/4 in. minimum width face for application of gypsum board.

11.5 Furring members should be installed vertically, spaced not over 24 in. o.c. Blocking or other backing as required for attachment and support of fixtures and furnishings should be provided.

11.6 Single ply gypsum board should be applied vertically with the long edges of the board located over furring members. Gypsum boards should be placed so that end joints are avoided.

11.6.1 Screw spacing for single application should be 12 in. o.c. maximum at board edges and in field of the board.

11.6.2 Nail spacing for wood should be as shown in Figure 2.

11.7 In double ply application, the base ply should be applied vertically. The face ply may be applied either vertically or horizontally. Edge joints of vertically applied face ply and end joints of horizontally applied face ply should be offset one furring member spacing from base ply edge joints.

11.7.1 Screw spacing for two ply application with vertically or horizontally applied face ply having no adhesive between the plies, should be 24 in. o.c. maximum for base ply and 12 in. o.c. for face ply.

11.7.2 Screw spacing for two ply application with vertically applied face ply and adhesive between plies should be 12 in. o.c. for base ply and 16 in o.c. maximum at top and bottom ends of face ply.

11.8 In single layer application, all joints between gypsum boards should be taped and finished. Vinyl faced gypsum board should not be adhesively applied over a rigid foam insulated wall.

11.9 In two ply application, face ply joints may be concealed or left exposed.

12. Gypsum Board Application Over Steel Framing and Furring

12.1 Steel framing and furring should comply with requirements of ASTM Standard Specification for Non-Load (Axial) Bearing Steel Studs, Runners (Track), and Rigid Furring Channels for Screw Application of Gypsum Board, ASTM C 645.

12.2 Screws for application to steel framing and furring should comply with requirements of ASTM C 1002 and should be a minimum of 3/8 in. longer than total thickness of gypsum board applied.

12.3 Screws should be applied as specified in 5.1.5.

12.4 Maximum spacing of steel framing and furring for screw application should be as specified in Table 1 for single ply gypsum board and as specified in Tables 3 and 4 for two ply gypsum board.

12.5 Screw Spacing

12.5.1 Maximum screw spacing for single ply gypsum board and face ply of two ply gypsum board with no adhesive is as specified in 5.3.3.

12.5.2 Maximum screw spacing for parallel applied base ply of two ply gypsum board over steel framing with no adhesives between plies is as specified in Table 5.

12.5.3 Maximum screw spacing for a perpendicularly applied base-ply of two ply gypsum board over steel framing with no adhesive between plies is one screw at each edge, stud or furring channel intersection and one screw midway between edges at each stud or channel.

12.5.4 Maximum screw spacing for perpendicular or parallel applied base ply of two-ply gypsum board over steel framing with adhesive between plies is as specified for single ply gypsum board in 5.3.3.

12.5.5 Maximum screw spacing for face ply of two ply gypsum board over steel framing with adhesive between plies on ceilings is 12 in. o.c. along perimeter and 16 in. o.c. in the field. On wall surfaces, only a sufficient number of screws should be used to hold gypsum board in place until adhesive, recommended and applied as specified by gypsum board manufacturer, develops adequate bond. On ceiling use fasteners as specified in Section 5.4.4.

13. Foil Backed Gypsum Board

13.1 Application of foil backed gypsum board should conform to the foregoing specifications for application of gypsum board. The reflective surface should be placed against face of framing members. Foil backed gypsum board should *not* be used in following areas:

(1) As a backing material for wall tile.
(2) For face ply on a two ply laminated system.
(3) For laminating directly to masonry.
(4) In conjunction with electric heating cables.

14. External Corners and Arches

14.1 External corners should be protected with a metal bead or other suitable type of corner protection which generally are attached to supporting construction with fasteners as required to maintain straightness. (See Section 3.12 and Figure 1.) Corner beads may also be attached with a crimping tool.

14.2 Where gypsum board is to be applied to soffit of arches, it should be carefully bent into place. (See Table 7.) If necessary, it first should either be dampened or scored approximately 1 in. o.c. on back side. In the latter case, after core has been broken at each cut, gypsum board should be applied to curved contour and anchored in place. At arrises of arch (exterior or interior "corners" formed at meeting of adjoining angle surfaces) joint compound and reinforcing tape or corner bead should be applied. Tape or corner bead should be snipped at intervals along one side to permit it to conform to curved contour.

14.2.1 To apply board, place a stop at one end of curve and then gently and gradually push on other end of board, forcing center against the framing until curve is complete.

14.2.2 By moistening face and back paper thoroughly and allowing water to soak well into core, board may be bent to still shorter radii. When board dries thoroughly, it will regain its original hardness.

Table 7
Bending Radii

Gypsum Board Thickness Inch	Bent Lengthwise Feet	Bent Widthwise Feet
1/4	5	15
3/8	7.5	25
1/2	*10	—

* Bending two 1/4 in. pieces successively permits radii shown for 1/4 in.

15. Application of Gypsum Board to Receive Adhesively Applied Ceramic or Plastic Wall Tile or Plastic Finished Wall Panels

15.1 Framing around tub enclosures and shower stalls should allow sufficient room so that inside lip of tub, prefabricated receptor or hot-mopped sub-pan will be properly aligned with face of gypsum board. (See Figures 9 and 10.) This may necessitate furring out from studs the thickness of gypsum board to be used 1/2 in. or 5/8 in. less thickness of the lip, on each wall abutting a tub, receptor, or sub-pan. Interior angles should be framed or blocked to provide solid backing for interior corners.

15.2 When framing is spaced more than 16 in. o.c., suitable blocking or backing should be located approximately 1 in. above top of tub or receptor and at gypsum board horizontal joints in area to receive tile. When surface finish is ceramic tile, spacing of studs 2-1/2 in. thick or less should not exceed 16 in. o.c. Studs 3-1/2 in. or more in thickness may be spaced 24 in. o.c. provided blocking described above is utilized.

Note to Specifier: Appropriate blocking, headers or supports should be provided to support tub, other plumbing fixtures and to receive soap dishes, grab bars, towel racks and similar items as may be required.

15.3 Water resistant gypsum board (ASTM C 630) should be used as a base for application of ceramic or plastic wall tile or plastic finished wall panels in wet areas such as tub and shower enclosures. Regular gypsum board may be used as a base for tile and wall panels

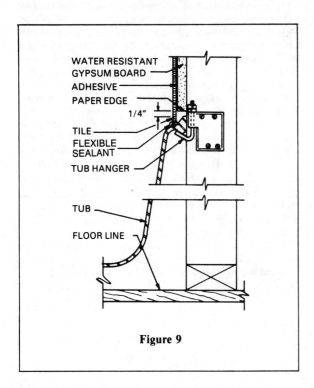

Figure 9

in dry areas. Gypsum board should not be used in extremely critical exposure areas such as saunas, steam rooms or gang shower rooms, and water resistant gypsum board should not be used on ceilings. Gypsum board used as a base for tile or wall panels in tub or shower enclosures, etc., should not be foil backed and **should not be applied over any vapor retarder.** Water resistant gypsum backing board should not be used on ceilings or soffits.

NOTE: A single layer of asphalt impregnated felt, 15 lb. or less, applied as part of the wall system, is not considered a vapor retarder.

15.3.1 Adhesive for application of surfacing material over gypsum board should be as recommended by manufacturers of surfacing material.

15.3.2 Waterproof receptors, pans or sub-pans should have an upstanding lip or flange which should be a minimum of 1 in. higher than water dam or threshold contained in entry way to shower.

15.4 Water resistant gypsum board should be applied perpendicularly with factory edge spaced a minimum of 1/4 in. above lip of receptor, tub or sub-pan. Shower pans, receptors or tubs should be installed prior to erection of gypsum board. FOR FIRE OR SOUND RATED CONSTRUCTION, A BASE PLY GYPSUM BOARD SHOULD EXTEND FULL HEIGHT, FLOOR TO CEILING. (See Figure 11.) If necessary, face ply gypsum board should be properly furred away from framing members as specified in 15.1 and installed so that inside surface of lip of tub, or upstanding leg of receptor or pan will be properly aligned with exposed surface of gypsum board.

15.4.1 In areas to receive tile, water resistant gypsum board should be attached by nails or screws spaced not more than 8 in. o.c.

15.4.2 Gypsum board applied with adhesive only should not be used as a base to receive tile.

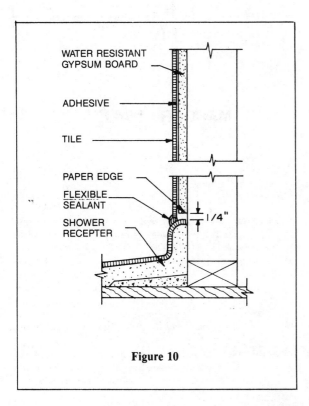

Figure 10

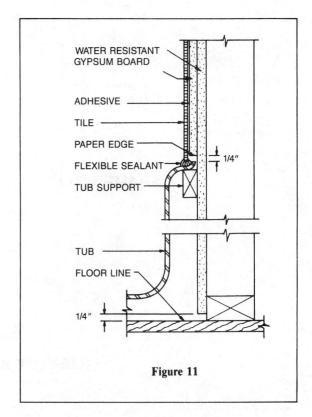

Figure 11

15.4.3 Tile, in combination with the bonding adhesive used, should be applied to protect the gypsum board and any water sensitive materials if present (such as joint compound) from penetration of water or moisture.

15.4.3.1 For application of ceramic, plastic or metal wall tile or plastic finished rigid wall panels or other types of surfacing materials over gypsum board in wet or dry areas, recommendations of manufacturer of tile, wall panels, or other surfacing materials should be followed. Responsibility for performance of completed installations shall rest with surfacing material manufacturer and/or surfacing material applicator.

15.4.3.2 For general recommendations see Appendix.

16. Exterior Application of Exterior Gypsum Soffit Board or Gypsum Wallboard

16.1 Exterior gypsum soffit board or gypsum wallboard 1/2 in. or 5/8 in. in thickness may be used for ceilings of carports, open walkways, porches and soffits of eaves that are horizontal or inclined downward away from building. Framing should be no more than 16 in. o.c. for 1/2 in. thick gypsum board nor more than 24 in. o.c. for 5/8 in. thick gypsum board and should be installed perpendicularly (horizontally) in accordance with foregoing specifications except as herein modified. For gypsum board lay-in panel application, consult panel manufacturer.

16.2 Suitable facia and molding should be provided around perimeter to protect gypsum board from direct exposure to water. Unless protected by metal or other water stops, edges of board should be placed not less than 1/4 in. away from abutting vertical surfaces. Joints and fastener heads should be treated as specified in Section 18.

16.3 Exposed surfaces of gypsum board should receive a prime coat and a finish coat of exterior paint.

16.4 Adequate ventilation should be provided for space immediately above installations as specified in Section 1.4 (See figure 12.)

16.5 When gypsum board is installed as a ceiling in covered walkways and canopies, adequate provision must be made against wind uplift of gypsum board panels.

16.6 Water resistant gypsum backing board should not be used on ceilings or soffits.

17. Electric Radiant Heating Systems for Gypsum Board Ceilings

17.1 Gypsum board should not be less than the minimum recommended for horizontal (perpendicular) application to the ceiling framing and in no case should the base layer be less than 1/2 in. thickness. Both layers should be applied horizontally to framing members. Application should be as required in Table 3 of Section 5.

17.2 Electric heating cables should be securely attached to gypsum board in accordance with recommendations of cable manufacturer and requirements of National Electrical Code.

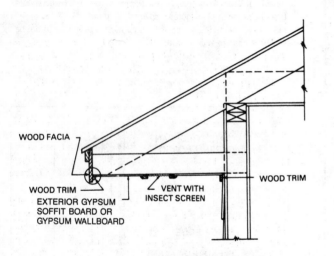

FRAME WALL

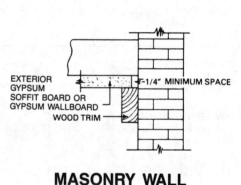

MASONRY WALL

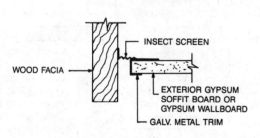

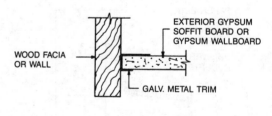

Figure 12

ALTERNATE FACIA DETAILS

17.3 Cables should be parallel to and between framing member with at least 1-1/4 in. clearance from center of framing member on each side so that at least 2-1/2 in. wide unobstructed strip is provided under each framing member.

17.4 Cables should cross framing members only at ends of ceiling 4 in. to 6 in. from wall. There should be at least a 4 in. space completely around perimeter of each ceiling which is clear of cables. Cables should be kept at least 8 in. clear of all openings.

17.5 All inspections and testing of heating system should be completed before application of a dense, low shrinkage filler.

17.6 Under operating conditions, gypsum board core should not be exposed to a temperature exceeding 125°F unless otherwise specifically recommended by gypsum board manufacturer.

17.7 Complete embedding of heating cable is necessary. Area between heating cables should be filled with a dense, low shrinkage filler and leveled.

17.8 Apply face layer of gypsum board, 1/2 in. maximum thickness, with nails or screws around perimeter of ceiling spaced 8 in. to 10 in. away from wall to prevent striking heating elements where they cross framing members. Field of the board should be applied as described in Section 5.3.1. Face layer should not be adhesively applied.

17.8.1 All joints and heads of fasteners should be finished in accordance with recommendations of gypsum board manufacturer. Allow a minimum one week with good drying conditions and two weeks in cold conditions before operating heating system.

17.9 If gypsum board is not used as a finish, a special spray texture finish specifically recommended for radiant heat ceilings may be applied. The single layer gypsum board and cables should be applied as recommended heretofore with floating ceiling angles per Section 5.5 and joint treatment as specified by the radiant heat finish manufacturer.

17.10 Design Data

17.10.1 Data in Table 8 is recommended for use in calculating heat transmission coefficient (U) for gypsum board constructions.

17.11 Decoration

17.11.1 Decoration of finished work, including sizing or sealing should not proceed until heating system has been test operated for a minimum of 24 hours.

17.11.2 The heating system should then be turned off and allowed to cool before decorating.

17.11.3 Where texture finishes are used, such finishes should be allowed to thoroughly set or dry prior to operation of heating system.

Table 8
Thermal Performance Values
of Gypsum Board

Thickness	Conductance "C"	Resistance "R"
Inch	Btu/(Ft²•H•°F)	°F•h•ft³/Btu
3/8	3.00	0.32
1/2	2.36	0.45
5/8	1.88	0.56
3/4	1.50	0.64
1	1.20	0.85

18. Finishing of Gypsum Board

18.1 Gypsum board should be finished with materials which meet ASTM C 475 or those materials recommended by gypsum board manufacturer. (See Section 3.4.)

18.2 Compounds for taping and finishing may be drying or setting type. Do not mix one type of compound with another unless recommended by joint compound manufacturer.

18.2.1 When applied, compounds should be of a chemical composition compatible with previous coats.

18.2.2 No finishing operation should be done until interior temperature has been maintained at a minimum of 50°F for a period of at least 48 hours and thereafter until compounds have completely dried.

18.2.3 Adequate and continuous ventilation should be provided to insure proper drying, setting or curing of taping and finishing compounds.

18.2.4 Gypsum board should be kept free of any dirt, oil or other foreign matter which could cause a lack of bond. All dents or gouges should be filled. Mechanical fasteners should be set below plane of board. All joints should be even and true. Board should be tight against framing members.

18.3 Taping and finishing should be done using proper hand tools such as broad knives or trowels with straight and true edges or mechanical tools designed for this purpose.

18.3.1 Tape should be properly applied either by applying compound to joint (buttering), pressing in tape, and wiping off excess compound, or by mechanical tool designed for this process.

18.3.2 Second coat should be applied with tools of sufficient width to extend beyond joint center approximately 3-1/2 inches. Compound should be drawn down to a smooth even plane. After drying or setting, treated surface should be sanded or otherwise smoothed as needed (See 18.10) to eliminate any high spots or excessive compound. Non-setting compounds should be allowed to dry thoroughly between coats before sanding (See 18.7) and application of next coat.

18.3.3 Third coat when required, should be applied with tools which will permit feathering of joint treatment edges approximately 6 in. from center of joint. After drying, final coat should be lightly sanded (See 18.10) to leave a smooth even surface covering the joint. Caution should be taken not to raise nap of paper when sanding. (See Section 18.10.)

18.4 Fastener heads should be covered with three coats each applied in a different direction except as noted in 18.3. Allow each coat to dry before applying subsequent coats. All cut-outs should be back-filled with compound used for taping or finishing so there is no opening larger than 1/4 in. between gypsum board and fixture or receptor.

18.5 All cut edges and openings around pipes and fixtures should be caulked flush with waterproof, nonhardening caulking compound.

18.6 Care must be taken to insure all tools and containers are kept clean and free from foreign materials. Only drinkable water should be used for mixing powder compounds or to thin premixed materials. Once setting materials have been mixed, no water should be added. Compounds should not be allowed to freeze.

18.7 Setting type compounds can receive additional coats as soon as material has set and before it dries completely.

18.8 For Gloss Paints. After all irregularities have been eliminated and the joint treatment surfaces sanded or sponged where required, a thin skim coat of joint compound shall be applied to the entire surface of the board. This will minimize any suction, porosity or other surface variations between the joint compound and the face paper surfaces. Caution should be taken to eliminate laps or tool marks in the skim coating operation. The wall or ceiling surface should be lightly sanded or sponged where required to assure a smooth and even surface. A high quality primer/sealer should be applied prior to decoration.

18.9 For Severe Lighting Conditions. Where the wall or ceiling surfaces will be subjected to severe natural or artificial side lighting, it is recommended that a thin skim coat of joint compound be applied to the entire board surface to improve fastener and joint concealment. It is also recommended that a high quality primer/sealer be applied prior to decoration.

18.10 APPROVED PROTECTIVE RESPIRATORS SHOULD BE WORN WHEN MIXING POWDER OR SANDING. MIXING SHOULD BE DONE ACCORDING TO MANUFACTURER'S DIRECTIONS. DRILL SPEEDS SHOULD NOT EXCEED THOSE RECOMMENDED BY JOINT COMPOUND MANUFACTURER.

19. Decoration of Gypsum Board

19.1 Decoration of gypsum board should not proceed until finishing materials, applied as recommended in Section 18 are dry.

19.2 Gypsum board surfaces to be decorated with paint, texture, wallpaper or vinyl covering should be primed or sealed prior to decoration and allowed to completely dry.

19.3 Gypsum board surfaces to be painted or textured should be primed or sealed with specific primer or sealer recommended by paint or texture manufacturer for application over gypsum board.

19.4 See section 18.8 for surface preparation when gloss paints will be used. See section 18.9 for surface preparation when severe lighting conditions will exist.

APPENDIX

A.1 Application of Ceramic or Plastic Wall Tile or Plastic Finished Wall Panels over Gypsum Board.

A.1.1 Application of gypsum board under this specification shall be in accordance with requirements of Gypsum Association Recommended Specifications for Application and Finishing of Gypsum Board, GA-216. Following are recommendations for application of surfacing materials, when not in conflict with recommendations of surfacing materials manufacturer.

A.1.1.1 All cut edges and openings around pipes and fixtures should be caulked flush with waterproof, flexible compound.

A.1.1.2 Surfacing material should be applied down to top surface or edge of finished shower floor, return, or tub and installed so as to overlap top lip of receptor, sub-pan or tub and should completely cover following areas:

(1) Over tubs without showerheads — 6 in. above rim of tub.
(2) Over tubs with showerheads — a minimum of 5 ft. above rim or 6 in. above height of showerhead, whichever is higher.

(3) Shower stalls — a minimum of 6 ft. above shower dam or 6 in. above showerhead, whichever is higher.
(4) All gypsum board window sills and jambs in shower or tub enclosures should be covered to a like height.
(5) Surfacing material should be applied to full specified height for a distance of at least 4 in. beyond external face to tub or receptor. Areas beyond an exterior corner are included.

A.1.1.3 Where plastic finished rigid wall panels are used as a surfacing material, following precautions should be taken:

(1) Type and shape of moldings recommended by manufacturer of surfacing material should be used. Recommended tub moldings should be used at base where surfacing materials abut tub, shower floor, or curb. Such moldings should be set in waterproof, flexible sealant.
(2) Joints should be filled in such a manner as to leave no voids for water penetration.
(3) A bead of adhesive should be applied as a dam between the back surface of finishing material and tub or receptor to prevent any leakage of water at joint.

A.2 Application of Water-based Spray Texture Finishes on Gypsum Wallboard Ceilings.

A.2.1 When water-base spray texture finishes are to be used on gypsum wallboard ceilings under this specification, the following conditions should be met:

A.2.1.1 For framing spaced 16″ o.c., 1/2″ thick gypsum wallboard may be used only if applied perpendicular to the framing.

A.2.1.2 For framing spaced 24″ o.c., 5/8″ thick gypsum wallboard may be used applied perpendicular to the framing.

A.2.2 A high quality pigmented alkyd or oil-based primer-sealer should be applied to the wallboard prior to the application of any water-based texture.

A.2.3 Failure to comply with Tables 1 and 3 of this specification can result in ceiling sag.

A.2.3.1 Textures, in conjunction with any one or more of the following conditions, may cause the gypsum wallboard ceiling to sag:

a) Unventilated building.
b) Use of vapor retarder(s) under certain conditions.
c) Prolonged high humidity due to either weather conditions or closed building units (poor drying conditions).
d) Inadequate framing support (can occur where framing changes direction).
e) Improper type or thickness of gypsum board.
f) Lack of proper primer-sealer.

Commonly Used Metric Conversions

Gypsum Board Thickness	Framing Spacing	Fastener Spacing	°F	°C	K•M²/W
1/4 in. — 6.4 mm	16 in. — 406 mm	2 in. — 51 mm	50 — 10		
3/8 in. — 9.5 mm	24 in. — 610 mm	2-1/2 in. — 63.5 mm	125 — 52		0.056
1/2 in. — 12.7 mm		7 in. — 178 mm			0.079
5/8 in. — 15.9 mm		8 in. — 203 mm			0.099
3/4 in. — 19.0 mm		12 in. — 305 mm			0.150
1 in. — 25.4 mm		16 in. — 406 mm			
		24 in. — 610 mm			

CHAPTER THREE-B

SPECIFICATIONS FOR DETENTION AND SECURITY WALLS OF LATH AND PLASTER FOR

JAILS AND PRISONS
COURT ROOMS
HOSPITALS
BANKS
VAULTS

LATH & PLASTER DETENTION & SECURITY WALLS

(Reproduced through courtesy of Western Conference of Lathing and Plastering Institutes Inc.)

GENERAL

For more than a half century, lath and plaster has been used to construct walls, partitions and ceilings in all kinds of security facilities. This includes jails, prisons, courtrooms, hospitals, sanitariums, banks and vaults. This brochure has been compiled to illustrate some of the recent systems used. Specific information on individual products is readily available from various manufacturers and your local lath and plaster information bureau.

COMPONENTS

STUDS, CHANNELS AND RODS

There are numerous steel sections which can be used to frame vertical and horizontal security elements. Steel studs are available in widths of 1-5/8" to 6" and in metal thickness of 25 gauge (0.0209") to 14 gauge (0.074"). Heavier gauge studs are preferred to facilitate welding and for producing stronger walls. Steel studs are available painted or fabricated from galvanized steel. Painted sections generally cost less and are more easily welded than galvanized steel. 16 gauge cold rolled channels are produced in 3/4, 1-1/2 and 2 inch widths. Hot rolled sections in 3/4", 1-1/2" and 2". Solid plaster partitions are sometimes framed with steel pencil rods or reinforcing bars, 3/8" to 3/4" in diameter. They can easily be wire tied or welded together to form a grid to which expanded metal security mesh and metal lath is attached.

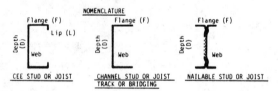

TYPES OF STUDS

METAL LATH

Expanded metal lath is available flat, self-furred and as riblath. The weight of expanded lath is measured in lbs. per sq. yard, such as 2.5, 2.75, 3.4 and 4.0. It can be fabricated from galvanized steel or coated with rust-resistant coating. Welded wire and woven wire laths are fabricated from galvanized wire into a grid or hexagonal pattern. The weight of wire lath can be 1.1 to 1.95 lbs. per square yard. Because of the unusually large openings in wire lath, this material is normally fabricated with a paper backing to aid plastering operations. Attachment of lath to steel framing is done with wire ties or self-tapping screws. Metallic laths are usually nailed or stapled to wood framing.

EXPANDED METAL LATH

EXPANDED METAL SECURITY MESH

Expanded metal is a rigid, non-raveling piece of metal which has been slit and drawn in a single operation. The expanded metal is stronger and more rigid than the original sheet, before expanding. It is too heavy and rigid to be used as a plaster base but provides a formidable barrier in the core of a plastered partition. Expanded metal security mesh is available in a standard or flattened pattern; carbon steel (plain or galvanized), stainless steel or aluminum in metal thickness of .030" to .119". The size (length and width) of the diamond openings designates the style. Usually the term SWD (Short Way of Design or Short Way of Diamond) is used along with the thickness of the metal to designate a style of mesh. Every manufacturer has a different variety of styles but generally speaking, diamond designs are available in SWD's of 3/16", 1/4", 1/2", 5/8", 3/4", 1" and 1-1/2".

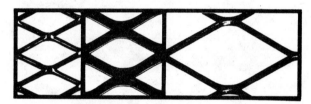

VARIOUS SIZES OF SECURITY MESH

PLASTER

Plaster for security wall construction must not only be strong but must also be crack resistant and economical to apply. Mixes prepared from portland cement and some gypsum cements offer these qualities. Authorities agree that a minimum 1400 psi plaster is desirable for security purposes.

Following are some compressive strength figures for different plaster mixes:

PLASTER MIX	COMPRESSIVE STRENGTH PSI
1. Portland Cement, Lime and Sand (1:3/4:6)	1500
2. Portland Cement, Plastic Cement & Sand (1/2:1/2:4)	1400
3. STRUCTOBASE Gypsum and Sand (1:2)	2800
STRUCTOBASE Gypsum and Sand (1:2-1/2)	1900
STRUCTOBASE Gypsum and Sand (1:3)	1400
4. Wood Fiber Gypsum without Aggregate (Neat)	1750
5. Wood Fiber Gypsum and Sand (1:1)	1400

Portland cement plaster is recommended in locations subject to frequent or severe wetting. It is, however, more difficult to work and generally costs more than gypsum plasters. It is difficult to get crack free, smooth troweled finishes in portland cement plaster — so sand floated finishes are suggested.

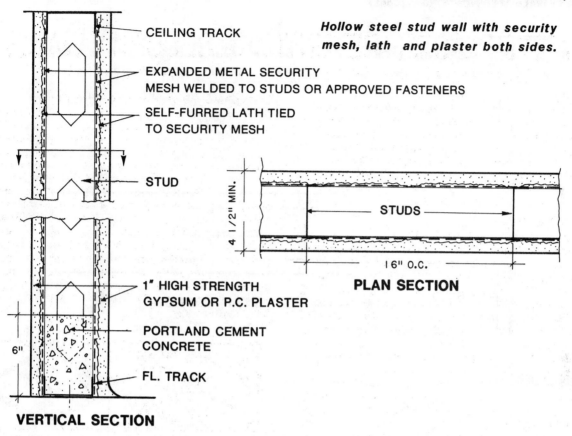

CEILING TRACK

Hollow steel stud wall with security mesh, lath and plaster both sides.

EXPANDED METAL SECURITY MESH WELDED TO STUDS OR APPROVED FASTENERS

SELF-FURRED LATH TIED TO SECURITY MESH

STUD

4 1/2" MIN.

STUDS

16" O.C.

PLAN SECTION

1" HIGH STRENGTH GYPSUM OR P.C. PLASTER

PORTLAND CEMENT CONCRETE

6"

FL. TRACK

VERTICAL SECTION

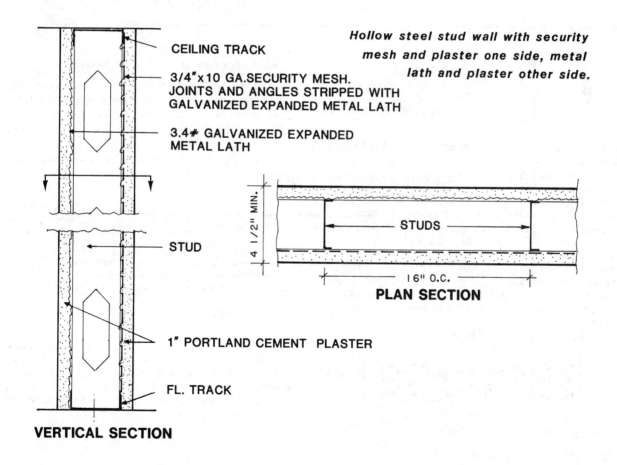

CEILING TRACK

Hollow steel stud wall with security mesh and plaster one side, metal lath and plaster other side.

3/4"x10 GA.SECURITY MESH. JOINTS AND ANGLES STRIPPED WITH GALVANIZED EXPANDED METAL LATH

3.4# GALVANIZED EXPANDED METAL LATH

4 1/2" MIN.

STUDS

16" O.C.

PLAN SECTION

STUD

1" PORTLAND CEMENT PLASTER

FL. TRACK

VERTICAL SECTION

255

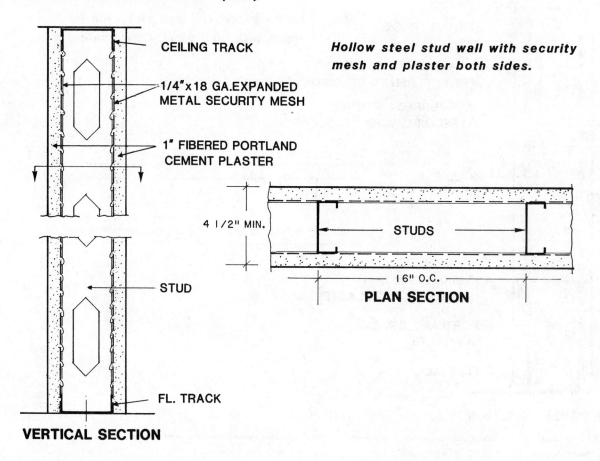

CEILING TRACK

1/4"x18 GA.EXPANDED METAL SECURITY MESH

1" FIBERED PORTLAND CEMENT PLASTER

Hollow steel stud wall with security mesh and plaster both sides.

4 1/2" MIN.

STUDS

16" O.C.

PLAN SECTION

STUD

FL. TRACK

VERTICAL SECTION

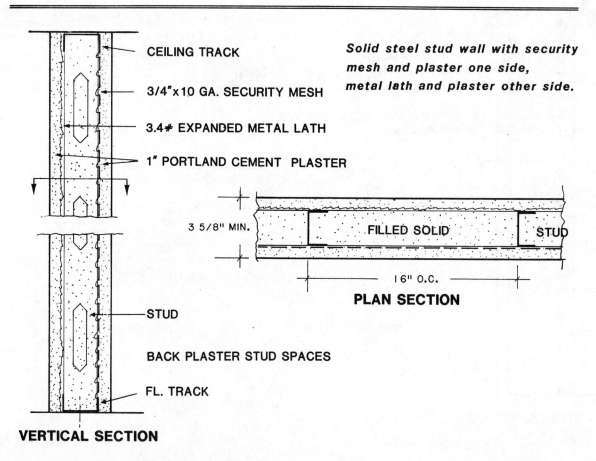

CEILING TRACK

3/4"x10 GA. SECURITY MESH

3.4# EXPANDED METAL LATH

1" PORTLAND CEMENT PLASTER

Solid steel stud wall with security mesh and plaster one side, metal lath and plaster other side.

3 5/8" MIN.

FILLED SOLID

STUD

16" O.C.

PLAN SECTION

STUD

BACK PLASTER STUD SPACES

FL. TRACK

VERTICAL SECTION

Channel framed, solid plaster wall with metal lath and security mesh .

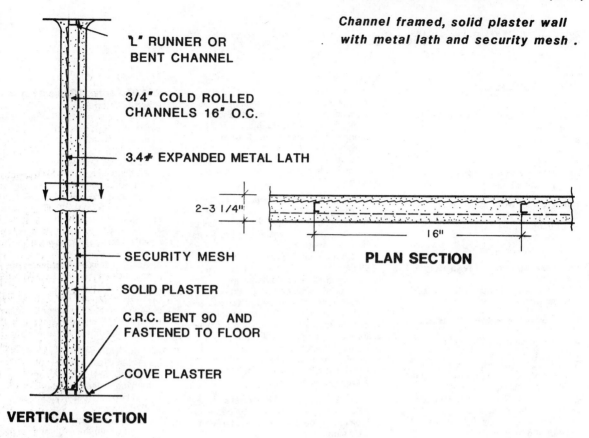

"L" RUNNER OR
BENT CHANNEL

3/4" COLD ROLLED
CHANNELS 16" O.C.

3.4# EXPANDED METAL LATH

2–3 1/4"

SECURITY MESH

SOLID PLASTER

C.R.C. BENT 90 AND
FASTENED TO FLOOR

COVE PLASTER

16"

PLAN SECTION

VERTICAL SECTION

Studless solid plaster partition .

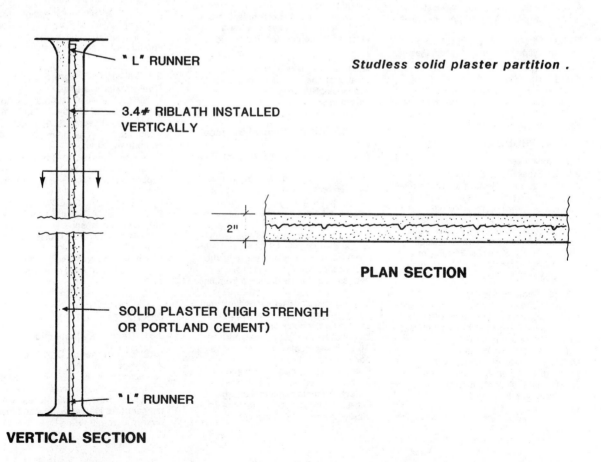

" L" RUNNER

3.4# RIBLATH INSTALLED
VERTICALLY

2"

SOLID PLASTER (HIGH STRENGTH
OR PORTLAND CEMENT)

" L" RUNNER

PLAN SECTION

VERTICAL SECTION

Security Wall Specifications

PART 1: GENERAL

1.01 Description
A. Steel Stud Framing, Lathing and Plastering.
B. Special items of work .
1. Steel studs
2. Security mesh
3. Metal lath and accessories
4. High strength plaster

1.02 Related work specified elsewhere
A. Concrete curbs and bases.
B. Wood framing.
C. Structural steel framing

1.03 Requirements of regulatory agencies
A. Comply with all applicable code and regulations.

1.04 Reference specifications and standards
A. Product manufacturers printed specifications.
B. Metal Lath/Steel Framing Manufacturers Specifications for Metal Lathing and Furring.
C. ANSI A42.3, Specification for Lathing and Furring for Portland Cement, Portland Cement-Lime Plastering, Exterior and Interior.
D. ASTMC841 Standard Specification for Installation of Interior Lathing and Furring
E. ASTM C926, Standard Specification for Application of Portland Cement Based Plaster
F. ASTM C842, Standard Specification for Application of Interior Gypsum Plaster

1.05 Submittals
A. Certification of conformance of materials with specification requirements.
B. Submit shop drawings and supporting calculations as required.

PART 2: PRODUCTS

2.01 Materials
2.01.01 Steel Studs & Joists
A. All studs and/or joists and accessories shall be of the type, size, gauge and spacing as shown on the drawings.
B. All studs and/or joists and accessories shall be coated with rust inhibitive paint, or be fabricated from galvanized steel, corresponding to the requirements of ASTM A446, Grade A, with a minimum yield of 33 ksi.

2.01.02 Security Mesh
A. Expanded steel security mesh shall be slit and drawn into a diamond patterned mesh from carbon steel and roll flattened to provide a finish thickness of not less than .048 inches (16 gauge).
B. The short way of the diamond openings shall not be more than 3/4".
C. Material shall conform to Military Specification MIL-M-17194C Type II Class 1.

2.01.03 Lath
A. Expanded metal lath shall be made from copper alloy steel sheets and given a protective coat of rust inhibitive paint after fabrication, or shall be made from galvanized steel.
B. Lath shall be self-furring design and weigh not less than 3.4 lbs. per square yard.

2.01.04 Lath Accessories
A. Casing beads, fabricated from not less than 26 gauge steel or zinc alloy, shall be installed wherever plaster stops or abuts dissimilar materials.
B. Corner reinforcement for Portland Cement Plaster. Plaster at all external corners shall be reinforced with galvanized, expanded wing corner reinforcements or welded galvanized wire corner reinforcements. Internal corners shall be reinforced with cornerite not less than 4" wide (2" on each surface).
C. Cornerbeads — for gypsum plaster. Small nose or bull nose cornerbeads shall be fabricated from minimum 26 gauge galvanized steel.

2.01.05 Mechanical Fasteners
A. Self-drilling screws shall be panhead type long enough to penetrate through steel studs at least 1/4 inch.
B. Nails (for attaching metal lath to wood supports) shall be not less than No. 11 gauge barbed roofing nails which provide penetration into supports at least 1-3/8" or power driven staples which will provide at least 3/4" penetration into supports.
C. Galvanized wire for securing metal lath to steel studs and security mesh shall be at least one loop of 16 gauge wire or two loops of 18 gauge wire. Wire for attaching accessories to lath shall be 18 gauge and annealed.

2.01.06 Portland Cement Basecoat Plaster Materials
A. Portland Cement, ASTM C150, Type I, II or III.
B. Plastic cement conforming to the requirements of ASTM C150, except in respect to the limitations on insoluable residue, air entrainment and additions subsequent to calcination.
C. Hydrated lime, ASTM C206, Type S.
D. Sand aggregate, ASTM C897. Where specification sand is not available, the best washed sand in terms of sieve analysis, cleanliness, freedom from fines, etc., may be used.
E. Alkali resistant natural or synthetic fibers, 1/4 to 1/2 inch in length shall be used in portland cement scratch and brown coats to improve crack resistance. Alkali resistant glass fiber shorts, or polypropylene fibers.
F. Water shall be clean, fresh and suitable for domestic consumption. It must be free of such amounts of mineral or organic substances which could affect the set, the plaster or any metal in the system.

2.01.07 Portland Cement Finish Coat Plaster Materials
A. Proprietary, portland cement based, acrylic modified finish coat plasters requiring only the addition of mixing liquid on the job.
B. Job mixed finish coats:
1. Portland cement — ASTM C-150, Type I, II, III, white cement where designated.
2. Hydrated lime, ASTM C-206, Type S.
3. Sand — ASTM C-897. Exceptions to sand gradation for finish coat shall be stipulated where texture surfaces are specified and aggregate gradation has a significant role.
4. Acrylic admix — Approved emulsified product developed for use with water in preparing mixing liquid.
C. Water — potable and clean and free of such amounts of mineral or organic substances as would affect the set, the plaster or any metal in the system.

2.01.08 High Strength Gypsum Basecoat Plaster Materials
A. STRUCTO-BASE Gypsum Plaster, complying with requirements of ASTM C-28 and Federal Specification SS-P-00402B, Type II.
B. Wood Fiber Gypsum Plaster, complying with the requirements of ASTM C-28 and Federal Specification SS-P-00402B, Type III.
C. Sand Aggregate, ASTM C-35, and shall be clean, sharp and free of materials which could aversely affect the plaster.
D. Water, potable and not contain impurities that affect the setting of gypsum.

2.01.09 High Strength Gypsum Finish Coat Plaster Materials
A. STRUCTO—GAUGE, high strength gypsum gauging plaster, ASTM C-28, Federal specification SSOP-00402B, Type V with the added requirement of 5000 psi compressive strength.
B. Keenes Cement, dead burned gypsum gauging, complying with ASTM C-61 and Federal Specification SS-P-00410, Type I or II.
C. Finish lime, ASTM C-206, Type S.
D. Sand, ASTM C-35, for float finishes shall be clean, washed, graded white silica sand passing a (30 mesh) (20 mesh) screen.
E. Water — potable, and not containing impurities that affect the setting of gypsum.

PART 3: EXECUTION

3.01 Preliminary Inspection

3.02 Environmental Conditions
A. Do not apply plaster to any base containing frost. Plaster mixes shall not contain frozen ingredients. Plastering operations must not take place when temperatures are below 40° F and must be protected from freezing for a period of not

Security Wall Specifications (Continued).

less than 48 hours after set has occurred.
B. Protect plaster from uneven and excessive evaporation during dry weather and hot dry winds.

3.03 Steel Stud Framing

3.03.01 Runner tracks shall be securely anchored to the supporting structure as shown on drawings.

A. Anchor floor track to concrete with washers and nuts fastened to 1/2″ diameter bolts set in concrete not over 24″ o.c. or power driven fasteners or expansion drivers providing equal holding power.
B. Attach track to steel with wire ties, bolts, screws or welds not over 24″ o.c.
C. Abutting sections of runner track shall be securely anchored to a common structural element or be butt welded or spliced together.

3.03.02 Attach studs to track by welding or screwing.

A. Studs shall be cut to proper length and installed plumb, without bows and evenly spaced.
B. Provision for structure vertical movement shall be provided as indicated on the drawings.
C. Install 1-1/2″ cold rolled channel stiffeners not over 4′-6″ apart. For partitions having unsupported height: from 8′ to 18′, place stiffeners on alternate interior sides; over 18′, place on each interior side. Wire tie stiffeners to each stud.

3.03.03 Expanded steel security mesh shall be fillet welded to steel studs not over 8″ o.c. or equivalent approved fasteners. Edge welds must be within 2 inches of edge.

A. End joints shall be butted and occur over a stud. Edge joints shall be butted and wire tied at mid point between supports.

3.03.04 Lath Installation

A. Attach expanded metal lath to security mesh with wire ties not over 6″ o.c. vertically and 16″ o.c. horizontally.
B. Stagger end laps. Return lath 4″ around corners or use cornerite. Carry lath over concrete foundations at least 2 inches.
C. Direct attachment to steel studs shall be made with wire ties or screws not over 6″ o.c.
D. Attach lath to wood supports at furring points and not over 6 inches o.c.

3.03.05 Lath Accessories

A. Install casing beads, control joints, corner reinforcement and other metal plaster accessories to plaster line (using shims if necessary). Attach accessories by wire-tying, nailing or stapling through wings or holes provided. Attachment shall be strong enough to hold accessory in place during plastering operations.

3.06 Basecoat Plaster

3.06.01 Proportions

A. Portland cement (common), lime and sand or common portland cement and plastic portland cement. In proportions set forth in ASTM specification C-926 with added requirement that 1 lb. of alkali resistant fibers be added per sack of cement.
B. STRUCTO BASE high strength gypsum, sanded 1:2 for scratch and 1:3 for brown.
C. Wood Fiber plaster neat or sanded up to 1:1 for scratch and 1:1 for brown.

3.06.02 Application — shall be done in accordance with ASTM C-926 or ASTM C-842, whichever is applicable.

3.07 Finish Coat Plasters

3.07.01 Proportions

A. Finish coat plasters for portland cement basecoats shall be proprietary acrylic modified portland cement based or job mixed finish coat plaster as described in ASTM C-926.
B. Finish coat plaster for STRUCTO-BASE high strength gypsum basecoat shall be prepared from STRUCTO—GAUGE gauging plaster and lime proportioned by dry weight 1:1.
C. Finish coat for wood fiber basecoat plasters shall be STRUCTO-GAUGE and lime proportioned by dry weight 1:2 or Keenes cement and lime proportioned by dry weight 4:1.

3.07.02 Application — shall be done in strict accordance with ASTM C-926 or ASTM C-842, whichever is most applicable.

CHAPTER THREE-C

FIREPROOFING
WITH
SPRAY-APPLIED
MATERIALS

SPECTEXT SPECIFICATION FOR

CELLULOSE FIREPROOFING
SECTION 07253

This section includes spray applied cellulosic fireproofing. A protective overcoat may be required to meet code imposed flame spread ratings. This insulation type should not be used in return air plenums unless the finished surface has been tamped, overcoated or sealed. This section includes performance, proprietary and descriptive type specifications. Edit to avoid conflicting requirements. Issued January 1986.

PART 1   GENERAL

1.01   SECTION INCLUDES

   A.   Cellulose fireproofing, spray applied.

   B.   Protective [overcoat.] [sealer.]

1.02   PRODUCTS INSTALLED BUT NOT FURNISHED UNDER THIS SECTION

   A.   Section 09206 - Metal Furring and Lathing: Metal lath over [structural members.] [_____.]

1.02 Use this Article when open web structural members require wrapping with metal furring to maximize the effectiveness of the insulation and minimize overspray. List sections which furnish products for installation under this section.

1.03   RELATED SECTIONS

   A.   Section 07218 - Sprayed Cellulose Insulation.

   B.   Section 07255 - Cementitious Fireproofing.

   C.   Section [____-_____]:   Intumescent coatings.

1.04   REFERENCES

   A.   ASTM E84 - Test for Surface Burning Characteristics of Building Materials.

   B.   ASTM E119 - Fire Tests of Building Construction and Materials.

   C.   ASTM E760 - Effect of Impact on Bonding of Sprayed Fire-Resistive Material Applied to Structural Members.

   D.   UL - Fire Hazard Classifications.

1.04 List only reference standards that are included within the text of this section. Refer to Section 01090 for assistance with abbreviations and addresses of standards agencies and associations. Edit the following as required for project conditions.

1.05   PERFORMANCE REQUIREMENTS

   A.   Installed materials to conform to UL Assembly No. [____] for a fire resistance rating of [_____] hours.

   B.   Bond Strength of Fireproofing: ASTM E760, tested to provide minimum bond strength twenty times weight of fireproofing materials.

   C.   Application: Verify fire test reports of fireproofing application to substrate materials similar to project conditions.

   D.   Verify reports from independant testing agencies, of product proposed for use, which indicate conformance to ASTM E119 and ASTM E84.

1.05 Include the UL Assembly number which applies to the application of this material. Coordinate specification text with adjacent floor or roof assembly (and ceiling assembly if applicable) in the paragraph below. If several assemblies or hour ratings are required, consider listing in the schedule at the end of this section.

1.06   SUBMITTALS

A.   Submit product data under provisions of Section [01300.] [01340.]

B.   Provide product data on specified products.

C.   Submit [two] [_____] copies of certified test reports of bond strength and product property tests.

D.   Submit manufacturer's installation instructions under provisions of Section [01300.] [01340.]

E.   Submit manufacturer's certificate under provisions of Section [01400] [01405] that [products] [_____] meet or exceed [specified fire resistance rating requirements.] [_____.]

1.07   QUALITY ASSURANCE

A.   Manufacturer: Company specializing in fireproofing materials with [three] [_____] years experience.

B.   Applicator: Company specializing in fireproofing materials [with [_____] years [documented] experience.] [approved by manufacturer.]

1.08   REGULATORY REQUIREMENTS

A.   Conform to [applicable] [_____] code for [flame/fuel/smoke ratings] [overcoating requirements] [_____] and fire resistance ratings.

1.09   MOCKUP

A.   Provide mockup of specified materials under provisions of Section [01400.] [01405.]

B.   Apply materials to representative substrates on site. Conform to requirements for fire ratings and finish texture.

C.   Apply materials to an area of [_____] sq ft ([_____] sq m).

D.   Comply with project requirements as to thickness, density of application, and fire rating.

E.   Examine installation within one hour of application to determine shrinkage and bond to substrate.

1.10   ENVIRONMENTAL REQUIREMENTS

A.   Provide ventilation in areas to receive fireproofing during and 24 hours after application to dry material. Maintain non-toxic, unpolluted working area.

B.   Provide temporary enclosure to prevent spray from contaminating air.

C.   Do not apply spray fireproofing when temperature of substrate material and surrounding air is below 40 degrees F (5 degrees C).

1.11   WARRANTY

A.   Provide [five] [____] year [manufacturer's] warranty under provisions of Section [01700.] [01740.]

B.   Warranty: Applied fireproofing to remain free from cracks, checking, dusting, flaking, spalling, separation, and blistering. Reinstall or repair such defects or failures.

1.06 Request product data and test reports only when specifying by reference standard where the manufacturer's names or product identity is unknown.

1.06C When manufacturer's instructions for specific installation requirements are referenced in Part 3, Execution, include the following request for submittal of those instructions. Edit the Part 3, Execution, statements to avoid conflict with those instructions.

1.08 Verify the fireproofing is acceptable to regulatory agencies for flame/fuel/smoke requirements. A cementitious overcoat may be required.

1.09 A mockup can demonstrate the capabilities of the installer, the ability of the insulation to bond to the substrate, and the overcoat bond to the insulation. Use this Article for full sized erected assemblies required for review of construction, coordination of work of several sections, and observation of operation.

1.09 Ensure the specified ambient and surface temperature requirements required by specified manufacturers are included in this Article.

1.11 This Article extends the warranty period beyond one year. Extended warranties add to the construction cost and may present difficulties to the owner in enforcing them. Specify with caution.

PART 2   PRODUCTS

2.01   MANUFACTURERS

A.   [_____.]

B.   [_____.]

C.   [_____.]

D.   Substitutions: Under provisions of Section [01600.]
      [01630.]

2.02   MATERIALS

A.   Cellulose Fiber Sprayed Fireproofing: Non-combustible
      cellulose with integral binder, conforming to the
      following:

      Fire hazard classification
      for flame/fuel/smoke   [20/10/0] [___/___/___]

      Thermal K (ksi) factor   [0.22] [____] ([0.032] [____])

      Density   [2.0] [____] lb/cu ft
                                   ([32] [____] kg/cu m)

B.   Primer Adhesive: Of type recommended by fireproofing
      manufacturer.

C.   [Overcoat] [and] [Sealer]: Comply with UL requirements
      and manufacturer's recommendations for specific site
      conditions.

                    ****** [OR] ******

C.   [Cementitious Overcoat: [_____] type, spray
      applied; flame/fuel/smoke rating of [____/____/____] in
      accordance with ASTM E84; manufactured by [_____].]

D.   Metal Lath: Expanded metal lath; galvanized.

E.   Water: Clean, potable.

PART 3   EXECUTION

3.01   EXAMINATION

A.   Verify that substrate surfaces are ready to receive work.

B.   Confirm compatibility of surfaces to receive fireproofing
      materials.

C.   Verify clips, hangers, supports, sleeves, and other items
      required to penetrate fireproofing are in place before
      application.

D.   Verify ducts, piping, equipment, or other items which
      would interfere with application of fireproofing are not
      positioned until fireproofing work is completed.

E.   Beginning of installation means acceptance of [existing
      conditions.] [substrate.]

3.02   PREPARATION

A.   Remove incompatible materials which affect bond.

B.   Fill voids and cracks in substrate, remove projections and
      level where sprayed fireproofing is exposed to view as
      finish material.

C.   Protect adjacent surfaces and equipment from damage by
      overspray fall-out, and dusting.

2.01 In this Article, list the product manufacturers acceptable for this project. If product substitution procedure is used, include Paragraph D.

2.02 Edit the following descriptive specifications for any conflicts with manufacturer's products specified above.

2.02 Select product criteria for project requirements when specifying by standards reference; minimize criteria if specifying by manufacturer's product name.

2.02C Use protective spray overcoat (or sealer) to prevent dusting of finished application, to protect fireproofing from contaminants (i.e. dust, oil, mist, moisture, etc.); where direct air flow, such as within a plenum, may erode fireproofing. A cementitious overcoat may be required by code to achieve the required fire rating. Consult with manufacturer. Edit Paragraph C as appropriate.

D. Close off and seal ductwork in areas where fireproofing is being applied.

03 APPLICATION

A. Mix and apply fireproofing in accordance with manufacturer's instructions.

B. [Install metal lath over structural members |and as detailed].]

C. Apply [primer adhesive and] fireproofing in sufficient thickness to achieve rating with as many passes as necessary to cover with monolithic blanket of uniform density and texture.

3.02C A primer adhesive is required by some manufacturers over certain substrate materials. Specify accordingly.

D. Tamp fireproofing after application to provide dense smooth surface.

E. Apply spray [overcoat] [sealer] on surfaces of fireproofing.

.04 FIELD QUALITY CONTROL

A. Field inspection [and testing] will be performed under provisions of Section [01400.] [01410.]

B. Inspections will be performed to verify compliance with requirements.

C. Correct unacceptable work and provide further inspection to verify compliance with requirements.

304 Usually field inspection relative to the application of sprayed fireproofing is limited to inspecting the substrate preparation, verifying the fireproofing (and overcoat) thickness.

.05 CLEANING

A. Remove excess material, overspray, droppings, and debris.

B. Remove fireproofing from materials and surfaces not specifically required to be fireproofed.

.06 PROTECTION

A. Protect applied sprayed fireproofing from damage under provisions of Section [01500.] [01535.]

.07 SCHEDULE

3.07 Provide a schedule when project requirements have a variety of surfaces. Identify and list required ratings by UL code number, and thickness of fireproofing.

hese Drawing Coordination Considerations describe Drawing
elated items that should be considered with SPECTEXT Section
7253.

Drawing Terms: The following generic terms should be used in Drawing notes to ensure definitions parallel the specification section. If more than one type of an item indicated below is required, indicate on Drawings and in the specification section as "Type A," "Type B," etc.

A. CELLULOSE FIREPROOFING: A wood fibered product, milled to a dry fibered powder consistency, then chemically treated to achieve non-combustible properties. This product is then blended with other cementitious materials to produce a spray applied mixture which air cures to a fire resistive product.

B. CEMENTITIOUS OVERCOAT: A Portland cement based product sometimes used to overcoat the spray applied cellulose to improve its fireproofing properties.

C. METAL LATH: An expanded (plaster) metal lathing product often used to wrap over structural steel framing members, such as open web steel joists, to assist the spray applied cellulose fireproofing to permanently bond to these framing members.

D. OVERCOAT or SEALER: A liquid product spray applied over cellulose fireproofing surface to seal the cellulose surface particles against being loosened by air movement.

E. PRIMER ADHESIVE: A chemical liquid, brush or spray applied to surfaces scheduled to receive cellulose fireproofing to improve bond to substrate.

A. Refer to UL and other applicable reference manuals for roof/ceiling, floor/ceiling, and wall assembly details. These details describe the thickness of fireproofing required, the requirements for adjacent materials, and the conditions to which the fireproofing can be installed. These details may also describe requirements for overcoat or sealer.

Coordination: The following items require special attention when coordinating Drawings and specifications, to ensure parallel information and references.

B. Other authorities having jurisdiction may supplement the UL or other agency assembly details. Determine the applicable agency or agencies and develop Drawing details accordingly.

A. Determine from the mechanical and electrical consultant if components in their design will require suspension or attachment to fireproofed surfaces. If so, require Drawing details to explicitly describe how these conditions will be achieved to ensure fireproofing integrity.

Caution: The following items indicate that caution should be observed prior to incorporating into Drawings or specifications.

B. Exterior wall cladding components such as curtain walls or precast concrete often require site welded attachments to structural spandrel beams. This welding process can ignite cellulose fireproofing under certain conditions. If possible, develop attachment details that could minimize this hazardous potential.

C. Use of this material may require isolation of floor areas of the project during application because of overspray particles.

SPECTEXT SPECIFICATION FOR

CEMENTITIOUS FIREPROOFING

SECTION 07255

This section includes cementitious fireproofing, spray applied to achieve fire protection. This section includes performance, proprietary and descriptive type specifications. **Edit** to avoid conflicting requirements. Issued January 1986.

PART 1   GENERAL

1.01   SECTION INCLUDES

   A.   Cementitious fireproofing, spray applied.

1.02   RELATED SECTIONS

   A.   Section [_____ - _____:] Building structural substrate surfaces.

   B.   Section 07218 - Sprayed Cellulose Insulation.

   C.   Section 07253 - Cellulose Fireproofing.

   D.   Section [_____ - _____:] Plaster fireproofing.

   E.   Section [_____ - _____:] Intumescent coatings.

1.03   REFERENCES

1.03   List reference standards that are included within the text of this section. Refer to Section 01090 for assistance with abbreviations and addresses of standards agencies and associations. Edit the following as required for project conditions. Refer to TAS Section 07255.

   A.   ASTM E84 - Test for Surface Burning Characteristics of Building Materials.

   B.   ASTM E119 - Fire Tests of Building Construction and Materials.

   C.   ASTM E605 - Thickness and Density of Sprayed Fire-Resistive Material Applied to Structural Members.

   D.   ASTM E736 - Cohesion/Adhesion of Sprayed Fire-Resistive Materials Applied to Structural Members.

   E.   ASTM E760 - Effect of Impact on Bonding of Sprayed Fire-Resistive Material Applied to Structural Members.

   F.   UL - Fire Hazard Classifications.

1.04   PERFORMANCE REQUIREMENTS

1.04   Use this Article carefully; restrict paragraph statements to identify system performance requirements or system function criteria only. If a schedule at the end of this section is used to list the assembly rating, delete this Article.

   A.   Cementitious fireproofing system to provide a fire rated assembly rating of [_____] hours for [roof] [_____] and [_____] hours for [typical floor] [_____] assembly.

1.05   SUBMITTALS

A. Submit product data under provisions of Section [01300.] [01340.]

B. Submit product data indicating product characteristics, performance and limitation criteria, [and] [_____.]

C. Submit manufacturer's installation instructions under provisions of Section [01300.] [01340.]

D. Submit manufacturer's certificate under provisions of Section [01400] [01405] that products meet or exceed the specified requirements.

E. Submit test reports under provisions of Section [01400.] [01405.]

F. Submit certified test reports indicating the following:
    1. Bond Strength of Fireproofing: ASTM E760, tested to provide minimum bond strength twenty times weight of fireproofing materials.
    2. Fire test reports of fireproofing application to substrate materials similar to project conditions.
    3. Reports from reputable independent testing agencies, of product proposed for use, which indicate conformance to ASTM E119 and ASTM E84.

1.06   QUALITY ASSURANCE

A. Manufacturer: Company specializing in manufacturing the products specified in this Section with minimum [three] [_____] years [documented] experience.

B. Applicator: Company specializing in applying the work of this Section [with minimum [_____] years [documented] experience.] [approved by manufacturer.]

1.07   REGULATORY REQUIREMENTS

A. Conform to [applicable] [_____] code for [fire resistance ratings.] [_____.]

B. Submit certification of acceptability of fireproofing materials to [authority having jurisdiction.] [[_____] agency.]

1.08   MOCKUP

A. Provide mockup of applied cementitious fireproofing under provisions of Section [01400.] [01405.]

B. Provide [testing] [and] [analysis] of mockup to specified requirements.

C. Apply sample section of [100] [_____] sq ft ([9] [_____] sq m) in size to representative substrates on site.

D. Comply with project requirements as to thickness, density of application, fire rating, and finish texture.

E. Examine installation within [one] [_____] hour of application to determine variance due to shrinkage, temperature, and humidity.

F. Where shrinkage and cracking are evident, adjust mixture and method of application as necessary.

G. If accepted, mockup will demonstrate minimum standard for the Work. Mockup may [not] remain as part of the Work.

1.05 Do not request submittals i Drawings sufficiently describe th products of this section or if pro prietary specifying techniques ar used. The review of submittals in creases the possibility of unin tended variations to Drawings thereby increasing the Specifier' liability.

1.05C When manufacturer's in structions for specific installation re quirements are referenced in Par 3, Execution, include the followin request for submittal of those in structions. Edit the Part 3 state ments to avoid conflict with thos instructions.

1.08 Use this Article for full sized erected assemblies required for re view of construction, coordination of work of several sections, testing or observation of operation.

1.08E Time duration for examina tion should be determined by an ticipated shrinkage, temperature and humidity conditions during dry ing time. Refer to manufacturers fo technical assistance.

.09   ENVIRONMENTAL REQUIREMENTS

   A.   Do not apply spray fireproofing when temperature of substrate material and surrounding air is below [40] [_____] degrees F ([5] [_____] degrees C).

   B.   Provide ventilation in areas to receive fireproofing during and 24 hours after application, to dry material.

   C.   Maintain non-toxic, unpolluted working area. Provide temporary enclosure to prevent spray from contaminating air.

1.10   SEQUENCING AND SCHEDULING

   A.   Sequence work under the provisions of Section [01005.] [_____.]

   B.   Sequence work in conjunction with placement of [ceiling hanger tabs,] [mechanical component hangers,] [and] [_____.]

1.11   WARRANTY

1.11   This Article extends the warranty period beyond one year. Extended warranties add to the construction cost and may present difficulties to the owner in enforcing them. Specify with caution.

   A.   Provide [two] [_____] year warranty under provisions of Section [01700.] [01740.]

   B.   Warranty: Fireproofing to remain free from cracking, checking, dusting, flaking, spalling, separation, and blistering. Reinstall or repair such defects or failures.

PART 2   PRODUCTS

2.01   MANUFACTURERS

2.01   In this Article, list the manufacturers acceptable for this project. If product substitution is allowed, include Paragraph D.

   A.   [_____] Type [_____.]

   B.   [_____] Type [_____.]

   C.   [_____] Type [_____.]

   D.   Substitutions: Under provisions of Section [01600.] [01630.]

2.02   MATERIALS

2.02   Edit the following descriptive specifications to identify project requirements and to eliminate any conflict with manufacturers' products specified above. Refer to TAS Section 07255.

   A.   Cementitious Spray Fireproofing: Factory mixed, asbestos free, cementitious material blended for uniform texture; non-fibrous materials; conforming to the following requirements:
     1.   Bond Strength: ANSI/ASTM E736, [200] [_____] lb/sq ft ([9.4] [_____] kPa) when set and dry.
     2.   Bond Impact: ASTM E760, no cracking, flaking, or delamination.
     3.   Dry Density: ASTM E605, minimum average density of [15] [_____] lb/cu ft ([240] [_____] kg/cu m), with minimum individual density of [14] [_____] lb/cu ft ([224] [_____] kg/cu m).
     4.   Compressive Strength: Minimum [500] [_____] lb/sq ft ([24] [_____] kPa).

   B.   Primer Adhesive: Of type recommended by fireproofing manufacturer.

2.02B   Confirm that the manufacturer has a primer for the specific substrate material and that incompatibility of materials does not exist.

C. [Overcoat:] [Sealer:] [_____.]

2.02C In certain applications, it may be appropriate to provide an overcoat or sealer to the cementitious fireproofing surfaces. The overcoat material will be a proprietary product associated with the fireproofing material. Specify accordingly.

D. [Metal Lath: Expanded metal lath; [3.4] [_____] lb/sq yd ([15] [_____] kg/sq m), [galvanized] [rust inhibitive primer] finish.]

E. Water: Clean, potable.

2.02D Include metal lath when fire rating requires it, or where existing building conditions will not tolerate excessive overspray.

PART 3   EXECUTION

3.01   EXAMINATION

A. Verify that surfaces are ready to receive work.

B. Verify that clips, hangers, supports, sleeves, and other items required to penetrate fireproofing are in place.

C. Verify ducts, piping, equipment, or other items which would interfere with application of fireproofing are not positioned until fireproofing work is complete.

D. Verify that voids and cracks in substrate are filled, and projections are removed where fireproofing is exposed to view as a finish material.

E. Beginning of installation means installer accepts existing [surfaces.] [substrates.]

3.02   PREPARATION

A. Clean substrate of dirt, dust, grease, oil, loose material, or other matter which may effect bond of fireproofing.

B. Remove incompatible materials which affect bond by scraping, brushing, scrubbing, or sandblasting.

C. [Install metal lath over structural members[.] [and as indicated on [shop drawings.] [Drawings.]]]

3.03   PROTECTION

A. Protect adjacent surfaces and equipment from damage by overspray, fall-out, and dusting.

B. Close off and seal ductwork in areas where fireproofing is being applied.

3.04   APPLICATION

A. Apply [primer adhesive,] fireproofing [and overcoat] [and sealer] in accordance with manufacturer's instructions.

B. Apply [primer adhesive and] fireproofing [in sufficient thickness to achieve rating,] [to a thickness of [_____] inches ([_____] mm),] with as many passes necessary to cover with monolithic blanket of uniform density and texture.

C. Apply [overcoat] [sealer] to a thickness of [_____] inches ([_____] mm).

3.05   FIELD QUALITY CONTROL

A. Field inspection [and testing] will be performed under provisions of Section [01400.] [01410.]

B. Inspections will be performed to verify compliance with requirements.

C. Reinspect the installed fireproofing for integrity of fire protection, prior to concealment of work.

D. Correct unacceptable work and provide further inspection to verify compliance with requirements, at no cost.

3.06 CLEANING

A. Clean work under provisions of [01700.] [01710.]

B. Remove excess material, overspray, droppings, and debris.

C. Remove fireproofing from materials and surfaces not specifically required to be fireproofed.

3.07 SCHEDULE

Location	Components	UL Rating
A. Mechanical Penthouse	Framing and underside of metal deck	2 hours
B. Main Roof	Underside of framing and metal deck	one hour
C. Mezzanine Boiler Room	All framing above 12 feet (3.7 m) above floor	2 hours

These Drawing Coordination Considerations describe Drawing related items that should be considered with SPECTEXT Section 07255.

A. METAL LATH: Expanded metal lath used to wrap structural members to assist bond of fireproofing material to component surface and to minimize overspray.

B. SPRAY FIREPROOFING: The cementitious fireproofing product of this specification section.

A. Do not dimension or indicate fireproofing thickness on Drawings. The thickness should be described in the specification section, along with the required fire resistance rating and/or the assembly rating. When developing Drawings, consult the Specifier for intended fireproofing thickness.

B. Do not make reference to assembly fire resistance ratings on Drawings or within Drawing 'general notes'.

C. Indicate metal lath on Drawing details. Do not indicate lath attachment method as the specification section will specify the type and extent of attachment by reference to the applicable fire resistance rating.

A. If cementitious fireproofing is required to be used in air plenums, determine if air velocity is sufficient to require an overcoat to the fireproofing surface. If a coating is required, it should be specified.

3.05C Include the following paragraph when work of other sections may occur after fireproofing installation. The integrity of protection may have been disrupted and require subsequent inspection prior to concealing fireproofing with other work.

3.07 Provide a schedule when a variety of surfaces, required ratings (by UL assembly number) and thicknesses of fireproofing area are required. The following examples may assist in developing such a schedule.

Drawing Terms: The following generic terms should be used in drawing notes to ensure definitions parallel the specification section. If more than one type of an item indicated below is required, indicate on drawings and in the specification section as "Type A," "Type B," etc.

Coordination: The following items require special attention when coordinating drawings and specifications, to ensure parallel information and references.

Caution: The following items indicate that caution should be observed prior to incorporating into drawings or specifications.

FACTS ABOUT PLASTER AND THERMAL INSULATION

In spite of advertising claims to the contrary, gypsum plaster and portland cement plaster are not particularly good insulation materials. Following are the thermal resistance (R) values of some typical plaster constructions:

A. ⅞″ portland cement-sand plaster on paper backed wire or expanded metal lath   0.17

B. ⅞″ portland cement-lightweight aggregate plaster on paper backed wire or expanded metal lath   0.23

C. ½″ gypsum-sand plaster on ⅜″ **plain** gypsum lath 1.41

D. ½″ gypsum-sand plaster on ⅜″ **foil** backed gypsum lath   2.72

E. ½″ gypsum-lightweight aggregate plaster on ⅜″ **plain** gypsum lath   0.64

F. ½″ gypsum-lightweight aggregate plaster on ⅜″ **foil** backed gypsum lath   2.95

For comparison here are the "R" values of some materials competitive to lath and plaster:

A. 8″ concrete block	1.11
B. 8″ lightweight concrete block	2.00
C. Wood shingles on backer board	1.40
D. 1 x 8 wood siding	.79
E. Tilt up concrete 6″ thick	.66
F. Brick masonry 4″ thick	.44

From this you can see virtually all conventional construction methods will require added insulation to meet the new energy standards. This insulation might be in the form of batt or sprayed insulation in the stud spaces, insulating sheathing under the lath or loose fill in the cores of masonry. Following are the resistance values for popular insulating materials per inch of thickness:

A. Fiberglass batts or blankets	4.00
B. Mineral wool batts or blankets	3.57
C. Sprayed mineral fiber insulation	3.93
D. Sprayed cellulosic insulation	3.70
E. Expanded polystyrene sheathing	4.00
F. Wood fiber sheathing	2.38

Modern energy conservation requirements have spurred development of a number of proprietary exterior portland cement plaster systems utilizing insulation as an integral component, with excellent thermal resistive results being achieved.

Detailed information on such systems available at the local area may be obtained by contacting local industry trade promotion bureaus.

CHAPTER THREE-D

PORTLAND
CEMENT
EXTERIOR
PLASTER

PORTLAND CEMENT PLASTER

Headquartered in Skokie, Illinois, the Portland Cement Association is a national organization of cement manufacturers with a primary mission to improve and extend the uses of portland cement and concrete through market development, engineering, research, education and public affairs work.

PCA's technical manuals are rated as the best in the industry. For many years, its publication, "Portland Cement Plaster (Stucco) Manual" has been regarded as one of the most informative treatises ever written on this subject. With permission of the publisher, the contents of this manual are reproduced herewith.

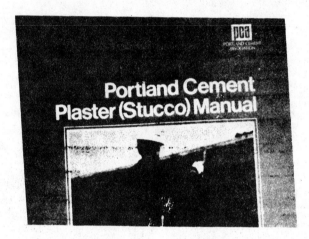

FOREWORD

This manual has been prepared as a guide to acceptable practices in designing, mixing, and applying portland cement plaster (stucco). It is intended as an authoritative reference manual for architects, specifiers, building officials, inspectors, contractors, plasterers, and apprentices. The newer materials, textures and finishes, and technology described provide ways to achieve attention-getting appearance and freedom of architectural form without sacrificing the utility and economy of plaster.

We gratefully acknowledge assistance in reviewing the text and valuable suggestions from many plastering contractors, as well as material and machinery representatives. Particular credit is given to the following:

California Lathing and Plastering Contractors Association

Chicago Plastering Institute

The Association of Wall and Ceiling Industries—International

Operative Plasterers' and Cement Masons' International Association of the United States and Canada

Stucco Manufacturers' Association

INTRODUCTION

Portland cement plaster is a versatile facing material that ca be applied to any flat or curved surface either inside or ou side of any structure or building. The serviceability of por land cement plaster has been proof-tested through long u and exposure in the cold temperatures of Alaska and Canad and the hot temperatures of the humid Gulf States and the d Southwestern States. The versatility of plaster extends b yond buildings to simulated mountains in amusement park rock grottos in zoos, ferrocement boats and ships, housing developing countries, and high-rise-building enclosures prefabricated panels that are built at ground level, then ele vated and fastened in their final positions. In the future th versatility of plaster will be extended to creative new uses.

Portland cement plaster is a combination of portland ce ment and aggregate mixed with a suitable amount of water form a plastic mass that will adhere to a surface and harder preserving any form and texture imposed on it while plastic The term "plaster" applies to the placed and hardened mix ture whether on an exterior or interior surface.

Portland cement stucco and portland cement plaster ar

ie same material. The term "stucco" is widely used to escribe the cement plaster used for coating exterior surfaces f buildings. In some localities "stucco" is used to refer only a factory-prepared finish coat mixture.

Throughout this text, the term "plaster" is used to refer to ie mixture of portland cement (or alternate cementitious iaterials or combinations), sand, and water, applied as either n interior or exterior facing material.

Portland cement plaster is usually applied in three coats ver metal reinforcement with or without solid backing. wo-coat work may be used over solid masonry or concrete irfaces that are reasonably true planes and provide adequate ond. Some backgrounds, such as smooth concrete, have vel surfaces and uniform suction that make thin, single- oat plastering possible. The cementitious material in plaster iay be either portland cement, air-entraining portland ce- ient, masonry cement, or plastic cement used singly; or it iay be portland cement used in combination with masonry ement, plastic cement, or lime. Pigments for integral color an be added while preparing the mixture and thus eliminate ie need for painting.

Plaster in the freshly mixed state is a plastic, workable iaterial. It can be applied by hand or machine to conform to ny shape, using two or three applications or coats. The final ppearance of the finish coat can be varied by changing the ze and shape of the aggregate, by adding color, by changing ie consistency of the finish mix, by the method or equipment sed for plastering, and by the plasterer's skill in manipulat- ig the finish coat.

Plaster in its hardened state is a desirable facing material: ard, strong, fire-resistant, rot- and fungus-resistant, color- etentive, and durable under weather cycles of wet and dry or eeze and thaw. Plaster has proved to be a durable wall cover i all climates from the very hot to the very cold. It has great ppeal as a surface finish because of its utility, low first cost, nd minimum need for maintenance.

Construction specifications setting forth the minimum re- uirements for plastering work required in a structure should e closely followed, whether stated in industrywide tandards and specifications or in local building codes and rdinances (see Appendix A). Construction specifications re written to establish the quality of materials and workman- hip that is acceptable for the work. Compliance with the onstruction specifications protects both the customer and ontractor.

This manual presents recommended procedures based on ccepted good practices that will produce high-quality port- and cement plaster with the desired beauty and serv- ceability regardless of exposure.

ORTLAND CEMENT PLASTERING

ortland cement plastering is simple or complex depending n the user's demands. In its simplest form portland cement laster may be any combination of cementitious material and ne aggregate. Experience, however, has taught us that the implest combination consists of one part portland cement, enerally with a small amount of lime, to approximately ree or four parts fine aggregate. This proportion will pro- ide necessary weather resistance and resistance to cracking.

When applied over frame construction, the two base coats should have a total thickness of approximately ¾ in. to pro- duce a solid base for any decorative finish coat of plaster.

In its more complex form, portland cement plastering can be as sophisticated as the craftsman's knowledge and experi- ence with the material will allow. The complex form is more commonly used than a simpler form because users want its uniformity, longevity, use of marginal aggregates, and economies of application. A native American plastering his hogan shames himself and causes his family to suffer if the plastering is inadequate. A plasterer applying a weather- resistant facing on a high-rise building degrades his craft and inconveniences many people if his work is unsatisfactory. In practice, simple plastering can be done by anyone. Complex plastering must be performed only by someone with a work- ing knowledge of the materials in plaster, how to combine them, and how the hardened plaster will perform in its par- ticular environment.

TOOLS FOR PLASTERING

Basic handtools

Hawk
Trowels: common, margin, pointing
Angle float, angle paddle
Brushes: browning, finishing, tool
Scarifier or scratcher
Floats: cork, rubber, sponge rubber, foam plaster, carpet
Half hatchet

Handtools for advanced work

Elastic knife
Steel square
Six-foot folding rule
Spring-loaded (push-pull) steel tape measure
Handsaw
Chalkline, plumb bob, level
Tin snips
Nippers
Angle plane

Specialized handtools

Pool trowel
Midget trowel
Pipe trowel
Inside- and outside-corner trowel
Angle plow
Lathing hatchet
Steel tape
Cement-stucco dash brush
Texture brushes: rice or stippling, wire texture
Rubber sponge

Ornamentation tools

Trowel and square
Leaf and square
Miter rod or joint rod
Rubber gloves
Bevel square

Larger tools

Rod
Featheredge
Darby
Slicker or shingle
Scaffold
Water level
Stilts

Machine equipment

Plaster-mortar mixer
Plastering machine or plaster gun
Glitter gun, aggregate gun
Texturing machines

Discoloration and other imperfections of the plaster surfa
may occur if tools are not properly cleaned after each us
Tool surfaces should be cleaned of any adhering plast
scoured with plasterer's sand, and wiped dry. Mixers shou
be thoroughly cleaned, especially if materials other th
portland cement plaster are to be mixed in the same mix
The use of sand is recommended for scouring interior s
faces of the mixer bowl.

Good tools and equipment that are clean, free of warpin
and without excessive wear are necessary to do a go
plastering job.

DESIRABLE PROPERTIES OF PLASTER

Portland cement plaster must possess certain properties
both fresh and hardened states to be totally satisfactory du
ing application and in service. The plaster as discharged fro
the mixer must have the properties necessary for either ha
or machine application. While being applied, the plast
must possess those properties that promote its adhesion to
solid base or a wire-mesh lath. It must be movable to allo
striking a smooth, plane surface for subsequent coats or f
surface texturing. After hardening, the plaster must posse
those properties that provide the desired serviceabilit
namely, hardness and strength, reasonable resistance
cracking, and resistance to weather. Fresh plaster must mai
tain its bond to the previous coat and with time develop t
desired hardened performance characteristics.

Fresh Plaster

Fresh plaster as discharged from the mixer should have t
following properties.

Adhesion. The capability to stick to an identical or simil
material is developed in portland cement plaster by the por
land cement paste, that is, the combination of cement, ai
and water. Air-entrained portland cements, masonry c
ments, and other types of cement of greater fineness gene
ally possess greater adhesion than do non-air-entraine
normal-fineness cements.

Cohesion. The capability of the plaster to hold together
cling to itself is also developed by the portland cement past
During application of plaster to metal-lath reinforcement, th
cohesion of the plaster governs its ability to remain fixed
place after the plaster envelops the metal reinforcement.
cohesive plaster remains where it is placed.

Workability. The ease with which the plaster is placed an
later rodded and floated is the measure of its workability
Workability involves adhesion and cohesion as well as un
weight and spreadability. The plasterer judges the degree c
workability during application. In the laboratory, degree c
workability is measured by various tests, including tests fc
flow, slump, unit weight, and penetrability. To give plaste
its best workability, aggregate with favorable shape and grac
ing, cement, and optimum amounts of air, water, and plas
ticizers must be properly proportioned and combined durin
mixing. Any of these ingredients incorporated into the mi

utside of optimum amounts will lessen the workability of e plaster. Plaster with less than optimum workability requires greater effort by the plasterer and may result in an ppreciable reduction in the desirable properties of the hardened plaster.

ardened Plaster

he finished, hardened plaster (stucco) should have the following performance characteristics.

eather Resistance. The ability of plaster to withstand eathering includes resistance to rain penetration; freeze-aw durability; resistance to extreme thermal and moisture nanges; and resistance to aggressive chemicals in the atosphere, such as acid rain. Plaster properly proportioned, ixed, applied, and cured will exhibit good durability and esistance to the natural environment.

Resistance to rain penetration is obtained by using properly cured portland cement mixtures that have been dequately densified during application. Freeze-thaw durality is achieved by using a plaster mixture that has been nproved with air-entrained cementitious materials or with n air-entraining admixture introduced during batching. Restance to excessive thermal and moisture changes is obined by using the maximum allowable aggregate content in ell-cured portland cement mixtures, properly densified uring application. Resistance to mildly aggressive chemials in the environment is achieved through the alkalinity of ortland cement plaster. In more aggressive environments, dditional resistance may require the use of special cement ypes such as Type II, its air-entrained equivalent, or Type V.

hemical Bond. The ability of the plaster to bond chemially with a base of identical or similar material is especially nportant when plaster is applied to a solid base. Separate upport, as provided by metal reinforcement, may be reuired to develop mechanical bond whenever chemical bond not attainable. Chemical bond is normally developed uring the portland cement hydration process. As cement ydrates, the resulting new chemical compounds develop rystals at and across the interface between the plaster and ne background material. The degree to which chemical ond is obtained depends on the area of contact, the vailability of water for cement hydration, and the hydration eriod.

reeze-Thaw Durability. The ability of plaster to resist the disruptive forces associated with the expansion of water reezing within saturated plaster is a measure of durability. Although plaster normally is positioned on a wall or ceiling vhere the water content of the hardened plaster is well below ritical saturation levels, the use of air-entrained plaster is specially beneficial where snow, slush, and deicer chemials may be splashed or sprayed against a wall surface.

ensile Strength. The ability of the plaster to resist tensile tresses, whether developed internally or externally, is losely related to the cracking characteristics of the plaster. Curing and hardening increase the tensile strength of the

plaster and its ability to resist cracking. When drying takes place, shrinkage stresses develop in the plaster. During early ages when moisture is available, the tensile strength of the plaster increases rapidly, thereby resisting the shrinkage stresses and lessening cracking. Proper curing of well-proportioned and densified plaster is compulsory to obtain maximum tensile strength.

Compressive Strength. The ability of the plaster to carry its own dead weight plus added service and wind loads is a physical property of interest. In itself, the development of high compressive strength as a property of plaster is really not important. The correlations, however, between compressive strength and other more important physical properties should be recognized. Compressive strength is easy to measure, but the other properties are not. These correlations permit compressive strength to be used as an indicator that the other properties have been attained. Tensile strength and compressive strength of plasters can be correlated, and compressive strength tests, being more easily performed, can be used to estimate the tensile strength of plaster. Similar relationships exist between compressive strength and completeness of hydration, permeability of plaster to rain and water, and other properties.

Compressive strength of plasters measured in the laboratory has resulted in the plaster-mixture recommendations set forth in this manual. Compressive strength of plaster, however, should not be specified as a quality control, field test, or criterion for acceptance or rejection of the finished work.

MATERIALS

The skilled plasterer knows how important the selection and use of proper materials is to the performance of the finished plaster segment of the structure. The choice of plastering materials is based on knowledge of the materials and their contribution to satisfactory appearance, serviceability, and durability. Proper materials are necessary to produce the desired results in both fresh and hardened plaster. All materials used in the plaster should meet the appropriate product requirements stated in the specifications. How to select good materials is described in the section "Mixtures."

Cement

The type of cement used should comply with current requirements of one of the following applicable product specifications:

Masonry cement	ASTM Designation: C91
Portland cement	ASTM Designation: C150, Types I, II, and III, or their air-entrained equivalents, Types IA, IIA, and IIIA
Plastic cement	No applicable specification; see text
White portland cement	No applicable specification. Cement can be tested in accordance with ASTM Designation: C150, Type I, IA, or III

Masonry cement, ASTM C91, is manufactured to produce a cementitious material that will be workable, achieve moderate strength, and be durable. Specifically manufactured for the masonry industry, this fine-grind, air-entraining, highly workable cement is ideally suited for the plaster industry.

Type I, normal, portland cement is a general-purpose cement recommended for most average conditions for which the special properties of the other types of cement are not required.

Type II is a modified portland cement that has slower heat liberation during hydration and improved resistance to sulfate attack. The improved sulfate resistance may be significant for plasters used in specific geographical areas.

Type III is a high-early-strength cement. Its chemical composition and somewhat greater fineness promote the development of high-early-age strength and greater heat liberation. These characteristics are especially beneficial when plastering in cold weather.

Type IA, IIA, and IIIA cements are similar to their Type I, II, and III counterparts but intentionally contain an air-entraining additive. These air-entraining cements produce minute, well-distributed air bubbles throughout the plaster during mixing. The entrained air bubbles improve the plasticity of the fresh plaster and increase the freeze-thaw durability of the hardened plaster.

Plastic cement is a special cement manufactured expressly for the plaster industry. The cement should meet the requirements of ASTM C150, Type I or II, except with respect to the limitations on insoluble residue, air entrainment, and additions subsequent to calcination. A plasticizing agent may be added to portland cement Types I or II in the manufacturing process, but not in excess of 12% of the total volume. When plastic cement is used, no lime or other plasticizer may be added to the plaster at the time of mixing.

White portland cement is a special cement used in finish-coat plasters. The cement should meet the requirements of ASTM C150, Types I, II, or III, or one of their air-entrained counterparts, Types IA, IIA, or IIIA.

Plasticizers

The function of a plasticizer is to improve workability. When plasticizing materials are added at the mixer they should be used only in combination with portland cement, and should not be added to other cementitious materials. The plasticizer should meet the requirements of the following ASTM standards:

Special finishing hydrated lime — ASTM C206, Type S
Hydrated lime for masonry purposes — ASTM C207, Type S

Water

Mixing water should be clean and fit to drink. It should also be free of harmful amounts of any mineral or organic substances that would affect the set or discolor the plaster. Water from a domestic water supply usually is suitable. Unclean or questionable water should be tested. Laboratory test mortars are made with the questionable water for comparison to simi-

lar mortars made with clean, pure, distilled water to obser the effect of water purity on time of setting (ASTM Te Method C191) and strength development (ASTM Te Method C109). Any significant increase in setting time any significant reduction in strength development when t questionable sample is compared to the sample containi distilled water justifies additional testing or the evaluation water from another source.

Water content of plaster should be as high as can be tole ated for either hand or machine application. Water conte when finishing operations have been completed should I low.

Aggregate (sand)

The amount and kind of aggregates used greatly affect th quality and performance of the plaster and the skills a methods required to apply the plaster. Aggregate for plaste ing should be well graded and clean, free from organic im purities, loam, clay, and vegetable matter. These foreig materials can interfere with the setting and hardenin (strength gain) of the cement paste and its adhesion to th aggregate particles. Aggregates used in plaster should co form to ASTM Specification C897, Aggregates for Jol Mixed Portland Cement Based Plasters.

To emphasize the role of a well-graded sand in plaste consider that 1 cu ft of plaster contains 0.97 cu ft (0.0275 m of sand and 0.25 to 0.33 cu ft (0.007 to 0.009 m³) of ceme titious material. The sand makes up practically the enti volume, whereas cement paste — a combination of cemen air, and water — fills all the voids or spaces between the san particles.

Since the sand occupies such a large percentage of th volume, the importance of exacting requirements for aggre gate cannot be overemphasized. Obtaining a clean, soun sand with the proper gradation of particles is of primar consideration.

The aggregates selected should be well graded to produc a proper plaster mixture. Large sand grains contain voi spaces that are filled by smaller sand grains; smaller voids a present that are filled with still smaller sand grains. The voi remaining after all sand particles have assumed their fina position will be filled with portland cement paste. The bes performing portland cement plaster is obtained when th sand grain particles are almost in contact with each other an the remaining void spaces are filled with cement paste.

The volume of void space within an aggregate is measure in the laboratory by filling completely a unit measure known volume with the aggregate and weighing it. A simpl calculation using this unit weight along with the relativ density (specific gravity) of the aggregate provides th aggregate's void content. As different methods — shoveling jigging, or rodding — can be used to fill the unit measure, th void content shown by each method will vary. The rodde unit weight value is considered a measure of the uniformit of sand-particle distribution. Slightly higher void conten will be noted in the plaster since cement paste envelops eac aggregate particle, thereby separating them. During actua plastering a slightly lower void content may result because (1) water lost by evaporation, (2) additional care taken i applying plaster, and (3) better consolidation obtained durin dding and finishing.

For each sand source there is a optimum paste content. This paste content will vary depending on the volume of voids in the aggregate. Once the paste content is established, either high-strength or low-strength plasters can be obtained the selection of cementitious material or combinations of cementitious materials.

In plaster mixes, the amount of paste required to carry the aggregate and produce a workable mixture governs the mix proportions. Because the paste content governs characteristics of the hardened plaster, the proper quantity of cement to be combined with a volume of aggregate should be determined.

Aggregate Grading

To reemphasize the importance of uniformly graded aggregates, the influence of void content on cement paste requirements and the workability characteristics of the plaster are largely governed by the shape and grading of the aggregates.

To produce high-quality base-coat plasters, the aggregate should meet the chemical and physical requirements of ASTM C897, Aggregates for Job-Mixed Portland Cement Based Plaster, except that the recommended gradation of the sand to be used in base coat portland cement plasters should be as given in Table 1.

Table 1. Sand Gradation for Base-Coat Plaster

Retained on U.S. Standard Sieve	Cumulative weight, percent retained	
	Minimum	Maximum
No. 4 (4.75 mm)	-	0
No. 8 (2.36 mm)	0	10
No. 16 (1.18 mm)	10	40
No. 30 (600 μm)	30	65
No. 50 (300 μm)	70	90
No. 100 (150 μm)	95	100
No. 200 (75 μm)	97	100

Many mortar sands may fall within the individual size and weight percentage limits, yet the resulting combined sizes may be difficult to use while plastering. A uniform gradation that avoids the boundry limits is desirable.

Uniformly graded aggregates are desirable because they produce plaster that is easy to apply. Certain outside-of-limits sands may be used for plaster if modifications are made while proportioning and if hand, not machine, placement is used to apply the plaster. Proof of any sand's acceptability is service performance in the hardened plaster. A uniformly graded specification sand improperly applied and consolidated during plastering is less desirable than a poorly graded, nonspecification sand properly proportioned in the mixture and properly consolidated during application.

Aggregates for finish coats need not meet the aggregate specification previously cited. Nonspecification aggregates of varied sizes and shapes can be evaluated with mockup or test panels to obtain special textures and finishes in the final surface. Test panels allow adjustment in finish-coat workability and application procedures to achieve the desired surface textures before starting the work. As a starting point, all aggregates for finish-coat plaster should be below a No. 16 (1.18-mm) sieve size and uniformly graded. Larger size particles may be added for finish appearance purposes.

Factory-Prepared Finish Coat

Factory-prepared products (commonly known as stucco finish) are available in some localities in many colors for the finish-coat plaster. These factory-prepared, colored, finish-coat mixes contain all the necessary ingredients except water. Their use generally improves batch-to-batch consistency and color uniformity for the type of texture specified.

Use of any stucco-finish product should closely follow the manufacturer's recommendations, and its effect should be appraised on a sample test panel.

Reinforcement

Metal lath or wire is used as a plaster base and reinforcement where plaster or stucco is applied to wood frame construction, steel stud partitions, and surfaces that in themselves will not provide adequate mechanical keying or chemical bond for the plaster. Examples of the latter are unsound, old stucco surfaces; painted surfaces; deteriorated or badly weathered brick and concrete masonry surfaces; and nonabsorbent concrete surfaces.

Metal reinforcement for plaster is used primarily to mechanically anchor the plaster to a base and provide support for the first coat of plaster as it is applied. This reinforcement is available in several different forms: expanded-metal lath, woven-wire-fabric lath, and welded-wire-fabric lath. Openings in the reinforcement should not exceed 4 sq. in. (250 mm²). The openings should be large enough to permit the first coat of plaster to be forced through the openings, form keys, and completely embed the reinforcement. Full embedment protects the metal lath or fabric against corrosion.

Expanded-metal lath—diamond-mesh, rib, or sheet lath, with or without paper backing—should be fabricated from copper-bearing steel sheets or should be fabricated from galvanized steel. The product should comply with Federal Specifications for Lath Metal (and Other Metal Plaster Bases), QQ-L-101c.

Expanded metal lath with diamond mesh should weigh not less than 2.5 lb per square yard (1.36 kg/m²) for use on vertical surfaces (walls) over solid substrates, and not less than 3.4 lb per square yard (1.84 kg/m²) for horizontal surfaces (ceilings). The weights of reinforcement needed for other types of metal bases are given in American National Standards Institute (ANSI) A 42.3-1971, Lathing and Furring for Portland Cement Plaster.

Recommended types of expanded metal for exterior surfaces are the flat-rib and the self-furring lath. A ⅜-in. (10-mm) V-rib lath is not recommended in locations exposed to moisture or weather because it is practically impossible to achieve adequate plaster penetration through the mesh and to completely embed the V-rib within the stucco or plaster.

Lacking full embedment, failure (evidenced by cracks) is possible from corrosion of the metal lath. The V-rib shape also forms a plane of weakness within the plaster, encouraging any plaster that is prone to excessive stress from drying shrinkage, climate changes, or building movements to crack over each V-rib.

V-rib metal lath is an option for areas where additional stiffness is needed. Locations such as soffits and ceiling areas protected from heavy rain and severe weather can be lathed with ⅜-in. (10-mm) V-rib lath weighing not less than 3.4 lb per square yard (1.84 kg/m²).

Hexagonal woven-wire-fabric lath (stucco netting) is available in rolls with 1½-in. (38-mm) openings of 17-gage galvanized wire or 1-in. (25-mm) hexagonal-mesh openings of 18-gage wire; some types include a paper backing and line wire. This mesh is nailed to studs or sheathing with furring nails. ANSI 42.3 permits but does not recommend woven-wire lath as a plaster base. The user should follow the manufacturer's recommendations.

Welded-wire-mesh lath is a grid made of copper-bearing, cold-drawn 16-gage (minimum) steel wires welded at their intersections or woven into squares or rectangles with openings not larger than 2x2 in. (50x50 mm). The wire should be galvanized steel. This welded-wire-mesh lath is available with or without paper backing and with or without furring crimps.

When paper or other backing is used on metal plaster bases where insulation or a moisture barrier is required, it should comply with manufacturer's specifications for building paper of the style and grade described for the intended use.

Building Paper

Building paper should comply with the current requirements of UU-B-790, Federal Specifications for Building Paper, Vegetable Fiber (Kraft, Waterproofed, Water Repellent, and Fire Resistant). For some building occupancies, such as hospitals, building paper must be flameproof and must comply with ASTM D777. Asphalt-saturated building felt should weigh not less than 14 lb per 108 sq ft (6.4 kg per 10 m²). Building paper should be free from holes and breaks.

Flashing

Flashing or membrane to redirect water to building exteriors should be properly placed and installed. Flashing is vital to the longevity of the plaster and the prevention of water seepage into the building. Flashings must be designed and installed so that any water getting behind the plaster will be channeled downward and outward toward the exterior of the building. Retention of water within the wall could result in breakup of the plaster. Such deterioration could be the result of rusting of the metal reinforcement, inserts, or attachments. Water collected within the wall may be responsible for disruptive expansion and deterioration of the plaster in a freeze-thaw environment. Only nonrusting flashings capable of directing water from the walls should be used. Metal reinforcement should be used when applying plaster over flashings.

Admixtures

Admixtures should not be used indiscriminately because th generally are not needed for portland cement plaster. Use an admixture should be approved only after it has been tri batch tested under conditions simulating its use and fi tested to establish that the results will be as expected.

Plaster mixtures are very sensitive to admixtures. If plas must be applied to absorptive substrates or during extrem dry conditions, an admixture that retards cement hydrati could be beneficial from the standpoint of workability. At same time, however, the same admixture may retard ceme hydration even though ample water is available and a lo strength plaster could result, because as time passes evapo tion will leave insufficient water in the mixture to compl hydration. An admixture that accelerates cement hydrati may cause a reduction in floating time, which may create problem. Proper curing of all admixtured plaster is importa to gain full benefit of the cement in the plaster.

Laboratory testing for acceptance of an admixture for pla ter should include a comparison of admixtured and nona mixtured plasters using time of setting and compressi strength test methods. Time of setting changes should be the direction intended (acceleration or retardation). Co pressive strength results should show that the admixtur plaster is equal to or greater than the nonadmixtured mortar both early and later ages. Compressive strength test spe mens should be cured as the plaster is to be cured in the fie either air-dried or moist-cured.

Air-Entraining Admixtures

As an alternative to the use of air-entraining portland c ments or masonry cements, an air-entraining admixture ma be introduced during mixing. The air-entraining admixtu will improve both the workability of the fresh plaster and t freeze-thaw durability of the hardened plaster. The im provement in workability of fresh plaster will be especial noticeable when using nonuniformly graded aggregates, aggregates deficient in the finer 100 (150 μm) and 200 (7 μm) mesh sizes.

During hand application, air-entrainment improves wor ability. During machine application air-entrainment eas pumping and reduces segregation during pumping. Althoug too much air can cause pump stoppages, the plaster (aft spray application to the surface) will not contain too mu entrained air, because the excess air will be dissipated durin the spraying.

The manufacturer's recommendations regarding dosag of air-entraining admixtures should be closely followed.

Accelerators

Admixtures capable of increasing the rate of strength d velopment may be beneficial under certain circumstance Accelerators for plaster modify the worktime needed b tween plaster application, rodding, and floating. The a celerator will shorten the time required between applicatio and completion of the finishing operation. Calcium chlorid is one accelerator capable of reducing the time betwee

ese operations and increasing the rate of strength development. Its use is not recommended however, because its presence in the plaster contributes to corrosion of the metal
reinforcement.

As an alternative to an accelerator, Type III portland cement can be used as part or all of the cementitious material.
Heating the water and sand will also increase the rate of
early-strength development.

Antifreeze Compounds

Products exist that are said to act as antifreeze agents. The
use of antifreeze compounds in portland cement plaster is not
recommended. Calcium chloride, alcohol, and ethylene
glycol have been considered for this purpose, but are not
adequate as an antifreeze in plaster because at the recommended safe dosages they produce only a slight lowering of
the freezing temperature of the plaster. Type III cement or
heated water and sand, or both, will be more satisfactory.

Water Repellents

Products are available that are said to reduce the water permeability of plaster. These formulations include water repellents, hydrophobic liquids, and inert fillers. Quality plaster
that is properly cured does not need water-repellent admixtures. Laboratory tests that measure time of setting and water
permeability of plaster show that water-repellent compounds
may actually retard cement hydration and increase
permeability.

Bonding Agents

Products are available to increase the adhesion of plaster to
substrate or of plaster to plaster. These organic compounds
are generally polyvinyl acetates or alcohols, cellulose derivatives, or acrylic resins. The polyvinyl alcohols and cellulose
derivatives do increase the adhesion of plaster to substrate
and are generally compatible with portland cement mixtures.
Polyvinyl acetates and alcohols require air curing and become less effective in moist environments. The acrylic resin
compounds also need air curing; when added to portland
cement mixtures acrylic resins play a significant role as the
main cementing compound. Bonding agents can be beneficial admixtures where conventional plasters cannot be
adequately moist-cured.

Color Pigments

A wide range of color pigments are available for obtaining
colored finish coats. The pigment should be carefully
selected to produce good dispersion and suspension in the
plastic mix and color permanence in the hardened state. The
amount of pigment required depends on the color of the
cement, the color of the fines in the sand, the color intensity
desired, and the water content of the plaster. Once the pigment dosage has been determined, sufficient coloring should
be preweighed for individual plaster mixes to ensure color
uniformity from batch to batch.

The effect of adding color pigments to the plaster should
be evaluated, using time of setting and strength test methods.

BASES

Portland cement plaster is applied to an array of bases, as can
be seen in Table 2. The successful application of plaster to a
base depends on the compatibility of the plaster and the base
material, the soundness of the base, and the application procedure. Plaster is very compatible with concrete masonry
and new concrete. Plaster may be compatible with old concrete or masonry, depending on the degree to which the
surface is contaminated. Where plaster is not compatible
with a particular base, the quality of the base must be up-
graded to gain both chemical and mechanical bond, or metal
reinforcement must be furred over the surface.

Table 2. **Bases for Portland Cement Plaster** (Stucco)

Materials of low or no suction[1]	Materials of medium or average suction	Materials of high suction
Glazed tile	Concrete block	Soft common brick
Hard-burnt brick (pavers)	Cinder block	Soft clay partition tile
Hard stone (granite)	Face or medium-hard brick	Aerated concrete
Plain and reinforced concrete	Medium-hard clay partition tile	Some lightweight concrete
Dense, hard concrete blocks	Good-grade common brick	
Hollow clay block, unscored	Soft stone Lightweight concrete block	

[1]Metal lath or wire and three-coat work are essential with a dense and
relatively impermeable mix if the first coat is to protect the reinforcement.

All bases should be sufficiently rigid to allow application of
the plaster without any movement of the base. A solid base,
such as old or new concrete or masonry, requires no special
attention to rigidity during application. Metal reinforcement
over open framing may require additional bracing during
plaster application. This additional bracing should be removed immediately after plaster application to eliminate or
reduce restraint during the subsequent drying period, when
the plaster is undergoing its greatest volume change.

Solid bases that are to receive plaster directly on the surface should have a sufficiently rough or scored texture to
ensure good mechanical bond and should be clean of any
surface contamination that may prevent good suction or
chemical bond. Mechanical bond is obtained by filling the
surface pores with plaster. Chemical bond is obtained
through cementing action across the substrate-plaster
interface.

The water intake (absorption or suction) of a masonry
surface (Table 2) is an indication of the chemical bond that
will be developed at the base-plaster interface. A normal to
highly absorptive surface will have greater attraction between the wet plaster and the base and develop good chemical bond. A low or nonabsorptive surface will develop little
or no chemical bond.

To establish the ability of any surface to absorb water, water can be dashed or sprinkled onto the surface. If the water forms beads and runs off the surface, a poor bond can be anticipated. If the water is rapidly absorbed, then a good chemical bond can be expected.

Under extreme drying conditions, highly absorptive surfaces must be premoistened to reduce absorption into the base. Premoistening prevents rapid loss of water from the plaster to the base. Highly absorptive surfaces should be moistened as heavily as possible without leaving a film of water on the surface when plastering starts. Prewetting is unnecessary when a plastering machine is to be used for spray application. The wet plaster that is used in machine application will have sufficient water to be applied to an absorptive base. If absorptive surfaces become wet or heavily moistened during humid weather, more time may be needed between plaster application and final floating. The plasterer has the option to either moisten the wall or adjust his work methods to compensate for wall moisture content and absorption characteristics.

High-gloss surfaces, or those with low absorption, can be made more receptive to plaster application by first applying a dash or bond coat. Dash-coat plasters are slurry mixtures rich in portland cement. These plasters are literally dashed against the base surface by hand with a brush, trowel or paddle, or by machine. Enough plaster should be dash-applied to cover most of the surface. The dash coat remains unfinished — that is, it is not rodded, floated, or troweled — and allowed to set "as is" before application of the next (first) coat of plaster. Alternatively, there are a number of special bonding materials that develop good bonding characteristics between plaster and low-suction bases. For example, products containing latex or acrylics will improve bond.

Cast-in-Place Concrete

Cast-in-place concrete that is to receive plaster should be straight, true to line and plane, and have a lightly scored or pitted rough surface texture. During forming, the concrete contractor should avoid misalignment of forms, using too much form-release agent, and using plastic-coated-plywood or metal forms. Close attention to these factors will reduce the amount of preplastering preparation needed to make the regular plaster coats adhere.

Misalignment in walls can be corrected with a plaster coating; however, the variable thickness of the plaster coat required to restore a true plane can result in differential shrinkage and subsequent cracking. Dense, smooth surfaces and those retaining an excessive amount of form-release agent on or absorbed into the concrete can cause delamination of the plaster from the concrete base. Preventive action for either situation can be taken by securing metal or wire lath to the concrete.

Newly placed concrete generally requires no special surface preparation, or at most a light sandblasting after the forms have been removed. Excess form oil can be removed by high-pressure waterblasting with hot water and fine silica sand.

Older concrete surfaces should be critically inspected to determine their alignment, surface texture, and absorption characteristics. Any painted or coated surface should have all

traces of the paint or coating removed. High-pressure-wa spray, without or with fine silica sand, will effectively move paint; a light-to-heavy sandblasting may be require remove coatings. The absorption characteristics of cleaned surface will dictate whether the base is satisfact for direct application or if metal reinforcement is needed.

A bonding agent painted on or spray applied to the surf of the base, may improve adhesion between the plaster c and the base. Bonding agents should be applied according manufacturers' recommendations.

Masonry

Masonry walls to be plastered may involve a wide variety building materials ranging from the high-absorption concr masonry or brick units to low or nonabsorptive hard bu brick and glass block. Plaster can be applied to all these uni Where poor or no bond is anticipated, the use of metal rei forcement anchored to the surface is recommended.

Masonry walls should be inspected for misalignment, u favorable surface conditions, and deterioration. The devi tion from true alignment of the wall will govern the plas thickness needed to restore a plane surface. If misalignme is excessive, furring and lathing prior to plaster application recommended. The surface should be examined for contam nation. Surface contamination, such as airborne so efflorescence, oils, paints, coatings, sealers, and mort droppings, must be removed. If the surface shows satisfa tory absorption after cleaning, plaster may be applied rectly to the masonry. If contamination remains aft cleaning, the use of metal reinforcement is recommended.

Before applying plaster to any masonry wall, assess t effectiveness of dampproof courses, flashings, and roof-w anchorages. All are intended to direct moisture away fro the interior of a wall toward the outside as rapidly as possib Make any necessary corrections to prevent water from ent ing at the top of the wall or at parapets, sills, and jamb sea before plastering.

Concrete masonry. Concrete masonry provides an exce lent base for plaster. When constructing concrete mason walls that are to be plastered, use an open-texture concr masonry unit. Also, flush (nontooled) masonry joints shou be specified. The open texture on the surface of the block a the joint promotes good mechanical bond. Because the por land cement in the block is the same cementitious material that in the plaster, the two have a great affinity.

Differential suction between the concrete masonry uni and the mortar joints may cause ghosting through of joi patterns in two-coat applications if the first coat was too thi This is most likely to occur if walls are plastered soon aft being built, and if the walls are of blocks that had becom very wet in the stack. It emphasizes the value of stori blocks under cover and of not plastering on wet walls.

New concrete masonry walls to be plastered should properly aligned and free from surface contaminatio Mortar droppings should be scraped off before plaster applied. The concrete masonry wall should have been pro erly cured and be carrying almost all of its design de load before plaster is applied.

Old concrete masonry walls should be critically inspecte

or alignment and surface condition prior to plastering. Old walls coated or moisture sealed with a product other than ortland cement paint should be restored to their original ondition by sandblasting the surface.

Clay masonry. When applying plaster to clay masonry walls, precautions should be taken similar to those for concrete masonry. Surfaces of hard and medium clay tile or unglazed clay brick usually are scored, have proper suction, and have textures rough enough to provide good mechanical bond. On old brick masonry where the mortar has become friable, the joints should be raked out and the entire surface cleaned before plastering. With old, disintegrating brick masonry, metal lathing should be used to provide a mechanical support for the plaster.

Metal Reinforcement*

Metal-lath reinforcement should be used wherever portland cement plaster is applied over one of the following:

Wood frame of open construction

Wood frame of sheathed construction

Metal framing

Flashings

Surfaces providing unsatisfactory bond

Chimneys

Metal reinforcement should be positioned and attached securely to the supporting structure, which may be a solid masonry wall or an open stud wall. When applied to solid concrete or masonry bases, the metal lath should be anchored directly to the base with hardened concrete nails or power-actuated fasteners. The location of control joints in the metal reinforcement should match those in the supporting wall. Where open-wood-frame or sheathed construction is encountered, the metal reinforcement must be firmly nailed to the wood support system. In metal-stud and runner-track wall systems, the metal reinforcement may be secured either to the studs or to the structure to provide rigid panels. In suspended-ceiling application, the metal reinforcement should be secured to the hanger and runner system in a manner that will produce free-floating panels. The floating-panel approach lessens shrinkage and settlement cracking within the plaster panels. When securing the metal reinforcement to the supporting frame, the metal reinforcement should be held out from the base or frame approximately ¼ in. (6 mm). The reinforcement must be stretched between supports to reduce the slack. Loosely attached reinforcement encourages an uneven thickness of plaster. The thick plaster between studs is stronger than the thinner plaster over the support stud, inducing cracking at stud locations.

Nails, wires, or other devices for attaching the reinforcement to the supports should be spaced closely to secure rigid support. Generally, nails and attachments should be spaced every 6 in. (150 mm). If the reinforcement is not self-furring,

the nails should have a self-furring device to keep the reinforcement at least ¼ in. (6 mm) away from the supports.

Metal reinforcement must form a continuous network of metal over the entire surface. Laps at ends and sides of large mesh reinforcement must be at least one full mesh and not less than 2 in. (50 mm). For small-mesh reinforcement, the laps should be at least 1 in. (25 mm). Laps should be wired securely and end laps should be staggered. In open-frame construction, joints and laps should be made at supports.

Paperbacked metal reinforcement should be lapped in the same way, being careful to maintain a metal to metal, paper to paper sequence to the lap joints. A metal to paper sequence leads to reduced plaster thicknesses and cracking at each lapped area.

Metal corner beads with solid metal noses should not be used on exterior surfaces. It is difficult or impossible to get sufficient plaster behind these beads to completely encase them and prevent corrosion. However, corner reinforcement built up with a couple of layers of wire can be used as long as the spacing of the wire permits complete encasement of the metal.

Wood Frame. Metal reinforcement is to be applied on all wood-frame structures, both open stud and sheathed, that will be plastered. The frame should be complete and carry the entire dead load of the structure prior to plaster application. Interior finishing may begin after exterior stucco work is complete, but vibrations or impacts against the exterior wall from the inside should be avoided.

In open-frame-wood construction, the structural frame should be properly braced and rigid. Soft, annealed-steel wire, No. 18 gage or heavier (often called line wire), is stretched horizontally across the face of the studs about 6 in. (150 mm) apart. The wire is stretched by nailing or stapling it to every fourth stud, and tightened by raising or lowering it and securing it to the center stud.

The wall surface is covered next by a waterproof building paper or felt that is nailed over the wire strands. The paper or felt should be lapped at least 2 in. (50 mm). The line wire provides a back support to minimize bulging of the paper or felt between the studs when plaster is applied and to help ensure a uniformly thick scratch coat.

Hexagonal-wire mesh (stucco netting) is applied over the waterproof building paper or felt, using furring nails that hold the mesh out ¼ to ⅜ in. (6 to 9 mm) from the framing. As the plaster is applied it is forced through the mesh openings and against the backing, completely embedding the metal lath in plaster.

Several types of metal reinforcement—diamond mesh, expanded rib, wire mesh—may be used over the line wire and waterproof paper backing. Some metal reinforcement is available that already has paper or polyethylene backing; this type can be used directly over the studs.

In open-frame-wood construction, when installing metal lath at internal or external corners, the lath should be started at least one full stud space away from the corner. The lath should be bent around the corner to avoid a junction or lap at the corner.

Wood-frame construction covered with sheathing eliminates the necessity for wrapping the horizontal strands of annealed line wire around the structure. Wood sheathing usually is applied horizontally or diagonally and securely

---

For application techniques, see Lathing and Furring for Portland Cement and Portland Cement-Lime Plastering, Exterior (Stucco) and Interior, A42.3, American National Standards Institute, Inc., New York, N.Y.

fastened to each stud. Plywood sheets and insulating boards may also be used as sheathing. Waterproof building paper or felt should always be placed over the sheathing, and metal reinforcement applied over the paper. Then the surface is ready for plastering. The lath and paper are lapped at the ends and sides for uninterrupted coverage of the surface. The metal reinforcement is fastened to the studs with furring nails. An alternate is to attach a waterproof, paperbacked, wire-fabric lath directly over the sheathing, nailing it to the studs.

Metal reinforcement should be installed with its long dimension at right angles to the stud supports. Nails or staples for attaching the metal reinforcement to the wood studs should be galvanized or rust resistant. Aluminum nails should not be used where they will be in direct contact with wet portland cement plaster in service. Portland cement will chemically react with the aluminum, causing corrosion and cracking.

Steel Frame. Studding or furring for steel channels, pressed-steel members, steel reinforcing rods, or metal studs should be properly spaced and securely fastened to the structural steel frame by wire-tying, clipping, bolting, or welding.

If welded, they should be separated into smaller panels and the individual panels tied together with wire. This allows for movement and prevents the incidence of extreme stresses within the portland cement plaster.

Where walls are rigidly fixed and restrain building movements, some deformation will occur, creating stresses in the plaster that can cause cracking. Therefore, control joints should be installed in large-area surfaces to relieve these stresses and prevent cracking. The smaller panel areas isolated by control joints can accommodate shrinkage stresses better than the same area without control joints.

Above. **Concrete masonry provides an excellent base for plastering. A water spray is used to premoisten the concrete blocks before applying the first coat of plaster.**

Below. **Metal-lath reinforcement is used where plaster is not compatible with a particular base.**

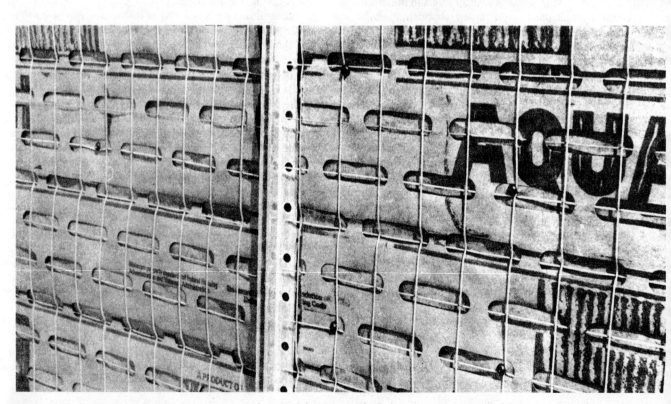

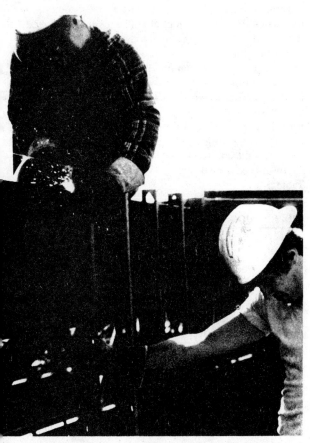

oove. **Metal studs should be properly spaced and ecurely fastened to the structural steel.**

elow. **Low-absorption (glazed) masonry walls are satisctory bases after a dash-coat plaster is applied to inease bond.**

MIXTURES

Portland cement plaster can be proportioned using a variety of cementitious materials and available aggregates to produce a hard surface that is resistant to abrasion and unaffected by dampness. Proportioning mixtures should be adapted to both the use and climatic conditions that will prevail at the jobsite. The merits of each cementitious material and the benefits of selecting aggregate on voids content and grading were explained in the section "Materials." The following paragraphs explain how the proper proportioning of the materials into a mixture will affect its performance in the fresh and hardened state, for application by hand and machine.

A properly proportioned plaster mixture can be recognized by its workability, ease of troweling, adhesiveness to bases, and resistance to sagging. Batch-to-batch uniformity in scratch-, brown-, and finish-coat mixtures will help assure uniform suction, color, and serviceability.

Proportioning of portland cement plaster mixtures is more an art than a science. Successful performance has been achieved by the combinations of cementitious materials and aggregates shown in Table 3. These proportions are recommended for the first and second coat applications. Designers and specifiers should explore the many special surface finishes available, or develop new ones, for the third or finish coat. Sample panels prepared prior to construction are an effective way to develop procedures, provide an acceptance sample, and establish performance characteristics. A test panel should be representative of the finished work in the structure and be at least 4 by 8 ft (1.2 by 2.4 m).

When selecting the plaster mix, basic considerations include: suction of the background, its surface irregularities, climate extremes, extent of surface exposure, type of base, and method of application. Tables 3 and 4 can help in the

Table 3. **Base-Coat Plasters** (Proportions, by volume[1])

Plaster type	Portland cement C	Lime L	Masonry cement M	Plastic cement P	First coat[4] (scratch)	Second coat[4] (brown)
	Cementitious materials[3]				Sand[2]	
C	1	0 to ½			2½ to 4	3 to 5
CM	1		1		2½ to 4	3 to 5
L	1	½ to 1¼			2½ to 4	3 to 5
M			1		2½ to 4	3 to 5
CP	1			1	2½ to 4	3 to 5
P				1	3 to 5	4 to 5

[1]A range of lime and sand contents allows for adjusting each mix to optimize workability using local materials.

[2]The same or a greater quantity of sand than that used in preparing the scratch coat should be used while preparing the brown coat. 21 No. 2 shovels equals approximately 3 cu ft (0.08 m³).

[3]The type of cement selected should be determined by weather conditions during plastering, availability of materials, and anticipated exposure.

[4]Volume of sand per sum (total amount) of cementitious materials used.

selection of plaster type and the recommended mixtures for the type selected.

These plaster mixtures can be prepared choosing among combinations of cementitious material and local sands. The sand content of the plaster is given as a range to allow the specifier to select high-strength, dense plaster where needed. A good rule is to select a mixture with the maximum amount of aggregate-to-cement ratio to reduce shrinkage and cracking. The use of an air-entrained plaster is recommended for geographical areas with freeze-thaw temperature cycles or where a deicer salt solution may be splashed against the plaster. For the brown-coat plaster that is applied over the scratch coat, the specifier should select the same plaster type (Table 4) with equal or larger aggregate-to-cement ratios. This design practice places the lower-strength mixtures with corresponding lower-shrinkage characteristics toward the exterior.

The plaster types recommended for application to specific bases (Table 4) lists the combinations considered satisfactory for bases with differing absorption characteristics and for application over metal lath. For economy and simplicity, it is better to select the same plaster type for both scratch- and brown-coat application. The proportions should be adjusted to allow for more sand in the brown coat.

Table 4. **Recommended Base Coat Plaster Types for Specific Bases**

Plaster base	First coat	Second coat
	Recommended plaster mixes	
Low absorption base, such as glazed and hard-burned brick or high-strength concretes	C CM CP	C or L or M or CM CM or M CP or P
High-absorption base, such as concrete masonry, absorptive brick, or tile	L M P	L M P
Metal reinforcement	C L CM M CP P	C or L or M or CM L CM or M M CP or P P

Lime or other plasticizers should not be added to the mixture when masonry cement or plastic cement are used, since these cements already contain plasticizers.

For the finish coat, a factory-prepared finish-coat mixture may be used. The manufacturer's instructions should be closely followed for factory-prepared finish-coat plaster mixtures. Alternatively, the finish coat may be proportioned and mixed at the jobsite. If the finish-coat plaster is job mixed, truer color and a more pleasing appearance can be obtained by using a white portland cement with a fine, graded, light-colored sand.

Job-mixed finish-coat plasters should be selected from Table 5.

Table 5. **Job-Mixed Finish-Coat Plaster**

Plaster type	Portland cement[2]	Lime	Masonry cement	Plastic cement	Aggregate
	Proportions, parts by volume[1]				
	Cementitious materials				
F[3]	1	½ to 1¼			3
FL	1	1¼ to 2			3
FP				1	1½
FM			1		1½

[1]Coloring compounds should be added by weight of portland cement and as addition to mixtures given.

[2]White or gray portland cement should be selected, using color of the hardened plaster as acceptance criterion.

[3]Surfaces subjected to abrasion should be plastered with plaster type F or FP.

[4]Volume of sand used per sum (total amount) of cementitious materials for finish coat. Quantity and gradation are dependent on surface texture desired.

The aggregate used for the finish-coat plaster need not comply with the grading requirements for base-coat plaster. Unless otherwise specified, the aggregate should all pass the No. 16 (1.18 mm) sieve and should be uniformly graded from coarse to fine. The sand should be selected for its ability to produce the desired surface texture.

Successful plastering depends on proper batching and mixing of the individual and combined materials. The plaster should possess both proper consistency and body to be either spread by hand or machine. Although batching by shovelful is still the most common procedure, shovelful batching should be checked daily by volume measures to establish both the required number of shovelfuls of each ingredient and the volume of mortar in the mixer when a batch is properly proportioned. Water additions should also be batched by volume with calibrated measures (containers of known volume), a quick-fill tank, or a watermeter.

Mixing should produce a uniform blend of all material

A. **Successful plastering depends upon proper batching and mixing to produce a uniform blend of all materials.**

B. **Machine mixing of portland cement plaster. Water is placed in mixer first.**

C. **Then 50% of the sand is added.**

D. **Next, the cement and any admixture are added.**

E. **Last, the balance of the sand is added and mixing is continued until the batch is uniform and of the right consistency. Three or four minutes usually is sufficient.**

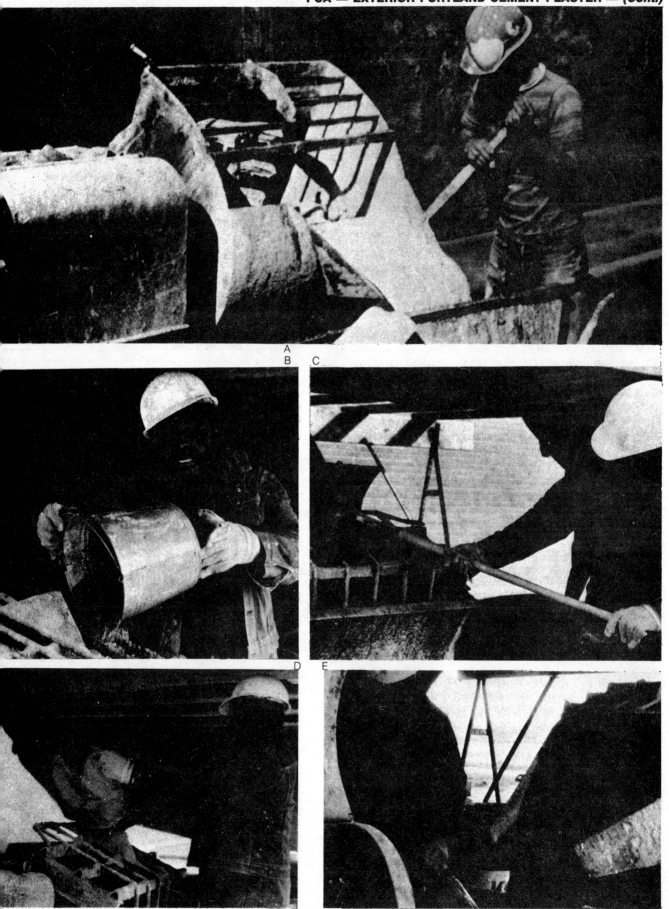

A

B   C

D   E

Ideally, mixing should be completed in a paddle-type mixer for 3 to 5 minutes after all ingredients are in the mixer. During machine application, this mixing period may slow the plastering operation so mixer speed (rpm) should be adjusted to allow for shorter mixing periods. The proof of adequate mixing is the uniformity of the plaster as received and judged by the plasterer who is applying the mixture.

The water content of the plaster should be determined by the plasterer. The plasterer can best judge the correct water content of the plaster by observing its ease of application and the degree of bonding to the substrate. Bonding to the substrate can easily be assessed by removing some of the applied plaster and observing the degree of wetting out of the substrate. Only water reductions or additions should be used during plaster mix adjustments.

APPLYING PLASTER

Whether plaster is applied by hand or machine, it must be applied with sufficient force to develop full adhesion between the plaster and the substrate and to put in place sufficient material to obtain the specified thickness of the coat. The applied plaster must be brought to the desired thickness, the walls made plumb, and the ceilings made level. The thickness of the coat of plaster is set by the use of screeds and grounds. When the area between the screeds and grounds is filled with mortar, a rod is used to even the surface. The rod can bear on the screeds or contact the grounds and be moved over the surface cutting off high spots and showing up the hollow places, which must be filled and rodded again. Additional manipulation of the surface is then required to prepare for the next coat. Scratch-coat plasters are scored or scratched to promote mechanical bond when the second or brown coat is applied. Brown-coat plasters are applied and floated to even the surface and bring it to proper plane, provide a uniform suction throughout the base-coat plaster, and provide a desirable surface for the finish coat. Brown-coat surfaces are floated with a wood float to improve bond with the final surface finish.

Both hand and machine application of plaster are common throughout the industry. Although procedures, tools, and equipment may differ, the method of application produces essentially the same plaster. For hand-applied plaster the water content is slightly less than for its machine-applied counterpart. For machine-applied plasters, the air content is slightly less than for its hand-applied counterpart, because entrained air within the plaster mixture is dissipated as the plaster is projected by the plastering machine toward the wall surface. Differences in the plaster because of different application procedures are further reduced during rodding and floating. After finishing, the two plasters are equal in performance; however, machine-applied plaster develops more intimate contact (bond) with a substrate. The thickness of the plaster coat should be in accordance with Table 6.

Hand Application

The operation of applying plaster by hand begins when the plaster for the scratch (first) coat is delivered to a mortar-

Plaster pump is placed adjacent to the mortar mixer deliver the plaster to the work area.

board. The plasterer verifies that the plaster is prope mixed by kneading the plaster on the board. He then ta some plaster from the board, puts it on his hawk and begin plaster the wall surface. Plastering can be done from bottom to the top of the work area, or from top to bottom. direction depends on the type of work being done and lo practice.

After lifting the plaster from the hawk onto the trowel, plasterer lays the plaster on the wall surface. He uses enou pressure to obtain good contact between plaster and b surface. This procedure continues until the entire wall plastered to the desired thickness. Excessive troweling movement of the scratch (first) coat must be avoided as being applied, because too much action will break the bo created between the plaster and the background, whetl masonry or metal lath.

The scratch coat should be scored in a horizontal directi shallow scratching is adequate.

The brown (second) coat is applied next in suffici thickness, usually ⅜ to ½ in. (10 to 13 mm), to bring surface to the proper plane. A few minutes after the plas has been applied, the surface is rodded to the desired pl with a darby. The plaster thickness is properly gaged w plaster screeds or wood slats of proper thickness as guides. After rodding, the plasterer floats the surface to g it the correct surface texture. Floating of the brown coat is most important part of plastering. Floating must be d only after the plaster has lost sufficient moisture so that surface sheen has disappeared and before the plaster become so rigid that it cannot be moved under the float. T interval is critical, since the degree of consolidation t occurs during floating influences the shrinkage-cracki characteristics of the plaster. The finish coat is applied t predamped, but still absorptive, base coat to a thickness about ⅛ in. (3 mm). The finish coat is applied from the down and the whole wall surface must be covered with joinings (laps or interruptions).

On this job, plaster was distributed by wheelbarrow from pump nozzle to mortarboards.

The plasterer takes plaster from the board, puts it on his hawk, and begins to plaster the surface.

Table 6. **Nominal Plaster Thickness for Three- and Two-Coat Work[1]**
(See "Glossary of Plastering Terms" for definition of "plaster thickness")

Base	Interior							
	Vertical[3]				Horizontal			
	1	2	3[7]	Total	1	2	3[7]	Total
Three-coat work:[2]								
Metal reinforcement	¼	¼	⅛	⅝	¼	¼	⅛	⅝
Solid base:[4]								
Unit masonry	¼	¼	⅛	⅝	(Use two-coat work.)			
Cast-in-place or precast concrete	¼	¼	⅛	⅝	—	—	—	⅜ max.[5,6]
Metal reinforcement over solid base	½	¼	⅛	⅞	½	¼	⅛	⅞
Two-coat work:[8]								
Solid base:[4]								
Unit masonry	⅜	¼	—	⅝	—	—	—	⅜ max.[5,6]
Cast-in-place or precast concrete	¼	⅛	—	⅜	—	—	—	⅜ max.[5,6]

Base	Exterior							
	Vertical[3]				Horizontal			
	1	2	3[7]	Total	1	2	3[7]	Total
Three-coat work:[2]								
Metal reinforcement	⅜	⅜	⅛	⅞	¼	¼	⅛	⅝
Solid base:[4]								
Unit masonry	¼	¼	⅛	⅝	(Use two-coat work.)			
Cast-in-place or precast concrete	¼	¼	⅛	⅝	—	—	—	⅜ max.[5,6]
Metal reinforcement over solid base	½	¼	⅛	⅞	½	¼	⅛	⅞
Two-coat work:[8]								
Solid base:[4]								
Unit masonry	⅜	¼	—	⅝	—	—	—	⅜ max.[5,6]
Cast-in-place or precast concrete	¼	⅛	—	⅜	—	—	—	⅜ max.[5,6]

[1] Where a fire rating is required, plaster thickness shall conform to the applicable building code or to an approved test assembly.

[2] Where three-coat work is required, a dash or brush coat of plaster materials will not be accepted as a required coat.

[3] Additional coats shall be applied to meet finished thickness specified for solid-plaster partitions.

[4] Where masonry and concrete surfaces vary in plane, plaster thickness required to produce level surfaces will not be uniform.

[5] On horizontal solid-base surfaces (such as ceilings or soffits) requiring more than ⅜-in. plaster thickness to obtain a level plane, metal reinforcement shall be attached to the concrete and the thickness specified for three-coat metal reinforcement over solid base shall apply.

[6] Where horizontal solid-base surfaces (such as ceilings or soffits) require ⅜-in. or less plaster thickness to level and decorate, and where there are no other requirements, a liquid bonding agent or dash-bond coat may be used.

[7] The finish-coat thickness may vary, provided that the total plaster thickness complies with Table 4 and is sufficient to achieve the texture specified. For exposed-aggregate finishes, the second (brown) coat may become the bedding coat and shall be of sufficient thickness to receive and hold the aggregate specified.

[8] Table 6 shows only the first and finish coats for vertical surfaces and only the total thickness on horizontal surfaces for two-coat work.

A

C D

B

The first (scratch) coat is applied with enough pres-
ure to obtain good contact between plaster and base
urface. The scratch coat is scored horizontally.

The second (brown) coat is applied in a thickness to
ing the surface to the proper plane.

A few minutes after brown coat is applied, the surface
rodded to the desired plane.

Finally, the finish coat is applied to a predamped, but
ill absorptive, brown coat and floated and textured to
e desired finish.

Machine Application

pplying plaster by machine requires preplanning so the
peration will go smoothly. The plaster pump should be
aced adjacent to the mortar mixer and sufficient hose at-
ched to the pump to allow quick, easy pumping of plaster
om the mixer to the surfaces to be plastered. Hose lengths
ould be relatively straight and no longer than necessary.
ormally, rigid pipe is used to accommodate the high pres-
res near the pump, and the pipe is coupled to a flexible
bber hose at the delivery or spray end. To ease the work
fort demanded of the nozzleman, a light, flexible whip-line
se is used between the flexible rubber hose and the nozzle.
he nozzle unit at the end of the pump line contains valves
at give the operator control of the pressurized-air discharge
d volume control of the pump.

Before pumping plaster, the hose should be prewetted and
bricated to ensure unimpeded flow of plaster through the
se. Enough water is pumped into the holding tank to par-
ally fill it. The pump is started and begins to move this water
rough the hose. The pump is stopped, the hose is discon-
ected at the pump, and a wet sponge is inserted in the hose.
he hose is reattached and the pump is started again. Water
der pressure forces the wet sponge through the hose. A
eat cement paste grout (cement and water without sand) now
poured into the nearly drained holding tank. As pumping
ntinues, the neat cement paste is forced through the hose,
bricating all hose surfaces. When the sponge emerges from
e hose at the discharge end, the nozzleman attaches the
zzle and sprays the remaining water and neat cement paste
nto the ground or into a waste container. The plaster mix is
aced in the holding tank, pumped through the hose, and
etected at the nozzle by the presence of sand. When the
ixture is uniform in appearance, it is applied to the surface
be plastered.

Good pumping practices require preconditioning, proper
aintenance during pumping, and good cleanup and preven-
ve maintenance of the hose at the end of the work. Hose
oppages may occur during work because of poor mixtures
r leakage at quick couplers located between pump and pipe,
ipe and hose, and hose and nozzle. A stoppage requires
mediate removal of the obstruction and, if necessary, re-
air or replacement of equipment. Excessive pressure in the
ose may cause a ruptured hose or a blowoff of the pump's
fety valve. Regular preventive maintenance of the pump
ould be established procedure. At the end of plastering, the

pump and hoses should be cleaned immediately by pumping
water through the hose and repeating the wet sponge operation. Water should be flushed through the hose using a second
sponge to free the hose of any residual plaster and cement
paste. When the nozzleman notices a marked change in the
plaster consistency, he must disconnect the nozzle, clean it,
and await the arrival of two sponges.

During pumping applications the nozzleman holds the
nozzle approximately 12 in. (0.305 m) from the surface.
Plaster is applied to the desired, or slightly greater, thickness.
The nozzleman can vary the spray pattern and pattern size by
adjusting the air pressure, changing the nozzle orifice size,
varying the distance between the end of the air stem and the
orifice, increasing or decreasing pump speed, or calling for a
change in water content to adjust the consistency of the plaster mix. Through proper selection of these options the nozzleman can control the plaster application.

Machine application eliminates lap and joint marks; produces a more uniform appearance in color and texture; and
produces, in colored finish coats, deeper, darker, and more
uniform colors than can be obtained by hand application.

For scratch and brown coats, the nozzle should be moved
with a steady, even stroke laying on the proper thickness with
one pass and overlapping successive strokes. The angle of the
nozzle to the wall should be uniform. Around door bucks and
window frames, the nozzle should be moved closer, to within
a few inches of the surface.

**Applying portland cement plaster by machine on high-
rise apartment building.**

Applying portland cement plaster by pump. Man in background is floating plaster and embedding the mesh.

The same rodding, floating, and finishing procedures used after hand application are used after machine application.

Manufacturers of plastering machines publish instructions regarding the proper use, care, and maintenance of their equipment. These instructions should be carefully followed.

Application to Bases

The first (scratch) coat of plaster over metal reinforcement should completely encase the metal reinforcement. After applying plaster to an entire area, the surface should be rodded and scratched to promote mechanical keying (bond) with a second coat. Scratching is normally done horizontally on a vertical wall. Horizontal scratches act as water dams and promote curing. Scratching vertically may promote cracking at studs when scratches or score marks are directly over and align with studs.

The second (brown) coat of plaster needs special attention after application and rodding. After rodding the plasterer should pause before starting to float the surface. This pause is related to the onset of cement hydration, evaporation of moisture, and the rate of absorption by the first coat of plaster. Floating with a large wood float must be done when the plaster is at the proper moisture content or the necessary reconsolidation (densification) of the plaster will not be obtained. The correct time to begin floating can be determined by placing a finishing trowel against the surface. The trowel should not adhere to unworked plaster but would adhere to worked plaster. Unless this condition (of the trowel not sticking to the plaster) exists at the beginning of the floating operation, the plaster will not be properly densified during floating.

During application of the first coat to a solid base or application of all second coats to first-coat plaster, the plasterer can determine the suitability of a plaster mix for developing intimate bond in the following way: Once the second coat has been applied using the pressure required to develop good contact and when the plaster has the required thickness, the plasterer removes the freshly applied plaster down to the previous coat or base. This is easily done with the square end of the trowel. If the underlying coat has not been completely wetted by the coat just removed, the plasterer must either apply more pressure during plaster application or increase the water content of the plaster. The same test can be used to assess the suitability of machine-applied plasters.

The use of a third (finish) coat of plaster permits variation in color and texture. Each texture requires special tools and techniques. The uniform appearance and texture of the finish-coat plaster will be influenced by the care taken during application of the first and second coats. Most color variations in the finish coat are traceable to base-coat variations. These color variations can be minimized by prewetting the base coat prior to applying the third coat. During the making of a test panel, any special finishing procedures and the correct timing and sequence should be documented for use later in the actual construction.

Color Variation

Color variation in finish-coat plasters can be traced to several causes, including the following:

- Differences in suction of the brown coat from one location to another
- Variations in thickness of the base coat
- Failure to maintain mix proportions of aggregate to cementitious materials
- Cold joints in the brown coat

Overfloating the brown coat

Nonuniform predampening of the brown coat prior to applying the finish coat

Color variation in exterior stucco can be corrected by fog spraying or brush coating. For brush coating, a factory-prepared brush coat is mixed to a milky consistency. The wall is dampened lightly and evenly before the brush coat is applied. Fog spraying should be applied to a *dry* surface, using as fine a spray as possible. Curing is necessary under hot, dry, or windy conditions. A light fog spray of water is applied the day following application

Delay Between Successive Coats

Within the past 10 years the time delay before applying the second coat has been reduced from seven days to one day or less. The full thickness of the base coats should be applied as rapidly as the two coats can be put in place. The second coat should be applied as soon as the first coat is sufficiently rigid to resist without cracking the pressures of second-coat application. Under certain conditions this may mean applying both first and second coats in a single day. The short delay, or even no delay, between the first and second coats promotes more intimate contact between them and more complete curing of the base coat. No stoppage of plaster is allowed within a panel.

Temperature Considerations

Temperature affects the speed of plastering by extending or reducing the time between consecutive operations. Cold weather lengthens the time between rodding and floating; hot weather shortens it. Dry weather has the same effect as hot weather. Dry or hot weather produces dry substrates and causes more rapid water loss from the plaster through both absorption and evaporation. Moderate changes in temperature and relative humidity can be overcome by heating materials during cool or cold weather and by prewetting during hot or dry weather to reduce the absorption of the base. Severely changing conditions require similar adjustments but to greater extremes. Artificial heat may be required during cold weather. Regardless of climatic conditions, when floating is completed at the proper time the base-coat plaster will perform satisfactorily.

Control Joints

Cracks may develop in plaster from a number of causes: drying shrinkage stresses; building movement; foundation settlement; restraints from lighting and plumbing fixtures that penetrate the plaster; intersecting walls, ceilings, pilasters and corners; weakened sections in a wall or ceiling from a reduction in surface areas or cross sections because of fenestration; severe thermal changes; and construction joints. Because the causes of cracking are so many, it is impossible to anticipate the precise location and direction where cracks may occur in a plaster surface. The strategic location of control joints helps to predetermine and prealign within the

joint most cracking caused by volume changes and building movements.

When plaster is applied to a concrete or masonry base without or with metal reinforcement, the control joints in the plaster coat should be installed directly over and aligned with any control joints existing in the base. Normally, cracking will not occur in plaster applied to uncracked concrete or masonry bases if the plaster bonds tightly to the base structure or is rigidly fixed to the base with metal lath. If excessive cracking does occur, the bond or mechanical anchorage is inadequate.

When plaster is applied over metal reinforcement, the metal reinforcement should be fabricated and attached in a way that allows free movement of the panel area between the control joints. Failure to allow free movement probably will result in cracking in the thinnest portion of the panel area. In determining the size of the free-moving panel, the total surface area should be divided into panels with control joints spaced 10 ft (3 m) apart, preferably to form a square of less than 150 sq ft (14.7 m²) enclosed within the control joint perimeter. Under ideal conditions these recommended limits may be exceeded.

Control joints can be installed with an array of plaster trim accessories. Examples of control joints are illustrated in Appendix C, drawing A. It is very important at a control joint that the metal reinforcement stop at each side of the joint and not continue through it.

The simple notched or ground control joint (Appendix C) is effective where plaster is applied uniformly and is of uniform thickness.

The accordion-pleat control joint seals out weather and is inserted where adjacent isolated panels are to receive control

The strategic location of control joints helps to predetermine and prealign within the joint most cracking caused by volume changes and building movements.

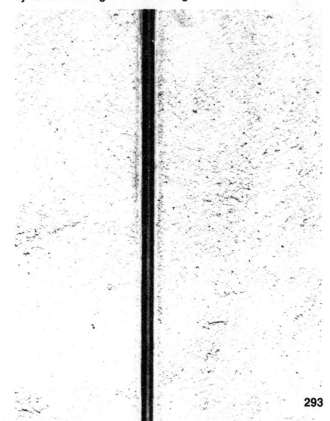

joints. Both legs of this joint must be attached to the isolated panels. If such a joint is used in open-frame construction, the control joint can be nailed through its center V-form or affixed to adjacent studs (double studs).

The control joint diaphragm should be closed, not open, when it is installed. When portland cement plaster hardens, it shrinks as it dries and the joint should open. As the weather changes from day to day, the joint will open further when the surface is cooling or drying and close partially when warming or being rewetted.

Curing

To obtain the best result from the cementitious materials in portland cement plaster, some moisture must be kept in the plaster for the first few days after application. Moist curing, under present plastering practice, usually is applied only to the base coat and continued during the delay before applying the finish coat. Generally, fogging the surface with water at the start and again at the end of the work day will suffice. If the relative humidity is high (more than 70%) the frequency of moistening the surface may be reduced. If it is hot, dry, and windy, the plaster surface should be moistened and covered with a single sheet of polyethylene plastic, weighted or taped down to prevent water loss through evaporation.

Immediately before finish-coat application, the base coat should be moistened. This moisture makes up the total curing of the finish-coat plaster — colored finish coats in particular — so that it is not necessary to further moist-cure the finish coat.

Cold-Weather Considerations

Although portland cement plaster can be applied during cold weather at temperatures below 40°F (4.4°C), better results can be achieved when plastering is done at temperatures above 40°F (4.4°C). The water content of plaster is sufficiently high that, upon freezing, the plaster will undergo disruptive expansion as the water within it freezes. A heated enclosure and use of heated materials are recommended when the daytime temperature is 40°F (4.4°C) and falling. The enclosure should be maintained above 40°F (4.4°C) for at least 24 hours after plastering, and preferably longer. When plaster materials are heated as a further protection from cold temperatures, the mixing water is logically the first material to be heated. Water heated to 130°-140°F (54°-60°C) and combined with cold but unfrozen sand will produce plaster in the 70°-100°F (21°-38°C) temperature range. Heating water to raise the mortar temperature has the effect of accelerating the plaster set, which protects against early freezing. When the plaster moisture content has been lowered by natural evaporation and reaction with portland cement (hydration), the plaster can withstand the lower ambient temperatures.

Interior portland cement plaster work should be maintained at or above 50°F (10°C) until the structure is occupied. Interior plaster such as is used in blast freezers or cold-storage rooms should be allowed to set, cure, and dry and should then be gradually brought down to the subfreezing working temperature.

SPECIAL APPLICATIONS

Surface Bonding of Masonry with Glass-Reinforced Portland Cement Plaster

A recent innovation in the plastering industry is the use of glass fibers as reinforcement in neat cement paste and mortar. These glass-reinforced cements and mortars are applied to dry-stacked masonry (with no mortar in the joints between units). Applying the surface bonding material to both faces of the masonry unit promotes wall stability and strength. Application is possible by hand or machine. The surface bonding material is applied in a 1/8-in. (3.2-mm) thickness with plastering tools. Rodding or floating operations are unnecessary with this thin, single coat. Textures are attainable as shown on page 1.

Commercial products are available with standard (E-glass) glass fibers, or alkali-resistant (AR-glass) glass fibers. Some of the mixtures include a fine 30-mesh (600-μm) silica sand. The mixtures generally contain glass fibers that are approximately 1/2 in. (12 mm) in length. Each fiber is a bundle of individual filaments held together and enveloped in a sizing material.

When the products first appeared there was concern about an alkali-glass reaction. Testing is still under way and will continue for some time to establish the effect of alkalies within the cement reacting with the high-silica glass. Present thinking is that any reaction that does occur is rapidly absorbed and dissipated by the low-modulus cement paste or mortar enveloping the individual fibers.

The structural capabilities of the glass-reinforced mixtures also are being incorporated into pleasing architectural and structural components.

Coolers and Walk-in Freezers

One of the successful uses of portland cement plaster with a hard, steel-trowel finish is on the interior surfaces of refrigeration or cooling rooms. In this construction a steel structure typically carries the roof load and concrete masonry units form the building's exterior walls. A vapor barrier is placed over the masonry and under two layers of insulation board which then receive a portland cement plaster hard-finish surface. Sometimes, particularly in remodeling work, it may be necessary first to plaster the rough masonry surface in order to provide a smooth plane on which to apply the insulation. This construction involves plastering over insulation board with or without metal reinforcement.

Portland cement plaster bonds tenaciously to most insulation, such as polyurethane, polystyrene, and cork. Each of these materials, although basically water resistant, possesses sufficient surface irregularities to allow mechanical keying of the plaster to the insulation material. Some materials, however, are too soft or too smooth to allow proper mechanical keying between plaster and insulation. Scoring the surface of the insulation may be sufficient, or it may be necessary to cover the surface of these materials with metal lath. A problem arises with any insulation because the applied plaster can lose water only by evaporation from the exposed interior surface. Freezing of any excess water n the plaster could cause delamination between the insulation and plaster.

Plastering is done in two coats. For the first coat, the plaster is applied with sufficient pressure to promote intimate contact with the insulation. After rodding the plaster to a true surface, it must be floated and refloated to promote water loss throughout its entire thickness before starting to trowel. These precautions are necessary because the plaster upon cooling and drying will undergo differential volume changes, that is, greater shrinkage on the interior surface than at the plaster and insulation interface. The warping effect can coincide with shrinkage cracking and will be most pronounced at the intersection of shrinkage cracks. Warping will curl the plaster away from the insulation. Consequently, after the surface has been troweled it must be cut into small square panels to prevent warping. A groover is used to score control joints into the plaster at short intervals, approximately 2 ft (0.61 m) apart on centers. Inserting calking compound in all control joints after shrinkage occurs will waterproof the joints.

Swimming pools

The white finish coat of a swimming pool is generally a white portland cement plaster that has been given a smooth, steel-trowel finish. The plaster usually is applied to the surface of a prepared base of cast-in-place concrete, concrete masonry, or shotcrete. These bases should be sufficiently roughened to develop bond with the plaster. A bonding agent of the type recommended for swimming pools may be necessary if the surface is too smooth.

Following is one method of plastering swimming pools: Proportions for the finish coat are

bag white portland cement

bag silica sand or marble dust

½ bags coarse, white marble dust

Sufficient water to achieve a trowelable consistency

Mortar should be mixed in a power mixer. Factory premixed finishes are available. Trowels used for pool work should be of stainless steel with rounded ends to prevent trowel burns.

Before applying the plaster, the base should be sprayed with clean water to reduce suction. The mix should be applied quickly to the walls in a thin coat, beginning at the deepest part of the pool. The plasterer then should double back for the second coat to a total thickness of approximately ⅜ in. (9.6 mm). When the second coat has stiffened sufficiently, troweling should begin and continue until all catfaces are eliminated and the surface properly compacted. Before troweling is completed, the trowel should literally ring as it is moved across the surface.

After the pool wall has been completely plastered, it must be cured immediately by filling the pool with water as rapidly as possible without causing any erosion of the paste from the surface of the plaster. Making arrangements with the local water department to use a convenient fire hydrant will ensure rapid filling if the pool is large. Small pools can be filled by a garden hose with a towel wrapped around the discharge end to prevent damage to the fresh plaster. The hose should be positioned at the deep end of the pool.

When the pool is in use, the owner must maintain a balance between alkaline and acid water conditions. An alkaline condition will promote the growth of crystals on the plaster surface, resulting in cuts and abrasions to swimmers. Conversely, an acid water condition will cause deterioration of the alkaline cement plaster and shorten the serviceability of the pool finish coat.

To protect against frost damage in cold weather, the pool owner may drain the pool and risk the need for replastering the following spring; keep the pool filled with warm water and use an insulating cover to maintain the water temperature above freezing; or place nonfreezable items (such as logs) in the pool to relieve expansive forces when the water freezes. Preventive maintenance considerations and economics may dictate completely enclosing the pool.

Old plaster pools in need of restoration can be repaired by first removing all unsound plaster and replastering these areas back to the original plane or contours. Unsound plaster, or plaster that has lost its bond to the base, can easily be detected by tapping the surface with a wooden mallet or metal rod. A dull, hollow sound indicates delamination or loss of bond; a sharp, solid sound indicates good adhesion between plaster and base. After cleaning the old plaster surface with a high pressure-velocity water-jet spray, the new plaster can be applied.

New plaster applied over existing plaster can be as thin as ⅛ in. (3.2 mm) if the plasterer can obtain sufficient bond with the mixture.

Discoloration of swimming pool plaster may be caused by deleterious substances in aggregates, overtroweling or burning the surface, contaminated mixing water, and improper use and maintenance. Deleterious substances include blast-furnace-slag contaminants (causing yellow, brown, green, blue, or purple spots) or graphite (causing gray specks), all of which can only be removed by coring discolored areas and replacing plaster. Burns from overtroweling can be removed by rubbing the affected area with an abrasive or by acid etching the entire area. Discoloration due to contaminated water sources can be remedied only by establishing the discoloration compound and either reducing or oxidizing the materials. Abuse in use of the pool may vary from improper chemical (acid) additions to the water, to suntan lotion, or to improper chemical balance. A white streak from pool top to the bottom drain indicates acids were added to pool in concentrated form. Rings around the pool surface indicate suntan lotions. An acid water condition in the pool may be indicated by bluish-white deposits leading from the water-filter return line to the pool drain.

Courts—Handball, Squash, Racquetball

The popularity of handball, squash, and racquetball as recreation is providing the plastering industry with a challenging application for smooth-trowel-finish surfaces on building interiors. Many court walls are constructed of concrete masonry, which serves as an excellent solid base for portland cement plaster.

Because of the frequent, high impact of the ball on the plaster, metal reinforcement is normally specified on the front wall surface. Side walls and ceilings also receive high impacts; however, the frequency of their occurrence is less than on the front wall and metal lath is not necessary. Metal lath should be used on the playing and sidewalls for racquetball courts.

Handball court construction requires that masonry walls be built to conventional masonry specifications and tolerances. Joints should be struck flush with the face of the masonry and not tooled. Although some courts have a vapor barrier at the interior concrete masonry surface, this practice is not recommended as it prevents the bonding of plaster to the walls.

An open metal reinforcement should be used on the front, high-impact wall surfaces. The metal reinforcement should be furred out ¼ in. (7 mm) from the wall using furring nails or self-furring metal lath. The lath should be affixed securely with nails to the concrete masonry wall; nails should be spaced approximately 16 in. (406 mm) apart.

Construction details must be planned carefully. The plaster should be separated from the finished flooring with casing beads installed ¼ in. (7 mm) above the finished floor, with the flooring underneath the base bead but not touching the basic wall construction. In the application procedure that follows, plaster can be placed on concrete masonry walls without control joints.

The first coat of portland cement plaster should be proportioned to contain 1 part of portland cement, 0 to ¼ part of hydrated lime, and 3 to 3¾ parts of sand, by volume, per sum of volume of cementitious materials. Immediately prior to plastering, the dry concrete masonry wall should be thoroughly moistened to reduce its absorption. Ideally, the plaster should be machine applied to obtain intimate contact and strong bond between the base and the plaster. Hand application of the first coat must be done with sufficient pressure to force the plaster through the metal lath and tightly against the base surface. After sufficient material has been applied, the surface should be scored horizontally.

The second coat of plaster should be applied either immediately after the first coat has become sufficiently rigid to allow application or the following day. This plaster should be proportioned to contain 1 part of portland cement, 0 to ¼ part of hydrated lime and 4 to 4½ parts of sand, by volume, per sum of volume of cementitious materials. The plaster should be applied with enough pressure to obtain tight contact with the first coat. It should be built up to the required thickness and rodded to plumb the surface and bring it to a smooth plane. After rodding the soft plaster, the plasterer should wait a few minutes for the plaster to harden before the surface is floated (reconsolidated) and brought to a good straight surface with sufficient roughness to develop mechanical bond with the third coat of plaster. On succeeding days the plaster should be moist-cured to develop full strength and hardness.

The third coat of plaster should be applied approximately seven days after the first coat. The third coat of plaster should be proportioned to contain 1 part of white portland cement, ½ to 1 part hydrated lime and 4 parts of minus 30-mesh (600 μm) white silica sand by volume per sum of cementitious material. The finish-coat plaster should be applied on the predampened base-coat plaster to a thickness of ⅛ in. (3.2 mm), and steel-troweled to obtain a smooth surface.

Throughout this construction sequence, the curing of plaster coats and the prevention of climatic effects, as recommended in this publication, should be practiced.

CARE AND MAINTENANCE OF PLASTER

Minimal care will keep a plaster (stucco) building attractive for many years. The simple act of washing — as with metal, wood, and masonry buildings — will keep the surface clean and the color bright.

Washing plaster wall surfaces is done in three steps:

1. Prewet the wall, saturating it. Start at the bottom and work to the top.

2. Use a garden hose to direct a high-pressure stream of water against the wall to loosen the dirt. Start at the top and wash the dirt down the wall to the bottom.

3. Flush remaining dirt off the wall with a followup stream.

Wash the walls from bottom to top. Start prewetting at the bottom and gradually wet the entire area to the top. Prewetting will overcome absorption and prevent dirty wash water from being absorbed and dulling the finish. A jet nozzle on a garden hose will clean effectively. Do not hold the nozzle too close to the surface because the high-pressure stream of water may erode the surface.

Chipped corners and small spalls can be patched. Premixed mortar will do the job. It requires only the addition of water, mixing to a doughy consistency, troweling into the space to be patched, and finishing to match the texture of the surrounding surface. The patch area should be prewetted before applying plaster.

A fresh, new look can be given to any exterior plaster wall by painting it with a cement-based paint or coating. These products are mixed with clean water to a brushable consistency and laid on heavily enough to fill and seal small cracks and holes. The surface should be dampened immediately ahead of application.

PROBLEMS—
THEIR CAUSES AND REMEDIES

Observed	Reason	Prevention
1. Cement floats on water during batching.	Waterproof cement is being used.	Mix with sand to proper proportions and workability.
2. Bagged cement and lime have lumps in them.	Materials were allowed to get moist during storage.	Screen bagged cement or lime to remove lumps. Mix plaster on the dry side (low water content) until lumps disappear.
3. Plaster froths in mixer.	Too much water has been used, or cold water and prolonged mixing.	Reduce amount of water. Use warmer water and mix for 3 to 5 minutes.
4. Plastering pump jams; but no leaks.	Air content of plaster is too high.	Review batching and mixing procedures. Check equipment.
5. Plaster pump jams; hose leaks at coupling; sand is jammed at hose coupling.	Coupling is not tight.	Check coupling and gasket. Repair or replace.
6. Plaster pump jams and nozzle is plugged. Sand and cement lumps are in plaster on wall.	Packed sand or lumps are impeding plaster and air flow.	Adjust mix cycle.
7. Plaster stiffens immediately after application to concrete masonry base.	Concrete masonry base is too dry or water retention of plaster is too low.	Moisten base prior to plaster application.
8. Plaster falls off metal reinforcement when applied by hand.	Plaster adhesion is low. Metal lath is on upside down.	Adjust mix to include air-entraining admixture or plasticizer. Check metal for proper installation.
9. Scratch-coat plaster is cracked over each wooden stud.	Metal reinforcement and backup were not tightened properly.	Check lathing procedures: line wire should be taut, paper tight, and metal lath tight.
10. Scratch-coat plaster shows white deposit seven days after application.	High water content was used in plaster; weather was cold; too much delay in applying coats.	Adjust water content and heat materials above 70°F (21°C). Apply brown coat.
11. The finished surface is blistering.	Mix was overly rich. Too much water was used. Surface was overtroweled.	Adjust mix with more sand, less water.

APPENDIX A

Industrywide Reference Standards and Specifications

Metal Lath/Steel Framing Association Specifications for Metal Lathing

ASTM C841 — Standard Specification for Installation of Interior Lathing and Furring, American Society for Testing and Materials

ASTM — Lathing and Furring for Portland Cement and Portland Cement-Lime Plastering Exterior (Stucco) and Interior, American National Standards Institute

Construction Specifications Institute 09110 — Specifying: Furring and Lathing

ASTM — Portland Cement and Portland Cement-Lime Plastering Exterior (Stucco) and Interior, American National Standards Institute

Construction Specifications Institute 09180 — Specifying: Cement Plaster

Department of Housing and Urban Development (HUD) Minimum Property Standards and Manual of Acceptable Practices

Requirements, Tests, and Standards

ASTM C897 — Standard Specification for Aggregates for Job-Mixed Portland Cement-Based Plasters

ASTM C150 — Standard Specification for Portland Cement

ANSI/ASTM C206 — Finishing Hydrated Lime

Codes and Ordinances

Uniform Building Code,

Uniform Building Code Standards.

BOCA Basic Building Code.

Standard Building Code (Southern).

APPENDIX B

Thickness of Portland Cement Plaster

Plaster base	Minimum finished thickness from face of lath, masonry, concrete
Expanded-metal lath	⅝ in.* (interior)[2]
	⅞ in. (exterior)[2]
Wire-fabric lath	⅝ in. (interior)[3]
	⅞ in. (exterior)[3]
Masonry Walls	½ in.
Concrete Walls	⅞ in. (maximum)[4]
Concrete Ceilings	½ in. (maximum)[5]

[1] Fire-resistant construction must conform to local building code.

[2] Thickness shall be ¾ in. measured from back plane of expanded-metal lath.

[3] Measured from face of support or backing.

[4] Because surfaces may not be true plane, thickness can vary.

[5] A 1/16-in. coat of approved skim-coat plaster may be applied directly to concrete.

*1 in. = 25.4 mm

APPENDIX C

A. **Beads, screeds, joints, accessories.**

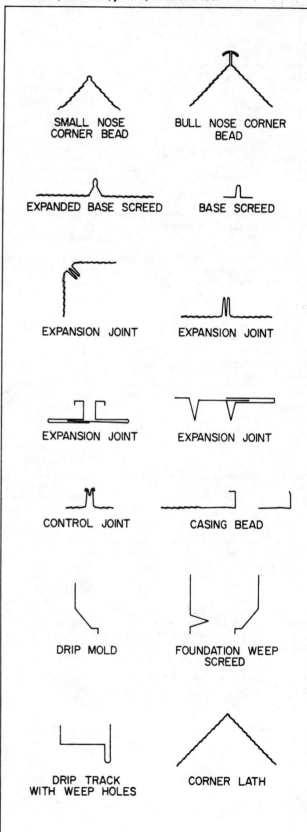

SMALL NOSE CORNER BEAD

BULL NOSE CORNER BEAD

EXPANDED BASE SCREED

BASE SCREED

EXPANSION JOINT

EXPANSION JOINT

EXPANSION JOINT

EXPANSION JOINT

CONTROL JOINT

CASING BEAD

DRIP MOLD

FOUNDATION WEEP SCREED

DRIP TRACK WITH WEEP HOLES

CORNER LATH

B. Portland cement plaster (stucco) on sheathed wood framing.

C. Portland cement plaster on open wood framing.

D. Portland cement plaster applied directly to concrete masonry.

E. Portland cement plaster at joining of unlike bases.

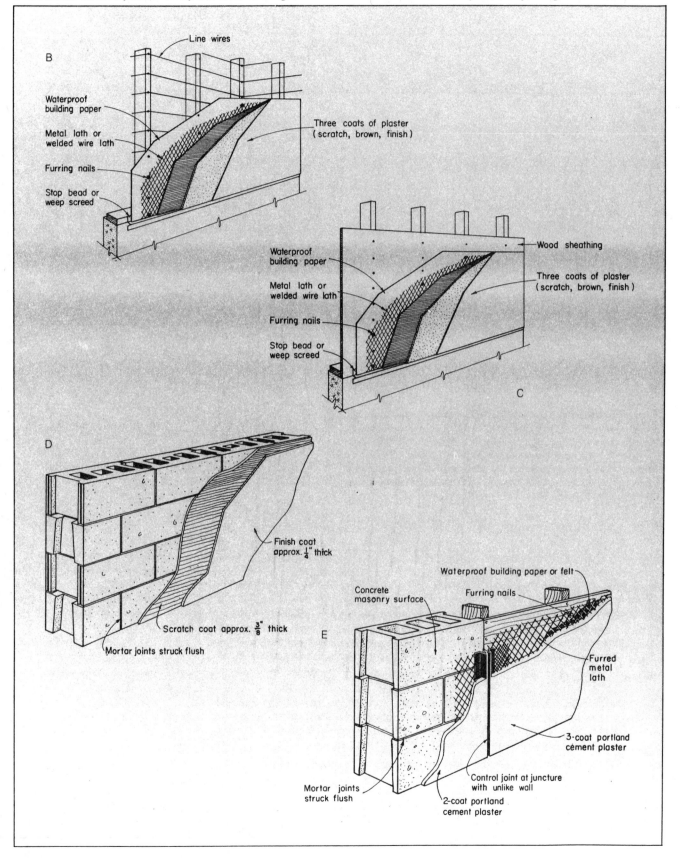

B
- Line wires
- Waterproof building paper
- Metal lath or welded wire lath
- Furring nails
- Stop bead or weep screed
- Three coats of plaster (scratch, brown, finish)

C
- Waterproof building paper
- Metal lath or welded wire lath
- Furring nails
- Stop bead or weep screed
- Wood sheathing
- Three coats of plaster (scratch, brown, finish)

D
- Finish coat approx. $\frac{1}{4}$" thick
- Scratch coat approx. $\frac{3}{8}$" thick
- Mortar joints struck flush

E
- Concrete masonry surface
- Waterproof building paper or felt
- Furring nails
- Furred metal lath
- 3-coat portland cement plaster
- Control joint at juncture with unlike wall
- 2-coat portland cement plaster
- Mortar joints struck flush

CHAPTER FOUR

REQUIREMENTS,
TESTS
AND
STANDARDS

INDUSTRY-WIDE REFERENCE STANDARDS

Following is a listing of industry-wide reference standards in most common use in the lathing and plastering industries. Those indicated by an asterisk are reproduced in full in this manual.

Abbreviations used in this listing are as follows: **ASTM,** for American Society for Testing and Materials; **ANSI,** for American National Standards Institute; **FS,** for Federal Specifications; and **GA,** for Gypsum Association.

FRAMING AND LATHING

FS QQ-W-46G — Specification for hanger and tie wire.
FS QQ-W-461G — Specification for hanger wire and staples.
FS QQ-L-101C — Specification for metal lath.
FS FF-N-015B — Specification for nails, wire, staples.
FS QQ-N-281C — Nickel-copper alloy, bar, plate, rod, sheet, strip, wire and special shaped sections.
FS W-461G — Wire, steel, carbon (round, bare and coated).
FS SSL-30A — Lath, sheathing, gypsum board.
FS TT-P-86 — Red lead oxide protection on steel studs.
FS UU-B-790a — Building paper.
FS HH-F-191a — Felt, asphalt saturated.
FS QQ-S-698 — Electro-galvanized metal studs, tracks, channels.
FS QQ-S-775E — Hot dip, galvanized metal studs, track, channels and trim items.
FS GGG-D-777a — Power actuated fastening devices.
FS GGG-D-780a
FS GGG-D-00570
ANSI A 97.2 Installation of steel framing to receive screw attached gypsum board and backing board.
ANSI A 42.3* — Lathing and furring for portland cement plaster exterior and interior.
SPR-R3-60 — Metal accessories.
ASTM A 90 — Weight of coating on zinc (galvanized), coated steel articles.
ASTM A 109M — Steel, carbon cold-rolled strip (metric).
ASTM A 112 — Zinc coated (galvanized) steel tie wire.
ASTM A 164 — Electro-deposited coatings of zinc on steel.
ASTM A 245 — Structural studs.
ASTM A 366 — Steel cold-rolled commercial quality.
ASTM A 390 — Zinc coated (galvanized) steel netting and woven wire.
ASTM A 446 — Steel sheet zinc coated (galvanized) by hot dip process structural quality. Grade A 33,000 psi yield, Grade D 50,000 psi yield.
ASTM A 525 Steel sheet, zinc-coated (galvanized) by hot-dip process general requirements G60.
ASTM A 526 — Steel sheet zinc coated (galvanized) hot dip commercial quality.
ASTM A 528 — Steel sheet zinc coated (galvanized) hot dip draw quality.
ASTM A 570 — Hot rolled carbon steel sheet and strip structural quality 50,000 psi yield.
ASTM A 591 — Commercial quality electrolytic zinc coated steel.
ASTM A 611 — Steel cold-rolled sheet carbon structural grade C 33,000 psi yield.
ASTM C 646 — Steel drill screw for gypsum sheet material to light gauge steel studs.
ASTM C 754 — Steel framing to receive screw attached gypsum base, backing board, or water resistant backing board.
ASTM C 847* — Metal lath.
ASTM C 933* — Standard specification for welded wire lath.

EXTERIOR AND INTERIOR

FS-SS-P-00402B — Plaster gypsum.
FS-SS-C-161A — Keene's cement gypsum.
FS-TT-P-0035 — Waterproof coatings.
FS-SS-J570B — Joint compounds and reinforcing tapes.
FS-SS-C-192g — Air entraining portland cement.
FS-SS-C-181e — Masonry cement.
FS-SS-C-185a — Natural cement.
FS-SS-C-192g — Portland cement.
FS-SS-A-III — Acoustical plaster.
FS-SS-S-118A — Acoustical materials.
FS-SS-L-0030 — Gypsum backing board.
FS-SS-L-30a — Gypsum lath.
FS-SS-P-00450 — Gypsum patching plaster.
FS-SS-L-30c — Gypsum sheathing board.

FS-SS-A-281b — Aggregate for portland cement plaster.
FS-MMM-A-00150a — Adhesives for acoustical materials.
FS-TT-F-001098a — Filler, surface styrene-butadienc filler for porous surface (stucco).
FS-TT-P-21 — Paint, cement water, powder, white and tints.
FS-TT-C-00555 — Coating system, textured for interior and exterior masonry surfaces.
FS-SS-L-0035a — Finishing hydrated lime.
FS-SS-L-351B — Finish lime.
ANSI A 42.2 — Portland cement plaster (stucco).
GA 150 — Application of gypsum base and veneer plaster.
GA 216 — Installation of steel door frames in steel stud gypsum base fire rated partitions.
GA 252 — Fire resistant gypsum sheathing.
FS-SS-W-110b — Water repellant, colorless, silicone resin base.
FS-TT-S-179e — Sealer surface pigmented oil plaster and gypsum board.
FS-TT-P-0033 — Paint, latex base, exterior (stucco).
ASTM C 6 — Normal finishing hydrated lime.
ASTM C 11 — Definitions of terms relating to ceilings and walls.
ASTM C 22 — Specification for gypsum.
ASTM C 28* — Specification for gypsum wood fiber.
ASTM C 35* — Organic aggregate, perlite, vermiculite.
ASTM C 37* — Gypsum lath.
ASTM C 52 — Gypsum partition tile and block.
ASTM C 59* — Gypsum coating and molding plaster.
ASTM C 61* — Gypsum Keene's cement.
ASTM C 79 — Gypsum sheathing board.
ASTM C 91 — Masonry cement.
ASTM C 136 — Sieve or screen analysis of fine and coarse aggregates.
ASTM C 144* — Aggregate for masonry mortar (portland cement).
ASTM C 150* — Portland cement.
ASTM C 175 — Air-entraining portland cement.
ASTM C 201 — Hydrated lime for masonry purposes.
ASTM C 206* — Special finishing hydrated lime.
ASTM C 226 — Air-entraining additions for use in portland cement.
ASTM C 318 — Gypsum formboard.
ASTM C 442 — Gypsum backing board.
ASTM C 474-5 — Joint treatment materials for gypsum board.
ASTM C 557 — Adhesives for gypsum board to wood framing.
ASTM C 587* — Gypsum veneer plaster.
ASTM C 588* — Gypsum base for veneer plaster.
ASTM C 630 — Water-resistant gypsum backing board.
ASTM C 631* — Bonding compounds for interior plastering.
ASTM C 778 — Standard sand.
ANSI/ASTM C 841* — Installation of interior lathing and furring.
ANSI/ASTM C 842* — Application of interior gypsum plaster.
ANSI/ASTM C 843* — Application of veneer plaster.
ANSI/ASTM C 844 — Application of gypsum base for veneer plaster.
ASTM C 897* — Aggregate for job mixed portland cement based plasters.
ASTM C 926* — Standard specification for application of portland cement based plaster.
ASTM C 932* — Standard specification for surface-applied bonding agents for exterior plastering.
ASTM D 1779 — Adhesive for acoustical materials.

INSULATION

FS-HH-I-521E — Insulation blanked, mineral fiber.
FS-LLL-1535A — Insulation blanked and thermal board.
FS-HH-I-558 — Rigid board type insulation.
FS-HH-I-001252 — Aluminum foil insulation.
FS-HH-I-515a — Insulation batt, blanket.

FS-HH-I-00529a — Insulation board.
FS-HH-I-524a — Insulation board polystyrene.
FS-HH-I-00530 — Insulation board urethane.
FS-L-P-00386 — Rigid urethane foam.
ASTM C 165 — Compressive properties of thermal insulation.
ASTM C 167 — Thickness and density of blanked or batt type thermal insulation material.
ASTM C 168 — Thermal insulation materials.
ASTM C 208 — Insulation board, structural and decorative.
ASTM C 303 — Density of block type building insulation.
ASTM C 519 — Density of fibrous loose fill building insulation.
ASTM C 553 — Mineral fiber blanket and felt insulation.
ASTM C 578 — Rigid block type cellular polystyrene insulation.
ASTM C 591 — Rigid performed cellular urethane thermal insulation.
ASTM C 653 — Thermal resistance of low density material fiber blanket building material.
ASTM C 665 — Mineral fiber blanket thermal insulation for wood frame and light construction.
ASTM C 720 — Spray applied fibrous insulation for elevated temperatures.
ASTM C 727 — Use of reflective insulation material in building construction.
ASTM C 739 — Celluosic fiber loose fill thermal insulation.
ASTM C 755 — Vapor barriers for thermal insulation.
ASTM C 762 — Spray applied fibrous thermal insulation.
ASTM C 764 — Mineral fibers loose fill thermal insulation.
ASTM D 2126 — Polystyrene insulation board.

TEST/PERFORMANCE/ METHODS

GA 600 — Fire resistance design manual.
ASTM C 67 — Absorption test of materials.
ASTM C 109 — Test for compressive strength of portland cement plaster.
ASTM C 190 — Tensile strength of materials.
ASTM C 355 — Test for water vapor transmission of thick materials.
ASTM C 448 — Exterior exposure test of finishes for thermal insulation.
ASTM C 471 — Chemical analysis of gypsum products.
ASTM C 472* — Standard methods for physical testing of gypsum plasters and gypsum concrete.
ASTM C 473 — Methods for physical testing of gypsum board and lath.
ASTM C 643 — Test for change in acoustical absorption of ceiling materials due to repainting.
ASTM D 696 — Thermal coefficient of linear expansion.
ASTM E 72 — Conducting strength test of panels for building construction.
ASTM E 84 — Test for surface burning characteristics of building materials.
ASTM E 90* — Recommended practice for lab measurement of airborne sound transmission loss of building partitions.
ASTM E 96 — Water vapor transmission of materials in sheet form.
ASTM E 119* — Fire test of building construction and materials.
ASTM E 152* — Fire test of door assemblies.
ASTM E 283 — Rate of air leakage through exterior curtain wall.
ASTM E 336 — Recommended practice for measurement of airborne sound insulation in buildings.
ASTM E 413 — Classification for determination of sound transmission class.
ASTM E 547 — Water penetration of exterior curtain walls.
ASTM E 662 — Sprayed fire resistance materials applied to structural members.

Designation: C 36 – 85

Standard Specification for
GYPSUM WALLBOARD[1]

This standard is issued under the fixed designation C 36; the number immediately following the designation indicates the year of original adoption or, in the case of revision, the year of last revision. A number in parentheses indicates the year of last reapproval. A superscript epsilon (ε) indicates an editorial change since the last revision or reapproval.

This specification has been approved for use by agencies of the Department of Defense and for listing in the DoD Index of Specifications and Standards.

1. Scope

1.1 This specification covers gypsum wallboard, that is designed to be used for walls, ceilings, or partitions and affords a surface suitable to receive decoration.

1.2 The values stated in inch-pound units are to be regarded as the standard. The metric equivalents of inch-pound units may be approximate.

2. Applicable Documents

2.1 *ASTM Standards:*
C 473 Methods for Physical Testing of Gypsum Board Products and Gypsum Lath[2]
E 84 Test Method for Surface Burning Characteristics of Building Materials[3]
E 96 Test Methods for Water Vapor Transmission of Materials[4]
E 119 Methods of Fire Tests of Building Construction and Materials[3]

3. Composition

3.1 Gypsum wallboard shall consist of an incombustible core, essentially gypsum, surfaced with paper firmly bonded to the core.

3.2 The back surface of foil-backed gypsum wallboard shall in addition be covered with aluminum foil.

3.3 Type X (special fire-resistant) designates gypsum wallboard complying with this specification that provides at least 1-h fire-retardant rating for boards ⅝ in. (16 mm) thick or ¾-h fire-resistance classification for boards ½ in. (13 mm) thick, applied to a partition in single-layer nail application on each face of loadbearing wood framing members, when tested in accordance with the requirements of Methods E 119.

3.4 Gypsum wallboard shall have a maximum flame-spread classification of 25 when tested in accordance with the requirements of Test Method E 84.

NOTE—Consult manufacturers for independent test data on assembly details and fire resistance classifications for other types of construction. See official fire test reports for assembly particulars, materials, and classifications.

4. Flexural Strength

4.1 Specimens shall be tested in accordance with Methods C 473. When tested face up and when tested face down the specimens shall have an average breaking load of not less than the following:

Thickness, in. (mm)	Load, lbf (N) Bearing Edges Across Fiber of Surfacing	Bearing Edges Parallel to Fiber of Surfacing
¼ (6.4)	50 (222)	20 (89)
5/16 (8)	65 (289)	25 (111)
⅜ (9.5)	80 (356)	30 (133)
½ (13)	110 (489)	40 (178)
⅝ (16)	150 (667)	50 (222)

Thickness, in. (mm)	Humidified Deflection, Eighths of an inch (mm)
¼ (6.4)	not applicable
5/16 (8)	not applicable
⅜ (9.5)	15 (48)
½ (13)	10 (32)
⅝ (16)	5 (16)

5. Humidified Deflection

5.1 When tested in accordance with Methods C 473, specimens taken from the gypsum wallboard shall have an average deflection of no more than the following

6. Core, End, and Edge Hardness

6.1 When tested in accordance with Methods C 473, specimens taken from the gypsum wallboard shall have an average hardness of not less than 15 lbf (67 N) for the core, ends, or edges.

7. Nail Pull Resistance

7.1 When tested in accordance with Methods C 473, specimens taken from the gypsum wallboard shall have an average nail pull resistance of not less than the following:

Thickness, in. (mm)	Nail Pull Resistance, lbf (N)
¼ (6.4)	40 (178)
5/16 (8)	50 (222)
⅜ (9.5)	60 (267)
½ (13)	80 (356)
⅝ (16)	90 (400)

8. Dimensions and Permissible Variations

8.1 *Thickness*—The nominal thickness of gypsum wallboard shall be ¼, 5/16, ⅜, ½, or ⅝ in. (6.4, 8, 9.5, 13, or 16 mm), with permissible variations in the nominal thickness of ±1/64 in. (0.4 mm) with permissible local variations of ±1/32 in. (0.8 mm) from the nominal thickness. Edges of ¼, 5/16, ⅜, ½, and ⅝-in. board may be either square, recessed, featured, tapered, or featured and tapered. The average thickness of the edge of recessed to tapered-edge wallboard shall be at least 0.015 in. (0.38 mm) but not more than 0.075 in. (1.90 mm) less than the average thickness of the wallboard as determined in accordance with Methods C 473.

8.2 *Width*—The nominal width of gypsum wallboard shall be up to 48 in. (1.22 m), with a permissible variation of 3/32 in. (2.4 mm) under the specified width.

8.3 *Length*—The nominal length of ¼-in. (6.4-mm) gypsum wallboard shall be from 4 to 12 ft (1.2 to 3.7 m), 5/16 in. (8-mm) gypsum wallboard shall be from 4 to 14 ft (1.2 to 4.3 m), and ⅜, ½, and ⅝-in. (9.5, 13, and 16 mm) wallboard shall be from 4 to 16 ft (1.2 to 4.9 m), inclusive, with permissible variations of ±¼ in. from the specified length.

9. Workmanship, Finish, and Appearance

9.1 The surfaces of gypsum wallboard shall be true and free from imperfections that would render the wallboard unfit for use with or without decoration. The edges and ends shall be straight. The corners shall be square with a permissible variation of ±⅛ in. (±3.2 mm) in the full width of the board.

10. Sampling

10.1 At least 0.25 % of the number of gypsum wallboards in a shipment, but not less than three boards, shall be so selected as to be representative of the shipment and shall constitute a sample for purpose of tests by the purchaser or user.

11. Additional Requirements for Foil-Backed Gypsum Wallboard

11.1 Foil-backed gypsum wallboard shall meet all the requirements specified above. In addition, the back surface shall be covered with pure bright-finished aluminum foil.

11.2 When tested in accordance with Test Methods E 96, the permeance of foil-backed gypsum wallboard shall not exceed 0.30 perm for the condition of 50 % relative humidity on Side I, the face of the board, and 0 % relative humidity on Side II, the foil-covered back of the board.

12. Packaging and Package Marking

12.1 Gypsum wallboard shall be shipped so as to be kept dry.

12.2 When shipped for resale, the name of the manufacturer or seller and the brand shall be legibly marked on each board or package.

[1] This specification is under the jurisdiction of ASTM Committee C-11 on Gypsum and Related Building Materials and Systems and is the direct responsibility of Subcommittee C 11.01 on Specifications and Test Methods for Gypsum Products.
Current edition approved Sept. 27, 1985. Published November 1985. Originally published as C 36 – 21 T. Last previous edition C 36 – 84a.
[2] *Annual Book of ASTM Standards*, Vol 04.01.
[3] *Annual Book of ASTM Standards*, Vol 04.07.
[4] *Annual Book of ASTM Standards*, Vols 04.06, 08.03, and 15.09.

 Designation: C 514 – 84

Standard Specification for

NAILS FOR THE APPLICATION OF GYPSUM WALLBOARD[1]

This standard is issued under the fixed designation C 514; the number immediately following the designation indicates the year of original adoption or, in the case of revision, the year of last revision. A number in parentheses indicates the year of last reapproval. A superscript epsilon (ε) indicates an editorial change since the last revision or reapproval.

1. Scope

1.1 This specification covers requirements for steel wire nails suitable for use in the application of gypsum wallboard and gypsum backing board.

NOTE 1—This specification does not necessarily cover nails for Type "X" gypsum wallboard where fire ratings are required. Consult manufacturers for independent data on assembly particulars, materials, and ratings.

1.2 The values stated in inch-pound units are to be regarded as the standard.

1.3 The following precautionary caveat pertains only to the test method portion, Sections 6 and 7, of this specification: *This standard may involve hazardous materials, operations, and equipment. This standard does not purport to address all of the safety problems associated with its use. It is the responsibility of whoever uses this standard to consult and establish appropriate safety and health practices and determine the applicability of regulatory limitations prior to use.*

2. Steel Wire

2.1 Steel wire, used in the manufacture of nails, shall be of hard-drawn low or medium-low-carbon steel, entirely suitable for the purpose intended. Before fabrication it shall be sufficiently ductile to withstand cold-bending, without fracture, through 180° over a radius not greater than the diameter of the wire (see Section 6).

3. Dimensions and Permissible Variations

3.1 *Heads*—The heads shall have a diameter of ¼ in. (6.4 mm) with a tolerance of +³/₆₄ or −¹/₃₂ in. (+1.2 or −0.8 mm), and shall be thin (approximately ¹/₆₄ in. (0.4 mm)) at the peripheral edge; they shall be uniformly tapered to a small fillet around the shank, they may be flat or concave, and shall be free from protrusions and sharp, irregular edges.

3.2 *Shanks*—The shank diameter may vary ± 0.003 in. (0.08 mm) for shank diameters 0.076 in. (1.93 mm) or larger.

3.3 *Points*—The nails shall have medium to long diamond or needle points.

3.4 *Clearance*—Should a deformation process produce other than a smooth shank nail, a clearance area (round and smooth), measured from the top of the head to the deformed section, shall be provided equal to the nominal thickness of gypsum wallboard for which the nails are specified.

4. Workmanship, Finish, and Appearance

4.1 The nails shall be bright, or chemically treated, or may be lightly coated with rust inhibiting material provided that such chemical treatment or rust inhibitor does not adversely affect the performance of the nail as specified in Section 7. The nails shall also be compatible with the joint compound and decoration. Nails shall be neatly formed and free from injurious defects or deformations.

5. Number of Tests

5.1 At least five nails from each lot of 100 individual containers shall be examined to determine conformance to the requirements of this specification.

6. Bend Test

6.1 When held in a vise and bent by means of a clamp or similar device attached to the free end of the test nail, the nail shall be sufficiently ductile to withstand cold bending, without fracture, through 90° over a radius not greater than the diameter of the nail.

7. Withdrawal Resistance Test

7.1 The minimum withdrawal resistance, immediate and delayed, shall be at least equal to that provided by a 12½-gage (0.099 ± 0.003 in. (2.515 ± 0.08 mm)) bright, smooth shank, nail with a medium diamond point, when tested as follows: Alternately use twenty of the above described nails and twenty of the nails being tested and hammer-drive to a depth of ⅞-in. (22.2 mm) penetration into the longitudinal center line of one edge and one side face of samples cut from the same pieces of nominal 2 by 6 in. Douglas fir, construction grade, containing at least 16 %, but not more than 19 %, free moisture as determined by a suitable moisture meter. Space nails approximately 3 in. (76 mm) apart.

7.2 Determine the comparative performance as follows:

7.2.1 *Immediate*: Withdraw alternate samples of each type of nail immediately from each face, ten of each nail, at the rate of 0.06 in. (1.5 mm)/min.

7.2.2 *Delayed*: Condition the lumber, into which the nails are driven, at 50 % relative humidity and 70 ± 5°F (21.1 ± 3°C) until the specimen has obtained a constant mass, after which withdraw the remaining nails.

7.2.3 The average withdrawal resistance, immediate and delayed, for the nail under test, shall be at least equal to that of the 12½-gage (0.099 in. (2.515 mm)) nail described in 7.1.

8. Certification

8.1 When specified in the purchase order a producer's or supplier's certification shall be furnished to the purchaser that the material is in compliance with this specification.

9. Packaging and Package Marking

9.1 Unless otherwise specified, nails shall be packaged in substantial commercial containers of the type, size, and kind commonly used for the purpose, so constructed as to preserve the contents in good condition and to ensure acceptance and safe delivery by common or other carriers, at the lowest rate, to the point of delivery. In addition, the containers shall be so made that the contents can be partially removed without destroying the container's ability to serve as a receptacle for the remainder of the contents.

9.2 Unless otherwise specified, individual packages and shipping containers shall be marked with the type and length of nail and the name of the manufacturer or distributor. Individual packages shall also be marked with the name of the manufacturer or distributor and the net mass.

[1] This specification is under the jurisdiction of ASTM Committee C-11 on Gypsum and Related Building Materials and Systems and is the direct responsibility of Subcommittee C11.03 on Specifications and Test Methods for Accessories and Related Products.

Current edition approved Aug. 31, 1984. Published October 1984. Originally published as C 514 – 63 T. Last previous edition C 514 – 83.

Designation: C 557 – 73 (Reapproved 1985)[ε1]

Standard Specification for

ADHESIVES FOR FASTENING GYPSUM WALLBOARD TO WOOD FRAMING[1]

This standard is issued under the fixed designation C 557; the number immediately following the designation indicates the year of original adoption or, in the case of revision, the year of last revision. A number in parentheses indicates the year of last reapproval. A superscript epsilon (ε) indicates an editorial change since the last revision or reapproval.

This specification has been approved for use by agencies of the Department of Defense and for listing in the DoD Index of Specifications and Standards.

[ε1] NOTE—Section 2 was added editorially and subsequent sections renumbered in October 1985.

Scope

1.1 This specification covers minimum standards for adhesives intended for bonding the back surface of gypsum wallboard to wood framing members.

1.2 This specification also covers test requirements and test methods for the adhesive used for the application of all thicknesses of gypsum wallboard.

Applicable Documents

2.1 *ASTM Standards:*

D 828 Test Method for Tensile Breaking Strength of Paper and Paperboard[2]

D 1779 Specification for Adhesive for Acoustical Materials[3]

Materials

3.1 *Adhesives*—The adhesives covered in this specification are organic adhesives.

NOTE 1—The term "organic adhesive" shall include any adhesive in which an organic material is used as the principal binding component.

3.2 *Workability*—The adhesive shall be essentially free of foreign matter and shall be of uniform consistency suitable for application in accordance with accepted commercial practice.

3.3 *Consistency*—The adhesive shall be of such uniform consistency so that when it is applied to the framing member in accordance with the manufacturer's instructions, it shall remain as applied until application of the gypsum wallboard, through an ambient temperature range of 40 to 110°F (4.4 to 43.3°C).

Detail Requirements

4.1 *Open Time*—Specimens tested in accordance with the method described in 6.2 shall show less than 75 % transfer (by area) to the back surface of the gypsum wallboard, or no less than 75 % paper failure.

4.2 *Wetting Characteristics*—The adhesive shall wet the plywood when tested in accordance with the method described in 6.3.

4.3 *Shear Strength:*

4.3.1 *Rate of Strength Development*—The shear strength of panels shall be as follows:

4.3.1.1 Not less than 10 psi (69 kPa) after 24 h when tested as specified in 6.4.1.

4.3.1.2 40 psi (276 kPa) min after 14 days when tested as specified in 6.4.2.

4.3.2 *Shear Strength After Cyclic Laboratory Exposure*—After cycling and testing specimens as specified in 6.4.3, the average shear strength of these specimens shall be not less than 80 % of the average actual value as determined in 4.3.1.2.

4.3.3 *Static Load in Shear*—Specimens tested as specified in 6.4.4 shall sustain a load of 40 lbf

(178 N) for 24 h at 73.4 ± 3.6°F (23 ± 2°C) and 50 ± 2 % relative humidity, and shall sustain a load of 20 lbf (89 N) for 24 h when tested at 100 ± 2°F (37.8 ± 1.1°C).

4.4 *Tensile Strength:*

4.4.1 *Rate of Strength Development*—The tensile strength of specimens tested shall be as follows:

4.4.1.1 Not less than 15 psi (103 kPa) after 24 h with 100 % paper tear when tested as specified in 6.5.1.

4.4.1.2 Not less than 25 psi (172 kPa) after 14 days when tested as specified in 6.5.2.

4.5 *Bridging Characteristics*—The adhesive shall be capable of maintaining a bridged condition between the framing member and gypsum wallboard after complete drying. When tested in accordance with 6.6 the bridged adhesive shall tear paper from the back of the wallboard in at least 90 % of the bonded area.

4.6 *Aging Properties*—The adhesive must not be embrittled, as evidenced by chipping or flaking, after 500 h exposure when tested in accordance with the method described in 6.7.

4.7 *Freeze-Thaw Stability*—After cycling and testing in accordance with the method described in 6.8, the adhesive must meet the consistency requirements of 3.3, and the shear strength must not be less than the minimum value required in 4.3.1.1.

4.8 *Vinyl-Covered Gypsum Board Compatibility*—When the adhesive is to be used to laminate vinyl-covered gypsum board, no blistering, vinyl film discoloration, or vinyl film bond damage shall occur when tested in accordance with 6.9.

4.9 *Adhesive Staining*—When the adhesive is to be used to laminate vinyl-covered gypsum board, no discoloration, swelling, or other damage shall be evident when tested in accordance with 6.10.

5. Sampling

5.1 Sampling shall be done in accordance with Specification D 1779.

6. Test Methods

6.1 Report the average value determined for the five specimens tested in each property category. If two of the values vary more than 15 % from the average of the five, disregard them and report the average of the remaining specimens. If more than two values vary more than 15 % from the average, disregard the results and repeat the test.

6.2 *Open Time:*

6.2.1 Using a template, form a uniform bead of adhesive ⅜ in. (9.5 mm) wide and ⅜ in. high onto the back surface of a piece of gypsum wall-

board that has been conditioned for 24 h at a temperature of 73.4 ± 2°F (23 ± 1°C) and 50 ± 2 % relative humidity.

6.2.2 After an open time of 30 min at standard conditions, position a 2 by 2-in. (50.8 by 50.8-mm) specimen of gypsum wallboard centrally over the extruded bead and place a 5-lb (2.27-kg) weight immediately on the assembly. Remove the weight after 30 min hold down.

6.2.3 After a period of 24 h at 73.4 ± 2°F and 50 ± 2 % relative humidity, pull the specimen apart and examine for the percentage of transfer and paper failure.

6.3 *Wetting Characteristics:*

6.3.1 Using a spatula, place a small amount of adhesive on the surface of douglas fir plywood, grade EXT-DFPA-AA, which has been conditioned 48 h at 73.4 ± 2°F (23 ± 1°C) and 50 ± 2 % relative humidity, and then, by reversing the pressure of the spatula, lift the adhesive from the surface.

6.3.2 Examine the surface of the plywood and the spatula to determine whether failure occurs in adhesion or cohesion. The adhesive is considered to have wetted the plywood if failure is in cohesion.

6.4 *Shear Strength (Rate of Strength Development):*

6.4.1 *Twenty-four-Hour Test:*

6.4.1.1 *Number of Test Specimens*—Test five specimens after exposure to 73.4 ± 2°F (23 ± 1°C) and 50 ± 2 % relative humidity for 24 h.

6.4.1.2 *Preparation of Test Specimens*—Prepare each shear test specimen individually from plywood and a gypsum board-plywood laminate. Condition the douglas fir plywood, grade EXT-DFPA-AA, and the gypsum board-plywood laminate at 73.4 ± 2°F (23 ± 1°C) and 50 ± 2 % relative humidity for a minimum of 48 h immediately preceding the preparation of specimens. Prepare each shear test specimen by bonding a 4 by 3½ by ¾-in. (102 by 88.9 by 19.0-mm) piece of the douglas fir plywood described to the previously prepared 4 by 3½ by 1¼-in. (102 by 88.9 by 31.8-mm) gypsum board-plywood laminate.

NOTE 2—It is necessary to laminate the gypsum board to a plywood backing to prevent the gypsum board from fracturing during the test.

6.4.1.3 The gypsum board-plywood assembly

---

[1] This specification is under the jurisdiction of ASTM Committee C-11 on Ceilings and Walls and is the direct responsibility of Subcommittee C11.02 on Specifications and Test Methods for Accessories and Related Products.

Current edition approved Jan. 29, 1973. Published March 1973. Originally published as C 557 – 65 T. Last previous edition C 557 – 67.

[2] *Annual Book of ASTM Standards*, Vol 15.04.

[3] *Annual Book of ASTM Standards*, Vol 15.06.

is made by laminating ½-in. (12.7-mm) gypsum wallboard to ¾-in. (19.0-mm) douglas fir plywood, grade EXT-DFPA-AA, with any commercially available poly(vinyl acetate) adhesive. The grain of the gypsum board facing paper shall run lengthwise (3½ in. dimension) of the specimen. Prepare the face of the ¾-in. thick plywood by sanding sufficiently to present a new wood, smooth surface with No. 120 grit 3/0 garnet paper. Wipe the sanded surface free of dust. Spread the adhesive on the sanded surface with a trowel having ³⁄₁₆-in. (4.8-mm) deep V-notches spaced ³⁄₁₆ in. on center. The adhesive ridges shall be parallel with the grain of wood. Hold the trowel at an angle of 90° to the surface during application to assure the deposition of a ridge of adhesive ³⁄₁₆ in. thick. Clean and dry the trowel thoroughly between the assembly of each specimen. Exercise care to assure that parts of the specimen are assembled perfectly square with each other (see Fig. 1 for a jig suitable to accomplish this). Allow the adhesive to remain exposed (open) exactly 30 s from completion of spreading, following which the gypsum board surface of the gypsum board-plywood laminate shall be positioned upon and overlapping the coated plywood exactly 2½ in. (63.5 mm), thus forming a 10-in.² (64.5-cm²) bonded area (Fig. 2). Insert six spacers, made from brass or bronze wire of No. 20 gage (American Standard or B & S) (0.812 mm) and at least 2 in. long, in the joint exactly 1 in. (25.4 mm). Position the spacers so that one is on the center line of the bonded area, and the others are 1 in. away from the center spacer and parallel to it (Figs. 1 and 2). Immediately following assembly of each specimen, compress it under a load of 15 lbf (67 N) for 3 min. The weight used for this purpose may be a 1-qt (1-L) metal can filled with sufficient No. 3 lead shot to give a total weight of shot and can of 15 lb (6.8 kg). Center the load over the bonded area. After the 3 min period, remove the load; wipe the excess adhesive from the bonded edges with a square-edged spatula; and then immediately withdraw the spacers, taking care to avoid disturbing alignment of the bonded pieces.

6.4.1.4 Store the specimens where the atmosphere is 73.4 ± 2°F (23 ± 1°C) and 50 ± 2 % relative humidity. Conduct shear tests in accordance with Test Method D 828, loading at a rate of 0.5 in./min (12.7 mm/min). Examine the specimen visually for squareness before insertion into the testing machine. Any specimens which are not perfectly square shall be adjusted in the machine with shims or pads to assure that the stress applied is parallel with the joint.

6.4.2 *Fourteen-Day Test:*

6.4.2.1 *Number of Test Specimens*—Test five samples after exposure to 73.4 ± 2°F (23 ± 1°C) and 50 ± 2 % relative humidity for 14 days.

6.4.2.2 *Preparation of Test Specimens*—Follow the procedures described in 6.4.1.2 through 6.4.1.4.

6.4.3 *Shear Strength After Cyclic Laboratory Exposure*—Age five specimens prepared as described in 6.4.1.2 through 6.4.1.4 two weeks at 73.4 ± 2°F (23 ± 1°C) and 50 ± 2 % relative humidity, then cycle as shown in Table 1.

6.4.3.1 After the specimens have been subjected to four complete aging cycles, condition them at 73.4 ± 2°F (23 ± 1°C) and 50 ± 2 %

relative humidity for 48 h, then test in accordance with 6.4.1.

6.4.3.2 Store specimens at 73.4 ± 2°F (23 ± 1°C) and 50 ± 2 % relative humidity over any weekend between cycles.

6.4.4 *Static Load in Shear*—Condition ten specimens 14 days at 73.4 ± 2°F (23 ± 1°C) and 50 ± 2 % relative humidity, then subject to static loads in shear at 73.4 ± 2°F (23 ± 1°C) and 100 ± 2°F (37.8 ± 1°C).

6.4.4.1 The load shall be 40 lbf (178 N) total at 73.4°F (23°C) and 20 lbf (89 N) total at 100°F (37.8°C) total, and it shall remain on the specimen without failure for 24 h.

6.4.4.2 Test five specimens at each temperature and normal room conditions. Clamp the top block to a rigid assembly for alignment.

6.5 *Tensile Strength (Rate of Strength Development):*

6.5.1 *Twenty-four-Hour Test:*

6.5.1.1 *Number of Test Specimens*—Test five samples after exposure to 73.4 ± 2°F (23 ± 1°C) and 50 ± 2 % relative humidity after 24 h.

6.5.1.2 *Preparation of Test Specimens*—Prepare each tensile test specimen individually. Condition the ½-in. gypsum wallboard and wood blocks for a min of 48 h immediately preceding the preparation of specimen, at 73.4 ± 2°F (23 ± 1°C) and 50 ± 2 % relative humidity. Prepare each tensile test specimen by bonding the back surface of a piece of ½ by 4 by 4-in. gypsum wallboard to a wood block 1⅝ by 1⅝ by 3⅝ in. cut from No. 1 straight-grain, knot-free douglas fir 2 by 4-in. studs (see Fig. 3). Drill a hole in the center of one end of the block and insert an eye hook.

6.5.1.3 Apply sufficient adhesive to the wood block to cause uniform squeeze-out of an excess of all sides when the bond is compressed to a glue-line thickness of approximately ¹⁄₃₂ in. (0.8 mm). Insert and position spacers made from brass or bronze wire of No. 20 gage (American Standard or B & S) and at least 2 in. long, as shown in Fig. 3. Compress each specimen immediately following assembly under a load of 15 lbf (67 N) and scrape all excess adhesive away from edges of board using a square-tipped spatula. Remove the weight and then immediately remove the spacers, taking care to avoid disturbing alignment. Condition these specimens at 73.4 ± 2°F (23 ± 1°C) and 50 ± 2 % relative humidity for 24 h. Test the specimens on a testing machine which can provide loading at the rate of 60 lbf/min (267 N/min).

6.5.2 *Fourteen-Day Test:*

6.5.2.1 *Number of Test Specimens*—Five samples are to be tested after exposure to 73.4 ± 2°F (23 ± 1°C) and 50 ± 2 % relative humidity after 14 days.

6.5.2.2 *Preparation of Test Specimens*—Prepare each tension test specimen individually. Condition the ½-in. plywood and No. 1 douglas fir wood blocks for a min of 48 h immediately preceding the preparation of specimens at 73.4 ± 2°F (23 ± 1°C) and 50 ± 2 % relative humidity. Prepare each tension test specimen by bonding a piece of ½ by 4 by 4-in. douglas fir plywood, grade EXT-DFPA-AA, to a wood block 1⅝ by 1⅝ by 3⅝ in. cut from No. 1 straight-grained knot-free douglas fir, 2 by 4-in. lumber. Prepare the bonding face of the plywood and the fir block

by sanding sufficiently to present a new woo smooth surface with No. 120 grit 3/0 garn paper. Wipe the sanded surface free of du (Similar to Fig. 3 except plywood is substitut for gypsum wallboard.) Drill a hole in the cent of one end of the block and insert an eye hook

6.5.2.3 Details of adhesive applicatio spacers, etc., should be as described in 6.5.1.

6.5.2.4 Condition the specimens at 73.4 ± 2 (23 ± 1°C) and 50 ± 2 % relative humidity f 14 days with the plywood resting on a table comparable stand. Strength tests will then be ru on a testing machine which can provide loadin at the rate of 60 lbf/min (267 N/min).

6.6 *Bridging Characteristics*—Construct test frame, 34 in. (864 mm) wide by 48 in. (12 mm) long, of No. 1 straight-grained, knot-free by 4-in. douglas fir lumber. Nail a stud 16 i (406 mm) on center, between the two outer stud but, recessed ¼ in. Condition this assembly 73.4 ± 2°F (23 ± 1°C) and 50 ± 2 % relativ humidity for 24 h prior to testing. Apply a ⅜ b ⅜-in. bead of adhesive to the center recessed stu See Fig. 4 for a description of a suitable templat for uniform bead application. After 15 min, na ½-in. gypsum wallboard, 34 in. wide and 42 i (1067 mm) long to the outside longitudinal stud using 10-in. (254 mm) nail spacing. Firmly pres the wallboard over the center recessed stud t ensure maximum deflection of the wallboar against the recessed stud. Then allow the wall board to spring back to its original position Condition the test panel for 48 h at 73.4 ± 2° (23 ± 1°C) and 50 ± 2 % relative humidity. Afte this time the adhesive shall form a bridge betwee the wallboard and framing member. To test th adhesive for bridging strength remove nails fro the outside studs. Grasp one edge of the wall board at points adjacent to each side of th displaced stud and pull outwardly at 90° to the stud. Draw over bead to ensure uniform size o ⅜ by ⅜ in.

6.7 *Aging Properties:*

6.7.1 *Adhesive Aging Test (Oven Test)*—The test specimen shall consist of a 1 by 4-in. by 10 to 15-mil (dry) adhesive strip deposited on a 2 by 6-in. strip of 0.032-in. (0.81 mm) aluminum. Dry the adhesive to constant weight before exposure. Maintain the test specimen in 158°F (70°C) environment (humidity uncontrolled) for 500 h. Upon completion of exposure allow specimen to cool at room temperature for 1 h. Then slowly bend the specimen around a ¼-in. steel mandrel with the adhesive side out.

6.8 *Freeze-Thaw Stability*—Samples to be tested shall consist of 4 oz of material in an 8-oz container. Place one sample at 0 ± 5°F (−17.8 ± 2.8°C) for 24 h, then at 73.4 ± 2°F (23 ± 1°C) for 24 h. After three cycles test the samples in accordance with 6.4.1.

6.9 *Vinyl-Covered Gypsum Board Compatibility*—Place 6 oz of adhesive into a clean, dry, open pint (16-oz) tin-lined can. Place the pint can into a gallon container. Seal a piece of vinyl-covered gypsum board face up on top of the gallon container using vapor impermeable duct tape. Place the assembly into an oven at 110°F for 24 h.

6.10 *Adhesive Staining*—Apply two dabs of adhesive approximately 2 in. (50.8 mm) in dimeter to the face surface of vinyl-covered gypsum board in two areas. Following the adhesive manufacturer's recommendations, clean both areas 1 h after application of the adhesive to the vinyl surface.

7. Packaging and Marking

7.1 *Packaging*—The adhesive shall be packaged in standard commercial containers. The containers shall be so constructed as to ensure acceptance by common or other carriers for safe transportation at the lowest rate to the point of delivery, unless otherwise specified in the contract or order.

7.2 *Marking*—Shipping containers shall be marked with the name of the adhesive, the quantity contained therein, the name of the manufacturer, and the batch number.

TABLE 1 Aging Cycles for Specimens

Period, h	Temperature, °F (°C)	Relative Humidity, %
4	140 (60.0)	85 ± 2
4	32 (0.0)	90 ± 2
Overnight (16)	140 (60.0)	10 (uncontrolled)
6	140 (60.0)	85 ± 2
Overnight (18)	140 (60.0)	10 (uncontrolled)

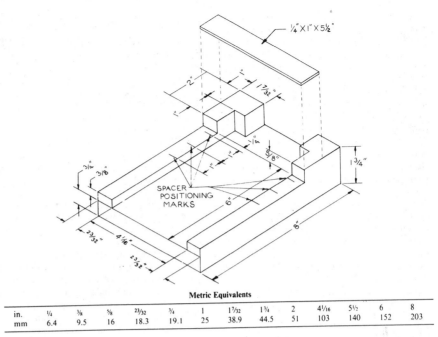

Metric Equivalents

in.	¼	⅜	⅝	23/32	¾	1	1 7/32	1¾	2	4 1/16	5½	6	8
mm	6.4	9.5	16	18.3	19.1	25	38.9	44.5	51	103	140	152	203

FIG. 1 Shear Strength Test Specimen Jig

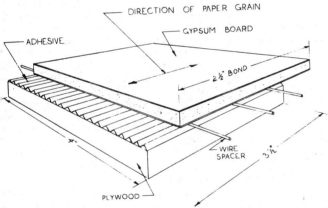

FIG. 2 Shear Strength Test Specimen After Assembly

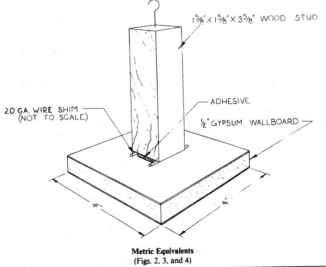

Metric Equivalents
(Figs. 2, 3, and 4)

in.	⅜	½	1⅝	2½	3½	3⅝	4
mm	9.5	13	41.3	64	89	92.1	102

FIG. 3 Tensile Strength Test Specimen

 **Designation: C 474 – 87**

Standard Test Methods for

JOINT TREATMENT MATERIALS FOR GYPSUM BOARD CONSTRUCTION[1]

This standard is issued under the fixed designation C 474; the number immediately following the designation indicates the year of original adoption or, in the case of revision, the year of last revision. A number in parentheses indicates the year of last reapproval. A superscript epsilon (ε) indicates an editorial change since the last revision or reapproval.

1. Scope*

1.1 These test methods cover the physical testing elements of a system for treating the joints of installed gypsum board with joint compound and joint reinforcing tape as classified in Specification C 475.

1.2 The values stated in inch-pound units are to be regarded as the standard. The values in parentheses are provided for information purposes only.

1.3 The test procedures appear in the following order:

	Sections
Joint Compound Tests (Note 1):	
Consistency	6 and 7
Shrinkage	8 to 10
Check Cracking of Joint Compound	11 and 12
Putrefaction	13
Joint Reinforcing Tape:	
Tensile Strength	14 and 15
Dimensional Stability	16
Width	17
Thickness	18
Bond of Joint Reinforcing Tape to Joint Compound	19 and 20
Cracking of Joint Compound at Tape Edges	21 and 22

NOTE 1—Ready-mixed joint compound, topping, or finishing compound prepared as a paste ready to apply, is tested in accordance with Sections 6 to 13. The ready-mixed joint compound shall be tested as received unless its viscosity, as determined in Section 7 is over 520, in which case the viscosity shall be adjusted by the addition of water until it is 500 ± 20.
Powdered topping or finishing compounds shall be tested in accordance with Section 6.

1.4 *This standard may involve hazardous materials, operations, and equipment. This standard does not purport to address all of the safety problems associated with its use. It is the responsibility of the user of this standard to establish appropriate safety and health practices and determine the applicability of regulatory limitations prior to use.*

2. Referenced Documents

2.1 *ASTM Standards:*
C 11 Definitions of Terms Relating to Gypsum and Related Building Materials and Systems[2]
C 36 Specification for Gypsum Wallboard[2]
C 475 Specification for Joint Treatment Materials for Gypsum Wallboard Construction[2]
D 685 Method for Conditioning Paper and Paper Products for Testing[3]
D 828 Test Method for Tensile Breaking Strength of Paper and Paperboard[3]
2.2 *TAPPI Standard:*
T 410 Grammage of Paper and Paperboard (Weight Per Unit Area)[4]

3. Definitions

3.1 *finishing compound*—a joint compound specifically designed for use as the covering or topping coat(s); not to be used to embed joint reinforcing tape.

3.2 *joint compound, powder*—A drying-type or setting-type cementitious material intended to be mixed with water and used in conjunction with joint reinforcing tape in covering gypsum board joints, for covering fastener heads, and with metal accessories.

3.3 *joint compound, ready-mixed*—a drying-tape cementitious material as defined in 3.2 except that it is factory mixed in ready-to-use form.

3.4 *steel reinforced broad knife*—a 5 to 8-in. (127 to 203-mm) drywall broad knife reinforced by a steel bar, 1 in. (25.4 mm) wide by ⅛-in. (3.2-mm) thick, by the knife width, attached to the back of the knife blade ¼ in. (6.4 mm) from the edge.

3.5 *topping compound*—a cementitious material; synonymous with Finishing Compound as defined in 3.1.

3.6 For additional definitions of terms see Definitions C 11.

4. Summary of Method

4.1 The test method for consistency of joint compound sets forth procedures to measure the viscosity (fluidity) of joint compound products under prescribed conditions. The other specified test methods are performed on conditioned samples prepared under prescribed procedures to determine compliance with Specification C 475.

5. Significance and Use

5.1 These test methods provide procedures for evaluating the physical properties and performance characteristics of joint treatment materials used in gypsum wallboard construction. These test methods are for use in determining compliance with Specification C 475. The degree of correlation between these tests and service performance has not been determined.

JOINT COMPOUND TEST

CONSISTENCY

6. Apparatus

6.1 *Viscosimeter*[5], adjusted to operate at 79 r/min, and with a 250 cm-g sensitivity cartridge.

6.2 *Viscosity Sample Container*, metal or plastic with an open top having an inside diameter of 2½ to 3 in. (64 to 76 mm) and a height of 2½ to 3 in. (64 to 76 mm).

6.3 *Viscosimeter Pin* having dimensions as follows:

	in. (mm)
Shaft diameter	0.187 ± 0.015 (4.75 ± 0.38)
Pin diameter	0.094 ± 0.015 (2.39 ± 0.38)
Immersion depth (from bottom of spindle)	1.625 ± 0.015 (41.3 ± 0.38)
Length of pin projecting from shaft	0.750 ± 0.015 (19.1 ± 0.38)
Upper pin from bottom of shaft	0.313 ± 0.015 (7.95 ± 0.38)
Lower pin from bottom of shaft	0.125 ± 0.015 (3.28 ± 0.38)

7. Procedure

7.1 Mix a minimum of 300 g of joint compound powder thoroughly with water to create a consistency suitable for normal hand application (viscosity of 500 ± 20). The temperature of the mixed sample should be 77 ± 2°F (25 ± 1°C). Allow the sample to stand for 30 min (Note 2) remix and adjust the temperature to 77 ± 20° (such as by placing the sample container in warm or cool water.)

NOTE 2—Setting type joint compounds should stand for one half of their established setting times, but not in excess of 30 min, prior to remixing.

7.2 Place the mixed sample in the viscosity sample container. Fill until level with the top of the container while puddling the joint-compound sample with a spatula to remove air bubbles. When the container is full, tap the bottom sharply several times on a hard flat surface to remove any remaining air bubbles.

7.3 Lock the filled container in the center of the viscosimeter spindle platform. Raise the platform until the level of the samples reaches the mark on the viscosimeter pin and lock the platform in place.

7.4 Start the viscosimeter and read the viscosity after the tracing pen starts to trace a straight line (usually within 1 min). If the tracing continues to be erratic, estimate the average viscosity reading.

7.5 The testing consistency is the amount of water measured in mL/100g of dry joint compound which yields a mixed sample having a viscosity of 500 ± 20.

SHRINKAGE

8. Apparatus

8.1 *Plastic or Rubber Film*, approximately 5 by 5 in. (130 by 130 mm). Any thin, flexible film that peels clean from the semi-dried patty may be used. Rubber dental dam dusted with talc, polyethylene, or PTFE films have been found satisfactory for this use.

8.2 *Balance*, having a sensitivity of 10 mg (Fig. 1 and 2).

9. Procedure

9.1 Mix approximately 60 mL of joint compound and water, allow to stand for 30 min, and remix to the testing consistency specified in 7.5.

9.2 Determine the specific gravity of the mixed sample (one way is to use a hydrometer)[5] and weigh out the equivalent of 25 mL. Spread

[1] These test methods are under the jurisdiction of ASTM Committee C-11 on Gypsum and Related Building Materials and Systems and are the direct responsibility of Subcommittee C11.03 on Specifications and Test Methods for Accessories and Related Products.
Current edition approved Feb. 27, 1987. Published April 1987. Originally published as C 474 – 61. Last previous edition C 474 – 85.
[2] *Annual Book of ASTM Standards,* Vol 04.01.
[3] *Annual Book of ASTM Standards,* Vol 15.09.
[4] The Brabender "Visco-Corder" Model VC-3, manufactured by C.W. Brabender Instruments Inc., South Hackensack, NJ has been found satisfactory.
[5] The Mudwate Hydrometer has been found satisfactory for this purpose.

the carefully measured volume of wet mix into an elongated patty, $3/16$ to $1/4$ in. (4.8 to 6.4 mm) thick, on the plastic or rubber film (see 8.1). Make two patties.

9.3 Dry the patties at a temperature of 100 to 120°F (38 to 49°C) for 16 to 24 h. Strip off the plastic or rubber film and dry for another 24 to 48 h to constant weight.

9.4 Determine the volume of the patties by the displacement method. First, immerse the patties in kerosine[6] for 3 to 4 h until they are saturated. The patties may be broken into two or three pieces to fit into the beaker without touching the sides. Remove the patties from the kerosine. Using a small cloth moistened with kerosine, lightly blot the excess kerosine from the surface of the patties. Weigh each patty in air on the pan of the balance, with the wire cradle suspended in the beaker of kerosine (Fig. 1). Reweigh each patty in the wire cradle and completely submerge in the kerosine (Fig. 2). The patties shall be completely submerged and freely suspended in the kerosine.

10. Calculation of Shrinkage

10.1 The difference between the weight of each saturated patty in the air and in the liquid, is the weight of the liquid having the same volume as the dried patty. To determine the volume of the dried patty, divide the weight difference by the specific gravity of the liquid used. (The specific gravity of the liquid shall be determined at room temperature).

10.2 Calculate the drying shrinkage as follows:

$$\% \text{ Shrinkage} = [(A - B)/A] \times 100$$

where:

A = average net volume, and
B = average dried volume.

10.3 Take the average of the two patties tested. If there is a difference between the volume of the two patties of more than $1\frac{1}{2}$ %, completely retest an additional two samples and take the average of the four.

CHECK CRACKING OF JOINT COMPOUND

11. Apparatus

11.1 *Rod,* metal or glass, $1/8$ in. (3.2 mm) in diameter by 7 in. (178 mm) in length.

11.2 *Steel-Reinforced Broad Knife,* as defined in 3.4.

11.3 *Gypsum Wallboard,* meeting the requirements of Specification C 36.

11.4 *Electric Fan,* capable of forcing a continuous current of air at a velocity of 350 to 450 ft/min (1.8 to 2.3 m/s) at a distance of 3 ft (1 m).

12. Procedure

12.1 Mix 100 g of joint compound powder with water in a 9-oz (270-mL) container as set forth in 7.1 to a viscosity of 500 ± 20 as specified in 7.5.

12.2 Place a rod $1/8$ in. (3.2 mm) in diameter by 7 in. (178 mm) in length on a piece of gypsum wallboard and place some of the paste next to the rod. Screed a $3\frac{1}{2}$ to 4-in. (89 to 102-mm) wide wedge with a 5 to 8-in. (127 to 203-mm) reinforced steel broad knife, using the rod on one side and the wallboard on the other side to guide the knife. When screeding the mixture, the broad knife shall be held at an angle less than 45° with

respect to the plane of the board, and screeded two to four times to leave the surface smooth. Remove the rod and adjust the wedge to a 5 in. (127 mm) length. Immediately dry the wedge-shaped sample by forcing a current of air 70 to 85°F (21 to 29°C) and 45 to 55 % relative humidity) at a velocity of 350 to 450 ft/min (1.8 to 2.3 m/s) over the surface of the wedge for 8 to 16 h.

12.3 Inspect the dried sample to determine the type and amount of cracking in both the thick half and the thin half of the wedge.

13. Putrefaction Procedure

13.1 *Joint Compound, Powder*—Mix 50 g of joint compound with water in a glass container as described in 7.1. Allow it to soak 30 min and then remix. Place a cover glass or piece of aluminum foil over the glass container and place in a humidity cabinet at 85 to 95°F (29 to 35°C) and 85 to 95 % relative humidity. Observe daily for odor of putrefaction or bacteriological growth.

13.2 *Joint Compound, Ready-mixed*—Remove 100 g of compound from a container that has not been opened and has not exceeded the manufacturer's specified shelf life and test in accordance with 13.1. If the material in the container has separated, mix thoroughly before removing the sample.

JOINT REINFORCING TAPE
TENSILE STRENGTH

14. Apparatus

14.1 The apparatus shall be as specified in Test Method D 828 except that the distance of the jaw spacing shall be reduced to $1/2 \pm 1/64$ in. (13 ± 0.4 mm).

15. Procedure

15.1 Test for cross-directional strength only.

15.2 Take ten samples at a minimum of 1 ft (0.3 m) intervals cut accurately to test width, preferably 1 in. (25.4 mm).

15.3 Condition the samples for a minimum of 24 h at 72 ± 4°F (22 ± 2°C) and 50 ± 2 % relatively humidity.

15.4 Test in accordance with Test Method D 828 with equipment as set forth in 14.1; the rate of loading shall be 5 ± 1.5 lb/s (2.3 ± 0.7 kg/s).

15.4 Results of the individual test samples shall be accepted or rejected in accordance with Test Method D 828. If any results are rejected, additional samples shall be tested. There shall be a minimum of ten individual test results for the evaluation of any unit of tape.

15.5 Report test results as pounds per inch (kilograms per millimetre) of width of sample.

DIMENSIONAL STABILITY

16. Procedure

16.1 Cut not less than three samples of tape 10 to 16 in. (254 to 406 mm) long from the roll to be tested, depending on the length of the rule to be used.

16.2 Suspend and condition the samples for a minimum of 16 h at 72 ± 4°F (22 ± 2°C) and 50 ± 2 % relative humidity.

16.3 Place the conditioned samples on a flat

surface and cut two $1/2$ in. (13 mm) long reference marks for the length-wise measurement crosswise of the tape with a sharp blade. Locate the reference marks about $1/2$ in. (13 mm) from each end of the tape. Place the rule on the tape so that one edge is centered lengthwise. To facilitate reading the rule a magnifying glass of 4 to 5X is recommended.

16.4 Move the rule so that its 1-in. (25.4-mm) mark coincides with the reference mark at one end of the tape. Take the reading at the other reference mark. Subtract one from this reading, to obtain the distance between the two reference marks, and record this figure. Read to the nearest 0.005 in. (0.13 mm).

16.5 Place the rule across the length of the tape and align the 1 in. (25.4-mm) mark of the rule with one edge of the tape. Take the reading at the opposite edge of the tape. Subtract one from this measurement to obtain the width of the tape at this point and record this figure. Mark the location of this crosswise measurement by drawing a $1\frac{1}{2}$ in. (38 mm) long pencil mark across the tape without injuring the edges of the tape.

16.6 Roll up the tape and place it in a container full of water at 100 ± 2°F (38 ± 1°C) so that the tape is entirely submerged. After 30 min, remove the tape from the water and roll it out on the flat surface again. Make the lengthwise and crosswise measurements in the same manner as the initial measurements were made. Determine the amount of expansion by subtracting the original measurement from the final measurement. Divide the expansion in inches or millimetres by the original reading and multiply by 100 to obtain the percentage expansion.

WIDTH

17. Procedure

17.1 Measure the width of the tape in ten places, at least 1 ft (0.3 m) apart, to the nearest $1/32$ in. (0.8 mm). Record the maximum, minimum and average.

THICKNESS

18. Thickness Procedure

18.1 Measure the thickness of the tape in 10 places, between the skived edges (if present), at least 1 ft. (0.3 m) apart, using a paper micrometer having circular faces of 0.25 to 0.33 in.^{2} (160 to 213 mm^2) in area. Faces shall be under steady pressure of 7 to 9 psi (48.2 to 62.1 kPa). For details see TAPPI T 410, except waive Section 7 and condition the sample for a minimum of 24 h as described in Method D 685.

18.2 Record the thickness to the nearest 0.001 in. (0.03 mm).

BOND OF JOINT REINFORCING
TAPE TO JOINT COMPOUND

19. Apparatus

19.1 *Feeler Gage Strips* (two required), each 12 in. (305 mm) long, $1/2$ in. (13 mm) wide, 0.025 in. (0.64 mm) thick with a small hole drilled in one end.

19.2 *Reinforced Broad Knife,* as defined in 3.4.

19.3 *Gypsum Wallboard,* two pieces, 6 by 14

---

[6] Mineral Spirits may be substituted for kerosine.

in. (152 by 356 mm) with the 14-in. length in the machine direction of the paper. See Specification C 36.

19.4 *Overlay Transparency Grid*—A transparent photo copy of 10 by 10 divisions/in. graph paper. An area 2 by 5 in. (51 by 127 mm) enclosing 1000 square divisions is outlined.

20. Procedure

20.1 For powder compounds, mix 200 g of joint compound powder to testing · iscosity of 500 ± 20 Brabender units (Section 2) using water at 70 to 80°F (21 to 27°C). Allow to stand 30 min and remix.

20.2 For ready mixed joint compounds, remix prior to testing to reincorporate any separated ingredients. Adjust the viscosity to 500 ± 20 Brabender units using water at 70 to 80°F (21 to 27°C).

20.3 Place the two 0.025-in. (0.64-mm) thick feeler gage strips parallel to each other about 4 in. (102 mm) apart and fasten to the face of the gypsum wallboard with a thumb tack through the hole in the end.

20.4 Apply a quantity of the joint compound paste to the board area between the feeler gage strips. With a few strokes of the reinforced broad knife, spread the paste evenly between the feeler gage strips slightly thicker than the strips. Place a 12-in. (305-mm) length of the paper tape in about the center of the paste. Press one end of the tape into the paste and hold it in place. With the reinforced broad knife embed the tape by applying two or three pressure strokes, wiping away from the end being held so the excess joint compound is squeezed out. The thickness of the joint compound plus the tape is about 0.025 in. (0.64 mm).

20.5 Allow the test assembly to reach constant weight by drying in an atmosphere of 75 ± 5°F (24 ± 2°C) and 50 ± 5 % relative humidity.

20.6 When the test assembly is dry, use a sharp knife to make a cut across and perpendicular to the tape 3½ in. (89 mm) from one end. Make a second cut 5 in. (127 mm) from and parallel to the first cut. Make two diagonal cuts across the tape connecting the opposite corners of the 5-in. section. With the tip of the knife, peel back the tabs formed by the "X" cuts and pull up sharply.

20.7 Using a sharp pencil, lightly outline the areas where fiber remains attached to the compound. Align the overlay transparency Grid so that the Grid outline matches the 2 by 5-in. (51 by 127-mm) sides of the tape bond area. Using the Overlay Transparency Grid count the number of squares that are more than half bare of fiber separated from the tape and outlined by pencil. Subtract this number from 1000 and divide by 1000 to determine the percent bond. Record the average of the two tests. A bond of 100 % is defined as the condition in which the tape delaminates within itself over the entire area.

CRACKING OF JOINT COMPOUND AT TAPE EDGES

21. Apparatus

21.1 *Feeler Gage Strips,* two required, each 12-in. (305 mm) long, ½ in. (13 mm) wide, 0.040 in. (1 mm) thick.

21.2 *Reinforced Broad Knife,* as defined in 3.4.

21.3 *Gypsum Wallboard* (C 36), approximately 6 by 12 in. (152 by 305 mm). See Specification C 36.

22. Procedure

22.1 Mix 200 g of joint compound powder with water as set forth in 7.1 to a viscosity of 500 ± 20 as specified in 7.5. Allow to stand 30 min and then remix.

22.2 Place the two 0.040-in. (1.02 mm) thick feeler gages on the face of the wallboard about 4 in. (102 mm) apart, fill the space between the gages with joint compound paste and screed off accurately to the thickness of the feeler gages.

22.3 After screeding to the correct thickness, center a strip of joint tape between the feeler gages. Using the broad knife, press the tape firmly into contact with the joint compound keeping the knife at about a 45° angle. A final flat stroke of the knife should be used to smooth the joint compound on both sides of the tape. Do not use enough pressure to squeeze the joint compound from under the tape, but be sure the tape is in firm contact with the joint compound. Remo the screeds.

22.4 Place this assembly in an atmosphere 100 ± 5°F (38 ± 2°C) and 25 ± 5 % relat humidity with an air current having a velocity 350 to 450 ft/min. (1.8 to 2.3 m/s). As an alt nate, in very dry air (under 20 % relative hum ity) at temperatures between 75 and 90°F (24 32°C) just provide an air current across the sa ple at a velocity of 350 to 450 ft/min. After 1 examine the sample for cracks along the edges the tape. For this examination, a magnifying gl of 4 to 5X is recommended. Report the perce age of cracking along the edges of the tape.

23. Precision and Bias

23.1 To obtain the best possible precision, t preparation of joint compound products shou be performed with strict adherence to the spe fied viscosity and conditioning requirements. I terlaboratory comparative studies have not be undertaken and no estimate of the precision bias is available.

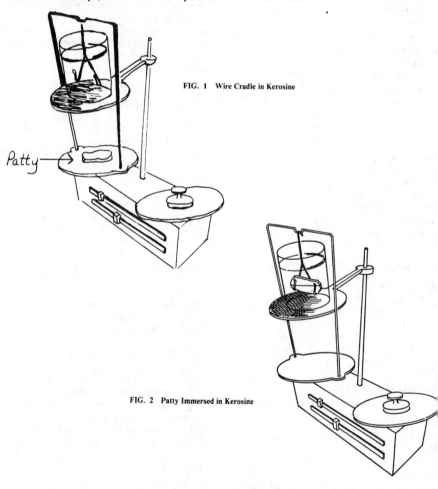

FIG. 1 Wire Cradle in Kerosine

FIG. 2 Patty Immersed in Kerosine

SUMMARY OF CHANGES

This section identifies the location of changes to these test methods that have been incorporated since the last issue. Committee C-11 has highlighted those changes that affect the technical interpretation or use of these test methods.

(1) Section 8.1 was revised.
(2) Sections 9.2 and 9.3 were revised.
(3) Section 16 was revised.
(4) Figure 1 was revised.

Designation: C 1002 - 83

Standard Specification for
STEEL DRILL SCREWS FOR THE APPLICATION OF GYPSUM BOARD[1]

This standard is issued under the fixed designation C 1002; the number immediately following the designation indicates the year of original adoption or, in the case of revision, the year of last revision. A number in parentheses indicates the year of last reapproval. A superscript epsilon (ϵ) indicates an editorial change since the last revision or reapproval.

1. Scope

1.1 This specification covers steel drill screws for use in fastening gypsum board to light-gage steel framing members, wood, and gypsum board.

1.2 The values stated in inch-pound units are to be regarded as the standard. The metric equivalents of inch-pound units may be approximate.

2. Applicable Documents

2.1 *ASTM Standards:*

A 548 Specification for Steel Wire, Carbon Cold-Heading Quality, for Tapping or Sheet Metal Screws[2]

C 36 Specification for Gypsum Wallboard[3]

C 442 Specification for Gypsum Backing Board and Coreboard[3]

C 630 Specification for Water-Resistant Gypsum Backing Board[3]

C 645 Specification for Non-Load (Axial) Bearing Steel Studs, Runners (Track), and Rigid Furring Channels for Screw Application of Gypsum Board[3]

3. Description of Terms Specific to This Standard

3.1 *gypsum board*—either gypsum wallboard (see Specification C 36), gypsum backing board (see Specification C 442), or water resistant gypsum backing board (see Specification C 640).

3.2 *steel framing member*—either steel studs, runners (track) or furring channels (see Specification C 645).

3.3 *wood member*—any wood member (stud, furring, plate, joist, rafter, or similar items) to which board is fastened.

4. Classification

4.1 Steel drill screws covered by this specification are of three types:

4.1.1 *Type G*—For the fastening of gypsum board to gypsum board,

4.1.2 *Type S*—For the fastening of gypsum board to steel framing members, and

4.1.3 *Type W*—For the fastening of gypsum board to wood members.

5. Materials and Manufacture

5.1 Screws shall be manufactured from Grade 1013 to 1022 steel wire manufactured in accordance with Specification A 548.

6. Physical Properties

6.1 *Hardness*—The surface of the screws shall be case hardened to a minimum case depth of 0.002 in. (0.05 mm) with case minimum hardness of 45 HRC.

7. Performance Requirements

7.1 *Type G Screws:*

7.1.1 Screws shall be able to self-drill and drive into gypsum board.

7.1.2 Screw threads shall be adequate to pull the head of the screw below the surface of the wallboard in the performance test in 7.1.3.

7.1.3 Performance tests shall be conducted by attaching ½ in. (12.7 mm) wallboard to ½ in. (12.7 mm) wallboard. The screw specimens will be driven at least ⅜ in. (9.5 mm) from the edge or end of the wallboard specimens during the tests. The wallboard must be rigidly supported. The drill motor propelling the screw shall turn at 2500 rpm and shall exert a total pressure of 20.0 lbf (89.0 N) (deadweight plus applied force).

7.2 *Type S Screws:*

7.2.1 Screws shall be able to self-drill and drive into the steel stud in less than 2 s.

7.2.2 Screw threads shall be adequate to pull the head of the screw below the surface of the wallboard through four layers of 0.010 in. (025 mm) thick kraft paper over ⅝ in. (15.9mm) Type X gypsum wallboard.

7.2.3 Make performance tests by attaching wallboard to the center of a flange of light-gage steel studs, as described in Specification C 645, with an approximate hardness of 52 HRB. The stud shall be rigidly supported. The drill propelling the screw shall turn 2500 rpm and shall exert a total pressure of 25 lbf (111 N) (deadweight plus applied force).

7.3 *Type W Screws:*

7.3.1 Screws shall be able to self-drill and drive into the wood stud.

7.3.2 Screw threads shall be adequate to pull the head of the screw below the surface of the wallboard through four layers of 0.010 in. (0.25 mm) thick kraft paper over ⅝ in. (15.9 mm) Type X gypsum wallboard.

7.3.3 Performance tests shall be conducted by attaching wallboard to the longitudinal center line of an edge of a wood stud. Wood stud specimens for the performance tests shall be cut from sample pieces of nominal 2 by 4 or 2 by 6 in. Douglas Fir, construction grade, containing at least 16 %, but not more than 19 %, free moisture determined by a suitable moisture monitor. The stud shall be rigidly supported. The drill propelling the screw shall turn 2500 rpm and shall exert a total pressure of 25 lbf (111 N).

7.3.4 *Ductility*—The screw must have sufficient ductility to avoid brittle failure.

7.3.5 *Ductility Test*—The sample screw will be held in a vise and bent by means of a clamp or similar device attached to the free end of the test specimen so a force can be exerted on that free end in order to produce a bend in the shank. The specimen shall be able to withstand a 15° bend without signs of fracture in the shank.

8. Dimensions, Mass, and Permissible Variations

8.1 Screws shall have the following shape and dimensions:

8.1.1 *Head Diameter*—The head of the screw shall be a minimum of 0.315 in. (8.00 mm) in diameter and may be out of round no more than 0.020 in. (0.51 mm).

8.1.2 *Head Contour*—The top of the head shall be flat. The outer flange thickness shall be 0.025 ±0.005 in. (0.64 ±0.13 mm). The contour beneath the flange head shall be such that the screw head can be pulled below the surface of the wallboard without cutting the paper and without leaving paper burrs.

8.1.3 *Screw Diameter*—The screw diameter of the threaded portion shall be such that the screw will pass the performance tests in Section 7.

8.1.4 *Threads*—The threads may be of single- or multiple-lead thread design. They shall be clean and smooth.

8.1.5 *Points*—The points shall be designed to be self-drilling and shall meet the performance tests in Section 7.

8.1.6 *Driving Recess*—The driving recess shall be a No. 2 "Phillips" design with a minimum depth of 0.104 in. (2.64 mm) as determined with a Phillips penetration depth gage or a recess of equal performance.

8.1.7 *Nominal Length*—The nominal length shall be minimum length. Type W screws shall be long enough to provide a minimum of ⅝ in. (15.9 mm) of fastener embedment in the wood member.

9. Workmanship and Finish

9.1 Screws shall be given a corrosion-resistant treatment. Any treatment shall not inhibit adhesion to joint finishing compounds and shall not bleed through decorative finishes.

9.2 Screws shall be straight, neatly formed, and free of defects such as burrs and deformations.

10. Sampling

10.1 Examine at least one screw from each of five containers with a minimum of five screws per 16 000 screws to determine conformance to the requirements of this specification. If the first sample lot fails, test 25 more screws. If two or more fail this second test, then the represented lot fails to meet this specification.

NOTE:
ASTM Designations C 646-78 *Standard Specification For Steel Drill Screws For The Application Of Gypsum Board To Lite-Gage Steel Studs* and C 893-78, *Standard Specification For Type G Steel Screws For The Application Of Gypsum Board To Gypsum Board,* were discontinued in 1983 and replaced by Specification C 1002, *Steel Drill Screws For The Application Of Gypsum Board.*

---

[1] This specification is under the jurisdiction of ASTM Committee C-11 on Gypsum and Related Building Materials and Systems and is the direct responsibility of Subcommittee C11.02 on Specifications and Test Methods for Accessories and Related Products.

Current edition approved Aug. 26, 1983. Published September 1983.

[2] *Annual Book of ASTM Standards*, Vol. 01.03.

[3] *Annual Book of ASTM Standards*, Vol 04.01.

Designation: C 645 – 83

Standard Specification for
NON-LOAD (AXIAL) BEARING STEEL STUDS, RUNNERS (TRACK), AND RIGID FURRING CHANNELS FOR SCREW APPLICATION OF GYPSUM BOARD[1]

This standard is issued under the fixed designation C 645; the number immediately following the designation indicates t′ year of original adoption or, in the case of revision, the year of last revision. A number in parentheses indicates the year of la reapproval. A superscript epsilon (ε) indicates an editorial change since the last revision or reapproval.

1. Scope

1.1 This specification covers steel studs, runners (track), rigid furring channels, and grid suspension systems for screw application of gypsum board in non-load bearing interior construction assemblies.

1.2 The values stated in inch-pound units are to be regarded as the standard.

2. Applicable Documents

2.1 *ASTM Standards:*
A 525 Specification for General Requirements for Steel Sheet, Zinc-Coated (Galvanized) by the Hot-Dip Process[2]
A 568 Specification for General Requirements for Steel Sheet, Carbon, and High-Strength, Low-Alloy, Hot-Rolled and Cold-Rolled Sheet[3]
C 36 Specification for Gypsum Wallboard[4]
C 646 Specification for Steel Drill Screws for the Application of Gypsum Board to Light-Gage Steel Studs[5]

2.2 *AISI Standard:*
Specifications for the Design of Cold-Formed Steel Structural Members[6]

3. Materials and Manufacture

3.1 Studs, runners (track), rigid furring channels, and grid suspension systems shall be manufactured from cold-rolled (coil or cut length) sheet complying with Specification A 568 or Specification A 525 with the additional requirement that the minimum thickness, individual measurement, shall be not less than 0.0179 in. (0.455 mm) before application of protective coating.

3.2 Studs, runners (track), rigid furring channels, and grid suspension systems shall have a protective coating to prevent corrosion in normal use.

4. Dimensions and Permissible Variations

4.1 Studs and rigid furring channels shall have a configuration and steel thickness such that the system in which they are used will carry the design transverse loads without exceeding either the allowable stress or a deflection of $L/240$ (grid suspension members $L/360$). Studs, furring channels, and grid suspension cross furring members shall be sufficiently rigid in the system to permit penetration of the screw. Studs, rigid furring channels, and grid suspension members shall be free of twist or camber of a degree that will prevent their use in the system for which they are intended. Minimum width of face to which gypsum board is screw-attached shall be not less than 1¼ in.(32 mm).

4.1.1 *Rigid Furring Channels*—Minimum depth shall be ¾ in. (19 mm). Minimum width of furring attachment flanges (see Fig. 2) shall be ½ in. (12.7 mm).

4.1.2 *Studs*—Length tolerance of studs shall be ±³⁄₁₆ in. (±4.8 mm).

4.1.3 Grid suspension systems include main beams and cross furring members which mechanically interlock to form a modular supporting network. Length tolerance for grid suspension members shall be ±¹⁄₁₆ in. (1.59 mm).

4.2 Runners (track) shall be formed in a U-

shaped configuration, having web depth compatible with those of the studs of the same nominal size. The runners (track) shall be designed such that when the studs are placed in both the top and bottom runners (track), they will be held by friction. Minimum height of flanges shall be 1 in. (25 mm).

5. Edges

5.1 Edges of studs, rigid furring channels, runners (track), and grid suspension members shall be manufactured in such a fashion as to minimize burrs and sharp edges.

6. Cutouts

6.1 Cutouts may be provided, but in no case shall they reduce the performance of the stud, runner (track), rigid furring channel, or grid suspension members in the gypsum board construction assembly below the specified performance requirements.

7. Sectional Properties

7.1 The sectional properties of studs, runners (track), rigid furring channels, and grid suspension members shall be computed in accordance

TABLE 1 Minimum Section Properties for Various Drywall Studs[A,B,C]

Size, in. (mm)	Area[D], in.² (mm²)	Moment of Inertia, I_x^D, in.⁴ (mm⁴)	Section Modulus, S_x^D, in.³ (mm³)
1⅝ (41.3)	0.074 (47.7)	0.035 (14568)	0.043 (705)
2½ (63.5)	0.090 (58.1)	0.092 (38293)	0.073 (1196)
3⅝ (92.1)	0.110 (71.0)	0.216 (89906)	0.119 (1950)

[A] Based on 0.0179 in. (0.46 mm) thickness.
[B] Properties of sections (area, moment of inertia, and section modulus) are determined in accordance with 1980 AISI Specification for the Design of Cold-Formed Steel Structural Members, section 2.3; allowable stress based on unstiffened flange element, section 3.2.
[C] See Appendix X1 for minimum cross section and corresponding calculations.
[D] Based on gross cross section area.

with AISI Specifications for the Design of Co Formed Steel Structural Members (See Tabl and Appendix X1).

8. Test Performance

8.1 Performance tests shall be made by taching ⅝-in. (16-mm) Type X gypsum wa board (complying with Specification C 36) the center of the attachment face of a minim 18-in (460-mm) long section of stud furri channel, or grid suspension cross furring me ber with steel drill screws (complying wi Specification C 646). Wallboard specimen sh be a minimum of 12 in. (305 mm) fr either edge or either end of a wallboard pan The member to be tested shall be rigidly su ported (see Fig. 1 for studs; Fig. 2 for furri channel). The drill propelling the screw sh turn at 2500 rpm (free spindle speed) and sh exert a total (dead weight and applied loa force of 25 lbf (111.2 N).

8.2 *Penetration Test*—Members shall be pable of pulling the head of the screw below surface of the Type X gypsum wallboard throu four layers of 0.010-in. (0.26-mm) thick kr paper in less than 2 s without spinout.

[1] This specification is under the jurisdiction of ASTM Co mittee C-11 on Gypsum and Related Building Materials a Systems and is the direct responsibility of Subcommittee C11. on Specifications and Test Methods for Accessories and Relat Products.
Current edition approved Aug. 26, 1983. Published Mar 1984. Originally published as C 645 – 70. Last previous editi C 645 – 81.
[2] Annual Book of ASTM Standards, Vol 01.06.
[3] Annual Book of ASTM Standards, Vol 01.03.
[4] Annual Book of ASTM Standards, Vol 04.01.
[5] Discontinued—Replaced by Specification C 1002, Ann Book of ASTM Standards, Vol 04.01.
[6] Available from the American Iron and Steel Institute, 10 16th St. N.W., Washington, DC 20036.

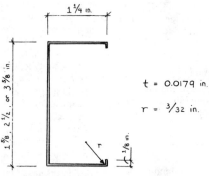

$t = 0.0179$ in.

$r = ³⁄₃₂$ in.

FIG. X2 Minimum Drywall Stud Cross Section

The American Society for Testing and Materials takes no position respecting the validity of any patent rights asserted in connection with any item mentioned in this standard. Users of this standard are expressly advised that determination of the validity of any such patent rights, and the risk of infringement of such rights, are entirely their own responsibility.

This standard is subject to revision at any time by the responsible technical committee and must be reviewed every five years and if not revised, either reapproved or withdrawn. Your comments are invited either for revision of this standard or for additional standards and should be addressed to ASTM Headquarters. Your comments will receive careful consideration at a meeting of the responsible technical committee, which you may attend. If you feel that your comments have not received a fair hearing you should make your views known to the ASTM Committee on Standards, 1916 Race St., Philadelphia, Pa. 19103.

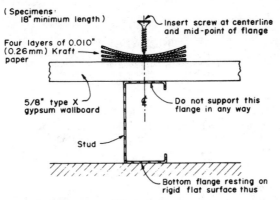

FIG. 1 Studs

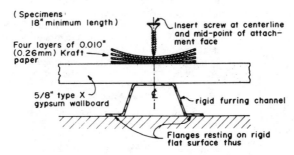

NOTE—Wallboard specimens shall be taken a minimum of 12-in. (305 mm) from either edge and either end of wallboard panel.

FIG. 2 Rigid Furring Channels

APPENDIX

(Nonmandatory Information)

X1. DRYWALL STUD MINIMUM SECTION PROPERTIES FOR 2½ IN. STUD

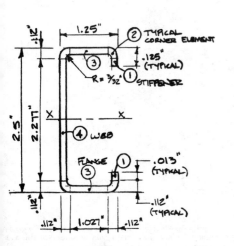

FIG. X1 Drywall Stud Cross Section of 2½ in. Stud

Material Thickness, $t = 0.0179$ in.
Distance from neutral (x-x) axis to farthest fibers, $C = 1.25$ in.
Properties of 90° Corners:
Radius to center of element, $r = R + t/2 = 0.094 + 0.0179/2 = 0.103$ in.

Length of element $= \frac{2\pi}{4} r = 1.57 (0.103) = 0.162$ in.

Distance of center of radius to center of gravity $= 0.637r = 0.066$ in.

Element	Element length, (nL)[A] in.	Distance of center of gravity of element x-x axis, Y, in.	$nL\,Y^2$ (in.³)	I_1 about own axis, $\left(\frac{h^3}{12}\right)$[B]
1	(2) 0.013 = 0.026	1.132	0.033	...
2	(4) 0.162 = 0.648	1.184	0.908	...
3	(2) 1.027 = 2.054	1.241	3.164	...
4	(1) 2.277 = 2.277	...	...	0.984
Sum (Σ)	5.005	...	4.105	0.984

[A] n = number of lengths.
[B] Only values of 10^{-5} and greater are shown.

$I'_x = \Sigma\, nL\, Y^2 + I'_1 = 4.105 + .984 = 5.089$ in.³
$I_x = I'_x (t) = 5.089 (0.0179) = 0.092$ in.⁴
$S_x = \dfrac{I_x}{c} = \dfrac{0.092}{1.25} = 0.073$ in.³
Area $= \Sigma\, L\, (t) = 5.005 (0.0179) = 0.090$ in.²

Designation: C 954 – 86

Standard Specification for

STEEL DRILL SCREWS FOR THE APPLICATION OF GYPSUM BOARD OR METAL PLASTER BASES TO STEEL STUDS FROM 0.033 in. (0.84 mm) TO 0.112 in. (2.84 mm) IN THICKNESS[1]

This standard is issued under the fixed designation C 954; the number immediately following the designation indicates the year of original adoption or, in the case of revision, the year of last revision. A number in parentheses indicates the year of last reapproval. A superscript epsilon (ε) indicates an editorial change since the last revision or reapproval.

1. Scope

1.1 This specification covers minimum requirements for steel drill screws for use in fastening gypsum board or metal plaster bases to steel members from 0.033 in. (0.84 mm) to 0.112 in. (2.84 mm) in thickness.

1.2 The values stated in inch-pound units are to be regarded as the standard. The metric equivalents of inch-pound units may be approximate.

2. Applicable Document

2.1 *ASTM Standard:*
A 510 Specification for General Requirements for Wire Rods and Coarse Round Wire, Carbon Steel[2]

3. Materials and Manufacture

3.1 Screws shall be manufactured from Grade 1018 to 1022 steel wire manufactured in accordance with Specification A 510.

4. Dimensions and Permissible Variations

4.1 Screws shall have the following shape and dimensions:

4.1.1 *Head Diameter*—The head of the screw shall be a minimum of 0.3145 in. (8.00 mm) in diameter for gypsum board and 0.437 in. (11.1 mm) for metal plaster bases. Heads may be out of round 0.020 in. (0.51 mm).

4.1.2 *Head Contour:*

4.1.2.1 *Gypsum Board*—The top of the head shall be flat. The outer flange thickness shall be 0.025 ± 0.005 in. (0.64 ± 0.12 mm). The contour beneath the flange head shall be such that the screw head can be pulled below the surface of the wallboard without cutting the paper, and free of paper burrs when used.

4.1.2.2 *Metal Plaster Bases*—The top of the head shall be flat or contoured. The underside of the head shall be flat. Threads shall extend to the underside of the head.

4.1.3 The *Screw Diameter* of the threaded portion shall be such that the screw will pass the performance tests in Section 8.

4.1.4 *Threads* may be of single- or double-lead thread design. They shall be clean and smooth.

4.1.5 *Points* shall provide for self-drilling into steel studs from 0.033 in. (0.84 mm) to 0.112 in. (2.84 mm) in thickness and meet the performance tests in Section 8.

4.1.6 The *Driving Recess* shall be a No. 2 "Phillips" design with a minimum depth of 0.104 in. (2.64 mm) as determined with a "Phillips" penetration depth gage or a recess or equal performance.

4.1.7 Nominal length shall be minimum length.

5. Metallurgical Requirements

5.1 The surface of the screws shall be carbonitrided to a depth of 0.004 to 0.008 in. (0.10 to 0.20 mm).

5.2 The core hardness of the screw shall be 32 to 40 HRC after being drawn at a minimum temperature of 625°F.

5.3 The screws shall have no band of free ferrite between the case and core nor shall the case contain appreciable amounts of retained austenite or other soft surface constituents.

5.4 Surface hardness shall be determined by a micro hardness instrument at "the root of the thread profile," as exposed by removal of enough material to form a flat along the length of the screw.

6. Finish and Workmanship

6.1 Screws shall be given a corrosion-resistant treatment. Any treatment shall not inhibit adhesion to joint-finishing compounds or plaster and shall not bleed through decorative finishes.

6.2 Screws shall be straight, neatly formed, and free of defects such as burrs and deformations.

7. Number of Tests and Retests

7.1 Test a minimum of one screw per container from each of 5 containers in a lot of 16 000 screws, but not less than 5 screws per 16 000 screws, for conformance to this specification. If any one screw of the sample of 5 screws fails to meet this specification, test a second sample of 25 additional screws. If two or more screws from the second sample fail, then the represented lot of 16 000 screws is considered to have failed to meet this specification.

8. Test Methods

8.1 *Performance Tests*—Make performance tests by attaching ⅝-in. (15.9-mm) Type X gypsum wallboard or 2.5-lb/yd² diamond mesh metal lath to the center of a flange of a steel stud. The stud shall be rigidly supported. The drill propelling the screw shall turn 2500 r/min (free spindle speed) and shall exert a total pressure of 30 lbf (133.5N) (deadweight plus applied force).

8.2 *Penetration Test:*

8.2.1 Screws shall self-drill and drive into a stud 0.0598 in. (1.5 mm) thick with an approximate hardness of 65 HRB in less than 4 s.

8.2.2 Screws shall also have the capability to drill into a similar test section 0.105 in. (2.7 mm) thick, with a hardness of approximately 80 HRB.

8.3 *Spin-Out Test:*

8.3.1 *Gypsum Wallboard*—Screw threads shall be adequate to pull the head of the screw below the surface of the wallboard through four layers of 0.010-in. (0.25-mm) thick kraft paper over ⅝-in. (15.9-mm) Type X gypsum wallboard and a steel stud 0.033 in. (0.84 mm) thick.

8.3.2 *Metal Plaster Bases*—Screw threads shall be adequate to pull the diamond mesh metal lath tightly enough against the flange of a steel stud 0.033 in. (0.84 mm) thick that when subjected to a steady pull the lath will tear and not slip out from under the screw head.

9. Packaging and Marking

9.1 Screws shall be packaged in substantial commercial containers constructed so as to preserve the contents in good condition and to ensure acceptance and safe delivery by common or other carriers, at the lowest rate, to the point of delivery.

9.2 The containers shall be so constructed that the contents can be partially removed without destroying the container's ability to serve as a receptacle for the remainder of the contents.

9.3 Individual packages and shipping containers shall be marked with the type and length of screw and the name of the manufacturer or distributor.

---

[1] This specification is under the jurisdiction of ASTM Committe C-11 on Gypsum and Related Building Materials and Systems and is the direct responsibility of Subcommittee C11.02 on Specifications and Test Methods for Accessories and Related Products.
Current edition approved Jan. 31, 1986. Published March 1986.
[2] *Annual Book of ASTM Standards*, Vol 01.03.

Designation: C 955 – 86

Standard Specification for
LOAD-BEARING (TRANSVERSE AND AXIAL) STEEL STUDS, RUNNERS (TRACK), AND BRACING OR BRIDGING, FOR SCREW APPLICATION OF GYPSUM BOARD AND METAL PLASTER BASES[1]

This standard is issued under the fixed designation C 955: the number immediately following the designation indicates the year of original adoption or, in the case of revision, the year of last revision. A number in parentheses indicates the year of last reapproval. A superscript epsilon (ε) indicates an editorial change since the last revision or reapproval.

1. Scope

1.1 This specification covers steel studs, runners (track), and bracing or bridging (within a base metal thickness range from 0.033 in. (0.84 mm) to 0.112 in. (2.84 mm) for screw application of gypsum board and metal plaster bases in load-bearing (transverse and axial) construction assemblies. Steel of lesser thickness may be used in additional engineered products.

1.2 The values stated in inch-pound units are to be regarded as the standard. The metric equivalents of inch-pound units may be approximate.

1.3 The following precautionary caveat pertains only to the test method portion, Section 9, of this specification: *This standard may involve hazardous materials, operations, and equipment. This standard does not purport to address all of the safety problems associated with its use. It is the responsibility of whoever uses this standard to consult and establish appropriate safety and health practices and determine the applicability of regulatory limitations prior to use.*

2. Applicable Documents

2.1 *ASTM Standards:*
A 446/A 446M Specification for Steel Sheet, Zinc-Coated (Galvanized) by the Hot-Dip Process, Structural (Physical) Quality[2]
A 525 Specification for General Requirements for Steel Sheet, Zinc-Coated (Galvanized) by the Hot-Dip Process[2]
C 36 Specification for Gypsum Wallboard[3]
C 1002 Specification for Steel Drill Screws for the Application of Gypsum Board[3]
2.2 *AISI Specification Publication:*
Specification for the Design of Cold-Formed Steel Structural Members[4]

3. Materials and Manufacture

3.1 The minimum steel thickness (base steel) shall be not less than 0.033 in. (0.84 mm).

3.2 Individual measurements before the application of protective coating shall be not less than 95 % of the intended design thickness.

3.3 Studs, runners (track), and bracing or bridging shall have a protective coating conforming to Specification A 525 or shall have a zinc chromate or iron oxide primer.

3.4 Edges of studs, runners (track), and bracing or bridging shall be manufactured to minimize burrs and sharp edges.

3.5 Punch-outs in webs may be provided, but in no case shall the punch-outs reduce the performance of the stud, runner (track), and bracing or bridging in the assembly below the specified performance requirements.

3.6 The sectional properties of studs, runners (track), and bracing or bridging shall be computed in accordance with the AISI Specifica-

tion for the Design of Cold-Formed Steel Structural Members.

4. Dimensions and Permissible Variations

4.1 Studs shall have a configuration and steel thickness such that the system in which they are used shall carry the transverse and axial design loads without exceeding the allowable stress of the steel or the allowable design deflection (Note). The manufacturer shall supply sufficient data for calculating design performance.

NOTE—Allowable deflection will vary depending on the cladding used and the lateral force from wind pressure, which depends on location. Detailed requirements shall be specified in application specifications.

4.2 Studs shall be free of twist or camber of a degree that will prevent use in the system for which intended, or that will reduce the intended design structural (load-bearing) capacity.

4.3 The minimum width of the face to which the gypsum board is screw-attached shall be a minimum of 1¼ in. (32 mm).

4.4 Stud length tolerance, as manufactured, shall be + ³⁄₁₆ in. – 0 (5 mm).

4.5 Runners (track) shall be formed in a U-shaped configuration, having a depth compatible with that of the studs of the same nominal size.

4.6 Minimum height of runner (track) flanges shall be ¾ in. (19 mm).

4.7 Bracing and bridging shall have configuration and steel thickness to provide secondary support for the studs in accordance with the AISI Specification for the Design of Cold-Formed Steel Structural Members.

5. Workmanship, Finish, and Appearance

5.1 The steel members shall be free of defects that interfere with the purpose for which they are intended, and shall be uniformly coated.

6. Sampling

6.1 One stud or runner, or both, shall be selected from each bundle or package but not more than ten from any one shipment for testing.

7. Number of Tests and Retests

7.1 Five sections of test stud shall be tested in the penetration test (9.2). If more than one test fails to meet the requirements, two more studs shall be chosen for retesting.

8. Specimen Preparation

8.1 Each member shall be cut into test specimens a minimum of 18 in. (457 mm) in length.

9. Test Methods

9.1 Make a performance test by attaching ⅝-in. (15.9-mm) Type X gypsum wallboard (conforming to Specification C 36) to the center of the attachment face of the test specimen obtained in 8.1 with steel drill screws (conforming to Specification C 1002). Take specimens a minimum of 12 in. (305 mm) from either edge or end of the gypsum wallboard panel. The member to be tested shall be rigidly supported (see Fig. 1). The drill propelling the screw shall turn at 2500 rpm (free spindle speed) and exert a total pressure (dead weight and applied load) of 30 lbf (135.5 N).

9.2 *Penetration Test*—Screws shall be able to self-drill and drive and lock into the stud. The screw used in this test shall conform to Specification C 1002.

10. Inspection

10.1 Inspection of the material shall be agreed upon between the purchaser and the supplier as part of the purchase order.

11. Rejection and Rehearing

11.1 When specified in the purchase order, material that fails to conform to the requirements of this specification may be rejected. Rejection should be reported to the producer or supplier promptly and in writing. In case of dissatisfaction with the results of the test, the producer or supplier may make claim for a rehearing.

12. Certification

12.1 When specified in the purchase order, a producer's or supplier's certification and report of the test results shall be furnished to the purchaser that the material was manufactured, sampled, tested, and inspected in accordance with this specification and has been found to meet the specified requirements.

12.2 When specified in the purchase order, the certification of an independent third party indicating conformance to the requirements of

[1] This specification is under the jurisdiction of ASTM Committee C-11 on Gypsum and Related Building Materials and Systems and is the direct responsibility of Subcommittee C11.02 on Specifications and Test Methods for Accessories and Related Products.
Current edition approved Jan. 31, 1986. Published March 1986. Originally published as C 955 – 81. Last previous edition C 955 – 83.
[2] *Annual Book of ASTM Standards*, Vol 01.06.
[3] *Annual Book of ASTM Standards*, Vol 04.01.
[4] Available from the American Iron and Steel Institute, 1000 16th St., N.W., Washington, DC 20036.

Designation: C 1047 – 85

Standard Specification for
ACCESSORIES FOR GYPSUM WALLBOARD AND GYPSUM VENEER BASE[1]

This standard is issued under the fixed designation C 1047; the number immediately following the designation indicates the year of original adoption or, in the case of revision, the year of last revision. A number in parentheses indicates the year of last reapproval. A superscript epsilon (ϵ) indicates an editorial change since the last revision or reapproval.

1. Scope

1.1 This specification covers accessories used in conjunction with assemblies of gypsum wallboard and gypsum veneer plaster to protect edges and corners and to provide architectural features. (See Fig. 1.)

1.2 The values stated in inch-pound units are to be regarded as the standard.

2. Applicable Documents

2.1 *ASTM Standards:*

A 525 Specification for General Requirements for Steel Sheet, Zinc-Coated (Galvanized) by the Hot-Dip Process[2]

A 591 Specification for Steel Sheet, Cold-Rolled, Electrolytic Zinc-Coated[2]

A 463 Specification for Steel Sheet, Cold-Rolled, Aluminum-Coated Type I and Type II[2]

B 69 Specification for Rolled Zinc[3]

B 117 Method of Salt Spray (Fog) Testing[4]

C 474 Test Methods for Testing Joint Treatment Materials for Gypsum Wallboard Construction[5]

C 475 Specification for Joint Treatment Materials for Gypsum Wallboard Construction[5]

C 587 Specification for Gypsum Veneer Plaster[5]

D 2092 Practices for Preparation of Zinc-Coated Steel Surfaces for Painting[6]

D 3678 Specification for Rigid Poly (Vinyl Chloride) (PVC) Interior-Profile Extrusions[7]

3. Terminology

3.1 *Description of Terms Specific to This Standard:*

3.1.1 *accessories*—cornerbeads, edge trims, and control joints, such as casing beads, bull noses, and stops.

3.1.2 *control joint*—a formed product used for designed or required separations between adjacent surfaces of gypsum boards or gypsum veneer base.

3.1.3 *cornerbead*—a formed metal, plastic, or metal and paper angle for outside corners of gypsum boards or gypsum veneer base.

3.1.4 *edge trim*—typically "J"- or "L"-shaped strip, as shown in Fig. 1, formed of metal, plastic, or metal and paper to cover exposed ends or edges of gypsum board or gypsum veneer base.

3.2 *Definition:*

3.2.1 *flange*—that part of the accessory extending out on the face of the gypsum board.

4. Materials and Manufacture

4.1 Steel accessories and steel components of accessories manufactured from steel and paper in combination shall be manufactured from zinc-coated cold-roll (coil or cut length) sheet steel not less than 0.012 in. (0.30 mm) thick before application of coating.

4.1.1 Sheet steel, zinc-coated by the hot-dip process, shall be in accordance with Specification A 525, minimized spangle, minimum G-30 coating.

4.1.2 Sheet steel, zinc-coated by the electrolytic process shall be in accordance with Specification A 591, minimum Class C coating.

4.1.3 Sheet steel, aluminum-coated, shall be in accordance with Specification A 463 minimum T1-40 coating.

4.1.4 Phosphatizing (as specified in Method A of Practices D 2092) or other surface treatments may be used to insure compatibility and bond as specified in Section 5.

4.2 Zinc accessories shall be manufactured from rolled zinc in accordance with Specification B 69, Type I, not less than 0.012 in. (0.30 mm) thick.

4.3 Plastic accessories shall be manufactured from rigid PVC plastic in accordance with Specification D 3678, Class II or III, not less than 0.02 in. (0.051 mm) thick.

4.4 Paper components of accessories manufactured from steel and paper in combination shall comply with requirements for thickness, tensile strength, dimensional stability, and bond of joint tape to joint compound as specified in Method C 474.

5. Physical Properties

5.1 *Compatibility and Bond*—Accessories shall be compatible with and provide a surface bond to the materials specified in Specifications C 475 and C 587.

5.2 *Test Performance*—Steel accessories and steel components of accessories manufactured from sheet metal, zinc-coated by the electrolytic process (see 4.1.2) or from aluminum-coated steel (see 4.1.3) or from rolled zinc (see 4.2) shall not show any red oxidation when tested as specified in Method B 117 for 120 h.

6. Dimensions and Permissible Variations

6.1 Cornerbeads shall have an interior angle between the flanges no greater than 89°.

6.2 Accessories shall be free of twist or camber of a degree that will prevent their use in the assembly for which they are intended.

6.3 Length tolerances shall be ±³⁄₁₆ in. (±4.8 mm).

6.4 The size of the edge trim shall suit the thickness of the gypsum board used.

6.5 The minimum width of the flange shall be ⅞ ± ¹⁄₃₂ in. (22 ± 0.8 mm).

7. Appearance

7.1 Edges of accessories shall be free of burrs and sharp edges.

8. Configuration

8.1 Flanges of accessories may have punch outs to accommodate fastening to framing members and may be either knurled, made of mesh, deformed, expanded or otherwise shaped to meet the performance requirements and intended use. Fig. 1 depicts the most commonly used shapes.

NOTE—Other types of accessories designed for special finish application (bull-nose, cove, etc.) shall be manufactured in accordance with the general requirements of this specification.

9. Sampling

9.1 At least 0.25 % of the number of each type of accessory in a shipment, but not less than three pieces, shall be selected to represent the shipment and shall constitute a sample for purpose of tests by the purchaser or user.

10. Inspection

10.1 Inspection of the accessories shall agreed upon between the purchaser and the s plier as part of the purchase order.

11. Rejection and Rehearing

11.1 Accessories that fail to conform to requirements of this specification may be jected. Rejection shall be reported to the p ducer or supplier in writing within 10 work days from receipt of shipment by the purchas The notice of rejection shall contain a spec statement of the respect in which the accesso have failed to conform to the requirements this specification. In case of dissatisfaction w the results of the test, at the request of the p ducer or supplier, such notice of rejection sh be supported by results of a test conducted by mutually agreeable independent laboratory.

12. Certification

12.1 When specified in the purchase order producer's or supplier's certification shall be f nished to the purchaser that the product me the requirements of this specification.

13. Packaging and Package Marking

13.1 Accessories shall be packaged to ens safe delivery by common or other carriers.

13.2 When shipped for resale, the name of manufacturer or seller and the brand and co tents shall be legibly marked on each package.

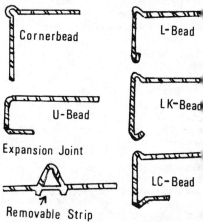

NOTE—Flanges may have punchouts, may be knurled, m deformed, expanded or otherwise shaped for intended use.

FIG. 1 Accessories for Gypsum Wallboard and Gypsur Veneer Base

[1] This specification is under the jurisdiction of ASTM Co mittee C-11 on Gypsum and Related Building Materials a Systems and is the direct responsibility of Subcommittee C11. on Specifications and Test Methods for Accessories and Rela Products.

Current edition approved Sept. 27, 1985. Published Octob 1985.

[2] *Annual Book of ASTM Standards*, Vol 01.06.
[3] *Annual Book of ASTM Standards*, Vol 02.04.
[4] *Annual Book of ASTM Standards*, Vol 03.02.
[5] *Annual Book of ASTM Standards*, Vol 04.01.
[6] *Annual Book of ASTM Standards*, Vol 06.01.
[7] *Annual Book of ASTM Standards*, Vol 08.04.

Designation: C 960 – 81

Standard Specification for
PREDECORATED GYPSUM BOARD[1]

This standard is issued under the fixed designation C 960; the number immediately following the designation indicates the year of original adoption or, in the case of revision, the year of last revision. A number in parentheses indicates the year of last reapproval. A superscript epsilon (ε) indicates an editorial change since the last revision or reapproval.

Scope

1.1 This specification covers predecorated gypsum board that may be used as the finished surfacing for walls or ceilings.

1.2 The values stated in inch-pound units are to be regarded as standard. The value given in parentheses are provided for information purposes only.

Applicable Documents

2.1 *ASTM Standards:*

C 11 Definitions of Terms Relating to Gypsum and Related Building Materials and Systems[2]

C 36 Specification for Gypsum Wallboard[2]

C 442 Specification for Gypsum Backing Board and Coreboard[2]

E 84 Test Method for Surface Burning Characteristics of Building Materials[3]

Composition

3.1 Predecorated gypsum board shall consist of a noncombustible core, essentially gypsum surfaced with paper firmly bonded to the core with the *face* covered with a decorative sheet, film, or coating.

Classification

4.1 Predecorated gypsum board is furnished in two classes and two types as follows:

4.1.1 Class I predecorated gypsum board shall have a decorative sheet or film laminated to the paper surface.

4.1.2 Class II predecorated gypsum board shall have a decorative coating applied to the paper surface.

4.1.3 The types are regular and Type X as specified in Specification C 36.

Terminology

5.1 *decorative sheet or film*—a plastic film backed or unbacked, plastic sheet, or combination of these that has been printed with or otherwise bears a decorative pattern.

5.2 *decorative coating*—a paint or other liquid material with or without aggregate, that has been applied by spraying, rolling, or other mechanical methods onto the paper surface of the gypsum board.

5.3 *face*—the side of the predecorated gypsum wallboard that is called the finish side.

Physical Requirements

6.1 Predecorated gypsum board shall meet the flexural strength, humidified deflection, core, end and edge hardness, and nail pull resistance requirements of Specification C 36.

6.2 Edges shall be either square or featured.

6.3 When Class I predecorated gypsum board is tested by the pick (tape) test described in Section 9, no part of the test surface shall be disturbed.

7. Dimensions and Permissible Variations

7.1 Predecorated gypsum board shall have the same dimensions and permissible variations after decoration as listed in Specification C 36. When heavy custom coatings, decorative sheet or film, are requested, the dimension specification may be waived when so specified on the order.

8. Sampling

8.1 At least 0.25 % of the number of gypsum boards in a shipment, but not less than three boards, shall be selected as to be representative of the shipment and shall constitute a sample for purpose of tests by the purchaser or user.

9. Test Method for Pick Resistance

9.1 A minimum of six tests shall be made, three in the machine direction and three perpendicular to the machine direction.

9.2 Apply a ½ to 9/16-in. (12.7 to 14.3-mm) wide strip of transparent adhesive tape[4] approximately 3 in. (76.2 mm) in length to the test surface, leaving 1 in. (25.4 mm) free. Press the tape firmly to the test surface using thumb pressure for a minimum of 5 s. Grasp the free end of tape and pull quickly at a 90° angle.

10. Rejection and Rehearing

10.1 Material that fails to conform to the requirements of this specification may be rejected. Rejection should be reported to the producer or supplier promptly and in writing. In case of dissatisfaction with the results of the test, the producer or supplier may make claim for a rehearing.

11. Certification

11.1 A producer's or supplier's certification shall be furnished upon request to the purchaser that the material was manufactured, sampled, tested, and inspected in accordance with this specification and has been found to meet the requirements. When specified in the purchase order or contract, a report of the test results shall be furnished.

12. Packaging

12.1 Predecorated gypsum board shall be shipped so as to be kept dry.

13. Marking

13.1 When shipped for resale, the name of the manufacturer or seller and the brand shall be legibly marked on each board, package, or pallet.

13.2 When required by the purchaser, the surface burning characteristic classification as tested in accordance with Method E 84 shall be legibly marked on each board, package, or pallet.

[1] This specification is under the jurisdiction of ASTM Committee C-11 on Gypsum and Related Building Materials and Systems and is the direct responsibility of Subcommittee C11.01 on Specifications and Test Methods for Gypsum Products.

Current edition approved Oct. 16, 1981. Published December 1981.

[2] *Annual Book of ASTM Standards*, Vol 04.01.

[3] *Annual Book of ASTM Standards*, Vol 04.07.

[4] "Scotch" brand transparent tape No. 600, manufactured by 3M Co., St. Paul, MN 55101, or equivalent, has been found satisfactory for this purpose.

Designation: C 931 – 85

Standard Specification for

EXTERIOR GYPSUM SOFFIT BOARD[1]

This standard is issued under the fixed designation C 931; the number immediately following the designation indicates the year of original adoption or, in the case of revision, the year of last revision. A number in parentheses indicates the year of last reapproval. A superscript epsilon (ε) indicates an editorial change since the last revision or reapproval.

1. Scope

1.1 This specification covers exterior gypsum soffit board, designed to be used for exterior soffits and carport ceilings that are completely protected from contact with liquid water.

1.2 The values stated in inch-pound units are to be regarded as the standard. The metric equivalents of the inch-pound units may be approximate.

2. Applicable Documents

2.1 *ASTM Standards:*

C 473 Test Methods for Physical Testing of Gypsum Board Products and Gypsum Lath[2]

E 84 Test Method for Surface Burning Characteristics of Building Materials[3]

E 119 Methods of Fire Tests of Building Construction and Materials[3]

3. Physical Properties

3.1 Gypsum soffit board shall consist of an incombustible core, essentially gypsum, surfaced with paper firmly bonded to the core.

3.2 Type "X" (special fire-resistant) designates gypsum soffit board complying with this specification and one that provides at least 1 h fire-resistance classification for boards ⅝ in. (15.9 mm) thick or ¾ h fire retardant rating for boards ½ in. (12.7 mm) thick applied to a partition in a single-layer nail application on each side of the load-bearing wood framing members, when tested in accordance with the requirements of Methods E 119.

3.3 Gypsum soffit board shall have a maximum flame-spread classification of 25 when tested in accordance with the requirements of Test Method E 84.

4. Flexural Strength

4.1 Specimens shall be tested in accordance with Methods C 473. When tested face up and when tested face down the specimens shall have an average breaking load of not less than the following:

Thickness, in. (mm)	Load, lbf (N) Bearing Edges Across Fiber of Surfacing	Bearing Edges Parallel to Fiber of Surfacing
½ (12.7)	110 (489)	40 (178)
⅝ (15.9)	150 (667)	50 (222)

5. Humidified Deflection

5.1 When tested in accordance with Methods C 473, specimens taken from the gypsum soffit board shall have an average deflection of no more than the following:

Thickness, in. (mm)	Humidified Deflection, Eighths of an in. (mm)
½ (12.7)	7 (22)
⅝ (15.9)	4 (13)

6. Core, End, and Edge Hardness

6.1 When tested in accordance with Methods C 473, specimens taken from the gypsum soffit board shall have an average hardness of not less than 15 lbf (67 N) for the core, ends, or edges.

7. Nail Pull Resistance

7.1 When tested in accordance with Methods C 473, specimens taken from the gypsum soffit board shall have an average nail pull resistance of not less than the following:

Thickness, in. (mm)	Nail Pull Resistance, lbf (N)
½ (12.7)	80 (356)
⅝ (15.9)	90 (400)

8. Dimensions and Permissible Variations

8.1 *Thickness, nominal*—½ or ⅝ in. (12.7 or 15.9 mm) with permissible variations in the nominal thickness of ¹⁄₆₄ in. (0.4 mm) and with permissible local variations of ±¹⁄₃₂ in. (0.8 mm) from the nominal thickness. Edges of ½ and ⅝ in. soffit board may be either plain, recessed, beveled, or tapered to receive a joint reinforcing strip.

8.2 *Width, nominal*—up to 48 in. (1.22 m), with a permissible variation of ³⁄₃₂ in. (2.4 mm) under the specified width.

8.3 *Length, nominal*—from 4 to 16 ft (1.2 to 4.9 m), inclusive, with permissible variations of ±¼ in. (6.35 mm) from the specified length.

9. Workmanship, Finish, and Appearance

9.1 The surfaces of gypsum soffit board shall be true and free of imperfections that would render the board unfit for use with or without decoration. The edges shall be square with a permissible variation of ±⅛ in. (±3.2 mm) in the full width of the board.

10. Sampling

10.1 Select at least 0.25 % of the number of gypsum soffit board in a shipment, but not less than three boards, to be representative of the shipment and to constitute a sample for the purpose of the tests by the purchaser or user.

11. Packaging and Package Marking

11.1 Ship the gypsum soffit board so as to be kept dry and free of moisture.

11.2 When shipped for resale, the name of the manufacturer or seller and the brand shall be legibly marked on each board or package.

---

[1] This specification is under the jurisdiction of ASTM Committee C-11 on Gypsum and Related Building Materials and Systems and is the direct responsibility of Subcommittee C11.0 on Specifications and Test Methods for Gypsum Products.

Current edition approved Sept. 27, 1985. Published November 1985. Originally published as C 930 – 80. Last previous edition C 930 – 84a.

[2] *Annual Book of ASTM Standards*, Vol 04.01.

[3] *Annual Book of ASTM Standards*, Vol 04.07.

Designation: C 79 – 87

Standard Specification for GYPSUM SHEATHING BOARD[1]

This standard is issued under the fixed designation C 79; the number immediately following the designation indicates the year of original adoption or, in the case of revision, the year of last revision. A number in parentheses indicates the year of last reapproval. A superscript epsilon (ε) indicates an editorial change since the last revision or reapproval.

This specification has been approved for use by agencies of the Department of Defense for listing in the DoD Index of Specifications and Standards.

Scope*

1.1 This specification covers gypsum sheath- g board, which is designed to be used as a eathing on buildings.

1.2 The values stated in inch-pound units are be regarded as the standard.

Referenced Documents

2.1 *ASTM Standards:*

C 473 Methods for Physical Testing of Gypsum Board Products and Gypsum Lath[2]

E 119 Methods of Fire Tests of Building Construction and Materials[3]

. Materials

3.1 Gypsum sheathing board shall consist of noncombustible core, essentially gypsum, sur- aced on both the face and back of the board ith water-repellent paper bonded to the core.

3.2 Water-resistant core gypsum sheathing hall have a water-resistant material incorporated the core.

3.3 Type X (special fire-resistant) gypsum heathing shall be gypsum sheathing that pro- ides at least 1-h fire-resistance rating for boards 8 in. (15.9 mm) thick, applied to a partition in ingle-layer nail application on each face of load- earing wood framing members, when tested in ccordance with the requirements of Methods 119.

NOTE—Consult manufacturers for independent test ata on assembly details and fire resistance classifica- ons for other types of construction. See official fire est reports for assembly particulars, materials, and lassifications.

. Physical Properties

4.1 *Flexural Strength*—Specimens shall be ested in accordance with Methods C 473. When ested face up and when tested face down the pecimens shall have an average breaking load of ot less than the values given in Table 1.

4.2 *Water Resistance of Core-Treated Gyp- um Sheathing*—The average water absorption f three specimens selected at random shall be ot more than 10 weight % after 2-h immersion hen tested for water resistance in accordance ith Methods C 473.

4.3 *Surface Water Absorption*—The average urface water absorption of each surface of the oard shall not be more than 1.6 g after 2 h of lapsed time when tested for surface water ab- orption in accordance with Methods C 473.

4.4 *Humidified Deflection*—Specimens shall ave an average deflection of no more than the alues given in Table 2 when tested for humidi- ied deflection in accordance with Methods 473.

4.5 *Core, End, and Edge Hardness*—Specimens shall have an average hardness of not less than 15 lbf (67 N) when tested for core, end, and edge hardness in accordance with Methods C 473.

4.6 *Nail-Pull Resistance*—Specimens shall have an average nail-pull resistance of not less than the values given in Table 3 when tested for nail pull resistance in accordance with Methods C 473.

5. Dimensions and Permissible Variations

5.1 Thickness, width, and length shall be determined in accordance with Methods C 473.

5.2 *Thickness*—The nominal thickness of gypsum sheathing board shall be 3/8, 4/10, 1/2, or 5/8 in. (9.5, 10.2, 12.7, or 15.9 mm), with permissible variations in the nominal thickness of ±1/32 in. (±0.8 mm), and with permissible local variations of ±1/16 in. (±1.6 mm) from the nominal thickness.

5.3 *Width*—The nominal width of gypsum sheathing board shall be 24 or 48 in. (610 or 1219 mm), with a permissible variation of ±1/8 in. (±3.2 mm) from the specified width.

5.4 *Length*—The nominal length of gypsum sheathing board shall be from 6 to 12 ft (1829 to 3658 mm) with permissible variations of ±1/4 in. (±6.4 mm) from the specified length.

5.5 *Edges and Ends*—Edges and ends of gypsum sheathing board shall be straight and solid. Edges shall be either square or V-tongue and groove. The corners shall be square with a permissible variation of ±1/8 in. (±3.2 mm) in the full width of the sheathing.

6. Workmanship, Finish, and Appearance

6.1 Gypsum sheathing board shall be free of any cracks and imperfections that would render it unfit for its intended use.

7. Sampling

7.1 At least 0.25 % of the number of gypsum sheathing boards in a shipment, but not less than three specimens shall be so selected as to be representative of the shipment and shall constitute a sample for purpose of tests by the purchaser or user.

8. Inspection

8.1 Inspection of the gypsum board shall be agreed upon between the purchaser and the supplier as part of the purchase order.

9. Rejection and Rehearing

9.1 Gypsum board that fails to conform to the requirements of this specification may be rejected. Rejection shall be reported to the producer or supplier in writing within 10 working days from receipt of shipment by the purchaser. The notice of rejection shall contain a specific statement of the respect in which the gypsum board has failed to conform to the requirements of this specification. In case of dissatisfaction of the producer or supplier with the results of the test, such notice of rejection shall be supported by results of a test conducted by a mutually agreeable independent laboratory.

10. Certification

10.1 When specified in the purchase order, a producer's or supplier's certification shall be furnished to the purchaser that the product meets the requirements of this specification.

11. Packaging and Package Marking

11.1 Gypsum sheathing board shall be shipped so as to be kept dry.

11.2 When shipped for resale, the name of the manufacturer or supplier and the brand shall be legibly marked on each board or package.

TABLE 1 Average Breaking Load

Thickness, in. (mm)	Load, lbf (N) Bearing Edges Across Fiber of Surfacing	Bearing Edges Parallel to Fiber of Surfacing
3/8 (9.5)	80 (356)	30 (133)
4/10 (10.2)	88 (391)	32 (142)
1/2 (12.7)	110 (489)	40 (178)
5/8 (15.9)	150 (667)	50 (222)

TABLE 2 Average Humidified Deflections

Thickness, in. (mm)	Humidified Deflection, in. (mm)
3/8 (9.5)	15/8 (48)
4/10 (10.2)	12/8 (38)
1/2 (12.7)	10/8 (32)
5/8 (15.9)	5/8 (16)

TABLE 3 Nail Pull Resistance

Thickness, in. (mm)	Nail Pull Resistance, lbf (N)
3/8 (9.5)	60 (267)
4/10 (10.2)	70 (311)
1/2 (12.7)	80 (356)
5/8 (15.9)	90 (400)

[1] This specification is under the jurisdiction of ASTM Committee C-11 on Gypsum and Related Building Materials and Systems and is the direct responsibility of Subcommittee C11.01 on Specifications and Test Methods for Gypsum Products.

Current edition approved Feb. 27, 1987. Published April 1987. Originally published as C 79 – 30 T. Last previous edition C 79 – 84a.

[2] *Annual Book of ASTM Standards*, Vol 04.01.

[3] *Annual Book of ASTM Standards*, Vol 04.07.

 **Designation: C 37 – 87**

Standard Specification for
GYPSUM LATH[1]

This standard is issued under the fixed designation C 37; the number immediately following the designation indicates the year original adoption or, in the case of revision, the year of last revision. A number in parentheses indicates the year of last reapprova A superscript epsilon (ε) indicates an editorial change since the last revision or reapproval.

This specification has been approved for use by agencies of the Department of Defense for listing in the DoD Index of Specification and Standards.

1. Scope*

1.1 This specification covers plain and foil-backed types of gypsum lath, designed to be used as a base for the application of gypsum plaster.

1.2 The values stated in inch-pound units are to be regarded as the standard.

2. Referenced Documents

2.1 *ASTM Standards·*
C 28 Specification for Gypsum Plasters[2]
C 473 Methods for Physical Testing of Gypsum Board Products and Gypsum Lath[2]
E 96 Test Methods for Water Vapor Transmission of Materials[3]

3. Materials and Manufacture

3.1 Plain gypsum lath shall consist of a noncombustible core, essentially gypsum, surfaced with paper firmly bonded to the core.

3.2 Foil-backed gypsum lath shall be plain gypsum lath with the back surface covered with aluminum foil.

NOTE 1—Foil-backed gypsum lath shall be installed with the foil surface against the framing members. The foil surface is not adapted to receive or retain plaster.

3.3 Type X (special fire-resistant) designates gypsum lath complying with this specification that provides at least 1 h fire-resistant classification rating for lath ⅜ in. (9.5 mm) thick, applied at right angles to each side of 2 ft by 4 ft wood studs 16 in. (406 mm) on center and nailed with blue lath nails 5 in. (127 mm) on center over which is applied ½ in. (12.7 mm) 1:2 gypsum-sand plaster.

4. Physical Properties

4.1 *Flexural Strength*—Specimens shall be tested in accordance with Methods C 473. When tested face up and when tested face down the specimens shall have an average breaking load of not less than the following:

	Load, lbf (N)	
Thickness, in. (mm)	Bearing Edges Across Fiber of Surfacing	Bearing Edges Parallel to Fiber of Surfacing
⅜ (9.5)	60 (267)	25 (111)
½ (12.7)	100 (445)	35 (156)

4.2 *Humidified Deflection*—Specimens taken from plain gypsum lath shall have an average deflection of no more than the following when tested in accordance with Methods C 473:

Thickness, in. (mm)	Humidified Deflection, in. (mm)
⅜ (9.5)	15/8 (48)
½ (12.7)	10/8 (32)

4.3 *Core, End, and Edge Hardness*—Specimens taken from gypsum lath shall have an average hardness of not less than 15 lbf (67 N) for the core, ends, or edges when tested in accordance with Methods C 473.

4.4 *Nail-Pull Resistance*—Specimens taken from the gypsum lath shall have an average nail-pull resistance of not less than the following when tested in accordance with Methods C 473:

Thickness, in. (mm)	Nail-Pull Resistance, lbf (N)
⅜ (9.5)	50 (222)
½ (12.7)	70 (312)

4.5 *Permeance*—Specimens taken from foil-backed gypsum lath shall have a permeance not to exceed 0.30 perm for the condition of 50 % relative humidity on Side I (the face of the lath) and 0 % relative humidity on Side II (the foil-covered back of the lath) when tested in accordance with the Desiccant Procedure of Test Methods E 96.

5. Dimensions and Permissible Variations

5.1 *Thickness*—The nominal thickness of gypsum lath shall be ⅜ or ½ in. (9.8 or 12.7 mm), with permissible variations in the nominal thickness of ±1/32 in. (±0.8 mm), and with permissible local variations ±1/16 in. (±1.6 mm) from the nominal thickness.

5.2 *Width*—The nominal width of gypsum lath shall be 16, 16⅓, 24, or 32 in. (406, 411, 610, or 813 mm), with permissible variations of 3/16 in. (4.8 mm) under and ⅛ in. (3.2 mm) over the specified width.

5.3 *Length*—The nominal length of gypsum lath shall be 32, 36, or 48 in. (813, 914, or 1219 mm), with permissible variations of ¼ in. (6.4 mm) under and ⅛ in. (3.2 mm) over the specified length. Lengths up to 12 ft (3.7 m) are available.

5.4 *Edges and Ends*—Edges and ends of gypsum lath shall be straight and solid. Edges shall be either round or square. The corners shall be square with a permissible variation of ±¼ in. (± 6.4 mm) in the full width of the lath.

6. Workmanship, Finish, and Appearance

6.1 The surfaces of gypsum lath shall be such that they will readily receive and retain gypsum plasters. Gypsum lath shall be free of any cracks and imperfections that would render it unfit for its intended use.

6.2 Gypsum lath having burred or broken corners shall be considered acceptable provided

that the burred or broken portion is not mo than 1 in. (25.4 mm) in either dimension.

7. Sampling

7.1 At least 0.25 % of the number of gypsu lath in a shipment, but not less than three lath shall be so selected as to be representative of th shipment and shall constitute a sample for pu pose of tests by the purchaser or user.

8. Inspection

8.1 Inspection of the material shall be agree upon between the purchaser and the supplier part of the purchase contract.

9. Rejection and Rehearing

9.1 When specified in the purchase order contract, material that fails to conform to th requirements of this specification may be re jected. Rejection should be reported to the pr ducer or supplier promptly and in writing. case of dissatisfaction with the results of the te the producer or supplier may make claim for rehearing.

10. Certification

10.1 When specified in the purchase order contract, a producer's or supplier's certificatio shall be furnished to the purchaser that the ma terial was manufactured, sampled, tested, an inspected in accordance with this specificatio and has been found to meet the specified require ments. When specified in the purchase order contract, a report of the test results shall b furnished.

11. Packaging and Package Marking

11.1 Gypsum lath shall be shipped so as to b kept dry and free of moisture.

11.2 When shipped for resale, the name of th manufacturer or supplier and the brand shall b legibly marked on each lath or package.

---

[1] This specification is under the jurisdiction of ASTM Com mittee C-11 on Gypsum and Related Building Materials an Systems and is the direct responsibility of Subcommittee C11.0 on Specifications and Test Methods for Gypsum Products.
Current edition approved May 15, 1987. Published Jun 1987. Originally published as C 37 – 21 T. Last previous editio C 37 – 84a.
[2] *Annual Book of ASTM Standards*, Vol 04.01.
[3] *Annual Book of ASTM Standards*, Vols 04.06, 08.03, and 15.09.

Summary of Changes

This section identifies the location of changes to this specification that have been incorporated since the last issue. Committee C-11 has highlighted those changes that affect the technical interpretation or use of this specification.

(1) Section 1.1, the word perforated was removed.   (2) Sections 3.4, 4.1.1, and 4.1.2 were deleted.

Designation: C 847 – 83

Standard Specification for
METAL LATH[1]

This standard is issued under the fixed designation C 847; the number immediately following the designation indicates the year of original adoption or, in the case of revision, the year of last revision. A number in parentheses indicates the year of last reapproval. A superscript epsilon (ε) indicates an editorial change since the last revision or reapproval.

This specification has been approved for use by agencies of the Department of Defense for listing in the DoD Index of Specifications and Standards.

1. Scope

1.1 This specification covers sheet lath, expanded metal lath, diamond mesh, flat and self-furring, and rib metal lath, ⅛, ⅜, and ¾-in. (3.2, 9.6, and 19.1-mm), all with or without backing and designed to be used as a base for gypsum or portland cement plaster.

2. Applicable Documents

2.1 *ASTM Standards:*
A 366/A 366M Specification for Steel, Sheet, Carbon, Cold-Rolled, Commercial Quality[2]
A 525 Specification for General Requirements for Steel Sheet, Zinc Coated (Galvanized) by the Hot-Dip Process[3]

3. Material

3.1 Metal lath shall be fabricated from cold-rolled carbon steel sheet of commercial quality conforming to Specification A 366/A 366M. Galvanized metal lath shall have a G60 coating in accordance with Specification A 525.

3.2 Backing shall be netting, film, paper, or felt attached to the lath sufficiently enough to prevent accidental removal during shipping, handling, or installation. Attachment of backing shall also allow lapping of metal to metal and backing to backing. 1 in. (25.4 mm) on the ends and ½ in. (12.7 mm) on the sides.

4. Dimensions, Weights, and Permissible Variations

4.1 *Thickness*—The nominal thickness of diamond mesh and flat rib metal lath shall be ⅛ in. (3.2 mm). The nominal thickness of other rib metal lath shall be as designated, ⅜ and ¾ in. (9.6 and 19.1 mm), respectively. The nominal thickness of self-furring diamond mesh shall be ⁵⁄₁₆ in. (7.9 mm).

4.2 *Width*—The nominal width of metal lath shall be 27 in. (686 mm) except for ¾-in. (19.1-mm) rib lath which shall be 24 in. (610 mm).

4.3 *Length*—The nominal length of metal lath shall be 96 in. (2438 mm).

4.4 *Weight*—The nominal weight of metal lath shall be as follows:

4.4.1 *U.S. Nominal Weights:*

Type:	Weight, lb/yd² (kg/m²)
Diamond mesh	2.5 (1.4); 3.4 (1.8)
Flat rib	2.75 (1.5); 3.4 (1.8)
⅜-in. rib	3.4 (1.8); 4.0 (2.1)
¾-in. rib	5.4 (2.9)
Sheet	4.5 (2.4)

4.4.2 *Canadian Nominal Weights:*

Type	Weight, lb/yd² (kg/m²)
Diamond mesh	2.5 (1.4); 3.0 (1.6); 3.4 (1.8)
Flat rib	2.5 (1.4); 3.0 (1.6)
⅜-in. rib	3.0 (1.5); 3.5 (1.9); 4.0 (2.1)
¾-in. rib	4.3 (2.3); 5.2 (2.8); 6.1 (3.3)

4.5 *Permissible Variations*—The permissible variations shall be as follows:
4.5.1 *Thickness,* ±¹⁄₆₄ in. (0.4 mm).
4.5.2 *Width,* ±³⁄₁₆ in. (4.8 mm).
4.5.3 *Length,* −0 in., +2½ in. (12.7 mm).
4.5.4 *Weight,* ±10 %

5. Finish

5.1 Metal lath shall be coated with a water barrier film such as asphalt or non-reemulsifiable water base paint unless fabricated from galvanized steel.

6. Inspection

6.1 Inspection of the material shall be agreed upon between the purchaser and supplier as part of the purchase contract.

7. Rejection and Rehearing

7.1 Any rejection shall be based upon the specific cause of failure to conform to the requirements of this specification, and shall be reported to the seller within 10 working days from the receipt of the shipment by the purchaser.

7.2 Claims for rehearing shall be valid only if made within 20 working days from receipt of notice of specific cause for rejection.

8. Certification

8.1 Where specified in the purchase order, a producer's or supplier's certification shall be furnished to the purchaser that the material was in accordance with this specification and has been found to meet the specified requirements.

9. Packaging and Package Marking

9.1 Metal lath shall be packaged ten sheets per bundle, except for ¾-in. (19.1-mm) rib lath which shall be six sheets per bundle and self-furring lath which shall be five or ten sheets per bundle (option of manufacturer).

9.2 When shipped for resale, the name of the manufacturer or the supplier and the brand shall be legibly marked on each lath or package.

[1] This specification is under the jurisdiction of ASTM Committee C-11 on Gypsum and Related Building Materials and Systems and is the direct responsibility of Subcommittee C11.02 on Specifications and Test Methods for Accessories and Related Products.
Current edition approved Aug. 26, 1983. Published October 1983. Originally published as C 847 – 77. Last previous edition C 847 – 82.
[2] *Annual Book of ASTM Standards,* Vol 01.03.
[3] *Annual Book of ASTM Standards,* Vol 01.06.

Designation: C 933 – 80 (Reapproved 1985)

Standard Specification for
WELDED WIRE LATH[1]

This standard is issued under the fixed designation C 933; the number immediately following the designation indicates the year of original adoption or, in the case of revision, the year of last revision. A number in parentheses indicates the year of last reapproval. A superscript epsilon (ε) indicates an editorial change since the last revision or reapproval.

1. Scope

1.1 This specification covers welded wire lath, flat or self-furring, with or without backing, designed for use as a base to receive portland cement-based interior plaster and exterior stucco.

2. Applicable Documents

2.1 *ASTM Standards*:

A 390 Specification for Zinc-Coated (Galvanized) Steel Poultry Netting (Hexagonal and Straight Line) and Woven Steel Poultry Fencing[2]

A 641 Specification for Zinc-Coated (Galvanized) Carbon Steel Wire[2]

2.2 *Federal Specifications:*[3]

UU-B-790 Building Paper, Vegetable Fiber: (Kraft, Waterproofed, Water Repellent and Fire Resistant)

3. Material

3.1 Welded wire lath shall be fabricated from not less than 0.0625 in. (1.588 mm), copper-bearing, cold-drawn, galvanized steel wire, conforming to Specification A 641.

3.2 Where specified, the backing shall be netting, film, paper or felt attached to the lath to prevent accidental removal during shipment, handling, or installation. Attachment of the backing shall allow lapping of wire to wire and backing to backing of one mesh at ends and edges and shall permit full embedment, in at least ⅛ in. (3.2 mm) of plaster, of at least one half of the total length and width of the wire. The backing shall be absorptive or water resistant as specified for intended use. The minimum bursting strength of backing shall be sufficient to maintain backing integrity under normal hand or machine application pressures.

4. Dimensions, Weights, and Permissible Variations

4.1 *Openings and Stiffening:*

4.1.1 Weld the welded wire lath at all intersections of wire to form openings not to exceed 2 in. (50.8 mm) and stiffen continuously, parallel to the long dimension of the lath at not more than 6-in. (152.4-mm) intervals with minimum 0.0720-in. (1.829-mm) thick wire.

4.1.2 Where lath is to be used over solid backing, self-furring crimps for welded wire laths shall provide a ¼-in. (6.4-mm) projection of the plane of the back of the lath from the plane of the solid backing.

4.2 *Thickness, nominal,* for welded wire laths shall be ⅛ in. (3.2-mm) for all types, exclusive of furring crimps.

4.3 *Width, nominal,* shall be 28 in. (0.71 mm).

4.4 *Length, nominal,* lath shall be 96 in. (2.44 m).

4.5 *Weight, nominal,* shall be 1.14 to 1.83 lb/yd² (0.618 to 0.993 kg/m²).

4.6 *Permissible Variations*—The permissible variations shall be as follows:

Thickness	+ 5/32 in. (4.0 mm)
	− 3/32 in. (2.4 mm)
Width	± ¾ in. (19.0 mm)
Length	± ¾ in. (19.0 mm)

NOTE— *Weight*—The allowable variations in percentage over and under the nominal weights of welded wire laths shall be in accordance with mill tolerances. Weights do not include weight of backing material.

5. Finish

5.1 The wire shall be zinc-coated (galvanized) in accordance with Specification A 390, Class 1.

6. Packaging and Marking

6.1 *Packaging*—Package in 8-ft (2.438 m) lengths with 12 sheets per bundle (25 yd²) (20.903 m²) or 4-ft (1.219 m) lengths with 19 sheets per bundle (20 yd²) (16.723 m²).

6.2 *Marking*—Lath with no backing shall have metal or paper tags securely attached to the bundles with the product description and name of manufacturer, or the paperbacked lath shall have this information imprinted on the backing of each sheet of lath.

---

[1] This specification is under the jurisdiction of ASTM Committee C-11 on Gypsum and Related Building Materials and Systems and is the direct responsibility of Subcommittee C 11.02 on Specifications and Test Methods for Accessories and Related Products.

Current edition approved March 28, 1980. Published May 1980.

[2] 1983 Annual Book of ASTM Standards, Vol 01.06.

[3] Available from Naval Publications and Forms Center, 5801 Tabor Ave., Philadelphia, PA 19120.

Designation: C 28 – 86

Standard Specification for
GYPSUM PLASTERS[1]

This standard is issued under the fixed designation C 28; the number immediately following the designation indicates the year of original adoption or, in the case of revision, the year of last revision. A number in parentheses indicates the year of last reapproval. A superscript epsilon (ε) indicates an editorial change since the last revision or reapproval.

1. Scope

1.1 This specification covers four gypsum plasters which are specified in the following order:

	Sections
Gypsum ready-mixed plaster	5
Gypsum neat plaster	6
Gypsum wood-fibered plaster	7
Gypsum gauging plaster for finish coat	8

1.2 The values stated in inch-pound units shall be regarded as the standard.

2. Referenced Documents

2.1 *ASTM Standards:*

C 11 Definitions of Terms Relating to Gypsum and Related Building Materials and Systems[2]

C 35 Specification for Inorganic Aggregates for Use in Gypsum Plaster[2]

C 471 Methods for Chemical Analysis of Gypsum and Gypsum Products[2]

C 472 Methods for Physical Testing of Gypsum Plasters and Gypsum Concrete[2]

C 778 Specification for Standard Sand[2]

C 842 Specification for Application of Interior Gypsum Plaster[2]

E 11 Specification for Wire-Cloth Sieves for Testing Purposes[3]

3. Terminology

3.1 *Definitions*—Definitions used in this standard shall be in accordance with Definitions C 11.

3.2 *Descriptions of Terms Used in This Standard:*

3.2.1 *gypsum gauging plaster for finish coat*—a calcined gypsum plaster designed to be mixed with lime putty for the finish coat.

3.2.2 *gypsum neat plaster*—calcined gypsum mixed at the mill with other ingredients to control working quality and setting time. Neat plaster may be fibered or unfibered. The addition of aggregate is required on the job.

3.2.3 *gypsum ready-mixed plaster*—calcined gypsum plaster, mixed at the mill with a mineral aggregate, designed to function as a base coat to receive various finish coats. It may contain other materials to control setting time and other desirable working properties.

NOTE 1—For determination of aggregate content of ready-mixed plasters see the Appendix.

3.2.4 *gypsum wood-fibered plaster*—A calcined gypsum plaster in which wood fiber is used as an aggregate.

4. Materials

4.1 *Calcined Gypsum*—Calcined gypsum shall have a purity of not less than 66.0 weight % of $CaSO_4.1/2H_2O$.

4.2 *Vermiculite Aggregate*, shall be in accordance with Specification C 35.

4.3 *Perlite Aggregate*, shall be in accordance with Specification C 35.

4.4 *Sand Aggregate*, shall be in accordance with Specification C 35.

4.5 *Ottawa Sand*, shall be in accordance with Specification C 778.

4.6 *Wood Fiber*, shall be in accordance with nonstaining wood fiber.

5. Gypsum Ready-Mixed Plaster

5.1 *Ready-Mixed Plaster With Vermiculite Aggregate For Use Over Lath Bases:*

5.1.1 *Composition*—It shall contain not more than 2 ft³ (0.057 m³) of aggregate per 100 lb (45 kg) of calcined gypsum plaster.

5.1.2 *Compressive Strength*, shall be not less than 450 psi (3.1 MPa).

5.2 *Ready-Mixed Plaster With Perlite Aggregate For Use Over Lath Bases:*

5.2.1 *Composition*—It shall contain not more than 2 ft³ (0.057 m³) of aggregate per 100 lb (45 kg) of calcined gypsum plaster.

5.2.2 *Compressive Strength*, shall be not less than 600 psi (4.1 MPa).

5.3 *Ready-Mixed Plaster With Sand Aggregate For Use Over Lath Bases:*

5.3.1 *Composition*—It shall contain not more than 2½ ft³ (0.071 m³) of aggregate per 100 lb (45 kg) of calcined gypsum plaster.

5.3.2 *Compressive Strength*, shall be not less than 700 psi (4.8 MPa).

5.4 *Ready-Mixed Plaster With Sand or Perlite Aggregate for Use Over Masonry Bases:*

5.4.1 *Composition*—It shall contain not more than 3 ft³ (0.085 m³) of aggregate per 100 lb (45 kg) of calcined gypsum plaster.

5.4.2 *Compressive Strength*, shall be not less than 400 psi (2.8 MPa).

5.5 *Read-Mixed Plaster With Vermiculite Aggregate For Use Over Masonry Bases:*

5.5.1 *Composition*—It shall contain not more than 3 ft³ (0.085 m³) of aggregate per 100 lb. (45 kg) of calcined gypsum plaster.

5.5.2 *Compressive Strength*, shall be not less than 325 psi (2.2 MPa).

5.6 *Time of Setting*—Gypsum ready-mixed plasters shall have a time set of not less than 1½ nor more than 8 h (Note 2).

NOTE 2—*Time of Setting*—Attention is directed to conditions affecting job set of gypsum plasters that are usually beyond the control of the producer. Materials added at the site of application such as water and aggregates affect job sets. In addition, the various bases with different absorptive values may cause a variation in job set performance. Attention is directed to the requirements of Specification C 842, 9.1.1, which states "Accelerate the plaster, if necessary, to provide a setting time on the jobs of not more than 4 h beginning with the addition of the mixing water to the batch of dry plaster."

6. Gypsum Neat Plaster

6.1 *Time of Setting*—When mixed with 3 parts by weight of standard Ottawa sand, the time of set shall be not less than 2 nor more than 16 h (Note 2).

6.2 *Compressive Strength*—When mixed with 2 parts by weight of Ottawa sand, the compressive strength shall be not less than 750 psi (5.2 MPa).

7. Gypsum Wood-Fibered Plaster

7.1 *Composition*—Gypsum wood-fibered plaster shall be calcined gypsum and wood fiber made from a nonstaining wood.

7.2 *Time of Setting*, shall be not less than 1½ nor more than 8 h (Note 2).

7.3 *Compressive Strength*, shall be not less than 1200 psi (8.3 MPa).

8. Gypsum Gauging Plaster for Finish Coat

8.1 *Composition*—Gypsum gauging plaster for finish coat shall be calcined gypsum, with or without materials to control setting time and working quality.

8.2 *Fineness*—Gypsum gauging plaster for finish coat all shall pass a No. 14 (1.4 mm) sieve and not less than 60% shall pass a No. 100 (150 μm) sieve.[4]

8.3 *Time of Setting*—When not retarded, the time of set shall be not less than 20 nor more than 40 min and, when retarded, not less than 40 min (Note 2).

8.4 *Compressive Strength*, shall be not less than 1200 psi (8.3 MPa).

9. Sampling

9.1 At least 1 % of the packages, but not less than 5 packages, shall be sampled. Packages to be sampled shall be selected at random. Samples shall be taken both from the outer portion and the center of each package. The materials so obtained shall be thoroughly mixed to provide a composite sample of not less than 15 lb (6.8 kg). This composite sample shall be placed immediately in a clean, dry, airtight container for delivery to the laboratory.

10. Test Methods

10.1 The chemical analysis and physical properties of gypsum plaster shall be determined in accordance with Methods C 471 and C 472.

11. Packaging and Package Marking

11.1 Gypsum plasters shall be dry and free of lumps, and shall be shipped in packages.

11.2 When shipped for resale, the following information shall be legibly marked on each package or on a tag of suitable size attached to the package:

11.2.1 Name of manufacturer or seller,

11.2.2 Brand, and

11.2.3 Net weight of the package.

12. Inspection

12.1 Inspection of the gypsum plaster shall be

---

[1] This specification is under the jurisdiction of ASTM Committee C-11 on Gypsum and Related Building Materials and Systems and is the direct responsibility of Subcommittee C11.01 on Specifications and Test Methods for Gypsum Products.

Current edition approved Sept. 26, 1986. Published November 1986. Originally published as C 28 – 20 T. Last previous edition C 28 – 80.

[2] *Annual Book of ASTM Standards*, Vol 04.01.

[3] *Annual Book of ASTM Standards*, Vol 04.01 and 14.02.

[4] Detailed requirements for these sieves are given in Specification E 11.

agreed upon between the purchaser and the supplier as part of the purchase order.

13. Rejection

13.1 Gypsum plaster that fails to conform to the requirements of this specification may be rejected. Rejection shall be reported to the producer or supplier in writing within 10 working days from the receipt of the shipment by the purchaser. The notice of rejection shall contain a specific statement as to the respects in which the plaster has failed to conform to the require-

ments of this specification. In case of dissatisfaction with the results of the test, at the request of the producer or supplier, such notice of rejection shall be supported by results of a test conducted by a mutually agreeable independent laboratory.

14. Certification

14.1 When specified in the purchase order, a producer's or supplier's certification shall be furnished to the purchaser that the product is in compliance with this specification.

APPENDIX

(Nonmandatory Information)

X1. DETERMINATION OF AGGREGATE CONTENT OF READY-MIXED PLASTER

X1.1 The determination of the aggregate content of ready-mixed plaster is normally not precise. Many gypsum plasters contain natural impurities that cannot be separated from the aggregate by either mechanical or chemical means. Specimens of the component parts of the ready-mixed plaster are required to accurately determine the aggregate content (Determination of Sand in Set Plaster of Methods see C 471).

X1.1.1 Reasonable estimates of the aggregate content may be determined by either mechanical or chemical separation. Two methods which may be used to estimate the aggregate content of the ready-mixed plaster follow.

X1.2 *Mechanical Separation:*

X1.2.1 It is assumed in this method that the aggregate meets the specifications of C 35 and that only a minor portion of the aggregate passes through a 100-mesh (150-μm) sieve.

X1.2.2 Sieve 100 g of the plaster through a 100-mesh (150-μm) sieve. The material retained on the sieve is assumed to be aggregate. Examine this portion to be sure small lumps or agglomerates of plaster are not retained.

X1.2.3 If the weight ratio of the aggregate and plaster is to be determined, weigh each portion and calculate the ratio as pounds of aggregate/100 lbs (45 kg) of plaster.

X1.2.4 If the volume of aggregate to the weight of plaster ratio is to be determined, measure the volume of aggregate retained on the sieve and weigh the material passing through the sieve. Calculate the ratio as cubic feet of aggregate/100 lbs (45 kg) of plaster.

X1.3 *Chemical Separation:*

X1.3.1 It is assumed in this method that all of the insoluble in ammonium acetate material is aggregate and the soluble material is plaster. The following procedure is used:

X1.3.2 *Reagents:*

X1.3.2.1 *Ammonium Acetate* (250 g/L)—dissolve 250 g of ammonium acetate ($NH_4C_2H_3O_2$) in water and dilute to 1 L.

X1.3.2.2 *Ammonium Hydroxide* (1 + 59)—Mix 1 volume of concentrated (NH_4OH) (sp gr 0.90) with 59 volumes of water.

X1.3.3 *Procedure:*

X1.3.3.1 Weigh accurately 40 ± 0.05 g of the calcined sample into a 1-L beaker. Add 600 to 700 mL of $NH_4C_2H_3O_2$ solution, that is slightly alkaline to litmus paper. If acidic, add a few millilitres of NH_4OH (1 + 59) to the stock $NH_4C_2H_3O_2$ solution to render it slightly alkaline prior to the addition to the test sample.

X1.3.3.2 Warm the suspension to a temperature of 70 ± 5°C and stir continuously for 20 to 30 min. Filter the warm suspension with the aid of suction through a small Buchner funnel or Gooch crucible in which filter paper has previously been formed, the funnel and mat having been dried at 110°C to a constant weight within 0.01 g. Refilter the first 100 mL of the filtrate. Wash the aggregate remaining in the beaker onto the filter with an additional 100 mL of warm ammonium acetate solution. Wash the beaker and residue with 200 to 300 mL of water, dry the funnel and aggregate at 100°C to constant weight. The weight of the residue is the weight of the aggregate.

X1.3.4 *Calculation:*

X1.3.4.1 *Weight to Weight Ratio*—The weight of the plaster is the original sample weight less the weight of the aggregate. Calculate the ratio as pound of aggregate/100 lbs (45 kg) of plaster.

X1.3.4.2 *Volume of Aggregate to Weight of Plaster Ratio*—The volume of aggregate recovered is determined. The weight of the plaster is the original sample weight less the weight of the aggregate. Calculate the ratio as cubic feet of aggregate/100 lbs (45 kg) of plaster.

Designation: C 472 – 84

Standard Methods for

PHYSICAL TESTING OF GYPSUM PLASTERS AND GYPSUM CONCRETE[1]

This standard is issued under the fixed designation C 472; the number immediately following the designation indicates the year of original adoption or, in the case of revision, the year of last revision. A number in parentheses indicates the year of last reapproval. A superscript epsilon (ε) indicates an editorial change since the last revision or reapproval.

These methods have been approved for use by agencies of the Department of Defense and for listing in the DoD Index of Specifications and Standards.

1. Scope

1.1 These methods cover the physical testing of gypsum plasters and gypsum concrete.

1.2 The methods appear in the following order:

	Sections
Precautions for Physical Tests	4
Free Water	5
Fineness	6
Plaster	7 and 8
Normal Consistency of Gypsum Concrete	9 and 10
Time of Setting	11 and 12
Time of Setting (Alternative IV Method)	13 and 14
Compressive Strength	15 to 18
Density	19

1.3 *This standard may involve hazardous materials, operations, and equipment. This standard does not purport to address all of the safety problems associated with its use. It is the responsibility of whoever uses this standard to consult and establish appropriate safety and health practices and determine the applicability of regulatory limitations prior to use. For a specific precautionary statement, see X1.2.1.*

2. Applicable Documents

2.1 *ASTM Standards:*
 C 11 Definitions of Terms Relating to Gypsum and Related Building Materials and Systems[2]
 E 11 Specification for Wire-Cloth Sieves for Testing Purposes[3]

3. Definitions

3.1 For useful definitions refer to Definition of Terms C 11.

4. Precautions for Physical Tests

4.1 Gypsum products are peculiar in that their properties are very greatly affected by the small amounts of impurities that may be introduced by careless laboratory manipulation. In order to obtain concordant results, it is, therefore, absolutely essential to observe the following precautions:

4.1.1 All apparatus shall be kept thoroughly clean. Especially, all traces of set plaster shall be removed. For mixing pastes and mortars, a 500-mL rubber dental bowl is a convenience.

4.1.2 Distilled water, free of chlorides and sulfates, shall be used exclusively for mixing putties and mortars. Tap water may be used in the tests for fineness, wood fiber, and water resistance.

4.1.3 Standard sand shall be used exclusively and shall consist of a natural silica sand from Ottawa, Ill., graded to pass a No. 20 (850-µm) sieve and to be retained on a No. 30 (600-µm) sieve. This sand shall be considered standard when not more than 15 g are retained on the No.

20 sieve and not more than 5 g pass the No. 30 sieve after 5 min of continuous sieving of a 100-g sample. Store in closed containers.

FREE WATER

5. Procedure

5.1 Weigh a sample of not less than 1 lb (0.45 kg) of the material as received and spread it out in a thin layer in a suitable-vessel. Place in an oven and dry at 113°F (45°C) for 2 h; then cool in an atmosphere free from moisture and weigh again. The loss of weight corresponds to the free water and shall be calculated as a percentage of the sample as received.

5.2 Retain the dried sample in an airtight container until used for the fineness test (Section 6).

FINENESS

6. Procedure

6.1 Determine fineness by sieving a known weight of the dried sample through sieves of the specified sizes (Notes 1 and 2). The size of the sample to be used in determining fineness depends upon the particle size of the material. If the material will pass a ¼-in. (6.4-mm) sieve,[4] a 100-g sample will be sufficient; if the largest particles are more than 1 in. (25.4 mm) in diameter, use at least a 1000-g sample. With these limitations the size of sample to be used is left to the discretion of the operator. Shake the sample through each sieve with as little abrasion as possible (Note 3). Weigh the amount of material retained on each sieve and calculate the fineness, expressed as a percentage of the weight of the original sample.

NOTE 1—For suggested method of sieving gypsum through a No. 325 (45-µm) sieve, see Appendix XI.
NOTE 2—The sizes of the sieves to be used are given in the specifications of ASTM covering the particular product in question.
NOTE 3—When sieving through a No. 100 sieve, use a lateral motion, and tap the side of the sieve with the palm of the hand. Continue without brushing until not more than 0.5 g passes through during 1 min of sieving. If the sieve openings become clogged, transfer the retained material temporarily to another vessel, invert the sieve over a sheet of paper on the table and tap it sharply against the table. Then transfer all the retained material back into the sieve and continue sieving.

NORMAL CONSISTENCY OF GYPSUM PLASTER

NOTE 4—Since accuracy in determining normal consistency is most important in the standardizing of physical methods for testing cementitious materials, it is essential that this test be performed with great care.

7. Apparatus

7.1 *Modified Vicat Apparatus*—The modified Vicat apparatus (Fig. 1)[5] shall consist of a bracket, *A*, bearing a movable brass rod, *B*, 6.3 mm in diameter and of suitable length to fit the Vicat

Bracket. On the lower end of the rod shall be attached a conical plunger, *C*, made of aluminum with an apex angle of 53° 08 min and a height of 45 mm. The total weight of the rod and plunger shall be 35 g. This total weight may be increased by means of a weight, *G*, screwed into the rod. The rod can be held in any desired position by a screw, *E*. The rod shall have a mark, *D*, midway between the ends which moves under a scale, *F*, graduated in millimetres, attached to the bracket, *A*.

7.2 *Mold*—The conical ring mold shall be made of a noncorroding, nonabsorbent material, and shall have an inside diameter of 60 mm at the base and 70 mm at the top and a height of 40 mm.

7.3 *Base Plate*—The base plate for supporting the ring mold shall be of plate glass and about 100 mm square.

8. Procedure

8.1 Clean the plunger, mold, and base plate of the modified Vicat apparatus. Apply a thin coat of petroleum jelly or other suitable lubricant to the upper surface of the base plate in order to prevent leaks during the test.

8.2 Sift a weighed quantity of the sample (200 to 300 g as required to fill the mold) into a known volume of water. If the plaster is unretarded, add to the mixing water 0.2 g of sodium citrate per 100 g of sample. After allowing the sample to soak for 2 min, stir the mixture for 1 min to an even fluidity. Pour this sample into the ring mold, work slightly to remove air bubbles, and then strike off flush with the top of the mold. Wet the plunger of the modified Vicat apparatus and lower it to the surface of the sample at approximately the center of the mold. Read the scale and release the plunger immediately. After the rod has settled, read the scale again. Readings are reproducible on a retarded mix, and therefore, in order to eliminate error, two or three determinations should be made on each mix, care being taken to have the mold completely filled and the plunger clean and wet.

8.3 Gypsum molding plaster and gypsum gauging plaster shall be considered of normal consistency when a penetration of 30 ± 2 mm is obtained when tested in accordance with 8.1 and

[1] These methods are under the jurisdiction of ASTM Committee C-11 on Gypsum and Related Building Materials and Systems and are the direct responsibility of Subcommittee C11.01 on Specifications and Test Methods for Gypsum Products.
Current edition approved Aug. 31, 1984. Published November 1984. Originally published as C 472 – 61. Last previous edition C 472 – 79ε¹.
[2] *Annual Book of ASTM Standards*, Vol 04.01.
[3] *Annual Book of ASTM Standards*, Vols 04.01 and 14.02.
[4] Detailed requirements for this sieve are given in Specification E 11.
[5] This method is described by Kuntze, R. A., "An Improved Method for the Normal Consistency of Gypsum Plasters," *ASTM Bulletin No. 246*, ASTM, May 1960, p. 35.

8.2, the weight of the rod and plunger for this determination to be 35 g. Normal consistency shall be expressed as the number of millilitres of water required to be added to 100 g of the gypsum.

8.4 All gypsum mixtures containing aggregates shall be considered of normal testing consistency when a penetration of 20 ± 3 mm is obtained when tested in accordance with 8.1 and 8.2, the weight of the rod and plunger for these determinations to be 50 g. Normal consistency shall be expressed as the number of millilitres of water required to be added to 100 g of the mixture.

NOTE 5—Gypsum neat plaster is tested for compressive strength (see 14.1) at a normal consistency determined after the plaster is first mixed with standard sand in the ratio of 200 g of sand to 100 g of plaster.

NORMAL CONSISTENCY OF GYPSUM CONCRETE

9. Apparatus

9.1 *Consistometer* (Fig. 2)—The consistometer consists of a conical vessel made of noncorroding, nonabsorbent material, and having an inside diameter of 9 in. (229 mm) at the top and 1¾ in. (44.5 mm) at the bottom, and a height of 5½ in. (139.7 mm). It shall be provided with a sliding gate at the bottom and supported so that the bottom is 4 in. (102 mm) above the base plate. The base plate shall be of plate glass, free of scratches and about 18 in. (457 mm) square.

10. Procedure

10.1 The consistometer and the base plate shall be clean and dry and the sliding gate shall be closed.

10.2 Sift 2000 g of the sample into a known volume of water to which 1.0 g of sodium citrate has previously been added. After allowing the sample to soak for 1 min, stir the mixture for 3 min to an even fluidity. Pour the mixture into the consistometer until level with the top. Then rapidly and completely open the sliding gate, allowing the mixture to run out freely upon the base plate. When the sliding gate is opened, take care to avoid jarring the consistometer.

10.3 Measure the resulting patty on the base plate along its major and minor axes and determine the average diameter.

10.4 Gypsum concrete shall be considered of normal consistency when a patty diameter of 15 ± ½ in. (380 ± 12.7 mm) is obtained when tested in accordance with 10.1 to 10.3. Normal consistency shall be expressed as the number of millilitres of water required to be added to 100 g of the gypsum concrete.

TIME OF SETTING

11. Apparatus

11.1 *Vicat Apparatus*—The Vicat apparatus (Fig. 3) shall consist of a frame, *A*, bearing a movable rod, *B*, weighing 300 g, one end, *C*, the plunger end, being 10 mm in diameter for a distance of at least 50 mm, the other end having a removable needle. *D*, 1 mm in diameter and 50 mm in length. The rod, *B*, shall be reversible, shall be able to be held in any desired position by a screw, *E*, and shall have an adjustable indicator, *F*, that moves over a scale (graduated in millimetres) attached to the frame, *A*. The paste shall be held in a rigid conical ring, *G*, resting on

a glass plate, *H*, about 100 mm square. The ring shall be made of a noncorroding, nonabsorbent material and shall have an inside diameter of 70 mm at the base and 60 mm at the top, and a height of 40 mm.

11.2 In addition, the Vicat apparatus shall conform to the following requirements:

Weight of plunger	300 ± 0.5 g
Diameter of larger end of plunger	10 ± 0.05 mm
Diameter of needle	1 ± 0.05 mm
Inside diameter of ring at bottom	70 ± 3 mm
Inside diameter of ring at top	60 ± 3 mm
Height of ring	40 ± 1 mm
Graduated scale	The graduated scale, when compared with a standard scale accurate to within 0.1 mm at all points, shall not show a deviation at any point greater than 0.25 mm.

12. Procedure

12.1 *Gypsum Concrete and All Gypsum Plasters, Except Gypsum Neat Plaster*—Starting the timing of the test approximately at the moment of contact of the dry material with the water, mix 200 g of the sample to make a paste of normal consistency. For the quantity of water and directions for mixing, see Sections 7 to 10, except that no retarder shall be added. When conducting the test, allow the needle to sink into the paste. After each penetration, wipe the needle clean, and move the paste slightly so that the needle will not strike the same place twice. The frequency of the penetration will depend upon the character of the material. Test the sample at such intervals as are necessary to determine whether it complies with the requirements for time of setting for the product tested. Set shall be considered complete when the needle no longer penetrates to the bottom of the paste. Until set, store the test specimens in a cabinet at a temperature of not less than 68°F (20°C) nor more than 72°F (22°C) in an atmosphere having a relative humidity of not less than 85 %. Record as the time of setting of the sample the elapsed time in minutes from the time when the sample was first added to the water to the time when set is complete.

12.2 *Gypsum Neat Plaster*—Test gypsum neat plaster for time of setting as mixed with three parts by weight of standard sand as described in 4.3. Mix dry a 100-g sample of the gypsum neat plaster and 300 g of the standard sand, and then add sufficient water to produce a stiff yet workable mortar. Stir for 1 min, to an even, lump-free consistency. Place the mortar in the conical ring and test for time of setting as described in 12.1.

12.3 *Frequency of Testing*—The frequency of testing of the materials shall be as follows:

Kind of Material	First Test	Frequency of Subsequent Tests
Molding plaster	15 min	5 min
Keene's cement:		
Standard	15 min	1 h
Quick set	15 min	5 min
Gypsum concrete	15 min	5 min
Ready mixed plaster	1½ h	1 h
Neat plaster	2 h	1 h
Wood-fibered plaster	1½ h	1 h
Bond plaster	1½ h	1 h
Gauging plaster:		
Slow set	40 min	2 h
Quick set	15 min	5 min

TIME OF SETTING
(Alternative IV Method)

13. Apparatus

13.1 *Potentiometer*—A single- or multiple-channel recording potentiometer or thermistor bridge shall be used to record the temperature change of the sample under test. The chart speed shall be at least 1 in. (25 mm)/h. Imprints recording the temperature shall not be longer than 1 min apart for each sample.

13.2 *Temperature Sensors*—Temperature changes may be indicated by either thermocouples or thermistors which may be movable or in a fixed position. Temperature sensing elements shall be of such capacity and sensitivity that, when connected to the recording potentiometer, a temperature change of 1°F (0.5°C) in the sample shall be recorded on the chart.

13.3 *Sample Containers and Test Conditions*—Waxed paper cups from 6 to 9-oz (178 to 268-mL) capacity shall be used. The cup containing the mixture under test shall be placed inside a matching paper cup held in an insulated block or beaker; the movable temperature sensor, in this case, shall be positioned ¼ to ⅓ the distance up from the bottom and between the inner and outer cup. Alternatively, the cup containing the test mixture may be positioned over a spring-loaded sensor to assure close contact with the bottom of the cup. Tests shall be made in a room or cabinet maintained at a temperature of 68°F (20°C) to 72°F (22°C). Materials and mixing water used for the test shall be at 68°F to 72°F.

NOTE 6—If a constant-temperature cabinet is not available, a constant-temperature water bath may be fitted with a cover which will admit the body of the cup holder but not its rim, so that the cup holder is in contact with the water in the bath.

14. Procedure

14.1 *Gypsum Concrete and All Gypsum Plaster, Except Gypsum Neat Plaster*—Mix 200 g of the dry sample to a paste of normal consistency in accordance with Sections 5 to 8 except that no retarder shall be added. Place the mixture in a clean dry waxed paper cup to about ¼ in. (19 mm) from the top. Place the cup into the empty cup in the cup holder and adjust the sensing element as required in 13.3, or place the cup upon a spring-loaded sensor. Cover the cup with a watch glass. Record as the time of setting of the sample the elapsed time in minutes from the time when the sample was first added to the water to the time of maximum temperature rise.

14.2 *Gypsum Neat Plaster*—Prepare the mixed plaster as required in 12.2. Fill a waxed paper cup and test as outlined in 14.1.

COMPRESSIVE STRENGTH

15. Specimen Molds

15.1 Molds for making test specimens shall be 2-in. (50.8 mm) split cube molds made of non-corrodible material and shall be sufficiently rigid to prevent spreading during molding. The molds shall have not more than three cube compartments and shall be separable into not more than two parts. When assembled, the parts of the molds shall be held firmly together, and dimensions shall conform to the following requirements: Interior faces shall be plane surfaces with a maximum variation of 0.001 in. (0.03 mm) for new molds and 0.002 in. (0.05 mm) for old molds; distance between opposite faces, and height of the molds, measured separately for each cube compartment, shall be 2 ± 0.005 in. (50.8 ± 0.13 mm) for new molds or 2 ± 0.020 in. (50.8 ± 0.51 mm) for old molds, angle between adjacent interior faces and between interior faces and top and bottom planes of the mold shall be 90 ± 0.5°, measured at points slightly removed from the intersection of the faces.

16. Test Specimens

16.1 Mix sufficient sample at normal consistency to produce not less than 1000 mL of mixed mortar and cast into six 2-in. (50.8-mm) split cube molds. Neat gypsum plaster shall be premixed dry with two parts by weight of standard sand (see 4.3). For the quantity of water, see Sections 7 to 10, except that no retarder shall be added. The temperature of the water shall be 70°F (21°C). Place the required amount of water in a clean 2- qt (2-L) mixing bowl. For all gypsum plasters except gypsum concrete, add the required amount of dried plaster and allow to soak for 2 min. Mix vigorously for 1 min with a metal spoon or stiff-bladed spatula to produce a mortar of uniform consistency. For gypsum concrete, soak for 1 min, and stir vigorously (about 150 complete circular strokes per minute) with a large metal spoon for 3 min. Setting time of the mortar shall be within the time limits shown in Table 1. If setting time is in excess of the maximum limit, the cubes shall be discarded; and the next mix shall be adjusted by adding freshly ground terra alba to the plaster, except for Keene's cement to which molding plaster shall be added to reduce setting time within the limits shown. The amount of terra alba or molding plaster used to accelerate set shall not exceed 1 % of the weight of the plaster or Keene's cement. Report dry density of the cubes.

16.2 Coat the molds with a thin film of mineral oil and place on an oiled glass or metal plate. Place a layer of mortar about 1 in. (25 mm) in depth in each mold and puddle ten times across the mold between each pair of opposite faces with a 1 in. wide metal spatula to remove air bubbles. Fill the molds to a point slightly above the tops of the molds, by the same filling and puddling procedure used for the first layer. Also fill the conical mold for the Vicat apparatus described in 11.1 and 11.2 and determine when the mortar or paste has set by allowing the needle to sink into the paste in the manner described in 12.1. After the mortar or paste has set, cut off the excess to a plane surface flush with the top of the mold, using a broad knife or similar implement. Place the filled molds in moist air (90 to 100 % relative humidity). The cubes may be removed from the molds as soon as thoroughly hardened, but must be retained in the moist air not less than a total of 24 h. Place the cubes in an oven provided with air circulation and adequate ventilation for removal of moisture, so that the air may be maintained at a temperature of 90 to 110°F (32 to 43°C) and a relative humidity not to exceed 50 %. Dry the cubes to constant weight as determined by weighing once each day but not to exceed 7 days, and then place in a desiccator over magnesium perchlorate for 24 h before testing. Test the cubes as soon as removed from the desiccator.

17. Procedure

17.1 As soon as the cube specimens have been dried (Section 16) determine their compressive strengths. Position the cubes in the testing machine so that the load is applied on surfaces formed by faces of the molds, not on top and bottom. Apply the load continuously and without shock, at a constant rate within the range 15 to 40 psi (103 to 276 kPa)/s. During application of the first half of the maximum load a higher rate of loading shall be permitted.

18. Report

18.1 Report the average compressive strength as the compressive strength of the material, except that if the strengths of one or two of the cubes vary more than 15 % from the average of the five, discard them and report the compressive strength as the average of the remaining specimens. In case the compressive strengths of three or more cubes vary more than 15 % from the average, discard the results and repeat the test.

DENSITY

19. Procedure

19.1 Determine the density of gypsum concrete, expressed in pounds per cubic foot, by multiplying by 0.095 the total weight in grams of the five compressive strength cube specimens, after drying (Section 16).

TABLE 1 Setting-Time Limits for Mortar

Kind of Material	Setting Time, min	
	min	max
Molding plaster	20	140
Keene's cement:		
Standard	40	120
Quick set	20	40
Gypsum concrete	20	40
Ready-mixed plaster	90	120
Neat plaster (with 2 parts sand)	120	150
Wood-fibered plaster	90	120
Bond plaster	120	150
Gauging plaster:		
Slow set	40	120
Quick set	20	40
Veneer plaster	30	90

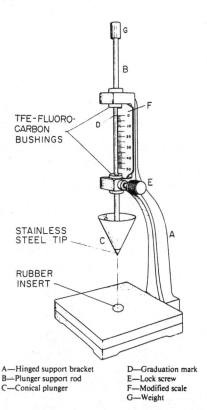

A—Hinged support bracket
B—Plunger support rod
C—Conical plunger
D—Graduation mark
E—Lock screw
F—Modified scale
G—Weight

FIG. 1 Modified Vicat Apparatus (Conical Plunger Method)

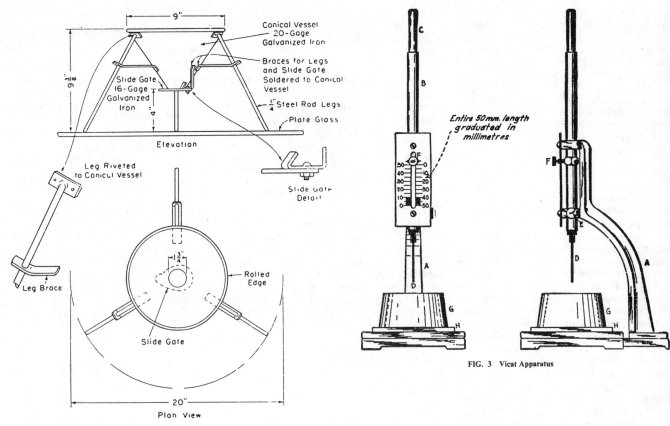

FIG. 2 Consistometer

FIG. 3 Vicat Apparatus

APPENDIX

(Nonmandatory Information)

X1. ALCOHOL WASH METHOD OF SIEVING GYPSUM AND GYPSUM PRODUCTS

X1.1 It is impracticable to sieve dry gypsum through a No. 325 (45-μm) sieve, and water cannot be used as a washing agent without introducing errors due to hydration or solution.

X1.2 Use isopropyl alcohol (99 %) as the washing agent (see X1.2.1). Place the alcohol in a vessel 1 to 2 in. (25.4 to 50.8 mm) larger in diameter than the sieve, to a depth above 2 in. Place 50 ± 0.01 g of the sample on a No. 325 (45-μm) sieve or, if desired, a coarser sieve up to and including the No. 100 (150-μm) sieve. Lower the sieve into the alcohol until the sample is wetted throughout. Lift the sieve out of the alcohol, with a swirling motion, permitting the alcohol to drain through the sample back into the vessel. Repeat this operation at least eight times, until the alcohol passes freely through the sieve and the sample is essentially free from fines. Wash the residue retained on the sieve with about 100 mL of clear alcohol, and then blot the bottom of the sieve with a soft, dry, lint-free cloth. Dry the sieve with the residue at 100 to 125°F (38 to 53°C). Shake the sieve on a mechanical shaker (Ro-Tap or equivalent) for 2 min ± 2 s. Weigh the material retained on the sieve. Multiply by 2 the weight of material retained after shaking in order to obtain percentage of the original sample. If desired, to permit simultaneous determinations of coarser fractions, the residue after drying and before shaking may be transferred to a series of coarser sieves up to and including the No. 100, nested in sequential order above the No. 325 sieve. The alcohol may be reused after decanting or filtering, provided its strength does not fall below 95 %.

X1.2.1 **Caution**—When using isopropyl alcohol, keep fire away, provide good ventilation, and avoid excessive inhalation of vapor.

Designation: C 631 – 81

Standard Specification for
BONDING COMPOUNDS FOR INTERIOR PLASTERING[1]

This standard is issued under the fixed designation C 631; the number immediately following the designation indicates the year of original adoption or, in the case of revision, the year of last revision. A number in parentheses indicates the year of last reapproval. A superscript epsilon (ε) indicates an editorial change since the last revision or reapproval.

1. Scope

1.1 This specification covers minimum standards for a bonding compound for adhering plaster to the surface of interior concrete, masonry, plaster, and any structurally sound surface such as surfaces painted with oil base paint, steel, tile, and similar material.

1.2 This specification also covers test requirements and test methods for bonding compound.

1.3 The values stated in inch-pound units are to be regarded as the standard.

2. Applicable Documents

2.1 *ASTM Standards:*
C 190 Test Method for Tensile Strength of Hydraulic Cement Mortars[2]
D 1779 Specification for Adhesive for Acoustical Materials[3]

3. Physical Requirements, General

3.1 *General Description*—The bonding compound shall be a film-forming, nonoxidizing, freeze-thaw, stable composition, suitable for application by brush, roller, or spray. It shall be tinted to show by visual inspection where it has been applied. The tinting material shall not bleed through the material being bonded. The compound shall be nontoxic and nonflammable.

3.2 *Consistency*—The bonding compound shall be free of foreign matter and shall be of such consistency as to be applied by brush, roller, or spray and when applied to surfaces in accordance with the manufacturer's recommendations shall cover evenly and dry to a uniform, smooth film.

NOTE 1—Surfaces to which the bonding compound is applied shall be prepared in accordance with the manufacturer's recommendations.

4. Physical Requirements, Detailed

4.1 *Film Characteristics*—The dry film of the bonding compound shall remain flexible and shall not be noticeably affected by alkalies or mild acids. It shall be capable of bonding cementitious materials immediately after application and drying, as recommended by the manufacturer, and up to 10 days after application when applied in accordance with the manufacturer's recommendations.

4.2 *Package Stability:*

4.2.1 The bonding compound shall be of a homogeneous nature with no settling of the solids.

4.2.2 The bonding compound shall stand exposure to storage temperatures up to 140°F when tested by the method specified in 6.1.

4.3 *Freeze-Thaw Stability*—When tested as specified in 6.2, the working and bonding properties shall not be adversely affected.

4.4 *Tensile Strength:*

4.4.1 *Tensile Strength of Fresh Sample*—When tested as specified in 6.3, a fresh sample

of bonding compound shall have a tensile strength of not less than 20 psi (138 kPa).

4.4.2 *Tensile Strength of 6-Month-Old Sample*—When tested as specified in 6.3, a 6-month-old sample of bonding compound shall have a minimum tensile strength of not less than 20 psi (138 kPa).

5. Sampling

5.1 Sampling shall be done in accordance with Specification D 1779.

6. Test Methods

6.1 *High-Temperature Test*—Place a 1-pt (400-cm³) container of bonding compound in an oven maintained at 140°F (60°C) for 30 days. Examine the contents for settling and separation at 15 and 30 days. The sample shall not separate to the extent that it cannot be remixed to uniform consistency. Exposure to 140°F for 30 days shall not affect its working and bonding properties. After such exposure the bonding compound shall be tested in accordance with 6.3.

6.2 *Freeze-Thaw Test*—Expose the bonding compound, in a closed, 1-pt (400-cm³) container, to a temperature of −10°F (−23.3°C) for 16 consecutive hours, then permit to thaw at room temperature. Repeat this procedure for a total of five cycles. After the fifth cycle test the bonding compound as specified in 6.3.

NOTE 2—Conditions conducive to dust accumulation, water, and moisture damage, etc., should be taken into account.

6.3 *Tension Test:*

6.3.1 Make briquets of neat wood-fibered plaster (CSA 82.22-1963) at normal consistency (CSA 82.20-1964). Add the plaster to the required amount of water and allow it to soak for ½ min. Mix for 1 min and cast into molds in accordance with Method C 190. Halve the briquets by placing a brass divider 0.010 in. (0.25 mm) thick in the center of the waistline of each mold. Add terra alba (gypsum) to the woodfibered plaster to provide a set in 2 to 3 h. Approximately 1 g of terra alba/100 g of plaster should give the desired set. Cast the plaster into the molds, puddle, and screed level, taking care not to disturb the brass divider. Place the cast briquets in a moist cabinet for at least 16 h, then remove the briquets from the molds. Separate them, and store them in an atmosphere of 70°F (21.1°C) and 50 % relative humidity for 48 h. Discard any briquets that have air holes or an excessively rough surface at this stage.

6.3.2 *Application of Plaster Bonding Agent*—Apply the plaster bonding agent as supplied, or diluted according to the manufacturer's directions, to the surface of the briquet by means of a 1-in. (25-mm) paint brush.

6.3.3 Replace the halved briquets with bonding agent in the molds, and fill the remaining

portion of the mold with neat wood-fibered plaster mixed and molded as in 5.2. Place the briquets in a moist closet for 16 h. Remove the briquets from the molds and test immediately.

6.3.4 Test the briquets for bond or tensile strength in accordance with Test Method C 190 except apply the load continuously at a rate of 0.05 in. (1.27 mm)/min.

6.4 *Report*—Test six briquets for bond strength and report the average. If any one result varies more than 15 % from the average, reject it, and report the average of five results. If the results from more than two samples vary over 15 % from the average, reject the series and retest.

7. Inspection

7.1 Inspection may be made at point of shipment or the point of delivery. The inspector shall have free access to the carrier being loaded.

8. Rejection

8.1 Any rejection shall be based upon a specific cause of failure to conform to the requirements of this specification and shall be reported to the seller within 20 working days from the receipt of the shipment by the purchaser.

9. Certification

9.1 When specified in the purchaser order, a producer's or supplier's certification and report of the test results shall be furnished to the purchaser that the material was manufactured, sampled, tested, and inspected in accordance with this specification and has been found to meet the specified requirements.

10. Packaging and Marking

10.1 *Packaging*—The bonding compound shall be packed in standard commercial containers. The containers shall be so constructed as to ensure acceptance by common or other carriers for safe transportation at the lowest rate to the point of delivery, unless otherwise specifed in the purchase order.

10.2 *Marking*—Shipping containers shall be marked with the name of the bonding compound, the quantity contained therein, the name, brand, or tradmark of the manufacturer or seller and period of time the material will be usable, the batch number, and ASTM designation.

---

[1] This specification is under the jurisdiction of ASTM Committee C-11 on Gypsum and Related Building Materials and Systems and is the direct responsibility of Subcommittee C11.02 on Specifications and Test Methods for Accessories and Related products.
Current edition approved June 26, 1981. Published August 1981. Originally published as C 631 – 68 T. Last previous edition C 631 – 70 (1975).
[2] Annual Book of ASTM Standards, Vol 04.01.
[3] Annual Book of ASTM Standards, Vol 15.06.

 **Designation: C 588 – 84a**

Standard Specification for
GYPSUM BASE FOR VENEER PLASTERS[1]

This standard is issued under the fixed designation C 588; the number immediately following the designation indicates the year of original adoption or, in the case of revision, the year of last revision. A number in parentheses indicates the year of last reapproval. A superscript epsilon (ε) indicates an editorial change since the last revision or reapproval.

This specification has been approved for use by agencies of the Department of Defense and for listing in the DoD Index of Specifications and Standards.

1. Scope

1.1 This specification covers types of gypsum base that are designed to be used as a base for application of veneer plaster.

1.2 The values stated in inch-pound units are to be regarded as the standard.

2. Applicable Documents

2.1 *ASTM Standards:*

C 473 Test Methods for Physical Testing of Gypsum Board Products and Gypsum Lath[2]

E 96 Test Methods for Water Vapor Transmission of Materials[3]

E 119 Methods of Fire Tests of Building Construction and Materials[4]

3. Composition

3.1 Gypsum base for veneer plaster shall consist of a noncombustible core, essentially gypsum, surfaced with paper firmly bonded to the core.

3.2 Foil-backed gypsum base shall, in addition, have the back surface covered with aluminum foil.

3.3 Type X (special fire-resistant) gypsum base for veneer plaster shall be gypsum base that provides at least a 1-h fire resistance rating for base ⅝ in. (15.9 mm) thick or ¾-h fire-resistance rating for base ½ in. (12.7 mm) thick, applied to a partition in a single-layer nail application on each face of load-bearing wood framing members, when tested in accordance with the requirements of Methods E 119.

NOTE—Consult manufacturers for independent test data on assembly details and fire resistance classifications for other types of construction. See official fire test reports for assembly particulars, materials, and classifications.

4. Physical Requirements

4.1 *Flexural Strength*—Specimens shall be tested in accordance with Methods C 473. When tested face up and when tested face down specimens shall have an average breaking load of not less than the following:

| | Load, lbf (N) | |
Thickness, in. (mm)	Bearing Edges Across Fiber of Surfacing	Bearing Edges Parallel to Fiber of Surfacing
⅜ (9.5)	80 (356)	30 (133)
½ (12.7)	110 (489)	40 (178)
⅝ (15.9)	150 (667)	50 (222)

4.2 *Humidified Deflection*—Specimens taken from the gypsum base for veneer plaster shall have an average deflection of no more than the following when tested in accordance with Methods C 473:

Thickness, in. (mm)	Humidified Deflection, Eighths of an inch (mm)
⅜ (9.5)	15 (48)
½ (12.7)	10 (32)
⅝ (15.9)	5 (16)

4.3 *Core, End, and Edge Hardness*—Specimens taken from the gypsum base for veneer plaster shall have an average hardness of not less than 15 lbf (67 N) for core, ends, or edges when tested in accordance with Methods C 473.

4.4 *Nail Pull Resistance*—Specimens taken from the gypsum base for veneer plaster shall have an average nail pull resistance of not less than the following when tested in accordance with Methods C 473:

Thickness, in. (mm)	Nail Pull Resistance, lbf (N)
⅜ (9.5)	60 (27.2)
½ (12.7)	80 (36.3)
⅝ (15.9)	90 (40.8)

4.5 *Permeance*—Foil-backed gypsum base for veneer plaster shall have a permeance of not more than 0.30 perm for the condition of 50 % relative humidity on Side I (the face of the board) and 0 % relative humidity on Side II (the foil-covered back of the board) when tested in accordance with the Desiccant Procedure of Test Methods E 96.

5. Dimensions and Permissible Variations

5.1 *Thickness*—The nominal thickness of gypsum base for veneer plaster shall be ⅜, ½, or ⅝ in. (9.5, 12.7, or 15.9 mm), with permissible variations in the nominal thickness of ±1/64 in. (± 0.4 mm), and with permissible local variations of ±1/32 in. (±0.8 mm) from the nominal thickness.

5.2 *Width*—The nominal width of gypsum base for veneer plaster shall be 48 in. (1.22 m) with permissible variations of ±3/32 in. (±2.4 mm) under the specified width.

5.3 *Length*—The nominal length of gypsum base for veneer plaster shall be from 6 to 16 ft (1.8 to 4.9 m) with permissible variations of ±¼ in. (±6.4 mm) from the specified length.

5.4 *Edges and Ends*—The edges and ends of gypsum base shall be straight and solid. Edges shall be square, rounded, or tapered. The average thickness of the tapered edge of gypsum base shall be at least 0.015 in. (0.38 mm) but not more than 0.075 in. (1.90 mm) less than the average thickness of the gypsum base as determined in accordance with Methods C 473. The corners shall be square with a permissible variation of ± ⅛ in. (±3.2 mm) in the full width of the base.

6. Workmanship, Finish, and Appearance

6.1 The surface of gypsum base for veneer plaster shall be free of any imperfections or foreign substances that would render it unfit for the intended use.

7. Sampling

7.1 At least 0.25 % of the number of gypsum base for veneer plaster in a shipment, but not less than three boards, shall be so selected as to be representative of the shipment and shall constitute a sample for purpose of tests by the purchaser or user.

8. Packaging and Package Marking

8.1 Gypsum base for veneer plaster shall be shipped so as to be kept dry and free of moisture.

8.2 When shipped for resale, the name of the manufacturer or supplier and the brand shall be legibly marked on each board or package.

[1] This specification is under the jurisdiction of ASTM Committee C-11 on Gypsum and Related Building Materials and Systems, and is the direct responsibility of Subcommittee C11.0 on Specifications and Test Methods for Gypsum Products.
Current edition approved Aug. 31, 1984. Published November 1984. Originally published as C 588 – 66 T. Last previous edition C 588 – 84.
[2] *Annual Book of ASTM Standards*, Vol 04.01.
[3] *Annual Book of ASTM Standards*, Vol 04.06.
[4] *Annual Book of ASTM Standards*, Vol 04.07.

The American Society for Testing and Materials takes no position respecting the validity of any patent rights asserted in connection with any item mentioned in this standard. Users of this standard are expressly advised that determination of the validity of any such patent rights, and the risk of infringement of such rights, are entirely their own responsibility.

This standard is subject to revision at any time by the responsible technical committee and must be reviewed every five years and if not revised, either reapproved or withdrawn. Your comments are invited either for revision of this standard or for additional standards and should be addressed to ASTM Headquarters. Your comments will receive careful consideration at a meeting of the responsible technical committee, which you may attend. If you feel that your comments have not received a fair hearing you should make your views known to the ASTM Committee on Standards, 1916 Race St., Philadelphia, Pa. 19103.

Designation: C 587 – 83

GYPSUM VENEER PLASTER[1]

This standard is issued under the fixed designation C 587; the number immediately following the designation indicates the year of original adoption or, in the case of revision, the year of last revision. A number in parentheses indicates the year of last reapproval. A superscript epsilon (ε) indicates an editorial change since the last revision or reapproval.

This specification has been approved for use by agencies of the Department of Defense and for listing in the DoD Index of Specifications and Standards.

1. Scope

1.1 This specification covers gypsum veneer plaster applied over a 4-ft (1.2-m) wide gypsum base material conforming to Specification C 588, designed to receive the plaster which may be applied in one or more coats to a maximum thickness of ¼ in. (6.4 mm).

2. Applicable Document

2.1 *ASTM Standard:*
C 588 Specification for Gypsum Base for Veneer Plasters[2]

3. Appearance

3.1 Gypsum veneer plaster constructions should provide an over-all monolithic trowel finish or textured surface free of surface defects. All angles should be cut sharply and be well detailed.

4. Joint Strength

4.1 Gypsum veneer plaster constructions applied in accordance with manufacturers' instructions shall exhibit joint strength between paperbound edges such that, when subjected to tensile test, specimens shall break within the field of the board and not at the joint leaving the plaster securely bonded thereto.

5. Bond

5.1 Gypsum veneer plaster shall exhibit sufficient bond strength between the plaster and the gypsum base and, where applicable, between the plaster base coat and finish coat to resist shock delamination when tested in the following manner:

5.1.1 Cut a 2-ft (0.6-m) square specimen from a panel of properly applied veneer plaster in accordance with manufacturers' instructions.

5.1.2 Allow the panel to thoroughly dry and come to equilibrium with the atmosphere of the test environment.

5.1.3 Raise the panel and slam the back side (face not treated with veneer plaster) on a bare concrete floor repeatedly until the panel is broken.

5.2 With this treatment, the bond of the plaster to the gypsum base shall not fail and individual coats of plaster if used in the system shall not delaminate.

6. Impact

6.1 Gypsum veneer plaster constructions shall exhibit sufficient impact strength such that when the surface is horizontally positioned on a firm background and struck by a polished steel ball, 1½ in. (38 mm) in diameter weighing 7.8 ± 0.1 oz (221 ± 3 g), dropped freely from a height of 36 in. (914 mm) there shall be no cracking or loss of bond beyond the impact area.

7. Flexure

7.1 Gypsum veneer plaster systems shall exhibit crack resistance within the field of the gypsum base such that when subjected to continuous, midpoint, line loading applied downward to the back of the panel (plaster surface in tension) across the 12-in. (305-mm) direction of 12 by 18-in. (305 by 457-mm) specimens supported over a clear span of 14 in. (356 mm), no radial or random cracking shall result at failure. Upon failure, for example, total rupture of the plasterboard, the veneer plaster shall remain securely bonded to the face paper exhibiting only a single break following the loading line.

8. Inspection

8.1 Inspection of the gypsum veneer plaster shall be agreed upon between the purchaser and the supplier as part of the purchase order.

9. Rejection and Rehearing

9.1 Gypsum veneer plaster that fails to conform to the requirements of this specification may be rejected. Rejection shall be reported to the producer or supplier in writing within 10 working days from receipt of shipment by the purchaser. The notice of rejection shall contain a specific statement of the respects in which the plaster has failed to conform to the requirements of this specification. In the case of dissatisfaction of the producer or supplier with the results of the test, such notice of rejection shall be supported by results of a test conducted by a mutually agreeable independent laboratory.

10. Certification

10.1 When specified in the purchase order, a producer's or supplier's certification shall be furnished to the purchaser that the product meets the requirements of this specification.

---

[1] This specification is under the jurisdiction of ASTM Committee C-11 on Gypsum and Related Building Materials and Systems and is the direct responsibility of Subcommittee C11.01 on Specifications and Test Methods for Gypsum Products.
Current edition approved Aug. 26, 1983. Published October 1983. Originally published as C 587 – 66. Last previous edition C 587 – 68 (1978).
[2] Annual Book of ASTM Standards, Vol 04.01.

Designation: C 35 – 76 (Reapproved 1981)[ε1]

Standard Specification for

INORGANIC AGGREGATES FOR USE IN GYPSUM PLASTER[1]

This standard is issued under the fixed designation C 35; the number immediately following the designation indicates the year of original adoption or, in the case of revision, the year of last revision. A number in parentheses indicates the year of last reapproval. A superscript epsilon (ε) indicates an editorial change since the last revision or reapproval.

[ε1] NOTE—Section 2 was added editorially and subsequent sections renumbered in May 1985.

1. Scope

1.1 This specification covers those aggregates most commonly used in gypsum plaster, which include perlite, sand (natural and manufactured), and vermiculite. Other aggregates may be employed, provided tests have demonstrated them to yield plaster of satisfactory quality.

2. Applicable Documents

2.1 *ASTM Standards:*
C 29 Test Method for Unit Weight and Voids in Aggregate[2]
C 40 Test Method for Organic Impurities in Fine Aggregates for Concrete[2]
C 136 Method for Sieve Analysis of Fine and Coarse Aggregates[2]
D 75 Practice for Sampling Aggregates[2]

3. Definitions

3.1 *perlite aggregate*—a siliceous volcanic glass properly expanded by heat.

3.2 *sand aggregate:*

3.2.1 *natural sand*—the fine granular material resulting from the natural disintegration of rock or from the crushing of friable sandstone.

3.2.2 *manufactured sand*—the fine material resulting from the crushing and classification by screening, or otherwise, of rock, gravel or blast furnace slag.

3.3 *vermiculite aggregate*—a micaceous mineral properly expanded by heat.

4. Grading

4.1 *Sieve Analysis*—The aggregate, except as provided in 4.2, shall be graded within the limits specified in Table 1.

4.2 For natural or manufactured sand, not more than 50 % shall be retained between any two consecutive sieves shown in 4.1, nor more than 25 % between the No. 50 (300-μm) and No. 100 (150-μm) sieves.

4.3 For natural or manufactured sand, the amount of material finer than a No. 200 (75-μm) sieve shall not exceed 5 %.

5. Weight of Lightweight Aggregates

5.1 The weight of perlite aggregate shall be not less than 6 nor more than 12 lb/ft³ (96 to 192 kg/m³).

5.2 The weight of vermiculite aggregate shall be not less than 6 more than 10 lb/ft³ (96 to 160 kg/m³).

6. Impurities

6.1 Water-soluble impurities in sand shall not exceed 0.15 weight % and sodium ion content shall not exceed 0.02 weight %.

6.2 Sand, when subjected to the colorimetric test for organic impurities, shall develop a color no darker than the standard, unless it is established by adequate tests that the impurities causing the color are not harmful in plaster.

7. Sampling

7.1 Samples of natural or manufactured sand shall be obtained in accordance with Practice D 75.

7.2 For bagged aggregates, at least one bag shall be taken at random for sampling from each 100 bags, but not less than 6 bags from each shipment of 100 bags or more, and for smaller shipments not less than 6 % of the bags shall be sampled. Representative portions from each bag selected shall be secured by means of a suitable sampling thief tube. The sampling tube shall be inserted the full distance between diagonally opposite corners of the bag with the bag lying in a horizontal position. The portions so obtained shall be combined to produce a composite sample having a volume of at least 1 ft³ (28 dm³). At least one composite sample shall be prepared and tested separately for each 2000 bags of aggregate used.

7.3 Samples secured in accordance with 7.1 or 7.2 shall be reduced by quartering or riffling to obtain samples of proper size for individual tests.

8. Test Methods

8.1 *Sieve Analysis*—Method C 136. For lightweight aggregates only the following modifications shall apply: The sample shall consist of approximately 500-mL bulk volume. Where a mechanical sieving device is used, the sieving time shall be 5 min. The volume of each sieve fraction shall be measured in a 250-mL graduated cylinder. The aggregate shall be poured loosely into the graduate without tamping or shaking, the surface leveled with a spatula, and the volume read within ±2 mL. The volumes of the individual sieve fractions shall be expressed as percentages of the sum of the volumes of all fractions. The percentage retained on each designated sieve shall be calculated by summing the individual percentages of all fractions larger than that sieve.

8.2 *Weight of Lightweight Aggregate*—Test Method C 29, using the shoveling procedure.

8.3 *Organic Impurities*—Test Method C 40.

8.4 *Water-Soluble Impurities*—Weigh accurately a 10-g sample of sand. Transfer to a 250-mL beaker. Add 100 mL of distilled water. Heat to boiling and allow to simmer on a hot plate for 5 min. Filter through a fine paper into a tared evaporating dish. Wash with hot distilled water until the volume of filtrate is about 125 mL. Evaporate to dryness. The temperature of the dish shall not exceed 250°F (121°C) during final drying. Cool in a desiccator, weigh, and calculate the weight of residue to percentage of water-soluble impurities.

9. Packaging and Package Marking

9.1 When lightweight aggregates covered by this specification are delivered in packages, the name of the manufacturer, type of aggregate, and minimum weight and approximate volume of the contents shall be plainly indicated thereon.

10. Rejection

10.1 The purchaser of materials covered by this specification shall have the option of evaluating these materials for rejection, by either minimum weight or approximate volume as stated.

10.2 Individual packages may be rejected when:

10.2.1 The contents, on a weight basis, are 5 % less than that indicated on the packages, or

10.2.2 The contents, on a volume basis, are 10 % less than that indicated on the package.

10.3 The entire shipment may be rejected:

10.3.1 On a weight basis when the average contents of two packages for each 100 but not less than six packages selected at random, in any one shipment, are less than that indicated on the package.

10.3.2 On a volume basis when the average contents of two packages for each 100 but not less than six packages selected at random, in any one shipment, differ by more than 5 % more or 10 % less from that indicated on the package.

10.4 The net weight of the contents shall be determined by weighing the package or packages and deducting the weight of the container.

10.5 The volume of the contents in the package shall be calculated by determining the weight of the contents of the package and then obtaining the weight per cubic foot of the aggregate, from an average weight package of the samples selected in 10.3.2, by the shoveling procedure given in Section 7 of Test Method C 29, and then dividing the weight of the contents of the bag by the weight per cubic foot of aggregate.

TABLE 1 Grading Requirements

| Sieve Size | Percentage Retained on Each Sieve, Cumulative | | | | | |
| | Perlite, by volume | | Vermiculite, by volume | | Sand, by weight | |
	Max	Min	Max	Min	Max	Min
No. 4 (4.75-mm)	0	...	0	...	0	...
No. 8 (2.36-mm)	5	0	10	0	5	0
No. 16 (1.18-mm)	60	5	75	40	30	5
No. 30 (600-μm)	95	45	95	65	65	30
No. 50 (300-μm)	98	75	98	75	95	65
No. 100 (150-μm)	100	85	100	90	100	90

[1] This specification is under the jurisdiction of ASTM Committee C-11 on Ceiling and Walls and is the direct responsibility of Subcommittee C 11.02 on Specifications and Test Methods for Accessories and Related Products.
Current edition approved Sept. 24, 1976. Published November 1976. Originally published as C 35 – 21 T. Last previous edition C 35 – 70 (1975).
[2] Annual Book of ASTM Standards, Vol 04.02.

Designation: C 61 – 76 (Reapproved 1981)

Standard Specification for
GYPSUM KEENE'S CEMENT[1]

This standard is issued under the fixed designation C 61: the number immediately following the designation indicates the year of original adoption or, in the case of revision, the year of last revision. A number in parentheses indicates the year of last reapproval. A superscript epsilon (ε) indicates an editorial change since the last revision or reapproval.

This specification has been approved for use by agencies of the Department of Defense and for listing in the DoD Index of Specifications and Standards.

1. Scope

1.1 This specification covers gypsum Keene's cement, anhydrous calcined gypsum the set of which is accelerated by the addition of other materials.

NOTE 1—Various grades of gypsum Keene's cement of different fineness and setting time are available. This specification is directly applicable to the grades manufactured for use in the base and finish coats of plastering. It is generally available as quick-setting and standard-setting gypsum Keene's cement. Grades intended for use in Scagliola, castings, and other special purposes should conform to the requirements of this specification in all respects except those for fineness and time of set, which will vary according to the particular use for which the grade is designed.

1.2 The values stated in inch-pound units are to be regarded as the standard.

2. Time of Setting

2.1 Gypsum Keene's cement shall set in not less than 20 min nor more than 6 h.

3. Compressive Strength

3.1 Gypsum Keene's cement shall have a compressive strength of not less than 2500 psi (17 MPa).

4. Fineness

4.1 The cement shall all pass a No. 14 (1.40-mm) sieve, not less than 98 percent shall pass a No. 40 (425-μm) sieve and not less than 80 percent shall pass a No. 100 (150-μm) sieve.[2]

5. Combined Water

5.1 The cement shall not show a combined water content of more than 2 percent.

6. Sampling

6.1 At least 3 percent of the packages shall be sampled and shall be so selected as to be representative of the shipment. Samples shall be taken from both the surface and the center of each package so selected. The material so obtained shall be thoroughly mixed and reduced by quartering to provide not less than a 15-lb (6.8-kg) sample for the laboratory.

7. Laboratory Samples

7.1 Each laboratory sample shall be placed immediately in an airtight container and shipped to the laboratory for test.

8. Test Methods

8.1 Determine the chemical analysis and physical properties of gypsum Keene's cement in accordance with ASTM Methods C 471, for Chemical Analysis of Gypsum and Gypsum Products,[3] and ASTM Methods C 472, for Physical Testing of Gypsum Plasters and Gypsum Concrete,[3] except that in the determination of the time of setting and compressive strength, the consistency used shall be such that a 35-g conical plunger described in Section 5 of Methods C 472, shall give a penetration of 30 ± 2 mm, 20 s after release, the measurement being made 5 min after the addition of the gypsum Keene's cement to the gauging water.

NOTE 2—A 20-min mixing and soaking period for the gypsum Keene's cement and water with occasional stirring to remove entrained air or gas bubbles is required before filling the molds in the strength determination. The use of 0.1 percent retarder in determining the testing consistency is necessary when working with quick-setting cements.

9. Packaging and Marking

9.1 Gypsum Keene's cement shall be dry and free from lumps and shall be shipped in packages.

9.2 When shipped for resale, the following information shall be legibly marked on each package or on a tag of suitable size attached thereto:

9.2.1 Name of manufacturer,
9.2.2 Brand,
9.2.3 Description, and
9.2.4 Net and gross weights of package.

10. Inspection

10.1 Inspection may be made either at the point of shipment or at the point of delivery. The inspector representing the purchaser shall have free access to the carriers being loaded for shipment to the purchaser. He shall be afforded all reasonable facilities for inspection and sampling, which shall be so conducted as not to interfere unnecessarily with the loading of the carriers.

11. Rejection

11.1 Any rejection shall be based upon the specific cause of failure to conform to the requirements of this specification and shall be reported to the seller within 10 working days from the receipt of the shipment by the purchaser.

12. Rehearing

12.1 Claims for rehearing shall be valid only if made within 20 working days from receipt of notice of specific cause for rejection.

ANNEX

A1. FIELD TEST FOR GYPSUM KEENE'S CEMENT

A1.1 In many instances it is desirable to have a simple test whereby gypsum Keene's cement may be identified. Such a test will not indicate the quality of the cement, which should be determined by the laboratory tests enumerated in the body of this specification, but is of especial value to plasterers, material dealers, and superintendents.

A1.1.1 The following procedure is satisfactory for this purpose:

A1.1.1.1 Take a cupful of the material, mix with water to the consistency of a thick paste, and pour upon a plate or piece of glass. Let stand until fairly firm and definite signs of set having begun are manifest. This will be some time less than 2 h, depending on climatic conditions.

A1.1.1.2 Divide the sample, allowing one half to remain undisturbed. Take the other half and break it down adding a little water, remix, and then allow it to "set up" again on the plate or glass.

A1.1.1.3 If the material is gypsum Keene's cement, the remixed portion will, within a few hours, become quite as hard and strong as the portion that was not remixed.

---

[1] This specification is under the jurisdiction of ASTM Committee C-11 on Gypsum and Related Building Materials and Systems and is the direct responsibility of Subcommittee C11.01 on Specifications and Test Methods for Gypsum Products.
Current edition approved Oct. 29, 1976. Published December 1976. Originally published as C 61 – 26 T. Last previous edition C 61 – 64 (1975).
[2] Detailed requirements for these sieves are given in ASTM Specification E 11, for Wire-Cloth Sieves for Testing Purposes see *Annual Book of ASTM Standards*, Vols 04.01 and 14.02.
[3] *Annual Book of ASTM Standards*, Vol 04.01.

 **Designation: C 206 – 84**

Standard Specification for
FINISHING HYDRATED LIME[1]

This standard is issued under the fixed designation C 206; the number immediately following the designation indicates the year of original adoption or, in the case of revision, the year of last revision. A number in parentheses indicates the year of last reapproval. A superscript epsilon (ϵ) indicates an editorial change since the last revision or reapproval.

This specification for Type S finishing hydrated lime has been approved for use by agencies of the Department of Defense and for listing in the DoD Index of Specifications and Standards.

1. Scope

1.1 This specification covers two types of finishing hydrated lime that are suitable for use in the scratch, brown, and finish coats of plaster, for stucco, for mortar, and as an addition to portland-cement concrete. The two types of lime sold under this specification shall be designated as follows:

1.1.1 *Type N*—Normal hydrated lime for finishing purposes, and

1.1.2 *Type S*—Special hydrated lime for finishing purposes.

NOTE—Type N, normal finishing hydrated lime, is differentiated from Type S, special finishing hydrated lime, in that no limitation on the amount of unhydrated oxides is specified for Type N hydrate, and the plasticity requirement for Type N hydrate shall be determined after soaking for 16 to 24 h.

2. Applicable Documents

2.1 *ASTM Standards:*

C 25 Methods for Chemical Analysis of Limestone, Quicklime, and Hydrated Lime[2]

C 50 Methods of Sampling, Inspection, Packing, and Marking of Lime and Limestone Products[2]

C 51 Definitions of Terms Relating to Lime and Limestone (As Used by the Industry)[2]

C 110 Methods of Physical Testing of Quicklime, Hydrated Lime, and Limestone[2]

C 842 Specification for Application of Interior Gypsum Plaster[2]

3. Definitions

3.1 For definitions of terms relating to hydrated lime, refer to Definitions C 51.

4. Chemical Requirements

4.1 Hydrated lime for finishing purposes shall conform to the following requirements as to chemical composition:

	Type N	Type S
Calcium and magnesium oxides (non-volatile basis), min, %	95	95
Carbon dioxide (as-received basis), max, %		
If sample is taken at the place of manufacture	5	5
If sample is taken at any other place	7	7
Unnydrated oxides (as-received basis), max, %	. . .	8

5. Residue

5.1 The percentage residue of finishing hydrated lime shall conform to the following requirements:

Residue retained on No. 30 (600-µm) sieve, max, %	0.5
Residue retained on No. 200 (75-µm) sieve, max, %	15

6. Popping and Pitting

6.1 Finishing hydrated lime shall show no pops or pits when tested in accordance with the method prescribed in 10.1.2.

7. Plasticity

7.1 The putty made from Type N, normal finishing hydrated lime, shall have a plasticity figure of not less than 200 when soaked for a period of not less than 16 h nor more than 24 h.

7.2 The putty made from Type S, special finishing hydrated lime, shall have a plasticity figure of not less than 200 when tested commencing within 30 min after mixing with water.

8. Application of Interior Gypsum Plaster

8.1 For recommended application procedures refer to Specification C 842.

9. Sampling, Inspection, etc.

9.1 The sampling, inspection, rejection, re-testing, packing, and marking shall be conducted in accordance with Methods C 50.

10. Test Methods

10.1 The properties enumerated in this specification shall be determined in accordance with the following methods:

10.1.1 *Chemical Analysis*—Methods C 25.

10.1.2 *Physical Tests*—Methods C 110.

11. Marking

11.1 Type N hydrated lime, in bags, conforming to this specification shall be soaked for a minimum of 16 h prior to use.

[1] This specification is under the jurisdiction of ASTM Committee C-7 on Lime and is the direct responsibility of Subcommittee C07.02 on Structural Lime.

Current edition approved May 25, 1984. Published July 198. Originally published as C 206 – 46. Replaces C 6 – 49 (1974). Last previous edition C 206 – 79.

[2] *Annual Book of ASTM Standards*, Vol 04.01.

Designation: C 59 – 83

Standard Specification for
GYPSUM CASTING AND MOLDING PLASTER[1]

This standard is issued under the fixed designation C 59; the number immediately following the designation indicates the year of original adoption or, in the case of revision, the year of last revision. A number in parentheses indicates the year of last reapproval. A superscript epsilon (ε) indicates an editorial change since the last revision or reapproval.

This specification has been approved for use by agencies of the Department of Defense and for listing in the DoD Index of Specifications and Standards.

1. Scope

1.1 This specification covers gypsum casting and molding plaster materials consisting essentially of calcined gypsum. Casting plaster is used to form objects, such as lamp bases, art ware, novelties, etc. Molding plaster is used in making interior embellishments and cornices, as gauging plaster, etc.

1.2 The values stated in the inch-pound units are to be regarded as the standard.

2. Applicable Documents

2.1 *ASTM Standards:*

C 471 Methods for Chemical Analysis of Gypsum and Gypsum Products[2]

C 472 Methods for Physical Testing of Gypsum Plasters and Gypsum Concrete[2]

3. Composition

3.1 Gypsum casting and molding plaster shall contain not less than 85 % rehydrated purity, $CaSO_4 \cdot 2H_2O$.

4. Time of Setting

4.1 Gypsum casting and molding plaster shall set in not less than 10 nor more than 50 min.

5. Compressive Strength

5.1 Gypsum casting and molding plaster shall have a compressive strength of not less than 1800 psi (12.4 MPa).

6. Fineness

6.1 Gypsum casting and molding plaster shall all pass a No. 30 (600-μm) sieve and not less than 90 % shall pass a No. 100 (150-μm) sieve.[3]

7. Sampling

7.1 At least 3 % of the packages shall be sampled and shall be so selected as to be representative of the shipment. Samples shall be taken from both the surface and the center of each package so selected. The material so obtained shall be thoroughly mixed and reduced by quartering to provide not less than a 15-lb (6.8-kg) sample for the laboratory.

8. Laboratory Samples

8.1 Each laboratory sample shall be placed immediately in an airtight container and shipped to the laboratory for test.

9. Test Methods

9.1 The chemical analysis and physical properties of gypsum casting and molding plasters shall be determined in accordance with Methods C 471 and C 472.

NOTE—Distilled water shall be used for conducting all tests.

10. Packaging and Package Marking

10.1 Gypsum casting and molding plasters shall be dry and free of lumps and shall be shipped in packages.

10.2 When shipped for resale, the following information shall be legibly marked on each package or on a tag of suitable size attached thereto:

10.2.1 Name of manufacturer or seller.

10.2.2 Brand,

10.2.3 Description, and

10.2.4 Net weight of package. Gross weight shall be shown where required.

[1] This specification is under the jurisdiction of ASTM Committee C-11 on Gypsum and Related Building Materials and Systems and is the direct responsibility of Subcommittee C11.01 on Specifications and Test Methods for Gypsum Products.

Current edition approved Aug. 26, 1983. Published October 1983. Originally published as C 59 – 26. Last previous edition C 59 – 81.

[2] *Annual Book of ASTM Standards*, Vol 04.01.

[3] Detailed requirements for these sieves are given in ASTM Specifications E 11, for Wire-Cloth Sieves for Testing Purposes, see *Annual Book of ASTM Standards*, Vol 14.02.

 **Designation: C 150 – 86**

Standard Specification for
PORTLAND CEMENT[1]

This standard is issued under the fixed designation C 150; the number immediately following the designation indicates the year of original adoption or, in the case of revision, the year of last revision. A number in parentheses indicates the year of last reapproval. A superscript epsilon (ε) indicates an editorial change since the last revision or reapproval.

This specification has been approved for use by agencies of the Department of Defense and for listing in the DoD Index of Specifications and Standards.

1. Scope

1.1 This specification covers eight types of portland cement, as follows (see Note):

1.1.1 *Type I*—For use when the special properties specified for any other type are not required.

1.1.2 *Type IA*—Air-entraining cement for the same uses as Type I, where air-entrainment is desired.

1.1.3 *Type II*—For general use, more especially when moderate sulfate resistance or moderate heat of hydration is desired.

1.1.4 *Type IIA*—Air-entraining cement for the same uses as Type II, where air-entrainment is desired.

1.1.5 *Type III*—For use when high early strength is desired.

1.1.6 *Type IIIA*—Air-entraining cement for the same use as Type III, where air-entrainment is desired.

1.1.7 *Type IV*—For use when a low heat of hydration is desired.

1.1.8 *Type V*—For use when high sulfate resistance is desired.

1.2 The values stated in inch-pound units are to be regarded as the standard.

2. Referenced Documents

2.1 *ASTM Standards:*

C 33 Specification for Concrete Aggregates[2]

C 109 Test Method for Compressive Strength of Hydraulic Cement Mortars (Using 2-in. or 50-mm Cube Specimens)[3]

C 114 Methods for Chemical Analysis of Hydraulic Cement[3]

C 115 Test Method for Fineness of Portland Cement by the Turbidimeter[3]

C 151 Test Method for Autoclave Expansion of Portland Cement[3]

C 183 Methods of Sampling and Acceptance of Hydraulic Cement[3]

C 185 Test Method for Air Content of Hydraulic Cement Mortar[3]

C 186 Test Method for Heat of Hydration of Hydraulic Cement[3]

C 191 Test Method for Time of Setting of Hydraulic Cement by Vicat Needle[3]

C 204 Test Method for Fineness of Portland Cement by Air Permeability Apparatus[3]

C 226 Specification for Air-Entraining Additions for Use in the Manufacture of Air-Entraining Portland Cement[3]

C 265 Test Method for Calcium Sulfate in Hydrated Portland Cement Mortar[3]

C 266 Test Method for Time of Setting of Hydraulic Cement by Gillmore Needles[3]

C 451 Test Method for Early Stiffening of Portland Cement (Paste Method)[3]

C 452 Test Method for Potential Expansion of Portland Cement Mortars Exposed to Sulfate[3]

C 465 Specification for Processing Additions for Use in the Manufacture of Hydraulic Cements[3]

C 563 Test Method for Optimum SO₃ in Portland Cement[3]

3. Definitions

3.1 *portland cement*—a hydraulic cement produced by pulverizing clinker consisting essentially of hydraulic calcium silicates, usually containing one or more of the forms of calcium sulfate as an interground addition.

3.2 *air-entraining portland cement*—a hydraulic cement produced by pulverizing clinker consisting essentially of hydraulic calcium silicates, usually containing one or more of the forms of calcium sulfate as an interground addition, and with which there has been interground an air-entraining addition.

4. Ordering Information

4.1 Orders for material under this specification shall include the following:

4.1.1 This specification number and date,

4.1.2 Type or types allowable. If no type is specified, Type I shall be supplied,

4.1.3 Any optional chemical requirements from Table 2, if desired,

4.1.4 Type of setting-time test required, Vicat or Gilmore. If not specified, the Vicat shall be used,

4.1.5 Any optional physical requirements from Table 4, if desired.

NOTE—Attention is called to the fact that cements conforming to the requirements for all types may not be carried in stock in some areas. In advance of specifying the use of other than Type I cement, it should be determined whether the proposed type of cement is or can be made available.

5. Additions

5.1 The cement covered by this specification shall contain no addition except as follows:

5.1.1 Water or calcium sulfate, or both, may be added in amounts such that the limits shown in Table 1 for sulfur trioxide and loss-on-ignition shall not be exceeded.

5.1.2 At the option of the manufacturer, processing additions may be used in the manufacture of the cement, provided such materials in the amounts used have been shown to meet the requirements of Specification C 465.

5.1.3 Air-entraining portland cement shall contain an interground addition conforming to the requirements of Specification C 226.

6. Chemical Composition

6.1 Portland cement of each of the eight types shown in Section 1 shall conform to the respective standard chemical requirements prescribed in Table 1. In addition, optional chemical requirements are shown in Table 2.

7. Physical Properties

7.1 Portland cement of each of the eight types shown in Section 1 shall conform to the respective standard physical requirements prescribed in Table 3. In addition, optional physical requirements are shown in Table 4.

8. Sampling

8.1 When the purchaser desires that the cement be sampled and tested to verify compliance with this specification, sampling and testing should be performed in accordance with Methods C 183.

8.2 Methods C 183 are not designed for manufacturing quality control and are not required for manufacturer's certification.

9. Test Methods

9.1 Determine the applicable properties enumerated in this specification in accordance with the following methods:

9.1.1 *Air Content of Mortar*—Test Method C 185.

9.1.2 *Chemical Analysis*—Methods C 114.

9.1.3 *Strength*—Test Method C 109.

9.1.4 *False Set*—Test Method C 451.

9.1.5 *Fineness by Air Permeability*—Method C 204.

9.1.6 *Fineness by Turbidimeter*—Test Method C 115.

9.1.7 *Heat of Hydration*—Test Method C 186.

9.1.8 *Autoclave Expansion*—Test Method C 151.

9.1.9 *Time of Setting by Gillmore Needles*—Test Method C 266.

9.1.10 *Time of Setting by Vicat Needles*—Test Method C 191.

9.1.11 *Sulfate Expansion*—Test Method C 452.

9.1.12 *Calcium Sulfate in Mortar*—Test Method C 265.

9.1.13 *Optimum SO₃*—Test Method C 563.

10. Inspection

10.1 Inspection of the material shall be made as agreed upon by the purchaser and the seller as part of the purchase contract.

11. Rejection

11.1 The cement may be rejected if it fails to meet any of the requirements of this specification.

11.2 Cement remaining in bulk storage at the mill, prior to shipment, for more than 6 months, or cement in bags in local storage in the hands of a vendor for more than 3 months, after com-

---

[1] This specification is under the jurisdiction of ASTM Committee C-1 on Cement and is the direct responsibility of Subcommittee C01.10 on Portland Cement.
Current edition approved Dec. 9, 1986. Published February 1987. Originally published as C 150 – 40 T. Last previous edition C 150 – 85a.
[2] *Annual Book of ASTM Standards*, Vol 04.02.
[3] *Annual Book of ASTM Standards*, Vol 04.01.

TABLE 1   Standard Chemical Requirements

Cement Type[A]	I and IA	II and IIA	III and IIIA	IV	V
Silicon dioxide (SiO$_2$), min. %	...	20.0	...	...	...
Aluminum oxide (Al$_2$O$_3$), max. %	...	6.0	...	...	...
Ferric oxide (Fe$_2$O$_3$), max. %	...	6.0	...	6.5	...
Magnesium oxide (MgO), max. %	6.0	6.0	6.0	6.0	6.0
Sulfur trioxide (SO$_3$),[B] max. %					
When (C$_3$A)[C] is 8 % or less	3.0	3.0	3.5	2.3	2.3
When (C$_3$A)[C] is more than 8 %	3.5	D	4.5	D	D
Loss on ignition, max. %	3.0	3.0	3.0	2.5	3.0
Insoluble residue, max. %	0.75	0.75	0.75	0.75	0.75
Tricalcium silicate (C$_3$S)[C] max. %	...	...	...	35[E]	...
Dicalcium silicate (C$_2$S)[C] min. %	...	...	...	40[E]	...
Tricalcium aluminate (C$_3$A)[C] max. %	...	8	15	7[E]	5[F]
Tetracalcium aluminoferrite plus twice the tricalcium aluminate[C] (C$_4$AF + 2(C$_3$A)), or solid solution (C$_4$AF + C$_2$F), as applicable, max. %	...	...	...	...	25[F]

[A] See Note.

[B] There are cases where optimum SO$_3$ for a particular cement is close to or in excess of the limit in this specification. When it has been demonstrated by Test Method C 563 that the optimum SO$_3$ exceeds a value 0.5 % less than the specification limit, an additional amount of SO$_3$ is permissible provided that, when the cement with the additional calcium sulfate is tested by Test Method C 265, the calcium sulfate in the hydrated mortar at 24 ± ¼ h expressed as SO$_3$ does not exceed 0.50 g/L. When the manufacturer supplies cement under this provision, he will, upon request, supply supporting data to the purchaser.

[C] The expressing of chemical limitations by means of calculated assumed compounds does not necessarily mean that the oxides are actually or entirely present as such compounds.

When expressing compounds, C = CaO, S = SiO$_2$, A = Al$_2$O$_3$, F = Fe$_2$O$_3$. For example, C$_3$A = 3CaO·Al$_2$O$_3$.

Titanium dioxide and phosphorus pentoxide (TiO$_2$ and P$_2$O$_5$) shall be included with the Al$_2$O$_3$ content. The value historically and traditionally used for Al$_2$O$_3$ in calculating potential compounds for specification purposes is the ammonium hydroxide group minus ferric oxide (R$_2$O$_3$ − Fe$_2$O$_3$) as obtained by classical wet chemical methods. This procedure includes as Al$_2$O$_3$ the TiO$_2$, P$_2$O$_5$ and other trace oxides which precipitate with the ammonium hydroxide group in the classical wet chemical methods. Many modern instrumental methods of cement analysis determine aluminum or aluminum oxide directly without the minor and trace oxides included by the classical method. Consequently, for consistency and to provide comparability with historic data and among various analytical methods, when calculating potential compounds for specification purposes, those using methods which determine Al or Al$_2$O$_3$ directly should add to the determined Al$_2$O$_3$ weight quantities of P$_2$O$_5$, TiO$_2$ and any other oxide except Fe$_2$O$_3$ which would precipitate with the ammonium hydroxide group when analyzed by the classical method and which is present in an amount of 0.05 weight % or greater. The weight percent of minor or trace oxides to be added to Al$_2$O$_3$ by those using direct methods may be obtained by actual analysis of those oxides in the sample being tested or estimated from historical data on those oxides on cements from the same source, provided that the estimated values are identified as such.

When the ratio of percentages of aluminum oxide to ferric oxide is 0.64 or more, the percentages of tricalcium silicate, dicalcium silicate, tricalcium aluminate, and tetracalcium aluminoferrite shall be calculated from the chemical analysis as follows:

Tricalcium silicate = (4.071 × % CaO) − (7.600 × % SiO$_2$) − (6.718 × % Al$_2$O$_3$) − (1.430 × % Fe$_2$O$_3$) − (2.852 × % SO$_3$)

Dicalcium silicate = (2.867 × % SiO$_2$) − (0.7544 × % C$_3$S)

Tricalcium aluminate = (2.650 × % Al$_2$O$_3$) − (1.692 × % Fe$_2$O$_3$)

Tetracalcium aluminoferrite = 3.043 × % Fe$_2$O$_3$

When the alumina-ferric oxide ratio is less than 0.64, a calcium aluminoferrite solid solution (expressed as ss(C$_4$AF + C$_2$F)) is formed. Contents of this solid solution and of tricalcium silicate shall be calculated by the following formulas:

ss(C$_4$AF + C$_2$F) = (2.100 × % Al$_2$O$_3$) + (1.702 × % Fe$_2$O$_3$)

Tricalcium silicate = (4.071 × % CaO) − (7.600 × % SiO$_2$) − (4.479 × % Al$_2$O$_3$) − (2.859 × % Fe$_2$O$_3$) − (2.852 × % SO$_3$).

No tricalcium aluminate will be present in cements of this composition. Dicalcium silicate shall be calculated as previously shown.

In the calculation of all compounds the oxides determined to the nearest 0.1 % shall be used.

All values calculated as described in this note shall be reported to the nearest 1 %.

[D] Not applicable.

[E] Does not apply when the heat of hydration limit in Table 4 is specified.

[F] Does not apply when the sulfate expansion limit in Table 4 is specified.

TABLE 4   Optional Physical Requirements[A]

Cement Type	I	IA	II	IIA	III	IIIA	IV	V
False set, final penetration, min, %	50	50	50	50	50	50	50	50
Heat of hydration:								
7 days, max, cal/g (kJ/kg)	...	...	70 (290)[B]	70 (290)[B]	...	...	60[C] (250)	...
28 days, max, cal/g (kJ/kg)	...	...	...	...	...	...	70[C] (290)	...
Strength, not less than the values shown:								
Compressive strength, psi (MPa)								
28 days	4000 (27.6)	3200 (22.1)	4000 (27.6) / 3200[B] (22.1)[B]	3200 (22.1) / 2560[B] (17.7)[B]	...	...	...	...
Sulfate expansion,[D] 14 days, max, %	...	...	...	...	...	...	...	0.040

[A] These optional requirements apply only if specifically requested. Availability should be verified. See Note in Section 4.

[B] The optional limit for the sum of the tricalcium silicate and tricalcium aluminate in Table 2 shall not be requested when this optional limit is requested. These strength requirements apply when either heat of hydration or the sum of tricalcium silicate and tricalcium aluminate requirements are requested.

[C] When the heat of hydration limit is specified, it shall be instead of the limits of C3S, C2S, and C3A listed in Table 1.

[D] When the sulfate expansion is specified, it shall be instead of the limits of C$_3$A and C$_4$AF + 2 C$_3$A listed in Table 1.

---

pletion of tests, may be retested before use and may be rejected if it fails to conform to any of the requirements of this specification.

11.3 Packages shall identify the weight as net weight. Packages more than 2 % below the weight marked thereon may be rejected; and if the average weight of packages in any shipment, as shown by weighing 50 packages taken at random, is less than that marked on the packages, the entire shipment may be rejected.

12. Manufacturer's Statement

12.1 At the request of the purchaser, the manufacturer shall state in writing the nature, amount, and identity of the air-entraining agent used, and of any processing addition that may have been used, and also, if requested, shall supply test data showing compliance of such air-entraining addition with the provisions of Specification C 226, and of any such processing addition with Specification C 465.

13. Packaging and Package Marking

13.1 When the cement is delivered in packages, the words "Portland Cement," the type of cement, the name and brand of the manufacturer, and the weight of the cement contained therein shall be plainly marked on each package. When the cement is an air-entraining type, the words "air-entraining" shall be plainly marked on each package. Similar information shall be provided in the shipping documents accompanying the shipment of packaged or bulk cement. All packages shall be in good condition at the time of inspection.

14. Storage

14.1 The cement shall be stored in such a manner as to permit easy access for proper inspection and identification of each shipment, and in a suitable weather-tight building that will protect the cement from dampness and minimize warehouse set.

15. Manufacturer's Certification

15.1 Upon request of the purchaser in the contract or order, a manufacturer's report shall be furnished at the time of shipment stating the results of tests made on samples of the material taken during production or transfer and certifying that the applicable requirements of this specification have been met.

TABLE 2 Optional Chemical Requirements[A]

Cement Type	I and IA	II and IIA	III and IIIA	IV	V	Remarks
Tricalcium aluminate (C₃A),[B] max, %	...	...	8	...	...	for moderate sulfate resistance
Tricalcium aluminate (C₃A),[B] max, %	...	..	5	...	...	for high sulfate resistance
Sum of tricalcium silicate and tricalcium aluminate,[B] max, %	...	58[C]	...	...	...	for moderate heat of hydration
Alkalies (Na₂O + 0.658K₂O), max, %	0.60[D]	0.60[D]	0.60[D]	0.60[D]	0.60[D]	low-alkali cement

[A] These optional requirements apply only if specifically requested. Availability should be verified. See note in Section 4.

[B] The expressing of chemical limitations by means of calculated assumed compounds does not necessarily mean that the oxides are actually or entirely present as such compounds.

When expressing compounds, C = CaO, S = SiO₂, A = Al₂O₃, F = Fe₂O₃. For example, C₃A = 3CaO·Al₂O₃.

Titanium dioxide and phosphorus pentoxide (TiO₂ and P₂O₅) shall be included with the Al₂O₃ content. The value historically and traditionally used for Al₂O₃ in calculating potential compounds for specification purposes is the ammonium hydroxide group minus ferric oxide (R₂O₃ − Fe₂O₃) as obtained by classical wet chemical methods. This procedure includes as Al₂O₃ the TiO₂, P₂O₅, and other trace oxides which precipitate with the ammonium hydroxide group in the classical wet chemical methods. Many modern instrumental methods of cement analysis determine aluminum or aluminum oxide directly without the minor and trace oxides included by the classical method. Consequently, for consistency and to provide comparability with historic data and among various analytical methods, when calculating potential compounds for specification purposes, those using methods which determine Al or Al₂O₃ directly should add to the determined Al₂O₃ weight quantities of P₂O₅, TiO₂ and any other oxide except Fe₂O₃ which would precipitate with the ammonium hydroxide group when analyzed by the classical method and which is present in an amount of 0.05 weight % or greater. The weight percent of minor or trace oxides to be added to Al₂O₃ by those using direct methods may be obtained by actual analysis of those oxides in the sample being tested or estimated from historical data on those oxides on cements from the same source, provided that the estimated values are identified as such.

When the ratio of percentages of aluminum oxide to ferric oxide is 0.64 or more, the percentages of tricalcium silicate, dicalcium silicate, tricalcium aluminate and tetracalcium aluminoferrite shall be calculated from the chemical analysis as follows:

Tricalcium silicate = (4.071 × % CaO) − (7.600 × % SiO₂) − (6.718 × % Al₂O₃) − (1.430 × % Fe₂O₃) − (2.852 × % SO₃)
Dicalcium silicate = (2.867 × % SiO₂) − (0.7544 × % C₃S)
Tricalcium aluminate = (2.650 × % Al₂O₃) − (1.692 × % Fe₂O₃)
Tetracalcium aluminoferrite = 3.043 × % Fe₂O₃

When the alumina-ferric oxide ratio is less than 0.64, a calcium aluminoferrite solid solution (expressed as ss (C₄AF + C₂F)) is formed. Contents of this solid solution and of tricalcium silicate shall be calculated by the following formulas:

ss(C₄AF + C₂F) = (2.100 × % Al₂O₃) + (1.702 × % Fe₂O₃)
Tricalcium silicate = (4.071 × % CaO) − (7.600 × % SiO₂) − (4.479 × % Al₂O₃) − (2.859 × % Fe₂O₃) − (2.852 × % SO₃).

No tricalcium aluminate will be present in cements of this composition. Dicalcium silicate shall be calculated as previously shown.

In the calculation of all compounds the oxides determined to the nearest 0.1 % shall be used.

All values calculated as described in this note shall be reported to the nearest 1 %.

[C] The optional limit for heat of hydration in Table 4 shall not be requested when this optional limit is requested.

[D] This limit may be specified when the cement is to be used in concrete with aggregates that may be deleteriously reactive. Reference should be made to Specification C 33 for suitable criteria of deleterious reactivity.

TABLE 3 Standard Physical Requirements

Cement Type[A]	I	IA	II	IIA	III	IIIA	IV	V
Air content of mortar,[B] volume %:								
max	12	22	12	22	12	22	12	12
min	...	16	...	16	...	16	...	...
Fineness,[C] specific surface, m²/kg (alternative methods):								
Turbidimeter test, min	160	160	160	160	...	...	160	160
Air permeability test, min	280	280	280	280	...	...	280	280
Autoclave expansion, max, %	0.80	0.80	0.80	0.80	0.80	0.80	0.80	0.80
Strength, not less than the values shown for the ages indicated below:[D]								
Compressive strength, psi (MPa):								
1 day	...	...	...	...	1800 (12.4)	1450 (10.0)	...	...
3 days	1800 (12.4)	1450 (10.0)	1500 (10.3) 1000 (6.9)[F]	1200 (8.3) 800 (5.5)[F]	3500 (24.1)	2800 (19.3)	...	1200 (8.3)
7 days	2800 (19.3)	2250 (15.5)	2500 (17.2) 1700 (11.7)[F]	2000 (13.8) 1350 (9.3)[F]	...	...	1000 (6.9)	2200 (15.2)
28 days	...	...	...	...	...	...	2500 (17.2)	3000 (20.7)
Time of setting (alternative methods):[E]								
Gillmore test:								
Initial set, min, not less than	60	60	60	60	60	60	60	60
Final set, min, not more than	600	600	600	600	600	600	600	600
Vicat test:[G]								
Time of setting, min, not less than	45	45	45	45	45	45	45	45
Time of setting, min, not more than	375	375	375	375	375	375	375	375

[A] See Note.

[B] Compliance with the requirements of this specification does not necessarily ensure that the desired air content will be obtained in concrete.

[C] Either of the two alternative fineness methods may be used at the option of the testing laboratory. However, when the sample fails to meet the requirements of the air-permeability test, the turbidimeter test shall be used, and the requirements in this table for the turbidimetric method shall govern.

[D] The strength at any specified test age shall be not less than that attained at any previous specified test age.

[E] The purchaser should specify the type of setting-time test required. In case he does not so specify, the requirements of the Vicat test only shall govern.

[F] When the optional heat of hydration or the chemical limit on the sum of the tricalcium silicate and tricalcium aluminate is specified.

[G] The time of setting is that described as initial setting time in Test Method C 191.

Standard Specification for

SURFACE-APPLIED BONDING AGENTS FOR EXTERIOR PLASTERING[1]

Designation: C 932 – 80 (Reapproved 1985)

This standard is issued under the fixed designation C 932; the number immediately following the designation indicates the year of original adoption or, in the case of revision, the year of last revision. A number in parentheses indicates the year of last reapproval. A superscript epsilon (ε) indicates an editorial change since the last revision or reapproval.

This specification has been approved for use by agencies of the Department of Defense and for listing in the DoD Index of Specifications and Standards

1. Scope

1.1 This specification establishes minimum standards for exterior surface-applied bonding agents for improving the adhesion of cementitious material to concrete or other masonry surface or any structurally sound surface. Included are specification limits and test methods.

2. Applicable Documents

2.1 *ASTM Standards:*
C 190 Test Method for Tensile Strength of Hydraulic Cement Mortars[2]
D 1779 Specification for Adhesive for Acoustical Materials[3]
2.2 *Federal Test Method:*[4]
Fed. Test Method Std. No. 141 Paint, Varnish, Lacquer, and Related Materials; Methods of Inspection, Sampling, and Testing
Method 2012.1 Preparation of Tin Panels
Method 6221 Flexibility

3. Physical Requirements

3.1 *Surface-Applied Bonding Agent*—A film-forming, freeze-thaw stable composition, suitable for brush, roller, or spray application. It shall be tinted to show by visual inspection where it has been applied. The tint shall not bleed through the material being bonded. The bonding agent shall be nontoxic and nonflammable.

NOTE 1—Inasmuch as there is no prescribed test to determine if a material is film-forming, this property shall be determined by visual inspection to determine the presence of a continuous film not broken by fisheyes, cracking, pull-back, or any other discontinuity in the film surface.

3.2 *Consistency*—The bonding agent shall be free of foreign matter and shall be of such uniform consistency that when applied in accordance with the manufacturer's recommendations by brush, roller, or spray to concrete, masonry, or structurally sound surface, the bonding agent shall flow on evenly and dry to a uniform, continuous smooth film.

NOTE 2—Determine the presence of foreign matter by visual inspection with the naked eye.

4. Detailed Physical Requirements

4.1 *Film Characteristics*—The dry film of the bonding agent shall remain flexible when tested in accordance with 6.1. It shall not be noticeably affected by alkalies or mild acids. It shall be capable of bonding cementitious materials immediately after application and drying of the bonding agent, as recommended by the manufacturer, and up to 10 days after application when applied in accordance with the manufacturer's recommendations.

4.2 *Package Stability*—The bonding agent shall be of a uniform consistency with no settling of the solids. Label the package with a date indicating the period of time in which the material shall be usable. When tested by the method specified in 6.2, the bonding agent shall stand exposure to storage temperatures up to 140°F (60°C).

4.3 *Freeze-Thaw Stability*—When tested as specified in 6.3, the bonding agent shall withstand five cycles of alternate freezing and thawing without adversely affecting its application and tensile bonding strength.

4.4 *Tensile Bond Strength:*

4.4.1 *Tensile Strength of Fresh Sample*—When tested in accordance with 6.4, a fresh sample of bonding agent shall have a minimum average tensile strength of 150 psi (1034 kPa).

4.4.2 *Tensile Strength of 6-Month-Old Sample*—When tested in accordance with 6.4, a 6-month-old sample of bonding agent shall meet the minimum average tensile strength of 150 psi (1034 kPa).

5. Sampling

5.1 Sample in accordance with Specification D 1779.

6. Test Methods

6.1 *Flexibility Test*—Use the bonding agent sampled in accordance with 5.1 to cast a continuous film in the following manner:

6.1.1 The bonding agent shall be brushed, rolled, sprayed, or applied by a film applicator on the surface of a tin panel conforming to the specifications in Federal Test Method Std. No. 141, Method 2012.1. The thickness of the wet film shall be 5 mils (0.13 mm). Allow the film to dry at room temperature for 10 days before flexing.

6.1.2 Condition the test panels for at least 24 h at 77 ± 2°F (25 ± 1°C) and 50 ± 5 % relative humidity, and test in the same environment or immediately upon removal therefrom.

6.1.3 Bend the test panels over a 1-in. (25 mm) diameter bar in accordance with Federal Test Method Std. No. 141, Method 6221.

6.1.4 A crack more than a ¼ in. (6.35 mm) long shall be cause for rejection. Disregard cracks less than ¼ in. (6.35 mm) at the ends.

6.2 *High-Temperature Test*—Place a 1-pt (473 mL) container of bonding agent in an oven maintained at 140°F (60°C) for 30 days. The sample shall not separate to the extent that it cannot be remixed to uniform consistency. Exposure to 140°F (60°C) for 30 days shall not affect its application and tensile bonding strength. After such exposure the bonding agent shall be tested in accordance with 6.4.

6.3 *Freeze-Thaw Test*—Expose the bonding agent in a closed 1-pt (400 mL) container to a temperature of −10°F (−23.3°C) for 16 consecutive hours, then permit to thaw at room temperature. Repeat this procedure for a total of five cycles. After the fifth cycle, test the bonding agent as specified in 6.4.

6.4 *Tensile Bond Strength:*

6.4.1 Make briquets according to Test Method C 190. Place the molded briquets in a moist cabinet for 48 h, remove, and then remove from the molds. Saw the cured briquets in half at the waistline and store in the atmosphere at 70°F (21.1°C) and 50 % relative humidity. Discard any briquets that have air holes or an excessively rough surface at the sawed face.

NOTE 3—Conditions conducive to dust accumulation, water, and moisture damage, etc., should be taken into account.

6.4.2 *Application of Bonding Agent*—Apply the bonding agent as supplied, or diluted in accordance with the manufacturer's directions, to the surface of the sawed face of the half briquets by means of a 1-in. (25-mm) paint brush and allow to air dry.

6.4.3 Place one end-coated half-briquet into each of twelve molds and fill the remaining half of the mold with hydraulic cement mortar in accordance with Test Method C 190. Remove the bonded briquets from the molds after 24 h. This time shall be considered as the first day of cure in an atmosphere at 75 ± 5°F (23.9 ± 2.8°C) and 50 % relative humidity.

6.4.4 Test the briquets for tensile bond strength in accordance with Test Method C 190. Air-cure three briquets for 4 days, soak in water for 2 days, air-cure for another 24 h, and test. Test three briquets after air-curing for 7 days. Allow three briquets to air-cure for 25 days, soak in water for 2 days, air-cure for another 24 h, and test. Air-cure three briquets for 28 days and test.

6.4.5 *Report*—Test each group of three briquets for bond strength and report the average of each group. If any one result varies more than 15 % from the average, reject it, and report the average of two results. If the results from more than one sample vary over 15 % from the average, reject the series and retest.

7. Packaging and Marking

7.1 *Packaging*—Pack the bonding agent in standard commercial containers constructed so as to ensure acceptance by common or other carriers for safe transportation at the lowest rate to the point of delivery, unless otherwise specified in the contract or order.

7.2 *Marking*—Mark the shipping containers with the name of the bonding agent, the quantity contained therein, the name of the manufacturer, and the batch number.

7.3 Label the package with a date indicating the period of time in which the material shall be usable.

---
[1] This specification is under the jurisdiction of ASTM Committee C-11 on Gypsum and Related Building Materials and Systems and is the direct responsibility of Subcommittee C11.02 on Specifications and Test Methods for Accessories and Related Products.
Current edition approved March 28, 1980. Published May 1980.
[2] *Annual Book of ASTM Standards*, Vol 04.01.
[3] *Annual Book of ASTM Standards*, Vol 15.06.
[4] Available from Naval Publications and Forms Center, 5801 Tabor Ave., Philadelphia, PA 19120.

 **Designation: C 897 – 83**

Standard Specification for

AGGREGATE FOR JOB-MIXED PORTLAND CEMENT-BASED PLASTERS[1]

This standard is issued under the fixed designation C 897; the number immediately following the designation indicates the year of original adoption or, in the case of revision, the year of last revision. A number in parentheses indicates the year of last reapproval. A superscript epsilon (ε) indicates an editorial change since the last revision or reapproval.

1. Scope

1.1 This specification covers natural or manufactured aggregate for use in job-mixed base and finish-coat portland cement, portland cement-lime and modified portland cement plasters.

2. Applicable Documents

2.1 *ASTM Standards:*

C 29 Test Method for Unit Weight and Voids in Aggregate[2]

C 35 Specification for Inorganic Aggregates for Use in Gypsum Plaster[2]

C 40 Test Method for Organic Impurities in Fine Aggregates for Concrete[2]

C 87 Test Method for Effect of Organic Impurities in Fine Aggregate on Strength of Mortar[2]

C 88 Test Method for Soundness of Aggregates by Use of Sodium Sulfate or Magnesium Sulfate[2]

C 117 Test Method for Material Finer Than 75-μm (No. 200) Sieve in Mineral Aggregates by Washing[2]

C 123 Test Method for Lightweight Pieces in Aggregate[2]

C 125 Definitions of Terms Relating to Concrete and Concrete Aggregates[2]

C 136 Method for Sieve Analysis of Fine and Coarse Aggregates[2]

C 142 Test Method for Clay Lumps and Friable Particles in Aggregates[2]

D 75 Practice for Sampling Aggregates[2]

E 11 Specification for Wire—Cloth Sieves for Testing Purposes[3]

2.2 *American National Standards Institute Standard:*

A42.2 Specification for Portland Cement and Portland Cement-Lime Plastering, Exterior (Stucco) and Interior

3. Definitions

3.1 *natural sand*—the fine granular material resulting from the natural disintegration of rock.

3.2 *manufactured sand*—the fine material resulting from the crushing and classification by screening, or otherwise, of rock, gravel, or blast furnace slag.

4. Materials and Manufacture

4.1 Manufactured aggregate shall be specially processed, to assure particle shape in addition to meeting the gradation requirements of Section 7.

NOTE 1—Fine sand having rounded particle shape, uniform in size, shall not be used for basecoat plastering.

5. Composition

5.1 *Deleterious Substances*—The amount of deleterious substances in aggregates, each determined on independent samples (see Test Method C 40) complying with the grading requirements of Section 7, shall not exceed the following:

Item	Maximum Permissible Weight, %
Friable particles	1.0
Light weight particles, floating on liquid having a specific gravity of 2.0	0.5

6. Physical Properties

6.1 *Soundness* (see 8.1.8):

6.1.1 Except as herein provided, aggregate subjected to five cycles of the soundness test shall show a loss, when weighed in accordance with the grading of a sampling complying with the limitations set forth in Section 7, not greater than 20 % when sodium sulfate is used, or 15 % when magnesium sulfate is used.

6.2.1 Aggregates failing to meet the requirements of 6.1.1 shall be accepted, provided there is evidence that plasters of comparable properties made from similar aggregates from the same source has been exposed to weathering, similar to that to be encountered, for a period of more than 5 years without appreciable disintegration.

7. Dimensions, Mass, and Permissible Variations

7.1 *Sand Aggregate for Base Coat*—Aggregate for use in base coat portland cement plasters shall be graded as follows:

U. S. Standard Sieve	Max %	Min %
No. 4 (4.75-mm)	0	—
No. 8 (2.36-mm)	10	0
No. 16 (1.18-mm)	40	10
No. 30 (600-μm)	65	30
No. 50 (300-μm)	90	70
No. 100 (150-μm)	100	95

7.1.1 Not more than 50 % shall be retained between any two consecutive sieves shown in the above table, nor more than 25 % between No. 50 (300-μm) and No. 100 (150-μm) sieves. The amount of material finer than a No. 200 (75-μm) sieve shall not exceed 3 %.

7.1.2 The fineness modulus shall fall between 2.05 and 3.05. If the fineness modulus varies by more than 0.25 from the value assumed in selecting proportions for the plaster, the aggregate shall be rejected unless adjustments are made in proportions to compensate for change in grading.

NOTE 2—If a change is made in gradation in the direction of smaller aggregate sizes, representing a greater specific surface area, this change shall be compensated by a change in the aggregate-cement ratio to a larger ratio. Conversely a change in the direction of larger aggregate sizes necessitates a smaller aggregate-cement ratio.

7.2 *Sand Aggregate for Finish Coat*—These plasters shall meet the requirements of 7.1, except that all aggregate shall pass the No. 8 (2.36-mm) sieve.

7.2.1 Job-mixed aggregates for special texture finishes require a graduation outside of the limits specified in 7.2. Where such special textures are specified, sample panels shall be required and special care taken to avoid unusual or excessive shrinkage or other detrimental characteristics.

7.3 *Lightweight Aggregate for Base and Finish Coat*—Perlite aggregate for base and finish coat shall meet the requirements for perlite in Specification C 35.

8. Sampling

8.1 The aggregate shall be sampled and tested in accordance with the following methods, except as otherwise specified in this specification:

8.1.1 *Sampling*, Practice D 75.

8.1.2 *Fineness Modulus*, Definitions C 125, is obtained by adding the total percentages shown by the sieve analysis to be retained on each of the sieves in 7.1 and dividing the sum by 100.

8.1.3 *Materials Finer than No. 200 (75-μm) Sieve*, Test Method C 117 by washing.

8.1.4 *Friable Particles*, Test Method C 142.

8.1.5 *Lightweight Constituents*, Test Method C 123.

8.1.6 *Organic Impurities*, Test Method C 40.

8.1.7 *Effect of Organic Impurities on Strength*, Test Method C 87.

8.1.8 *Soundness*, Test Method C 88.

8.1.9 *Sieve Analysis*, Test Method C 136.

8.1.10 *Unit Weight*, Test Method C 29.

9. Inspection

9.1 Inspection of the material shall be agreed upon between the purchaser and supplier as part of the purchase contract.

10. Rejection and Rehearing

10.1 Any rejection shall be based upon the specific cause of failure to conform to the requirements of this specification, and shall be reported to the seller within ten working days from the receipt of the shipment by the purchaser.

10.2 Claims for rehearing shall be valid only if made within 20 working days from receipt of notice of specific cause for rejection.

11. Certification

11.1 When specified in the purchase order a producer's or supplier's certification shall be furnished to the purchaser that the material was prepared or manufactured in accordance with this specification and has been found to meet the specified requirements.

12. Packaging and Package Marking

12.1 Where packaged, the materials shall be packaged in substantial commercial containers of the type, size, and kind commonly used for the purpose, so constructed as to preserve the contents in good condition, and to ensure acceptance and safe delivery by common or other carriers, at the lowest rate, to the point of delivery.

12.2 The containers shall be so made that the contents can be partially removed without destroying the container's ability to serve as a receptacle for the remainder of the contents.

12.3 The containers shall be labeled to show the content, weight, and the name of the manufacturer or supplier.

12.4 Where delivered in bulk quantities, the bill of lading shall show the quantity by weight, or cubic yards, and the name of the manufacturer or supplier.

12.5 Each package or container shall be marked with the name of manufacturer, description of contents, and amount.

[1] This specification is under the jurisdiction of ASTM Committee C-11 on Gypsum and Related Building Materials and Systems, and is the direct responsibility of Subcommittee C 11.02 on Specifications and Test Methods for Accessories and Related Products.

Current edition approved Aug. 26, 1983. Published October 1983. Originally published as C 897 – 78. Last previous edition C 897 – 78.

[2] Annual Book of ASTM Standards. Vol 04.02.

[3] Annual Book of ASTM Standards, Vol 14.02.

Designation: C 144 – 84

Standard Specification for
AGGREGATE FOR MASONRY MORTAR[1]

This standard is issued under the fixed designation C 144; the number immediately following the designation indicates the year of original adoption or, in the case of revision, the year of last revision. A number in parentheses indicates the year of last reapproval. A superscript epsilon (ε) indicates an editorial change since the last revision or reapproval.

This specification has been approved for use by agencies of the Department of Defense and for listing in the DoD Index of Specifications and Standards.

Scope

1.1 This specification covers aggregate for in masonry mortar.

Applicable Documents

2.1 *ASTM Standards:*

C 40 Test Method for Organic Impurities in Fine Aggregates for Concrete[2]

C 87 Test Method for Effect of Organic Impurities in Fine Aggregate on Strength of Mortar[2]

C 88 Test Method for Soundness of Aggregates by Use of Sodium Sulfate or Magnesium Sulfate[2]

C 117 Test Method for Materials Finer than 75-μm (No. 200) Sieve in Mineral Aggregates by Washing[2]

C 123 Test Method for Lightweight Pieces in Aggregate[2]

C 125 Definitions of Terms Relating to Concrete and Concrete Aggregates[2]

C 136 Method for Sieve Analysis of Fine and Coarse Aggregates[2]

C 142 Test Method for Clay Lumps and Friable Particles in Aggregates[2]

C 270 Specification for Mortar for Unit Masonry[3]

C 404 Specification for Aggregates for Masonry Grout[2,3]

D 75 Practice for Sampling Aggregates[4]

Material

3.1 Aggregate for use in masonry mortar shall consist of natural sand or manufactured sand. Manufactured sand is the product obtained by crushing stone, gravel, or air-cooled iron blast-furnace slag specially processed to assure suitable particle shape as well as gradation.

4. Grading

4.1 Aggregate for use in masonry mortar shall be graded within the following limits, depending upon whether natural sand or manufactured sand is to be used:

	Percent Passing	
Sieve Size	Natural Sand	Manufactured Sand
No. 4 (4.75-mm)	100	100
No. 8 (2.36-mm)	95 to 100	95 to 100
No. 16 (1.18-mm)	70 to 100	70 to 100
No. 30 (600-μm)	40 to 75	40 to 75
No. 50 (300-μm)	10 to 35	20 to 40
No. 100 (150-μm)	2 to 15	10 to 25
No. 200 (75-μm)	...	0 to 10

4.2 The aggregate shall not have more than 50 % retained between any two consecutive sieves of those listed in 3.1 nor more than 25 % between No. 50 (300-μm) and the No. 100 (150-μm) sieve.

4.3 If the fineness modulus varies by more than 0.20 from the value assumed in selecting proportions for the mortar, the aggregate shall be rejected unless suitable adjustments are made in proportions to compensate for the change in grading.

NOTE 1—For heavy construction employing joints thicker than ½ in. (13 mm), a coarser aggregate may be desirable; for such work a fine aggregate conforming to Specification C 404 is satisfactory.

4.4 When an aggregate fails the gradation limits specified in 4.1 and 4.2, it may be used provided the mortar can be prepared to comply with the aggregate ratio, water retention, and compressive strength requirements of the property specifications of Specification C 270.

5. Composition

5.1 *Deleterious Substances*—The amount of deleterious substances in aggregate for masonry mortar, each determined on independent samples complying with the grading requirements of Section 4, shall not exceed the following:

Item	Maximum Permissible Weight Percent
Friable particles	1.0
Lightweight particles, floating on liquid having a specific gravity of 2.0	0.5[A]

[A] This requirement does not apply to blast-furnace slag aggregate.

5.2 *Organic Impurities:*

5.2.1 The aggregate shall be free of injurious amounts of organic impurities. Except as herein provided, aggregates subjected to the test for organic impurities and producing a color darker than the standard shall be rejected.

5.2.2 Aggregate failing in the test may be used, provided that the discoloration is due principally to the presence of small quantities of coal, lignite, or similar discrete particles.

5.2.3 Aggregate failing in the test may be used provided that, when tested for the effect of organic impurities on strength of mortar, the relative strength at 7 days calculated in accordance with Section 10 of Test Method C 87, is not less than 95 %.

6. Soundness

6.1 Except as herein provided, aggregate subjected to five cycles of the soundness test shall show a loss, weighted in accordance with the grading of a sample complying with the limitations set forth in Section 4, not greater than 10 % when sodium sulfate is used or 15 % when magnesium sulfate is used.

6.2 Aggregate failing to meet the requirements of 6.1 may be accepted, provided that mortar of comparable properties made from similar aggregates from the same source has been exposed to weathering, similar to that to be encountered, for a period of more than 5 years without appreciable disintegration.

7. Methods of Sampling and Testing

7.1 Sample and test the aggregate in accordance with the following ASTM methods, except as otherwise provided in this specification:

7.1.1 *Sampling*—Practice D 75, except that the fine aggregate portion only of Table 1 shall be used for approximate minimum mass of field samples. Sampling from stockpiles shall be allowed if the sampling plan specified in 3.3.3 is a hollow tube with an inside diameter of approximately 2 in. (50 mm) to be inserted or driven into the stockpile to obtain samples from the interior of the pile.

7.1.2 *Sieve Analysis*—Method C 136.

7.1.3 *Amount of Material Finer Than No. 200 (75-μm) Sieve*—Test Method C 117.

7.1.4 *Organic Impurities*—Test Method C 40.

7.1.5 *Effect of Organic Impurities on Strength*—Test Method C 87.

7.1.6 *Friable Particles*—Test Method C 142.

7.1.7 *Lightweight Constituents*—Test Method C 123.

7.1.8 *Fineness Modulus*—The fineness modulus, as defined in Definitions C 125 is obtained by adding the total percentages shown by the sieve analysis to be retained on each of the following sieves, and dividing the sum by 100:

No. 100 (150-μm)
No. 50 (300-μm)
No. 30 (600-μm)
No. 16 (1.18-mm)
No. 8 (2.36-mm)
No. 4 (4.75-mm)

7.1.9 *Soundness*—Test Method C 88.

[1] This specification is under the jurisdiction of ASTM Committee C-12 on Mortars for Unit Masonry and is the direct responsibility of Subcommittee C12.04 on Specifications for Aggregates for Mortar.

Current edition approved Nov. 30, 1984. Published January 1985. Originally published as C 144–39 T. Last previous edition C 144–81.

[2] *Annual Book of ASTM Standards*, Vol 04.02.

[3] *Annual Book of ASTM Standards*, Vol 04.05.

[4] *Annual Book of ASTM Standards*, Vols 04.02 and 04.03.

Designation: E 119 – 83ᵉ¹

Standard Methods of
FIRE TESTS OF BUILDING CONSTRUCTION AND MATERIALS[1]

This standard is issued under the fixed designation E 119; the number immediately following the designation indicates the year of original adoption or, in the case of revision, the year of last revision. A number in parentheses indicates the year of last reapproval.

These methods have been approved for use by agencies of the Department of Defense and for listing in the DoD Index of Specifications and Standards.

ᵉ¹ NOTE—SI equivalents in 9.4 and X5.9.1 were corrected editorially in July 1984.

INTRODUCTION

The performance of walls, columns, floors, and other building members under fire exposure conditions is an item of major importance in securing constructions that are safe, and that are not a menace to neighboring structures nor to the public. Recognition of this is registered in the codes of many authorities, municipal and other. It is important to secure balance of the many units in a single building, and of buildings of like character and use in a community; and also to promote uniformity in requirements of various authorities throughout the country. To do this it is necessary that the fire-resistive properties of materials and assemblies be measured and specified according to a common standard expressed in terms that are applicable alike to a wide variety of materials, situations, and conditions of exposure.

Such a standard is found in the methods that follow. They prescribe a standard exposing fire of controlled extent and severity. Performance is defined as the period of resistance to standard exposure elapsing before the first critical point in behavior is observed. Results are reported in units in which field exposures can be judged and expressed.

The methods may be cited as the "Standard Fire Tests," and the performance or exposure shall be expressed as "2-h," "6-h," "½-h," etc.

When a factor of safety exceeding that inherent in the test conditions is desired, a proportional increase should be made in the specified time-classification period.

1. Scope

1.1 These methods are applicable to assemblies of masonry units and to composite assemblies of structural materials for buildings, including bearing and other walls and partitions, columns, girders, beams, slabs, and composite slab and beam assemblies for floors and roofs. They are also applicable to other assemblies and structural units that constitute permanent integral parts of a finished building.

1.2 It is the intent that classifications shall register performance during the period of exposure and shall not be construed as having determined suitability for use after fire exposure.

1.3 *This standard should be used to measure and describe the properties of materials, products, or assemblies in response to heat and flame under controlled laboratory conditions and should not be used to describe or appraise the fire hazard or fire risk of materials, products, or assemblies under actual fire conditons. However, results of this test may be used as elements of a fire risk assessment which takes into account all of the factors which are pertinent to an assessment of the fire hazard of a particular end use.*

NOTE 1—A method of fire hazard classification based on rate of flame spread is covered in ASTM Method E 84, Test for Surface Burning Characteristics of Building Materials.[2]

1.4 The results of these tests are one factor in assessing fire performance of building construction and assemblies. These methods prescribe a standard fire exposure for comparing the performance of building construction assemblies. Application of these test results to predict the performance of actual building construction requires careful evaluation of test conditions.

2. Significance

2.1 This standard is intended to evaluate the duration for which the types of assemblies noted in 1.1 will contain a fire, or retain their structural integrity or exhibit both properties dependent upon the type of assembly involved during a predetermined test exposure.

2.2 The test exposes a specimen to a *standard fire exposure* controlled to achieve specified temperatures throughout a specified time period. In some instances, the *fire exposure may be* followed by the application of a *specified standard* fire hose stream. The exposure, however, may not be representative of all fire conditions which may vary with changes in the amount, nature and distribution of fire loading, ventilation, compartment size and configuration, and heat sink characteristics of the compartment. It does, however, provide a relative measure of fire performance of comparable assemblies under these specified fire exposure conditions. Any variation from the construction or conditions (that is, size, method of assembly, and materials) that are tested may substantially change the performance characteristics of the assembly.

2.3 The test standard provides for the following:

2.3.1 In walls, partitions, and floor or roof assemblies:

2.3.1.1 Measurement of the transmission of heat.

2.3.1.2 Measurement of the transmission of hot gases through the assembly, sufficient to ignite cotton waste.

2.3.1.3 For load bearing elements, measurement of the load carrying ability of the *test specimen* during the test exposure.

2.3.2 For individual load bearing assemblies such as beams and columns: Measurement of

the load carrying ability under the test exposure with some consideration for the end support conditions (that is, restrained or not restrained).

2.4 The test standard does not provide the following:

2.4.1 Full information as to performance of assemblies constructed with components or lengths other than those tested.

2.4.2 Evaluation of the degree by which the assembly contributes to the fire hazard by generation of smoke, toxic gases, or other products of combustion.

2.4.3 Measurement of the degree of control or limitation of *the passage of* smoke or products of combustion through the assembly.

2.4.4 Simulation of the fire behavior of joints between building elements such as floor-wall or wall-wall, etc., connections.

2.4.5 Measurement of flame spread over surface of tested element.

2.4.6 The effect of fire endurance of conventional openings in the assembly, that is, electrical receptacle outlets, plumbing pipe, etc., unless specifically provided for in the construction tested.

CONTROL OF FIRE TESTS

3. Time-Temperature Curve

3.1 The conduct of fire tests of materials and construction shall be controlled by the standard time-temperature curve shown in Fig. 1. The points on the curve that determine its character are:

1000°F (538°C)	at 5 min
1300°F (704°C)	at 10 min
1550°F (843°C)	at 30 min
1700°F (927°C)	at 1 h
1850°F (1010°C)	at 2 h
2000°F (1093°C)	at 4 h
2300°F (1260°C)	at 8 h or over

3.2 For a closer definition of the time-temperature curve, see Appendix X1.

NOTE 2—*Recommendations for Recording Fuel Flow to Furnace Burners*—The following provides guidance on the desired characteristics of instrumentation for recording the flow of fuel to the furnace burners. Fuel flow data may be useful for a furnace heat balance analysis, for measuring the effect of furnace or control changes, and for comparing the per-

---

[1] These methods are under the jurisdiction of ASTM Committee E-5 on Fire Standards and are the direct responsibility of Subcommittee E05.11 on Building Construction.

Current edition approved March 25, 1983. Published July 1983. Originally published as C 19–1917 T. Last previous edition E 119–82.

These methods, of which the present standard represents a revision, were prepared by Sectional Committee A2 on Fire Tests of Materials and Construction, under the joint sponsorship of the National Bureau of Standards, the ANSI Fire Protection Group, and the American Society for Testing and Materials, functioning under the procedure of the American National Standards Institute.

[2] *1983 Annual Book of ASTM Standards*, Vol 04.07.

formance of assemblies of different properties in the fire endurance test.[3]

Record the integrated (cumulative) flow of gas (or other fuel) to the furnace burners at 10 min, 20 min, 30 min, and every 30 min thereafter or more frequently. Total gas consumed during the total test period is also to be determined. A recording flow meter has advantages over periodic readings on an instantaneous or totalizing flow meter. Select a measuring and recording system to provide flow rate readings accurate to within ± 5 %.

Report the type of fuel, its higher (gross) heating value, and the fuel flow (corrected to standard conditions of 60°F (16°C) and 30.0 in. Hg) as in function of time.

4. Furnace Temperatures

4.1 The temperature fixed by the curve shall be deemed to be the average temperature obtained from the readings of not less than nine thermocouples for a floor, roof, wall, or partition and not less than eight thermocouples for a structural column symmetrically disposed and distributed to show the temperature near all parts of the sample, the thermocouples being enclosed in protection tubes of such materials and dimensions that the time constant of the protected thermocouple assembly lies within the range from 5.0 to 7.2 min (Note 3). The exposed length of the pyrometer tube and thermocouple in the furnace chamber shall be not less than 12 in. (305 mm). Other types of protecting tubes or pyrometers may be used that, under test conditions, give the same indications as the above standard within the limit of accuracy that applies for furnace-temperature measurements. For floors and columns, the junction of the thermocouples shall be placed 12 in. away from the exposed face of the sample at the beginning of the test and, during the test, shall not touch the sample as a result of its deflection. In the case of walls and partitions, the thermocouples shall be placed 6 in. (152 mm) away from the exposed face of the sample at the beginning of the test, and shall not touch the sample during the test, in the event of deflecton.

NOTE 3—A typical thermocouple assembly meeting these time constant requirements may be fabricated by fusion-welding the twisted ends of No. 18 gage Chromel-Alumel wires, mounting the leads in porcelain insulators and inserting the assembly so the thermocouple bead is ½ in. from the sealed end of a standard weight nominal ½ in. iron, steel, or Inconel pipe. The time constant for this and for several other thermocouple assemblies was measured in 1976. The time constant may also be calculated from knowledge of its physical and thermal properties.[4]

4.2 The temperatures shall be read at intervals not exceeding 5 min during the first 2 h, and thereafter the intervals may be increased to not more than 10 min.

4.3 The accuracy of the furnace control shall be such that the area under the time-temperature curve, obtained by averaging the results from the pyrometer readings, is within 10 % of the corresponding area under the standard time-temperature curve shown in Fig. 1 for fire tests of 1 h or less duration, within 7.5 % for those over 1 h and not more than 2 h, and within 5 % for tests exceeding 2 h in duration.

5. Temperatures of Unexposed Surfaces of Floors, Roofs, Walls, and Partitions

5.1 Temperatures of unexposed surfaces shall be measured with thermocouples or thermometers (Note 5) placed under dry, felted

pads meeting the requirements listed in Annex A1. The wire leads of the thermocouple or the stem of the thermometer shall have an immersion under the pad and be in contact with the unexposed surface for not less than 3½ in. (89 mm). The hot junction of the thermocouple or the bulb of the thermometer shall be placed approximately under the center of the pad. The outside diameter of protecting or insulating tubes, and of thermometer stems, shall be not more than 5⁄16 in. (8 mm). The pad shall be held firmly against the surface, and shall fit closely about the thermocouples or thermometer stems. Thermometers shall be of the partial-immersion type, with a length of stem, between the end of the bulb and the immersion mark, of 3 in. (76 mm). The wires for the thermocouple in the length covered by the pad shall be not heavier than No. 18 B & S gage (0.04 in.) (1.02 mm) and shall be electrically insulated with heat-resistant and moisture-resistant coatings.

NOTE 4—For the purpose of testing roof assemblies, the unexposed surface shall be defined as the surface exposed to ambient air.
NOTE 5—Under certain conditions it may be unsafe or impracticable to use thermometers.

5.2 Temperature readings shall be taken at not less than nine points on the surface. Five of these shall be symmetrically disposed, one to be approximately at the center of the specimen, and four at approximately the center of its quarter sections. The other four shall be located at the discretion of the testing authority to obtain representative information on the performance of the construction under test. None of the thermocouples shall be located nearer to the edges of the test specimen than one and one-half times the thickness of the construction, or 12 in. (305 mm). An exception can be made in those cases where there is an element of the construction that is not otherwise represented in the remainder of the test specimen. None of the thermocouples shall be located opposite or on top of beams, girders, pilasters, or other structural members if temperatures at such points will obviously be lower than at more representative locations. None of the thermocouples shall be located opposite or on top of fasteners such as screws, nails, or staples that will be obviously higher or lower in temperature than at more representative locations if the aggregate area of any part of such fasteners projected to the unexposed surface is less than 0.8 percent of the area within any 5-in. (127-mm) square. Such fasteners shall not extend through the assembly.

5.3 Temperature readings shall be taken at intervals not exceeding 15 min until a reading exceeding 212°F (100°C) has been obtained at any one point. Thereafter the readings may be taken more frequently at the discretion of the testing body, but the intervals need not be less than 5 min.

5.4 Where the conditions of acceptance place a limitation on the rise of temperature of the unexposed surface, the temperature end point of the fire endurance period shall be determined by the average of the measurements taken at individual points; except that if a temperature rise 30 % in excess of the specified limit occurs at any one of these points, the remainder shall be ignored and the fire endurance period judged as ended.

CLASSIFICATION AS DETERMINED BY TEST

6. Report of Results

6.1 Results shall be reported in accordance with the performance in the tests prescribed in these methods. They shall be expressed in time periods of resistance, to the nearest integral minute. Reports shall include observations of significant details of the behavior of the material or construction during the test and after the furnace fire is cut off, including information on deformation, spalling, cracking, burning of the specimen or its component parts, continuance of flaming, and production of smoke.

6.2 Reports of tests involving wall, floor, beam, or ceiling constructions in which restraint is provided against expansion, contraction, or rotation of the construction shall describe the method used to provide this restraint.

6.3 Reports of tests which other than maximum load conditions are imposed shall fully define the conditions of loading used in the test and shall be designated in the title of the report of the test as a restricted load condition.

6.4 When the indicated resistance period is ½ h or over, determined by the average or maximum temperature rise on the unexposed surface or within the test sample, or by failure under load, a correction shall be applied for variation of the furnace exposure from that prescribed, where it will affect the classification, by multiplying the indicated period by two thirds of the difference in area between the curve of average furnace temperature and the standard curve for the first three fourths of the period and dividing the product by the area between the standard curve and a base line of 68°F (20°C) for the same part of the indicated period, the latter area increased by 54°F·h or 30°C·h (3240°F·min or 1800°C·min) to compensate for the thermal lag of the furnace thermocouples during the first part of the test. For fire exposure in the test higher than standard, the indicated resistance period shall be increased by the amount of the correction and be

CONDUCT OF FIRE TESTS

8. Fire Endurance Test

8.1 Continue the fire endurance test on the specimen with its applied load, if any, until failure occurs, or until the specimen has withstood the test conditions for a period equal to that herein specified in the conditions of acceptance for the given type of construction.

8.2 For the purpose of obtaining additional performance data, the test may be continued beyond the time the fire endurance classification is determined.

9. Hose Stream Test

9.1 Where required by the conditions of acceptance, subject a duplicate specimen to a fire exposure test for a period equal to one half of

[3] Harmathy, T. Z., "Design of Fire Test Furnaces," *Fire Technology*, Vol. 5, No. 2, May 1969, pp. 146–150; Seigel, L. G., "Effects of Furnace Design on Fire Endurance Test Results," *Fire Test Performance, ASTM STP 464*, Am. Soc. Testing Mats., 1970, pp. 57–67; and Williamson, R. B., and Buchanan, A. H., "A Heat Balance Analysis of the Standard Fire Endurance Test."
[4] Supporting data are available on loan from ASTM Headquarters, 1916 Race St., Philadelphia, Pa. 19103. Request RR: E05–1001.

that indicated as the resistance period in the fire endurance test, but not for more than 1 h, immediately after which subject the specimen to the impact, erosion, and cooling effects of a hose stream directed first at the middle and then at all parts of the exposed face. changes in direction being made slowly.

9.2 *Exemption*—The hose stream test shall not be required in the case of constructions having a resistance period, indicated in the fire endurance test, of less than 1 h.

9.3 *Optional Program*—The submitter may elect, with the advice and consent of the testing body, to have the hose stream test made on the specimen subjected to the fire endurance test and immediately following the expiration of the fire endurance test.

9.4 *Stream Equipment and Details*—The stream shall be delivered through a 2½-in. (64-mm) hose discharging through a National Standard Playpipe of corresponding size equipped with a 1⅛-in. (28.5-mm) discharge tip of the standard-taper smooth-bore pattern without shoulder at the orifice. The water pressure and duration of application shall be as prescribed in Table 1.

9.5 *Nozzle Distance*—The nozzle orifice shall be 20 ft (6 m) from the center of the exposed surface of the test specimen if the nozzle is so located that when directed at the center its axis is normal to the surface of the test specimen. If otherwise located, its distance from the center shall be less than 20 ft by an amount equal to 1 ft (305 mm) for each 10 deg of deviation from the normal.
similarly decreased for fire exposure below standard.

NOTE 6—The correction can be expressed by the following equation:

$$C = 2I(A - A_s)/3(A_s + L)$$

where:
C = correction in the same units as I,
I = indicated fire-resistance period,
A = area under the curve of indicated average furnace temperature for the first three fourths of the indicated period,
A_s = area under the standard furnace curve for the same part of the indicated period, and
L = lag correction in the same units as A and A_s (54°F·h or 30°C·h (3240°F·min or 1800°C·min)).

6.5 Unsymmetrical wall assemblies may be tested with either side exposed to the fire, and the report shall indicate the side so exposed. Both sides may be tested, and the report then shall so indicate the fire endurance classification applicable to each side.

TEST SPECIMEN

7. Test Specimen

7.1 The test specimen shall be truly representative of the construction for which classification is desired, as to materials, workmanship, and details such as dimensions of parts, and shall be built under conditions representative of those obtaining as practically applied in building construction and operation. The physical properties of the materials and ingredients used in the test specimen shall be determined and recorded.

7.2 The size and dimensions of the test specimen specified herein are intended to apply for rating constructions of dimensions within the usual general range employed in buildings. If the conditions of use limit the construction to smaller dimensions, a proportionate reduction may be made in the dimensions of the specimens for a test qualifying them for such restricted use.

7.3 When it is desired to include a built-up roof covering, the test specimen shall have a roof covering of 3-ply, 15-lb (6.8-kg) type felt not in excess of 120 lb (54 kg) per square (100 ft² (9 m²)) of hot mopping asphalt without gravel surfacing. Tests of assemblies with this covering do not preclude the field use of other built-up roof coverings.

10. Protection and Conditioning of Test Specimen

10.1 Protect the test specimen during and after fabrication to assure normality of its quality and condition at the time of test. It shall not be tested until a large portion of its final strength has been attained, and, if it contains moisture, until the excess has been removed to achieve an air-dry condition in accordance with the requirements given in 10.1.1 through 10.1.3. Protect the testing equipment and sample undergoing the fire test from any condition of wind or weather, that might lead to abnormal results. The ambient air temperature at the beginning of the test shall be within the range of 50 to 90°F (10 to 32°C). The velocity of air across the unexposed surface of the sample, measured just before the test begins, shall not exceed 4.4 ft (1.3 m)/s, as determined by an anemometer placed at right angles to the unexposed surface. If mechanical ventilation is employed during the test, an air stream shall not be directed across the surface of the specimen.

10.1.1 Prior to fire test, condition constructions with the objective of providing, within a reasonable time, a moisture condition within the specimen approximately representative of that likely to exist in similar construction in buildings. For purposes of standardization, this condition is to be considered as that which would be established at equilibrium resulting from drying in an ambient atmosphere of 50 % relative humidity at 73°F (Note 7). However, with some constructions, it may be difficult or impossible to achieve such uniformity within a reasonable period of time. Accordingly, where this is the case, specimens may be tested when the dampest portion of the structure, the portion at 6-in. (152-mm) depth below the surface of massive constructions, has achieved a moisture content corresponding to drying to equilibrium with air in the range of 50 to 75 % relative humidity at 73 ± 5°F (23 ± 3°C). In the event that specimens dried in a heated building fail to meet these requirements after a 12-month conditioning period, or in the event that the nature of the construction is such that it is evident that drying of the specimen interior will be prevented by hermetic sealing, these requirements may be waived, except as to attainment of a large portion of final strength, and the specimen tested in the condition in which it then exists.

10.1.2 If, during the conditioning of the specimen it appears desirable or is necessary to use accelerated drying techniques, it is the responsibility of the laboratory conducting the test to avoid procedures which will significantly alter the structural or fire endurance characteristics of the specimen or both from those produced as the result of drying in accordance with procedures given in 10.1.1.

10.1.3 Within 72 h prior to the fire test information on the actual moisture content and distribution within the specimen shall be obtained. Include this information in the test report (Note 8).

NOTE 7—A recommended method for determining the relative humidity within a hardened concrete specimen with electric sensing elements is described in Appendix I of the paper by Menzel, C. A., "A Method for Determining the Moisture Condition of Hardened Concrete in Terms of Relative Humidity," *Proceedings*, Am. Soc. Testing Mats., ASTEA, Vol 55, 1955, p. 1085. A similar procedure with electric sensing elements can be used to determine the relative humidity within fire test specimens made with other materials.

With wood constructions, the moisture meter based on the electrical resistance method can be used, when appropriate, as an alternative to the relative humidity method to indicate when wood has attained the proper moisture content. Electrical methods are described on pages 320 and 321 of the 1955 edition of the *Wood Handbook of the Forest Products Laboratory*, U.S. Department of Agriculture. The relationships between relative humidity and moisture content are given by the graphs in Fig. 23 on p. 327. They indicate that wood has a moisture content of 13 % at a relative humidity of 70 % for a temperature of 70 to 80°F (21 to 27°C).

NOTE 8—If the moisture condition of the fire test assembly is likely to change drastically from the 72-h sampling time prior to test, the sampling should be made not later than 24 h prior to the test.

11. Precision[5]

11.1 No comprehensive test program has been conducted to develop data on which to derive statistical measures of repeatability (within-laboratory variability) and reproducibility (among-laboratory variability). The limited data suggest that there is a degree of repeatability and reproducibility for some types of assemblies. Results depend on factors such as the type of assembly and materials being tested, the characteristics of the furnace, the type and level of applied load, the nature of the boundary conditions (restraint and end fixity), and details of workmanship during assembly.

TESTS OF BEARING WALLS AND PARTITIONS

12. Size of Sample

12.1 The area exposed to fire shall be not less than 100 ft² (9 m²), with neither dimension less than 9 ft (2.7 m). The test specimen shall not be restrained on its vertical edges.

13. Loading

13.1 Throughout the fire endurance and fire and hose stream tests apply a constant superimposed load to simulate a maximum load condition. The applied load shall be as nearly as practicable the maximum load allowed by design under nationally recognized structural design criteria. The tests may also be conducted by applying to the specimen a load less than the maximum. Such tests shall be identified in the test report as having been conducted under restricted load conditions. The applied load, and the ap-

[5] Supporting data on repeatability and reproducibility are available from ASTM Headquarters, 1916 Race St., Philadelphia, PA 19103. Request RR: E05-1003.

plied load expressed as a percentage of the maximum allowable design load, shall be included in the report. A double wall assembly shall be loaded during the test to simulate field use conditions, with either side loaded separately or both sides together (Note 9). The method used shall be reported.

NOTE 9—The choice depends on the intended use, and whether the load on the exposed side, after it has failed, will be transferred to the unexposed side. If, in the intended use, the load from the structure above is supported by both walls as a unit and would be or is transferred to the unexposed side in case of collapse of the exposed side, both walls should be loaded in the test by a single unit. If, in the intended use the load from the structure above each wall is supported by each wall separately, the walls should be loaded separately in the test by separate load sources. If the intended use of the construction system being tested involved situations of both loading conditions described above, the walls should be loaded separately in the test by separate load sources. In tests conducted with the walls loaded separately the condition of acceptance requiring the walls to maintain the applied load shall be based on the time at which the first of either of the walls fail to sustain the load.

14. Conditions of Acceptance

14.1 Regard the test as successful if the following conditions are met:

14.1.1 The wall or partition shall have sustained the applied load during the fire endurance test without passage of flame or gases hot enough to ignite cotton waste, for a period equal to that for which classification is desired.

14.1.2 The wall or partition shall have sustained the applied load during the fire and hose stream test as specified in Section 9, without passage of flame, of gases hot enough to ignite cotton waste, or of the hose stream. The assembly shall be considered to have failed the hose stream test if an opening develops that permits a projection of water from the stream beyond the unexposed surface during the time of the hose stream test.

14.1.3 Transmission of heat through the wall or partition during the fire endurance test shall not have been such as to raise the temperature on its unexposed surface more than 250°F (139°C) above its initial temperature.

TESTS OF NONBEARING WALLS AND PARTITIONS

15. Size of Sample

15.1 The area exposed to fire shall be not less than 100 ft² (9 m²), with neither dimension less than 9 ft (2.7 m). Restrain the test specimen on all four edges.

16. Conditions of Acceptance

16.1 Regard the test as successful if the following conditions are met:

16.1.1 The wall or partition shall have withstood the fire endurance test without passage of flame or gases enough to ignite cotton waste, for a period equal to that for which classification is desired.

16.1.2 The wall or partition shall have withstood the fire and hose stream test as specified in Section 9, without passage of flame, of gases hot enough to ignite cotton waste, or of the hose stream. The assembly shall be considered to have failed the hose stream test if an opening develops that permits a projection of water from the stream beyond the unexposed surface during the time of the hose stream test.

16.1.3 Transmission of heat through the wall or partition during the fire endurance test shall not have been such as to raise the temperature on its unexposed surface more than 250°F (139°C) above its initial temperature.

TESTS OF COLUMNS

17. Size of Sample

17.1 The length of the column exposed to fire shall, when practicable, approximate the maximum clear length contemplated by the design, and for building columns shall be not less than 9 ft (2.7 m). Apply the contemplated details of connections, and their protection if any, according to the methods of acceptable field practice.

18. Loading

18.1 Throughout the fire endurance test expose the column to fire on all sides and load it in a manner calculated to develop theoretically, as nearly as practicable, the working stresses contemplated by the design. Make provision for transmitting the load to the exposed portion of the column without unduly increasing the effective column length.

18.2 If the submitter and the testing body jointly so decide, the column may be subjected to 1¾ times its designed working load before the fire endurance test is undertaken. The fact that such a test has been made shall not be construed as having had a deleterious effect on the fire endurance test performance.

19. Condition of Acceptance

19.1 Regard the test as successful if the column sustains the applied load during the fire endurance test for a period equal to that for which classification is desired.

ALTERNATIVE TEST OF PROTECTION FOR STRUCTURAL STEEL COLUMNS

20. Application

20.1 This test procedure does not require column loading at any time and may be used at the discretion of the testing laboratory to evaluate steel column protections that are not required by design to carry any of the column load.

21. Size and Character of Sample

21.1 The size of the steel column used shall be such as to provide a test specimen that is truly representative of the design, materials, and workmanship for which classification is desired. Apply the protection according to the methods of acceptable field practice. The length of the protected column shall be at least 8 ft (2.4 m). The column shall be vertical during application of the protection and during the fire exposure.

21.2 Restrain the applied protection against longitudinal temperature expansion greater than that of the steel column by rigid steel plates or reinforced concrete attached to the ends of the steel column before the protection is applied. The size of the plates or amount of concrete shall be adequate to provide direct bearing for the entire transverse area of the protection.

21.3 Provide the ends of the specimen, including the means for restraint with sufficient thermal insulation to prevent appreciable direct heat transfer from the furnace.

22. Temperature Measurement

22.1 Measure the temperature of the steel in the column by at least three thermocouples located at each of four levels. The upper and lower

levels shall be 2 ft (0.6 m) from the ends of the steel column, and the two intermediate levels shall be equally spaced. So place the thermocouples at each level as to measure significant temperatures of the component elements of the steel section.

23. Exposure to Fire

23.1 Throughout the fire endurance test expose the specimen to fire on all sides for its full length.

24. Conditions of Acceptance

24.1 Regard the test as successful if the transmission of heat through the protection during the period of fire exposure for which classification is desired does not raise the average (arithmetical) temperature of the steel at any one of the four levels above 1000°F (538°C), or does not raise the temperature above 1200°F (649°C) at any one of the measured points.

TESTS OF FLOORS AND ROOFS

25. Application

25.1 This procedure is applicable to floor and roof assemblies with or without attached, furred, or suspended ceilings and requires application of fire exposure to the underside of the specimen under test.

25.2 Two fire endurance classifications shall be developed for assemblies restrained against thermal expansion; a restrained assembly classification based upon the conditions of acceptance specified in 29.1.1 and 29.1.2 in addition to 29.1.3, 29.1.4, or 29.1.5; and an unrestrained assembly classification based upon the conditions of acceptance specified in 30.1.1 and 30.1.2 in addition to 30.1.3, 30.1.4, 30.1.5 or 30.1.6.

NOTE 10—See Appendix X3, which is intended as a guide for assisting the user of this method in determining the conditions of thermal restraint applicable to floor and roof constructions and individual beams in actual building construction.

25.3 One fire endurance classification shall be developed from tests of assemblies not restrained against thermal expansion based upon the conditions of acceptance specified in 30.1.1 and 30.1.2.

25.4 Individual unrestrained classifications may be developed for beams tested in accordance with this test method using the conditions of acceptance specified in 38.1.1, 38.1.2, or 38.1.3.

26. Size and Characteristics of Specimen

26.1 The area exposed to fire shall be not less than 180 ft² (16 m²) with neither dimension less than 12 ft (3.7 m). Structural members, if a part of the construction under test, shall lie within the combustion chamber and have a side clearance of not less than 8 in. (203 mm) from its walls.

26.2 The specimen shall be installed in accordance with recommended fabrication procedures for the type of construction and shall be representative of the design for which classification is desired. Where a restrained classification is desired, specimens representing forms of construction in which restraint to thermal expansion occurs shall be reasonably restrained in the furnace.

27. Loading

27.1 Throughout the fire endurance test apply a superimposed load to the specimen to simulate a maximum load condition. The maximum load condition shall be as nearly as practicable the maximum load allowed by the limiting condition of design under nationally recognized structural design criteria. A fire endurance test may be conducted applying a restricted load condition to the specimen which shall be identified for a specific load condition other than the maximum allowed load condition.

28. Temperature Measurement

28.1 For specimens employing structural members (beams, open-web steel joists, etc.) spaced at more than 4 ft (1.2 m) on centers, measure the temperature of the steel in these structural members by thermocouples at three or more sections spaced along the length of the members with one section preferably located at midspan except that in cases where the cover thickness is not uniform along the specimen length, at least one of the sections at which temperatures are measured shall include the point of minimum cover.

28.2 For specimens employing structural members (beams, open-web steel joists, etc.) spaced at 4 ft (1.2 m) on center or less, measure the temperature of the steel in these structural members by four thermocouples placed on each member, except that no more than four members shall be so instrumented. Place the thermocouples at significant locations, such as at midspan, over joints in the ceiling, and over light fixtures, etc.

28.3 For steel structural members there shall be four thermocouples at each section except that where only four thermocouples are required on a member, the thermocouples may be distributed along the member at significant locations as provided for in 28.2; locate two on the bottom of the bottom flange or chord, one on the web at the center, and one on the top flange or chord. The recommended thermocouple distribution at each section is shown in Fig. 2.

28.4 For reinforced or prestressed concrete structural members, locate thermocouples on each of the tension reinforcing elements, unless there are more than eight such elements, in which case place thermocouples on eight elements selected in such a manner as to obtain representative temperatures of all the elements.

28.5 For steel floor or roof units locate four thermocouples on each section (a section to comprise the width of one unit), one on the bottom plane of the unit at an edge joint, one on the bottom plane of the unit remote from the edge, one on a side wall of the unit, and one on the top plane of the unit. The thermocouples should be applied, where practicable, to the surface of the units remote from fire and spaced across the width of the unit. No more than four nor less than two sections need be so instrumented in each representative span. Locate the groups of four thermocouples in representative locations. Typical thermocouple locations for a unit section are shown in Fig. 3.

29. Conditions of Acceptance—Restrained Assembly

29.1 In obtaining a restrained assembly classification, the following conditions shall be met:

29.1.1 The specimen shall have sustained the applied load during the classification period without developing unexposed surface conditions which will ignite cotton waste.

29.1.2 Transmission of heat through the specimen during the classification period shall not have been such as to raise the average temperature on its unexposed surface more than 250°F (139°C) above its initial temperature.

29.1.3 For specimens employing steel structural members (beams, open-web steel joists, etc.) spaced more than 4 ft (1.2 m) on centers, the assembly shall achieve a fire endurance classification on the basis of the temperature criteria specified in 30.1.3 for assembly classifications up to and including 1 h. For classifications greater than 1 h, the above temperature criteria shall apply for a period of one half the classification of the assembly or 1 h, whichever is the greater.

29.1.4 For specimens employing steel structural members (beam, open-web steel joists, etc.) spaced 4 ft (1.2 m) or less on centers, the assembly shall achieve a fire endurance classification on the basis of the temperature criteria specified in 30.1.4 for assembly classifications up to and including 1 h. For classifications greater than 1 h, the above temperature criteria shall apply for a period of one half the classification of the assembly or 1 h, whichever is the greater.

29.1.5 For specimens employing conventionally designed concrete beams, spaced more than 4 ft (1.2 m) on centers, the assembly shall achieve a fire endurance classification on the basis of the temperature criteria specified in 30.1.5 for assembly classifications up to and including 1 h. For classifications greater than 1 h, the above temperature criteria shall apply for a period of one half the classification of the assembly or 1 h, whichever is the greater.

30. Conditions of Acceptance—Unrestrained Assembly

30.1 In obtaining an unrestrained assembly classification, the following conditions shall be met:

30.1.1 The specimen shall have sustained the applied load during the classification period without developing unexposed surface conditions which will ignite cotton waste.

30.1.2 The transmission of heat through the specimen during the classification period shall not have been such as to raise the average temperature on its unexposed surface more than 250°F (139°C) above its initial temperature.

30.1.3 For specimens employing steel structural members (beams, open-web steel joists, etc.), spaced more than 4 ft (1.2 m) on centers, the temperature of the steel shall not have exceeded 1300°F (704°C) at any location during the classification period nor shall the average temperature recorded by four thermocouples at any section have exceeded 1100°F (593°C) during the classification period.

30.1.4 For specimens employing steel structural members (beams, open-web steel joists, etc.), spaced 4 ft (1.2 m) or less on center, the average temperature recorded by all joist or beam thermocouples shall not have exceeded 1100°F (593°C) during the classification period.

30.1.5 For specimens employing conventionally designed concrete structural members (excluding cast-in-place concrete roof or floor slabs having spans equal to or less than those tested), the average temperature of the tension steel at any section shall not have exceeded 800°F (427°C) for cold-drawn prestressing steel or 1100°F (593°C) for reinforcing steel during the classification period.

30.1.6 For specimens employing steel floor or roof units intended for use in spans greater than those tested, the average temperature recorded by all thermocouples located on any one span of the floor or roof units shall not have exceeded 1100°F (593°C) during the classification period.

31. Report of Results

31.1 The fire endurance classification of a restrained assembly shall be reported as that developed by applying the conditions of acceptance specified in 29.1.1, 29.1.2, and 29.1.3.

31.2 The fire endurance classification of an unrestrained assembly shall be reported as that developed by applying the conditions of acceptance specified in 30.1.1 and 30.1.2 and, where applicable, 30.1.3, 30.1.4, and 30.1.5 to a specimen tested in accordance with this test procedure.

TESTS OF LOADED RESTRAINED BEAMS

32. Application

32.1 An individual classification of a restrained beam may be obtained by this procedure and based upon the conditions of acceptance specified in Section 35. The fire endurance classification so derived shall be applicable to the beam when used with a floor or roof construction which has a comparable or greater capacity for heat dissipation from the beam than the floor or roof with which it was tested. The fire endurance classification developed by this method shall not be applicable to sizes of beams smaller than those tested.

33. Size and Characteristics of Specimen

33.1 Install the test specimen in accordance with recommended fabrication procedures for the type of construction. It shall be representative of the design for which classification is desired. The length of beam exposed to the fire shall be not less than 12 ft (3.7 m) and the member shall be tested in its normal horizontal position. A section of a representative floor or roof construction not more than 7 ft (2.1 m) wide, symmetrically located with reference to the beam, may be included with the test specimen and exposed to the fire from below. Restrain the beam including that part of the floor or roof element forming the complete beam as designed (such as composite steel or concrete construction) against longitudinal thermal expansion in a manner simulating the restraint in the construction represented. Do not support or restrain the perimeter of the floor or roof element of the specimen, except that part which forms part of a beam as designed.

34. Loading

34.1 Throughout the fire endurance tests apply a superimposed load to the specimen. This load, together with the weight of the specimen, shall be as nearly as practicable the maximum theoretical dead and live loads permitted by nationally recognized design standards.

35. Conditions of Acceptance

35.1 The following conditions shall be met:

35.1.1 The specimen shall have sustained the applied load during the classification period.

35.1.2 The specimen shall have achieved a fire endurance classification on the basis of the temperature criteria specified in 30.1.3 or 30.1.4 of one half the classification of the assembly or 1 h, whichever is the greater.

ALTERNATIVE CLASSIFICATION PROCEDURE FOR LOADED BEAMS

36. Application

36.1 Individual unrestrained classifications may be developed for beams tested as part of a floor or roof assembly as described in Sections 25 through 28 (except 25.3) or for restrained beams tested in accordance with the procedure described in Sections 32 through 34. The fire endurance classification so derived shall be applicable to beams when used with a floor or roof construction which has a comparable or greater capacity for heat dissipation from the beam than the floor or roof with which it was tested. The fire endurance classification developed by this method shall not be applicable to sizes of beams smaller than those tested.

37. Temperature Measurement

37.1 Measure the temperature of the steel in structural members by thermocouples at three or more sections spaced along the length of the members with one section preferably located at midspan, except that in cases where the cover thickness is not uniform along the specimen length, at least one of the sections at which temperatures are measured shall include the point of minimum cover.

37.2 For steel beams, there shall be four thermocouples at each section; locate two on the bottom of the bottom flange, one on the web at the center, and one on the bottom of the top flange.

37.3 For reinforced or prestressed concrete structural members, locate thermocouples on each of the tension reinforcing elements unless there are more than eight such elements, in which case place thermocouples on eight elements selected in such a manner as to obtain representative temperatures of all the elements.

38. Conditions of Acceptance

38.1 In obtaining an unrestrained beam classification the following conditions shall be met:

38.1.1 The specimen shall have sustained the applied load during the classification period.

38.1.2 For steel beams the temperature of the steel shall not have exceeded 1300°F (704°C) at any location during the classification period nor shall the average temperature recorded by four thermocouples at any section have exceeded 1100°F (593°C) during this period.

38.1.3 For conventionally designed concrete beams, the average temperature of the tension steel at any section shall not have exceeded 800°F (427°C) for cold-drawn prestressing steel or 1100°F for reinforcing steel during the classification period.

ALTERNATIVE TEST OF PROTECTION FOR SOLID STRUCTURAL STEEL BEAMS AND GIRDERS

39. Application

39.1 Where the loading required in Section 27 is not feasible, this alternative test procedure may be used to evaluate the protection of steel beams and girders without application of design load, provided that the protection is not required by design to function structurally in resisting applied loads. The conditions of acceptance of this alternative test are not applicable to tests made under design load as provided under tests for floors and roofs in Sections 26, 29, and 30.

40. Size and Character of Sample

40.1 The size of the steel beam or girder shall be such as to provide a test specimen that is truly representative of the design, materials, and workmanship for which classification is desired. Apply the protection according to the methods of acceptable field practice and the projection below the ceiling, if any, shall be representative of the conditions of intended use. The length of beam or girder exposed to the fire shall be not less than 12 ft (3.7 m) and the member shall be tested in a horizontal position. A section of a representative floor construction not less than 5 ft (1.5 m) wide, symmetrically located with reference to the beam or girder and extending its full length, shall be included in the test assembly and exposed to fire from below. The rating of performance shall not be applicable to sizes smaller than those tested.

40.2 Restrain the applied protection against longitudinal expansion greater than that of the steel beam or girder by rigid steel plates or reinforced concrete attached to the ends of the member before the protection is applied. The ends of the member, including the means for restraint, shall be given sufficient thermal insulation to prevent appreciable direct heat transfer from the furnace to the unexposed ends of the member or from the ends of the member to the outside of the furnace.

41. Temperature Measurement

41.1 Measure the temperature of the steel in the beam or girder with not less than four thermocouples at each of four sections equally spaced along the length of the beam and symmetrically disposed and not nearer than 2 ft (0.6 m) from the inside face of the furnace. Symmetrically place the thermocouples at each section so as to measure significant temperatures of the component elements of the steel section.

42. Conditions of Acceptance

42.1 Regard the test as successful if the transmission of heat through the protection during the period of fire exposure for which classification is desired does not raise the average (arithmetical) temperature of the steel at any one of the four sections above 1000°F (538°C), or does not raise the temperature above 1200°F (649°C) at any one of the measured points.

PERFORMANCE OF PROTECTIVE MEMBRANES IN WALL, PARTITION, FLOOR, OR ROOF ASSEMBLIES

43. Application

43.1 When it is desired to determine the thermal protection afforded by membrane elements in wall, partition, floor, or roof assemblies, the nonstructural performance of protective membranes shall be obtained by this procedure. The performance of protective membranes is supplementary information only and is not a substitute for the Fire Endurance Classification determined by Sections 12 through 42 of this method.

44. Characteristics and Size of Sample

44.1 The characteristics of the sample shall conform to 7.1.

44.2 The size of the sample shall conform to 12.1 for bearing walls and partitions, 15.1 for nonbearing walls and partitions, or 26.1 for floors or roofs.

45. Temperature Performance of Protective Membranes

45.1 The temperature performance of protective membranes shall be measured with thermocouples, the measuring junctions of which are in intimate contact with the exposed surface of the elements being protected. The diameter of the wires used to form the thermo-junction shall not be greater than the thickness of sheet metal framing or panel members to which they are attached and in no case greater than No. 18 B&S gage (0.040 in.) (1.02 mm). The lead shall be electrically insulated with heat-resistant and moisture-resistant coatings.

45.2 For each class of elements being protected, temperature readings shall be taken at not less than five representative points. None of the thermocouples shall be located nearer to the edges of the test assembly than 12 in. (305 mm). An exception can be made in those cases where there is an element or feature of the construction that is not otherwise represented in the test assembly. None of the thermocouples shall be located opposite, on top of or adjacent to fasteners such as screws, nails, or staples when such locations are excluded for thermocouple placement on the unexposed surface of the test assembly in 5.2.

45.3 Thermocouples shall be located to obtain representative information on the temperature of the interface between the exposed membrane and the substrate or element being protected.

45.4 Temperature readings shall be taken at intervals not exceeding 5 min, but the intervals need not be less than 2 min.

46. Conditions of Performance

46.1 Unless otherwise specified, the performance of protective membranes shall be determined as the time at which following conditions occur:

46.1.1 The average temperature rise of any set of thermocouples for each class of element being produced is more than 250°F (139°C) above the initial temperature, or

46.1.2 The temperature rise of any one thermocouple of the set for each class of element being protected is more than 325°F (181°C) above the initial temperature.

47. Report of Results

47.1 The Protective Membrane Performance, for each class of element being protected, shall be reported to the nearest integral minute.

47.2 The test report shall identify each class of elements being protected and shall show the location of each thermocouple.

47.3 The test report shall show the time-temperature data recorded for each thermocouple and the average temperature for the set of thermocouples on each element being protected.

47.4 The test report shall state any visual observations recorded that are pertinent to the performance of the protective membrane.

TABLE 1 Conditions For Hose Stream Test

Resistance Period	Water Pressure at Base of Nozzle, psi (kPa)	Duration of Application, min/100 ft² (9 m²) exposed area
8 h and over	45 (310)	6
4 h and over if less than 8 h	45 (310)	5
2 h and over if less than 4 h	30 (207)	2½
1½ h and over if less than 2 h	30 (207)	1½
1 h and over if less than 1½ h	30 (207)	1
Less than 1 h, if desired	30 (207)	1

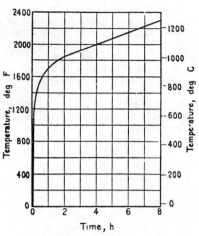

FIG. 1 Time-Temperature Curve

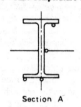

Section A

Section B

FIG. 2 Recommended Thermocouple Distributions

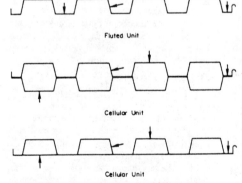

FIG. 3 Typical Location of Thermocouples

ANNEX

A1. REQUIREMENTS FOR THERMOCOUPLE PADS

A1.1 *Asbestos Pads*—Asbestos pads used in measurements of temperature of unexposed surfaces of specimens shall be of felted amosite asbestos free of organic additives and shall exhibit the following properties:

A1.1.1 Length and width, 6 ± ⅛ in. (152 ± 3 mm).

A1.1.2 Thickness, 0.40 ± 0.05 in. (10.2 ± 1.3 mm). The thickness measurement shall be made under the light load of a ½-in. (13-mm) diameter pad of a dial micrometer gage.

A1.1.3 Dry weight, 0.260 ± 0.026 lb (0.12 ± 0.01 kg).

A1.1.4 Thermal conductivity (at 150°F (66°C)), 0.38 ± 0.027 Btu·in./h·ft².°F (0.055 ± 0.003 W/m·K).

A1.1.5 Hardness indentation shall be 0.157 ± 0.07 in. (4.0 ± 1.8 mm) or 10–25 (modified Brinell). Indentation shall be determined in accordance with ASTM Method C 569, Test for Indentation Hardness of Preformed Thermal Insulations.[2] Modified Brinell values of hardness are obtained by the relationship

$$\text{Hardness} = \frac{2.24}{y}$$

where y = the measured indentation in inches.

A1.1.6 The pads shall be sufficiently soft so that, without breaking, they may be shaped to contact over the whole surface against which they are placed.

A1.2 *Refractory Fiber Pads*—Comparative fire tests have demonstrated that a refractory fiber material manufactured by Johns-Manville, designated Ceraform 126, placed with the softer surfaces in contact with the thermocouple, may be substituted for the asbestos pad when distortion of the unexposed face of the sample is minimal. The pads are relatively rigid and shall not be used on surfaces subject to sharp distortions or discontinuities during the test. Properties of Ceraform 126 material shall be as follows:

A1.2.1 Length and width, 6 ± ⅛ in. (152 ± 3 mm).

A1.2.2 Thickness, 0.375 ± 0.063 in. (9.5 ± 1. mm) (see A1.2).

A1.2.3 Dry weight, 0.147 ± 0.053 lb (67 ± 24 g)

A1.2.4 Thermal conductivity (at 150°F (66°C)), 0.37 ± 0.03 Btu·in./h·ft².°F (0.053 ± 0.004 W/m·K).

A1.2.5 Hardness (see A1.1.5) indentation on soft face shall be 0.075 ± 0.025 in. (1.9 ± 0.6 mm).

A1.2.6 The pads may be shaped by wetting, forming, and then drying to constant weight to provide complete contact on sharply contoured surfaces.

[6] Supporting data are available on loan from ASTM Headquarters, 1916 Race St., Philadelphia, Pa. 19103. Request RR: E05-1004.

APPENDIXES

X1. STANDARD TIME-TEMPERATURE CURVE FOR CONTROL OF FIRE TESTS

Time	Temperature,	Area Above 68°F Base		Temperature,	Area Above 20°C Base	
h:min	°F	°F·min	°F·h	°C	°C·min	°C·h
0:00	68	00	0	20	00	0
0:05	1 000	2 330	39	538	1 290	22
0:10	1 300	7 740	129	704	4 300	72
0:15	1 399	14 150	236	760	7 860	131
0:20	1 462	20 970	350	795	11 650	194
0:25	1 510	28 050	468	821	15 590	260
0:30	1 550	35 360	589	843	19 650	328
0:35	1 584	42 860	714	862	23 810	397
0:40	1 613	50 510	842	878	28 060	468
0:45	1 638	58 300	971	892	32 390	540
0:50	1 661	66 200	1 103	905	36 780	613
0:55	1 681	74 220	1 237	916	41 230	687
1:00	1 700	82 330	1 372	927	45 740	762
1:05	1 718	90 540	1 509	937	50 300	838
1:10	1 735	98 830	1 647	946	54 910	915
1:15	1 750	107 200	1 787	955	59 560	993
1:20	1 765	115 650	1 928	963	64 250	1 071
1:25	1 779	124 180	2 070	971	68 990	1 150
1:30	1 792	132 760	2 213	978	73 760	1 229
1:35	1 804	141 420	2 357	985	78 560	1 309
1:40	1 815	150 120	2 502	991	83 400	1 390
1:45	1 826	158 890	2 648	996	88 280	1 471
1:50	1 835	167 700	2 795	1 001	93 170	1 553
1:55	1 843	176 550	2 942	1 006	98 080	1 635
2:00	1 850	185 440	3 091	1 010	103 020	1 717
2:10	1 862	203 330	3 389	1 017	112 960	1 882
2:20	1 875	221 330	3 689	1 024	122 960	2 049
2:30	1 888	239 470	3 991	1 031	133 040	2 217
2:40	1 900	257 720	4 295	1 038	143 180	2 386
2:50	1 912	276 110	4 602	1 045	153 390	2 556
3:00	1 925	294 610	4 910	1 052	163 670	2 728
3:10	1 938	313 250	5 221	1 059	174 030	2 900
3:20	1 950	332 000	5 533	1 066	184 450	3 074
3:30	1 962	350 890	5 848	1 072	194 940	3 249
3:40	1 975	369 890	6 165	1 079	205 500	3 425
3:50	1 988	389 030	6 484	1 086	216 130	3 602
4:00	2 000	408 280	6 805	1 093	226 820	3 780
4:10	2 012	427 670	7 128	1 100	237 590	3 960
4:20	2 025	447 180	7 453	1 107	248 430	4 140
4:30	2 038	466 810	7 780	1 114	259 340	4 322
4:40	2 050	486 560	8 110	1 121	270 310	4 505
4:50	2 062	506 450	8 441	1 128	281 360	4 689
5:00	2 075	526 450	8 774	1 135	292 470	4 874
5:10	2 088	546 580	9 110	1 142	303 660	5 061
5:20	2 100	566 840	9 447	1 149	314 910	5 248
5:30	2 112	587 220	9 787	1 156	326 240	5 437
5:40	2 125	607 730	10 129	1 163	337 630	5 627
5:50	2 138	628 360	10 473	1 170	349 090	5 818
6:00	2 150	649 120	10 819	1 177	360 620	6 010
6:10	2 162	670 000	11 167	1 184	372 230	6 204
6:20	2 175	691 010	11 517	1 191	383 900	6 398
6:30	2 188	712 140	11 869	1 198	395 640	6 594
6:40	2 200	733 400	12 223	1 204	407 450	6 791
6:50	2 212	754 780	12 580	1 211	419 330	6 989
7:00	2 225	776 290	12 938	1 218	431 270	7 188
7:10	2 238	797 920	13 299	1 225	443 290	7 388
7:20	2 250	819 680	13 661	1 232	455 380	7 590
7:30	2 262	841 560	14 026	1 239	467 540	7 792
7:40	2 275	863 570	14 393	1 246	479 760	7 996
7:50	2 288	885 700	14 762	1 253	492 060	8 201
8:00	2 300	907 960	15 133	1 260	504 420	8 407

X4.5 Illustrative Example

X4.5.1 A wall made from normal weight concrete having 23.2 volume % of paste, was conditioned at 200°F (93°C) and 5 % *RH* until the *RH* at its mid-depth was reduced to 70 %. It had a 2.90-h fire endurance. Determine the adjusted fire endurance.

X4.5.1.1 *Step 1*—Calculate m_a as follows:
For 70 % *RH*,

$m_e = 0.225$ (see Table X4.1)

For 200°F (93°C) and 5 % *RH* conditioning, for normal weight concrete

$A = 0.45$ (see Table X4.2)

$\therefore m_c = 0.45 \times 0.225 = 0.101$ (see X4.3.2)

For $v = 0.232$

$m_a = 0.232 \times 0.101 = 0.0234$ (see X4.3.3)

that is, the concrete contains 2.34 volume % moisture at time of test.

X4.5.1.2 *Step 2*—Calculate m_s as follows:
As an example, if the standard moisture level is assumed to correspond to a middepth *RH* of 75 %, then $m_c = 0.24$

$m_s = 0.232 \times 0.24 = 0.0557$ (see X4.3.4)

that is, the standard moisture level is 5.57 volume %.

X4.5.1.3 *Step 3*—Calculate b_m as follows:
$b = 5.5$ see Table X4.3
$bm_a = 5.5 \times 0.0234 = 0.129$
$bm_s = 5.5 \times 0.0557 = 0.306$

X4.5.1.4 *Step 4*—Draw lines on the nomogram from point *R* to bm_a and bm_s (see Fig. X4.1).

X4.5.1.5 *Step 5*—Draw a line from the *FE* ordinate, 2.90, parallel to line *R*-bm_a to intersect the curve.

X4.5.1.6 *Step 6*—Draw a line parallel to *R*-bm_s from a point on the curve, to intersect the *FE* ordinate scale. The value of *FE* = 3.19 is the adjusted fire endurance; that is, the fire endurance that would have resulted if the specimen had been tested at the standard moisture level, here assumed to correspond to 75 % RH at middepth.

X5. COMMENTARY

X5.1 Introduction

X5.1.1 This commentary has been prepared to provide the user of Methods E 119 with background information on the development of the standard and its application in fire protection of buildings. It also provides guidance in the planning and performance of fire tests and in the reporting of results. No attempt has been made to incorporate all the available information on fire testing in this commentary. The serious student of fire testing is strongly urged to peruse the referenced documents for a better appreciation of the history of fire-resistant design (**1, 2**)[3] and the intricate problems associated with testing and with interpretation of test results.

X5.1.2 Floors and walls designed as fire separations have been recognized for many years as efficient tools in restricting fires to the area of origin, or limiting their spread (**3, 4, 5, 6, 7, 8, 9, 10, 11**). Prior to 1900, relative fire safety was achieved by mandating specific materials. By the year 1900, the appearance of a multitude of new materials and innovative designs and constructions accelerated the demand for performance standards. The British Fire Prevention Committee, established in 1894, was the first to produce tables listing fire resisting floors, ceilings, doors and partitions (**5**). Test furnaces in the United States were constructed shortly after 1900 at the Underwriters Laboratories Inc., Columbia University, and the National Bureau of Standards (NBS) (**1, 12**). These early furnaces eventually led to the development of Methods E 119.

X5.2 Historical Aspects

X5.2.1 Methods E 119 was first published by ASTM as C 19 in 1918. A number of refinements have been made in the standard since that time. However, several provisions, including the temperature-time curve, the major apparatus, and the acceptance criteria remain essentially unchanged. The roots of fire testing as we define it today, can be traced back to about 1800. A comprehensive review of early fire testing has been published (**1**).

X5.3 Fire-Load Concept

X5.3.1 Specifications for fire resistance in regulatory documents continue to be based largely on the fire-load concept developed by NBS in the 1920s and reported in the 1928 NFPA Quarterly by Ingberg. The concept incorporates the premise that the duration of a fire is proportional to the fire loading, that is, the mass of combustible materials per unit floor area. The relationship between the mass of combustible materials and fire duration was established on the basis of burnout tests in structures incorporating materials having calorific or potential heat values equivalent to wood and paper, that is, 7000 to 8000 BtU/lb (16.3 to 18.6 MJ/kg). The fire-load of non-cellulosic materials such as oils, waxes, and flammable liquids were interpreted on the basis of their equivalent calorific content (**5, 13, 14, 15**). In the simplest terms, the above premise states that 10 lb of combustible materials per square foot (50 kg/m²) of floor area will produce a fire of 1 h duration.

X5.3.2 Increasing sophistication in the knowledge of materials and the fire process has resulted from numerous research activities (**9, 11, 13, 14, 15, 16, 17, 18, 19, 20, 21, 22, 23, 24, 25, 26, 27**). It is now generally conceded that fire severity as well as the temperature-time relationship of a fire depends on several factors, including;*if0*

1. *Fire load--amount and type.*
2. *Distribution of this fire load.*
3. *Specific surface characteristics of the fire load* (**5, 27**).
4. Ventilation, as determined by the size and shape of openings (**17, 18, 19, 21, 27, 28, 29**).
5. Geometry of the fire compartment--size and shape.
6. Thermal characteristics of the enclosure boundaries.
7. Relative humidity of the atmosphere.

For the purposes of this commentary, fire severity is defined as a measure of the fire intensity (temperature) and fire duration. It is expressed in terms of minutes or hours of fire exposure and in Methods E 119 is assumed to be equivalent to that defined by the standard temperature-time (*T-t*) Curve, that is, the area under the *T-t* curve (**27**).

X5.4 Scope and Significance

X5.4.1 Methods E 119 is intended to evaluate in terms of endurance time, the ability of an assembly to contain a fire, or to retain its structural integrity, or both during the test conditions imposed by the standard. It also contains standard conditions for measuring heat transfer through membrane elements protecting combustible framing or surfaces.

X5.4.2 The end-point criteria by which the test result is assessed are related to:

1. Transmission of heat through the test assembly.
2. Ability of the test assembly to withstand the transmission of flames or gases hot enough to ignite combustible material.
3. Ability of the assembly to carry the load and withstand restraining forces during the fire test period.
4. Temperatures of the steel under some conditions.

X5.4.3 It is the intent that classifications shall register performance during the period of exposure and shall not be construed as having determined suitability for use after the exposure.

X5.4.4 The standard, although being specific about the assembly to be tested, enables the testing laboratory to determine whether the specimen is "truly representative" of the assembly intended for evaluation. This is necessary because of the wide variation in assemblies. For instance, wall test specimens generally do not contain electric switches and outlets, that in some designs may affect test results. Floor test specimens may or may not contain electrical raceways and outlets or pull boxes for power and communication wiring. Cover plates over trench headers are also present in some designs. The testing laboratory is in the best position to judge the effects of such items.

X5.5 Test Furnaces

X5.5.1 Methods E 119 does not provide specific construction details of the furnace. Readers are urged to consult reference documents for a more comprehensive review of furnace design and performance (**25**).

X5.6 Temperature - Time Curve

X5.6.1 A specific temperature - time relationship for the test fire is defined in the standard and in Appendix X1. The actual recorded temperatures in the furnace is required to be within specified percentages of those of the standard curve. Accuracy in measuring temperature is generally easier to achieve after 1 h due to stabilizing of the furnace and the slope of the *T-t* curve. The number and type of temperature-measuring devices are outlined in the standard. Specific standard practices for location and use of these temperature-measuring devices are also outlined in the standard. However, no uniformity of the temperatures within the fire chamber is specified.

X5.6.2 The standard *T-t* curve used in Methods E 119 is considered to represent a severe building fire (**5**). The curve was adopted in 1918 as a result of several conferences by eleven technical organizations, including testing laboratories, insurance underwriters, fire protection associations, and technical societies (**1, 16, 30**). The *T-t* relationship of these test methods represents only one fire situation. Data are available to evaluate the performance of assemblies under fire exposure conditions that may be more representative of particular fire situations, that is, using different *T-t* relationships to simulate specific fire conditions (**9, 11, 16, 19, 22, 23, 27, 29, 31, 32**).

X5.6.3 Furnace pressure is not specified and is generally slightly negative. The pressure may have an effect on the test results, and the test conditions should always be carefully controlled.

X5.7 Test Specimen

X5.7.1 The test specimen is required to represent as closely as possible the actual construction in the field, subject to the limits imposed by the test facilities.

X5.7.2 All specimens are required to be conditioned so as to attain a moisture content comparable to that in the field prior to testing. For uniformity, the standard moisture content is defined as that in equilibrium with an atmosphere of 50 % relative humidity at 73°F (23°C). Massive concrete units which may require unusually long drying periods may be fire tested after a 12-month conditioning period. Appendix X4 describes how the test result should be corrected to account for any variation from the standard moisture condition (**33**).

X5.7.3 With few exceptions, only the interior face of exterior wall assemblies and the ceiling portion or underside of floor or roof assemblies are exposed to the standard fire (**24, 25**). This practice is rationalized on the assumption that the outside face of exterior walls is not usually subjected to the same fire as the interior face and that the fire exposure of the upper side of a floor or roof assembly is seldom as intense as that of the underside.

X5.7.4 Although the standard does not contain specific criteria for judging the impact of through joints nor "poke-through" devices, such as electrical or telephone outlets, it should be recognized that these components should be evaluated with respect to structural-performance and temperature-rise criteria if they constitute a significant part of the tested assembly.

X5.7.5 For obvious reasons, symmetrical walls and partitions are tested only on one side. Assymmetrical walls and partitions may be required to be tested with either or both sides individually exposed to the fire. If both sides are exposed, the report should indicate the fire endurance classification for each case.

X5.8 Loading

X5.8.1 Floors and roofs are required to be loaded during test to provide a maximum load condition determined by the applicable nationally recognized design criteria. This practice allows for more confidence in extrapolating testing results. For instance, the maximum length of a floor specimen in most test facilities is 16 ft (4.9 m). It is, therefore, necessary to extrapolate developed fire-endurance ratings to much longer spans.

X5.8.2 When a floor or roof assembly is designed for a specific use, such as used in prefabricated housing units, the assembly may be tested with a restricted load condition. The loading condition used for such tests shall be defined in the test report. The standard does not require specific loading devices. Some laboratories use large containers of water; oth-

[3] The boldface numbers in parentheses refer to the list of references at the end of this appendix.

TABLE X3.1   Construction Classification, Restrained and Unrestrained

I. Wall bearing:
Single span and simply supported end spans of multiple bays:[a]
  (1) Open-web steel joists or steel beams, supporting concrete slab, precast units, or metal decking .... unrestrained
  (2) Concrete slabs, precast units, or metal decking .... unrestrained
Interior spans of multiple bays:
  (1) Open-web steel joists, steel beams or metal decking, supporting continuous concrete slab .... restrained
  (2) Open-web steel joists or steel beams, supporting precast units or metal decking .... unrestrained
  (3) Cast-in-place concrete slab systems .... restrained
  (4) Precast concrete where the potential thermal expansion is resisted by adjacent construction[b] .... restrained
II. Steel framing:
  (1) Steel beams welded, riveted, or bolted to the framing members .... restrained
  (2) All types of cast-in-place floor and roof systems (such as beam-and-slabs, flat slabs, pan joists, and waffle slabs) where the floor or roof system is secured to the framing members .... restrained
  (3) All types of prefabricated floor or roof systems where the structural members are secured to the framing members and the potential thermal expansion of the floor or roof system is resisted by the framing system or the adjoining floor or roof construction[b] .... restrained
III. Concrete framing:
  (1) Beams securely fastened to the framing members .... restrained
  (2) All types of cast-in-place floor or roof systems (such as beam-and-slabs, flat slabs, pan joists, and waffle slabs) where the floor system is cast with the framing members .... restrained
  (3) Interior and exterior spans of precast systems with cast-in-place joints resulting in restraint equivalent to that which would exist in condition III (1) .... restrained
  (4) All types of prefabricated floor or roof systems where the structural members are secured to such systems and the potential thermal expansion of the floor or roof systems is resisted by the framing system or the adjoining floor or roof construction[b] .... restrained
IV. Wood construction:
All types .... unrestrained

[a] Floor and roof systems can be considered restrained when they are tied into walls with or without tie beams, the walls being designed and detailed to resist thermal thrust from the floor or roof system.
[b] For example, resistance to potential thermal expansion is considered to be achieved when:
  (1) Continuous structural concrete topping is used,
  (2) The space between the ends of precast units or between the ends of units and the vertical face of supports is filled with concrete or mortar, or
  (3) The space between the ends of precast units and the vertical faces of supports, or between the ends of solid or hollow core slab units does not exceed 0.25 % of the length for normal weight concrete members or 0.1 % of the length for structural lightweight concrete members.

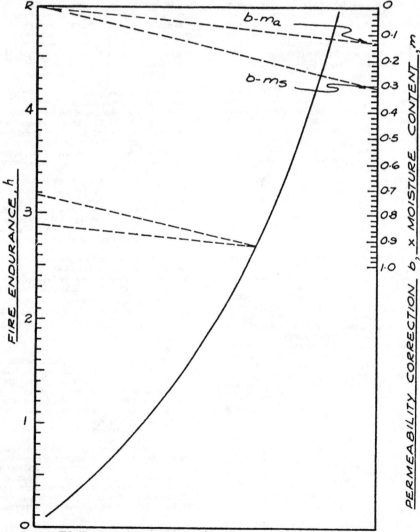

FIG. X4.1   Nomogram for Correcting Fire Endurance for Nonstandard Moisture Content.

X4. METHOD OF CORRECTING FIRE ENDURANCE FOR CONCRETE SLABS DETERMINED BY UNEXPOSED SURFACE TEMPERATURE RISE FOR NONSTANDARD MOISTURE CONTENT

X4.1 Scope

X4.1.1 The standard fire endurance is the time determined by unexposed surface temperature rise of a test specimen at a standard moisture level.

X4.1.2 This appendix gives a procedure to correct the fire endurance of unprotected vertical or horizontal slabs (solid or hollow), made from essentially inorganic building materials; and conditioned on both sides, when moisture content at the time of test is other than at a standard moisture level.

X4.1.3 From among the common inorganic building materials, only the hydrated portland cement products can hold (after due conditioning in accordance with Section 10) sufficient amount of moisture to affect noticeably the result of the fire test. Consequently, correcting the experimental fire endurance of constructions containing less than 5 volume % of portland cement paste is not necessary.

X4.2 Symbols

X4.2.1 The symbols used in this Appendix are defined as follows:
A = factor characterizing the drying conditions (see Table X4.2),
b = factor characterizing the permeability of the specimen (see Table X4.3),
FE = fire endurance of specimen, h,
RH = relative humidity,
m = moisture content, volume fraction ft^3/ft^3 or cm^3/cm^3,
m_a = average moisture content of test specimen,
m_c = average moisture content of cement paste,
m_e = nominal equilibrium moisture content of cement paste for a given RH (see Table X4.1),
m_{ea} = equilibrium moisture content of cement paste at the standard RH level (see Table X4.1).
m_s = average moisture content of a standard conditioned concrete specimen of same concrete and cement paste volume as the test specimen, and
v = volume fraction of cement paste, ft^3/ft^3 or cm^3/cm^3.

X4.3 Calculation of Moisture Content

X4.3.1 The average moisture content, m_a, is the volume fraction of moisture (ft^3/ft^3 (or cm^3/cm^3)) in the material relative to its dry condition; where dry condition is defined as that resulting when the material is heated in an oven at $221 \pm 1°F$ ($105 \pm 0.5°C$) until no further weight loss occurs.

X4.3.2 The average moisture content of the cement paste can be estimated from the known value of RH at middepth (assuming the material has never been subject to rewetting) by calculating first the moisture content in the cement paste as follows:

$$m_c = A \cdot m_e$$

X4.3.3 The average moisture content of the test specimen is then calculated as follows:

$$m_a = v \cdot m_c$$

X4.3.4 Calculate the average moisture content of a standard conditioned specimen as follows:

$$m_s = v \cdot m_{ea}$$

where m_{ea} is the value of m_e in Table X4.1 pertaining to the standard RH level.

X4.4 Correction Procedure

X4.4.1 The correction procedure starts with the selection of an empirical factor to reflect the permeability of the material as suggested in Table X4.3. With known values of m_a and m_s calculate the products bm_a and bm_s. On the nomogram (Fig. X4.1) draw lines from point R to values of bm_a and bm_s on the right-hand scale. From the point representing the actual fire endurance time (FE) on the left-hand scale draw a line parallel to R-bm_a to intersect the curve. From this point on the curve draw a line parallel to R-bm_s and find the corrected fire endurance on the FE scale.

X2. SUGGESTED REPORT FORM

ASTM E 119

TITLE PAGE

(Preferably Cover)

Laboratory _____

Project Number _____

ASTM E 119 (Year)

STANDARD FIRE ENDURANCE TEST

Fire Endurance Time _____

Construction _____

Date Tested _____

Sponsor _____

Material _____

Maximum Load Conditions, or Restricted Load Conditions (as the conditions of the test dictate) _____

(Identify if test is part of a research program)

(Add—Table of Contents)

X2.1 *Description of Laboratory Test Facility*—Furnace, restraining frame, details of end conditions, including wedges, bearing, etc.

X2.1.1 If construction is to be tested under load indicate how the load is applied and controlled. (Give loading diagram.) Indicate whether the load is a maximum load condition or a restricted load condition and for either condition, report the specific loads and the basis for limitation, such as bending stress, shear, etc. A restricted load condition shall be reported as a percentage of the maximum load condition.

X2.1.2 If construction is to be tested as nonload bearing indicate whether frame is rigid or moves in test, or whether test is of temperature rise only.

X2.2 *Description of all Materials*—Type, size, class, strength, densities, trade name, and any additional data necessary to define materials. The testing laboratory should indicate whether materials meet ASTM Standards by markings, or by statement of sponsor, or by physical or chemical test by the testing laboratory.

X2.3 *Description of Test Assembly*:

X2.3.1 Give size of test specimen.

X2.3.2 Give details of structural design, including safety factors of all structural members in test assembly.

X2.3.3 Include plan, elevation, principal cross section, plus other sections as needed for clarity.

X2.3.4 Give details of attachment of test panel in frame.

X2.3.5 Location of thermocouples, deflection points, and other items for test.

X2.3.6 Describe general ambient conditions at:

X2.3.6.1 Time of construction,

X2.3.6.2 During curing (time from construction to test), and

X2.3.6.3 Time of test.

X2.4 *Description of Test*:

X2.4.1 Report temperature at beginning and every 5 min. If charts are included in report, clearly indicate time and temperature:

X2.4.1.1 In furnace space,

X2.4.1.2 On unexposed surface, and

X2.4.1.3 On protected framing members as stipulated in Standard.

NOTE X2.1—It is recommended that temperature observations not required by the standard, but useful, be reported in the Appendix to the report. These include temperatures on the face of framing members in back of protection and others that may be required by various Building Codes.

X2.4.2 Report deflections every 5 min for the first 15 min of test and the last hour. In between, every 10 min.

X2.4.3 Report appearance of exposed face:

X2.4.3.1 Every 15 min,

X2.4.3.2 At any noticeable development, give details and time, that is, cracks, buckling, flaming, smoke, loss of material, etc., and

X2.4.3.3 At end of the test include the amount of drop out, condition of fasteners, sag, etc.

X2.4.4 Report appearance of unexposed face:

X2.4.4.1 Every 15 min,

X2.4.4.2 At any noticeable development including

face exposed to fire with comments on fire resistance from the opposite side.

X2.5.3 Fire test.

X2.6 *Summarize Results, include*:

X2.6.1 Endurance time,

X2.6.2 Nature of failure, and

X2.6.3 Hose stream test results.

X2.7 *List Official Observers*—Signatures of responsible persons.

X2.8 *Appendix*—Include all data not specifically required by test standard, but useful to better understanding of test results. Special observations for Building Code approvals should be in appendix.

X2.9 *Pictures*—All taken to show what cannot be covered in the report or to clarify.

X2.9.1 Assembly in construction.

X2.9.2 Exposed face prior to fire test.

X2.9.3 Unexposed face at start of endurance test; include recording equipment when possible.

X2.9.4 Unexposed face at end of fire endurance test.

X2.9.5 Exposed face at end of fire endurance test.

X2.9.6 Unexposed face at end of fire exposure before hose test.

X2.9.7 Exposed face at end of fire exposure before hose test.

X2.9.8 Exposed face after hose stream test.

X2.9.9 Unexposed face after hose stream test.

X2.10 It is essential to have the following:

X2.10.1 Detailed drawing of test assembly.

X2.10.2 Pictures (X2.9.1, X2.9.4, X2.9.8, and X2.9.9) for every test report.

cracking, smoking, buckling, give details and time, and

X2.4.4.3 At the end of test.

X2.4.5 Report time of failure by:

X2.4.5.1 Temperature rise,

X2.4.5.2 Failure to carry load, and

X2.4.5.3 Passage of flame-heat-smoke.

X2.4.6 If a hose stream test is required repeat necessary parts of X2.1 and X2.3. If failure occurs in hose stream test—describe!

X2.5 *Official Comments on*:

X2.5.1 Included shall be a statement to the effect that the construction truly represents field construction. If the construction does not represent typical field construction, then the deviations shall be noted.

X2.5.2 If construction is unsymmetrical (has different details on each face) be sure to indicate the

X3. GUIDE FOR DETERMINING CONDITIONS OF RESTRAINT FOR FLOOR AND ROOF ASSEMBLIES AND FOR INDIVIDUAL BEAMS

X3.1 The revisions adopted in 1970 were introduced, for the first time in the history of the standard, the concept of fire endurance classifications based on two conditions of support: restrained and unrestrained. As a result, most specimens will be fire tested in such a manner as to derive these two classifications.

X3.2 A restrained condition in fire tests, as used in this method, is one in which expansion at the supports of a load carrying element resulting from the effects of the fire is resisted by forces external to the element. An unrestrained condition is one in which the load carrying element is free to expand and rotate at its supports.

X3.3 Some difficulty is recognized in determining the condition of restraint that may be anticipated at

elevated temperatures in actual structures. Until a more satisfactory method is developed, this guide recommends that all constructions be temporarily classified as either restrained or unrestrained. This classification will enable the architect, engineer, or building official to correlate the fire endurance classification, based on conditions of restraint, with the construction type under consideration.

X3.4 For the purpose of this guide, restraint in buildings is defined as follows: "Floor and roof assemblies and individual beams in buildings shall be considered restrained when the surrounding or supporting structure is capable of resisting substantial thermal expansion throughout the range of anticipated elevated temperatures. Construction not complying with this definition are assumed to be free to rotate and expand and shall therefore be considered as unrestrained."

X3.5 This definition requires the exercise of engineering judgment to determine what constitutes restraint to "substantial thermal expansion." Restraint may be provided by the lateral stiffness of supports for floor and roof assemblies and intermediate beams forming part of the assembly. In order to develop restraint, connections must adequately transfer thermal thrusts to such supports. The rigidity of adjoining panels or structures should be considered in assessing the capability of a structure to resist thermal expansion. Continuity, such as that occurring in beams acting continuously over more than two supports, will induce rotational restraint which will usually add to the fire resistance of structural members.

X3.6 In Table X3.1 only the common types of constructions are listed. Having these examples in mind as well as the philosophy expressed in the preamble, the user should be able to rationalize the less common types of construction.

X3.7 Committee E-5 considers the foregoing methods of establishing the presence or absence of restraint according to type and detail of construction to be a temporary expedient, necessary to the initiation of dual fire endurance classifications. It is anticipated that methods for realistically predetermining the degree of restraint applicable to a particular fire endurance classification will be developed in the near future.

TABLE X4.1 Equilibrium Moisture Content (Desorption) of Cement Paste at Given Relative Humidity

RH at middepth, %	m_o
90	0.30
85	0.275
80	0.255
75	0.24
70	0.225
65	0.21
60	0.195
55	0.185
50	0.175
45	0.16
40	0.15

TABLE X4.2 Factor Characterizing Drying Conditions

Conditioning Environment	Middepth RH of Test Specimen, %	Factor A for Portland Cement	
		Normal Weight Concrete	Light-Weight Concrete
60 to 80°F (15.6 to 26.7°C) atmospheric conditions	any	1.0	1.0
120 to 160°F (48.9 to 71.1°C) 20 to 35 % RH	70 to 75	0.7	0.7
190 to 200°F (87.8 to 93.3°C) 0 to 5 % RH	70 to 75	0.45	0
120 to 200°F (48.9 to 93.3°C) 5 to 35 % RH	less than 70	0	0

TABLE X4.3 Factor Characterizing Permeability of Test Specimen

Material	b
Normal weight and gun-applied concrete (dry unit weight greater than 135 lb/ft³ (2162 kg/m³))	5.5
Lightweight concrete (dry unit weight between 85 and 115 lb/ft³ (1361 and 1841 kg/m³))	8.0
Lightweight insulating concrete (dry unit weight less than 50 lb/ft³ (801 kg/m³))	10.0

ers use a system of hydraulic rams for floor and roof assemblies. When uniformly distributed load is simulated by point-loading (several, small-area loads), it is recommended that the load at any such area not exceed 25 % of the total load and that the individual have a width at least equal to the depth of the floor. Wall furnaces are generally equipped with hydraulic rams.

X5.8.3 The standard requires that load-bearing walls and partitions sustain the applied test load during the fire endurance and hose stream tests. A former requirement that load-bearing walls and partitions also sustain twice the specified superimposed test load after cooling but with 72 h of the test period has been deleted from the method as being unrealistic. Non-bearing walls and partitions are not loaded during the test but are restrained on all sides. This restraint may impose more stress than a load on top. Committee E-5 has several times reviewed the loading procedures for framed walls and partitions. It was the committee's unanimous decision that such a wall be tested either with calculated maximum design load or with a load expected to occur in practice. The method used to compute the design loads must be reported.

X5.8.4 Some important stresses, such as those caused by creep and shrinkage in the wall itself and its supporting frame must be present, and the designer should recognize these stresses in his analysis. Committee E-5 has investigated the possibility of openings occurring at a joint at corners of non-load-bearing enclosures due to differential movement. While the possibility exists that this will occur, the committee has not found it feasible to amend the test based on data available.

X5.8.5 Double walls pose a unique problem as to load application. Which wall should be loaded? Or should both walls be loaded simultaneously? Committee E-5 has devoted considerable time to debating this problem, and recommends the decision be made by the user after an analysis of the loading conditions anticipated in service both before and after a fire. Such loading conditions are to be reported.

X5.9 Integrity

X5.9.1 All walls and partitions that qualify for a fire endurance classification of 1 h or more are required to be subjected to the cooling impact and erosion effects of a stream of water from a 2-½-in. (63.5-mm) hose discharging through a standard playpipe equipped with a 1⅛-in. (28.5-mm) tip under specified pressures. In this hose stream test, the ability of the construction to resist disintegration under adverse conditions is examined. The requirement for a hose stream test was earlier removed from the test procedure for columns and floor or roof assemblies because of impracticality and the possibility of excessive damage to the furnace.

X5.10 Conditions of Tests

X5.10.1 Columns are generally tested with all four sides exposed to the test fire. However, it is possible to test a column with three sides exposed (fourth side against a wall). The standard requires that specimens be tested under conditions contemplated in the design. The former general practice of testing columns with pin connection at top and bottom to simulate the most critical condition is no longer a criterion.

X5.10.2 Columns are required to sustain successfully the design load during the test period. The standard also permits columns to be loaded up to 1-¾ times the design load prior to the fire test if desired by the submitter. Such loading, however, shall not be construed as having had a deleterious effect on the fire endurance test performance. Instead of loading, steel columns, whose protective covering does not carry load, may be assigned a fire-resistance classification on the basis of temperature of the steel only. With such columns, the protective cover shall be restrained against longitudinal expansion. Wood columns are tested for load-carrying ability only.

X5.10.3 From test results, it has been established that variations of restraint conditions can considerably influence the time of fire resistance for a structure or a structural element. Restraints are generally beneficial to fire resistance; however, there are conditions where restraint can have a detrimental effect on the performance of a specimen during a fire-resistance test (34, 35). The users of test results are advised to study the reference documents as well as

Appendix X3 and Table X3.1.

X5.10.4 An unrestrained classification for a steel beam or a reinforced concrete beam used as part of an assembly tested in restrained condition can be assessed from the temperature records obtained for the steel or the reinforcing steel respectively (Sections 36–38). It is also possible to evaluate the protective cover of steel beams by measuring the temperature of the steel that is protected (Sections 39–42). The fire endurance classification derived under the provisions of Sections 36–38 is only applicable to beams used with a floor or roof construction that has a comparable or greater capacity for heat dissipation from the beam than the floor or roof with which it is tested.

X5.11 Other Observations

X5.11.1 No limitation is imposed on the deformation of the specimen during or after the test period. It is assumed that the deflection or deformation of an assembly is limited only by its ability to stay in place (under load where specified) during the test period.

X5.11.2 A complete record of deformation during the endurance test may be helpful in the application of test results, and shall be reported.

X5.11.3 Other observations, such as the evolution of unusual quantities of visible smoke, vapors, or gases that may affect the proper decision regarding use of the test results, should be reported.

X5.12 Protective Membranes

X5.12.1 The standard provides criteria for evaluating the protection that membrane elements can offer to combustible framing and paneling, for example, joists, wall studs, and paneling or boards on the unexposed side of an assembly, and other combustible materials. The results of these tests are reported as protective membrane ratings.

X5.13 Future

X5.13.1 ASTM Committee E-5 on Fire Standards through Subcommittee E05.11 is continually striving to update Methods E 119. Users of the test method are encouraged to contact the committee for further information.

REFERENCES

(1) Babrauskas, Vytenis; Williamson, Robert Brady, Historical Basis of Fire Resistance Testing, Part I and Part II, *Fire Technology*, Vol 14, No. 3 and No. 4, 1978, p. 184–194, 304–316.

(2) Shoub, Harry, "Early History of Fire Endurance Testing in the United States," *Symposium on Fire Test Methods, ASTM STP 301*, Am Soc. Testing Mats., 1961, pp. 1–9

(3) Dilam, C. H., Modern Building Inspection, et al, R. C. Colling and Associates, Los Angeles, Calif. 1942.

(4) *Facts About Fires*, National Fire Protection Assn., 1971.

(5) Bird and Docking, *Fire in Buildings*, D. VanNostrand Co., Inc., New York, 1949.

(6) Ferguson, R. S., *Principles of Fire Protection - National Bldg. Code of Canada* - Technical Paper No. 272, Div. of Bldg. Research - National Research Council of Canada, Ottawa, March 1970.

(7) Konicek L., and Lie, T. T., "Temperature Tables for Ventilation Controlled Fires," Bldg. Research Note No. 94, National Research Council of Canada, September 1974.

(8) Gordon, C. C., *Considerations of Life Safety and Bldg. Use*, DBR Paper No. 699, Bldg. Research Div., National Research Council of Canada, Ottawa, January 1977.

(9) Shorter, G. W., *Fire Protection Engineer and Modern Building Design, Fire Technology*, Vol 4, No. 3, August 1968, p. 206–213.

(10) Harmathy, T. Z., *Performance of Building Elements in Spreading Fire*, DBR Paper No. 752, National Research Council of Canada, NRCC 16437, *Fire Research* Vol 1, 1977/78, pp. 119–132.

(11) Harmathy, T. Z., "*Design Approach to Fire Safety in Buildings*" Progressive Architecture, NRCC 14076, April 1974, pp. 82–87.

(12) Rule 508 - Industrial Code, New York State Labor Law. See p. 513, New York City Building Code, 1934 Edition.

(13) Robertson, A. F., and Gross, D., "Fire Load, Fire Severity, and Fire Endurance," *Fire Test*

Performance, ASTM STP 464, Am. Soc. Testing Mats., 1970, pp. 3–29.

(14) *Building Materials and Standards*, BMS 92, National Bureau of Standards, Washington, D.C. October 1942.

(15) Ingberg, S. H., et al, "Combustible Contents in Buildings," BMS 149, National Bureau of Standards, Washington, D. C., July 1957.

(16) Seigel, L. G., *The Severity of Fires in Steel Framed Buildings* - Her Majesty's Stationary Office, 1968, London, Proceedings of the Symposium held at the Fire Research Station, Boreham Woods, Herts (England) on Jan. 24, 1967, pp. 59–63.

(17) Gross, D., *Field Burnout Tests of Apartment Dwelling Units*, Bldg. Science Series 10, U. S. Dept. of Commerce, National Bureau of Standards, Sept. 29, 1967.

(18) Law, Margaret, *Radiation from Fires in a Compartment*, Fire Research Technical Paper No. 20, Her Majesty's Stationary Office, London, 1968.

(19) Heselden, A. J. M., *Parameters Determining the Severity of Fire*, Symposium No. 2 Her Majesty's Stationary Office, 1968, London, Proceedings of the Symposium held at the Fire Research Station, Boreham Woods, Herts (England) on Jan. 24, 1967, pp. 19–27.

(20) Sfintesco, D., *Furnace Tests and Fire Resistance ibid* 17.

(21) Gross, D., and Robertson, A. F., *Experimental Fires in Enclosures*, Tenth Symposium (International) on Combustion, The Combustion Institute, 1965, pp. 931–942.

(22) Odeen, Kai, *Theoretical Study of Fire Characteristics in Enclosed Spaces*, Div. of Bldg. Construction, Royal Institute of Technology, Stockholm, Sweden, 1965.

(23) Ryan, J. E., "*Perspective on Methods of Assessing Fire Hazards in Buildings*," Ignition, Heat Release, and Noncombustibility of Materials ASTM STP 502, Am. Soc. Testing Mats., 1972, pp. 11–23.

(24) Harmathy, T. Z., *Performance of Building Elements in Spreading Fire*, DBR Paper No. 752, National Research Council of Canada, NRCC 16437, *Fire Research* Vol 1, 1977/78, pp. 119–132.

(25) Harmathy, T. Z., "Design of Fire Test Furnaces," *Fire Technology*, Vol 5, No. 2, May 1969, pp. 146–150.

(26) Harmathy, T. Z., "Fire Resistance versus Flame Spread Resistance," *Fire Technology*, Vol 12, No. 4, November 1976, pp. 290–302 and 330.

(27) Harmathy, T. Z., "A New Look at Compartment Fires, Part I and Part II," *Fire Technology*, Vol 8, No. 3 and No. 4, August and November 1972, pp. 196–217; 326–351.

(28) Satsberg, F., Illinois Institute of Technology Research Institute Limited release on research data conducted for U. S. Dept. of Civil Defense.

(29) Harmathy, T. Z., *Designers Option: Fire Resistance or Ventilation*, Technical Paper No. 436, Division of Building Research, National Research Council of Canada, Ottawa, NRCC 14746.

(30) *Fire Protection Handbook*, Revised Thirteenth Edition, National Fire Protection Assn., Boston, Mass., 1969.

(31) ISO/TE-WG5 Report, March 9–10, 1967, at Copenhagen, Denmark (Sweden-B)-Exhibit 14. Preliminary report on some theoretical studies for structural elements of the effect on their fire resistance of variations of T-t curve for cooling down period. - Magusson, Sven-Erik and Pettersson, Ove.

(32) Ryan, J. V., and Robertson, A. F., "Proposed Criteria for Defining Load Failure of Beams, Floors, and Roof Construction During Fire Tests," *Journal of Research of the National Bureau of Standards*, Washington, D. C., Vol 63C, No. 2, 1959.

(33) Harmathy, T. Z., "Experimental Study on Moisture and Fire-Endurance," *Fire Technology*, Vol 2, No. 1, February 1966.

(34) Carlson, C. C., Selvaggio, S. L., Gustaferro, A. H., *A Review of Studies of the Effects of Restraint on the Fire-Resistance of Prestressed Concrete*, Feuerwider-stansfahigkeit von Spannbeton, Ergebnisse einer Tagung der F.I.P. in Braunschweig, Juni 1965. Wiesbaden-Berlin, 1966.

Designation: C 475 – 81

Standard Specification for

JOINT TREATMENT MATERIALS FOR GYPSUM WALLBOARD CONSTRUCTION[1]

This standard is issued under the fixed designation C 475; the number immediately following the designation indicates the year of original adoption or, in the case of revision, the year of last revision. A number in parentheses indicates the year of last reapproval. A superscript epsilon (ε) indicates an editorial change since the last revision or reapproval.

This specification has been approved for use by agencies of the Department of Defense and for listing in the DoD Index of Specifications and Standards.

1. Scope

1.1 This specification covers joint compound (an adhesive), reinforcing joint tape, and the combined use of the two materials.

1.2 The values stated in inch-pound units are to be regarded as the standard.

NOTE 1—Decoration of the wallboard and the joint treatment is not covered in this specification.

2. Applicable Documents

2.1 *ASTM Standards:*
C 474 Test Methods for Testing Joint Treatment Materials for Gypsum Wallboard Construction[2]

3. Material

3.1 *Joint Compound (An Adhesive)*, an adhesive with or without fillers (Note 2).

NOTE 2—Caution—This product may contain ingredients that have been judged hazardous. Approved protective measures should be observed during its use.

3.2 *Joint Tape*, a strip of reinforcing material designed to be embedded in the joint compound to reinforce the joints in gypsum wallboard construction.

3.3 *Finishing or Topping Compound (An Adhesive)*, an adhesive with or without fillers (Note 2).

4. Physical Requirements for Joint Compound and Finishing Compound

4.1 *Check Cracking Test*—There shall be no cracks in the thinner half of the wedge, and there shall be no deep fissure cracks in the thicker half of the wedge, when tested as specified in the section on Check Cracking of Joint Compound and Finishing Compound in Methods C 474.

4.2 *Putrefaction Test*—There shall be no putrefaction in less than 4 days, when tested as specified in the section on Putrefaction Test in Methods C 474.

4.3 *Shrinkage*—The shrinkage shall be not more than 35 %.

5. Requirements for Reinforcing Joint Tape

5.1 *Width*—The width shall be not less than 7/8 in. (47.6 mm) nor more than 2¼ in. (57.2 mm). Width of the tape in individual rolls shall not vary more than ±1/32 in. (0.8 mm).

5.2 *Thickness*—The thickness shall be not more than 0.012 in. (0.30 mm).

5.3 *Tensile Strength*—The tensile strength in cross direction shall be not less than 30 lbf (133 N)/in. (25.4 mm).

5.4 *Dimensional Stability*—The expansion shall be no more than 0.40 % lengthwise and not more than 2.5 % crosswise.

6. Requirements for Joint Compound and Reinforcing Joint Tape

6.1 *Bond of Joint Tape to Joint Compound*—The bond of tape to the joint compound shall be not less than 90 %.

5.2 *Edge Cracking*—Edge cracking of joint compound shall not exceed 10 %.

NOTE 3—Requirements of the Sampling section do not apply to finishing or topping compound.

7. Sampling

7.1 *Sampling of Joint Compound and Finishing or Topping Compound*—At least 1 %, but not less than one package, shall be sampled and shall be so selected as to be representative of the shipment. The material so obtained shall be thoroughly blended and reduced by quartering to provide not less than one 20-lb (9.1-kg) sample for the test.

7.2 *Sampling of Joint Tape*—At least one roll for every ten cases or fraction thereof, but not more than ten rolls per shipment shall constitute a sample.

8. Test Methods

8.1 The properties of the joint compound, joint tape, and finishing compound shall be determined in accordance with Methods C 474.

9. Inspection

9.1 Inspection may be made at point of shipment or the point of delivery. The inspector shall have free access to the carrier being loaded.

10. Rejection

10.1 Any rejection shall be based upon the specific cause of failure to conform to the requirements of this specification, and shall be reported to the seller within 20 working days from the receipt of the shipment by the purchaser.

NOTE 4—In the event that combined use tests fail when joint compound and tape are made by different manufacturers, then the manufacturers of the joint compound and the tape shall be contacted for their recommendations on types of materials to be used with their respective products.

11. Certification

11.1 When specified in the purchase order, a producer's or supplier's certification and report of the test results shall be furnished to the purchaser that the material was manufactured, sampled, tested, and inspected in accordance with this specification and has been found to meet the specified requirements.

12. Packaging and Marking

12.1 Joint compound, joint tape, and finishing compound shall be shipped in packages marked with the name, brand or trademark of the manufacturer or seller, size or net weight of contents, ASTM designation, and if the product contains ingredients that have been judged hazardous, the approved protective measures that should be observed during its use.

---

[1] This specification is under the jurisdiction of ASTM Committee C-11 on Gypsum and Related Building Materials and Systems and is the direct responsibility of Subcommittee C11.02 on Specifications and Test Methods for Accessories and Related Products.

Current edition approved June 26, 1981. Published August 1981. Originally published as C 475 – 61 T. Last previous edition C 475 – 64 (1975).

[2] *Annual Book of ASTM Standards*, Vol 04.01.

 **Designation: E 152 – 81a'[1]**

Standard Methods of
FIRE TESTS OF DOOR ASSEMBLIES[1]

This standard is issued under the fixed designation E 152; the number immediately following the designation indicates the year of original adoption or, in the case of revision, the year of last revision. A number in parentheses indicates the year of last reapproval. A superscript epsilon (ϵ) indicates an editorial change since the last revision or reapproval.

These methods have been approved for use by agencies of the Department of Defense and for listing in the DoD Index of Specifications and Standards.

---

[1] NOTE—Paragraphs 10.2 and X1.13.1 were changed editorially and all references were renumbered in July 1984.

---

1. Scope

1.1 These methods of fire test are applicable to door assemblies of various materials and types of construction, for use in wall openings to retard the passage of fire (see commentary in Appendix).

1.2 Tests made in conformity with these test methods will register performance during the test exposure; but such tests shall not be construed as determining suitability for use after exposure to fire.

1.3 It is the intent that tests made in conformity with these test methods will develop data to enable regulatory bodies to determine the suitability of door assemblies for use in locations where fire resistance of a specified duration is required.

1.4 *This standard should be used to measure and describe the properties of materials, products, or assemblies in response to heat and flame under controlled laboratory conditions and should not be used to describe or appraise the fire hazard or fire risk of materials, products, or assemblies under actual fire conditions. However, results of this test may be used as elemens of a fire risk assessment which takes into account all of the factors which are pertinent in an assessment of the fire hazard of a particular end use.*

2. Significance

2.1 These methods are intended to evaluate the ability of a door assembly to remain in an opening during a predetermined test exposure.

2.2 The tests expose a specimen to a standard fire exposure controlled to achieve specified temperatures throughout a specified time period, followed by the application of a specified standard fire host stream. The exposure, however, may not be representative of all fire conditions, which may vary with changes in the amount, nature, and distribution of fire loading, ventilation, compartment size and configuration, and heat sink characteristics of the compartment. It does, however, provide a relative measure of fire performance of door assemblies under these specified fire exposure conditions.

2.3 Any variation from the construction or conditions that are tested may substantially change the performance characteristics of the assembly.

2.4 The methods do not provide the following:

2.4.1 Full information as to performance of all door assemblies in walls constructed of materials other than that tested.

2.4.2 Evaluation of the degree by which the door assembly contributes to the fire hazard by generation of smoke, toxic gases, or other products of combustion.

2.4.3 A specific requirement that the unexposed surface temperatures be reported although the temperature measurement procedure is described.

2.4.4 A limit on the number of openings allowed in glazed areas or of the number and size of lateral openings between the door and frame.

2.4.5 Measurement of the degree of control or limitation of the passage of smoke or products of combustion through the door assembly.

CONTROL OF FIRE TESTS

3. Time - Temperature Curve

3.1 The fire exposure of door assemblies shall be controlled to conform to the applicable portion of the standard time-temperature curve shown in Fig. 1. The points on the curve that determine its character are:

1000°F (538°C)	at 5 min
1300°F (704°C)	at 10 min
1550°F (843°C)	at 30 min
1700°F (926°C)	at 1 h
1850°F (1010°C)	at 2 h
2000°F (1093°C)	at 4 h
2300°F (1260°C)	at 8 h or over

3.1.1 For a closer definition of the time-temperature curve, see Table A1.1.

4. Furnace Temperatures

4.1 The temperatures of the test exposure shall be deemed to be the average temperature obtained from the readings of not less than nine thermocouples symmetrically disposed and distributed to show the temperature near all parts of the test assembly. The thermocouples shall be protected by sealed porcelain tubes having ¾-in. (19-mm) outside diameter and ⅛-in. (3-mm) wall thickness, or, as an alternative, in the case of base metal thermocouples, protected by ½-in. (13-mm) wrought steel or wrought iron pipe of standard weight. The junction of the thermocouples shall be 6 in. (152 mm) from the exposed face of the test assembly or from the masonry in which the assembly is installed, during the entire test exposure.

4.2 The temperatures shall be read at intervals not exceeding 5 min during the first 2 h, and thereafter the intervals may be increased to not more than 10 min.

4.3 The accuracy of the furnace control shall be such that the area under the time-temperature curve, obtained by averaging the results from the thermocouple readings, is within 10 % of the corresponding area under the standard time-temperature curve for fire tests of 1 h or less duration, within 7.5 % for those over 1 h and not more than 2 h, and within 5 % for tests exceeding 2 h in duration.

5. Unexposed Surface Temperatures

5.1 Unexposed surface temperatures shall be recorded and shall be determined in the following manner:

5.1.1 Unexposed surface temperatures shall be taken at not less than three points with at least one thermocouple in each 16 ft² (1.5 m²) area of the door. Thermocouples shall not be located over reinforcements extending through the door, over vision panels, or nearer than 12 in. (305 mm) from the edge of the door.

5.1.2 Unexposed surface temperatures sha be measured with thermocouples placed und flexible, oven-dry, felted asbestos pads 6 i (152 mm) square, 0.4 in. (10 mm) in thicknes and weighing not less than 1.0 nor more tha 1.4 lb/ft² (4.88 to 6.83 kg/m²). The pads sha be held firmly against the surface of the do and fit closely about the thermocouples. Th thermocouple leads shall be immersed unde the pad for a distance of not less than 3½ i (89 mm) with the hot junction under the cent of the pad. The thermocouple leads under th pads shall be not heavier than No. 18 B & gage (0.04 in.) (1.02 mm) and shall be electr cally insulated with heat-resistant and mois ture-resistant coatings.

5.1.3 Unexposed surface temperatures shal be read at intervals not exceeding 5 min for th first 30 min of the test.

TEST ASSEMBLIES

6. Construction and Size

6.1 The construction and size of the test doc assembly, consisting of single doors, doors i pairs, special-purpose doors (such as Dutc doors, double-egress doors, etc.), or multisec tion doors, shall be representative of that fo which classification or rating is desired.

6.2 A floor structure shall be provided a part of the opening to be protected, excep where such floor interferes with the operatio of the door. The floor segment shall be c noncombustible material and shall project int the furnace approximately twice the thicknes of the test door, or to the limit of the fram whichever is greater.

7. Mounting for Test

7.1 Swinging doors shall be mounted so a to open into the furnace chamber. Sliding an rolling doors, except horizontal slide-type ele vator shaft doors, shall be mounted on th exposed side of the opening in the wall closin the furnace chamber. Horizontal slide-type e evator shaft doors shall be mounted on th unexposed side of the opening in the wall clos ing the furnace chamber. Access-type doo and chute-type doors and frame assemblie shall be mounted so as to have one assembl open into the furnace chamber and anothe assembly open away from the furnace chambe Dumb-waiter and service-counter doors an frame assemblies shall be mounted on the ex posed side of the opening in the wall.

7.2 The mounting of all doors shall be suc that they fit snugly within the frame, agains the wall surfaces, or in guides, but such mount ing shall not prevent free and easy operation o the test door.

---

[1] These methods are under the jurisdiction of ASTM Committee E-5 on Fire Standards.
Current edition approved Sept. 25 and Nov. 10, 1981. Published March 1982. Originally published as C 152 – 40 T. Redesignated E 152 in 1941. Last previous edition E 152 – 80.

7.2.1 Clearances for swinging doors shall be as follows: With a minus ¹⁄₁₆-in. (1.6-mm) tolerance: ⅛ in. (3.2 mm) along the top, ⅛ in. along the hinge and latch jambs, ⅛ in. along the meeting edge of doors, in pairs, and ⅜ in. (9.5 mm) at the bottom edge of a single swinging door, and ¼ in. (6.3 mm) at the bottom of a pair of doors.

7.2.2 Clearances of horizontal sliding doors not mounted within guides shall be as follows: With a minus ⅛-in. (3.2-mm) tolerance: ½ in. (12.7 mm) between the door and wall surfaces, ⅜ in. (9.5 mm) between the door and floor structure and ¼ in. (6.3 mm) between the meeting edges of center-parting doors. A maximum lap of 4 in. (102 mm) of the door over the wall opening at sides and top shall be provided.

7.2.3 Clearances of vertical sliding doors moving within guides shall be as follows: With a minus ⅛-in. (3.2-mm) tolerance: ½ in. (12.7 mm) between the door and wall surfaces along the top and/or the bottom door edges with guides mounted directly to the wall surfaces and ³⁄₁₆ in. (4.8 mm) between the meeting edges of bi-parting doors or ³⁄₁₆ in. between the door and floor structure or the sill.

7.2.4 Clearances for horizontal slide type elevator doors shall be as follows: With a minus ⅛-in. (3.2-mm) tolerance: ⅜ in. (9.5 mm) between the door and wall surfaces, ⅜ in. between the multisection door panels, and ⅜ in. from the bottom of a panel to the sill. Multisection door panels shall overlap ¾ in. (19.0 mm). Door panels shall lap the wall opening ¾ in. at the sides and top.

CONDUCT OF TESTS

8. Time of Testing

8.1 *Time of Testing*—Masonry shall have sufficient strength to retain the assembly securely in position throughout the fire and hose stream test.

9. Fire Endurance Test

9.1 Maintain the pressure in the furnace chamber as nearly equal to the atmospheric pressure as possible.

9.2 Continue the test until the exposure period of the desired classification or rating is reached unless the conditions of acceptance set forth in Section 12 are exceeded in a shorter period.

10. Hose Stream Test

10.1 Immediately following the fire endurance test, subject the test assembly to the impact, erosion, and cooling effects of a hose stream directed first at the middle and then at all parts of the exposed surface, making changes in direction slowly.

10.2 Deliver the hose stream through a 2½-in. (64-mm) hose discharging through a National Standard Playpipe of corresponding size equipped with a 1⅛-in. (28.5-mm) discharge tip of the standard-taper smooth-bore pattern without shoulder at the orifice. The water pressure at the base of the nozzle and duration of application in s/ft² (m²) of exposed area shall be as prescribed in Table 1.

10.3 The tip of the nozzle shall be located 20 ft (6 m) from and on a line normal to the center of the test door. If impossible to be so located, the nozzle may be on a line deviating not to exceed 30° from the line normal to the center of the test door. When so located the distance from the center shall be less than 20 ft by an amount equal to 1 ft (0.3 m) for each 10° of deviation from the normal.

11. Report

11.1 Report results in accordance with the performance in the tests prescribed in these test methods. The report shall show:

11.1.1 The performance under the desired exposure period chosen from the following: 20 min, 30 min, ¾ h, 1 h, 1½ h, or 3 h.

11.1.2 The temperature measurements of the furnace.

11.1.3 The temperature measurement of the unexposed side.

11.1.4 All observations having a bearing on the performance of the test assembly.

11.1.5 Flaming, if any, on the unexposed surface of the door leaf during the first 20 min of the fire test.

11.1.6 The amount of movement of any portion of the edges of the door adjacent to the door frame from the original position (see Section 12).

11.1.7 The materials and the construction of the door and frame, and the details of the installation, hardware, hangers, guides, trim, finish, and clearance or lap shall be recorded or appropriately referenced to ensure positive identification or duplication in all respects.

11.1.8 Pressure measurements made in the furnace and their relationship to the top of the door.

CONDITIONS OF ACCEPTANCE

12. Conditions of Acceptance

12.1 A door assembly shall be considered as meeting the requirements for acceptable performance when it remains in the opening during the fire endurance test and host-stream test within the following limitations:

12.1.1 The movement of swinging doors shall not permit any portion of the edges to move from the original position more than the thickness of the door, during the first half of the classification period, nor more than 1½ times the thickness during the entire classification period, nor more than 1½ times the thickness immediately following the hose stream test.

12.1.2 An assembly consisting of a pair of swinging doors shall not separate more than ¾ in. (19 mm) or equal to the throw of the latch bolt at the latch location.

12.1.3 An assembly consisting of a single swinging door shall not separate more than ½ in. (13 mm) at the latch location.

12.1.4 The lap edges of passenger (A17.1 horizontal slide-type) elevator doors, including the lap edges of multisection doors, shall not move from the wall or adjacent panel surfaces sufficiently to develop a separation of more than 2⅞ in. (73.0 mm) during the entire classification period, or immediately following the hose stream test. The meeting edges of center-parting elevator door assemblies, for a fire and hose stream exposure of 1½ h or less, shall not move apart more than 1¼ in. (31.7 mm) as measured in any horizontal plane during the entire classification period or immediately following the hose stream test.

12.1.5 Doors mounted in guides shall not release from guides and guides shall not loosen from fastenings.

12.1.6 The test assembly shall have withstood the fire endurance test and hose-stream test, without developing openings anywhere through the assembly, except that small portions of glass dislodged by the hose stream shall not be considered a weakness.

13. Precision and Bias

13.1 Precision and bias data are not available at this time; however, a task group of Subcommittee E05.12 has been established to investigate the subject and prepare a statement.

TABLE 1 Water Pressure at Base of Nozzle and Duration of Application[A]

Desired Rating	Water Pressure at Base of Nozzle, psi (kPa)	Duration of Application, s/ft² (0.09 m²) exposed area
3 h	45 (310)	3
1½ h and over, if less than 3 h	30 (207)	1.5
1 h and over, if less than 1½ h	30 (207)	0.9
Less than 1 h	30 (207)	0.6

[A] The exposed area may be calculated using the outside dimensions of the test specimen, including a frame, hangers, tracks, or other parts of the assembly if provided, but normally not including the wall into which the specimen is mounted. Where multiple test specimens are mounted in the same wall, the rectangular or square wall area encompassing all of the specimens will have to be considered as the exposed area since the hose stream must traverse this area during its application.

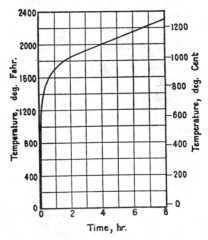

FIG. 1 Time - Temperature Curve

ANNEX

(Mandatory Information)

TABLE A1.1 **Standard Time-Temperature Curve for Control of Fire Tests**

Time	Temperature, °F	Area Above 68°F Base		Temperature, °C	Area Above 20°C Base	
h:min		°F-min	°F-h		°C-min	°C-h
0:00	68	00	0	20	00	0
0:05	1 000	2 330	39	538	1 290	22
0:10	1 300	7 740	129	704	4 300	72
0:15	1 399	14 150	236	760	7 860	131
0:20	1 462	20 970	350	795	11 650	194
0:25	1 510	28 050	468	821	15 590	260
0:30	1 550	35 360	589	843	19 650	328
0:35	1 584	42 860	714	862	23 810	397
0:40	1 613	50 510	842	878	28 060	468
0:45	1 638	58 300	971	892	32 390	540
0:50	1 661	66 200	1 103	905	36 780	613
0:55	1 681	74 220	1 237	916	41 230	687
1:00	1 700	82 330	1 372	927	45 740	762
1:05	1 718	90 540	1 509	937	50 300	838
1:10	1 735	98 830	1 647	946	54 910	915
1:15	1 750	107 200	1 787	955	59 560	993
1:20	1 765	115 650	1 928	963	64 250	1 071
1:25	1 779	124 180	2 070	971	68 990	1 150
1:30	1 792	132 760	2 213	978	73 760	1 229
1:35	1 804	141 420	2 357	985	78 560	1 309
1:40	1 815	150 120	2 502	991	83 400	1 390
1:45	1 826	158 890	2 648	996	88 280	1 471
1:50	1 835	167 700	2 795	1 001	93 170	1 553
1:55	1 843	176 550	2 942	1 006	98 080	1 635
2:00	1 850	185 440	3 091	1 010	103 020	1 717
2:10	1 862	203 330	3 389	1 017	112 960	1 882
2:20	1 875	221 330	3 689	1 024	122 960	2 049
2:30	1 888	239 470	3 991	1 031	133 040	2 217
2:40	1 900	257 720	4 295	1 038	143 180	2 386
2:50	1 912	276 110	4 602	1 045	153 390	2 556
3:00	1 925	294 610	4 910	1 052	163 670	2 728
3:10	1 938	313 250	5 221	1 059	174 030	2 900
3:20	1 950	332 000	5 533	1 066	184 450	3 074
3:30	1 962	350 890	5 848	1 072	194 940	3 249
3:40	1 975	369 890	6 165	1 079	205 500	3 425
3:50	1 988	389 030	6 484	1 086	216 130	3 602
4:00	2 000	408 280	6 805	1 093	226 820	3 780
4:10	2 012	427 670	7 128	1 100	237 590	3 960
4:20	2 025	447 180	7 453	1 107	248 430	4 140
4:30	2 038	466 810	7 780	1 114	259 340	4 322
4:40	2 050	486 560	8 110	1 121	270 310	4 505
4:50	2 062	506 450	8 441	1 128	281 360	4 689
5:00	2 075	526 450	8 774	1 135	292 470	4 874
5:10	2 088	546 580	9 110	1 142	303 660	5 061
5:20	2 100	566 840	9 447	1 149	314 910	5 248
5:30	2 112	587 220	9 787	1 156	326 240	5 437
5:40	2 125	607 730	10 129	1 163	337 630	5 627
5:50	2 138	628 360	10 473	1 170	349 090	5 818
6:00	2 150	649 120	10 819	1 177	360 620	6 010
6:10	2 162	670 000	11 167	1 184	372 230	6 204
6:20	2 175	691 010	11 517	1 191	383 900	6 398
6:30	2 188	712 140	11 869	1 198	395 640	6 594
6:40	2 200	733 400	12 223	1 204	407 450	6 791
6:50	2 212	754 780	12 580	1 211	419 330	6 989
7:00	2 225	776 290	12 938	1 218	431 270	7 188
7:10	2 238	797 920	13 299	1 225	443 290	7 388
7:20	2 250	819 680	13 661	1 232	455 380	7 590
7:30	2 262	841 560	14 026	1 239	467 540	7 792
7:40	2 275	863 570	14 393	1 246	479 760	7 996
7:50	2 288	885 700	14 762	1 253	492 060	8 201
8:00	2 300	907 960	15 133	1 260	504 420	8 407

APPENDIX

(Nonmandatory Information)

X1. COMMENTARY

X1.1 Introduction

X1.1.1 This commentary has been prepared to provide the user of Methods E 152 with background information on the development of the standard and its application in fire protection of buildings. It also provides guidance in the planning and performance of fire tests and in the reporting of results. No attempt has been made to incorporate all of the available information on fire testing in this commentary. The serious student of fire testing is strongly urged to peruse the referenced documents for a better appreciation of the history of fire-resistant design and the intricate problems associated with testing and with interpretation of test results.

X1.2 Application

X1.2.1 Compartmentation of buildings by fire-resistive walls has been recognized for many years as an efficient method of restricting fires to the area of origin (**1, 2, 3, 4, 5, 6, 7, 8, 9**)[2] or limiting their spread. The functional use of buildings however, demands a reasonable amount of communication between compartments necessitating openings in these fire-resistive walls. Fire door assemblies are utilized to protect these openings and maintain the integrity of the fire barrier (**10**). Openings in walls have been classified by fire protection standards (**6, 11, 12**) and building codes in accordance with the location and purpose of the wall in which the opening occurs, and these standards and codes specify the fire rating of the assembly required to protect the openings.

X1.2.2 These fire protection standards and building codes permit labeled wire glass panels and other penetrations, such as labeled ventilation louvres, in some rated doors. The reader is referred to the model building codes, NFPA Standard No. 80, and the specific fire door manufacturer's label service for information on the types and sizes of these openings.

X1.2.3 Fire doors must also be properly installed to maintain their fire rating. Again, NFPA Standard No. 80 and the specific fire door manufacturer's label service should be consulted for details on the installation of fire door assemblies and for limitations on the application of specific labeled fire doors.

X1.3 Historical Aspects

X1.3.1 The first effort to test fire doors is reported in a series of tests conducted in Germany in 1893 (**13, 14, 15**). The British Fire Prevention Committee began testing in 1899 and produced a Standard Table of Fire Resisting Elements including Fire Resisting Doors (**1**). Underwriters Laboratories Inc. was involved in testing and listing fire doors shortly after 1900 using their own standards. ASTM adopted Methods E 152 on fire door assembly tests in 1941.

X1.4 Scope and Significance

X1.4.1 Methods E 152 are intended to provide methods for measuring the relative performance of fire door assemblies when exposed to predetermined standard fire conditions. The standard provides for testing of several classifications, types, and methods and sectional doors (**6**). Since the effectiveness of the opening protection is dependent upon the entire assembly, proper attention must be paid to the installation as a unit. Accordingly, fire door assemblies are required to be tested as an assembly of all necessary elements and equipment, including the door frame and hardware.

X1.4.2 Fire protection ratings are assigned to indicate that the assembly has continued to perform as required for periods of 3, 1½, 1, ¾, ½ or ⅓ h. Labels on assemblies also carry the lettered designations of A, B, C, D, or E. These letter designations are not a part of the E 152 standard classification system but are used to designate the class of opening for which the door is designed as determined by other standards (**6, 11**).

X1.4.3 The ⅓-h, or 20-min fire-rated door is relatively new. Concern about the uniform adequacy of the 1¾ in. (44.5-mm) solid bonded wood core construction and the difficulty of determining equivalency of other types of doors, led to a voluntary consensus to test such doors for 20 min in the E 152 furnace using the same acceptance criteria as specified for door assemblies traditionally tested for a longer period of time except that the hose stream test is required by the test method but may not be required by regulatory codes.

X1.4.4 It is usual for fire door to have a fire protection rating lower than the wall in which it is installed, for example, a 1½-h fire door in a wall having a fire-resistance rating of 2 h. This is justified by the fact that under normal conditions of use the potential fire exposure in the vicinity of a door opening is lessened since there will be a clear space on both sides of the opening for traffic purposes. If the opening is not used, combustibles may be piled against the door, and the assumed enclosure protection will not be maintained. In these instances, the openings should be made equal to the rating of the wall or precautions taken to prevent storage of combustibles against the doors (**2, 6**).

X1.5 Limitations

X1.5.1 Methods intend that the door be tested until the conditions of acceptance are met for the desired exposure period unless the conditions of ac-

[2] The boldface numbers in parentheses refer to the list of references at the end of this appendix.

eptance are exceeded in a shorter period. It is not ntended that a fire door subjected to a building fire vill be satisfactory for reuse after the fire.

X1.5.2 The variations in material performance reclude any prediction of an assembly's performnce in walls other than those types used in the test. he standard also makes no provisions for measuring e generation of smoke and gases or other products f combustion from the unexposed side of the door. emperature measurements on the unexposed side. vhen recorded, are stopped after 30 min.

1.6 Furnace

X1.6.1 The methods provide details on the operting characteristics and temperature-measurement equirements of the test furnace. The walls of the urnace should be typically of furnace refractory naterials and should be sufficiently rugged to mainain the overall integrity of the furnace during the re-exposure period.

X1.6.2 The thermocouples in the furnace are loated 6 in. (152 mm) from the face of the door or the vall in which the door is installed. Otherwise no urnace depth is specified. A depth of 8 to 18 in. (203) 457 mm) has been considered desirable by most aboratories. The reader is urged to consult reference ocuments for a more comprehensive review of furnace esign and performance (**16, 17**).

1.7 Temperature – Time Curve

X1.7.1 A specific temperature – time relationship or the test fire is defined in the standard and in the Annex. The actual recorded temperature – time conlition obtained in the furnace is required to be within specified percentages of those of the standard curve. The number and type of temperature-measuring deices are outlined in the standard. Specific standard ractices for location and use of these temperatureneasuring devices are also outlined in the standard.

X1.7.2 The standard temperature – time $(T - t)$ urve used in E 152 is considered to represent a severe uilding fire (**3**). The curve was adopted in 1918 as a esult of several conferences by eleven technical orgaizations, including testing laboratories, insurance unerwriters, fire protection associations, and technical ocieties. It should be recognized that the $T - t$ relaionship of these test methods represent only one real re situation (**7, 8, 9, 18, 19, 20, 21, 22, 23, 24, 25, 26, 27**).

X1.8 Furnace Control

X1.8.1 The standard contains specific instruction or measuring temperatures in the furnace and for election of the required thermocouples. Thermocoules of the design specified are sufficiently rugged to etain accuracy throughout anticipated test periods. lowever, their massive construction results in a sigificant time delay in response to temperature change nd results in temperatures exceeding the indicated emperatures during the early stages of the test period vhen the temperature rises rapidly. The iron or porelain tubes surrounding the junction and leads of he thermocouple provide a shield against degradaion of the junction and increase the thermal inertia. t is customary for laboratories to replace furnace hermocouples after three or four accumulated hours f use.

1.9 Unexposed Surface Temperature

X1.9.1 Conditions of acceptance for fire-resistive valls specify that the temperature increase on the inexposed side of the wall not exceed an average of 250°F (139°C) above ambient, and that there be no assage of flames or gases hot enough to ignite ombustibles. It is obvious that the necessity of mainaining some clearances for efficient operation of the loor and the possibility of warping preclude comletely any attempt to restrict escape of gases and ninor flames on the periphery of doors.

X1.9.2 The standard describes a standard proceure for measuring the unexposed surface temperaures. However, unexposed surface temperatures are ot a condition of acceptance for E 152. Building egulations do restrict temperature transmission for ome wall-opening protectives (**6, 11**). For instance, it s usual for codes to limit the temperature rise on the inexposed side of fire doors protecting exit stairways o 450°F (250°C) during the first 30 min of test. This riterion assumes that a higher temperature would proide enough radiant heat to discourage if not prevent ccupants from passing by the door during an emerency. It is present practice for testing laboratories to rovide labels on fire doors indicating that the maximum transmitted temperature on the unexposed side is 250°F, 450°F, or 650°F (139°C, 250°C, or 361°C) above ambient. It not indicated on the label, the temperature rise during the first 30 min may or may not be in excess of 650°F (361°C). Temperature rise on the unexposed side of glass panels and louvres is not measured.

X1.9.3 Information on the properties of pads used to cover the thermocouples on the unexposed surfaces may be found in Appendix A2 of ASTM Standard E 119.

X1.10 Test Assemblies

X1.10.1 Standard E 152 provides a relative measure of performance for door assemblies. In order to establish confidence that the tested doors will perform in a building as expected, the tested assembly and its installation in the test frame must be representative of actual use conditions. Therefore, the National Fire Protection Assn. Standard No. 80 (**6**) or such other standards or specifications should be consulted before testing an assembly.

X1.10.2 Methods E 152 provide additional minimum requirements including direction of door swing, location in relation to the exposed side of the wall, and specific clearance between the door and its frame or wall, or both. Regardless of other specifications, these instructions must be followed in order to make a comparative judgment on test results.

X1.11 Conduct of Tests

X1.11.1 The test frame or wall in which a door assembly is installed should be rugged enough to endure the exposed fire during the time period, without affecting the door assembly. Traditionally, this wall has been of masonry construction. Today, fire doors are installed in other than masonry walls and have been tested in walls framed with metal and wood studs covered with a number of materials.

X1.12 Furnace Pressures

X1.12.1 A fire in a building compartment will create both negative and positive pressures on door assemblies depending upon atmospheric conditions, height above ground, wind conditions, and ventilation of the compartment at the beginning and during the fire.

X1.12.2 Methods E 152 specify that the pressure in the furnace be maintained as nearly equal to atmospheric pressure as possible. Experience has shown this practice to be acceptable. The pressure in the furnace is required to be reported but the method of measuring it is optional with the laboratory.

X1.13 Hose Stream Test

X1.13.1 Immediately following a fire test, the test frame is removed from the furnace and the door assembly is subjected to the impact, erosion, and cooling effects of a stream of water from a 2½-in. (63.5-mm) hose discharging through a standard playpipe equipped with a 1⅛-in. (28.5-mm) tip under specified pressures (see 10.2). The application of water produces stresses in the assembly and provides a measure of its structural capability. Weights were once used to provide a measure of the ability of the assembly to withstand impact. The hose stream is considered to be an improvement in uniformity and accuracy over the weights.

X1.14 Conditions of Acceptance

X1.14.1 The standard provides a specific set of conditions by which the performance of the door is measured, the most important being that it remain in place during both the fire test and the hose stream test. Instructions for conducting the hose stream test are detailed in the standard.

X1.15 Additional Information

X1.15.1 Inquiries concerning Methods E 152 should be addressed to ASTM Subcommittee E05.12.

REFERENCES

(1) Bird and Docking, *Fire in Buildings*, D. Van-Nostrand Co., Inc., New York, 1949.

(2) Ferguson, R. S., *Principles of Fire Protection, National Bldg. Code of Canada Technical Paper No. 272, Div. of Bldg. Research, National Research Council of Canada, Ottawa, March 1970.*

(3) Gordon, C., "Considerations of Life Safety and Building Use," DBR Paper No. 699, Division of Building Research, National Research Council of Canada, Ottawa, January, 1977.

(4) Gross, D., *Field Burnout Tests of Apartment Dwelling Units*, Bldg. Science Series 10, U.S. Dept. of Commerce, National Bureau of Standards, Sept. 29, 1967.

(5) Law, Margaret, "Radiation from Fires in a Compartment," Fire Research Technical Paper No. 20, Her Majesty's Stationary Office, London 1968.

(6) NFPA 80, "Fire Doors and Windows," National Fire Protection Assn.

(7) Harmathy, T. Z., "Designers Option: Fire Resistance or Ventilation," Technical Paper No. 436, Division of Building Research, National Research Council of Canada, Ottawa, NRCC 14746.

(8) Harmathy, T. Z., "Design Approach to Fire Safety in Buildings," *Progressive Architecture*, April 1974, pp. 82–87, NRCC 14076.

(9) Harmathy, T. Z., "A New Look of Compartment Fires Part I and Part II," *Fire Technology*, Vol 8, No. 3 and No. 4, 1972, pp. 196–217; 326–351.

(10) Shoub, H., and Gross, D., *Doors as Barriers to Fire and Smoke*, Building Science Series 3, National Bureau of Standards, March 25, 1966.

(11) Model Building Codes: *Basic Building Code*-Building Officials & Code Administrators International Inc. *Uniform Building Code*-International Conference of Building Officials Inc. *Standard Building Code*-Southern Building Code Congress International *National Building Code*-American Insurance Assn.

(12) *Fire Protection Handbook*, Revised Fourteenth Edition, National Fire Protection Assn., Boston, 1978.

(13) Shoub, Harry, "*Early History of Fire Endurance Testing in the United States*," *Symposium on Fire Test Methods, ASTM STP 301*, Am. Soc. Testing Mats., 1961.

(14) Konicek, L., and Lie, T. T., *Temperature Tables for Ventilation Controlled Fires*, Building Research Note No. 94, National Research Council of Canada, September, 1974.

(15) Babrauskas, Vytenis; Williamson, Robert Brady, "Historical Basis of Fire Resistance Testing, Part I and Part II," *Fire Technology*, Vol 14, No. 3 and No. 4, 1978, pp. 184–194, 304–316.

(16) Seigel, L. G., "Effects of Furnace Design on Fire Endurance Test Results," *Fire Test Performance, ASTM STP 464*, Am. Soc. Testing Mats., 1970, pp. 57–67.

(17) Harmathy, T. Z., "Design of Fire Test Furnaces," *Fire Technology*, Vol. 5, No. 2, May 1969, pp. 146–150.

(18) Seigel, L. G., "The Severity of Fires in Steel-Framed Buildings," Symposium No. 2, Her Majesty's Stationery Office, 1968, London, Proceedings of the Symposium held at the Fire Research Station, Boreham Woods, Herts (England), January 1967.

(19) Odeen, Kai, "Theoretical Study of Fire Characteristics in Enclosed Spaces," *Bulletin No. 10*, Royal Institute of Technology, Division of Bldg. Construction, Stockholm, 1963.

(20) Shorter, G. W., "Fire Protection Engineer and Modern Building Design," *NFPA Fire Technology*, Vol. 4, No. 3, August 1968, pp. 206–213.

(21) *Wall Street Journal*, Dec. 8, 1970, "Danger-Flammable," by-line Richard Stone.

(22) Ryan, J. E., "Assessment of Fire Hazards in Buildings, *Ignition, Heat Release, and Noncombustibility of Materials, ASTM STP 502*, Am. Soc. Testing Mats., 1972.

(23) Robertson, A. F., and Gross, Daniel, "Fire Load, Fire Severity, and Fire Endurance," *Fire Test Performances ASTM, STP 464*, Am. Soc. Testing Mats., 1970.

(24) Heselden, A. J. M., *Parameters Determining the Severity of Fire*, Symposium No. 2, Her Majesty's Stationery Office, 1968, London, Proceedings of the Symposium held at the Fire Research Station, Boreham Woods, Herts (England) on January, 1967.

(25) Gross, Daniel, and Robertson, A. F., *Experimental Fires in Enclosures*, Tenth Symposium (International) on Combustion, The Combustion Institute, 1965, pp. 731–942.

(26) Ingberg, et al., *Combustible Contents in Buildings*, National Bureau of Standards, BMS 149, July 1957.

(27) Harmathy, T. Z., "Performance of Building Elements in Spreading Fire," DBR Paper No. 752, National Research Council of Canada, NRCC 16437, *Fire Research*, Vol 1, 1977/78, pp. 119–132.

Designation: E 90 – 85

Standard Method for

LABORATORY MEASUREMENT OF AIRBORNE SOUND TRANSMISSION LOSS OF BUILDING PARTITIONS[1]

This standard is issued under the fixed designation E 90; the number immediately following the designation indicates the year of original adoption or, in the case of revision, the year of last revision. A number in parentheses indicates the year of last reapproval. A superscript epsilon (ϵ) indicates an editorial change since the last revision or reapproval.

This method has been approved for use by agencies of the Department of Defense and for listing in the DoD Index of Specifications and Standards.

1. Scope

1.1 This method covers the laboratory measurement of airborne sound transmission loss of building partitions such as walls of all kinds, floor-ceiling assemblies, doors, windows, roofs, and other space-dividing elements.

1.2 *Laboratory Accreditation*—A procedure for accrediting a laboratory for purposes of this test is given in Annex A2.

1.3 *This standard may involve hazardous materials, operations, and equipment. This standard does not purport to address all of the safety problems associated with its use. It is the responsibility of whoever uses this standard to consult and establish appropriate safety and health practices and determine the applicability of regulatory limitations prior to use.*

2. Applicable Documents

2.1 *ASTM Standards:*

C 423 Test Method for Sound Absorption and Sound Absorption Coefficients by the Reverberation Room Method[2]

C 634 Definitions of Terms Relating To Environmental Acoustics[2]

E 336 Test Method for Measurement of Airborne Sound Insulation in Buildings[2]

2.2 *ANSI Standards:*

S1.11 Specification for Octave, Half-Octave and Third-Octave Band Filter Sets[3]

S1.32-1980 Precision Methods for the Determination of Sound Power Levels of Discrete-Frequency and Narrow-Band Noise Sources in Reverberation Rooms[3]

2.3 *ISO Standard:*

ISO 3741-1975 Acoustics—Determination of Sound Power Level of Noise Sources—Precision Methods for Broad-Band Sources in Reverberation Rooms[3]

3. Definitions

3.1 The acoustical terms used in this standard are defined in Definitions C 634. A few definitions of special relevance are repeated here.

3.1.1 *diffuse sound field*—the sound in a region where the intensity is the same in all directions and at every point.

3.1.2 *direct sound field*—the sound that arrives directly from a source without reflection.

3.1.3 *flanking transmission*—transmission of sound from the source to a receiving location by a path other than that under consideration (in this case other than, through the test partition).

3.1.4 *reverberation room*—a room so designed that the reverberant sound field closely approximates a diffuse sound field, both in the steady state when the sound source is on, and during decay after the source of sound has stopped.

3.1.5 *reverberant sound field*—the sound in an enclosed or partially enclosed space that has

been reflected repeatedly or continuously from the boundaries.

3.1.6 *sound absorption, A; [L²]*; metric sabin—*in a specified frequency band*, the measure of the magnitude of the absorptive property of a material, an object, or a structure such as a room.

3.1.7 *sound transmission coefficient, τ; [dimensionless]*—*of a partition, in a specified frequency band*, the fraction of the airborne sound power incident on the partition that is transmitted by the partition and radiated on the other side. See Note 2.

NOTE 1—The sound transmission coefficient and sound transmission loss are related through the equation:

$$TL = 10 \log (1/\tau) \qquad (2)$$

3.1.8 *sound transmission loss, TL*—*of a partition, in a specified frequency band*, ten times the common logarithm of the ratio of the airborne sound power incident on the partition to the sound power transmitted by the partition and radiated on the other side. The quantity so obtained is expressed in decibels.

NOTE 2—Unless qualified, the term denotes the sound transmission loss obtained when the specimen is exposed to a diffuse sound field as approximated, for example, in reverberation rooms meeting the requirements of this standard.

4. Significance and Use

4.1 Sound transmission loss as usually defined refers to the response of specimens exposed to a diffuse incident sound field, and this is the test condition approached by this laboratory test method. The test results are therefore most directly relevant to the performance of specimens exposed to similar sound fields, but they also provide a useful general measure of performance for the variety of sound fields to which a partition may typically be exposed.

4.2 This method is not appropriate for determining the sound-insulation performance of a partition exposed to a sound field that contains only a small range of angles of incidence, nor is it applicable to sounds produced by direct mechanical contact or impact with the partition.

4.3 This method is not intended for field tests. Field tests should be performed in accordance with Test Method E 336.

5. Summary of Method

5.1 Two adjacent reverberation rooms are arranged with an opening between them in which the test partition is installed, care being taken that the only significant sound transmission path between rooms is by way of the test partition. An approximately diffuse sound field is produced in one room, the source room, and the resulting space average sound pressure levels in the two rooms are determined. In addition, with the test specimen in place, the sound absorption in the

receiving room is determined. Then the sound transmission loss is calculated from

$$TL = L_1 - L_2 + 10 \log S - 10 \log A_2 \qquad (1$$

where:

TL = transmission loss, dB,

L_1 = average sound pressure level in the source room, dB,

L_2 = average sound pressure level in the receiving room, dB,

S = area of test specimen common to both rooms, m², and

A_2 = sound absorption of the receiving room with test specimen in place, metric sabin (1).[4]

Since transmission loss is a function of frequency measurements are made in a series of frequency bands.

NOTE 3—The convention used in this standard is that if X is the symbol for a physical quantity, log X denotes the common logarithm of the numerical value of the quantity.

NOTE 4—The difference $L_1 - L_2$ in Eq 1 is by definition the noise reduction, *NR*.

6. Test Rooms

NOTE 5—This section describes the physical features of test rooms that are found to facilitate meeting the acoustical requirements of the test method (see 12.3 and Annex A2).

6.1 *Flanking Transmission*—The test room shall be constructed and arranged so as to minimize the possibility of transmission by paths other than through the test partition. Such flanking transmission should be at least 10 dB lower than the power transmitted into the receiving room by the test partition.

6.1.1 The limit on the transmission loss measurement introduced by this condition should be investigated by building in the test opening a partition having a very high transmission loss and measuring the transmission loss. When significant additions to the test wall no longer increase the measured transmission loss, then the residual sound transmission can be ascribed to flanking paths. Alternatively, the transmission loss along a particular flanking path may be increased to determine if the measured transmission loss increases.

NOTE 6—A potential flanking path is through the perimeter of the partition, including the mounting frame if any (14). It is therefore important that the partition mounting arrangement used in determining the transmission loss limit be the same as is used for normal testing operations.

6.1.2 If the specimen is rigidly connected to

[1] This method is under the jurisdiction of ASTM Committee E-33 on Environmental Acoustics and is the direct responsibility of Subcommittee E33.03 on Sound Transmission.

Current edition approved July 26, 1985. Published September 1985. Originally published as E 90 – 50 T. Last previous edition E 90 – 83.

[2] *Annual Book of ASTM Standards*, Vol 04.06.

[3] Available from American National Standards Institute, 1430 Broadway, New York, NY 10018.

[4] The boldface numbers in parentheses refer to the list of references at the end of this standard.

the source-room structure, there is some danger that, in addition to the incident airborne sound, extra energy may be imparted to the specimen at the edges because of vibration of the source-room structure. A structural discontinuity is therefore recommended between the source-room walls and the structure in which the test specimen is mounted.

NOTE 7—An extraneous signal similar in effect to flanking transmission may be produced by electrical pickup between source and microphone systems. This possibility should be checked, whenever the systems are changed, by measuring the residual signals when the loudspeaker is replaced by an equivalent electrical load.

6.2 *Room Size and Shape*—To produce an acceptable approximation to the assumed diffuse sound fields, especially in the lowest test frequency band, the rooms should meet the following requirements:

6.2.1 *Minimum Volume*—The volume of each room should be no less than 80 m³. Laboratories with room volumes less than 80 m³ shall report the room volume in the test report. For meaningful measurements at frequencies lower than 125 Hz, larger room volumes may be necessary to ensure an acceptably diffuse sound field. For example, for measurements down to 100 Hz, a minimum room volume of 125 m³ is recommended.

NOTE 8—The minimum room volume is derived by assuming that a minimum of approximately 18 modes in the 125 Hz one-third octave band will ensure a satisfactorily diffuse sound field in the room. The recommendation for 100 Hz is obtained by assuming that the same mean modal separation, required for measurements at 125 Hz, will be adequate for measurements at 100 Hz.

6.2.2 *Room Shapes*—No two dimensions of the pair of rooms should be the same or in the ratio of small whole numbers. The ratio of largest to smallest dimension of either room should be less than two.

NOTE 9—Theoretical studies of rectangular rooms (6, 7, 15) suggest that the proportions $1:2^{1/3}:2^{2/3}$ provide an optimum distribution of modes in the lowest bands. Minor deviations in construction, or the presence of diffusers, will alter the actual distribution.

6.3 *Diffusion*—Even in rooms meeting the requirements of 6.2 the results for the lower test bands are likely to depend critically on arbitrary features of the test geometry such as positioning of the sound source and individual microphones. These effects can be ameliorated by one or more of the following techniques:

6.3.1 *Stationary Diffusing Panels*—Each test room should be fitted with a set of diffusing panels, suspended at random orientations throughout the room space. The minimum width of a panel should be one quarter-wavelength at the lowest test band. The appropriate number and distribution of panels should be determined experimentally.

6.3.2 *Rotating Diffusers*—One or more rotating panels, set at oblique angles relative to the room surfaces, should be installed in either or both rooms. The diffusers should be large enough that during rotation they produce a significant variation in the sound field, yet small enough that they do not effectively partition the room at any point in their rotation. Determination of average sound pressure levels should be made preferably during an integral number of diffuser rotations.

6.3.3 *Multiple Sound Sources*—Two or more

sound source positions may be used, either in sequence or simultaneously if driven by incoherently related signals.

6.4 *Room Absorption*—The sound absorption in each of the rooms should be made as low as possible in order to achieve the best possible simulation of the ideal diffuse field condition, and in order to keep the region dominated by the direct field (of the source or of the test partition) as small as possible (see 8.4.1). Within the frequency limits described below, the room absorption for each room (as furnished with diffusers) should be no greater than

$$A = V^{2/3}/3 \qquad (4)$$

where:
V = room volume, m³, and
A = room absorption in metric sabins.
For frequencies below $f = 2000/^3\sqrt{V}$ (where the number 2000 is an empirical constant with the units seconds per metre), somewhat higher absorption may be desirable to accommodate other test requirements (for example, ANSI S1.32, ISO 3741); in any case, the absorption should be no greater than three times the value given by Eq 4. For frequencies above 2000 Hz, atmospheric absorption may make it impossible to avoid a slightly higher value than that given by Eq 4.

NOTE 10—To minimize errors related to atmospheric absorption, the temperature and humidity in the receiving room should be kept the same during both the transmission and absorption measurements; for monitoring purposes, temperature and humidity should be measured and recorded during each day of testing.

7. Test Specimens

7.1 *Size and Mounting*—The test specimen that is to typify a wall or floor shall be large enough to include all the essential constructional elements in their normal size, and in a proportion typical of actual use. The smallest dimension (excluding thickness) shall be not less than 2.4 m, except that specimens of doors and other smaller building elements shall be their customary size. Preformed panel structures should include at least two complete modules (panels plus edge mounting elements). In all cases the test specimen shall be installed in a manner as similar as possible to actual construction, with a careful simulation of normal constraint and sealing conditions at the perimeter and at joints within the specimen. Detailed procedures for particular types of building separation elements are recommended in Annex A1.

7.1.1 The test frame or opening in which the specimen is installed should be arranged so that in each room the depth from the surface of the room proper to the specimen surface is small compared to the lateral dimensions of the specimen. An alternative is to have the specimen form a whole room surface (wall or floor) (8).

7.2 *Aging of Specimens*—Test specimens that incorporate materials for which there is a curing process (for example, adhesives, plasters, concrete, mortar, damping compound) should age for a sufficient interval before testing. Aging periods for certain common materials are recommended in Annex A1. In the case of materials whose aging characteristics are not known, repeat tests over a reasonable time shall be made on at least one specimen to determine an appropriate aging period.

NOTE 11—A suggested procedure is to make tests at intervals in the series 1, 2, 4, 7, 14, and 28 days from

date of construction, until no change is observed, to the precision of the measurement, between successive tests. A sufficient number of measurements should be made to achieve somewhat greater precision than is otherwise specified in this method. The minimum aging period should be the interval beyond which no change greater than the precision values of 9.7 is observed.

7.3 *Testing of Specimens Smaller Than Test Opening*—When the test opening is reduced by means of a filler element to accommodate a small test specimen, it is necessary to consider whether a significant portion of the sound is transmitted by way of the filler element. It may be necessary to measure the transmission loss of the filler element (filling the entire test opening with a construction similar to the filler element). The following derivation may be made when the transmission losses of the complete filler element and of the composite construction formed by the test specimen and filler element are known.

NOTE 12—When a filler element is used, care should be exercised to ensure that sound does not flank the test specimen through the joint between the filler element and the test specimen. Such flanking can easily occur when the filler element is thicker than the test specimen.

7.3.1 The transmission coefficients τ_c and τ_f, for the composite partition and filler element, respectively, can be calculated from the corresponding transmission losses by means of Eq 2. These may be substituted in the following expression for the sound transmission through the composite wall:

$$\tau_c S_c = \tau_s S_s + \tau_f S_f \qquad (5)$$

or

$$\tau_s = (\tau_c S_c - \tau_f S_f)/S_s \qquad (6)$$

where:
S_c, S_s, S_f = areas of the composite construction, test specimen, and filler element, respectively, ($S_c = S_s + S_f$), and
τ_s = transmission coefficient of the test specimen.

Then the transmission loss of the test specimen may be calculated from Eq 2.

NOTE 13—It is assumed in the above expressions that the two parts of the composite construction react to the sound field independently of each other, and that the transmission coefficient of a complete filler element is the same as that of the portion surrounding the test specimen. These simplifications are acceptable as long as $\tau_f S_f$ is small compared to $\tau_c S_c$. Another obvious limitation of the accuracy of the calculation is the accuracy of the measurements from which the difference $(\tau_c S_c - \tau_f S_f)$ is derived; again, the calculation is most reliable if $\tau_f S_f$ is much smaller than $\tau_c S_c$. The filler element selected should be such that $\tau_f S_f$ is likely to be less than one third of $\tau_c S_c$.

7.3.2 When a small specimen such as a door or window assembly is mounted in a filler element, the considerations discussed in 7.1.1 should apply to the filler element.

8. Test Signal

8.1 *Signal Spectrum*—The sound signals used for these tests shall form a series of bands of random noise containing an essentially continuous distribution of frequencies over each test band.

8.2 *Bandwidth*—The measurement bandwidth shall be one third octave. Specifically, the overall frequency response of the electrical system, including the filter or filters in the source and microphone sections, shall for each test band conform to the specifications given in ANSI Specification S1.11 for a Class III filter.

8.2.1 Filtering may be done either in the source or measurement system or partly in both, provided that the required overall characteristic is achieved. Apart from defining the one-third-octave bands of test signals, filters in the microphone system serve to filter out extraneous noise lying outside the test bands, including possible distortion in the source system; a filter in the source system serves to concentrate the available power in the test band or bands.

NOTE 14—Sections 8.1, 8.2, and 8.2.1 are intended to describe the effective results, rather than specific instrumentation. Any system that achieves the specified results is acceptable.

8.3 *Standard Test Frequencies* The minimum range of measurements shall be a series of contiguous one-third-octave bands with geometric center frequencies from 125 to 4000 Hz. It is desirable that the range be extended to include at least the 100 and 5000-Hz bands. Note that larger room volumes are recommended when extending to lower frequencies (see 6.2.2).

8.4 *Location of Sound Sources*—The sound source positions shall be selected so as to minimize spatial fluctuations in the reverberant field in the source room. In addition, they shall be far enough away from the test partition that the direct field reaching the latter is negligible compared to the reverberant field (see 8.4.1).

8.4.1 *Direct Field of Sound Source(s)*—The minimum distance from the source to the test partition, or from the source to the nearest measurement point, shall be such that

$$r \geq 0.63\, A^{1/2} \qquad (7)$$

where:
A = room absorption, metric sabins, and
r = distance from the source, m.
The minimum source distance shall be determined for the absorption conditions in the source room.

8.4.2 *Sound Power of Sources*—The sound power requirement is dependent on the room absorptions, the nature of the test specimen, and the background noise in the receiving room; the power must be sufficient to meet the requirements of 9.4.

9. Measurement of Average Sound Pressure Levels $\bar{L}_1$ and $\bar{L}_2$

9.1 One of the two measurement procedures implied by Eq 1 is the determination of the average sound pressure levels $\bar{L}_1$, and $\bar{L}_2$ produced in the two rooms by the sound source in the source room. The measurement process must account for variations with measurement position, microphone sensitivity, revolution of rotating diffusers, and possible changes in the spectrum of the source. The process must be repeated for each test frequency band. Various systems of data collection and data processing are possible, ranging from a single microphone placed in sequence at the several measurement positions to a multiplicity of microphones making simultaneous measurements. The system adopted shall meet the precision requirements of Section 12. To achieve this end the points discussed in 9.2 through 9.6 should be considered.

9.2 *Location of Microphone Positions* Mi-

crophone positions shall be located so as to sample adequately the reverberant sound field in each room space, with the following restrictions:

9.2.1 The perpendicular distance from any microphone position to any major extended surface shall be no less than half a wavelength or 1 m, whichever is less. The same limit applies relative to any fixed diffuser, if a perpendicular can be drawn to the diffuser surface (excluding edges) and relative to any possible position of a rotating diffuser.

9.2.2 In the source room, no microphone shall be so close to the source as to be affected significantly by its direct field (see 8.4.1).

9.2.3 In the receiving room, no microphone shall be closer to the test partition than the direct-field limit given in 8.4.1.

9.2.4 For rooms and test signals that conform to this method, the sound pressure level will be nearly the same at all positions within the restricted space delineated above. The number and location of microphone positions required for a space average to a given precision (see Section 11) should be determined initially by a detailed survey of the spaces, especially at the lowest test frequencies. A reduced number of locations, that yield the same average result as the detailed survey and that meet the precision requirements of Section 11, may be selected for day-to-day measurements.

9.2.5 If the estimates of precision of average sound pressure levels are to be reliable, the observation points should be sufficiently far apart to provide independent samples of the sound field. This requires, if fixed microphone positions are used, that they be spaced at least half a wavelength apart. (See Appendix X1 and Ref 3.) (For a moving microphone see 9.7.)

9.3 *Averaging Time*—For each sampling position, the averaging time shall be sufficient to yield an accurate estimate of the time-average level: for example, if a rotating diffuser is used, the level shall be averaged (by appropriate techniques or instrumentation) over a time interval long enough to average out variations with diffuser position.

9.4 *Background Noise*—Measurements of background noise levels shall routinely be made to ensure that the observations are not affected by extraneous sound such as flanking transmission, electrical noise in the receiving system, or electrical cross-talk between source and receiving systems. At each measurement position the background level should be at least 5 dB below the level of signal and background combined, and corrections must be made unless the background level is more than 10 dB below the combined level. If the background level is between 5 and 10 dB below the combined level due to signal and background, the adjusted value of the signal level is given by

$$L_s = 10 \log (10^{L_{sb}/10} - 10^{L_b/10})$$

where:
L_b = background noise level,
L_{sb} = level of signal and background combined, and
L_s = adjusted signal level.

9.5 *Microphone Calibration*—Each microphone shall be calibrated at regular intervals as detailed below, and a record shall be kept of the dates of such calibration. If microphone sensitivities differ among themselves by more

than 0.5 dB in any of the specified frequency bands the individual readings shall be adjusted to correspond to the same sensitivity. Calibration over the whole range of test frequencies shall be done annually, and calibration checks for at least one frequency shall be made at least monthly.

9.6 *Determination of Space-Average Levels*—Following the procedures of 9.2 and 9.3, two sets of sound pressure levels, corresponding to the sampling positions in the two rooms, will be obtained. The space-average level corresponding to each set is given by

$$\bar{L} = 10 \log \left(\frac{1}{n} \sum_{i=1}^{n} 10^{L_i/10} \right) \qquad (8)$$

where L_i are one set of time-average levels taken at n locations.

NOTE 15—If the range of values L_i is no more than 4 dB, an arithmetic mean may be used.

9.6.1 *Number and Precision of Measurements*—The number of measurements of L_1 and L_2 shall be sufficient to meet the precision requirements of Section 12.

9.7 A single microphone continuously moving along a defined traverse may be used instead of stationary microphone positions provided that the restrictions given in 9.2 are met. For purposes of verifying the measurement precision, the adequacy of the sampling provided by the traverse may be studied by examining the data obtained from stationary positions along the traverse path approximately half a wavelength apart at the lowest test frequency. These data may be compared with those of a complete survey as described in 9.2.4. Note that if both moving microphones and moving vanes (6.3.2) are used, their periods of rotation and displacement should be chosen so that each observation adequately samples the possible combinations of microphone and vane positions.

10. Determination of Receiving Room Absorption

10.1 The sound absorption of the receiving room, A_2, is determined by measuring the rate of decay of sound pressure level in the room, following essentially the measurement procedures of Test Method C 423. Note, however, that the precision requirement is less stringent in this application than in Test Method C 423 since the quantity entering into the transmission loss calculation is only $10 \log A_2$.

10.1.1 Following Test Method C 423, the room absorption is given by the Sabine equation:

$$A = (0.9210\, Vd)/(c) \qquad (9)$$

where:
A = room absorption, metric sabins,
c = speed of sound in the medium, m/s,
V = volume of room, m³, and
d = rate of decay of sound pressure level in the room, dB/s.

NOTE 16—This is the only absolute method for determining A_2. For determining changes in A_2, an alternative method, based on steady-state sound power considerations, is under study, but because of precision questions it is not permitted for purposes of this standard.

10.1.1.1 Since the speed of sound changes with temperature, it should be calculated for the onditions existing at the time of test from the equation:

$$c = 20.047 \sqrt{273.15 + t} \ \text{m/s}$$

where:
= receiving room temperature, measured to nearest degree, °C.

10.1.2 *Room Condition*—The determination of A_2 shall be made with the receiving room in the same condition as for the measurement of noise reduction (Section 9). Specifically, the test specimen shall remain in place so that its effective absorption (which includes transmission back to the source room) is included; the loudspeakers needed for measuring A_2 shall be present during the measurement of L_2, so that their absorption is present during both measurements.

10.1.3 *Room Coupling*—Because the two rooms are coupled by the test specimen, it is possible that the reverberation measurements in the receiving room will be influenced by energy transmitted into the source room and then back again during the decay process (11). The effect will be small if τS is small compared to A_1 and A_2, or if d_1/d_2, the ratio of decay rates times in the two rooms, is sufficiently large. The latter requirement may be met by adding absorption to the source room until no further effect is observed on the measured value of d_2.

NOTE 17—The additional absorption in the source room is required only during the measurement of the receiving room absorption. In some cases, it may be inappropriate to retain it during the measurement of noise reduction.

10.2 *Number and Precision of Measurements*—A sufficient number of determinations of A_2 shall be taken to ensure that the precision requirements given in Section 12 are met.

11. Report

11.1 The report shall include the following items:

11.1.1 A statement, if true in every respect, that the tests were conducted in accordance with the provisions of this method. Conformance to the relevant sections of Annex A1 shall also be reported when applicable.

11.1.2 A description of the test specimen. It is desirable that the description be sufficiently detailed to identify the specimen, at least in terms of the elements that may affect its sound transmission loss. The specimen size, including thickness, and the average weight per square metre shall always be reported. Wherever possible, the testing laboratory should observed and report the materials, dimensions, weight, and other relevant physical properties of the major components and the manner in which they are combined, including a description of fastening elements. A designation and description furnished by the sponsor of the test may be included in the report provided that they are attributed to the sponsor. The curing period, if any, and the final condition of the specimen (shrinkage, cracks, etc.) shall be reported.

11.1.3 The method of installation of the specimen in the test opening, including the location of framing members relative to the edges, and the treatment of the junction with the test opening. The use and type of caulking, gaskets, tape, or other sealant on perimeter or interior joints shall be carefully described. Clearances around movable elements such as doors shall be reported.

11.1.4 Sound transmission losses to the nearest decibel for the individual test bands as detailed in 8.3.

NOTE 18—If results are presented in graphical form, it is recommended that the abscissa length for a 10:1 frequency ratio equal the ordinate length for 25 dB. Whenever practicable, the scales should be 50 mm for a 10:1 frequency ratio and 20 mm for ten decibels, and the ordinate scale should start at zero decibels.

11.1.5 The precision of the *TL* data, as derived from the precisions of the two main measurements. When a traveling microphone is used, quantity entering into the transmission loss calculation is only 10 log A_2.

10.1.1 Following Test Method C 423, the room absorption is given by the Sabine equation:

$$A = (0.9210\, Vd)/(c) \qquad (9)$$

where:
A = room absorption, metric sabins,
c = speed of sound in the medium, m/s,
V = volume of room, m³, and
d = rate of decay of sound pressure level in the room, dB/s.

NOTE 16—This is the only absolute method for determining A_2. For determining changes in A_2, an alternative method, based on steady-state sound power considerations, is under study, but because of precision questions it is not permitted for purposes of this standard.

10.1.1.1 Since the speed of sound changes with temperature, it should be calculated for the conditions existing at the time of test from the equation:

$$c = 20.047 \sqrt{273.15 + t} \ \text{m/s}$$

where:
t = receiving room temperature, measured to nearest degree, °C.

10.1.2 *Room Condition*—The determination of A_2 shall be made with the receiving room in the same condition as for the measurement of noise reduction (Section 9). Specifically, the test specimen shall remain in place so that its effective absorption (which includes transmission back to the source room) is included; the loudspeakers needed for measuring A_2 shall be present during the measurement of L_2, so that their absorption is present during both measurements.

10.1.3 *Room Coupling*—Because the two rooms are coupled by the test specimen, it is possible that the reverberation measurements in the receiving room will be influenced by energy transmitted into the source room and then back again during the decay process (11). The effect will be small if τS is small compared to A_1 and A_2, or if d_1/d_2, the ratio of decay rates times in the two rooms, is sufficiently large. The latter requirement may be met by adding absorption to the source room until no further effect is observed on the measured value of d_2.

NOTE 17—The additional absorption in the source room is required only during the measurement of the receiving room absorption. In some cases, it may be inappropriate to retain it during the measurement of noise reduction.

10.2 *Number and Precision of Measurements*—A sufficient number of determinations of A_2 shall be taken to ensure that the precision requirements given in Section 12 are met.

11. Report

11.1 The report shall include the following items:

11.1.1 A statement, if true in every respect, that the tests were conducted in accordance with the provisions of this method. Conformance to the relevant sections of Annex A1 shall also be reported when applicable.

11.1.2 A description of the test specimen. It is desirable that the description be sufficiently detailed to identify the specimen, at least in terms of the elements that may affect its sound transmission loss. The specimen size, including thickness, and the average weight per square metre shall always be reported. Wherever possible, the testing laboratory should observed and report the materials, dimensions, weight, and other relevant physical properties of the major components and the manner in which they are combined, including a description of fastening elements. A designation and description furnished by the sponsor of the test may be included in the report provided that they are attributed to the sponsor. The curing period, if any, and the final condition of the specimen (shrinkage, cracks, etc.) shall be reported.

11.1.3 The method of installation of the specimen in the test opening, including the location of framing members relative to the edges, and the treatment of the junction with the test opening. The use and type of caulking, gaskets, tape, or other sealant on perimeter or interior joints shall be carefully described. Clearances around movable elements such as doors shall be reported.

11.1.4 Sound transmission losses to the nearest decibel for the individual test bands as detailed in 8.3.

NOTE 18—If results are presented in graphical form, it is recommended that the abscissa length for a 10:1 frequency ratio equal the ordinate length for 25 dB. Whenever practicable, the scales should be 50 mm for a 10:1 frequency ratio and 20 mm for ten decibels, and the ordinate scale should start at zero decibels.

11.1.5 The precision of the *TL* data, as derived from the precisions of the two main measurements. When a traveling microphone is used, the precision as inferred from periodic studies as described in 9.7 may be reported, but details of such studies, including the dates shall be included.

11.1.6 The room temperature and relative humidities in both the source and receiving rooms.

11.1.7 The volumes of the test rooms if either one is less than the limit in 6.2.1.

12. Precision

12.1 The precision of a sound transmission loss determination is derived from the precisions of the individual quantities in Eq 1, namely, $\bar{L}_1$, $\bar{L}_2$, S, and A_2. The procedure is to calculate the 95 % uncertainties for the four quantities and then to combine them to calculate the 95 % uncertainty for *TL*. The method, described in *ASTM STP 15 D* (12), is summarized in 12.2.

12.2 Calculation of 95 % Uncertainty of a Transmission Loss Determination:
12.2.1 Calculate the standard deviation for

the set of determinations of each quantity from the expression:

$$s = \left(\frac{\sum_{i=1}^{n} (X_i - \bar{X})^2}{n-1} \right)^{1/2} \qquad (10)$$

where:

s = standard deviation,

X_i = an individual determination (of L_1, L_2, 10 log s, or 10 log A_2, as the case may be),

$\bar{X}$ = arithmetic mean of the set of individual determinations, and

n = number of determinations.

12.2.2 Calculate the 95 % uncertainties from:

$$\Delta X = as \qquad (11)$$

where the factor a, which depends on the number of measurements, is given in Table 1.

12.2.3 The uncertainty for the transmission loss is the combined uncertainty of the measured quantities, namely,

$$\Delta^2 TL = \Delta^2 L_1 + \Delta^2 L_2 + \Delta^2(10 \log S) \\ + \Delta^2(10 \log A_2) \qquad (12)$$

The last two terms of Eq 12 may be simplified for calculation purposes, using $(10 \log e)^2 = 18.86$.

$$\Delta^2 TL = \Delta^2 L_1 + \Delta^2 L_2 + 18.86 \ (\Delta S / \bar{S})^2 \\ + 18.86 \ (\Delta A_2 / \bar{A}_2)^2$$

Generally the uncertainty of the measurement of area S is taken to be zero. If the absorption of the receiving room has been calculated from the average decay rate $\bar{d}$, 18.86 $(\Delta d / \bar{d})^2$ may be substituted for the last term of Eq 12.

NOTE 19—A more exact general procedure for combining uncertainties is described in Ref 16.

12.2.4 The 95 % confidence limits for the transmission loss are given by:

$$TL = \overline{TL} + \Delta TL \ \text{and} \ TL = \overline{TL} - \Delta TL \qquad (14)$$

12.3 *Precision Requirement*—It is required that the transmission loss uncertainty be no greater than 3 dB for the one-third-octave bands centered on 125 and 160 Hz, 2 dB for bands centered on 200 and 250 Hz, and 1 dB for the bands centered in the range 315 to 4000 Hz.

TABLE 1 Factors for 95 % Confidence Limits for Averages

Number of Measurements	Confidence Limits[A], $\bar{X} + as$
n	a
4	1.591
5	1.241
6	1.050
7	0.925
8	0.836
9	0.769
10	0.715
11	0.672
12	0.635
13	0.604
14	0.577
15	0.554
16	0.533
17	0.514
18	0.497
19	0.482
20	0.468
21	0.455
22	0.443
23	0.432
24	0.422
25	0.413

[A] Limits that may be expected to include the "true" average, $\bar{X}$, 95 times in 100 in a series of problems, each involving a single sample of observations.

ANNEXES
(Mandatory Information)
A1. PREPARATION AND DESCRIPTION OF TEST SPECIMENS

A1.1 Scope

A1.1.1 This annex constitutes an interpretation and elaboration, for certain generic types of construction, of the general requirements given in 7.1, 7.2, 11.1.2, and 11.1.3. The types of construction are dealt with in separate sections, largely independent of each other but not independent of the main method; thus, the appropriate sections should be considered as a supplement to the main method. Special details are spelled out relating to the preparation, installation, and aging of test specimens and the reporting of such matters.

NOTE A1.1 If the recommended aging periods seem inappropriate for a particular construction the repeat test procedure described in 7.2 may be used.

A1.1.2 The following generic types of partition are considered:

A1.1.2.1 Concrete and masonry walls (A1.3),

A1.1.2.2 Plaster partitions (A1.4).

A1.1.2.3 Wall-board partitions (A1.5).

A1.1.2.4 Demountable modular wall panel systems (A1.6).

A1.1.2.5 Operable (folding or sliding) walls and doors (A1.7), and

A1.1.2.6 Swinging doors (A1.8).

A1.2 Composite Construction

A1.2.1 If a partition includes components corresponding to more than one of the listed types, the appropriate requirements of each type shall apply. For example, a block wall to which plaster is applied shall meet the relevant requirements of masonry walls for the basic wall and plaster partitions for the plaster layer.

A1.3 Concrete and Masonry Walls

A1.3.1 *Materials*—The material, dimensions, and average weight of an individual masonry unit, and the material and thickness of mortar shall be determined and reported. The type and density of concrete shall be reported. The weight per unit area of the completed wall shall be determined by weighing a representative portion of the wall after test.

A1.3.2 *Construction*—The wall shall be built in accordance with usual construction practice except that extra control procedures may be desirable to ensure maintenance of the specified dimensions. The construction procedures should be reported in detail (see also 11.1.2 and 11.1.3).

A1.3.3 *Aging*—Following construction, the specimen shall be allowed to age a minimum of 28 days before testing. A temperature of 18 to 24°C (65 to 75°F) and 30 to 55 % relative humidity are recommended. For 24 h immediately prior to a test, the specimen should be conditioned at a temperature of 21 to 24°C (70 to 75°F) and 45 to 55 % relative humidity.

A1.4 Plaster Partitions

A1.4.1 *Materials*—The following information shall be reported:

A1.4.1.1 *Studs*—Material (state grade of wood, if used), true as well as nominal dimensions, spacing in test opening, end fastening conditions, and weight per unit length.

A1.4.1.2 *Fillers*—Materials, weight per unit area of wall, location, and method of fastening.

A1.4.1.3 *Lathing*—Material, dimensions of individual sections and orientation in test specimen, weight per unit area of wall, number and location of fasteners (Note A1.2), and treatment of edges of specimen.

NOTE A1.2—If a resilient fastening device is used, an accurate description of its dimensions and material, preferably in the form of a drawing, shall be reported.

A1.4.1.4 *Plaster*—Materials, thickness of each layer, and the method of application. The weight per unit area of the completed wall (including studs, other framing members, and filler materials) should be determined by weighing representative sections after test.

A1.4.2 *Construction*—The test specimen may

either be built into a suitable frame, which is then inserted in the test opening, or built into the opening itself. The type of installation and the steps in constructing the specimen (for example, plastering techniques) should be reported in detail. The actual thickness of plaster layers should be determined, for example, by inspection of representative sections after test (see also 11.1.2 and 11.1.3).

A1.4.3 *Aging*—Thick coats of plaster, forming significant contributions to the structure, shall age at least 28 days before testing (see Note A1.1). Superficial coats (3 mm thick or less) shall age at least [...] days. A temperature of 18 to 24°C (65 to 75°F) and a relative humidity of 30 to 55 % are recommended.

A1.5 Wall-Board Partitions

A1.5.1 *Materials*:

A1.5.1.1 *Studs*—Material (state grade of wood used), true as well as nominal dimensions, spacing in test opening, end fastening conditions, and weight per unit length.

A1.5.1.2 *Fillers*—Materials, weight per unit area of wall location, and method of fastening.

A1.5.1.3 *Wallboard*—Materials, weight per unit area of material, number of layers and thickness of each orientation of individual panels in test specimen location and treatment of joints, and number and type of fasteners (Note A1.2).

A1.5.1.4 *Laminating Adhesives*—Type of adhesive, method of application, and thickness.

A1.5.2 *Construction*—See 11.1.2 and 11.1.3.

A1.5.3 *Aging*—If laminating adhesives are used the specimen shall age before testing a minimum of 14 days for water-base adhesives and 3 days for nonwater-base adhesives. If no laminating adhesives are used, but joints are finished with typical joint and finishing compounds, the minimum aging period shall be 12 h.

A1.6 Demountable Modular Wall Panel Systems

A1.6.1 *Materials and Construction*—The testing laboratory shall report as much physical information as can be determined about the materials and methods of assembly of all components of the partition including weights and dimensions of the component parts and the average weight per unit area of the completed partition (see 11.1.2).

A1.6.2 *Installation*—Installation of the test specimen shall be carried out or observed by the testing laboratory and reported in detail (see also 11.1.3).

A1.6.3 *Aging*—If significant quantities of caulking or adhesive materials are required, appropriate aging procedures shall be used (see, for example, 7.2, A1.5.3).

A1.7 Operable (Folding or Sliding) Walls and Doors

A1.7.1 *Materials and Construction*—The testing laboratory shall report as much physical information as can be determined about the materials and methods of assembly of all components of the partition including weights and dimensions of the component parts and the average weight per unit area of the completed partition. If the specimen consists of an assembly of panels, the number and dimensions of panels comprising the specimen shall be reported. If the specimen is an accordion-type partition, the number of volutes, their spacing and width when extended, shall be reported. Header construction and dimensions shall be reported. Weights of header and the hanging portion of the door shall be given. Latching and sealing devices shall be fully described.

A1.7.2 *Installation*—Installation of the test specimen shall be carried out or observed by the testing laboratory and reported in detail (see also 11.1.3). Clearances at the perimeter between nondeformable portions of door and frame shall be measured and reported. In particular, any features of the installation that require dimensional control closer than 6 mm (¼ in.) on the height or width or the test specimen shall be reported.

A1.7.3 *Operation*—The specimen shall not be designated an operable wall unless it opens and closes in a normal manner. It shall be fully opened and closed at least five times after installation is completed, and tested without further adjustments.

A1.7.4 *Aging*—If significant quantities of caulking or cementing materials are required, appropriate aging procedures shall be used (see, for example, 7.2, A1.5.3).

A1.7.5 The specimen area should include the

der and other framing elements if these constitute ortion of the separating partition in a typical allation.

8 Swinging Doors

\1.8.1 *Materials and Construction*—The testing oratory shall report as much physical information an be determined about the materials and method ssembly of all components of the door including ghts and dimensions of the component parts and average weight per unit area of the completed r. Latching and sealing devices shall be fully cribed. The recommended size for a single door cimen shall be 0.9 by 2 m (3 by 7 ft) (inside ne); departures from this shall be noted.

\1.8.2 *Installation*—Installation of the test speci- n shall be carried out or observed by the testing oratory and reported in detail (see also 11.1.3). arances at the perimeter between nondeformable tions of door and frame shall be measured and orted. Less than 6 mm (¼ in.) clearance on overall ght and 3 mm (⅛ in.) on overall width shall be arded as special.

\1.8.3 *Operation*—The specimen shall not be des- ated a door unless it opens and closes in a normal nner. It shall be fully opened and closed at least times after installation is completed, and tested hout further adjustments.

A2. LABORATORY ACCREDITATION

1 Scope

A2.1.1 This annex describes procedures for ac- diting an acoustical testing laboratory to perform s in conformance with this test method.

2 Applicable Documents

548 Practice for Generic Criteria for Use in the Evaluation of Testing and Inspection Agencies[5]
717 Guide for Preparation of Accreditation Sec- tion of Acoustical Test Standards[2]

3 General Requirements

A2.3.1 The testing agency shall make available to accrediting authority the information required by tions 4 to 7 of Practice E 548.

4 Requirements Specific to This Method

A2.4.1 *Physical Facilities*—The testing agency ll provide information demonstrating compliance h the following provisions of this method:

.1 Flanking transmission,
.2 Room size and shape,
.4 Room absorption,
.1 Size and mounting of test specimens,
.1 Test signal spectrum,
.4 Measurement bandwidth, and
.4 Location of sound sources.

A2.4.2 *Procedures*—The agency shall furnish a ple report of a complete test (including raw data), wing compliance with the following provisions of s method:

.1 Determination of average sound pressure levels,
.2 Location of microphone positions,
.3 Averaging time,
.4 Background noise,
.5 Microphone calibration,
.6 Averaging procedure,
.7 Continuous traverse of a single microphone (if applicable),
10. Determination of receiving room absorption, and
12. Calculation of precision.

A2.4.3 *Repeatability and Reproducibility*—Results repeated tests made on a particular reference ecimen shall be reported as a demonstration of the ng-term repeatability of the test procedures. Each st should include installation of the reference spec- ien. The record should include for each test the und transmission losses for all standard frequency nds, together with the precision of each value llowing Section 12 of this method.

A2.4.3.1 *Reference Specimen*—The reference ecimen should be such that all details of construc- n and installation that may affect its sound trans- ission loss can be specified and controlled.

A2.4.3.2 To provide evidence of reproducibility in comparison with other testing laboratories it is pref- erable to use a reference specimen for which test data are available from several laboratories.

NOTE A2.1—Guidance in the selection of suitable reference specimens can be found in Ref. **(14)**.

APPENDIX

(Nonmandatory Information)

X1. ACCURACY AND PRECISION OF LABORATORY MEASUREMENT OF TRANSMISSION LOSS

X1.1 Accuracy

X1.1.1 The accuracy of a transmission loss mea- surement depends on how well the experimental conditions approximate the theoretical model on which the method is based. In the laboratory test, an important objective is to achieve the diffuse sound fields assumed in the theory. This is important in two respects:

X1.1.1.1 Transmission loss is defined in terms of the ratio of incident to transmitted *sound power*. If diffuse sound fields exist in the two rooms, then the desired sound power ratio may be determined from measurements of average sound pressure levels in the two rooms, and Eq 1 applies.

X1.1.1.2 The value of transmission loss obtained depends on the distribution of angles of incidence in the ensemble of incident sound waves. If a diffuse sound field exists in the source room, then the sound incident on the partition will include waves of all angles of incidence, up to grazing incidence. Even in the laboratory, in finite rooms, the diffusion condi- tion cannot quite be achieved, especially with respect to the grazing incidence condition. This results in measured transmission loss values significantly

higher (about 5 dB) than the theoretical value for the full range of incident angles.

X1.1.2 Another important question relating to laboratory tests is the degree to which the test speci- men exemplifies actual construction in buildings. Apart from such matters as differences in construc- tion and flanking transmission (discussed in 6.1 and Annex A1) the specimen should be large enough to reduce the importance of edge constraint conditions and to ensure that it behaves in the same way as in a field installation, especially under obliquely inci- dent sound. Note that at frequencies where the lateral dimensions of the specimens are not large in com- parison with the wavelength for flexural waves in the specimen, the test results may be significantly af- fected by the impedance discontinuity and damping at the edge of the specimen. This may be particularly true for lightweight rigid specimens with low internal damping **(8)**. In addition, the installation in the test opening should be so arranged that the specimen is coupled in the assumed manner to the sound fields in the two rooms (see 7.1.1).

X1.1.3 It will be recognized that the accuracy of measurements on a given specimen, as distinguished from the precision of a given set of measurements, cannot be estimated from analytical considerations. Experimental or theoretical evidence may, however, be used to give some indication of the validity of test procedures. The reproducibility of tests on the same construction in different laboratories provides an overall estimate of precision and accuracy, from which inferences may be drawn. A more useful test is to measure the transmission of a consensus refer- ence specimen whose performance may be consid- ered predictable **(14)**.

[5] *Annual Book of ASTM Standards*, Vols 03.01 and 14.02.

REFERENCES

(1) Buckingham, E., "Theory and Interpretation of Experiments on the Transmission of Sound Through Partition Walls," *Scientific Papers*, Na- tional Bureau of Standards, Vol 20, s506, 1925, p. 193.

(2) Chrisler, V. L., and Snyder, W. F., "Recent Sound-Transmission Measurements at the Na- tional Bureau of Standards," *Journal of Re- search of the National Bureau of Standards*, Vol 14, RP800, 1935, p. 749.

(3) Lubman, D., "Precision of Reverberant Sound Power Measurements," *Journal of the Acoustical Society of America*, Vol 56, No. 2, 1974, p. 523– 33.

(4) Bowers, H. D., and Lubman, D., "Decibel Av- eraging in Reverberant Rooms," LTV Research Center (now Advanced Technology Center, Inc.), *Technical Report* 0-71200/8 TR-130, 1968.

(5) Maling, G., "Guidelines for the Determination of the Average Sound Power Radiation by Dis- crete Frequency Sources in a Reverberation Room," *Journal of the Acoustical Society of America*, Vol 53, No. 4, 1973, p. 1064–1069.

(6) Sepmeyer, L. W., "Computed Frequency and Angular Distribution of the Normal Modes of Vibration in Rectangular Rooms," *Journal of the Acoustical Society of America*, Vol 37, 1965, pp. 413–423.

(7) Morse, P. M., and Bolt, R. H., "Sound Waves in Rooms," *Reviews of Modern Physics*, Vol 16, No. 2, April 1944.

(8) Kihlman, T., and Nilsson, A. C., "The Effects

of Some Laboratory Designs and Mounting Conditions on Reduction Index Measure- ments," *Journal of Sound and Vibration*, Vol 24, No. 3, 1972, pp. 349–364.

(9) Higginson, R. F., "A Study of Measuring Tech- niques for Airborne Sound Insulation in Build- ings," *Journal of Sound and Vibration*, Vol 21, No. 4, 1972, pp. 405–429.

(10) Waterhouse, R. V., "Interference Patterns in Reverberant Sound Fields," *Journal of the Acoustical Society of America*, Vol 27, 1955, p. 247.

(11) Mariner, T., "Critique of the Reverberant Room Method of Measuring Air-Borne Sound Trans- mission Loss," *Journal of the Acoustical Society of America*, Vol 33, 1961, p. 1131.

(12) *Manual on Presentation of Data and Control Chart Analysis, ASTM STP 15D*, Am. Soc. Test- ing Mats., 1976, p. 56.

(13) Sharp, B. H., "Prediction Methods for the Sound Transmission of Building Elements," *Noise Control Engineering*, Vol 11 No. 2, Sep- tember–October 1978.

(14) Jones, R. E., "Intercomparisons of Laboratory Determinations of Airborne Sound Transmis- sion Loss," *Journal of the Acoustical Society of America*, Vol 66, July 1979.

(15) Donato, R. J., "Angular Distributions of Lower Room Modes," *Journal of the Acoustical Society of America*, Vol 41, 1967, pp. 1496–1499.

(16) Hoel, P. G., *Introduction to Mathematical Statis- tics*, John Wiley, New York, NY, 1962, p. 275 ff.

FIRE AND SOUND RATINGS

WOOD FRAMED INTERIOR PARTITIONS/CONVENTIONAL PLASTER

#	MIN. THICKNESS	CONSTRUCTION	FIRE	*STC	TESTING AGENCY FIRE	TESTING AGENCY SOUND	ICBO	SFM	HCD	L.A.	S.F.	SOURCE
1	5-1/2"	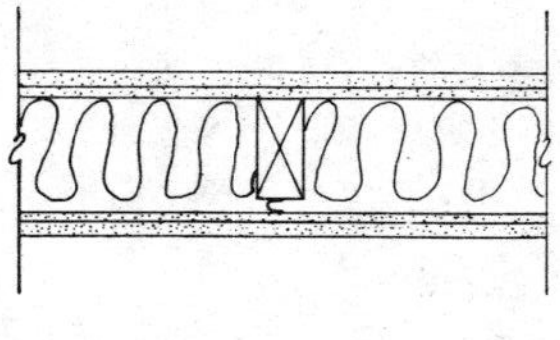 A. ½" Gypsum Sanded Plaster B. Approved Nails or Staples (Direct Attachment) C. ⅜" Type X Gypsum Lath D. 3" Mineral Wool E. 2" x 4" Wood Studs 16" O/C F. Resilient Clips G. ⅜" Type X Gypsum Lath H. ½" Gypsum Sanded Plaster NOTE: Perimeter Caulked	1 hr	56		USG 118-FT (Geiger & Hamme)						USG Code Selector #20 (1972) USG A-1362 (1972)
2	5-1/2"	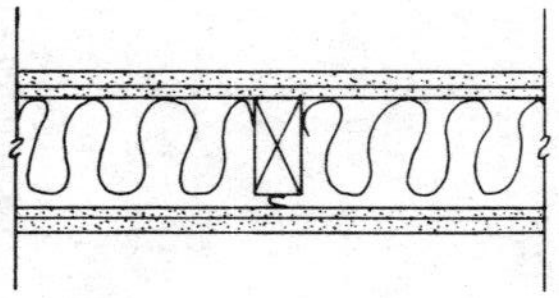 A. ½" Gypsum Sanded Plaster B. Approved Nails or Staples (Direct Attachment) C. ⅜" Type X Gypsum Lath D. 3" Mineral Wool E. 2" x 4" Wood Studs 16" O/C F. Resilient Clips G. ⅜" Type X Gypsum Lath H. ½" Gypsum Sanded Plaster	1 hr	50	Ohio State University T-1329	USG 118-FT (Geiger & Hamme)						USG Code Selector #20 (1972) USG A-1362 (1972)
3	5-3/4"	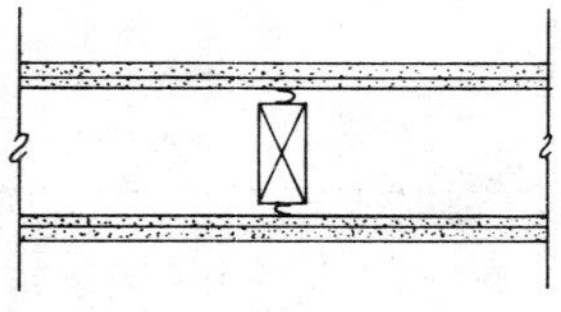 A. ½" Gypsum Sanded or Gypsum Perlite Plaster B. ⅜" Type X Gypsum Lath C. Resilient Clips D. 2" x 4" Wood Studs 16" O/C E. Resilient Clips F. ⅜" Type X Gypsum Lath G. ½" Gypsum Sanded or Gypsum Perlite Plaster	1 hr	51	Ohio State University T-1329	National Bureau Standards 167	R.R.# 1174	W & P Design No.59- 1 hr.				Sound: Kaiser Gypsum Fire: USG A-1362 (1972)
4	5-3/4"	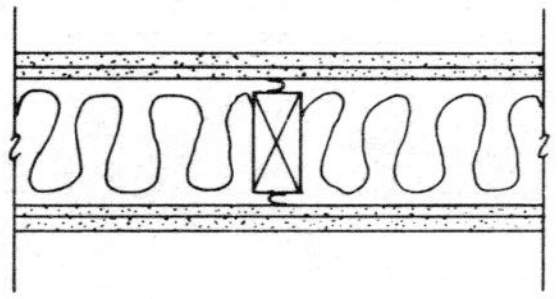 A. ½" Gypsum Sanded or Gypsum Perlite Plaster B. ⅜" Type X Gypsum Lath C. Resilient Clips D. 3" Mineral Wool E. 2" x 4" Wood Studs 16" O/C F. Resilient Clips G. ⅜" Type X Gypsum Lath H. ½" Gypsum Sanded or Gypsum Perlite Plaster	1 hr	57	Ohio State University T-1329	National Bureau Standards 168	R.R.# 1174	W & P Design No.59- 1 hr.				Sound: Kaiser Gypsum Fire: USG A-1362 (1972)
5	6-1/4"	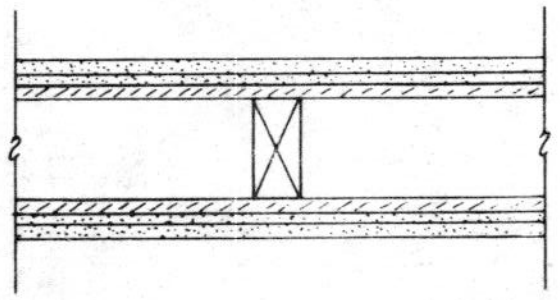 A. ½" Gypsum Sanded Plaster B. Approved Nails or Staples (Direct Attachment) C. ⅜" Type X Gypsum Lath D. ½" Wood Fiber Sound Deadening Board E. 2" x 4" Wood Studs 16" O/C F. ½" Wood Fiber Sound Deadening Board G. Approved Nails or Staples (Direct Attachment) H. ⅜" Type X Gypsum Lath I. ½" Gypsum Sanded Plaster NOTE: Perimeter Caulked	1 hr	50 est.	SUBSTITUTED 7/8" LATH & PLASTER FOR 1/2" WALLBOARD		R.R.# 1602 R.R.# 2741	W & P Design No.90- 1 hr.		R.R. #192U1.8 2198		USG Code Selector #41 (1972)

			Fire	*Sound		
6	6"	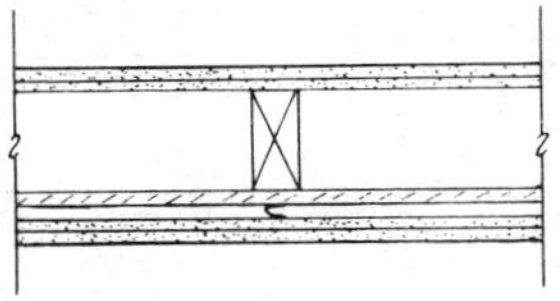 A. ½" Gypsum Sanded Plaster B. Approved Nails or Staples (Direct Attachment) C. ⅜" Type X Gypsum Lath D. 2" x 4" Wood Studs 16" O/C E. ½" Sound Deadening Board F. Resilient Clips (Surface Mounted) G. ⅜" Type X Gypsum Lath H. ½" Gypsum Sanded Plaster	1 hr est.	54	USG 119-FT (Geiger & Hamme)	USG A-1377
7	6-1/8"	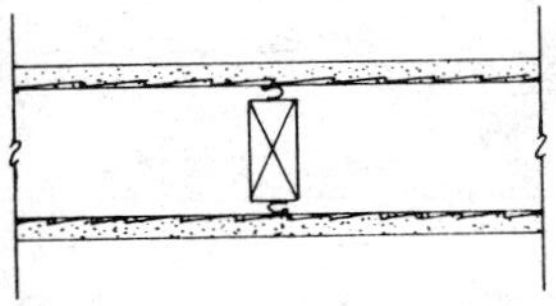 A. ¾" Gypsum Sanded Plaster B. 3.4# Expanded Metal Lath C. ¼" Pencil Rods D. Resilient Clips E. 2" x 4" Wood Studs 16" O/C F. Resilient Clips G. ¼" Pencil Rods H. 3.4# Expanded Metal Lath Wire Tied I. ¾" Gypsum Sanded Plaster	1 hr	50	National Bureau Standards 425	Sound: Metal Lath Assn. Tech. Bulletin #151 (1972) Fire: USG Code Sel. #16 (1972)

*Sound Transmission Class

(Continued)

WOOD FRAMED INT. PARTITIONS STAGGERED/DOUBLE STUDS CONVENTIONAL PLASTER

FIRE AND SOUND RATINGS

#	MIN. THICK-NESS	CONSTRUCTION	RATINGS FIRE	RATINGS *STC	TESTING AGENCY FIRE	TESTING AGENCY SOUND	ICBO	SFM	HCD	L.A.	S.F.	SOURCE
8	7-3/8"	A. 1/2" Gypsum Sanded Plaster B. Approved Nails or Staples (Direct Attachment) C. 3/8" Type X Gypsum Lath D. 2" Mineral Wool, One Side E. 2" x 4" Wood Studs 16" O/C Each Side (Staggered)—6" Common Plate F. Approved Nails or Staples (Direct Attachment) G. 3/8" Type X Gypsum Lath H. 1/2" Gypsum Sanded Plaster	1 hr	50	Ohio State University T-1329	Riverbank Acoustical Lab. 58-64	43-B	43-B		43-B	43-B	USG Code Selector #14 (1972)
9	7-3/8"	A. 1/2" Gypsum Sanded Plaster B. Approved Nails or Staples (Direct Attachment) C. 3/8" Type X Gypsum Lath D. 2" Mineral Wool, Each Side E. 2" x 4" Wood Studs 16" O/C Each Side (Staggered)—6" Common Plate F. Approved Nails or Staples (Direct Attachment) G. 3/8" Type X Gypsum Lath H. 1/2" Gypsum Sanded Plaster	1 hr	51 est.	SUBSTITUTED 7/8" LATH & PLASTER FOR 1/2" WALLBOARD							Fiberglas 1-BL-6066-B (1974) Pg.16 Item A
10	7-1/2"	A. 1/2" Gypsum Sanded Plaster B. Approved Nails or Staples (Direct Attachment) C. 3/8" Type X Gypsum Lath D. 2" x 4" Wood Studs 16" O/C Each Side (Staggered)—6" Common Plate E. 1/2" Sound Board F. Approved Nails or Staples (Direct Attachment) G. 3/8" Type X Gypsum Lath H. 1/2" Gypsum Sanded Plaster	1 hr	53								Kaiser Gypsum KGC-258-51-64
11	8"	A. 1/2" Gypsum Sanded Plaster B. Approved Nails or Staples C. 3/8" Type X Gypsum Lath (Direct Attachment) D. 1/2" Sound Board E. 2" x 4" Wood Studs 16" O/C Each Side (Staggered)—6" Common Plate F. 1/2" Sound Board G. Approved Nails or Staples (Direct Attachment) H. 3/8" Type X Gypsum Lath I. 1/2" Gypsum Sanded Plaster	1 hr	57								Kaiser Gypsum KGC-254-205-64
12	7"	A. 3/4" Gypsum Sanded Plaster B. Approved Nails or Staples (Direct Attachment) C. 3.4# Expanded Metal Lath D. 2" x 4" Wood Studs 16" O/C Turned Flat (Staggered) E. 2" x 4" Wood Studs 16" O/C (Staggered)—6" Common Plate Resting on 1" Cork F. Approved Nails or Staples (Direct Attachment) G. 3.4# Expanded Metal Lath H. 3/4" Gypsum Sanded Plaster	1 hr	50	Building Materials & Structures 92	National Bureau Standards 175						Sound: Metal Lath Assn. Tech. Bulletin #151 (1972) Fire: MLA Technical Bulletin #18 (1958)

#	Thickness	Fire Rating	STC*	Construction	Test / Reference
13	10-3/4"	1 hr	56	A. ½" Gypsum Sanded Plaster B. ⅜" Type X Gypsum Lath C. Approved Nails or Staples (Direct Attachment) D. ½" Sound Board (Installed Vertically) E. 2" x 4" Wood Studs 16" O/C on Separate Plate F. 2" x 4" Wood Studs 16" O/C on Separate Plate G. ½" Sound Board (Installed Vertically) H. Approved Nails or Staples (Direct Attachment) I. ⅜" Type X Gypsum Lath J. ½" Gypsum Sanded Plaster NOTE: Continuous Sub-Floor & Ceiling	Kaiser Gypsum KG-36FT (Geiger & Hamme) Kaiser Gypsum Tech. #475 (1964) Kaiser Gypsum KGC-258-51-64
14	10-3/4"	1 hr	64	A. ½" Gypsum Sanded Plaster B. ⅜" Type X Gypsum Lath C. Approved Nails or Staples (Direct Attachment) D. ½" Sound Board (Installed Vertically) E. 2" x 4" Wood Studs 16" O/C on Separate Plate F. ½" Sound Board (Installed Vertically) G. Approved Nails or Staples (Direct Attachment) I. ⅜" Type X Gypsum Lath J. ½" Gypsum Sanded Plaster NOTE: Discontinuous Sub-Floor & Ceiling	Kaiser Gypsum KG-36 Lab Kaiser Gypsum Tech. #475 (1964) Kaiser Gypsum KGC-258-51-64
15	9"	1 hr	53 est.	A. ½" Gypsum Sanded Plaster B. ⅜" Type X Gypsum Lath C. Approved Nails or Staples (Direct Attachment) D. ½" Sound Board E. 2" x 3" Wood Studs 16" O/C on Separate Plate F. 2" x 3" Wood Studs 16" O/C on Separate Plate G. ½" Sound Board H. Approved Nails or Staples (Direct Attachment) I. ⅜" Type X Gypsum Lath J. ½" Gypsum Sanded Plaster	R.R.# 1602 R.R.# 2741 W & P Design No. 90- 1 hr. R.R.# 2198 192U1.8 USG Code Selector #37 (1972)

*Sound Transmission Class

(Continued)

FIRE AND SOUND RATINGS
WOOD FRAMED INTERIOR PARTITIONS VENEER PLASTER

#	MIN. THICK-NESS	CONSTRUCTION	RATINGS FIRE	RATINGS *STC	TESTING AGENCY FIRE	TESTING AGENCY SOUND	ICBO	SFM	HCD	L.A.	S.F.	SOURCE
16	5-1/2"	A. 1/16" Minimum Veneer Plaster B. 5/8" Type X Veneer Plaster Base Applied at Right Angles to Studs With Vertical Joints Staggered 48" C. 1" Type S Screws 12" O/C (Veneer Base Attachment) D. Resilient Channel 24" O/C Maximum (Installed Horizontally) E. 3" Mineral Fiber F. 2" x 4" Wood Studs 16" or 24" O/C G. 1-1/4" Type W Screws 12" O/C (Veneer Base Attachment) H. 5/8" Type X Veneer Plaster Base Applied at Right Angles to Studs With Vertical Joints Staggered 48" I. 1/16" Minimum Veneer Plaster	1 hr	50		USG 111-FT (Geiger & Hamme)	R.R.# 1602 R.R.# 2410 R.R.# 2741	W & P Design No.94- 1 hr.	WP 3230	R.R.# 22535	198U1.1	USG Code Selector #12 (1972) USG A-1337 (1967)
17	6-1/8"	A. 1/16" Minimum Veneer Plaster B. 5/8" Type X Veneer Plaster Base Applied Parallel to Resilient Channels. End Joints Back Blocked with Resilient Channels. C. 1" Type S Screws 12" O/C (Veneer Base Attachment) D. Resilient Channels 24" O/C E. 3" Mineral Fiber F. 2" x 4" Wood Studs 16" O/C G. Base Layer Attached With 6D Coated Nails 1-5/8" or Approved Staples H. Face Layer Attached With 8D Coated Nails 2-3/8" long (Vertical Joints Staggered From Base Layer Joints) I. 2 Layers 5/8" Type X Veneer Plaster Base (Direct Attachment) J. 1/16" Minimum Veneer Plaster	1 hr	53		Cedar Knolls 654-38						USG A-1337 (1967)
18	5-1/2"	A. 1/16" Minimum Veneer Plaster B. 5/8" Type X Veneer Plaster Base Applied at Right Angles to Studs with Vertical Joints Staggered 48" C. 1" Type S Screws 12" O/C (Veneer Base Attachment) D. Resilient Channel 24" O/C Maximum (Installed Horizontally) E. 3" Mineral Fiber F. 2" x 4" Wood Studs 16" or 24" O/C G. 1-1/4" Type W Screws 12" O/C (Veneer Base Attachment) H. 5/8" Type X Veneer Plaster Base Applied at Right Angles to Studs with Vertical Joints Staggered 48" I. 1/16" Minimum Veneer Plaster NOTE: Perimeter Caulking	1 hr	52						R.R.# 2198	192U1.9	USG Code Selector #38 (1972)
19	6"	A. 1/16" Minimum Veneer Plaster B. 5/8" Type X Veneer Plaster Base Approved Nails or Staples (Direct Attachment) C. 5/8" Type X Veneer Plaster Base D. 1/2" Wood Fiber Sound Deadening Board E. 2" x 4" Wood Studs 16" O/C F. 1/2" Wood Fiber Sound Deadening Board G. 5/8" Type X Veneer Plaster Base H. Approved Nails or Staples (Direct Attachment) I. 1/16" Minimum Veneer Plaster NOTE: Perimeter Caulking	1 hr	50			R.R.# 1602 R.R.# 2741	W & P Design No.90- 1 hr.		R.R.# 2198	192U1.8	USG Code Selector #41 (1972)

No.	Thickness	Diagram	Components	Fire Rating	STC*	Test Design	Test Authority	Reference
20	6"		A. 1/16" Minimum Veneer Plaster B. 5/8" Type X Veneer Plaster Base C. 1/4" Adhesive Bead Attachment D. 1/4" Gypsum Sound Deadening Board E. Approved Nails or Staples (Direct Attachment) F. 2" x 4" Wood Studs 16" O/C G. 1½" Insulation H. Approved Nails or Staples (Direct Attachment) I. 1/4" Gypsum Sound Deadening Board J. 1/4" Adhesive Bead Attachment K. 5/8" Type X Veneer Plaster Base L. 1/16" Minimum Veneer Plaster	1 hr	50	Underwriters Laboratory Design #U312	Geiger & Hamme	Georgia Pacific Brochure Assembly 10-4438G
21	6-1/8"		A. 1/16" Minimum Veneer Plaster B. Approved Nails or Staples (Direct Attachment) C. 2 Layer 1/2" Type X Veneer Plaster Base D. 3" Glass Fiber Batt E. 2" x 4" Wood Studs 16" O/C F. Resilient Furring Channel (Installed Horizontally) G. Approved Screw Attachment H. 1/2" Type X Veneer Plaster Base I. 1/4" Adhesive Bead Attachment J. 1/2" Type X Veneer Plaster Base K. 1/16" Minimum Veneer Plaster	1 hr	56			Fiberglas 1-BL-6066-B (1974)
22	6-1/8"		A. 1/16" Minimum Veneer Plaster B. Approved Nails or Staples (Direct Attachment) C. 2 Layer 1/2" Type X Veneer Plaster Base D. 2" x 4" Wood Studs 16" O/C E. Resilient Furring Channel (Installed Horizontally) F. Approved Screw Attachment G. 1/2" Type X Veneer Plaster Base H. 1/4" Adhesive Bead Attachment I. 1/2" Type X Veneer Plaster Base J. 1/16" Minimum Veneer Plaster	1 hr	52			Fiberglas 1-BL-6066-B (1974)
23	6-7/8"		A. 1/16" Minimum Veneer Plaster B. 3/8" Type X Veneer Plaster Base C. 3/4" Daubs of Adhesive 12" O/C D. 8D Nails 12" O/C E. 1/2" Type X Veneer Plaster Base F. 5D Coated Nails 32" O/C G. 5/8" Type X Veneer Plaster Base H. 2" Mineral Batts I. 2" x 4" Wood Stud 16" O/C J. Resilient Channel 24" O/C (Installed Horizontally) K. Screw Attached 12" O/C (Veneer Base Attachment) L. 5/8" Type X Veneer Plaster Base M. 3/4" Daubs of Adhesive 12" O/C N. 5/8" Type X Face Layer Veneer Plaster Base O. 1/16" Minimum Veneer Plaster	1 hr	60	Underwriters Laboratory Design #U313	Riverbank Acoustical Laboratory TL69-117	

*Sound Transmission Class

(Continued)

FIRE AND SOUND RATINGS
WOOD FRAMED INTERIOR PARTITIONS (STAGGERED STUDS) VENEER PLASTER

	MIN. THICK-NESS	CONSTRUCTION	RATINGS FIRE	RATINGS *STC	TESTING AGENCY FIRE	TESTING AGENCY SOUND	ICBO	SFM	HCD	L.A.	S.F.	SOURCE
24	6-7/8"	A. 1/16" Minimum Veneer Plaster B. Approved Nails or Staples (Direct Attachment) C. ⅝" Type X Veneer Plaster Base D. 6" Common Plate E. 2" x 4" Wood Studs 16" O/C (Staggered) F. 2" Mineral Batts G. ⅝" Type X Veneer Plaster Base H. Approved Nails or Staples (Direct Attachment) I. 1/16" Minimum Veneer Plaster	1 hr	50						Rule of Gen. Appl. 2-74		
25	6-7/8"	A. 1/16" Minimum Veneer Plaster B. Approved Nails or Staples (Direct Attachment) C. ½" Type X Veneer Plaster Base D. 6" Common Plate E. 2" x 4" Wood Studs 16" O/C (Staggered) F. 2" Mineral Batts, Each Side G. ½" Type X Veneer Plaster Base H. Approved Nails or Staples (Direct Attachment) I. 1/16" Minimum Veneer Plaster	1 hr	51						Rule of Gen. Appl. 2-74		
26	7-5/8"	A. 1/16" Minimum Veneer Plaster B. Approved Nails or Staples (Direct Attachment) C. 2 Layers ½" Type X Veneer Plaster Base D. 3" Glass Fiber Insulation E. 6" Common Plate F. 2" x 4" Wood Studs 24" O/C (Staggered) G. 2 Layers ½" Type X Veneer Plaster Base H. Approved Nails or Staples (Direct Attachment) I. 1/16" Minimum Veneer Plaster Double Layer ⅝" Each Side	1 hr 2 hr	53 58		Fiberglas					Fiberglas 1-BL-6066-B (1974)

No.	Construction	Fire	STC / IIC	Test	Source
27	A. 14 Oz. Face Weight Carpet B. 1½" Lightweight Cellular Concrete C. 15# Felt D. 7/16" Resilient Underlayment E. 5/8" Wood Sub-Floor F. R-11 Mineral Wool G. 2" x 10" Wood Joists 16" O/C H. 3/8" Type X Gypsum Lath I. Approved Metal or Wire Lath J. Approved Nails or Staples (Direct Attachment) K. ½" Gypsum Sanded Plaster NOTE: Vinyl or Linoleum Floor Covering With 1/8" Min. Resilient Backing May Be Used in Lieu of Carpeting.	1 hr	51 on bare floor **IIC 53 w/carpet no pad 50 with linoleum	Building Materials & Structures 92 1974 – Bolt, Beranek & Newman Field Test	43-C
28	A. 14 Oz. Face Weight Carpet B. 1½" Lightweight Cellular Concrete C. 15# Felt D. 5/8" Wood Sub-Floor E. R-11 Mineral Wool F. 2" x 10" Wood Joists 16" O/C G. Approved Resilient Clips H. 3/8" Type X Gypsum Lath I. 18 Ga. Line Wire Across Lath, Under Resilient Clip Tabs J. ½" Gypsum Sanded Plaster Applied in Two Coats NOTE: Vinyl or Linoleum Floor Covering with 1/8" Min. Resilient Backing May Be Used in Lieu of Carpeting.	1 hr	55 on bare floor est. **IIC 60	1974 – Bolt, Beranek & Newman Field Test Gypsum Assn. Manual	43-C
29	A. Carpet With Pad B. 15# Felt C. 1" Nominal Sub-Floor D. 3" Mineral Fiber E. 2" x 10" Wood Joist 16" O/C F. Resilient Channels Spaced 16" O/C Attached to Wood Joists With 6D Coated Nails (One Per Bearing) G. 3/8" Type X Gypsum Lath (16" Width) H. 3" Wide Approved Metal or Wire Lath Stripping Over Longitudinal Joints I. 1" Type S Screws, Three Per Lath Per Bearing J. ½" Gypsum Sanded Plaster	1 hr	52 Flint-kote **IIC 68 Gyp. Assn. Man.	Cedar Knolls 6712-5 Gypsum Assn. Manual	FC5110 Gypsum Assn. Manual '73 – '74 Pgs. 86-87 FC5110
30	A. 1" Nominal Sub And Finish Floor B. 2" x 10" Wood Joist 16" O/C C. Resilient ¼" Pencil Rod Clip (Attached to Each Joist With Min. 1⅛" by 13 Ga. Nail) D. ¼" Pencil Rod Spaced 12" O/C E. 18 Ga. Tie Wire (Attaching Metal Lath to Pencil Rod) F. 3.4# Expanded Metal Lath G. ¾" Gypsum Sanded Plaster	1 hr	52	National Bureau Standards 710	USG A-1356

*Sound Transmission Class **Impact Insulation Class

(Continued)

FIRE AND SOUND RATINGS

WOOD FRAME CONSTRUCTION FLOOR/CEILING ASSEMBLIES

#	CONSTRUCTION	FIRE	*STC	TESTING AGENCY — FIRE	TESTING AGENCY — SOUND	ICBO	SFM	HCD	L.A.	S.F.	SOURCE
31	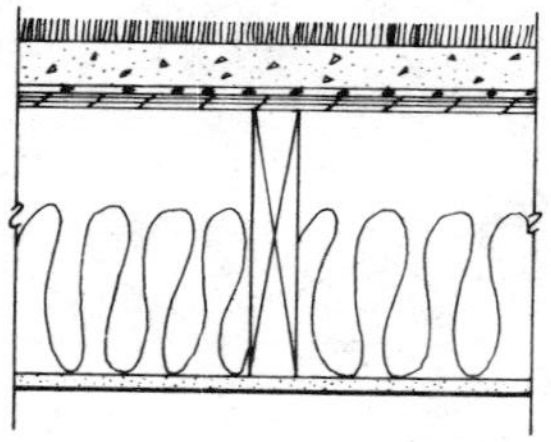 A. 14 Oz. Face Weight Carpet B. 1½" Lightweight Cellular Concrete C. 15# Felt D. 7/16" Resilient Underlayment E. ⅝" Wood Sub-Floor F. R-11 Mineral Wool G. 2" x 10" Wood Joist 16" O/C H. ½" Type X Veneer Plaster Base (Direct Attachment) I. Approved Nails or Staples 7" O/C J. 1/16" Minimum Veneer Plaster NOTE: Vinyl or Linoleum Floor Covering With ⅛" Min. Resilient Backing May Be Used in Lieu of Carpeting.	1 hr	51 on bare floor **IIC 50 w/ linoleum 53 w/ carpet no pad		1974 – Bolt, Beranek & Newman Field Test	43-C (Gypsum Assn. Manual Ref.)			43-C (Gyp. Assn. Man. Ref.)		
32	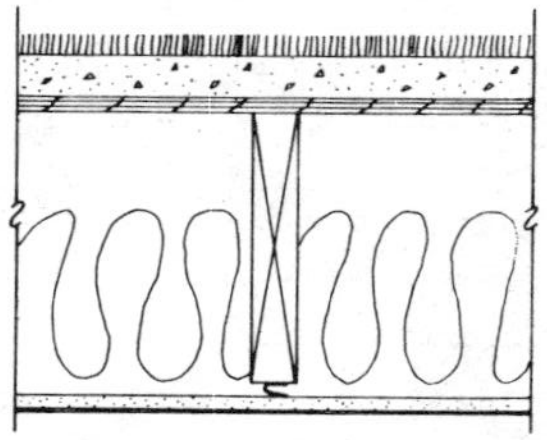 A. Standard Carpet And Pad B. 1½" Lightweight Cellular Concrete C. 15# Felt D. ⅝" Plywood Sub-Floor F. R-11 Fiberglas Insulation G. 2" x 10" Wood Joist 16" O/C Resilient Channels Spaced 24" O/C Attached to Wood Joist With 6D Coated Nails or 1¼" Type W Screw, One Attachment Per Bearing H. 1" Type S Screws 12" O/C (Veneer Base Attachment) I. ½" Type X Veneer Plaster Base J. 1/16" Minimum Veneer Plaster	1 hr	58 **IIC 74	Underwriters Laboratory Design #L502							Fiberglas 1-BL-6066-B (1974)
33	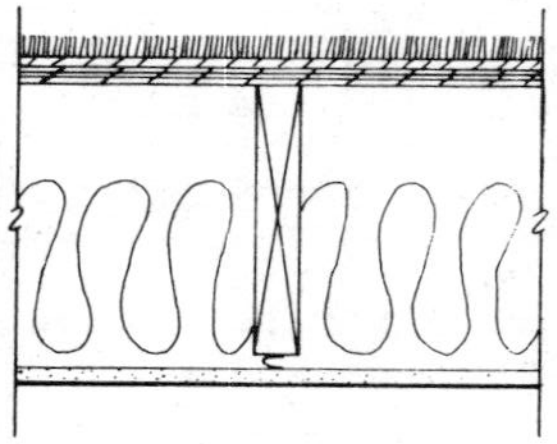 A. Standard Carpet And Pad B. ⅜" Particle Underlayment Board C. ⅝" Plywood Sub-Floor D. R-11 Fiberglas Insulation E. 2" x 10" Wood Joist 16" O/C F. Resilient Channels Spaced 24" O/C Attached to Wood Joist With 6D Coated Nails or 1¼" Type W Screws, One Attachment Per Bearing G. 1" Type S Screws 12" O/C (Veneer Base Attachment) H. ½" Type X Veneer Plaster Base I. 1/16" Minimum Veneer Plaster	1 hr	53 **IIC 73	Underwriters Laboratory Design #L502		R.R.# 1602 R.R.# 2014 R.R.# 2741			R.R.# 22047 R.R.# 22535		Fiberglas 1-BL-6066-B (1974)
34	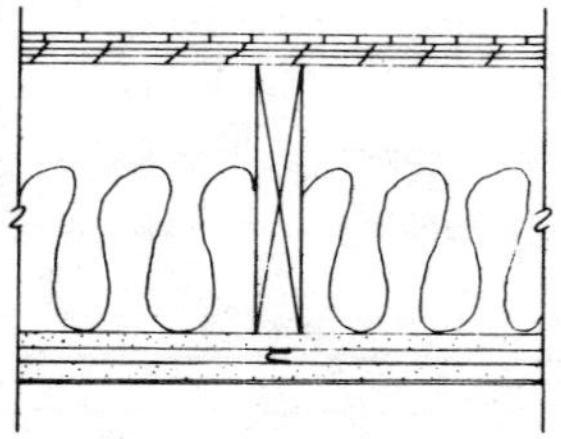 A. 1" Nominal Sub And Finish Floor B. R-11 Mineral Wool C. 2" x 10" Wood Joist 16" ... O/C D. ⅝" Type X Veneer Pla... ...se E. 8D Coated Nails 7" O/C F. Resilient Channels Spaced 2... O/C G. 1⅞" Type S Screws 16" O/L H. ⅝" Type X Veneer Plaster Base I. 1" Type S Screws 12" O/C J. 1/16" Minimum Veneer Plaster	2 hr	51 est.								Gypsum Assn. Manual '75 – '76 Pg. 93 FC5710

FIRE AND SOUND RATINGS

EXTERIOR LATH & PLASTER WALL ASSEMBLIES

MIN. THICK-NESS	CONSTRUCTION	RATINGS		TESTING AGENCY		ICBO	THERMAL RESISTANCE	L.A.	S.F.	SOURCE
		FIRE	*STC	FIRE	SOUND					
1 6"	A. 7/8" Portland Cement Plaster B. 1½" X 17 Ga. Wire Lath C. Weather Resistive Barrier Paper D. Approved Nails or Staples (Direct Attachment) E. 2" x 4" Wood Studs 16" O/C F. 3" Rockwool Insulation G. Resilient Clips H. 3/8" Type X Gypsum Lath I. ½" Sanded Gypsum Plaster	1 hr	47		Riverbank Acoustical Laboratory TL 77-40		$R_T = 12.43$ $U = \dfrac{1}{R_T}$ $U = .080$			Fire: Chapter 43B 1976 UBC & RR# 1174 Sound: WCLP1 12/3/76
	Same As (#1) Above With Perimeter Caulked		53		RAL TL 77-45					Sound:WCLP1 12/7/76
2 6-1/4"	A. 7/8" Portland Cement Plaster B. 1½" x 17 Ga. Wire Lath C. Weather Resistive Barrier Paper D. Approved Nails or Staples (Direct Attachment) E. 2" x 4" Wood Studs 16" O/C F. 3" Rockwool Insulation G. Resilient Channel 24" O/C H. ½" Type X Veneer Base (Vertically Applied) I. 1" Type S Screw 12" O/C (Veneer Base Attachment) J. 1/16" Min. Veneer Plaster	1 hr (est)	50		Riverbank Acoustical Laboratory TL 77-38		$R_T = 12.47$ $U = \dfrac{1}{R_T}$ $U = .080$			Fire: Chapter 43B 1976 UBC & Gypsum Assn. Manual '75-'76 WP 3620 Sound:WCLP1 12/3/76
	Same As (#2) Above With Perimeter Caulked		56		RAL TL 77-67					Sound:WCLP1 1/12/77
3 5-1/2"	A. 7/8" Portland Cement Plaster B. 1½" x 17 Ga. Wire Lath C. Weather Resistive Barrier Paper D. Approved Nails or Staples (Direct Attachment) E. 2" x 4" Wood Studs 16" O/C F. 3" Rockwool Insulation G. 3/8" Type X Gypsum Lath H. Approved Nails or Staples (Direct Attachment) I. ½" Gypsum Sanded Plaster	1 hr	41		Riverbank Acoustical Laboratory TL 77-43		$R_T = 12.43$ $U = \dfrac{1}{R_T}$ $U = .080$			Fire: Chapter 43B 1976 UBC Sound:WCLP1 12/7/76

(Continued)

FIRE AND SOUND RATINGS

	Thickness	Detail	Fire Rating	STC*	Test	Acoustical Test	U-Value	References
4	5-1/4"	A. 7/8" Portland Cement Plaster B. 1½" x 17 Ga. Wire Lath C. Weather Resistive Barrier Paper D. Approved Nails or Staples (Direct Attachment) E. 2" x 4" Wood Studs 16" O/C F. 3" Rockwool Insulation G. ½" Type X Veneer Base H. 6d Nails 7" O/C (Direct Attachment) I. 1/16" Min. Veneer Plaster	1 hr	40		Riverbank Acoustical Laboratory TL 77-41	$R_T = 12.47$ $U = \dfrac{1}{R_T}$ $U = .080$	Fire: Chapter 43B 1976 UBC & Gypsum Assn. Manual '75-'76 WP 3620 Sound: WCLPI 12/3/76
5	5-5/8"	A. 1" Portland Cement Plaster B. 3.4 Expanded Metal Lath C. 1¼" Type S-12 Screws 8" O/C D. ½" Gypsum Sheathing E. 35/8" 20 Ga. Steel Stud 16" O/C F. 3" Rockwool Insulation G. ½" Type X Veneer Base H. 1" Type S-12 Screws, 12" O/C in Field, 8" O/C at Perimeters (Direct Attachment) I. 1/16" Min. Veneer Plaster NOTE: Perimeter Caulked	2 hr	54	Ohio State University T-4851 6/70	Riverbank Acoustical Laboratory TL 77-66	$R_T = 12.95$ $U = \dfrac{1}{R_T}$ $U = .077$	Fire: Chapter 43B 1976 UBC & Gypsum Assn. Manual '75-'76 WP 8320 Sound: WCLPI 1/5/77

*Sound Transmission Class

FIRE AND SOUND RATINGS

STEEL STUD FRAMED INTERIOR PARTITIONS/CONVENTIONAL PLASTER

#	MIN. THICK-NESS	CONSTRUCTION	RATINGS FIRE	RATINGS *STC	TESTING AGENCY FIRE	TESTING AGENCY SOUND	ICBO	SFM	HCD	L.A.	S.F.	SOURCE
1	5-1/2"	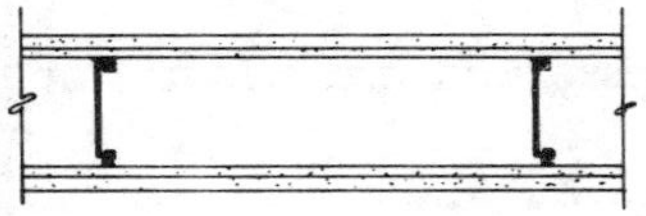 A. ½" Gypsum Sanded Plaster B. ⅜" Type X Gypsum Lath C. Attachment Clips D. 3¼" Steel Studs 16" O/C E. Resilient Clips F. ⅜" Type X Gypsum Lath G. ½" Gypsum Sanded Plaster NOTE: Perimeter Caulked	1 hr est.	55	Extrapolated:1976 UBC Table 43-B #61 & 1975 Underwriter's Lab. U408	USG 104-FT (Geiger & Hamme)	R.R.# 1562				112U1.4	USG A-1186 (1966)
2	5-3/4"	 A. ½" Gypsum Sanded Plaster B. ⅜" Type X Gypsum Lath C. Resilient Clips, Staggered 16" O/C D. 2½" Steel Studs 8" O/C E. 2" Thermofiber Batts F. Resilient Clips, Staggered 16" O/C G. ⅜" Type X Gypsum Lath H. ½" Gypsum Sanded Plaster	1 hr est.	56	Underwriter's Lab. U408	USG 133-FT (Geiger & Hamme)						USG A-1186 (1966)
3	5"	 A. ½" Gypsum Sanded Plaster B. ⅜" Type X Gypsum Lath C. ⅜" Veneer Base D. 2½" Steel Studs 24" O/C E. ⅜" Veneer Base F. ⅜" Type X Gypsum Lath G. ½" Gypsum Sanded Plaster	1 hr	52 54 with economy batts	American Society for Testing & Materials E-11q (test at Kaiser)	Kaiser Gypsum Co. Acoustical Laboratory						Kaiser Gypsum Co. Technical Bulletin Plate 477-2
4	6-1/8"	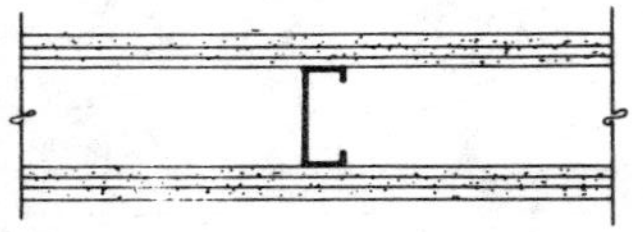 A. ½" Gypsum Sanded Plaster B. ⅜" Type X Gypsum Lath C. ⅜" Veneer Base D. 3⅝" Steel Studs 24" O/C E. ⅜" Veneer Base F. ⅜" Type X Gypsum Lath G. ½" Gypsum Sanded Plaster	1 hr	53 56 with thick batts	American Society for Testing & Materials E-119 (test at Kaiser)	Kaiser Gypsum Co. Acoustical Laboratory						Kaiser Gypsum Co. Technical Bulletin Plate 477-3
5	5-3/4" 6-1/2" 8-1/2"	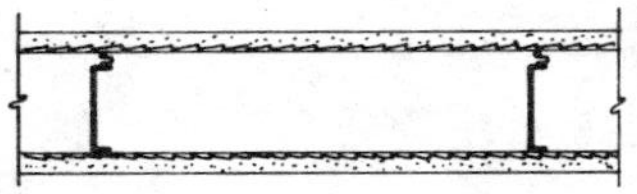 A. ¾" Gypsum Sanded Plaster B. 3.4# Expanded Metal Lath C. ¼" Pencil Rods D. Resilient Clips E. 3¼" Steel Studs 16" O/C F. 3.4# Expanded Metal Lath G. ¾" Gypsum Sanded Plaster	1 hr	50 51 est. (4" studs) 52 est. (6" studs)		National Bureau of Standards Cedar Knolls 664-20						Wheeling 303-5M T-866-24 BMS 144 NBS Metal Lath Assn. Tech. Bulletin #151

*Sound Transmission Class

(Continued)

375

FIRE AND SOUND RATINGS

DOUBLE PARTITIONS, COLD ROLLED CHANNELS OR STEEL STUDS/CONV. PLASTER

#	MIN. THICK-NESS	CONSTRUCTION	RATINGS FIRE	*STC	TESTING AGENCY FIRE	TESTING AGENCY SOUND	ICBO	SFM	HCD	L.A.	S.F.	SOURCE
6	4-3/8" 5-1/2" 7" 8-1/2" 10"	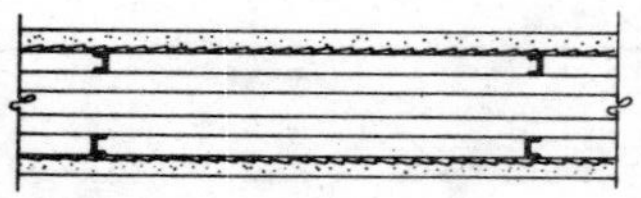 A. ¾" Portland Cement Plaster B. 3.4# Expanded Metal Lath C. Cold Rolled Channels or Studs D. ¾" Cold Rolled Channels E. ¾" Cold Rolled Channels F. Cold Rolled Channels or Studs G. 3.4# Expanded Metal Lath H. ¾" Portland Cement Plaster NOTE: Resting on 1" Cork	1 hr	50 51 52 53 54	Building Materials & Structures 92-Table 31	NBS 160E NBS 160D NBS 160C NBS 160B NBS 160A						Metal Lath Assn. Tech. Bulletins #141 (fire) #151 (sound)
7	4-3/16"	A. 3/32" Veneer Plaster B. ½" Veneer Base C. ¼" Sound Board D. 2½" Steel Studs 24" O/C E. 2" Batt Insulation F. ¼" Sound Board G. ½" Veneer Base H. 3/32" Veneer Plaster NOTE: Perimeter Caulked Same As (#7) Above With ⅝" Veneer Base	1 hr 1 hr	50 54	WP 1051	Geiger & Hamme Cedar Knolls 684.14	UBC UBC					Georgia Pacific DC-25-1/70 Gypsum Assn. Manual '75-76 WP1015
8	4-11/16"	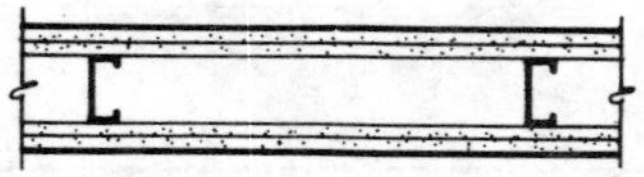 A. 3/32" Veneer Plaster B. 2 Layers ½" Veneer Base C. 2½" Steel Studs 16" O/C D. 2 Layers ½" Veneer Base E. 3/32" Veneer Plaster	2 hr	50								Georgia Pac. DC-1/73-5M-44-38G;P.12
9	5-1/8"	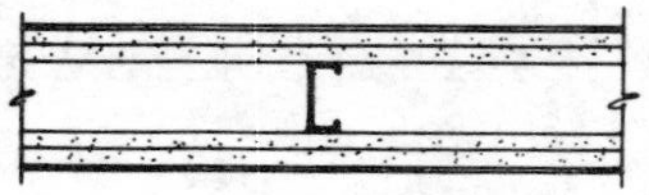 A. 1/16" Veneer Plaster B. 2 Layers ⅝" Veneer Base C. 2½" Steel Studs 24" O/C D. 2 Layers ⅝" Veneer Base E. 1/16" Veneer Plaster NOTE: Perimeter Caulked	2 hr	50	U.L. #U411	TL-63-177						USG Imperial P-456C(1974) USG SA912 USG A-1141
10	4-5/8"	 Same As (#9) Above With ½" Veneer Base and 1" Batt Insulation	2 hr	53	U.L. #U303	Cedar Knolls 65466	Table 43-B R.R.# 1530			R.R.# 23961		USG SA912

CHAPTER FIVE

CODES
AND
ORDINANCES

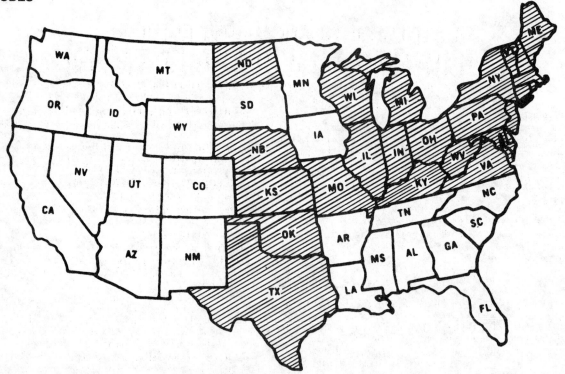

MAP SHOWS breakout by states in which three national model building codes have been adopted. Western states, including Alaska and Hawaii, indicated by white background, follow the Uniform Building Code. Central states, indicated by shading, follow the BOCA Basic National Building Code. Southern states, shown with white background in bottom, right area of map, follow Standard Building Code.

LATH, PLASTER, FIREPROOFING AND DRYWALL INSTALLATIONS UNDER THE MODEL BUILDING CODES

In the past half century there have been an inception and development of three regional model building codes (also often described as "uniform" codes), which collectively have replaced local city, county and state building codes, bringing a welcome degree of uniformity to the design and construction of buildings in three respective regions.

The three model code organizations are non-profit institutions whose memberships consist primarily of building officials, who conceive the building laws after in-depth study and continue to research and update them to conform to the changing technology of the construction industry and the public need for safety from fire and earthquakes.

Most recently, in a new movement toward uniformity, a new **national** code for dwellings has been developed and published (latest edition, 1986) under the title of the "CABO One and Two Family Dwelling Code."

None of the model codes, which are all published triennially and updated annually, have the effect of law until adopted by the legislative bodies of local cities, counties and states.

Lath, plaster, fireproofing and drywall installations are fully covered in appropriate sections in the four model codes, extracts from which follow on the ensuing pages of this manual.

In adopting the model codes by reference, the local authorities have the option to modify or amend any section of the rules where it is felt that local conditions might require such a change. In the case of lath, plaster, fireproofing and drywall, the deviations away from the model codes have been very minimal.

Following is a brief description of the four model code writing organizations:

In the *West* it's the International Conference of Building Officials, headquartered at 5360 South Workman Mill Road, Whittier, California 90601. The latest edition of its *Uniform Building Code*, popularly known as the "UBC", is dated 1988.

In the *Midwest* it's the Building Officials & Code Administrators International, Inc., headquartered at 4051 West Flossmoor Road, Country Club Hills, Illinois 60477. The latest edition of its BOCA National Building Code is 1987.

In the *South* it's the Southern Building Code Congress International, Inc., headquartered at 900 Montclair Road, Birmingham, Alabama 35213. The latest edition of its Standard Building Code, popularly known as the "SBBC", is dated 1988.

The CABO One and Two Family Dwelling Code is written and promulgated by the Council of American Building Officials, 5203 Leesburg Pike, Suite 708, Falls Church, Virginia 22041.

UNIFORM BUILDING CODE

Weather Protection

Sec. 1707. (a) **Weather-resistive Barriers.** All weather-exposed surfaces shall have a weather-resistive barrier to protect the interior wall covering. Such barrier shall be equal to that provided for in U.B.C. Standard No. 17-1 for kraft waterproof building paper or U.B.C. Standard No. 32-1 for asphalt-saturated rag felt. Building paper and felt shall be free from holes and breaks other than those created by fasteners and construction system due to attaching of the building paper, and shall be applied over studs or sheathing of all exterior walls. Such felt or paper shall be applied weatherboard fashion, lapped not less than 2 inches at horizontal joints and not less than 6 inches at vertical joints.

Weather-protected barrier may be omitted in the following cases:

1. When exterior covering is of approved weatherproof panels.
2. In back-plastered construction.
3. When there is no human occupancy.
4. Over water-repellent panel sheathing.
5. Under approved paperback metal or wire fabric lath.
6. Behind lath and portland cement plaster applied to the underside of roof and eave projections.

(b) **Flashing and Counterflashing.** Exterior openings exposed to the weather shall be flashed in such a manner as to make them weatherproof.

All parapets shall be provided with coping of approved materials. All flashing, counterflashing and coping, when of metal, shall be of not less than No. 26 U.S. gauge corrosion-resistant metal.

(c) **Waterproofing Weather-exposed Areas.** Balconies, landings, exterior stairways and similar surfaces exposed to the weather and sealed underneath shall be waterproofed.

Chapter 43

FIRE-RESISTIVE STANDARDS

General

Sec. 4301. In addition to all the other requirements of this code, fire-resistive materials shall meet the requirements for fire-resistive construction given in this chapter.

Fire-resistive Materials and Systems

Sec. 4302. (a) **General.** Materials and systems used for fire-resistive purposes shall be limited to those specified in this chapter unless accepted under the procedure given in Section 4302 (b) or 4302 (c). For standards referred to in this chapter, see Chapter 60.

The materials and details of construction for the fire-resistive systems described in this chapter shall be in accordance with all other provisions of this code except as modified herein.

For the purpose of determining the degree of fire resistance afforded, the materials of construction listed in this chapter shall be assumed to have the fire-resistance rating indicated in Table No. 43-A, 43-B or 43-C.

As an alternate to Tables Nos. 43-A, B and C, fire-resistive construction may be approved by the building official on the basis of evidence submitted by the person responsible for the structural design showing that the construction meets the required fire-resistive classification.

(b) **Qualification by Testing.** Material or assembly of materials of construction tested in accordance with the requirements set forth in U.B.C. Standard No. 43-1 shall be rated for fire resistance in accordance with the results and conditions of such tests.

> **EXCEPTION:** The acceptance criteria of U.B.C. Standard No. 43-1 for exterior bearing walls shall not be required to be greater with respect to heat transmission and passage of flame or hot gases than would be required of a nonbearing wall in the same building with the same distance to the property line. The fire exposure time period, water pressure and duration of application for the hose stream test shall be based upon the fire-resistive rating determined by this exception.

Fire-resistive assemblies tested under U.B.C. Standard No. 43-1 shall not be considered to be restrained unless evidence satisfactory to the building official is furnished by the person responsible for the structural design showing that the construction qualifies for a restrained classification in accordance with U.B.C. Standard No. 43-1. Restrained construction shall be identified on the plans.

(c) **Calculating Fire Resistance.** The fire-resistive rating of a material or assembly may be established by calculations. The procedures used for such calculations shall be in accordance with U.B.C. Standard No. 43-9.

CARBONATE AGGREGATE CONCRETE is concrete made with aggregates consisting mainly of calcium or magnesium carbonate, e.g., limestone or dolomite, and containing 40 percent or less quartz, chert or flint.

LIGHTWEIGHT AGGREGATE CONCRETE is concrete made with aggregates of expanded clay, shale, slag or slate or sintered fly ash and weighting 85 to 115 pcf.

SAND-LIGHTWEIGHT CONCRETE is concrete made with a combination of expanded clay, shale, slag or slate or sintered fly ash and natural sand. Its unit weight is generally between 105 and 120 pcf.

SILICEOUS AGGREGATE CONCRETE is concrete made with normal-weight aggregates consisting mainly of silica or compounds other than calcium or magnesium carbonate, and may contain more than 40 percent quartz, chert or flint.

(e) **Shotcrete.** Shotcrete without coarse aggregate shall be classified in accordance with the aggregate used.

Protection of Structural Members

Sec. 4303. (a) **General.** Structural members having the fire-resistive protection set forth in Table No. 43-A shall be assumed to have the fire-resistance ratings set forth therein.

(b) **Protective Coverings. 1. Thickness of protection.** The thickness of fire-resistive materials required for protection of structural members shall be not less than set forth in Table No. 43-A, except as modified in this section. The figures shown shall be the net thickness of the protecting materials and shall not include any hollow space back of the protection.

2. **Unit masonry protection.** Where required, metal ties shall be embedded in transverse joints of unit masonry for protection of steel columns. Such ties shall be as set forth in Table No. 43-A or be equivalent thereto.

3. **Reinforcement for cast-in-place concrete column protection.** Cast-in-place concrete protection for steel columns shall be reinforced at the edges of such members with wire ties of not less than .18 inch in diameter wound spirally around the columns on a pitch of not more than 8 inches or by equivalent reinforcement.

4. **Embedment of pipes.** Conduits and pipes shall not be embedded in required fire protection of structural members.

5. **Column jacketing.** Where the fire-resistive covering on columns is exposed to injury from moving vehicles, the handling of merchandise or other means, it shall be protected in an approved manner.

6. **Ceiling protection.** Where a ceiling forms the protective membrane for fire-resistive assemblies, the constructions and their supporting horizontal structural members need not be individually fire protected except where such members support directly applied loads from more than one floor or roof. The required fire resistance shall be not less than that required for individual protection of members.

Ceilings shall form continuous fire-resistive membranes but may have openings for copper, sheet steel or ferrous plumbing pipes, ducts and electrical outlet boxes, provided the areas of such openings through the ceiling aggregate not more than 100 square inches for any 100 square feet of ceiling area. Regardless of size, duct openings in such ceilings shall be protected by approved ceiling fire dampers. Access doors installed in such ceilings shall be approved horizontal access door assemblies listed for such purpose.

> **EXCEPTIONS:** 1. Larger openings than permitted above may be installed where such openings and the assemblies in which they are utilized are in accordance with the results of tests pursuant to the provisions of Section 4302 (b).
>
> 2. Ceiling fire dampers may be omitted from duct openings where fire-resistive tests have shown that fire dampers are not necessary to maintain the fire resistance of the assembly.

Individual electrical outlet boxes shall be of steel and not greater than 16 square inches in area.

7. **Plaster application.** Plaster protective coatings may be applied with the finish coat omitted when they comply with the design mix and thickness requirements of Tables Nos. 43-A, 43-B and 43-C.

Floor-Ceilings or Roof-Ceilings

Sec. 4305. (a) **General.** Fire-resistive floor-ceiling or roof-ceiling construction systems shall be assumed to have the fire-resistance ratings set forth in Table No. 43-C. Penetrations in floors and ceilings requiring protected openings shall be fire stopped. Firestopping shall be of an approved material, securely installed and capable of maintaining its integrity when subjected to the time-temperature curve of U.B.C. Standard No. 43-1 for the specific floor-ceiling or roof-ceiling construction.

> **EXCEPTION:** Where penetrations are protected as shaft enclosures as required in Section 1706.

When materials are incorporated into an otherwise fire-resistive assembly which may change the capacity for heat dissipation, fire test results or other substantiating data shall be made available to the building official to show that the required fire-resistive time period is not reduced.

TABLE 43-A—MINIMUM PROTECTION OF STRUCTURAL PARTS BASED ON TIME PERIODS FOR VARIOUS NONCOMBUSTIBLE INSULATING MATERIALS—(Continued)

STRUCTURAL PARTS TO BE PROTECTED	ITEM NUMBER	INSULATING MATERIAL USED	MINIMUM THICKNESS OF INSULATING MATERIAL FOR FOLLOWING FIRE-RESISTIVE PERIODS (In Inches)			
			4 Hr.	3 Hr.	2 Hr.	1 Hr.
1. Steel Columns and All Members of Primary Trusses	1-1.1	Carbonate, lightweight and sand-lightweight aggregate concrete, members 6" by 6" or greater (not including sandstone, granite and siliceous gravel).[1]	2½	2	1½	1
	1-1.2	Carbonate, lightweight and sand-lightweight aggregate concrete, members 8" by 8" or greater (not including sandstone, granite and siliceous gravel).[1]	2	1½	1	1
	1-1.3	Carbonate, lightweight and sand-lightweight aggregate concrete, members 12" by 12" or greater (not including sandstone, granite and siliceous gravel).[1]	1½	1	1	1
	1-1.4	Siliceous aggregate concrete and concrete excluded in Item No. 1-1.1, members 6" by 6" or greater.[1]	3	2	1½	1
	1-1.5	Siliceous aggregate concrete and concrete excluded in Item No. 1-1.1, members 8" by 8" or greater.[1]	2½	2	1	1
	1-1.6	Siliceous aggregate concrete and concrete excluded in Item No. 1-1.1, members 12" by 12" or greater.[1]	2	1	1	1
	1-2.1	Clay or shale brick with brick and mortar fill.[1]	3¾			2¼
	1-3.1	4" Hollow clay tile in two 2" layers; ½" mortar between tile and column; ⅜" metal mesh (.046" wire diameter) in horizontal joints; tile fill.[1]	4			
	1-3.2	2" Hollow clay tile; ¾" mortar between tile and column; ⅜" metal mesh (.046" wire diameter) in horizontal joints; limestone concrete fill;[1] plastered with ¾" gypsum plaster.	3			
	1-3.3	2" Hollow clay tile with outside wire ties (.08" diameter) at each course of tile or ⅜" metal mesh (.046" diameter wire) in horizontal joints; limestone or trap-rock concrete fill[1] extending 1" outside column on all sides.			3	
1. Steel Columns and All Members of Primary Trusses	1-4.1	Portland cement plaster over metal lath wire tied to ¾" cold-rolled vertical channels with No. 18 gauge wire ties spaced 3" to 6" on center. Plaster mixed 1:2½ by volume, cement to sand.			2½[2]	⅞
	1-5.1	Vermiculite concrete, 1:4 mix by volume over paperbacked wire fabric lath wrapped directly around column witn additional 2" by 2" No. 16/16 gauge wire fabric placed ¾" from outer concrete surface. Wire fabric tied with No. 18 gauge wire spaced 6" on center for inner layer and 2" on center for outer layer.	2			
	1-6.1	Perlite or vermiculite gypsum plaster over metal lath wrapped around column and furred 1¼" from column flanges. Sheets lapped at ends and tied at 6" intervals with No. 18 gauge tie wire. Plaster pushed through to flanges.	1½	1		
	1-6.2	Perlite or vermiculite gypsum plaster over self-furring metal lath wrapped directly around column, lapped 1" and tied at 6" intervals with No. 18 gauge wire.	1¾	1⅜	1	
	1-6.3	Perlite or vermiculite gypsum plaster on metal lath applied to ¾" cold-rolled channels spaced 24 inches apart vertically and wrapped flatwise around column.	1½			
1. Steel Columns and All Members of Primary Trusses	1-6.4	Perlite or vermiculite gypsum plaster over 2 layers of ½" plain full-length gypsum lath applied tight to column flanges. Lath wrapped with 1" hexagonal mesh of No. 20 gauge wire and tied with doubled No. 18 gauge wire ties spaced 23" on center. For three-coat work the plaster mix for the second coat shall not exceed 100 pounds of gypsum to 2½ cubic feet of aggregate for the three-hour system.	2½	2		
	1-6.5	Perlite or vermiculite gypsum plaster over one layer of ½" plain full-length gypsum lath applied tight to column flanges. Lath tied with doubled No. 18 gauge wire ties spaced 23" on center and scratch coat wrapped with 1" hexagonal mesh No. 20 gauge wire fabric. For three-coat work the plaster mix for the second coat shall not exceed 100 pounds of gypsum to 2½ cubic feet of aggregate.		2		
	1-7.1	Multiple layers of ½" gypsum wallboard[3] adhesively[4] secured to column flanges and successive layers. Wallboard applied without horizontal joints. Corner edges of each layer staggered. Wallboard layer below outer layer secured to column with doubled No. 18 gauge wire ties spaced 15" on center. Exposed corners taped and treated.			2	1

(Continued)

TABLE 43-A—MINIMUM PROTECTION OF STRUCTURAL PARTS BASED ON TIME PERIODS FOR VARIOUS NONCOMBUSTIBLE INSULATING MATERIALS—(Continued)

STRUCTURAL PARTS TO BE PROTECTED	ITEM NUMBER	INSULATING MATERIAL USED	MINIMUM THICKNESS OF INSULATING MATERIAL FOR FOLLOWING FIRE-RESISTIVE PERIODS (In Inches)			
			4 Hr.	3 Hr.	2 Hr.	1 Hr.
1. Steel Columns and All Members of Primary Trusses	1-7.2	Three layers of 5/8″ Type X gypsum wallboard.³ First and second layer held in place by 1/8″ diameter by 1 3/8″ long ring shank nails with 5/16″ diameter heads spaced 24″ on center at corners. Middle layer also secured with metal straps at mid-height and 18″ from each end, and by metal corner bead at each corner held by the metal straps. Third layer attached to corner bead with 1″ long gypsum wallboard screws spaced 12″ on center.			1 7/8	
	1-7.3	Three layers of 5/8″ Type X gypsum wallboard,³ each layer screw attached to 1 5/8″ steel studs (No. 25 gauge) at each corner of column. Middle layer also secured with No. 18 gauge double strand tie wire, 24″ on center for inner layer, No. 6 by 1 5/8″ spaced 12″ on center for middle layer and No. 8 by 2 1/4″ spaced 12″ on center for outer layer.		1 7/8		
	1-8.1	Wood-fibered gypsum plaster mixed 1:1 by weight gypsum to sand aggregate applied over metal lath. Lath lapped 1″ and tied 6″ on center at all ends, edges and spacers with No. 18 gauge tie wire. Lath applied over 1/2″ spacers made of 3/4″ furring channel with 2″ legs bent around each corner. Spacers located 1″ from top and bottom of member and a maximum of 40″ on center and wire tied with a single strand of No. 18 gauge wire. Corner bead tied to the lath at 6″ on center along each corner to provide plaster thickness.			1 5/8	
	2-2.1	Portland cement plaster on metal lath attached to 3/4″ cold-rolled channels with No. 18 gauge wire ties spaced 3″ to 6″ on center. Plaster mixed 1:2 1/2 by volume, cement to sand.			2 1/2²	7/8
	2-3.1	Vermiculite gypsum plaster on a metal lath cage, wire tied to No. 8 steel wire hangers wrapped around beam and spaced 16″ on center. Metal lath ties spaced approximately 5″ on center at cage sides and bottom.		7/8		
2. Webs or Flanges of Steel Beams and Girders	2-4.1	Two layers of 5/8″ Type X gypsum wallboard³ are attached to U-shaped brackets spaced 24″ on center. No. 25 gauge 1 5/8″ deep by 1″ galvanized steel runner channels are first installed parallel to and on each side of the top beam flange to provide a 1/2″ clearance to the flange. The channel runners are attached to steel deck or concrete floor construction with approved fasteners spaced 12″ on center. U-shaped brackets are formed from members identical to the channel runners. At the bent portion of the U-shaped bracket, the webs of the channel are cut out so that 1 5/8″ deep corner channels can be inserted without attachment parallel to each side of the lower flange. As an alternate No. 24 gauge 1″ by 2″ runner and corner angles may be used in lieu of channels and the web cutouts in the U-shaped brackets may be omitted. Each angle is attached to the bracket with 1/2″ long No. 8 self-drilling screws. The vertical legs of the U-shaped bracket are attached to the runners with one 1/2″ long No. 8 self-drilling screw. The completed steel framing provides a 2 1/8″ and 1 1/2″ space between the inner layer of wallboard and the sides and bottom of the steel beam respectively. The inner layer of wallboard is attached to the top runners and bottom corner channels or corner angles with 1 1/4″ long No. 6 self-drilling screws spaced 16″ on center. The outer layer of wallboard is applied with 1 3/4″ long No. 6 self-drilling screws spaced 8″ on center. The bottom corners are reinforced with metal corner beads.			1 1/4	
2. Webs or Flanges of Steel Beams and Girders	2-4.2	Three layers of 5/8″ Type X gypsum wallboard³ attached to a steel suspension system as described immediately above utilizing the No. 25 gauge 1″ by 2″ lower corner angles. The framing is located so that a 2 1/8″ and 2″ space is provided between the inner layer of wallboard and the sides and bottom of the beam respectively. The first two layers of wallboard are attached as described immediately above. A layer of No. 20 gauge 1″ hexagonal galvanized wire mesh is applied under the soffit of the middle layer and up the sides approximately 2″. The mesh is held in position with the No. 6 1 5/8″ long screws installed in the vertical leg of the bottom corner angles. The outer layer of wallboard is attached with No. 6 2 1/4″ long screws spaced 8″ on center. One screw is also installed at the mid-depth of the bracket in each layer. Bottom corners are finished as described above.			1 7/8	

ªGeneric fire-resistance ratings (those not designated by a company code letter) as listed in the Fire Resistance Design Manual, Eleventh Edition, dated October, 1984, as published by the Gypsum Association, may be accepted as if herein listed.

¹Reentrant parts of protected members to be filled solidly.

²Two layers of equal thickness with a 3/4-inch air space between.

³For all of the construction with gypsum wallboard described in Table No. 43-A, gypsum base for veneer plaster of the same size, thickness and core type may be substituted for gypsum wallboard, provided attachment is identical to that specified for the wallboard and the joints on the face layer are reinforced and the entire surface is covered with a minimum of 1/16-inch gypsum veneer plaster. The gypsum base for veneer plaster and the veneer plaster shall comply with U.B.C. Standard No. 47-15.

TABLE NO. 43-B—RATED FIRE-RESISTIVE PERIODS FOR VARIOUS WALLS AND PARTITIONS[a 1]
—(Continued)

MATERIAL	ITEM NUMBER	CONSTRUCTION	MINIMUM FINISHED THICKNESS FACE-TO-FACE[2] (In Inches)			
			4 Hr.	3 Hr.	2 Hr.	1 Hr.
1. Brick of Clay or Shale	1-1.2	Solid units plastered each side with ⅝" gypsum or portland cement plaster. Portland cement plaster mixed 1:2½ by weight, cement to sand.			4¾[4]	
	1-2.2	Hollow brick units[5] at least 71 percent solid, plastered each side with ⅝" gypsum plaster.	8¾			
	1-3.2	Hollow (rowlock[6]) plastered each side with ⅝" gypsum or portland cement plaster. Portland cement plaster mixed 1:2½ by weight, cement to sand.	9			
	1-6.1	4" nominal thick units at least 75 percent solid backed with a hat-shaped metal furring channel ¾ inch thick formed from 0.021-inch sheet metal attached to the brick wall on 24-inch centers with approved fasteners; and ½ inch Type X gypsum wallboard[7] attached to the metal furring strips with 1-inch-long Type S screws spaced 8 inches on center.			5[4]	
2. Hollow Clay Tile, Non-load-bearing (End or Side Construction)	2-1.1	One cell in wall thickness, units at least 50 percent solid, plastered each side with ⅝" gypsum plaster.				4¼
	2-1.2	Two cells in wall thickness, units at least 45 percent solid.				6
	2-1.3	Two cells in wall thickness, units at least 45 percent solid. Plastered each side with ⅝" gypsum plaster.			7	
	2-1.4	Two cells in wall thickness, units at least 60 percent solid. Plastered each side with ⅝" gypsum plaster.			5	
	3-1.2	Two cells in wall thickness, units at least 40 percent solid. Plastered one side with ⅝" gypsum plaster.			8½	
	3-2.1	Two units and three cells in wall thickness, units at least 40 percent solid. Plastered one side with ⅝" gypsum plaster.	12½			
	3-2.2	Three cells in wall thickness, units at least 43 percent solid. Plastered one side with ⅝" gypsum plaster.		8½		
3. Hollow Clay Tile, Load-bearing (End or Side Construction)	3-2.3	Two cells in wall thickness, units at least 40 percent solid. Plastered each side with ⅝" gypsum plaster.		9		
	3-2.4	Three cells in wall thickness, units at least 43 percent solid. Plastered each side with ⅝" gypsum plaster.	9			
	3-2.5	Three cells in wall thickness, units at least 40 percent solid. Plastered each side with ⅝" gypsum plaster.	13			
	3-3.1	Hollow cavity wall consisting of two 4" nominal clay tile units (at least 40 percent solid) with air space between. Plastered one side (exterior) with ¾" portland cement plaster and other side with ⅝" gypsum plaster. Portland cement plaster mixed 1:3 by volume, cement to sand.	10			
4. Combination of Clay Brick and Load-bearing Hollow Clay Tile	4-1.3	4" brick and 4" tile plastered on the tile side with ⅝" gypsum plaster.	8½			
6. Solid Concrete[8 15]	6-1.1	Siliceous Aggregate Concrete.	7.0	6.2	5.0	3.5
		Carbonate Aggregate Concrete.	6.6	5.7	4.6	3.2
		Sand-lightweight Concrete.	5.4	4.6	3.8	2.7
		Lightweight Concrete.	5.1	4.4	3.6	2.5
7. Glazed or Unglazed Facing Tile, Nonload-bearing	7-1.1	One 2" unit cored 15 percent maximum and one 4" unit cored 25 percent maximum with ¾" mortar filled collar joint. Unit positions reversed in alternate courses.		6⅜		
	7-1.2	One 2" unit cored 15 percent maximum and one 4" unit cored 40 percent maximum with ⅜" mortar filled collar joint. Plastered one side with ¾" gypsum plaster. Two wythes tied together every fourth course with No. 22 gauge corrugated metal ties.		6¾		
	7-1.3	One unit with three cells in wall thickness, cored 29 percent maximum.			6	
	7-1.4	One 2" unit cored 22 percent maximum and one 4" unit cored 41 percent maximum with ¼" mortar filled collar joint. Two wythes tied together every third course with No. 22 gauge corrugated metal ties.			6	
	7-1.5	One 4" unit cord 25 percent maximum with ¾" gypsum plaster on one side.			4¾	
	7-1.6	One 4" unit with two cells in wall thickness, cored 22 percent maximum.				4

TABLE NO. 43-B—RATED FIRE-RESISTIVE PERIODS FOR VARIOUS WALLS AND PARTITIONS[a][1]

MATERIAL	ITEM NUMBER	CONSTRUCTION	MINIMUM FINISHED THICKNESS FACE-TO-FACE[2] (In Inches)			
			4 Hr.	3 Hr.	2 Hr.	1 Hr.
8. Solid Gypsum Plaster	8-1.1	¾″ by No. 16 gauge vertical cold-rolled channels, 16″ on center with 2.5-pound flat metal lath applied to one face and tied with No. 18 gauge wire at 6″ spacing. Gypsum plaster each side mixed 1:2 by weight, gypsum to sand aggregate.				2[4]
	8-1.2	¾″ by No. 16 gauge cold-rolled channels 16″ on center with metal lath applied to one face and tied with No. 18 gauge wire at 6″ spacing. Perlite or vermiculite gypsum plaster each side. For three-coat work the plaster mix for the second coat shall not exceed 100 pounds of gypsum to 2½ cubic feet of aggregate for the one-hour system.			2½[4]	2[4]
	8-1.3	¾″ by No. 16 gauge vertical cold-rolled channels, 16″ on center, with ⅜″ gypsum lath applied to one face and attached with sheet metal clips. Gypsum plaster each side mixed 1:2 by weight, gypsum to sand aggregate.				2[4]
	8-2.1	Studless with ½″ full-length plain gypsum lath and gypsum plaster each side. Plaster mixed 1:1 for scratch coat and 1:2 for brown coat, by weight, gypsum to sand aggregate.				2[4]
	8-2.2	Studless with ½″ full-length plain gypsum lath and perlite or vermiculite gypsum plaster each side.			2½[4]	2[4]
	8-2.3	Studless partition with ⅜″ rib metal lath installed vertically, adjacent edges tied 6″ on center with No. 18 gauge wire ties, gypsum plaster each side mixed 1:2 by weight, gypsum to sand aggregate.				2[4]
9. Solid Perlite and Portland Cement	9-1.1	Perlite mixed in the ratio of 3 cubic feet to 100 pounds of portland cement and machine applied to stud side of 1½″ mesh by No. 17 gauge paper-backed woven wire fabric lath wire-tied to 4″ deep steel trussed wire[9] studs 16″ on center. Wire ties of 18 gauge galvanized steel wire 6″ on center vertically.			3⅛[4]	
10. Solid Neat Wood Fibered Gypsum Plaster	10-1.1	¾″ by No. 16 gauge cold-rolled channels, 12″ on center with 2.5-pound flat metal lath applied to one face and tied with No. 18 gauge wire at 6″ spacing. Neat gypsum plaster applied each side.			2[4]	
11. Solid Gypsum Wallboard Partition	11-1.1	One full-length layer ½″ Type X gypsum wallboard[7] laminated to each side of 1″ full-length V-edge gypsum coreboard with approved laminating compound. Vertical joints of face layer and coreboard staggered at least 3″.			2[4]	
12. Hollow (Studless) Gypsum Wallboard Partition	12-1.1	One full-length layer of ⅝″ Type X gypsum wallboard[7] attached to both sides of wood or metal top and bottom runners laminated to each side of 1″ x 6″ full-length gypsum coreboard ribs spaced 24″ on center with approved laminating compound. Ribs centered at vertical joints of face plies and joints staggered 24″ in opposing faces. Ribs may be recessed 6″ from the top and bottom.				2¼[4]
12. Hollow (Studless) Gypsum Wallboard Partition	12-1.2	1″ regular gypsum V-edge full-length backing board attached to both sides of wood or metal top and bottom runners with nails or 1⅝″ drywall screws at 24″ on center. Minimum width of runners 1⅝″. Face layer of ½″ regular full-length gypsum wallboard laminated to outer faces of backing board with approved laminating compound.			4⅝[4]	
13. Noncombustible Studs—Interior Partition with Plaster Each Side	13-1.1	3¼″ by No. 18 gauge steel studs spaced 24″ on center. ⅝″ gypsum plaster on metal lath each side mixed 1:2 by weight, gypsum to sand aggregate.				4¾[4]
	13-1.2	3⅝″ No. 16 gauge approved nailable[10] studs spaced 24″ on center. ⅝″ neat gypsum wood fibered plaster each side over ⅜″ rib metal lath nailed to studs with 6d common nails, 8″ on center. Nails driven 1¼″ and bent over.			5⅝	
	13-1.3	4″ No. 18 gauge channel-shaped steel studs at 16″ on center. On each side approved resilient clips pressed onto stud flange at 16″ vertical spacing, ¼″ pencil rods snapped into or wire-tied onto outer loop of clips, metal lath wire-tied to pencil rods at 6″ intervals, 1″ perlite gypsum plaster, each side.		7⅝[4]		
	13-1.4	2½″ No. 18 gauge steel studs spaced 16″ on center. Wood fibered gypsum plaster mixed 1:1 by weight gypsum to sand aggregate applied on 3.4 pound metal lath wire tied to studs, each side. ¾″ plaster applied over each face, including finish coat.			4¼[4]	

TABLE NO. 43-B—RATED FIRE-RESISTIVE PERIODS FOR VARIOUS WALLS AND PARTITIONS[a][1]
—(Continued)

MATERIAL	ITEM NUMBER	CONSTRUCTION	MINIMUM FINISHED THICKNESS FACE-TO-FACE[2] (In Inches)			
			4 Hr.	3 Hr.	2 Hr.	1 Hr.
14. Wood Studs Interior Partition with Plaster Each Side	14-1.1[11][16]	2" x 4" wood studs 16" on center with ⁵/₈" gypsum plaster on metal lath. Lath attached by 4d common nails bent over or No. 14 gauge by 1¼" x ¾" crown width staples spaced 6" on center. Plaster mixed 1:1½ for scratch coat and 1:3 for brown coat, by weight, gypsum to sand aggregate.				5⅛
	14-1.2[11]	2" x 4" wood studs 16" on center with metal lath and ⁷/₈" neat wood fibered gypsum plaster each side. Lath attached by 6d common nails, 7" on center. Nails driven 1¼" and bent over.			5½[4]	
	14-1.3[11][16]	2" x 4" wood studs 16" on center with ³/₈" perforated or plain gypsum lath and ½" gypsum plaster each side. Lath nailed with 1⅛" by No. 13 gauge by ¹⁹/₆₄" head plasterboard blued nails, 4" on center. Plaster mixed 1:2 by weight, gypsum to sand aggregate.				5¼
	14-1.4[11][16]	2" x 4" wood studs 16" on center with ³/₈" Type X gypsum lath and ½" gypsum plaster each side. Lath nailed with 1⅛" by No. 13 gauge by ¹⁹/₆₄" head plasterboard blued nails, 5" on center. Plaster mixed 1:2 by weight, gypsum to sand aggregate.				5¼
15. Noncombustible Studs— Interior Partition with Gypsum Wallboard Each Side	15-1.1	No. 25 gauge channel-shaped studs 24" on center with one full-length layer of ⁵/₈" Type X gypsum wallboard[7] applied vertically attached with 1" long No. 6 drywall screws to each stud. Screws are 8" on center around the perimeter and 12" on center on the intermediate stud. The wallboard may be applied horizontally when attached to 3⅝" studs and the horizontal joints are staggered with those on the opposite side. Screws for the horizontal application shall be 8" on center at vertical edges and 12" on center at intermediate studs.				2⅞[4]
	15-1.2	No. 25 gauge channel-shaped studs 24" on center with two full-length layers of ½" Type X gypsum wallboard[7] applied vertically each side. First layer attached with 1" long, No. 6 drywall screws, 8" on center around the perimeter and 12" on center on the intermediate stud. Second layer applied with vertical joints offset one stud space from first layer using 1⅝" long, No. 6 drywall screws spaced 9" on center along vertical joints, 12" on center at intermediate studs and 24" on center along top and bottom runners.			3⅝[4]	
15. Noncombustible Studs— Interior Partition with Gypsum Wallboard Each Side	15-1.3	No. 16 gauge approved nailable metal studs[10] 24" on center with full-length ⁵/₈" Type X gypsum wallboard[7] applied vertically and nailed 7" on center with 6d cement-coated common nails. Approved metal fastener grips used with nails at vertical butt joints along studs.				4⅞
16. Wood Studs— Interior Partition with Gypsum Wallboard Each Side	16-1.1[11][16]	2" x 4" wood studs 16" on center with two layers of ³/₈" regular gypsum wallboard[7] each side, 4d cooler[12] or wallboard[12] nails at 8" on center first layer, 5d cooler[12] or wallboard[12] nails at 8" on center second layer with laminating compound between layers. Joints staggered. First layer applied full length vertically, second layer applied horizontally or vertically.				5
	16-1.2[11][16]	2" x 4" wood studs 16" on center with two layers ½" regular gypsum wallboard[7] applied vertically or horizontally each side, joints staggered. Nail base layer with 5d cooler[12] or wallboard[12] nails at 8" on center, face layer with 8d cooler[12] or wallboard[12] nails at 8" on center.				5½
	16-1.3[11][16]	2" x 4" wood studs 24" on center with ⁵/₈" Type X gypsum wallboard[7] applied vertically or horizontally nailed with 6d cooler[12] or wallboard[12] nails at 7" on center with end joints on nailing members.				4¾
	16-1.4[11]	2" x 4" fire-retardant-treated wood studs spaced 24" on center with one layer of ⁵/₈" thick Type X gypsum wallboard[7] applied with face paper grain (long dimension) parallel to studs. Wallboard attached with 6d cooler[12] or wallboard[12] nails at 7" on center.				4¾[4]
16. Wood Studs— Interior Partition With Gypsum Wallboard Each Side	16-1.5[11][16]	2" x 4" wood studs 16" on center with two layers ⁵/₈" Type X gypsum wallboard[7] each side. Base layers applied vertically and nailed with 6d cooler[12] or wallboard[12] nails at 9" on center. Face layer applied vertically or horizontally and nailed with 8d cooler[12] or wallboard[12] nails at 7" on center. For nail-adhesive application, base layers are nailed 6" on center. Face layers applied with coating of approved wallboard adhesive and nailed 12" on center.			6	
	16-1.6[11]	2" x 3" fire-retardant-treated wood studs spaced 24" on center with one layer of ⁵/₈" thick Type X gypsum wallboard[7] applied with face paper grain (long dimension) at right angles to studs. Wallboard attached with 6d cement-coated box nails spaced 7" on center.				3⅝[4]

TABLE NO. 43-B—RATED FIRE-RESISTIVE PERIODS FOR VARIOUS WALLS AND PARTITIONS[a][1] —(Continued)

MATERIAL	ITEM NUMBER	CONSTRUCTION	MINIMUM FINISHED THICKNESS FACE-TO-FACE[2] (In inches)			
			4 Hr.	3 Hr.	2 Hr.	1 Hr.
17. Exterior or Interior Walls	17-1.2[11][16]	2″ x 4″ wood studs 16″ on center with metal lath and ³/₄″ exterior cement plaster on each side. Lath attached with 6d common nails 7″ on center driven to 1″ minimum penetration and bent over. Plaster mix 1:4 for scratch coat and 1:5 for brown coat, by volume, cement to sand.				5³/₈
	17-1.3[11][16]	2″ x 4″ wood studs 16″ on center with ⁷/₈″ exterior cement plaster (measured from the face of studs) on the exterior surface with interior surface treatment as required for interior wood stud partitions in this table. Plaster mix 1:4 for scratch coat and 1:5 for brown coat, by volume, cement to sand.				Varies
	17-1.4	3⁵/₈″ No. 16 gauge noncombustible studs 16″ on center with ⁷/₈″ exterior cement plaster (measured from the face of the studs) on the exterior surface with interior surface treatment as required for interior, nonbearing, noncombustible stud partitions in this table. Plaster mix 1:4 for scratch coat and 1:5 for brown coat, by volume, cement to sand.				Varies[4]
	17-1.5[16]	2¹/₄″ x 3³/₄″ clay face brick with cored holes over ¹/₂″ gypsum sheathing on exterior surface of 2″ x 4″ wood studs at 16″ on center and two layers ⁵/₈″ Type X gypsum wallboard[7] on interior surface. Sheathing placed horizontally or vertically with vertical joints over studs nailed 6″ on center with 1³/₄″ by No. 11 gauge by ⁷/₁₆″ head galvanized nails. Inner layer of wallboard placed horizontally or vertically and nailed 8″ on center with 6d cooler[12] or wallboard[12] nails. Outer layer of wallboard placed horizontally or vertically and nailed 8″ on center with 8d cooler[12] or wallboard[12] nails. All joints staggered with vertical joints over studs. Outer layer joints taped and finished with compound. Nailheads covered with joint compound. No. 20 gauge corrugated galvanized steel wall ties ³/₄″ by 6⁵/₈″ attached to each stud with two 8d cooler[12] or wallboard[12] nails every sixth course of bricks.			10	
17. Exterior or Interior Walls	17-1.6[11][16]	2″ x 6″ fire-retardant-treated wood studs 16″ on center. Interior face has two layers of ⁵/₈″ Type X gypsum wallboard[7] with the base layer placed vertically and attached with 6d box nails 12″ on center. The face layer is placed horizontally and attached with 8d box nails 8″ on center at joints and 12″ on center elsewhere. The exterior face has a base layer of ⁵/₈″ Type X gypsum wallboard placed vertically with 6d box nails 8″ on center at joints and 12″ on center elsewhere. An approved building paper is next applied, followed by self-furred exterior lath attached with 2¹/₂″, No. 12 gauge galvanized roofing nails with a ³/₈″ diameter head and spaced 6″ on center along each stud. Exterior cement plaster consisting of a ¹/₂″ brown coat is then applied. The scratch coat is mixed in the proportion of 1:3 by weight, cement to sand with 10 pounds of hydrated lime and 3 pounds of approved additives or admixtures per sack of cement. The brown coat is mixed in the proportion of 1:4 by weight, cement to sand with the same amounts of hydrated lime and approved additives or admixtures used in the scratch coat.			8¹/₄	
	17-1.7[11][16]	2″ x 6″ wood studs 16″ on center. The exterior face has a layer of ⁵/₈″ Type X gypsum wallboard[7] placed vertically with 6d box nails 8″ on center at joints and 12″ on center elsewhere. An approved building paper is next applied, followed by i″ by No. 18 gauge self-furred exterior lath attached with 8d by 2¹/₂″ long galvanized roofing nails spaced 6″ on center along each stud. Exterior cement plaster consisting of a ¹/₂″ scratch coat, a bonding agent and a ¹/₂″ brown coat and a finish coat is then applied. The scratch coat is mixed in the proportion of 1:3 by weight, cement to sand with 10 pounds of hydrated lime and 3 pounds of approved additives or admixtures per sack of cement. The brown coat is mixed in the proportion of 1:4 by weight, cement to sand with the same amounts of hydrated lime and approved additives or admixtures used in the scratch coat. The interior is covered with ³/₈″ gypsum lath with 1″ hexagonal mesh of No. 20 gauge woven wire lath furred out ⁵/₁₆″ and 1″ perlite or vermiculite gypsum plaster. Lath nailed with 1¹/₈″ by No. 13 gauge by ¹⁹/₆₄″ head plasterboard blued nails spaced 5″ on center. Mesh attached by 1³/₄″ by No. 12 gauge by ³/₈″ head nails with ³/₈″ furrings, spaced 8″ on center. The plaster mix shall not exceed 100 pounds of gypsum to 2¹/₂ cubic feet of aggregate.			8³/₈	

(Continued)

TABLE NO. 43-B—RATED FIRE-RESISTIVE PERIODS FOR VARIOUS WALLS AND PARTITIONS[a 1] —(Continued)

MATERIAL	ITEM NUMBER	CONSTRUCTION	MINIMUM FINISHED THICKNESS FACE-TO-FACE[2] (In Inches)			
			4 Hr.	3 Hr.	2 Hr.	1 Hr.
17. Exterior or Interior Walls	17-1.8[11 16]	2″ x 6″ wood studs 16″ on center. The exterior face has a layer of ⁵⁄₈″ Type X gypsum wallboard[7] placed vertically with 6d box nails 8″ on center at joints and 12″ on center elsewhere. An approved building paper is next applied, followed by 1¹⁄₂″ by No. 17 gauge self-furred exterior lath attached with 8d by 2¹⁄₂″ long galvanized roofing nails spaced 6″ on center along each stud. Exterior cement plaster consisting of a ¹⁄₂″ scratch coat, and a ¹⁄₂″ brown coat is then applied. The plaster may be placed by machine. The scratch coat is mixed in the proportion of 1:4 by weight, plastic cement to sand. The brown coat is mixed in the proportion of 1:5 by weight, plastic cement to sand. The interior is covered with ³⁄₈″ gypsum lath with 1″ hexagonal mesh of No. 20 gauge woven wire lath furred out ⁵⁄₁₆″ and 1″ perlite or vermiculite gypsum plaster. Lath nailed with 1¹⁄₈″ by No. 13 gauge by ¹⁹⁄₆₄″ head plasterboard blued nails spaced 5″ on center. Mesh attached by 1³⁄₄″ by No. 12 gauge by ³⁄₈″ head nails with ³⁄₈″ furrings, spaced 8″ on center. The plaster mix shall not exceed 100 pounds of gypsum to 2¹⁄₂ cubic feet of aggregate.			8³⁄₈	
	17-1.9	4″ No. 18 gauge, nonload-bearing metal studs, 16″ on center, with 1″ portland cement lime plaster (measured from the back side of the 3.4# expanded metal lath) on the exterior surface. Interior surface to be covered with 1″ of gypsum plaster on 3.4# expanded metal lath proportioned by weight—1:2 for scratch coat, 1:3 for brown, gypsum to sand. Lath on one side of the partition fastened to ¹⁄₄″ diameter pencil rods supported by No. 20 gauge metal clips, located 16″ on center vertically, on each stud. 3″ thick mineral fiber insulating batts friction fitted between the studs.			6¹⁄₂[4]	

[a]Generic fire-resistance ratings (those not designated by company code letter) as listed in the Fire Resistance Design Manual, Eleventh Edition, dated October, 1984, as published by the Gypsum Association, may be accepted as if herein listed.

[1]Staples with equivalent holding power and penetration may be used as alternate fasteners to nails for attachment to wood framing.

[2]Thickness shown for brick and clay tile are nominal thicknesses unless plastered, in which case thicknesses are net. Thickness shown for concrete masonry is equivalent thickness defined as the average thickness of solid material in the wall and is represented by the formula:

$$T_E = \frac{V_n}{L \times H}$$

WHERE:

T_E = Equivalent thickness, in inches

V_n = Net volume (gross volume less volume of voids), in cubic inches

L = Length of block, in inches

H = Height of block, in inches

Thickness includes plaster, lath and gypsum wallboard, where mentioned, and grout when all cells are solid grouted.

[3]Single-wythe brick.

[4]Shall be used for nonbearing purposes only.

[5]Hollow brick units 4-inch by 8-inch by 12-inch nominal with two interior cells having a 1¹⁄₂-inch web thickness between cells and 1³⁄₄-inch-thick face shells.

[6]Rowlock design employs clay brick with all or part of bricks laid on edge with the bond broken vertically.

[7]For all of the construction with gypsum wallboard described in Table No. 43-B, gypsum base for veneer plaster of the same size, thickness and core type may be substituted for gypsum wallboard, provided attachment is identical to that specified for the wallboard and the joints on the face layer are reinforced and the entire surface is covered with a minimum of ¹⁄₁₆-inch gypsum veneer plaster. The gypsum base for veneer plaster and the veneer plaster shall comply with U.B.C. Standard No. 47-15.

[8]See also Footnote 2. The equivalent thickness may include the thickness of portland cement plaster or 1.5 times the thickness of gypsum plaster applied in accordance with the requirements of Chapter 47 of the code.

[9]Studs are welded truss wire studs with No. 7 gauge flange wire and No. 7 gauge truss wires.

[10]Nailable metal studs consist of two channel studs spot welded back-to-back with a crimped web forming a nailing groove.

[11]Plywood may be installed between the fire protection and the wood studs on either the interior or exterior side of the wood frame assemblies in this table, provided the length of the fasteners used to attach the fire protection are increased by an amount at least equal to the thickness of the plywood.

[12]For properties of cooler or wallboard nails, see U.B.C. Standard No. 25-17, Table No. 25-17-I.

[13]The fire-resistive time period for concrete masonry unit construction meeting the equivalent thicknesses required for a two-hour fire-resistive rating in Items 30, 31, 32 and 33, and having a minimum thickness of not less than 7⁵⁄₈ inches is four-hour when cores which are not grouted are filled with silicone-treated perlite loose-fill insulation conforming to U.B.C. Standard No. 43-10; vermiculite loose-fill insulation conforming to U.B.C. Standard No. 43-11; or expanded clay, shale or slate lightweight aggregate conforming to U.B.C. Standard No. 26-3; sand or slag having a maximum particle size of ³⁄₈ inch.

[14]For determining equivalent thickness of concrete masonry units made from unblended aggregates, see Footnote 2. The equivalent thickness of units composed of blends of two or more aggregate categories shall be determined by interpolating between the equivalent thickness values specified in Table No. 43-B in proportion to the percent by volume of each aggregate.

The equivalent thickness required to provide a desired fire-resistive time period for concrete masonry composed of units manufactured with fine aggregates passing a No. 4 sieve listed in Item 5-1.4 of Table No. 43-B blended with aggregates listed in Item 5-1.1 or 5-1.2 shall be determined by interpolating between the equivalent thickness values specified in Table No. 43-B in proportion to the percent by volume of each aggregate as follows:

1. $ET_{required} = ET_{1.4} \times V_{1.4} + ET_{1.1} \times V_{1.1}$

2. $ET_{required} = ET_{1.4} \times V_{1.4} + ET_{1.2} \times V_{1.2}$

The required equivalent thickness of concrete masonry units manufactured with aggregates listed in Items 5-1.1, 5-1.2 and 5-1.4 of Table No. 43-B shall be determined as follows:

3. $ET_{required} = ET_{1.4} \times V_{1.4} + ET_{1.2} \times (V_{1.1} + V_{1.2})$

WHERE:

$ET_{1.1}, ET_{1.2}, ET_{1.4}$ = specified equivalent thickness for Items 5-1.1, 5-1.2 and 5-1.4 of Table No. 43-B.

$V_{1.1}, V_{1.2}, V_{1.4}$ = volume of aggregates expressed as a percentage of the total aggregate volume for Item 5-1.1, 5-1.2 or 5-1.4 of Table No. 43-B.

[15]Concrete walls shall be reinforced with horizontal and vertical temperature reinforcement as required by Subsections 2614 (d) 2 and 3.

[16]The design stress of studs shall be reduced to 78 percent of allowable F'_c with the maximum not greater than 78 percent of the calculated stress with studs having a slenderness ratio $\dfrac{le}{d}$ of 33.

TABLE NO. 43 C—MINIMUM PROTECTION FOR FLOOR AND ROOF SYSTEMS[a 1]

FLOOR OR ROOF CONSTRUCTION	ITEM NUMBER	CEILING CONSTRUCTION	THICKNESS OF FLOOR OR ROOF SLAB (In Inches)				MINIMUM THICKNESS OF CEILING (In Inches)			
			4 Hr.	3 Hr.	2 Hr.	1 Hr.	4 Hr.	3 Hr.	2 Hr.	1 Hr.
5. Reinforced Concrete Joists	5-1.1	Slab with suspended ceiling of vermiculite gypsum plaster over metal lath attached to $^3/_4$" cold-rolled channels spaced 12" on center. Ceiling located 6" minimum below joists.	3	2			1	$^3/_4$		
5. Reinforced Concrete Joists	5-2.1	$^3/_8$ Type X gypsum wallboard[3] attached to No. 25 gauge by $^7/_8$" deep by $2^5/_8$" hat-shaped galvanized steel channels with 1" long No. 6 screws. The channels are spaced 24" on center, span 35" and are supported along their length at 35" intervals by No. 21 gauge galvanized steel flat strap hangers having formed edges which engage the lips of the channel. The strap hangers are attached to the side of the concrete joists with $^5/_{32}$" by $1^1/_4$" long powder-driven fasteners. The wallboard is installed with the long dimension perpendicular to the channels. All end joints occur on channels and supplementary channels and supplementary channels are installed parallel to the main channels, 12" each side, at end joint occurrences. The finish ceiling is located approximately 12" below the soffit of the floor slab.			$2^1/_2$				$^5/_8$	

TABLE NO. 43 C—MINIMUM PROTECTION FOR FLOOR AND ROOF SYSTEMS[a][1]—(Continued)

FLOOR OR ROOF CONSTRUCTION	ITEM NUMBER	CEILING CONSTRUCTION	THICKNESS OF FLOOR OR ROOF SLAB (In Inches)				MINIMUM THICKNESS OF CEILING (In Inches)			
			4 Hr.	3 Hr.	2 Hr.	1 Hr.	4 Hr.	3 Hr.	2 Hr.	1 Hr.
6. Steel Joists Constructed with a Poured Reinforced Concrete Slab on Metal Lath Forms or Steel Form Units[4]	6-1.1	Gypsum plaster on metal lath attached to the bottom cord with single No. 16 gauge or doubled No. 18 gauge wire ties spaced 6" on center. Plaster mixed 1:2 for scratch coat, 1:3 for brown coat, by weight, gypsum to sand aggregate for two-hour system. For three-hour system plaster is neat.		$2^1/_2$	$2^1/_4$			$3/_4$	$5/_8$	
	6-2.1	Vermiculite gypsum plaster on metal lath attached to the bottom chord with single No, 16 gauge or doubled No. 18 gauge wire ties 6" on center.		2				$5/_8$		
	6-3.1	Portland cement plaster over metal lath attached to the bottom chord of joists with single No. 16 gauge or doubled No. 18 gauge wire ties spaced 6" on center. Plaster mixed 1:2 for scratch coat, 1:3 for brown coat for one-hour system and 1:1 for scratch coat, $1:1^1/_2$ for brown coat for two-hour system, by weight, cement to sand.			$2^1/_4$	2				$5/_8$[5]
6. Steel Joists Constructed with a Poured Reinforced Concrete Slab on Metal Lath Forms or Steel Form Units[4]	6-4.1	Ceiling of $5/_8$" Type X wallboard[3] attached to $7/_8$" deep by $2^5/_8$" by No. 25 gauge hat-shaped furring channels 12" on center with 1" long No. 6 wallboard screws at 8" on center. Channels wire tied to bottom chord of joists with doubled No. 18 gauge wire or suspended below joists on wire hangers.[9]			$2^1/_2$				$5/_8$	
	6-5.1	Wood-fibered gypsum plaster mixed 1:1 by weight gypsum to sand aggregate applied over metal lath. Lath tied 6" on center to $3/_4$" channels spaced $13^1/_2$" on center. Channels secured to joists at each intersection with two strands of No. 18 gauge galvanized wire.			$2^1/_2$				$3/_4$	
7. Reinforced Concrete Slab and Joists with Hollow Clay Tile Fillers Laid End to End in Rows $2^1/_2$" or More Apart; Reinforcement Placed Between Rows and Concrete Cast Around and Over Tile	7-1.1	$5/_8$" gypsum plaster on bottom of floor or roof construction.			8^{10}				$5/_8$	
	7-1.2	None.				$5^1/_2$[11]				
8. Steel Joists Constructed with a Reinforced Concrete Slab on Top Poured on a $1/_2$" Deep Steel Deck.	8-1.1	Vermiculite gypsum plaster on metal lath attached to $3/_4$" cold-rolled channels with No. 18 gauge wire ties spaced 6" on center.	$2^1/_2$[12]					$3/_4$		

TABLE NO. 43 C—MINIMUM PROTECTION FOR FLOOR AND ROOF SYSTEMS[a 1]—(Continued)

FLOOR OR ROOF CONSTRUCTION	ITEM NUMBER	CEILING CONSTRUCTION	THICKNESS OF FLOOR OR ROOF SLAB (In Inches)				MINIMUM THICKNESS OF CEILING (In Inches)			
			4 Hr.	3 Hr.	2 Hr.	1 Hr.	4 Hr.	3 Hr.	2 Hr.	1 Hr.
9. 3″ Deep Cellular Steel Deck with Concrete Slab on Top. Slab Thickness Measured to Top of Cells.	9-1.1	Suspended ceiling of vermiculite gypsum plaster base coat and vermiculite acoustical plaster on metal lath attached at 6″ intervals to 3/4″ cold-rolled channels spaced 12″ on center and secured to 1 1/2″ cold-rolled channels spaced 36″ on center with No. 16 gauge wire. 1 1/2″ channels supported by No. 8 gauge wire hangers at 36″ on center. Beams within envelope and with a 2 1/2″ air space between beam soffit and lath have a 4-hour rating.	2 1/2				1 1/8[13]			
10. 1 1/2″ Deep Steel Roof Deck on Steel Framing. Insulation Board, 30 lbs. per Cubic Foot Density, Composed of Wood Fibers with Cement Binders of Thickness Shown Bonded to Deck with Unfilled Asphalt Adhesive. Covered with a Fire-retardant Roof Covering	10-1.1	Ceiling of gypsum plaster on metal lath. Lath attached to 3/4″ furring channels with No. 18 gauge wire ties spaced 6″ on center. 3/4″ channel saddled-tied to 2″ channels with doubled No. 16 gauge wire ties. 2″ channels spaced 36″ on center suspended 2″ below steel framing and saddle-tied with No. 8 gauge wire. Plaster mixed 1:2 by weight, gypsum to sand aggregate.			1 7/8	1			3/4[7]	3/4[7]
11. 1 1/2″ Deep Steel Roof Deck on Steel Framing Wood Fiber Insulation Board, 17.5 lbs. per Cubic Foot Density on Top Applied Over a 15-lb. Asphalt Saturated Felt. Fire-retardant Roof Covering.	11-1.1	Ceiling of gypsum plaster on metal lath. Lath attached to 3/4″ furring channels with No. 18 gauge wire ties spaced 6″ on center. 3/4″ channels saddle tied to 2″ channels with doubled No. 16 gauge wire ties. 2″ channels spaced 36″ on center suspended 2″ below steel framing and saddle tied with No. 8 gauge wire. Plaster mixed 1:2 for scratch coat and 1:3 for brown coat, by weight, gypsum to sand aggregate for one-hour system. For two-hour system plaster mix is 1:2 by weight, gypsum to sand aggregate.			1 1/2	1			7/8[7]	3/4[7]
12. 1 1/2″ Deep Steel Roof Deck on Steel Framing Insulation of Rigid Board Consisting of Expanded Perlite and Fibers Impregnated With Integral Asphalt Waterproofing; Density 9 to 12 Lbs./Cu. Ft. Secured to Metal Roof Deck by 1/2″ Wide Ribbons of Waterproof, Cold-process Liquid Adhesive Spaced 6″ Apart. Steel Joist or Light Steel Construction with Metal Roof Deck, Insulation, and Built-up Fire-retardant Roof Covering.	12-1.1	Gypsum-vermiculite plaster on metal lath wire tied at 6″ intervals to 3/4″ furring channels spaced 12″ on center and wire tied to 2″ runner channels spaced 32″ on center. Runners wire tied to bottom chord of steel joists.			1				7/8	

(Continued)

TABLE NO. 43-C—MINIMUM PROTECTION FOR FLOOR AND ROOF SYSTEMS[a][1]—(Continued)

FLOOR OR ROOF CONSTRUCTION	ITEM NUMBER	CEILING CONSTRUCTION	THICKNESS OF FLOOR OR ROOF SLAB (In Inches)				MINIMUM THICKNESS OF CEILING (In Inches)			
			4 Hr.	3 Hr.	2 Hr.	1 Hr.	4 Hr.	3 Hr.	2 Hr.	1 Hr.
13. Double Wood Floor Over Wood Joists Spaced 16" on Center[14][15]	13-1.1	Gypsum plaster over 3/8" Type X gypsum lath. Lath initially applied with not less than four 1 1/8" by No. 13 gauge by 19/64" head plasterboard blued nails per bearing. Continuous stripping over lath along all joist lines. Stripping consists of 3" wide strips of metal lath attached by 1 1/2" by No. 11 gauge by 1/2" head roofing nails spaced 6" on center. Alternate stripping consists of 3" wide .049" diameter wire stripping weighing one pound per sq. yd. and attached by No. 16 gauge by 1 1/2" by 3/4" crown width staples, spaced 4" on center. Where alternate stripping is used the lath nailing may consist of two nails at each end and one nail at each intermediate bearing. Plaster mixed 1:2 by weight, gypsum to sand aggregate.								7/8
	13-1.2 13-1.2	Portland cement or gypsum plaster on metal lath. Lath fastened with 1 1/2" by No. 11 gauge by 7/16" head barbed shank roofing nails spaced 5" on center. Plaster mixed 1:2 for scratch coat and 1:3 for brown coat, by weight, cement to sand aggregate.								5/8
13. Double Wood Floor Over Wood Joists Spaced 16" On Center[14][15]	13-1.3	Perlite or vermiculite gypsum plaster on metal lath secured to joists with 1 1/2" by No. 11 gauge by 7/16" head barbed shank roofing nails spaced 5" on center.								5/8
	13-1.4	1/2" Type X gypsum wallboard[3] nailed to joists with 5d cooler[12] or wallboard[12] nails at 6" on center. End joints of wallboard centered on joists.								1/2
14. Plywood Stressed Skin Panels Consisting of 5/8" Thick Interior C-D (Exterior Glue) Top Stressed Skin on 2" by 6" Nominal (Minimum) Stringers. Adjacent Panel Edges Joined with 8d Common Wire Nails Spaced 6" on Center. Stringers Spaced 12" Maximum on Center.	14-1.1	1/2" thick wood fiberboard weighing 15 to 18 lbs. per cu. ft. installed with long dimension parallel to stringers or 3/8" C-D (exterior glue) plywood glued and/or nailed to stringers. Nailing to be with 5d cooler[12] or wallboard[12] nails at 12" on center. Second layer of 1/2" Type X gypsum wallboard[3] applied with long dimension perpendicular to joists and attached with 8d cooler[12] or wallboard[12] nails at 6" on center at end joints and 8" on center elsewhere. Wallboard joints staggered with respect to fiberboard joints.								1

(Continued)

TABLE NO. 43 C—MINIMUM PROTECTION FOR FLOOR AND ROOF SYSTEMS[a][1]—(Continued)

FLOOR OR ROOF CONSTRUCTION	ITEM NUMBER	CEILING CONSTRUCTION	THICKNESS OF FLOOR OR ROOF SLAB (In Inches)				MINIMUM THICKNESS OF CEILING (In Inches)			
			4 Hr.	3 Hr.	2 Hr.	1 Hr.	4 Hr.	3 Hr.	2 Hr.	1 Hr.
17. Perlite Concrete Slab Proportioned 1:6 (Portland Cement to Perlite Aggregate) on a $^9/_{16}$" Deep Steel Deck Supported by Steel Joists 4' on Center. Fire-retardant roof covering on top.	17-1.1	Perlite gypsum plaster on metal lath wire tied to $^3/_4$" furring channels attached with No. 16 gauge wire ties to lower chord of joists.		2[17]	2[17]			$^7/_8$	$^3/_4$	
19. Floor and Beam Construction Consisting of 3" Deep Cellular Steel Floor Units Mounted on Steel Members with 1:4 (Proportion of Portland Cement to Perlite Aggregate) Perlite-concrete floor slab on top.	19-1.1	Suspended envelope ceiling of perlite gypsum plaster on metal lath attached to $^3/_4$" cold-rolled channels, secured to $1^1/_2$" cold-rolled channels spaced 42" on center supported by No. 6 wire 36" on center. Beams in envelope with 3" minimum air space between beam soffit and lath have a 4-hour rating.	2[17]				1"			
21. Wood joist, floor trusses and roof trusses spaced 24" o.c. with $^1/_2$" plywood with exterior glue applied at right angles to top of joist or truss with 8d nails. The plywood thickness shall not be less than $^1/_2$", nor less than required by Chapter 25.	21-1.1	Base layer $^5/_8$" Type X gypsum wallboard applied at right angles to joist or truss 24" o.c. with $1^1/_4$" Type S or Type W drywall screws 24" o.c. Face layer $^5/_8$" Type X gypsum wallboard or veneer base applied at right angles to joist or truss through base layer with $1^7/_8$" Type S or Type W drywall screws 12" o.c. at joints and intermediate joist or truss. Face layer joints offset 24" from base layer joints, $1^1/_2$" Type G drywall screws placed 2" back on either side of face layer end joints, 12" o.c.					Varies			$1^1/_4$

FOOTNOTES FOR TABLE NO. 43-C

[a]Generic fire-resistance ratings (those not designated by company code letter) as listed in the Fire Resistance Design Manual, Eleventh Edition, dated October, 1984, as published by the Gypsum Association, may be accepted as if herein listed.

[1]Staples with equivalent holding power and penetration may be used as alternate fasteners to nails for attachment to wood framing.

[2]The thickness may be reduced to 3 inches where limestone aggregate is used.

[3]For all of the construction with gypsum wallboard described in Table No. 43-C, gypsum base for veneer plaster of the same size, thickness and core type may be substituted for gypsum wallboard, provided attachment is identical to that specified for the wallboard and the joints on the face layer are reinforced and the entire surface is covered with a minimum of $^1/_{16}$-inch gypsum veneer plaster. The gypsum base for veneer plaster and the veneer plaster shall comply with U.B.C. Standard No. 47-15.

[4]Slab thickness over steel joists measured at the joists for metal lath form and at the top of the form for steel form units.

[5]Portland cement plaster with 15 pounds of hydrated lime and 3 pounds of approved additives or admixtures per bag of cement.

[6]One-inch by No. 20 gauge hexagonal wire mesh installed below lath and tied to each furring channel at joints between lath.

[7]Furring channels spaced 12 inches on center.

[8]No. 14 gauge wires spaced 11.3 inches on center or 10 inches on center (for channel spacing of 16 inches and 12 inches, respectively) installed below lath sheets in a diagonal pattern. Wires tied to furring channels or clips at lath edges.

[9]Gypsum wallboard ceilings attached to steel framing may be suspended with $1^1/_2$-inch cold-formed carrying channels spaced 48 inches on center which are suspended with No. 8 SWG galvanized wire hangers spaced 48 inches on center. Cross-furring channels are tied to the carrying channels with No. 18 SWG galvanized wire (double strand) and spaced as required for direct attachment to the framing. This alternative is also applicable to those assemblies recognized under Footnote a.

[10]Six-inch hollow clay tile with 2-inch concrete slab above.

[11]Four-inch hollow clay tile with $1^1/_2$-inch concrete slab above.

[12]Thickness measured to bottom of steel form units.

[13]Five-eighths inch of vermiculite gypsum plaster plus $^1/_2$ inch of approved vermiculite acoustical plastic.

[14]Double wood floor may be either of the following:
(a) Subfloor of 1-inch nominal boarding, a layer of asbestos paper weighing not less than 14 pounds per 100 square feet and a layer of 1-inch nominal tongue-and-groove finish flooring; or
(b) Subfloor of 1-inch nominal tongue-and-groove boarding or $^{15}/_{32}$-inch interior-type plywood with exterior glue and a layer of 1-inch nominal tongue-and-groove finish flooring or $^{19}/_{32}$-inch interior-type plywood finish flooring or a layer of Type I Grade M-1 particleboard not less than $^5/_8$ inch thick.

[15]The ceiling may be omitted over unusable space, and flooring may be omitted where unusable space occurs above.

[16]For properties of cooler or wallboard nails, see U.B.C. Standard No. 25-17, Table 25-17-1.

[17]Thickness measured on top of steel deck unit.

[18]When the slab is in an unrestrained condition, minimum reinforcement cover shall not be less than $1^1/_4$ inches for three-hour; 1 inch for two-hour (siliceous aggregate only); and $^3/_4$ inch for all other restrained or unrestrained conditions.

Part IX

WALL AND CEILING COVERINGS

Chapter 47

INSTALLATION OF WALL AND CEILING COVERINGS

Scope

Sec. 4701. (a) **General.** The installation of lath, plaster and gypsum board shall be done in a manner and with materials as specified in this chapter and, when required for fire-resistive construction, also shall conform with the provisions of Chapter 43.

Other approved wall or ceiling coverings may be installed in accordance with the recommendations of the manufacturer and the conditions of approval.

(b) **Inspection.** No lath or gypsum board or their attachments shall be covered or finished until it has been inspected and approved by the building official in accordance with Section 305 (e).

(c) **Tests.** The building official may require tests to be made in accordance with approved standards to determine compliance with the provisions of this chapter, provided the permit holder has been notified 24 hours in advance of the time of making such tests.

The testing of gypsum and gypsum products shall conform with U.B.C. Standard No. 47-17.

(d) **Definitions.** For purposes of this chapter, certain terms are defined as follows:

CORNER BEAD is a rigid formed unit or shape used at projecting or external angles to define and reinforce the corners of interior surfaces.

CORNERITE is a shaped reinforcing unit of expanded metal or wire fabric used for angle reinforcing and having minimum outstanding legs of not less than 2 inches.

CORROSION-RESISTANT MATERIALS are materials that are inherently rust resistant or materials to which an approved rust-resistive coating has been applied either before or after forming or fabrication.

EXTERIOR SURFACES are weather-exposed surfaces as defined in Section 424.

EXTERNAL CORNER REINFORCEMENT is a shaped reinforcing unit for external corner reinforcement for portland cement plaster formed to ensure mechanical bond and a solid plaster corner.

INTERIOR SURFACES are surfaces other than weather-exposed surfaces.

MOIST CURING is any method employed to retain sufficient moisture for hydration of portland cement plaster.

PORTLAND CEMENT PLASTER is a mixture of portland cement or portland cement and lime and aggregate and other approved materials as specified in this code.

STEEL STUDS, LOAD-BEARING AND NONLOAD-BEARING, are prefabricated channel shapes, welded wire or combination wire and steel angle types, galvanized or coated with rust-resistive material.

STRIPPING is flat reinforcing units of expanded metal or wire fabric or other materials not less than 3 inches wide to be installed as required over joints of gypsum lath.

TIE WIRE is wire for securing together metal framing or supports, for tying metal and wire fabric lath and gypsum lath and wallboard together and for securing accessories.

WIRE BACKING is horizontal strands of tautened wire attached to surfaces of vertical wood supports which, when covered with building paper, provide a backing for portland cement plaster.

(e) **Suspended Acoustical Ceiling Systems.** Suspended acoustical ceiling systems shall be installed in accordance with U.B.C. Standard No. 47-18.

(f) **Gypsum Veneer Plaster Systems.** Gypsum veneer base and gypsum veneer plaster shall be installed in accordance with U.B.C. Standard No. 47-19.

Materials

Sec. 4702. Lathing, plastering, wallboard materials, ceiling suspension systems and plywood paneling shall conform to the applicable standards listed in Chapter 60.

Vertical Assemblies

Sec. 4703. (a) **General.** In addition to the requirements of this section, vertical assemblies of plaster or gypsum board shall be designed to resist the loads specified in Chapter 23 of this code. For wood framing, see Chapter 25. For metal framing, see Chapter 27.

EXCEPTION: Wood-framed assemblies meeting the requirements of Section 2517 need not be designed.

(b) **Wood Framing.** Wood supports for lath or gypsum board shall be not less than 2 inches nominal in least dimension. Wood stripping or furring shall be not less than 2 inches nominal thickness in the least dimension except that furring strips not less than 1-inch by 2-inch nominal dimension may be used over solid backing.

(c) **Studless Partitions.** The minimum thickness of vertically erected studless solid plaster partitions of 3/8-inch and 3/4-inch rib metal lath or 1/2-inch-thick long length gypsum lath and gypsum board partitions shall be 2 inches. The installation of metal lath used in studless partitions shall conform with the provisions of U.B.C. Standard No. 47-4.

Horizontal Assemblies

Sec. 4704. (a) **General.** In addition to the requirements of this section, supports for horizontal assemblies of plaster or gypsum board shall be designed to support all loads as specified in Chapter 23 of this code.

EXCEPTION: Wood-framed assemblies meeting the requirements of Section 2517 need not be designed.

(b) **Wood Framing.** Wood stripping or suspended wood systems, where used, shall be not less than 2 inches nominal thickness in the least dimension, except that furring strips not less than 1-inch by 2-inch nominal dimension may be used over solid backing.

(c) **Hangers.** Hangers for suspended ceilings shall be not less than the sizes set forth in Table No. 47-A, fastened to or embedded in the structural framing, masonry or concrete.

Hangers shall be saddle-tied around main runners to develop the full strength of the hangers. Lower ends of flat hangers shall be bolted with 3/8-inch bolts to runner channels or bent tightly around runners and bolted to the main part of the hanger.

(d) **Runners and Furring.** The main runner and cross-furring shall be not less than the sizes set forth in Table No. 47-A, except that other steel sections of equivalent strength may be substituted for those set forth in this table. Cross-furring shall be securely attached to the main runner by saddle-tying with not less than one strand of No. 16 or two strands of No. 18 U.S. gauge tie wire or approved equivalent attachments.

Interior Lath

Sec. 4705. (a) **General.** Gypsum lath shall not be installed until weather protection for the installation is provided. Where wood frame walls and partitions are covered on the interior with portland cement plaster or tile of similar material and are subject to water splash, the framing shall be protected with an approved moisture barrier.

Showers and public toilet walls shall conform to Section 510 (b).

(b) **Application of Gypsum Lath.** The thickness, spacing of supports and the method of attachment of gypsum lath shall be as set forth in Tables No. 47-B and No. 47-C. Approved wire and sheet metal attachment clips may be used.

Gypsum lath shall be applied with the long dimension perpendicular to supports and with end joints staggered in successive courses. End joints may occur at one support when stripping is applied the full length of the joints.

Where electrical radiant heat cables are installed on ceilings, the stripping, if conductive, may be omitted a distance not to exceed 12 inches from the walls.

Where lath edges are not in moderate contact and have joint gaps exceeding 3/8 inch, the joint gaps shall be covered with stripping or cornerite. Stripping or cornerite may be omitted when the entire surface is reinforced with not less than 1 inch No. 20 U.S. gauge woven wire. When lath is secured to horizontal or vertical supports not used as structural diaphragms, end joints may occur between supports when lath ends are secured together with approved fasteners. Vertical assemblies also shall conform with Section 2309 (b).

Cornerite shall be installed so as to retain position during plastering at all internal corners. Cornerite may be omitted when plaster is not continuous from one plane to an adjacent plane.

(c) **Application of Metal Plaster Bases.** The type and weight of metal lath and the gauge and spacing of wire in welded or woven lath, the spacing of supports, and the methods of attachment to wood supports shall be as set forth in Tables No. 47-B and No. 47-C.

Metal lath shall be attached to metal supports with not less than No. 18 U.S. gauge tie wire spaced not more than 6 inches apart or with approved equivalent attachments.

Metal lath or wire fabric lath shall be applied with the long dimension of the sheets perpendicular to supports.

Metal lath shall be lapped not less than 1/2 inch at sides and 1 inch at ends. Wire fabric lath shall be lapped not less than one mesh at sides and ends, but not less than 1 inch. Rib metal lath with edge ribs greater than 1/8 inch shall be lapped at sides by nesting outside ribs. When edge ribs are 1/8 inch or less, rib metal lath may be lapped 1/2 inch at sides, or outside ribs may be nested. Where end laps of sheets do not occur over supports, they shall be securely tied together with not less than No. 18 U.S. gauge wire.

Cornerite shall be installed in all internal corners to retain position during plastering. Cornerite may be omitted when lath is continuous or when plaster is continuous from one plane to an adjacent plane.

Exterior Lath

Sec. 4706. (a) General. Exterior surfaces are weather-exposed surfaces as defined in Section 424. For eave overhangs required to be fire resistive, see Section 1710.

(b) Corrosion Resistance. All lath and lath attachments shall be of corrosion-resistant material. See Section 4701 (d).

(c) Backing. Backing or a lath shall provide sufficient rigidity to permit plaster application.

Where lath on vertical surfaces extends between rafters or other similar projecting members, solid backing shall be installed to provide support for lath and attachments.

Gypsum lath or gypsum board shall not be used, except that on horizontal supports of ceilings or roof soffits it may be used as backing for metal lath or wire fabric lath and portland cement plaster.

Backing is not required under metal lath or paperbacked wire fabric lath.

(d) Weather-resistive Barriers. Weather-resistive barriers shall be installed as required in Section 1707 (a) and, when applied over wood base sheathing, shall include two layers of Grade D paper.

(e) Application of Metal Plaster Bases. The application of metal lath or wire fabric lath shall be as specified in Section 4705 (c) and they shall be furred out from vertical supports or backing not less than 1/4 inch except as set forth in Footnote No. 2, Table No. 47-B.

Where no external corner reinforcement is used, lath shall be furred out and carried around corners at least one support on frame construction.

A minimum 0.021-inch (No. 26 gauge) corrosion-resistant weep screed with a minimum vertical attachment flange of 3 1/2 inches shall be provided at or below the foundation plate line on all exterior stud walls. The screed shall be placed a minimum of 4 inches above grade and shall be of a type which will allow trapped water to drain to the exterior of the building. The weather-resistive barrier and exterior lath shall cover and terminate on the attachment flange of the screed.

Interior Plaster

Sec. 4707. (a) General. Plastering with gypsum plaster or portland cement plaster shall be not less than three coats when applied over metal lath or wire fabric lath and shall be not less than two coats when applied over other bases permitted by this chapter. Showers and public toilet walls shall conform to Section 510 (b).

Plaster shall not be applied directly to fiber insulation board. Portland cement plaster shall not be applied directly to gypsum lath, gypsum masonry or gypsum plaster except as specified in Section 4706 (c).

When installed, grounds shall assure the minimum thickness of plaster as set forth in Table No. 47-D. Plaster thickness shall be measured from the face of lath and other bases.

(b) Base Coat Proportions. Proportions of aggregate to cementitious materials shall not exceed the volume set forth in Table No. 47-E for gypsum plaster and Table No. 47-F for portland cement and portland cement-lime plaster.

(c) Base Coat Application. Base coats shall be applied with sufficient material and pressure to form a complete key or bond.

1. **Gypsum plaster.** For two-coat work, the first coat shall be brought out to grounds and straightened to a true surface, leaving the surface rough to receive the finish coat. For three-coat work, the surface of the first coat shall be scored sufficiently to provide adequate bond for the second coat and shall be permitted to harden and set before the second coat is applied. The second coat shall be brought out to grounds and straightened to a true surface, leaving the surface rough to receive the finish coat.

2. **Portland cement plaster.** The first two coats shall be as required for the first coats of exterior plaster, except that the moist-curing time period between the first and second coats shall be not less than 24 hours and the thickness shall be as set forth in Table No. 47-D. Moist curing shall not be required where job and weather conditions are favorable to the retention of moisture in the portland cement plaster for the required time period.

(d) Finish Coat Application. Finish coats shall be applied with sufficient material and pressure to form a complete bond. Finish coats shall be proportioned and mixed in an approved manner. Gypsum and lime and other interior finish coats shall be applied over gypsum base coats which have hardened and set. Thicknesses shall be not less than 1/16 inch.

Portland cement and lime finish coats may be applied over interior portland cement base coats which have been in place not less than 48 hours.

Approved acoustical finish plaster may be applied over any base coat plaster, over lean masonry or concrete, or other approved surfaces.

(e) Interior Masonry or Concrete. Condition of surfaces shall be as specified in Section 4708 (h). Approved specially prepared gypsum plaster designed

for application to concrete surfaces or approved acoustical plaster may be used. The total thickness of base coat plaster applied to concrete ceilings shall be as set forth in Table No. 47-D. Should ceiling surfaces require more than the maximum thickness permitted in Table No. 47-D, metal lath or wire fabric lath shall be installed on such surfaces before plastering.

Exterior Plaster

Sec. 4708. (a) General. Plastering with portland cement plaster shall be not less than three coats when applied over metal lath or wire fabric lath and shall be not less than two coats when applied over masonry, concrete or gypsum backing as specified in Section 4706 (c). If plaster surface is completely covered by veneer or other facing material, or is completely concealed by another wall, plaster application need be only two coats, provided the total thickness is as set forth in Table No. 47-F.

On wood frame or metal stud construction with an on-grade concrete floor slab system, exterior plaster shall be applied in such a manner as to cover, but not extend below, lath and paper. See Section 4706 (e) for the application of paper and lath, and flashing or drip screeds.

Only approved plasticity agents and approved amounts thereof may be added to portland cement. When plastic cement is used, no additional lime or plasticizers shall be added. Hydrated lime or the equivalent amount of lime putty used as a plasticizer may be added to portland cement plaster in an amount not to exceed that set forth in Table No. 47-F.

Gypsum plaster shall not be used on exterior surfaces. See Section 424.

(b) Base Coat Proportions. The proportion of aggregate to cementitious materials shall be as set forth in Table No. 47-F.

(c) Base Coat Application. The first coat shall be applied with sufficient material and pressure to fill solidly all openings in the lath. The surface shall be scored horizontally sufficiently rough to provide adequate bond to receive the second coat.

The second coat shall be brought out to proper thickness, rodded and floated sufficiently rough to provide adequate bond for finish coat. The second coat shall have no variation greater than 1/4 inch in any direction under a 5-foot straight edge.

(d) Environmental Conditions. Portland cement based plaster shall not be applied to frozen base or those containing frost. Plaster mixes shall not contain frozen ingredients. Plaster coats shall be protected from freezing for a period of not less than 24 hours after set has occurred.

(e) Curing and Interval. First and second coats of plaster shall be applied and moist cured as set forth in Table No. 47-F.

When applied over gypsum backing as specified in Section 4706 (c) or directly to unit masonry surfaces, the second coat may be applied as soon as the first coat has attained sufficient hardness.

(f) Alternate Method of Application. As an alternate method of application, the second coat may be applied as soon as the first coat has attained sufficient rigidity to receive the second coat.

When using this method of application, calcium aluminate cement up to 15 percent of the weight of the portland cement may be added to the mix.

Curing of the first coat may be omitted and the second coat shall be cured as set forth in Table No. 47-F.

(g) Finish Coats. Finish coats shall be proportioned and mixed in an approved manner and in accordance with Table No. 47-F.

Portland cement and lime finish coats shall be applied over base coats which have been in place for the time periods set forth in Table No. 47-F. The third or finish coat shall be applied with sufficient material and pressure to bond to and to cover the brown coat and shall be of sufficient thickness to conceal the brown coat.

(h) Preparation of Masonry and Concrete. Surfaces shall be clean, free from efflorescence, sufficiently damp and rough to assure proper bond. If surface is insufficiently rough, approved bonding agents or a portland cement dash bond coat mixed in the proportions of 1 1/2 cubic feet of sand to 1 cubic foot of portland cement shall be applied. Approved bonding agents shall conform with the provisions of U.B.C. Standard No. 47-1. Dash bond coat shall be left undisturbed and shall be moist cured not less than 24 hours. When dash bond is applied, first coat of base coat plaster may be omitted. See Table No. 47-D for thickness.

Exposed Aggregate Plaster

Sec. 4709. (a) General. Exposed natural or integrally colored aggregate may be partially embedded in a natural or colored bedding coat of portland cement or gypsum plaster, subject to the provisions of this section.

(b) Aggregate. The aggregate may be applied manually or mechanically and shall consist of marble chips, pebbles or similar durable, nonreactive materials, moderately hard (three or more on the MOH scale).

(c) Bedding Coat Proportions. The exterior bedding coat shall be composed of one part portland cement, one part Type S lime and a maximum three parts of graded white or natural sand by volume. The interior bedding coat shall be composed of 100 pounds neat gypsum plaster and a maximum 200 pounds of graded white sand, or exterior or interior may be a factory-prepared bedding coat.

The exterior bedding coat shall have a minimum compressive strength of 1000 pounds per square inch.

(d) **Application.** The bedding coat may be applied directly over the first (scratch) coat of plaster, provided the ultimate overall thickness is a minimum of $7/8$ inch including lath. Over concrete or masonry surfaces the overall thickness shall be a minimum of $1/2$ inch.

(e) **Bases.** Exposed aggregate plaster may be applied over concrete, masonry, portland cement plaster base coats or gypsum plaster base coats.

(f) **Preparation of Masonry and Concrete.** Masonry and concrete surfaces shall be prepared in accordance with the provisions of Section 4708 (h).

(g) **Curing.** Portland cement base coats shall be cured in accordance with Table No. 47-F. Portland cement bedding coat shall retain sufficient moisture for hydration (hardening) for 24 hours minimum or, where necessary, shall be kept damp for 24 hours by light water spraying.

Gypsum Wallboard

Sec. 4711. (a) **General.** All gypsum wallboard shall conform to U.B.C. Standard No. 47-11 and shall be installed in accordance with the provisions of this section. Gypsum wallboard shall not be installed on exterior surfaces. See Section 424. For use as backing under stucco, see Section 4706 (c).

Gypsum wallboard shall not be installed until weather protection for the installation is provided.

(b) **Supports.** Supports shall be spaced not to exceed the spacing set forth in Table No. 47-G for single-ply application and Table No. 47-H for two-ply application. Vertical assemblies shall conform with Section 4703. Horizontal assemblies shall comply with Section 4704.

(c) **Single-ply Application.** All edges and ends of gypsum wallboard shall occur on the framing members, except those edges and ends which are perpendicular to the framing members. All edges and ends of gypsum wallboard shall be in moderate contact except in concealed spaces where fire-resistive construction or diaphragm action is not required.

The size and spacing of fasteners shall conform with Table No. 47-G except where modified by fire-resistive construction meeting the requirements of Section 4302 (b). Fasteners shall be spaced not less than $3/8$ inch from edges and ends of gypsum wallboard. Fasteners at the top and bottom plates of vertical assemblies, or the edges and ends of horizontal assemblies perpendicular to supports, and at the wall line may be omitted except on shear-resisting elements or fire-resistive assemblies. Fasteners shall be applied in such a manner as not to fracture the face paper with the fastener head.

Gypsum wallboard may be applied to wood framing members with an approved adhesive conforming with U.B.C. Standard No. 47-2. A continuous bead of the adhesive shall be applied to the face of all framing members, except top and bottom plates, of sufficient size as to spread to an average width of 1 inch and thickness of $1/16$ inch when the gypsum wallboard is applied. Where the edges or ends of two pieces of gypsum wallboard occur on the same framing member, two continuous parallel beads of adhesive shall be applied to the framing member. Fasteners shall be used with adhesive application in accordance with Table No. 47-G.

(d) **Two-ply Application.** The base of gypsum wallboard shall be applied with fasteners of the type and size as required for the nonadhesive application of single-ply gypsum wallboard. Fastener spacings shall be in accordance with Table No. 47-H except where modified by fire-resistive construction meeting the requirements of Section 4302 (b).

The face ply of gypsum wallboard may be applied with gypsum wallboard joint compound or approved adhesive furnishing full coverage between the plies or with fasteners in accordance with Table No. 47-H. When the face ply is installed with joint compound or adhesive, the joints of the face ply need not occur on supports. Temporary nails or shoring shall be used to hold face ply in position until the joint compound or adhesive develops adequate bond.

(e) **Joint Treatment.** Gypsum wallboard single-layer fire-rated assemblies shall have joints treated.

EXCEPTIONS: Joint treatment need not be provided when any of the following conditions occur:

1. Where the wallboard is to receive a decorative finish such as wood panel, battens, acoustical finishes or any similar application which would be equivalent joint treatment.

2. Joints occur over wood-framing members.

3. Assemblies tested without joint treatment.

Gypsum wallboard tape and joint compound shall conform with the provision of U.B.C. Standard No. 47-6.

Use of Gypsum in Showers and Water Closets

Sec. 4712. When gypsum is used as a base for tile or wall panels for tub shower enclosures or water closet compartment walls, water-resistant gypsum backing board complying with U.B.C. Standard No. 47-14 shall be used, except that water-resistant gypsum board shall not be used in the following locations:

1. Over a vapor retarder.

2. In areas subject to continuous high humidity, such as saunas, steam rooms, gang shower rooms.

3. On ceilings.

Softwood Plywood Paneling

Sec. 4713. All softwood plywood paneling shall conform with the provisions of Chapters 25 and 42 and shall be installed in accordance with Table No. 47-.

Shear-resisting Construction with Wood Frame

Sec. 4714. (a) **General.** Portland cement plaster, gypsum lath and plaster, gypsum veneer base, gypsum sheathing board and gypsum wallboard may be used on wood studs for vertical diaphragms if applied in accordance with this section. Shear-resisting values shall not exceed those set forth in Table No. 47-I. The effects of overturning on vertical diaphragms shall be investigated in accordance with Section 2303 (b) 3.

The shear values tabulated shall not be cumulative with the shear value of other materials applied to the same wall. The shear values may be additive when identical materials applied as specified in this section are applied to both sides of the wall.

(b) **Masonry and Concrete Construction.** Portland cement plaster, gypsum lath and plaster, gypsum veneer base, gypsum sheathing board and gypsum wallboard shall not be used in vertical diaphragms to resist forces imposed by masonry or concrete construction.

(c) **Wall Framing.** Framing for vertical diaphragms shall conform with Section 2517 (g) for bearing walls, and studs shall be spaced not farther apart than inches center to center. Sills, plates and marginal studs shall be adequately connected to framing elements located above and below to resist all design forces.

(d) **Height-to-Length Ratio.** The maximum allowable height-to-length ratio for the construction in this section shall be 2 to 1. Wall sections having height-length ratios in excess of $1 1/2$ to 1 shall be blocked.

(e) **Application.** End joints of adjacent courses of gypsum lath, gypsum veneer base, gypsum sheathing board or gypsum wallboard sheets shall not occur over the same stud.

Where required in Table No. 47-I, blocking having the same cross-sectional dimensions as the studs shall be provided at all joints that are perpendicular to studs.

The size and spacing of nails shall be as set forth in Table No. 47-I. Nails shall be spaced not less than $3/8$ inch from edges and ends of gypsum lath, gypsum veneer base, gypsum sheathing board, gypsum wallboard or sides of studs, blocking and top and bottom plates.

1. **Gypsum lath.** Gypsum lath shall be applied perpendicular to the studs. Maximum allowable shear values shall be as set forth in Table No. 47-I.

2. **Gypsum sheathing board.** Four-foot-wide pieces may be applied parallel or perpendicular to studs. Two-foot-wide pieces shall be applied perpendicular to the studs. Maximum allowable shear values shall be as set forth in Table No. 47-I.

3. **Gypsum wallboard or veneer base.** Gypsum wallboard or veneer base may be applied parallel or perpendicular to studs. Maximum allowable shear values shall be as set forth in Table No. 47-I.

TABLE NO. 47-A—SUSPENDED AND FURRED CEILINGS[1]
(For Support of Ceilings Weighing Not More than 10 Pounds per Square Foot)

Minimum Sizes for Wire and Rigid Hangers				
SIZE AND TYPE			MAXIMUM AREA SUPPORTED (In Square Feet)	SIZE
Hangers for Suspended Ceilings			12.5	No. 9 gauge wire
			16	No. 8 gauge wire
			18	3/16" diameter, mild steel rod[2]
			20	7/32" diameter, mild steel rod[2]
			22.5	1/4" diameter, mild steel rod[2]
			25.0	1" x 3/16" mild steel flats[3]
Hangers for Attaching Runners and Furring Directly to Beams and Joists	For Supporting Runners	Single Hangers Between Beams[4]	8	No. 12 gauge wire
			12	No. 10 gauge wire
			16	No. 8 gauge wire
		Double Wire Loops at Beams or Joists[3]	8	No. 14 gauge wire
			12	No. 12 gauge wire
			16	No. 11 gauge wire
	For Supporting Furring without Runners[4] (Wire Loops at Supports)	Type of Support: Concrete Steel Wood	8	No. 14 gauge wire No. 16 gauge wire (2 loops)[5] No. 16 gauge wire (2 loops)[5]

Minimum Sizes and Maximum Spans for Main Runners [6] [7]		
SIZE AND TYPE	MAXIMUM SPACING OF HANGERS OR SUPPORTS (ALONG RUNNERS)	MAXIMUM SPACING OF RUNNERS (TRANSVERSE)
3/4"— .3 pound per foot, cold- or hot-rolled channel	2'0"	3'0"
1 1/2"— .475 pound per foot, cold-rolled channel	3'0"	4'0"
1 1/2"— .475 pound per foot, cold-rolled channel	3'6"	3'6"
1 1/2"— .475 pound per foot, cold-rolled channel	4'0"	3'0"
1 1/2"—1.12 pounds per foot, hot-rolled channel	4'0"	5'0"
2 "—1.26 pounds per foot, hot-rolled channel	5'0"	5'0"
2 "— .59 pound per foot, cold-rolled channel	5'0"	3'6"
1 1/2" x 1 1/2" x 3/16" angle	5'0"	3'6"

Minimum Sizes and Maximum Spans for Cross Furring [6] [7]		
SIZE AND TYPE OF CROSS FURRING	MAXIMUM SPACING OF RUNNERS OR SUPPORTS	MAXIMUM SPACING OF CROSS FURRING MEMBERS (TRANSVERSE)
1/4" diameter pencil rods	2'0"	12"
3/8" diameter pencil rods	2'0"	19"
3/8" diameter pencil rods	2'6"	12"
3/4"—.3 pound per foot, cold- or hot-rolled channel	3'0"	24"
	3'6"	16"
	4'0"	12"
1 "—.410 pound per foot, hot-rolled channel	4'0"	24"
	4'6"	19"
	5'0"	12"

[1]Metal suspension systems for acoustical tile and lay-in panel ceiling systems weighing not more than 4 pounds per square foot, including light fixtures and all ceiling-supported equipment and conforming to U.B.C. Standard No. 47-18, are exempt from Table No. 47-A.

For furred and suspended ceilings with metal lath construction, see U.B.C. Standard No. 47-4.

[2]All rod hangers shall be protected with a zinc or cadmium coating or with a rust-inhibitive paint.

[3]All flat hangers shall be protected with a zinc or cadmium coating or with a rust-inhibitive paint.

[4]Inserts, special clips or other devices of equal strength may be substituted for those specified.

[5]Two loops of No. 18 gauge wire may be substituted for each loop of No. 16 gauge wire for attaching steel furring to steel or wood joists.

[6]Spans are based on webs of channels being erected vertically.

[7]Other sections of hot- or cold-rolled members of equivalent strength may be substituted for those specified.

TABLE NO. 47-B[1]—TYPES OF LATH—MAXIMUM SPACING OF SUPPORTS

TYPE OF LATH[2]		MINIMUM WEIGHT (Per Square Yard) GAUGE AND MESH SIZE	VERTICAL (In Inches)			HORIZONTAL (In Inches)	
				Metal			
			Wood	Solid Plaster Partitions	Other	Wood or Concrete	Metal
1. Expanded Metal Lath (Diamond Mesh)		2.5 3.4	16[3] 16[3]	16[3] 16[3]	12 16	12 16	12 16
2. Flat Rib Expanded Metal Lath		2.75 3.4	16 19	16 24	16 19	16 19	16 19
3. Stucco Mesh Expanded Metal Lath		1.8 and 3.6	16[4]	—	—	—	—
4. 3/8" Rib Expanded Metal Lath		3.4 4.0	24 24	24[5] 24[5]	24 24	24 24	24 24
5. Sheet Lath		4.5	24	[5]	24	24	24
6. Wire Fabric Lath	Welded	1.95 pounds, No. 11 gauge, 2" × 2" 1.16 pounds, No. 16 gauge, 2" × 2" 1.4 pounds, No. 18 gauge, 1" × 1"[6]	24 16 16[4]	24 16 —	24 16 —	24 16 —	24 16 —
	Woven[4]	1.1 pounds, No. 18 gauge, 1 1/2" Hexagonal[6] 1.4 pounds, No. 17 gauge, 1 1/2"Hexagonal[6] 1.4 pounds, No. 18 gauge, 1" Hexagonal[6]	24 24 24	16 16 16	16 16 16	24 24 24	16 16 16
7. 3/8" Gypsum Lath (plain)			16	—	16[7]	16	16
8. 1/2" Gypsum Lath (plain)			24	—	24	24	24

[1]For fire-resistive construction, see Tables No. 43-A, No. 43-B and No. 43-C. For shear-resisting elements, see Table No. 47-I. Metal lath, wire lath, wire fabric lath and metal accessories shall conform with the provisions of U.B.C. Standard No. 47-4. Gypsum lath shall conform with the provisions of U.B.C. Standard No. 47-8.

[2]Metal lath and wire fabric lath used as reinforcement for portland cement plaster shall be furred out away from vertical supports at least 1/4 inch. Self-furring lath meets furring requirements. Exception: Furring of expanded metal lath is not required on supports having a bearing surface width of 1-5/8 inches or less.

[3]Span may be increased to 24 inches with self-furred metal lath over solid sheathing assemblies approved for this use.

[4]Wire backing required on open vertical frame construction except under expanded metal lath and paperbacked wire fabric lath.

[5]May be used for studless solid partitions.

[6]Woven wire or welded wire fabric lath, not to be used as base for gypsum plaster without absorbent paperbacking or slot-perforated separator.

[7]Span may be increased to 24 inches on vertical screw or approved nailable assemblies.

TABLE NO. 47-C—TYPES OF LATH—ATTACHMENT TO WOOD AND METAL[1] SUPPORTS

TYPE OF LATH	NAILS[2][3] TYPE AND SIZE	MAXIMUM SPACING[5] Vertical (In Inches)	MAXIMUM SPACING[5] Horizontal (In Inches)	SCREWS[3][6] MAX. SPACING[5] Vertical (In Inches)	SCREWS[3][6] MAX. SPACING[5] Horizontal (In Inches)	STAPLES[3][4] Wire Gauge No.	STAPLES Crown	STAPLES Leg[7]	STAPLES MAX. SPACING[5][6] Vertical (In Inches)	STAPLES MAX. SPACING[5][6] Horizontal (In Inches)
1. Diamond Mesh Expanded Metal Lath and Flat Rib Metal Lath	4d blued smooth box 1 1/2″[11] No. 14 gauge 7/32″ head (clinched)[8] 1″ No. 11 gauge 7/16″ head, barbed 1 1/2″ No. 11 gauge 7/16″ head, barbed	6 6 6	— — 6	6	6	16	3/4	7/8	6	6
2. 3/8″ Rib Metal Lath and Sheet Lath	1 1/2″ No. 11 gauge 7/16″ head, barbed	6	6	6	6	16	3/4	1 1/4	At Ribs	At Ribs
3. 3/4″ Rib Metal Lath	4d common 1 1/2″ No. 12 1/2 gauge 1/4″ head 2″ No. 11 gauge 7/16″ head, barbed	At Ribs	— At Ribs	At Ribs	At Ribs	16	3/4	1 5/8	At Ribs	At Ribs
4. Wire Fabric Lath [9]	4d blued smooth box (clinched)[8] 1″ No. 11 gauge 7/16″ head, barbed	6 6	— —	6	6	16	3/4	7/8	6	6
	1 1/2″ No. 11 gauge 7/16″ head, barbed 1 1/4″ No. 12 gauge 3/8″ head, furring 1″ No. 12 gauge 3/8″ head	6 6 6	6 6			16	7/16[9]	7/8	6	6
5. 3/8″ Gypsum Lath	1 1/8″ No. 13 gauge 19/64″ head, blued	8[10]	8[10]	8[10]	8[10]	16	3/4	7/8	8[10]	8[10]
6. 1/2″ Gypsum Lath	1 1/4″ No. 13 gauge 19/64″ head, blued	8	8[10] 6[11]	8[10]	8[10] 6[11]	16	3/4	1 1/8	8[10]	8[10] 6[11]

[1]Metal lath, wire lath, wire fabric lath and metal accessories shall conform with the provisions of U.B.C. Standard No. 47-4.

[2]For nailable nonload-bearing metal supports, use annular threaded nails or approved staples.

[3]For fire-resistive construction, see Tables No. 43-B and No. 43-C. For shear-resisting elements, see Table No. 47-I. Approved wire and sheet metal attachment clips may be used.

[4]With chisel or divergent points.

[5]Maximum spacing of attachments from longitudinal edges shall not exceed 2 inches.

[6]Screws shall be an approved type long enough to penetrate into wood framing not less than 5/8 inch and through metal supports adaptable for screw attachment not less than 1/4 inch.

[7]When lath and stripping are stapled simultaneously, increase leg length of staple 1/8 inch.

[8]For interiors only.

[9]Attach self-furring wire fabric lath to supports at furring device.

[10]Three attachments per 16-inch-wide lath per bearing. Four attachments per 24-inch-wide lath per bearing.

[11]Supports spaced 24 inches o.c. Four attachments per 16-inch-wide lath per bearing. Five attachments per 24-inch-wide lath per bearing.

TABLE NO. 47-D—THICKNESS OF PLASTER[1]

| PLASTER BASE | FINISHED THICKNESS OF PLASTER FROM FACE OF LATH, MASONRY, CONCRETE | |
	Gypsum Plaster	Portland Cement Plaster
1. Expanded Metal Lath	5/8″ minimum[2]	5/8″ minimum[2]
2. Wire Fabric Lath	5/8″ minimum[2]	3/4″ minimum (interior)[3]
		7/8″ minimum (exterior)[3]
3. Gypsum Lath	1/2″ minimum	
4. Masonry Walls[4]	1/2″ minimum	1/2″ minimum
5. Monolithic Concrete Walls[4][5]	5/8″ maximum[8]	7/8″ maximum[8]
6. Monolithic Concrete Ceilings[4][5]	3/8″ maximum[6][7][8]	1/2″ maximum[7][8]

[1]For fire-resistive construction, see Tables Nos. 43-A, 43-B and 43-C.

[2]When measured from back plane of expanded metal lath, exclusive of ribs, or self-furring lath, plaster thickness shall be 3/4-inch minimum.

[3]When measured from face of support or backing.

[4]Because masonry and concrete surfaces may vary in plane, thickness of plaster need not be uniform.

[5]When applied over a liquid bonding agent, finish coat may be applied directly to concrete surface.

[6]Approved acoustical plaster may be applied directly to concrete, or over base coat plaster, beyond the maximum plaster thickness shown.

[7]On concrete ceilings, where the base coat plaster thickness exceeds the maximum thickness shown, metal lath or wire fabric lath shall be attached to the concrete.

[8]An approved skim-coat plaster 1/16 inch thick may be applied directly to concrete.

TABLE NO. 47-E—GYPSUM PLASTER PROPORTIONS[1]

| NUMBER | COAT | PLASTER BASE OR LATH | MAXIMUM VOLUME AGGREGATE PER 100 POUNDS NEAT PLASTER[2][3] (Cubic Feet) | |
			Damp Loose Sand[4]	Perlite or Vermiculite[4]
1. Two-coat Work	Base Coat	Gypsum Lath	2½	2
	Base Coat	Masonry	3	3
2. Three-coat Work	First Coat	Lath	2[5]	2
	Second Coat	Lath	3[5]	2[6]
	First and Second Coats	Masonry	3	3

[1]Wood-fibered gypsum plaster may be mixed in the proportions of 100 pounds of gypsum to not more than 1 cubic foot of sand where applied on masonry or concrete.

Gypsum plasters shall conform with the provisions of U.B.C. Standard No. 47-9.

[2]For fire-resistive construction, see Tables No. 43-A, No. 43-B and No. 43-C.

[3]When determining the amount of aggregate in set plaster, a tolerance of 10 percent shall be allowed.

[4]Combinations of sand and lightweight aggregate may be used, provided the volume and weight relationship of the combined aggregate to gypsum plaster is maintained. Sand and lightweight aggregate shall conform with U.B.C. Standard No. 47-3.

[5]If used for both first and second coats, the volume of aggregate may be 2½ cubic feet.

[6]Where plaster is 1 inch or more in total thickness, the proportions for the second coat may be increased to 3 cubic feet.

TABLE NO. 47-F—PORTLAND CEMENT PLASTERS[1]

PORTLAND CEMENT PLASTER						
COAT	**VOLUME CEMENT**	**MAXIMUM WEIGHT (OR VOLUME) LIME PER VOLUME CEMENT**[2]	**MAXIMUM VOLUME SAND PER VOLUME CEMENT**[3]	**APPROXIMATE MINIMUM THICKNESS**[4]	**MINIMUM PERIOD MOIST CURING**	**MINIMUM INTERVAL BETWEEN COATS**
First	1	20 lbs.	4	⅜″[5]	48[6] Hours	48[7] Hours
Second	1	20 lbs.	5	1st and 2nd Coats total ¾″	48 hours	7 Days[8]
Finish	1	1[9]	3	1st, 2nd and Finish Coats ⅞″	—	[8]
PORTLAND CEMENT-LIME PLASTER[10]						
COAT	**VOLUME CEMENT**[11]	**MAXIMUM VOLUME LIME PER VOLUME CEMENT**	**MAXIMUM VOLUME SAND PER COMBINED VOLUMES CEMENT AND LIME**	**APPROXIMATE MINIMUM THICKNESS**[4]	**MINIMUM PERIOD MOIST CURING**	**MINIMUM INTERVAL BETWEEN COATS**
First	1	1	4	⅜″[5]	48[6] Hours	48[7] Hours
Second	1	1	4½	1st and 2nd Coats total ¾″	48 hours	7 Days[8]
Finish	1	1[9]	3	1st, 2nd and Finish Coats ⅞″	—	[8]

[1]Exposed aggregate plaster shall be applied in accordance with Section 4709. Minimum overall thickness shall be ¾ inch.

[2]Up to 20 pounds of dry hydrated lime (or an equivalent amount of lime putty) may be used as a plasticizing agent in proportion to each sack (cubic foot) of Type I and Type II standard portland cement in first and second coats of plaster. See Section 4708 (a) for use of plastic cement.

[3]When determining the amount of sand in set plaster, a tolerance of 10 percent may be allowed.

[4]See Table No. 47-D.

[5]Measured from face of support of backing to crest of scored plaster.

[6]See Section 4707 (c) 2.

[7]Twenty-four hours minimum interval between coats of interior portland cement plaster. For alternate method of application, see Section 4708 (e).

[8]Finish coat plaster may be applied to interior portland cement base coats after a 48-hour period.

[9]For finish coat plaster, up to an equal part of dry hydrated lime by weight (or an equivalent volume of lime putty) may be added to Types I, II and III standard portland cement.

[10]No additions of plasticizing agents shall be made.

[11]Type I, II or III standard portland cement. See Section 4708 (a) for use of plastic cement.

TABLE NO. 47-G—APPLICATION OF SINGLE-PLY GYPSUM WALLBOARD

THICKNESS OF GYPSUM WALL-BOARD (Inch)	PLANE OF FRAMING SURFACE	LONG DIMENSION OF GYPSUM WALLBOARD SHEETS IN RELATION TO DIRECTION OF FRAMING MEMBERS	MAXIMUM SPACING OF FRAMING MEMBER[1] (Center to Center) (In Inches)	MAXIMUM SPACING OF FASTENERS[1] (Center to Center) (In Inches)		NAILS[2]—TO WOOD
				Nails[3]	Screws[4]	
1/2	Vertical	Either direction	16	8	16	No. 13 gauge, 1 3/8″ long, 19/64″ head; 0.098″ diameter, 1 1/4″ long, annular ringed; 5d, cooler or wallboard[5] nail (0.086″ dia., 1 5/8″ long, 15/64″ head).
	Horizontal	Either direction	16	7	12	
	Horizontal	Perpendicular	24	7	12	
	Vertical	Either direction	24	8	12	
5/8	Vertical	Either direction	16	8	16	No. 13 gauge, 1 5/8″ long, 19/64″ head; 0.098″ diameter, 1 3/8″ long, annular ringed; 6d, cooler or wallboard[5] nail (0.092″ dia., 1 7/8″ long, 1/4″ head).
	Horizontal	Either direction	16	7	12	
	Horizontal	Perpendicular	24	7	12	
	Vertical	Either direction	24	8	12	
Nail or Screw Fastenings With Adhesives (Maximum Center to Center in Inches)						
(Column headings as above)				End	Edges	Field
1/2 or 5/8	Horizontal	Either direction	16	16	16	24	As required for 1/2″ and 5/8″ gypsum wallboard, see above.
		Perpendicular	24	16	24	24	
	Vertical	Either direction	24	16	24	6	

[1]For fire-resistive construction, see Tables Nos. 43-B and 43-C. For shear-resisting elements, see Table No. 47-I.

[2]Where the metal framing has a clinching design formed to receive the nails by two edges of metal, the nails shall be not less than 5/8 inch longer than the wallboard thickness, and shall have ringed shanks. Where the metal framing has a nailing groove formed to receive the nails, the nails shall have barbed shanks or be 5d, No. 13 1/2 gauge, 1 5/8 inch long, 15/64-inch head for 1/2-inch gypsum wallboard; 6d, No. 13 gauge, 1 7/8-inch long, 15/64-inch head for 5/8-inch gypsum wallboard.

[3]Two nails spaced 2 inches to 2 1/2 inches apart may be used where the pairs are spaced 12 inches on center except around the perimeter of the sheets.

[4]Screws shall conform with U.B.C. Standard No. 47-5 and be long enough to penetrate into wood framing not less than 5/8 inch and through metal framing not less than 1/4 inch.

[5]For properties of cooler or wallboard nails, see U.B.C. Standard No. 25-17, Table No. 25-17-I.

[6]Not required.

TABLE NO. 47-H—APPLICATION OF TWO-PLY GYPSUM WALLBOARD [1]

				MAXIMUM SPACING OF FASTENERS (Center to Center) (In Inches)				
FASTENERS ONLY								
THICKNESS OF GYPSUM WALLBOARD (Each Ply) (Inch)	PLANE OF FRAMING SURFACE	LONG DIMENSION OF GYPSUM WALLBOARD SHEETS	MAXIMUM SPACING OF FRAMING MEMBERS (Center to Center) (In Inches)	Base Ply			Face Ply	
				Nails[2]	Screws[3]	Staples[4]	Nails[2]	Screws[3]
3/8	Horizontal	Perpendicular only	16	16	24	16	7	12
	Vertical	Either Direction	16				8	
1/2	Horizontal	Perpendicular only	24				7	
	Vertical	Either Direction	24				8	
5/8	Horizontal	Perpendicular only	24				7	
	Vertical	Either Direction	24				8	
Fasteners and Adhesives								
3/8 Base Ply	Horizontal	Perpendicular only	16	7	12	5	Temporary Nailing or Shoring to Comply with Section 4711 (d)	
	Vertical	Either Direction	24	8		7		
1/2 Base Ply	Horizontal	Perpendicular only	24	7		5		
	Vertical	Either Direction	24	8		7		
5/8 Base Ply	Horizontal	Perpendicular only	24	7		5		
	Vertical	Either Direction	24	8		7		

[1]For fire-resistive construction, see Tables Nos. 43-B and 43-C. For shear-resisting elements, see Table No. 47-I.

[2]Nails for wood framing shall be long enough to penetrate into wood members not less than 3/4 inch and the sizes shall conform with the provisions of Table No. 47-G. For nails not included in Table No. 47-G, use the appropriate size cooler or wallboard nails as set forth in Table No. 25-17-I of U.B.C. Standard No. 25-17. Nails for metal framing shall conform with the provisions of Table No. 47-G.

[3]Screws shall conform with the provisions of Table No. 47-G.

[4]Staples shall be not less than No. 16 gauge by 3/4-inch crown width with leg length of 7/8 inch, 1 1/8 inch and 1 3/8 inch for gypsum wallboard thicknesses of 3/8 inch, 1/2 inch and 5/8 inch, respectively.

TABLE NO. 47-I—ALLOWABLE SHEAR FOR WIND OR SEISMIC FORCES IN POUNDS PER FOOT FOR VERTICAL DIAPHRAGMS OF LATH AND PLASTER OR GYPSUM BOARD FRAME WALL ASSEMBLIES[1]

TYPE OF MATERIAL	THICKNESS OF MATERIAL	WALL CONSTRUCTION	NAIL SPACING[2] MAXIMUM (In Inches)	SHEAR VALUE	MINIMUM NAIL SIZE[3]
1. Expanded metal, or woven wire lath and portland cement plaster	$7/8''$	Unblocked	6	180	No. 11 gauge, $1\,1/2''$ long, $7/16''$ head No. 16 gauge staple, $7/8''$ legs
2. Gypsum lath	$3/8''$ Lath and $1/2''$ Plaster	Unblocked	5	100	No. 13 gauge, $1\,1/8''$ long, $19/64''$ head, plasterboard blued nail
3. Gypsum sheathing board	$1/2'' \times 2' \times 8'$	Unblocked	4	75	No. 11 gauge, $1\,3/4''$ long, $7/16''$ head, diamond-point, galvanized
	$1/2'' \times 4'$	Blocked	4	175	
	$1/2'' \times 4'$	Unblocked	7	100	
4. Gypsum wallboard or veneer base.	$1/2''$	Unblocked	7	100	5d cooler or wallboard
			4	125	
		Blocked	7	125	
			4	150	
	$5/8''$	Unblocked	7	115	6d cooler or wallboard
			4	145	
		Blocked	7	145	
			4	175	
		Blocked Two ply	Base ply 9 Face ply 7	250	Base ply—6d cooler or wallboard Face ply—8d cooler or wallboard

[1]These vertical diaphragms shall not be used to resist loads imposed by masonry or concrete construction. See Section 4714 (b). Values are for short term loading due to wind. Values must be reduced 25 percent for normal loading. The values for gypsum products must be reduced 50 percent for dynamic loading due to earthquake in Seismic Zones Nos. 3 and 4.

[2]Applies to nailing at all studs, top and bottom plates and blocking.

[3]Alternate nails may be used if their dimensions are not less than the specified dimensions.

[4]For properties of cooler or wallboard nails, see U. B. C. Standard No. 25-17-I

UNIFORM BUILDING CODE STANDARDS

The Uniform Building Code Standards, 1988 Edition, is a corollary publication to the 1988 Uniform Building Code. Both are published by the International Conference of Building Officials, 5360 South Workman Mill Road, Whittier, California 90601. (Copies of this book may be purchased for $74.75, shipping included, from Building News, Inc., 3055 Overland Ave., Los Angeles, CA 90034.)

CHAPTER 17

17-1; 1707 (a)
Kraft Waterproof Building Paper. Federal Specification UU-B-790a, February 5, 1968.

CHAPTER 43

43-1; 4302 (b), 4304 (d), 4304 (e), 4305 (a), Table No. 43-A
Fire Tests of Building Construction and Materials. Standard Methods E119-83 of the ASTM.

43-3; 4306 (e)
Tinclad Fire Doors. American National Standards Institute/Underwriters Laboratories Inc. 10A-1979 (R 1985).

43-4; 4306 (e), 4306 (i)
Fire Tests of Window Assemblies. Standard Methods E163-76 of the ASTM.

43-7; 503 (c), 1706 (b), 4306 (d), 4306 (e), 4306 (j)
Fire Dampers. Test Standards of the International Conference of Building Officials.

CHAPTER 47

47-1; 4708 (h)
Plaster Bonding Agents. U.S. Government Military Specification MIL-B-19235 (Docks) (December 12, 1965), and Standard Specifications of the California Lathing and Plastering Contractors Association (1965), and C631-780 of the ASTM, and Recommendations of the Gypsum Association.

47-2; 4711 (c)
Adhesives for Fastening Gypsum Wallboard to Wood Framing. Standard Specification C557-73 of the ASTM.

47-3; Table No. 47-E
Perlite, Vermiculite and Sand Aggregates for Gypsum Plaster. Standard Specification C35-70 of the ASTM.

47-4; 4703 (c), Tables Nos. 47-A, 47-B and 47-C
Metal Lath, Wire Lath, Wire Fabric Lath and Metal Accessories. Standard Specification A42.4-1955 of the American National Standards Institute, Inc., and Specification 2.6.73 of the California Lathing and Plastering Contractors Association.

47-5; Table No. 47-G
Drill Screws. Standard Specification C646-76 of the ASTM.

47-6; 4711 (e)
Gypsum Wallboard Tape and Joint Compound. Standard Specification C475-70 and Standard Methods of Testing C474-73 of the ASTM.

47-7; 4702
Gypsum Backing Board. Standard Specification C442-72 of the ASTM.

47-8; Table No. 47-B
Gypsum Lath. Standard Specification C37-69 of the ASTM.

47-9; Table No. 47-E
Gypsum Plasters. Standard Specification C28-76a of the ASTM.

47-10; 4702
Gypsum Sheathing Board. Standard Specification C79-82 of the ASTM.

47-11; 4711 (a)
Gypsum Wallboard. Standard Specification C36-76a of the ASTM.

47-12
Keene's Cement. Standard Specification C61-70 of the ASTM.

47-14; 4712
Water-resistant Gypsum Backing Board. Standard Specification C630-76 the ASTM.

47-15; Tables Nos. 43-A, 43-B, 43-C
Gypsum Base for Veneer Plaster and Gypsum Veneer Plaster. Standard Specifications C588-68 and C587-73 of the ASTM.

47-16
Lime. Standard Specifications C6-49 (1968) and C206-49 (1968) of the ASTM.

47-17; 2627 (a), 4701 (c)
Testing Gypsum and Gypsum Products. Standard Specification C22- (R74) and Standard Methods C472-73 and C473-76a of the ASTM.

47-19; 4701 (f)
Application of Gypsum Base for Veneer Plaster and Gypsum Veneer Plaster. Standard Specifications C843-76 and C844-79 of the ASTM.

47-20
Nails for the Application of Gypsum Wallboard, Gypsum Backing Board and Gypsum Veneer Base. Standard Specification C514-77 of the ASTM.

EXCERPTS FROM 1987 EDITION

BOCA BASIC BUILDING CODE

The BOCA Basic Code, published by Building Officials & Code Administrators International, Inc., 4051 W. Flossmoor Road, Country Club Hills, Illinois 60477, is a model code which is updated and reproduced every three years. This code is the official building code of the majority of building authorities in the midwest and eastern part of the country. Following excerpts are reproduced with permission of the publisher. Copies of this code may be purchased from Building News, Inc., 3055 Overland Ave., Los Angeles, Calif. 90034 at a cost of $40.00, plus sales tax (if applicable) and $2.95 shipping charge.

SECTION 1600.0 GENERAL

1600.1 Scope: The provisions of this article shall govern the materials, design, construction and quality of gypsum and plaster.

SECTION 1601.0 INTERIOR LATHING AND GYPSUM PLASTERING

1601.1 General: All lathing and gypsum plaster materials and accessories shall be marked with appropriate standards referenced in this section and stored in such manner as to protect them from the weather.

1601.2 Standards: All interior lathing and gypsum plastering materials shall conform to the standards listed in Table 1601 and Appendix A and, when required for fire protection, shall also conform to the provisions of Article 9.

1601.3 Installation: Installation of these materials shall conform to Section 1604.0.

Table 1601
PLASTER MATERIALS AND ACCESSORIES

Material	Standard
Exterior plaster bonding compounds	ASTM C932
Gypsum base for veneer plasters	ASTM C588
Gypsum casting and molding plaster	ASTM C59
Gypsum Keene's cement	ASTM C61
Gypsum lath	ASTM C37
Gypsum plaster	ASTM C28
Gypsum veneer plaster	ASTM C587
Interior bonding compounds, gypsum	ASTM C631
Lime plasters	ASTM C5; C206
Metal lath	ASTM C847
Plaster aggregates	
Sand	ASTM C35; C897
Perlite	ASTM C35
Vermiculite	ASTM C35
Portland cement	ASTM C150
Steel studs and track	ASTM C645; C955
Steel screws	ASTM C1002; C954
Welded wire lath	ASTM C933

SECTION 1602.0 PORTLAND CEMENT STUCCO LATH AND PLASTER

1602.1 General: All exterior and interior portland cement stucco lathing and plastering shall be done with appropriate materials listed in Table 1601 and Appendix A.

1602.2 Weather protection: All materials shall be stored in such a manner as to protect them from the weather.

1602.3 Installation: Installation of these materials shall be in conformance with ASTM C926 listed in Appendix A and Section 1602.4.

1602.4 Protection after application: At all times during application and for a period of not less than 48 hours after application of each coat, provisions shall be made to keep stucco work above 40 degrees F. (4 degrees C.).

SECTION 1603.0 GYPSUM BOARD MATERIALS

1603.1 General: All gypsum board materials and accessories shall be marked with appropriate standards referenced in this section and stored so as to protect them from the weather.

1603.2 Standards: All gypsum board materials shall conform to the appropriate standards listed in Table 1603 and Appendix A.

1603.3 Installation: Installation of these materials shall conform to Section 1604.0 and, when required for fireresistance, to Article 9.

Table 1603
GYPSUM BOARD MATERIALS AND ACCESSORIES

Material	Standard
Gypsum sheathing	ASTM C79
Gypsum wallboard	ASTM C36
Joint reinforcing tape and compound	ASTM C474; C475
Nails for gypsum boards	ASTM C514
Steel screws	ASTM C1002; C954
Steel studs, nonloadbearing	ASTM C645
Water-resistant gypsum backing board	ASTM C630

1603.4 Water-resistant gypsum backer board: In all areas subjected to repeated damp conditions and moisture accumulation such as bathtub and shower compartments, water-resistant gypsum backer board complying with ASTM C630 listed in Appendix A shall be used as a substratum unless protected with a moisture-proof and vapor-proof covering.

SECTION 1604.0 GYPSUM CONSTRUCTION

1604.1 General: Gypsum board and plaster construction shall be of the materials listed in Table 1603. These materials shall be assembled and installed in conformance with appropriate standards listed in Table 1604 and Appendix A.

1604.2 Limitations: Gypsum construction shall not be used in any exterior location where it would be directly exposed to the weather.

1604.3 Inspection: The code official shall be notified not less than 24 hours in advance of all plastering work or installation of any gypsum board except gypsum lath. Plaster shall not be applied until after the lathing or other plaster base has been inspected and approved by the code official.

1604.4 Weather protection: When plastering work is in progress, the building or structure shall be enclosed and conditioned to provide proper ventilation and temperatures not less than 40 degrees F. (4 degrees C.) nor more than 80 degrees F. (27 degrees C.) from one week prior to the plastering operation and until one week following or until the plaster is dry.

Table 1604
INSTALLATION OF GYPSUM CONSTRUCTION

Material	Standard
Gypsum plaster	ASTM C842
Gypsum veneer base	ASTM C844
Gypsum veneer plaster	ASTM C843
Gypsum wallboard	GA-216
Interior lathing and furring	ASTM C841
Steel framing for gypsum boards	ASTM C754; C1007

EXCERPTS FROM 1988 EDITION

STANDARD BUILDING CODE

The STANDARD BUILDING CODE, published by Southern Building Code Congress, International, Inc., 900 Montclair Road, Birmingham, Alabama 35213-1206, is a model code which is updated and republished every three years. This code is the official building code of the majority of building authorities in the southern part of the country. Following excerpts are reproduced with permission of the publisher. Copies of this code may be purchased from Building News, Inc., 3055 Overland Avenue, Los Angeles 90034 at a cost of $45.00, plus sales tax (to California destinations, 6%; L.A. County 6½%) and $2.95 shipping charge.

CHAPTER 10
FIRE RESISTANCE STANDARDS FOR MATERIALS AND CONSTRUCTION

1001 GENERAL

1001.1 Tests

1001.1.1 Fire protection requirements of this Code are based on fire resistance ratings. Materials, thicknesses, and assemblies which have successfully performed under tests made by a recognized laboratory in accordance with the requirements of ASTM E 119 or based on calculations and accepted engineering practice as set forth in Chapter 31 shall be accepted by the Building Official for specific ratings.

EXCEPTION: In determining the fire resistance rating of exterior bearing walls, compliance with the ASTM E 119 criteria for unexposed surface temperature rise and ignition of cotton waste due to passage of flame or gases, is required only for a period of time corresponding to the required fire resistance rating of an exterior, nonbearing wall with the same horizontal separation distance, and in a building of the same type of construction. When the fire resistance rating determined in accordance with this exception exceeds the fire resistance rating determined in accordance with ASTM E119, the fire exposure time period, water pressure and application duration criteria for the hose stream test of ASTM E119 shall be based upon the fire resistance rating determined in accordance with this exception.

1001.1.2 When insulation or other materials which may change the capacity for heat dissipation are added to or subtracted from fire resistant roof or ceiling assemblies whose fire ratings are listed in this Code or listed in reference documents, fire test results or other substantiating data shall be submitted to the Building Official to show that the required fire resistance time period is not reduced.

1001.1.3 Thicknesses established by fire tests shall be construed as establishing minimum requirements for fire resistance only, and shall not preclude the application of other requirements of this Code where consideration of strength, durability or stability require greater thicknesses.

1001.1.4 Combustible materials shall not enter into the construction of assemblies except as provided in the foregoing prescribed tests.

1001.2 Opening Protection

Fire doors, curtains, shutters, windows, or other protection required for openings in fire resistive walls, shall be in accordance with the requirements of 703.

1001.3 Penetrations

1001.3.1 Penetration of fire resistance rated walls, floors, or floor/ceiling assemblies for electrical, telephone, plumbing, ducts, intercommunication systems or similar facilities shall not be permitted.

EXCEPTIONS:

1. Penetrations which are included in assemblies tested in accordance with ASTM E 119.

2. Penetrations by noncombustible pipe and conduit when all openings around the pipe or conduit are firestopped in accordance with 705.1.4.

3. Penetrations by pipe, conduit and cables when the penetration is protected with a system which has an F rating equal to or greater than the assembly in which the penetration occurs, when tested in accordance with ASTM E 814 conducted with a minimum positive differential pressure of 0.03-inch water column. A T rating equal to one-half of the required fire resistance rating of the floor shall be required for floor penetrations which are outside of a shaft enclosure.

4. Duct penetrations in accordance with 703.3.

1001.3.2 When walls, floors and partitions are required to have a minimum 1-hour or greater fire resistance rating, cabinets, bathroom componer lighting and other fixtures shall be so installed such that the required resistance will not be reduced.

EXCEPTION: Fixtures which are listed for such installation are permitt

1001.4 Column Protection

Where columns require a fire resistance rating, the entire column, includi its connections to beams or girders, shall be protected. Where the colu extends through a ceiling, fire protection of the column shall be continu from the top of the floor through the ceiling space to the underside of floor deck above, except as provided in Table 600, Note f or other su provisions of this Code.

1002 MATERIALS FOR FIRE PROTECTION

1002.1 Scope

Materials prescribed herein for fire resistance and fire protection shall confo with the requirements of this Chapter.

1002.2 Brick

Brick shall be laid in Type M, S, N or O mortar. Solid clay and sh brick shall conform to ASTM C 216 or ASTM C 62. Hollow clay a shale brick shall conform to ASTM C 652. Concrete brick shall confo to ASTM C 55. Sand-lime brick shall conform to ASTM C 73. Cera glazed structural facing tile and facing brick shall conform to ASTM C 12

1002.3 Clay or Shale Tile

Hollow clay or shale tile shall be laid in Type M, S, N, O or gypsum mort Clay or shale tile used in nonbearing partitions and for fire protection sh meet the requirements of ASTM C 56. Clay or shale tile used in exter walls and in all loadbearing walls shall comply with the requirements ASTM C 34 and ASTM C 212.

1002.4 Gypsum

1002.4.1 Poured gypsum used for fire protection and floor and roof constr tion shall contain not more than 12 1/2% of wood chips, shavings or fib measured in a dry condition, as a percentage by weight of the dry m Gypsum mortar shall be composed of one part gypsum and not more th three parts clean, sharp, well-graded sand, by weight.

1002.4.2 Fibered plaster may be used where unsanded or neat gypsum plas is prescribed.

1002.4.3 All plaster mixes for sanded gypsum plasters shall be measu by dry weight.

1002.4.4 When gypsum plaster is used with an aggregate, the proportio shall be as required in 1803.1.

1002.5 Gypsum Lath, Wallboard and Sheathing Board

1002.5.1 Gypsum lath shall comply with the provisions of ASTM C 37.

1002.5.2 Gypsum lath shall be nailed to wood studs or joists in all constru ions required to be fire resistant, with 1 1/8-inch, 13 ga, 19/64-inch head blued nails at intervals not exceeding 4 inches on centers (five na per lath for support of 16-inch lath) or equivalent attachment.

1002.5.3 Gypsum wallboard shall comply with the provisions of AST C 36.

1002.5.4 Gypsum sheathing board shall comply with the provisions of AST C 79.

1002.5.5 Gypsum veneer base shall comply with the provisions of AST C 588.

1002.5.6 Gypsum veneer plaster shall comply with the provisions of AST C 587.

1002.5.7 Exterior gypsum soffit board shall comply with the provisions ASTM C 931.

002.5.8 Water resistant gypsum backing board shall comply with the provisions of ASTM C 630.

002.6 Metal or Wire Lath

002.6.1 Metal lath shall comply with the provisions of ASTM C 847. Wherever metal lath or wire lath and plaster are used as required protection against the spread of fire, the weight of lath shall be not less than 1/2 lb per sq yd when used in vertical position, and not less than 3/4 lb per sq yd when used in horizontal position. Wire lath shall not be lighter than 2 1/2 meshes per inch, or equivalent.

002.6.2 Weight tags shall be left on all metal lath or wire lath until inspected and approved by the Building Official.

002.6.3 Metal lath for ceilings below wood joists in construction which is required to be fire resistant shall be attached with 1 1/2-inch, 11 ga, 7/16-inch head barbed roofing nails spaced at intervals not to exceed 6 inches on centers, or equivalent attachment.

002.6.4 Welded wire lath shall comply with the provisions of ASTM C 933.

002.6.5 Woven wire lath shall comply with ASTM C 1032.

002.8 Vermiculite

Vermiculite, when used as an aggregate with plaster, shall conform in particle size to ASTM C 35. The weight of vermiculite shall be not less than 6 nor more than 10 pcf as determined by measurement in a cubic-foot box, using the shoveling procedure as outlined in ASTM C 29.

002.9 Perlite

Perlite, when used as an aggregate with plaster, shall conform in particle size to ASTM C 35. The weight of perlite shall be not less than 7 1/2 nor more than 15 pcf, as determined by measurement in a cubic-foot box, using the shoveling procedure as outlined in ASTM C 29.

002.11 Sprayed Fire Resistant Materials

002.11.1 Sprayed fibrous and cementitious materials used for structural fire resistance and fire protection shall provide the fire resistance ratings set forth in this Code. The density and thickness shall be determined in accordance with 1002.12.2 and 1002.12.3.

002.11.2 Thickness measurement and acceptance criteria:

1. 25% of the structural frame, columns and beams in each story shall be inspected for thickness determination in accordance with ASTM E 605.

2. 10% of beams (other than structural frame members) on each floor shall be selected at random and shall be measured for thickness as required by these methods.

3. Floor thickness measurements, where required, shall be taken on a random basis for each 10,000 sq ft of area.

4. The average thickness as determined by this procedure shall not be less than that specified in inches subject to a tolerance of ± 1/8 inch. The acceptance of measurements with a minus tolerance greater than 1/8 inch shall not be permitted. Measurements greater than 1/8 inch above the required shall not be used to determine the thickness average.

5. Where thicknesses are less than that required, the condition shall be corrected. The location of any uncorrected areas shall be reported to the Building Official.

002.11.3 Density measurement and acceptance criteria:

1. There shall be density test specimens taken from a column, a beam and a deck for each 10,000 sq ft of floor area or fraction thereof or from each floor if the floor area is smaller than 10,000 sq ft in accordance with ASTM E 605.

2. No density sample shall have a density less than 5% below the specified density. Where the density is less than the 5% tolerance allowed above, the work shall be corrected to the satisfaction of the Building Official.

1003 FIRE RESISTANCE REFERENCES

1003.1 Reference Tables

1003.1.1 To meet the fire resistance requirements of this Code, it shall be determined that materials, constructions and assemblies of construction materials have successfully performed under accepted tests as prescribed in 1001.

1003.1.2 Appropriate fire resistant materials, constructions and assemblies of constructions as listed in Appendix B and the following publications may be accepted as if herein listed:

FM Specification Tested Products Guide, 1986.
GA Fire Resistance Design Manual.
UL Fire Resistance Directory.

1003.1.3 Other fire resistance ratings may be accepted by the Building Official for fire protection on evidence of compliance with 1001.

CHAPTER 18
LATHING, PLASTERING, AND GYPSUM CONSTRUCTION

1801 GENERAL

1801.1 Lathing, plastering and gypsum construction shall be done in the manner and with the materials specified in this chapter, and when required for fire protection shall also comply with the provisions of Chapter 10.

1801.2 No plaster shall be applied until the lathing has been inspected and approved by the Building Official.

1801.3 The Building Official may require that test holes be made in the wall for the purpose of determining the thickness and proportioning of the plaster, provided the permit holder has been notified 24 hours in advance of the time of making such tests.

1801.4 Joint treatment of gypsum wallboard shall not be applied until the wallboard application has been approved by the Building Official.

1802 MATERIALS

Materials	Designation
Sand—Shall be washed and when used with portland cement for scratch coat plastering the amount of sand retained on a No. 16 sieve shall not be less than 10% nor more than 40%	ASTM C 35
Perlite	ASTM C 35
Vermiculite	ASTM C 35
Gypsum Plasters	ASTM C 28
Gypsum Veneer Plaster	ASTM C 587
Gypsum Veneer Base	ASTM C 588
Water Resistant Gypsum Backing Board	ASTM C 630
Bonding Compounds for Interior Plastering	ASTM C 631
Lime-Special Finishing Hydrated Lime Type "S"	ASTM C 206
Quicklime for structural purposes (Lime putty shall be made from quicklime or hydrated lime and shall be prepared in an approved manner.)	ASTM C 5
Keene's Cement	ASTM C 61
Portland Cement Type I, II, or III	ASTM C 150
Type I-A, II-A, or III-A EXCEPTION: Approved types of plasticizing agents may be added to portland cement Type I or II in the manufacturing process, but not in excess of 12% of the total volume. Plastic or waterproofed cements so manufactured shall meet the requirements for portland cement as specified in ASTM C 150 except in respect to the limitation on insoluble residue, air-entrainment and additions subsequent to calcination.	
Masonry Cement Type II	ASTM C 91
Portland Blast Furnace Slag Cement	ASTM C 595 Type IS-A
Gypsum Lath	ASTM C 37
Metal Lath	ASTM C 847
Exterior Soffit Board	ASTM C 931
Gypsum Wallboard	ASTM C 36
Gypsum Backing Board	ASTM C 442
Joint Reinforcing Tape and Adhesive Materials	ASTM C 474 ASTM C 475
Exterior Soffit Board	ASTM C 931
Steel Studs (for use with Gypsum Boards)	ASTM C 645
Steel Studs, Loadbearing (for use with Gypsum Boards)	ASTM C 955
Screws (for use with Framing covered with gypsum boards; types G, S and W)	ASTM C 1002
Screws (for Loadbearing Steel Framing)	ASTM C 954

1803 APPLICATION

1803.1 Interior Lathing And Plastering

1803.1.1 Installation of interior gypsum lathing and furring shall comply with ASTM C 841.

1803.1.2 Interior gypsum plastering shall comply with ASTM C 842.

1803.1.3 Portland cement plaster shall comply with ASTM C 926.

1803.2 Exterior Lathing And Plastering

1803.2.1 Exterior use of portland cement plaster shall comply with the application requirements of ASTM C 926.

1803.2.2 Installation of exterior lathing and framing shall comply with ASTM C 1047.

1803.3 Pneumatically Placed Portland Cement Plaster

1803.3.1 Pneumatically placed portland cement plaster shall be a mixture of portland cement and aggregate conveyed by air through a pipe or flexible tube, and deposited by air pressure in its final position.

1803.3.2 Rebound material may be screened and reused as aggregate in an amount not greater than 25% of the total sand in any batch.

1803.3.3 Pneumatically placed portland cement plaster shall consist of a mixture of one part cement to not more than five parts of aggregate. Plasticity agents may be used as specified elsewhere in this chapter. Except when applied to concrete or masonry, such plaster shall be applied in not less than two coats to a minimum total thickness of 7/8 inch.

1803.4 Application Of Gypsum Wallboard

1803.4.1 Interior and exterior applications and finishing of gypsum board, other than gypsum veneer base and plaster, shall be done in accordance with 1003, 1805 or GA 216.

1803.4.2 Gypsum veneer base and veneer plaster shall be applied and finished in compliance with 1003, 1805, or ASTM C 844 and ASTM C 843.

1803.5 Application Of Steel Studs

1803.5.1 Nonloadbearing steel framing shall be installed in compliance with the provisions of ASTM C 754.

1803.5.2 Loadbearing (Transverse and Axial) Steel Studs and Related Accessories shall be installed in compliance with the provisions of ASTM C 1007.

1804 ALLOWABLE PARTITION HEIGHTS

Composite partitions of gypsum wallboard and steel studs shall be limited in height in accordance with Table 1804.

1805 VERTICAL GYPSUM BOARD DIAPHRAGMS

1805.1 General

1805.1.1 Gypsum wallboard, gypsum sheathing and gypsum veneer base may be used on wood studs for vertical diaphragms if applied in accordance with 1805. Shear-resisting values shall not exceed those set forth in Table 1805.

1805.1.2 The shear values tabulated shall not be cumulative with the shear value of other materials applied to the same wall. The shear values may be doubled when identical materials applied as specified in 1805.3 are applied to both sides of the wall.

TABLE 1804 — ALLOWABLE PARTITION HEIGHTS BASED ON WALLBOARD AND NO. 25 GAGE STUDS ACTING AS A COMPOSITE SECTION[1] [2]

Stud Spacing (in)	Facing On Each Side	Stud Depth (in)					
		1 5/8	2 1/2	3 1/4	3 5/8	4	6
16	1/2"-one ply	11'0"	14'8"	17'10"	19'5"	20'8"	18'10
24	1/2"-one ply	10'0"	13'5"	16'0"	17'3"	18'5"	17'8
24	1/2"-two ply	12'4"	15'10"	18'3"	19'5"	20'8"	19'0

1. The tabulated stud heights are based on 0.0179" uncoated thickness (25 ga) steel studs manufactured in compliance with ASTM C 754 for installation of screw type steel framing members to receive gypsum boards.
2. Gypsum board product must be 1/2" minimum thick and may be applied vertical or horizontally.

1805.2 Wall Framing

1805.2.1 Framing for vertical diaphragms shall comply with 1707.2 for bearing walls. Studs shall be spaced no farther apart than 16 inches center to center. Marginal studs and plates shall be anchored to resist all design forces.

1805.2.2 The maximum allowable height to length ratio for the construction shall be 1 1/2:1.

1805.3 Application

1805.3.1 End joints of adjacent courses of gypsum board sheets shall not occur over the same stud.

1805.3.2 Where required in Table 1805, blocking having the same cross sectional dimensions as the studs shall be provided at all joints that are perpendicular to the studs.

1805.3.3 The size and spacing of nails shall be as set forth in Table 1805. Nails shall be spaced not less than 3/8 inch from edges and ends of gypsum boards or sides of studs, blocking and top and bottom plates.

1805.3.4 Gypsum sheathing 4 ft wide may be applied parallel or perpendicular to studs. Pieces 2 ft wide shall be as set forth in Table 1805.

1805.3.5 Gypsum wallboard or veneer base may be applied parallel or perpendicular to studs. Maximum allowable shear values shall be as set forth in Table 1805.

1805.4 Masonry And Concrete Construction

Gypsum board shall not be used in vertical diaphragms to resist forces imposed by masonry or concrete construction.

TABLE 1805
ALLOWABLE SHEAR FOR WIND OR SEISMIC FORCES IN POUNDS PER FOOT FOR GYPSUM BOARD VERTICAL DIAPHRAGMS[1]

Material	Size	Wall Construction	Nail Spacing[2] (in)	Shear Value	Minimum Nail Size
Gypsum Sheathing Board	1/2" x 2' x 8'	Unblocked	4	75	No. 11 gage, 13/4" long 7/16" head, diamond-point, galvanized
	1/2" x 4'	Blocked	4	175	
	1/2" x 4'	Unblocked	7	100	
Gypsum Wallboard or Veneer Base	1/2"	Unblocked	7	100	5d cooler nails
		Unblocked	4	125	
		Blocked	7	125	
		Blocked	4	150	
	5/8"	Blocked	4	175	6d cooler nails
		Blocked	Base Ply 9	250	Base Ply - 6d cooler nails
		Two-Ply	Face Ply 7		Face Ply - 8d cooler nails

1. These vertical diaphragms shall not be used to resist loads imposed by masonry or concrete walls. Values are for short-time loading due to wind or earthquake and must be reduced 25% for normal loading.
2. Applies to nailing at all studs, top and bottom plates, and blocking.

EXCERPTS FROM 1986 EDITION

CABO ONE AND TWO FAMILY DWELLING CODE

Note: The CABO Code is a national building code authored by the Council of American Building Officials, a coalition of the three regional model code writing organizations listed in the copyright notice below.

Chapter 5
WALL COVERING

Section R-501—General

R-501.1—Application. The provisions of this chapter shall control the design and construction of the interior and exterior wall covering for all buildings. Conformity with the applicable material, test and construction standards specified in this chapter and Section S-26.501 shall be evidence that the coverings applied in accordance with the provisions of this chapter are reasonably safe to persons and property. The use of other materials or methods of construction accomplishing the purposes intended by this code and approved by the building official in accordance with Section R-108 shall be accepted as complying with this code.

R-501.2—Installation. Products sensitive to adverse weather shall not be installed until adequate weather protection for the installation is provided. Exterior sheathing shall be dry before applying exterior cover.

Section R-502—Interior Covering

R-502.1—General. Interior coverings shall be installed in accordance with this chapter and Tables Nos. R-502.1.1A, R-502.1.1B, R-502.1.2, R-502.1.3 and R-502.1.4.

R-502.2—Vertical Assemblies. Vertical support for lath or gypsum wallboard shall be not less than 2 inches nominal in least dimension. Wood stripping for furring shall not be less than 2 inches nominal thickness in the least dimension except that furring strips not less than 1-inch by 2-inch nominal dimension may be used over solid backing.

R-502.3—Interior. Where wood-framed walls and partitions are covered in the interior with portland cement plaster or tile or similar material and subject to water splash, the framing shall be protected with an approved moisture barrier. Vapor barriers shall not be used behind water-resistant gypsum backing board.

Gypsum board may be applied at right angles or parallel to framing members, except gypsum lath shall be applied at right angles with end joints staggered.

Support spacing for gypsum and metal lath shall conform with Table No. R-502.1.1A, and fastener spacing shall conform to Table No. R-502.1.1B.

R-502.4—Interior Plaster. Plastering with gypsum plaster or portland cement plaster shall be not less than two coats when applied over other bases permitted by this section, except that veneer plaster may be applied in one coat not to exceed 3/16-inch thickness.

R-502.5—Gypsum Wallboard. All gypsum wallboard shall be installed in accordance with the provisions of this section. Gypsum wallboard shall not be installed until weather protection is provided.

All edges and ends of gypsum wallboard shall occur on the framing members, except those edges and ends which are perpendicular to the framing members.

Support spacing and the size and spacing of fasteners shall comply with Table No. R-502.1.2.

R-502.8—Finishes and Materials. Interior finishes and materials shall conform to requirements of Section R-218.

Section R-503—Exterior Covering

R-503.1—General. All exterior walls shall be covered with approved materials designed and installed to provide a barrier against the weather and insects to enable environmental control of the interior spaces. The exterior coverings contained in this section shall be installed in the specified manner unless otherwise approved.

R-503.2—Exterior Lath. All lath and lath attachments shall be of corrosion-resistant materials and shall conform to Table No. R-502.1.1.

Backing for vertical surfaces shall consist of sheathing or of not less than No. 18 U.S. gauge steel wire stretched taut horizontally and spaced not more than 6 inches apart vertically.

Where lath on vertical surfaces extends between rafters or other similar projecting members, solid backing shall be installed to provide support for lath and attachments.

Gypsum lath shall not be used, except that on horizontal supports of ceilings or roof soffits it may be used as backing for metal lath or wire lath and portland cement plaster.

Backing is not required under expanded metal lath or paperbacked wire lath.

R-503.3—Exterior Plaster. Plastering with portland cement plaster shall be not less than three coats when applied over metal lath or wire lath and shall be not less than two coats when applied over masonry, concrete or gypsum backing. If plaster surface is completely covered by veneer or other facing material or is completely concealed, plaster application need be only two coats, provided the total thickness is as set forth in Table No. R-502.1.3.

On wood-frame construction with an on-grade concrete floor slab system, exterior plaster shall be applied in such a manner as to cover, but not to extend below, lath, paper and screed.

Only approved plasticity agents and approved amounts thereof may be added to portland cement. When plastic cement is used, no additional lime or plasticizers shall be added. Hydrated lime or the equivalent amount of lime putty used as a plasticizer, may be added to standard portland cement in an amount not to exceed 20 percent by weight of the portland cement.

The proportion of aggregate to cementitious materials shall be as set forth in Tables Nos. R-502.1.4 and R-502.1.5.

Table No. R-502.1.1 A
MAXIMUM SPACING OF SUPPORTS FOR LATH

TYPE OF LATH	MINIMUM WEIGHT (Per Square Yard) GAUGE AND MESH SIZE	VERTICAL (In Inches)			HORIZONTAL (In Inches)	
		Wood	Metal — Solid Plaster Partitions	Other	Wood or Concrete	Metal
Expanded metal lath (diamond mesh)	2.5	16	16	12	—	—
	3.4	16	16	16	16	13¹/₂
Flat rib expanded metal lath	2.75	16	16	16	16	12
	3.4	19	24	19	19	19
Stucco mesh expanded metal lath	1.8 and 3.6	16²	—	—	—	—
³/₈" rib expanded metal lath	3.4	24	24	24	24	24
	4.0	24	24	24	24	24
Sheet lath	4.5	24	—	24	24	24
³/₄" rib expanded metal lath	5.4	—	—	—	36³	36³
Welded	1.95 pounds, No. 11 gauge, 2" × 2"	24	24	24	24	24
	1.4 pounds, No. 16 gauge, 2" × 2"	16	16	16	16	16
Wire lath	1.4 pounds, No. 18 gauge, 1" × 1"⁴	16	—	—	—	—
Woven	1.4 pounds, No. 17 gauge, 1¹/₂" hexagonal⁴	16	—	—	—	—
	1.4 pounds, No. 18 gauge, 1" hexagonal⁴	16	—	—	—	—
³/₈" Gypsum lath (perforated)		16	—	16	16	16
³/₈" Gypsum lath (plain)		16	—	16	16	16
¹/₂" Gypsum lath (perforated)		16	—	16	16	16
¹/₂" Gypsum lath (plain)		24	—	24	24	24

¹Metal lath and wire lath as reinforcement for portland cement plaster shall be furred out away from vertical supports at least ¹/₄ inch. Self-furring lath meets furring requirements.
²Wire backing required on open vertical frame construction except under expanded metal lath and paperback wire lath.
³Contact or furred ceilings only. May not be used in suspended ceilings.
⁴Stucco netting, not to be used as a base for gypsum plaster.

Table No. R-502.1.1 B
MAXIMUM SPACING OF FASTENERS FOR SUPPORT OF LATH

TYPE OF LATH	TYPE (NAILS)	Maximum Spacing Vertical (In inches)	Maximum Spacing Horizontal (In inches)	Leg⁶	Wire Gauge No.	Minimum Crown Width (In inches)	Maximum Spacing Vertical	Maximum Spacing Horizontal
Diamond mesh, expanded metal lath and flat rib metal lath	4d blued box (clinched)¹; 1" No. 11 Ga., ⁷/₁₆ head, barbed; 1¹/₂" No. 11 Ga., ⁷/₁₆", barbed	6; 6; 6	—; —; 6	⁷/₈	16	⁷/₁₆	6	6
³/₈" rib metal lath and sheet lath	1¹/₂" No. 11 Ga., ⁷/₁₆ head, barbed	6	6	1¹/₄	16	⁷/₁₆	6	6
³/₄" rib metal lath	4d Common; 2" No. 11 Ga., ⁷/₁₆ head, barbed	At ribs; —	At ribs	1³/₄; —	16; —	⁷/₁₆; —	At ribs; —	At ribs; —
Wire lath⁴	4d blued box (clinched)³; 1" No. 11 Ga., ⁷/₁₆ head, barbed; 1¹/₂" No. 11 Ga., ⁷/₁₆, barbed; 1¹/₄" No. 12 Ga., ³/₈ head, furring	6; 6; 6; 6	—; —; 6; —	⁷/₈	16	⁷/₁₆	6	6
³/₈" gypsum lath³	1¹/₈" No. 13 Ga., ¹⁹/₆₄ head, blued	5	5	⁷/₈	16	⁷/₁₆	5	5
¹/₂" gypsum lath³	1¹/₄" No. 13 Ga., ¹⁹/₆₄ head, blued	5⁵ 4⁶	5⁵ 4⁶	1¹/₈	16	⁷/₁₆	4	4

¹With divergent points and semiflattered round wire for gypsum lath.
²When lath and stripping are stapled simultaneously, increase leg length of staple ¹/₈ inch.
³For interior only.
⁴Attach self-furring wire fabric lath to supports at furring device.
⁵Perforated lath.
⁶Plain lath.

Table No. R-502.1.2
APPLICATION AND MINIMUM THICKNESSES OF GYPSUM WALLBOARD

THICK-NESS OF GYPSUM WALLBOARD (Inch)	PLANE OF FRAMING SURFACE	LONG DIMENSION OF GYPSUM WALLBOARD SHEETS IN RELATION TO DIRECTION OF FRAMING MEMBERS	MAXIMUM SPACING OF FRAMING MEMBERS (center-to-center, in inches)	Nails¹²	Screws³	NAILS¹—TO WOOD
fastening		**Fastening required without adhesive application.**				
³/₈	Horizontal⁴	Perpendicular	16	7	12	No. 13 gauge, 1¹/₂" long, ¹⁹/₆₄ head, .098" diameter, 1¹/₂" long, annular-ringed 4d cooler nail
	Vertical	Either direction	16	8	12	
¹/₂	Horizontal	Either direction	16	7	12	No. 13 gauge, 1³/₈" long, ¹⁹/₆₄ head, .098" diameter, 1¹/₄" long, annular-ringed 5d cooler nail
	Horizontal	Perpendicular	24	7	12	
	Vertical	Either direction	24	8	12	
⁵/₈	Horizontal	Either direction	16	7	12	No. 13 gauge, 1⁵/₈" long, ¹⁹/₆₄ head, .098" diameter, 1³/₈" long, annular-ringed 6d cooler nail
	Horizontal	Perpendicular	24	7	12	
	Vertical	Either direction	24	8	12	
		With adhesive application.				
³/₈	Horizontal⁴	Perpendicular	16	16	16	Same as above for ³/₈"
	Vertical	Either direction	16	16	24	
¹/₂ or ⁵/₈	Horizontal	Either direction	16	16	16	As required for ¹/₂" and ⁵/₈" gypsum wallboard, see above
	Horizontal	Perpendicular	24	12	16	
	Vertical	Either direction	24	24	24	
2—³/₈ layers	Horizontal	Perpendicular	24	16	16	Base ply nailed as required for ¹/₂" gypsum wallboard and face ply placed with adhesive
	Vertical	Either direction	24	24	24	

¹Where the metal framing has a clinching design formed to receive the nails by two edges of metal, the nails shall be not less than ⁵/₈ inch longer than the wallboard thickness and shall have ringed shanks. Where the metal framing has a nailing groove formed to receive the nails, the nails shall have barbed shanks or be 5d, No. 13 1/2 gauge, 1⁵/₈ inches long, ¹⁵/₆₄-inch head for ¹/₂-inch gypsum wallboard; 6d, No. 13 gauge, 1⁷/₈ inches long, ¹⁵/₆₄-inch head for ⁵/₈-inch gypsum wallboard.
²Two nails spaced not less than 2 inches apart, nor more than 2¹/₂ inches apart, and pairs of nails spaced not more than 12 inches center to center may be used.
³Screw shall be Type S or W per ASTM C 1002 and long enough to penetrate wood framing not less than ⁵/₈ inch and metal framing not less than ¹/₄ inch.
⁴Three-eighths-inch single-ply gypsum board shall not be installed if water-based spray-textured finish is applied nor to support insulation above a ceiling.

Table No. R-502.1.3
THICKNESS OF PLASTER

PLASTER BASE	Gypsum Plaster	Portland Cement Mortar
Expanded metal lath	⁵/₈" minimum¹	⁵/₈" minimum¹
Wire lath	⁵/₈" minimum¹	³/₄" minimum (interior)²; ⁷/₈" minimum (exterior)²
Gypsum lath	¹/₂" minimum	
Masonry walls³	¹/₂" minimum	¹/₂" minimum
Monolithic concrete walls³ ⁴	⁵/₈" maximum	⁷/₈" maximum
Monolithic concrete ceilings³ ⁴	³/₈" maximum⁵	¹/₂" maximum
Gypsum veneer base⁶	¹/₁₆" minimum	

¹When measured from back plane of expanded metal lath, exclusive of ribs, or self-furring lath, plaster thickness shall be ³/₄ inch minimum.
²When measured from face of support or backing.
³Because masonry and concrete surfaces may vary in plane, thickness of plaster need not be uniform.
⁴When applied over a liquid bonding agent, finish coat may be applied directly to concrete surface.
⁵Approved acoustical plaster may be applied directly to concrete or over base coat plaster, beyond the maximum plaster thickness shown.
⁶Attachment shall be in accordance with Table No. 502.1.2.

Table No. R-502.1.4
GYPSUM PLASTER PROPORTIONS¹

NUMBER	COAT	PLASTER BASE OR LATH	Damp Loose Sand¹	Perlite or Vermiculite³
Two-coat work	Base coat	Gypsum lath	2¹/₂	2
	Base coat	Masonry	3	3
Three-coat work	First coat	Lath	2⁴	2
	Second coat	Lath	3⁴	2⁵
	First and second coats	Masonry	3	3

¹Wood-fibered gypsum plaster may be mixed in the proportions of 100 pounds of gypsum to not more than 1 cubic foot of sand where applied on masonry or concrete.
²When determining the amount of aggregate in set plaster, a tolerance of 10 percent shall be allowed.
³Combinations of sand and lightweight aggregate may be used, provided the volume and weight relationship of the combined aggregate to gypsum plaster is maintained.
⁴If used for both first and second coats, the volume of aggregate may be 2¹/₂ cubic feet.
⁵Where plaster is 1 inch or more in total thickness the proportions for the second coat may be increased to 3 cubic feet.

Table No. R-502.1.5
PORTLAND CEMENT PLASTER

COAT	Portland Cement Plaster² Maximum Volume Aggregate per Volume Cement	Maximum Volume Lime per Volume Cement	Maximum Volume Sand per Volume Cement and Lime	Approximate Minimum Thickness⁴	MINIMUM PERIOD MOIST Curing	MINIMUM INTERVAL BETWEEN COATS
First	4	³/₄	4	³/₈⁵	48⁶ Hours	48⁷ Hours
Second	5	³/₄	5	First and second coats	48 Hours	7 Days⁸
Finished	3⁹	—	3⁹	¹/₈"	—	—⁸

¹When determining the amount of aggregate in set plaster, a tolerance of 10 percent may be allowed.
²From 10 to 20 pounds of dry hydrated lime (or an equivalent amount of lime putty) may be added as a plasticizing agent to each sack of Type I and Type II standard portland cement in base coat plaster.
³No additions of plasticizing agents shall be made.
⁴See Table No. R-501.1.3.
⁵Measured from face of support or backing to crest of scored plaster.
⁶Twenty-four-hour minimum period for moist curing of interior portland cement plaster.
⁷Twenty-four-hour minimum interval between coats of interior portland cement plaster.
⁸Finish coat plaster may be applied to interior portland cement base coats after a 48-hour period.
⁹For finish coat, plaster up to an equal part of dry hydrated lime by weight (or an equivalent volume of lime putty) may be added to Type I, Type II and Type III standard portland cement.

EXCERPTS FROM

CALIFORNIA ADMINISTRATIVE CODE

Lathing, plastering and drywall covering of walls and ceilings in hospitals and public schools in the State of California is governed by rules and standards promulgated and enforced by the Office of the State Architect (for schools) and the Office of Statewide Health, Planning and Development (for hospitals) and as approved by the State of Building Standards Commission (to be renamed California Building Standards Commission in 1989).

Basically these state agencies follow Chapter 47 of the 1979, 1982 and 1985 editions of the Uniform Building Code, which code is adopted by reference as the primary building code of the state, with certain modifications. Adoption of the 1988 edition the UBC is scheduled to take place and become effective about January, 1989.

Following are excerpts from the current State of California Building Code (Part 2, Title 24, of the California Administrative Code) which relate to lath, plaster and drywall.

2-4702. Materials. Lathing and plastering shall conform to UBC Standards listed in Chapters 47 and 60, UBC, except as modified in this Chapter.

2-4703. Vertical Assemblies.

(a) **General.** In addition to the requirements of this section, vertical assemblies of plaster or gypsum board shall be designed to resist the loads specified in Chapter 2-33 of this Code.

> Note No. 1: For wood framing, see Chapter 2-25.
> Note No. 2: For metal framing, see Chapter 2-27.
> *EXCEPTION: Wood-framed assemblies meeting the requirements of Section 2-2518 need not be designed.*

2-4704. Horizontal Assemblies.

(a) **General.** In addition to the requirements of this section, supports for horizontal assemblies of plaster or gypsum board shall be designed to support all loads as specified in Chapter 2-23 of this code.

> *EXCEPTION: Wood-framed assemblies meeting the requirements of Section 2-2518 need not be designed.*

(b) **Wood Framing.** Wood stripping or suspended wood systems, where used, shall be not less than two inches nominal thickness in the least dimension, except that furring strips not less than one inch × two inches nominal dimension may be used over solid backing.

Wood furring strips for ceilings fastened to floor or ceiling joists shall be nailed at each bearing with two common wire nails, one of which shall be a slant nail and the other a face nail, or by one nail having spirally grooved or annular grooved shanks approved by the Office of the State Architect for this purpose. All stripping nails shall penetrate not less than 1¾ inches into the member receiving the point. Holes in stripping at joints shall be subdrilled to prevent splitting.

Where common wire nails are used to support horizontal wood stripping for plaster ceilings, such stripping shall be wire tied to the joists four feet on center with two strands of No. 18 W & M gauge galvanized annealed wire to an 8d common wire nail driven into each side of the joist two inches above the bottom of the joist or to each end of a 16d common wire nail driven horizontally through the joist two inches above the bottom of the joist, and the ends of the wire secured together with three twists of the wire.

2-4705. Interior Lath.

(a) **Application of Gypsum Lath.** The thickness, spacing of supports and the method of attachment of gypsum lath shall be as set forth in Tables No. 47-B and No. 47-C, UBC. Approved wire and sheet metal attachment clips may be used.

Gypsum lath shall be applied with the long dimension perpendicular to supports and with end joints staggered in successive courses. End joints may occur on one support when stripping is applied the full length of the joints. Where electrical radiant heat cables are installed on ceilings, the stripping, if conductive, may be omitted a distance not to exceed 12 inches from the walls. Where lath edges are not in moderate contact and have joint gaps exceeding ⅜ inch, the joint gaps shall be covered with stripping or cornerite. Stripping or cornerite may be omitted when the entire surface is reinforced with not less than one inch No. 20 U.S. gauge woven wire. When lath is secured to horizontal or vertical supports not used as structural diaphragms, end joints may occur between supports when lath ends are secured together with approved fasteners. Vertical assemblies also shall conform with Subsection 2-2309(b). Cornerite shall be installed so as to retain position during plastering at all internal corners. Cornerite may be omitted when plaster is not continuous from one plane to an adjacent plane.

(c) **Application of Metal Plaster Bases.** The type and weight of metal lath, and the gauge and spacing of wire in welded or woven lath, the spacing of supports, and the methods of attachment to wood supports shall be as set forth in Tables No. 47-B and No. 47-C, UBC.

Where interior lath is attached to horizontal wood supports, either of the following attachments shall be used in addition to the methods of attachment set forth in Table No. 47-C, UBC.

1. Secure lath to alternate supports with ties consisting of a double strand of No. 18 W & M gauge galvanized annealed wire at one edge of each sheet of lath. Wire ties shall be installed not less than three inches back from the edge of each sheet and shall be looped around stripping, or attached to an 8d common wire nail driven into each side of the joists two inches above the bottom of the joist or to each end of a 16d common wire nail driven horizontally through the joist two inches above the bottom of the joist and the ends of the wire secured together with three twists of the wire.

2. Secure lath to each support with ½ inch wide, 1½ inches long No. 9 W & M gauge, ring shank, hook staple

placed around a 10d common nail laid flat under the surface of the lath not more than three inches from edge of each sheet. Such staples may be placed over ribs of ⅜-inch rib lath or over back wire of welded wire fabric or other approved lath, omitting the 10d nails.

Metal lath shall be attached to metal supports with not less than No. 18 U.S. gauge tie wire spaced not more than six inches apart or with approved equivalent attachments.

Metal lath or wire fabric lath shall be applied with the long dimension of the sheets perpendicular to supports.

Metal lath shall be lapped not less than ½ inch at sides and one inch at ends. Wire fabric lath shall be lapped not less than one mesh at sides and ends, but not less than one inch. Rib metal lath with edge ribs greater than ⅛ inch shall be lapped at sides by nesting outside ribs. When edge ribs are ⅛ inch or less, rib metal lath may be lapped ½ inch at sides, or outside ribs may be nested. Where end laps of sheets do not occur over supports, they shall be securely tied together with not less than No. 18 U.S. gauge wire.

Cornerite shall be installed in all internal corners to retain position during plastering. Cornerite may be omitted when lath is continuous or when plaster is not continuous from one plane to an adjacent plane.

2-4706. Exterior Lath.

(a) **Application of Metal Plaster Bases.** The application of metal lath or wire fabric lath shall be as specified in Subsection 2-4705(c) and they shall be furred out from vertical supports or backing not less than ¼ inch, except as set forth in footnote No. 2, Table No. 47-B, UBC.

Where exterior lath is attached to horizontal wood supports, either of the following attachments shall be used in addition to the methods of attachment set forth in Table No. 47-C, UBC:

1. Secure lath to alternate supports with ties consisting of a double strand of No. 18 W & M gauge galvanized annealed wire at one edge of each sheet of lath. Wire ties shall be installed not less than three inches back from the edge of each sheet and shall be looped around stripping, or attached to an 8d common wire nail driven into each side of the joist two inches above the bottom of the joist or to each end of a 16d common wire nail driven horizontally through the joist two inches above the bottom of the joist and the ends of the wire secured together with three twists of the wire.

2. Secure lath to each support with ½-inch wide, 1½-inches long No. 9 W & M gauge ring shank, hook staple placed around a 10d common nail laid flat under the surface of the lath not more than three inches from edge of each sheet. Such staples may be placed over ribs of ⅜-inch rib lath or over back wire of welded wire fabric or other approved lath, omitting the 10d nails.

Where no external corner reinforcement is used, lath shall be furred out and carried around corners at least one support on frame construction.

A weep screed shall be provided at the foundation plate line on all exterior stud walls. The screed shall be of a type which will allow trapped water to drain to the exterior of the building.

2-4708. Exterior Plaster.

(a) **General.** Plastering with portland cement plaster shall be not less than three coats when applied over metal lath or wire fabric lath and shall be not less than two coats when applied over masonry, concrete or gypsum backing as specified in Section 4706(c), UBC. If plaster surface is completely covered by veneer or other facing material, or is completely concealed by another wall, plaster application need be only two coats, provided the total thickness is as set forth in Table No. 47-F, UBC.

On wood frame or metal stud construction with an on-grade concrete floor slab system, exterior plaster shall be applied in such a manner as to cover, but not extend below, lath and paper.

Note: See Subsection 2-4706(e) for the application of paper and lath, and flashing or drip screeds.

Only approved plasticity agents and approved amounts thereof may be added to portland cement. When plastic cement is used, no additional lime or plasticizers shall be added. Hydrated lime or the equivalent amount of lime putty used as a plasticizer may be added to portland cement plaster in an amount not to exceed that set forth in Table No. 47-F, UBC.

For machine-placed plasters, approved fiber may be added to portland cement plaster in approved amounts. Approved portland cement plaster containing approved fiber, blended at the time of manufacture, and so labeled, may be used. Gypsum plaster shall not be used on exterior surfaces.

Note: See Section 424, UBC.

2-4710. Pneumatically Placed Plaster (Gunite).
Pneumatically placed portland cement plaster shall be a mixture of portland cement and sand, mixed dry, conveyed by air through a pipe or flexible tube, hydrated at the nozzle at the end of the conveyor and deposited by air pressure in its final position.

Rebound material may be screened and reused as sand in an amount not greater than 25 percent of the total sand in any batch.

Pneumatically placed portland cement plaster shall consist of a mixture of one part cement to not more than five parts sand. Plasticity agents may be used as specified in Subsection 2-4708(a). Except when applied to concrete or masonry, such plaster shall be applied in not less than two coats to a minimum total thickness of ⅞ inch. The first coat shall be rodded as specified in Section 4708(c), UBC, for the second coat. The curing period and time interval shall be as set forth in Table No. 47-F, UBC.

2-4711. Gypsum Wallboard.

(b) **Supports.** Supports shall be spaced not to exceed the spacing set forth in Table No. 47-G, UBC, for singly-ply application and Table No. 47-H, UBC, for two-ply application. Vertical assemblies shall conform with Section 2-4703. Horizontal assemblies shall comply with Section 2-4704.

CHAPTER SIX

INSPECTION

EXCERPTS FROM

CONSTRUCTION INSPECTION MANUAL

Construction Inspection Manual is a widely used handbook of inspection, published by Building News, Inc., 3055 Overland Avenue, Los Angeles, Calif. 90034, on behalf of the Construction Industry Advancement Fund. Following excerpts from this manual provides an inspection checklist for lath and plaster and drywall construction.

DIVISION 9 — FINISHES
GYPSUM DRYWALL 09250

STANDARDS: GA (45) UL (93)

1. Refer to 3.2 GENERAL ITEMS CHECKLIST where applicable to this section.
2. Material is stored in dry location and does not overload floor systems.

FRAMING SYSTEM

3. Materials are galvanized where exposed to exterior and damp conditions and where otherwise required.
4. Stud spacing is as required. Studs are doubled-up at jambs. Special reinforcement and heavy gauge studs are provided as required.
5. Studs are set to allow for vertical movement such as shrinkage, slab deflection, etc.
6. Studs are securely anchored to walls and columns and friction fit or fastened as required to securely anchored plates. Sound proofing, such as caulking beads, is provided at floors, walls, etc., if required.
7. Locations, layout, and plumbness are verified.
8. Channel stiffeners are provided as required.
9. Special field conditions of fastening and connection are observed for accuracy.
10. Anchorages, blocking, plates, etc., required for other equipment support and fastening are provided and installed.
11. Cut studs for cut-outs and openings are properly framed.
12. Observe size, gauge, spacing, and fastening of runner and furring channels.
13. Hangers of proper type, size, and gauge are provided, and are saddletied, bolted, or clipped as required.
14. Tie wire material and size for connection of channels to runners is provided and properly tied.
15. Elevation and layout of furring is understood. Installation provides a true plane surface, plumb or level as required.
16. Corner beads, expansion devices, casings, trim, and other accessories are provided and properly installed. Long lengths or single lengths are provided. Control and contraction joint type, installation, and method is understood.
17. Frames are provided for access panels as required.
18. Connections and provisions are made at corners or adjoining surfaces of different materials.
19. Holes in metal studs are in alignment.
20. Perimeter sealing or treatment is provided as required for sound or thermal isolation.
21. Wood studs are in alignment, and out-of-line members are corrected. Spacing and construction are as specified.
22. Blocking, bracing, nailers, and back-up to attach wallboard are provided. Verify whether all edges of wallboard require continuous blocking. Provisions are made for required anchorage and support of other equipment.
23. Wood materials are sufficiently dry to avoid "nail popping" due to shrinkage.

WALLBOARD

24. Agency inspection is performed before "closing-in" if required.
25. Type, thickness, length, and edges are as required. Verify if horizontal or vertical application is required.
26. Type of nail or fastener, gauge, length, and spacing are provided as required. Verify whether special nailing is required.
27. Installation complies with manufacturer's recommendations or other requirements.
28. Wallboard is not erected until building is closed-in (depends on weather). Ventilation for air circulation is provided, with adequate dry heat.
29. Fire-rated wallboard and installation system is provided where required by type of construction, occupancy, or otherwise.
30. If fire-rated, all recesses over 16″ square are boxed in, and all penetrations are tight and sealed and otherwise as required by codes.
31. Special type of wallboard is used for damp or other special locations if required. Observe that cut edges and cut outs of moisture resistant edges are properly sealed.
32. Special lengths are provided if required.
33. Special installations required for soundproofing are provided. Observe method and see that isolation is achieved. See that rigid jointing is tight to obstruct passage of sound.
34. Correct sizing for cut-outs, outlet boxes, etc., is performed to avoid patching, sound passage, and thermal loss. Require sawing and do not allow scoring and knock-out.
35. Wallboard is held up from floor 3/8″ minimum.
36. Wallboard is installed with staggered application — back-to-back staggering, wall-to-ceiling staggering, double-layer staggering, etc.
37. Vertical joints are aligned with door jambs.
38. Nailing or fastening is performed as required. Nailing from center outward, paper surfaces not broken, sheets not driven together, etc. Observe that non-metallic cable, plastic, or copper piping is not close to surfaces or damaged.
39. Observe that excessive piecing or jointing is not provided. Damaged sheets are not used or are removed.
40. Taping system is of type, compound, and method required.
41. Number of coatings required is provided, equipment and tools are suitable, sanding between coats is performed, feathering is out 12″ to 16″, and joints will be unnoticeable after finish is applied. Observe that curing time is adequate, and check for bubbles and dimples.
42. Type of texture or finish is understood.
43. Types of internal and external metal corners are provided as required. Wallboard accessories are of type required.
44. Clean-up at intervals is performed and complete clean-up and debris disposal is performed at end of operations. Do not allow excessive debris to accumulate, and see that precautions are taken to prevent tracking to other areas.

SECTIONS ON INSPECTION — EXCERPTS FROM

CONSTRUCTION SPECIFICATIONS INSTITUTE SPECIFICATIONS

Note: The following excellent suggestiones on the subject of inspecting plaster installations appeared as appendices to specifications authored and promulgated by the Construction Specifications Institute prior to the merger of these specifications into the industry-wide Spectext program.

FURRING AND LATHING

Before Installation: Check materials for conformance to specifications.

Check to assure materials are in properly labeled bundles and undamaged bags.

Assure that materials are stored in a dry area protected from weather and accidental damage by other trades.

During Installation: Check that all framing and furring members, and lathing, are installed in accordance with specifications and partitions are located accurately.

After Installation: Assure that all openings in framing are in accordance with specifications and are properly finished, including corner beads and partition intersections.

GYPSUM PLASTER

Before Installation: Check materials for conformance to specifications.

Check to assure materials are in properly labeled and undamaged containers.

Assure that materials are stored in a dry area protected from weather and accidental damage by other trades.

During Installation: Assure that ventilation of area being plastered is maintained during installation.

Check to assure that surfaces adjacent to areas receiving plaster are protected during plaster operations.

Check to assure recommended application procedures are followed for application of bonding agent where required.

Check proportions of mix, method of mixing, and quality of plaster coats.

Check thickness and finish of plaster coats for conformance to specifications.

Assure that scratch coat plaster has been properly dried and aged prior to application of brown coat.

Assure that brown coat plaster has been properly dried and aged prior to application of finish coat.

Check proportions of mix, method of mixing, and quality of applied finish coat material.

Assure that all areas scheduled to receive plaster have been properly finished and dried.

After Installation: Assure that all protective material has been removed and areas in which plaster work was performed have been cleared of waste materials and equipment.

Assure that all openings in plaster are in accordance with specifications; all plastered areas are properly finished.

CEMENT PLASTER

Before Application: 1. Verify that materials are delivered to job site in original containers with labels intact and legible.

2. Assure that cementitious materials are kept dry and stored off ground.

3. Check that finished surfaces which have been installed prior to plastering are protected from contact with wet plaster.

During Installation: 1. Verify that materials are proportioned and mixed in accordance with specification requirements.

2. Check that batches are sized for complete use within one hour after mixing.

3. Check that plaster is applied to required thicknesses and tolerances.

4. Assure that partially hydrated materials are not retempered.

After Application: 1. Verify that cement plaster is cured as required.

2. Assure that cement plaster finish matches approved sample.

3. Check that defects in cement plaster are patched and finished to match surrounding area.

4. Verify that debris is removed from job site.

SPRAYED FIREPROOFING

Verify the materials furnished and methods of installation proposed meet requirements established by testing.

Assure that the thickness required meets the hourly design rating before application of any work is started.

Verify that substrate materials are clean and free of dirt, grease, rust or other materials that would affect bond of fireproofing materials and the installation.

Check the installation of sprayed fireproofing that conditions meet approved sample installation as to substrate preparation, thickness, density, texture and sealer coat if applied.

To assure field quality control, have the sprayed fireproofing installation tested to verify the thickness and density of the actual installation.

Assure that damaged work is patched and corrected and restored to the desired fire rating requirements.

APPENDIX

CONTRACTOR ASSOCIATIONS

ASSOCIATED INTERIOR AND EXTERIOR CONTRACTORS OF OREGON AND SOUTHWEST WASHINGTON, INC., 3420 SW Macadam Avenue, Portland, Oregon 97201. (503) 295-0333.

ASSOCIATION OF PLASTERING AND LATHING CONTRACTORS, P.O. Box 19753, San Diego, California 92119. (619) 464-7681.

ASSOCIATION OF THE WALL & CEILING INDUSTRIES — INTERNATIONAL, 1600 Cameron Street, Alexandria, Virginia 22314. (703) 684-AWCI.

ATLANTA DRYWALL AND PLASTERING CONTRACTORS' ASSOCIATION, 600 Virginia Avenue, N.E., Atlanta, Georgia 30306. (404) 872-4726.

BAKERSFIELD CHAPTER, WESTERN LATH/PLASTER/DRYWALL INDUSTRIES ASSOCIATION, 6420 Midwick Court, Bakersfield, California 93306. (805) 871-2593.

CALIFORNIA ASSOCIATION OF LATHING CONTRACTORS, 14039 Sherman Way, Van Nuys, California 91405. (213) 782-2012.

CEILING AND INTERIOR SYSTEMS CONTRACTORS' ASSOCIATION, 104 Wilmot Road, #201, Deerfield, Illinois 60015. (312) 940-8800.

CONTRACTING PLASTERERS' ASSOCIATION OF SOUTHERN CALIFORNIA, INC. 1280 West Lambert Road, Suite J2, Brea, California 92621. (213) 690-5721; (213) 691-9531; (213) 691-8061.

GREATER CINCINNATI BUREAU FOR LATH AND PLASTER, 3537 Epley Road, Cincinnati, Ohio. (513) 385-2600.

INLAND EMPIRE LATHING AND PLASTERING CONTRACTORS ASSOCIATION, 950 West Second Street, San Bernardino, California 92410. (714) 885-5769.

LATHING AND PLASTERING INSTITUTE OF NORTHERN CALIFORNIA, 1043 #2, Stewart Street, Lafayette, California 94549. (415) 283-5160.

MASTER PLASTERING AND LATHING CONTRACTORS/FRESNO AREA, 5889 Midwick Lane, Fresno, California 93727. (209) 255-5617.

MINNEAPOLIS MASTER PLASTERERS' ASSOCIATION, 3433 Broadway #250, N.E. Minneapolis, Minnesota 55413. (612) 378-9526.

MONTEREY BAY AREA LATHING AND PLASTERING CONTRACTORS ASSOCIATION, 13540 Blackie Road, Castroville, California 95012. (408) 633-3315.

NEW JERSEY CONTRACTING LATHERS AND PLASTERERS ASSOCIATION, 40 Brunswick Avenue, Edison, New Jersey 08817.

NORTHWEST LATH & PLASTER TRUST, 4621 S.W. Kelly Avenue, Portland, Oregon 97201. (503) 224-8226.

ORANGE COUNTY LATHING AND PLASTERING CONTRACTORS ASSOCIATION, 12680 Hoover Street, Garden Grove, California 92641. (714) 892-2014.

PLASTER INFORMATION CENTER/SOUTH BAY ASSOCIATION, 4960 Hamilton Avenue, Suite 100, San Jose, California 95130. (408) 866-8318.

REDWOOD PLASTERING CONTRACTORS ASSOCIATION, 47 De Luca Place, San Rafael, California 94901. (415) 459-4707.

WESTERN LATH/PLASTER/DRYWALL INDUSTRIES ASSOCIATION, 8635 Navajo Road, San Diego, California 92119. (619) 466-9070.

INDUSTRY TRADE PROMOTION BUREAUS

BUREAU FOR LATHING AND PLASTERING OF BATON ROUGE, 2350 Wooddale Boulevard, Baton Rouge, Louisiana 70806. (504) 926-0783.

CHICAGO PLASTERING INSTITUTE, 5859 West Fullerton Ave., Chicago, Illinois 60639. (312) 237-6910.

GYPSUM ASSOCIATION, 1603 Orrington Avenue, Evanston, Illinois 60201. (312) 491-1744.

INFORMATION BUREAU FOR LATH, PLASTER AND DRYWALL, 3127 Los Feliz Boulevard, Los Angeles, California 90059. (213) 663-2213.

INTERNATIONAL INSTITUTE FOR LATH AND PLASTER, 795 Raymond Avenue, St. Paul, Minnesota 55114. (612) 645-0208.

LATHING AND PLASTERING INSTITUTE OF NORTHERN CALIFORNIA, P.O. Box 1053, Lafayette, California 94549. (415) 357-6410.

METAL LATH/STEEL FRAMING DIVISION of National Association of Architectural Metal Manufacturers, 600 So. Federal Street, Chicago, Illinois 60605. (312) 922-6222.

MINNESOTA LATHING AND PLASTERING BUREAU, 795 Raymond Avenue, St. Paul, Minnesota 55114. (612) 645-0208.

NATIONAL LIME ASSOCIATION, 3601 North Fairfax Drive, Arlington, Virginia 22201. (703) 243-LIME.

NORTHWEST WALL AND CEILING BUREAU, INC., 325 Second Avenue West, Seattle, Washington 98119. (206) 284-7160 — (208) 284-4380 and 284-7160.

PACIFIC BUREAU FOR LATHING AND PLASTERING, 905 Umi Street, Room 303, Honolulu, Hawaii 96819. (808) 847-4321 — (800) 847-5607.

PERLITE INSTITUTE, 600 So. Federal Street, Chicago, Illinois 60605. (312) 922-6222.

PLASTERING INDUSTRY BUREAU OF SAN FRAN-CISCO AND SAN MATEO COUNTIES, 2010 Ocean Avenue, Suite D, San Francisco, California 94127. (415) 239-1422.

PLASTERING INFORMATION BUREAU — A DIVISION OF SOUTHERN CALIFORNIA PLASTERING INSTITUTE, INC., 3127 Los Feliz Boulevard, Los Angeles, California 90039. (213) 663-2213.

PORTLAND CEMENT ASSOCIATION, 5420 Old Orchard Road, Skokie, Illinois 60077. (312) 966-6200.

SACRAMENTO VALLEY BUREAU FOR LATH AND PLASTER, INC., And CHAPTER OF WESTERN LATH/PLASTER/DRYWALL INDUSTRIES ASSOCIATION, 1400 "S" Street, Suite 203, Sacramento, California 95814. (916) 444-2397.

SCAFFOLD INDUSTRY ASSOCIATION, 14039 Sherman Way, Van Nuys, California 91405. (213) 782-2012.

SCAFFOLDING, SHORING & FORMING INSTITUTE, INC., 1230 Keith Building, Cleveland, Ohio 44115. (216) 241-7333.

SPRAYED MINERAL FIBER MANUFACTURERS ASSOCIATION, INC., 1 Wall Street, Suite 2400, New York, New York 10005.

STUCCO MANUFACTURERS ASSOCIATION, 14006 Ventura Boulevard, Sherman Oaks, California 91403. (213) 789-8733.

TEXAS BUREAU FOR LATHING & PLASTERING, 6500 Greenville Avenue, Suite 460, Dallas, Texas 75206. (214) 363-6747.

VERMICULITE ASSOCIATION, 52 Executive Park, South Atlanta, Georgia 30329. (404) 631-5621.

WESTERN CONFERENCE OF LATHING AND PLASTERING INSTITUTES, P.O. Box 6468, Santa Ana, California 92706. (714) 531-1278.

LABOR UNIONS

BUILDING TRADES DEPARTMENT OF THE AFL-CIO, 815 16th Street, N.W., Washington, D.C. 20006. (202) 347-1461.

INTERNATIONAL UNION OF BRICKLAYERS AND ALLIED CRAFTSMEN, 815 15th Street, N.W., Washington, D.C. 20005. (202) 783-3788.

LABORERS INTERNATIONAL UNION OF NORTH AMERICA, 905 16th Street, N.W., Washington, D.C. 20006. (203) 737-8320.

OPERATIVE PLASTERERS AND CEMENT MASONS INTERNATIONAL ASSOCIATION, 1125 17th Avenue, N.W., Washington, D.C. 20036. (202) 393-6569.

UNITED BROTHERWOOD OF CARPENTERS & JOINERS OF AMERICA, 101 Constitution Avenue, N.W., Washington, D.C. 20001. (202) 546-6206.

TRADE MAGAZINES

CONSTRUCTION DIMENSIONS, 1600 Cameron Street, Alexandria, Virginia 22314. (703) 684-AWCI.

WALLS AND CEILINGS, 8602 North 40th Street, Tampa, Florida 33604. (813) 989-9300.